T0321025

Building Electrical Systems
and Distribution Networks

Nano and Energy Series

Series Editor: Sohail Anwar

Pennsylvania State University, Altoona College, USA

For more information about this series, please visit: https://www.crcpress.com/Nano-and-Energy/book-series/NANANDENE

Building Electrical Systems and Distribution Networks

An Introduction

Radian Belu

CRC Press
Taylor & Francis Group
Boca Raton London New York

CRC Press is an imprint of the
Taylor & Francis Group, an **informa** business

CRC Press
Taylor & Francis Group
6000 Broken Sound Parkway NW, Suite 300
Boca Raton, FL 33487-2742

© 2020 by Taylor & Francis Group, LLC
CRC Press is an imprint of Taylor & Francis Group, an Informa business

No claim to original U.S. Government works

Printed on acid-free paper

International Standard Book Number-13: 978-1-4822-6351-0 (Hardback)

Visit the Taylor & Francis Web site at
http://www.taylorandfrancis.com

and the CRC Press Web site at
http://www.crcpress.com

Dedication

To my wife Paulina Belu, my best friend and partner in life, for her continuous encouragement, patience, and support.

To my children Alexandru, Mirela, and Maria-Ruxandra, and to my grandchildren Stefan-Ovidiu and Ana-Victoria—our greatest joy in life and hope for the future.

To all my teachers, professors, and mentors.

This book is also dedicated to the memory of my parents Grigore and Gheorghita Belu, my first and best teachers and mentors.

Contents

Preface

This book covers the main topics of building energy systems and industrial power distribution, such as load calculation, electric services, wire devices, branch circuits, transformers, electric motors, load and motor centers, conduits and raceways, lighting, heat, ventilation, and air condition (HVAC), building automation, control, and safety. Topics related to building-integrated renewable energy systems, microgrids, DC nanogrids, energy economics, management, or energy audit methods are also included and discussed in detail. Buildings, industrial, and commercial facilities share a significant portion of the overall energy consumption of any country. In recent decades, there has been increased interest in energy conservation; use of energy-efficient equipment, devices, or systems; security of the energy supply; energy portfolio diversification; sustainability; and pollution reduction and control. Consequently, a large number of universities, colleges, and schools have included in their curriculum energy engineering, building energy systems, or industrial power distribution courses, which include topics such as energy management, building automation, and lighting. In addition, the ABET (Accreditation Board for Engineering and Technology) criteria in the United States encourages the development and changes of curricula that underline broad engineering education, contemporary engineering topics, and impacts of engineering solutions on society, industry, environment, or economic contexts. All these requirements can be in part met by updating and restructuring the energy courses, making them relevant to most of the engineering students. This is one of the book's main purposes: covering such issues and problems related to building electrical energy systems and industrial power distribution. The goals and objectives of this book are to provide engineering students, as well as engineers and technicians interested in building energy systems and industrial power distribution, with essential knowledge of the major energy system technologies, fundamental principles, characteristics, how they work, and how they are evaluated so these professionals can properly select the optimum system or equipment. It is written to assist schools, programs, and instructors who wish to establish a course in building energy systems while maintaining in-depth coverage of topics. The book covers major concepts such as power system structure, components, power engineering basics, motors and transformers, building and industrial power distribution, lighting, electric load characteristics and calculations, load and motor centers, building electrical systems, building thermal envelopes, thermal load-calculation methods, HVAC, noise control, building automation and control, distributed generation, microgrid concepts, architecture, structure and types, DC nanogrids, building-integrated renewable systems, energy management, and energy audit concepts, methods, and applications. Chapter 1 through Chapter 3 gives a comprehensive review of power systems and electric circuits. Chapter 4 and Chapter 5 discuss electric system design, electric load calculations, wiring devices, electric services, branch circuits, feeders, faults, conducts, cables, voltage drop calculations, wiring devices, protection circuits and methods, load and motor centers, raceways and conduits. Chapter 6 presents, in great detail, lighting concepts and systems, luminaires, and indoor and outdoor lighting calculations and design, while Chapter 7 focuses on transformers and electric motors. Chapter 8 gives an in-depth presentation of heat transfer concepts, theory, and types, building energy envelope, thermal resistance concept, heat pumps, and HVAC systems. Chapter 9 and Chapter 10 discuss building-integrated renewable energy sources, distributed generation, building applications of microgrids and DC nanogrids, and thermal energy storage, related concepts, architecture, structures, and types. We conclude with Chapter 11 and Chapter 12 presenting issues and topics related to building automation, control, and management systems, HVAC control methods, building noise-control methods, energy economics concepts and calculations, energy management, and energy audit concepts, theory, planning, methods, and applications. This book originates from the author's research and industrial experience and from courses that the author taught in the areas of energy and power engineering, renewable energy, industrial power distribution, and building electrical systems. The technical content and presentations are independent of any specific technology, being neutral to any specific technology, while still adhering to several premises of building or industrial energy systems. Likewise, this book assumes no specific knowledge of power and energy engineering; it guides the reader through basic understanding from topic to topic and inside of each topic. The topics of each chapter are self-contained and well-structured, requiring only basic and general engineering, physics, and mathematics knowledge with almost no relation to topics presented in other book chapters. Essential references for interested readers, instructors, and students are included at the end of each chapter, as well as questions and problems. The book is intended both as a textbook and a reference book for students, instructors, engineers, and professionals interested in industrial energy systems, renewable energy, and electrical systems design. Each chapter also contains several solved examples to help readers to understand and instructors to demonstrate the material. Support materials for interested instructors are also provided. The author is indebted to students, colleagues, and co-professionals for their feedback and suggestions over the years, and last but not least to the editorial technical staff for support and help.

About the Author

Radian Belu, PhD, is an associate professor in the Electrical Engineering Department at Southern University and A&M College, Baton Rouge, Louisiana. He has a PhD in power engineering and a PhD in physics. Before joining Southern University, Dr. Belu held faculty and research positions at universities and research institutes in Romania, Canada, and the United States. He also worked for several years in industry as a project manager, senior engineer, and consultant. His research focuses on energy conversion, renewable energy, microgrids, power electronics, climate, and extreme event impacts on power systems. He has taught and developed courses in power engineering, renewable energy, smart grids, control, electric machines, and environmental physics. His research interests include energy management and engineering education. Dr. Belu has published three books, fifteen book chapters, and more than 200 papers in referred journals and in conference proceedings. He has been principal investigator or co-principal investigator for various research projects in the United States and abroad.

1 Review, Power Systems, Energy, and Economics

1.1 INTRODUCTION TO MODERN POWER SYSTEMS

The electric grid is a vast physical and human network connecting thousands of electricity generators to millions of consumers, consisting of public and private enterprises operating within a web of government institutions: federal, regional, state, and municipal. The electric power system is a complex technical system with the main function of delivering electricity between generation, consumption, and storage. A power system serves one important function: to supply customers with electricity as economically and as reliably as possible. It is an interconnected network with components converting non-electrical energy continuously into the electrical form and transporting electrical energy from generating sources to loads/users. Electric power systems are real-time energy-delivery systems, meaning that power is generated, transported, and supplied the moment we turn on the light switch. Electric power systems are not storage systems like water systems and gas systems. Instead, generators produce the energy as the demand calls for it. The system starts with generation, by which electrical energy is produced in power plants and then transformed in power stations to high-voltage electrical energy that is more suitable for efficient long-distance transportation. The power plants transform other energy sources in the process of producing electrical energy. For example, heat, mechanical, hydraulic, chemical, solar, wind, geothermal, nuclear, and other energy sources are used in electrical energy production. High-voltage (HV) power lines in the transmission portion of the electric power system efficiently transport electrical energy over long distances to users. Finally, substations transform this HV electrical energy into lower-voltage (LV) energy that is transmitted over distribution power lines, which is more suitable for the distribution of electrical energy to its destination, where it is again transformed for residential, commercial, and industrial consumption. Power systems can be divided into four subsystems:

1. *Generation*—Generating and/or sources of electrical energy.
2. *Transmission*—Transporting electrical energy from its sources to load centers with high voltages (115 kV and above) to reduce transmission losses.
3. *Distribution*—Distributing electrical energy from substations (44 kV ~ 12 kV) to end users/customers.
4. *Utilization subsystem*—Using electrical energy in residential, commercial, and industrial facilities.

Planning of the power systems and the power distribution network is an essential task, intended to ensure that the required demand can be met based on various forecast loading figures, supply security, and reliability. Planning goals are to provide service at low cost, extended services, and higher reliability. Planning requires a mix of geographic, engineering, and economic analysis skills. New circuits (or other solutions) must be integrated into the existing power distribution system within a variety of economic, political, environmental, electrical, and geographic constraints. The three main categories of planning involved in the power systems are: the long-term, network, and construction planning. Long-term planning determines the most optimum network arrangements and the associated investment with consideration of future developments. Stage-by-stage development must be in line with the forecasted load growth so that electricity demands can be met in a timely manner. Network and power distribution planning is based on meeting the predicted energy demand and load forecasting trough new or upgraded sub-transmission lines, substations, and power distribution networks, while improving the reliability and power quality to the customers. Construction planning or design is the actual design and engineering work when the required circuits and substations have been planned and adopted. The planning of electric power distribution in buildings and infrastructure facilities is subject to constant transformation. The search for an assignment-compliant, dependable solution should fulfill those usual requirements placed on cost optimization, efficiency, and time needs. At the same time, technical development innovations and findings from the practical world are constantly seeping into the planning process. Opportunities for improving the functioning and reliability of the grid arise from technological developments in sensing, communications, control, and power electronics. These technologies can enhance efficiency and reliability, increase capacity utilization, enable more rapid response to remediate contingencies, and increase flexibility in controlling power flows on transmission lines. If properly deployed and accompanied by appropriate policies, the technologies can effectively overcome power distribution challenges, facilitate the integration of large volumes of renewable and distributed generation, provide greater visibility of the instantaneous state of the grid, and make possible the engagement of demand as a resource. Increasingly greater demands are placed on modern building energy systems. In order to exploit the full potential of economic efficiencies and fulfill technical demands, the earlier planning stages must take into account the demands for high levels of safety and flexibility throughout the entire life cycle, low pollution levels, integration of renewable energies, and low costs. A special challenge is the coordination of individual installations. The main building installations are heating, ventilation, air-conditioning, refrigeration, fire protection, safety, building control and monitoring systems, and electric power distribution. With innovative planning, the requirements are not simply broken down into the individual installations; rather, the efforts are coordinated. Electricity networks that are developed on a more holistic basis reflecting system-wide planning and more objective cost-benefit analyses can enable more efficient, timely, and cost-effective investments. Improved efficiency and transparency can help to build general acceptance toward new investments, while also helping to foster local acceptance. Transparent and consultative network infrastructure planning processes can build local community understanding, allowing proponents to draw on specific local knowledge to support appropriate power system developments. Benefit-cost analysis needs to take account of all applicable costs to the greatest extent possible, including environmental and distribution costs. Unaccounted costs can become a significant driver for local resistance during the siting process.

1.1.1 Power System Structure

The electric power industry is undergoing fundamental changes since the deregulation of telecommunication and other energy industries, the advent of new computing, information technology, communication, smart sensor and monitoring, and smart grid technologies. The growth in size and complexity of power plants and in higher-voltage equipment and devices was accompanied by interconnections of generating facilities, to decrease the probability of service interruptions, making the most possible reductions in generation costs, and reserve capacities. At the same time, in order to cope with the increased system complexity and to meet load requirements, safety, reliability, and reduced costs, a central control and operation, and the use of advanced computing and information technology were introduced. In order to reduce the transmission losses, extra-higher-voltage (EHV) has become dominant in electric transmission over large distances. The trend is also motivated by the economy due to higher transmission capacity possible, more efficient use of right-of-way, and reduced environmental impacts. The power industry faces new challenges and problems associated with smart grids, such as the interactions of the power system entities, distributed and dispersed generation, cyber security, demand-side management, consumer participation, and environmental issues, while striving to achieve the highest level of human welfare.

Generators convert non-electrical energy into electrical energy. Transformers are the devices connecting generators to the transmission system and from the transmission system to the distribution system. Their main functions are stepping up the lower-generation voltage to the higher-transmission voltage and stepping down the higher-transmission voltage to the lower-distribution voltage. The main advantage of having higher voltage in the transmission system is to reduce losses in the grid. Because transformers operate at constant power, when the voltage is higher, then the current has a lower value. Therefore, the losses, a function of the current square, will be lower at a higher voltage. A primary distribution substation is the connection point of a distribution system to the transmission or sub-transmission networks. Electric power distribution is the portion of the power delivery infrastructure that takes the electricity from the highly meshed, high-voltage transmission circuits and delivers it to customers. Outgoing feeders from a primary distribution substation are typically feeding secondary distribution substations and bigger, most often industrial type, consumers directly. What is considered to be the voltage level for a primary distribution substation varies country by country and depends on the whole electricity network structure and extent and historical and organizational issues. Primary distribution lines are medium-voltage networks, normally thought of as 600 V to 35 kV. At distribution substations, a substation transformer takes the incoming transmission-level voltage (35 to 230 kV) and steps it down to several distribution primary networks, originating from the substation. Close to each end user, a distribution transformer takes the primary-distribution voltage and steps it down to a low-voltage secondary circuit (commonly 120/240 V; other utilization voltages are used as well). From the distribution transformer, the secondary distribution circuits connect to the end user where the connection is made at the service entrance. Figure 1.1 shows the basic building blocks of an electric power system, from generation to the delivery infrastructure, and where distribution fits in (Von Mayer, 2006; Patrick and Fardo, 2008). Functionally, distribution circuits are those that feed customers. Some also think of distribution as anything that is radial or anything that is below 35 kV. The distribution infrastructure is extensive; after all, electricity has to be delivered to customers concentrated in cities, customers in the suburbs, and customers in very remote regions; very few places in developed countries do not have electricity from a distribution system readily available.

The *generation subsystem* of an electrical grid includes generators and transformers. Generators, an essential component of power systems, are in the vast majority the three-phase AC synchronous generators or alternators. They have two synchronously rotating electromagnetic fields, one produced by the rotor circuit driven at synchronous speed of rotation and excited by DC current, supplied by the excitation circuits, and the other generated by the three-phase armature currents in the

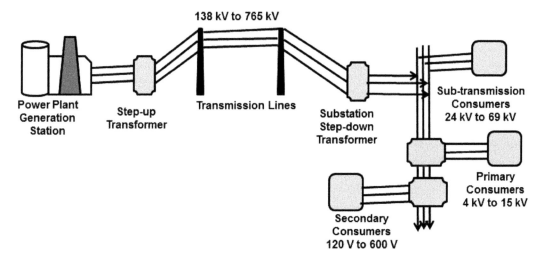

FIGURE 1.1　Power system structure.

stator windings. The alternator excitation system maintains and controls the voltage and the reactive power flow. Synchronous generators are able to operate at high power and high voltage, typically 30 kV, due to lack of commutator and the fact the armature circuits are on the stator. The source of mechanical power, the prime mover, may be hydraulic turbines, operating at a lower speed of rotation; steam turbines whose energy is coming from burning coal, gas, or nuclear fuel; gas turbines; or occasionally internal combustion engines. Steam turbines operate at high RPM, 1800 or 3600, coupled with four-pole and two-pole generators, while hydropower turbines operate at much lower speeds of rotation. In power plants, several alternators are operated in parallel with the power grid to provide the needed power, being connected to a common point, called the *bus*. Due to environmental concerns and conservation of fossil fuels, alternate energy sources, such as wind, solar, wave, tide, or geothermal energy, are considered for power generation. Transformers, the second component of the generating subsystems, transfer power from one voltage level to another with very high efficacy, above 90%. Insulation constraints and design requirements limit the generated voltage to low values, usually 30 kV. In order to reduce transmission losses, making electricity transfer over large distances, step-up transformers are used for power transmission. The electricity in a power system may undergo four or five transformations between generator and user.

Transmission and sub-transmission subsystems consist of overhead transmission lines and transformers designed to transfer power from generating units to the distribution systems, which ultimately supply the loads. Transmission lines also interconnect the neighboring utilities to allow economic power dispatch during normal operation conditions or power transfer between regions during emergencies. Standard transmission voltage, established in the United States by the American National Standards Institute (ANSI), is standardized at 60 kV, 115 kV, 138 kV, 161 kV, 230 kV, 345 kV, 500 kV, and 765 kV line-to-line voltage. High-voltage transmission lines are terminated in substations, called *high-voltage substations*, *receiving substations*, or *primary substations*. At the primary substations, voltages are stepped down to values more suitable for the distribution subsections, which are the next part of the power flow to the loads. Some substations are designed to function by switching circuits in and out service. Very large industrial customers are connected directly from the transmission subsystems. The section of the transmission subsystem that connects high-voltage substations to distribution substations is called the sub-transmission network. Typically a sub-transmission voltage ranges from 69 to 138 kV. However, there is no clear distinction between transmission and sub-transmission voltage levels. Some large industrial or commercial facilities may be served directly by sub-transmission networks. Capacitor and reactor banks are usually installed in substations for maintaining transmission voltage levels.

The primary purpose of an electricity distribution system is to meet customers' demands for energy after receiving bulk electrical energy from generation stations, through transmission or sub-transmission substation subsystems. Power distribution circuits or networks are found along most secondary roads and streets. Urban construction is mainly underground, while rural networks are mainly overhead. Suburban structures are a mix, with a good deal of new construction going underground. There are basically two major types of distribution substations: primary substation and customer substation. The primary substation serves as a load center and the customer substation interfaces to the low-voltage network. Depending on the geographical location, the power distribution network can be in the form of overhead lines or underground cables. Different network configurations are possible in order to meet the required supply reliability. Protection, control, and monitoring equipment are provided to enable effective operation of the distribution network. Distribution networks are located at the end of the generation and transmission, and directly face to the customers. With the rapid development of the economy, power demand keeps increasing, and distribution network structures become more complex. Standards by which customers judge the power supply have increased from a demand for quantity to the satisfaction of quality. Major problems in distribution network transformation and planning include: facing the grid crises; load information changing over time; a wide variety of power equipment; ensuring the scientific standard, efficiency, security, and reliability; reducing network losses; and improving the security and economy of running power distribution equipment (Cogdell, 1999; Chapman, 2002; Schavemaker and van der Sluis, 2008; El-Hawary, 2008; El-Sharkawi, 2009).

The *distribution subsystem* is divided into *primary distribution* and *secondary distribution* sections. The division between transmission and distribution is defined in terms of voltage levels, though individual utilities have different customs for where, exactly, to draw the line. Physically the boundary between transmission and distribution subsystems is demarked by transformers grouped at distribution substations along with other equipment such as circuit breakers and monitoring instrumentation. Primary distribution lines, with voltage ranges from 4 to 34.5 kV, supply the loads in a well-defined geographical area. Some small commercial and industrial customers are served directly from primary distribution feeders. Secondary distribution networks step down the voltage at levels appropriate for commercial and residential users. Lines and cables not exceeding a few hundred feet (or less than 100 m) in length deliver power to the individual consumers. The secondary distribution networks serve the consumers at voltage levels of 240/120 V, single-phase, three-wire; 208/120 V, three-phase, four-wire; or 480/277 V, three-phase, four-wire configurations. Distribution subsystems utilize both *overhead* and *underground* conductors, with underground being installed in most urban and suburban areas, and overhead more common for rural areas. The power for a typical residence is taken from a transformer that reduces the primary feed voltage to 240/120 V using a three-wire line. Organizationally, most utilities have separate divisions responsible for the operation, maintenance, and management of transmission and distribution subsystems, while substations often fall under the jurisdiction of power distribution.

The last section of the power system is the *load subsystems*. The loads are divided into industrial, commercial, and residential loads. Industrial and some commercial loads are composite loads, consisting mainly of induction motors, as well as power electronics circuits and other equipment and devices. These composite loads are a function of voltage and frequency and form a major part of these loads. Residential and most commercial loads consist largely of lighting, cooking, heating, and cooling, being independent of frequency, consuming very little reactive power. However, regardless of type, all loads vary during the day, and the power needs to be available on demand. The utility daily load curve is made of various user classes. The maximum value of this curve is called the *peak* or *maximum demand*. The *load factor*, assessing the usefulness of the generating plant, is defined as the ratio of the average load over a specific period of time to the peak load occurring during that period, such as a day, a month, or a year, and is expressed as:

$$Daily\ (Monthly,\ Yearly)\ \text{L.F.} = \frac{\text{Average Load}}{\text{Peak Load}} \tag{1.1}$$

The annual or yearly load factor is the most useful because a year represents a full cycle of time. In the above expression, by multiplying numerator and denominator by 24 hours, we obtain:

$$Daily\ (Monthly,\ Yearly)\ \text{L.F.} = \frac{\text{Average Load} \times 24\ \text{hr}}{\text{Peak Load} \times 24\ \text{hr}} = \frac{\text{Energy consumed during 24 hr}}{\text{Peak Load} \times 24\ \text{hr}} \tag{1.2}$$

The annual load factor is expressed as:

$$\text{Annual L.F.} = \frac{\text{Total Annula Energy}}{\text{Peak Load} \times 8760\ \text{hr}} \tag{1.3}$$

In order for a power plant to operate economically, a high load factor is required. In general, the diversity in the peak load between different classes tends to improve the load factor. Typical system load factors range from 55% to 75%. Load forecasting at different levels is critical in the operation, management, and planning of an electrical system. Other devices, equipment, and systems for the satisfactory operation and protection of power systems include: instrument transformers, circuit breakers, disconnect switches, fuses, and lightning arresters. This associated control and protection equipment is needed to de-energize either for normal or fault occurrences. For reliable and eco-nomic power system operation, it is necessary to monitor and control the entire power system in a so-called control center, equipped with online computers and data acquisition to perform data analysis, signal processing, and visualization in normal and emergency conditions in order to make the best operating decisions.

1.1.2 Industrial Power Distribution and Energy Systems

Power systems used in oil, gas, manufacturing, mining, chemical industries, or large industrial facilities have characteristics significantly different from those found in large-scale power genera-tion and long-distance transmission systems operated by public utility industries. One important difference is the common use of self-contained generating facilities, with little or no reliance upon connections to the public utility. This necessitates special consideration being given to installing spare and reserve equipment and to their interconnection configurations. These systems often have very large induction motors that require being started direct-on-line. Their large size would not be permitted if they were to be supplied from a public utility network. Therefore, the system design must ensure that they can be started without unduly disturbing other consumers. If the industrial facility is small enough to be fed from a single distribution transformer, the utility may serve the facility at utilization voltage, defined as the voltage at the line terminals of utilization equipment. The voltage is often 480Y/277 V, four-wire wye-connected service provided with a line-to-line volt-age of 480 V and a line-to-neutral voltage of 277 V. Single-phase 120 V and three- or single-phase 208 V can be provided using dry-type transformers rated 480 – 208Y/120 V. For large industrial facility increases, it is necessary to supply the facility at higher voltage levels to limit excessive voltage drops and to reduce transmission losses, minimizing the operation costs. Large industrial facility distribution transformers and associated high- and low-voltage switching equipment are known as *unit substations*. The unit substations, fed from the utility distribution tap, are supplying power to the facility. Distribution feeders are typically in the 2.4–15 kV range, with a trend toward the higher part of that range. Some utilities use distribution voltages higher than 15 kV, particularly if the load density is high or the feeders are very long. Equipment in the 25 kV class can be used on 22 and 24.9 kV systems, while 35 kV class equipment can be used on 33 and 34.5 kV systems.

1.1.2.1 Single-Line Diagrams

A one-line diagram is a type of diagram that uses single lines and graphic symbols to indicate the path and components of an electrical circuit. One-line diagrams are used when information about a circuit is required but detail of the actual wire connections and operation of the circuit is not strictly needed. A line diagram is used to show the relationship between circuits and their components but not the actual location of components. Line diagrams provide a fast, easy understanding of the connections and use of components. The one-line or single-line diagram is an important type of drawing used by power engineers to convey topological information about a power system. Most industrial power systems are three-phase systems, where each of the three phases is very similar to if not identical to the other two. This symmetry can be exploited by showing only one of the three phases, thus the term one-line diagram. Any asymmetries can be noted on the diagram. Some of the common power system ANSI symbols are shown in Figure 1.2. Special forms of one-line diagrams can be produced to convey specific information such as protection schemes or switching procedures. A single-line diagram is designed to show in a simple way the key elements of a power, energy, or distribution system, such as: sources of power (e.g., generators, utility intakes, the main switchboard and the interconnections to the subsidiary or secondary switchboards) and important equipment and components (e.g., power transformers, busbars, busbar section circuit breakers, incoming and interconnecting circuit breakers, large items of equipment such as high-voltage induction motors, series reactors for fault current limitation, and connections to old or existing equipment if these are relevant, and the main earthing arrangements). The key single-line diagram must show, at least,

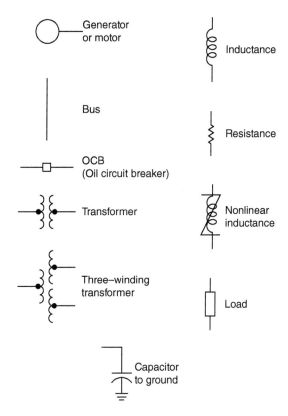

FIGURE 1.2 Some ANSI single-line diagram symbols.

the various voltage levels, system frequency, and power or volt-ampere capacity of main items or system components, such as generators, motors, and transformers, switchboard fault current levels, the vector group for each power transformer, and the identification names and unique tag numbers of the main equipment. The set of single-line diagrams forms the basis of all the electrical work carried out in a project, so it must be regularly reviewed and updated throughout the project duration. Similar to a schematic diagram, the one-line diagram uses fairly standardized symbols to represent system components.

1.2 ENERGY CONVERSION, HEAT TRANSFER, THERMODYNAMICS, AND FLUID MECHANICS ESSENTIALS

Energy has a large number of different forms, and there are relationships for each of them, as well as for conversion from one form to another. Among the energy forms are: gravitational energy, kinetic energy, heat energy, elastic energy, electrical energy, chemical energy, radiant energy, nuclear energy, and mass energy. If we total up the formulas for each of these contributions, it will not change except for energy going in and out. It is important to realize that in today's physics, we do not have a full knowledge and understanding of what energy is. However, this is beyond the scope of this book. In energy engineering, we are focusing on understanding energy transformation, conversion, transfer, and conservation. In order to do that, a basic understanding and knowledge of mechanics, thermal energy, heat transfer, and fluid mechanics is needed. Three very important physical quantities are *energy, force*, and *power*. Energy is a universal concept that bridges all engineering and science disciplines, being a unifying concept in the physical sciences. The understating of these physical quantities is critical for understanding of energy conversion and transfer processes, energy system operation, characteristics, and performances (Bueche, 1975; Petrecca, 2014; Belu, 2018). The subject of thermodynamics stems from the notions of temperature, heat, and work; thermodynamics as a science deals with matter as continuous rather than as discrete. Its study is crucial to understand the operation and characteristics of energy equipment or devices or to rigorously assess the efficiency of devices and equipment or industrial processes described later and provide a quantitative means of comparing competing technologies. A brief presentation of mechanics and thermodynamics concepts, thermal energy, and energy conversion processes is given in the following paragraphs.

If a force F moves a mass m through distance Δs in the force direction, the work done by the force is:

$$W = F \cdot \Delta s \tag{1.4}$$

Or more accurately, the work is computed by taking the integral of the force vs. distance:

$$W = \int F \cdot ds$$

If the time taken for this process is Δt, then the power, i.e., the energy per unit time, is defined as:

$$P = \frac{W}{\Delta t} = F \frac{\Delta s}{\Delta t} = F \cdot v \tag{1.5}$$

Here, v is the velocity. In the case of rotation, when a torque, τ (Nm) is acting on a body with momentum of inertia I (kg·m^2), rotating with an angular velocity, ω (rad/s), then Equation (1.5) is changed to:

$$P = \tau \cdot \omega \tag{1.6}$$

Example 1.1

What is the work done if a force of 25 N moves an object through a distance of 4 m? What is the average speed of the object if this displacement is carried out in 8 s? What is the power?

SOLUTION

Work done is $W = F \times d = 25 \times 4 = 100$ J W. The average speed is $v = d/t = 4/8 = 0.5$ m/s. Power input is computed with Equation (1.5), as:

$$P = F \cdot v = 25 \times 0.5 = 12.5 \text{ W}$$

The mechanical energy of a body, E, which is a scalar, can be kinetic, due to its motion, or potential, due to its mechanical state, for example due to its height above a reference or the compression of a spring. Kinetic energy (E_{KE}) of a body is associated with its velocity $\bar{v}$ (m/s) and can be evaluated in terms of the work required to change the velocity of the body, as follows:

$$E_{KE} = \int F \cdot dx = \int m \cdot a \cdot dx = \int m \frac{d\bar{v}}{dt} dx$$

Because the velocity is the derivative of the distance, integrating from v_1 to v_2, yields to:

$$E_{KE} = \int_{v_1}^{v_2} mv \cdot dv = \frac{1}{2} m \cdot \left(v_2^2 - v_1^2 \right) \tag{1.7}$$

Example 1.2

What is the potential energy of a coiled spring with a spring constant of 100 N/m that has been compressed a distance of 5 cm?

SOLUTION

By using Equation (1.9), the potential energy of the spring is:

$$PE = \frac{1}{2}100 \cdot \left(5 \times 10^{-2}\right)^2 = 0.125 \text{ J}$$

We introduce the various forms of energy of interest to us in terms of a solid body having a mass m. These include not only kinetic energy, but also potential energy and how we will discuss later internal energy. Potential energy (E_{PE}) is associated with the elevation of the body or with compression or extension of a spring. In the first case, it can be evaluated in terms of the work done to lift the body from one datum level to another under a constant acceleration due to gravity g (m/s^2), as follows:

$$E_{PE} = \int_{h_1}^{h_2} mg \cdot dz = mg \cdot (h_2 - h_1) = mg \cdot \Delta h \tag{1.8}$$

While in the second case of elastic potential energy, when the elastic force, $F = k \cdot x$, the above relationship is expressed as:

$$E_{PE} = \int_{x_1}^{x_2} F \cdot dx = \int_{x_1}^{x_2} kx \cdot dx = \frac{1}{2}k \cdot \left(x_2^2 - x_1^2\right) \tag{1.9}$$

In these two equations, h_1, h_2, x_1, and x_2 refer to the initial and final position of the body and spring, respectively. The total mechanical energy is the sum of the kinetic and potential energies:

$$E_{mech} = E_{KE} + E_{PE} \tag{1.10}$$

For rectilinear motion, work is done by the application of a force. Multiplying Newton's second law ($F = mdv/dt$) by velocity for linear motion by the force F gives the power involved:

$$P = F \cdot v = F \cdot a \cdot dt = \frac{d}{dt}\left(\frac{1}{2}mv^2\right) \tag{1.11}$$

which is a mechanical energy equation. The right side is the rate of increase of the kinetic energy, which is:

$$E_{KE} = \frac{1}{2}mv^2 \tag{1.12}$$

Likewise, for angular rotation, the mechanical power equation for the rotating systems is given by:

$$P = \tau \cdot \omega = \tau \cdot a_{rot} \times dt = \frac{d}{dt}\left(\frac{1}{2}I\omega^2\right) \tag{1.13}$$

Now, the kinetic energy for the rotating systems is expressed as:

$$E_{KE} = \frac{1}{2}I\omega^2 \tag{1.14}$$

Example 1.3

How much kinetic energy is stored in a flywheel of moment of inertia 10 kg·m² and rotational speed 1200 rpm?

SOLUTION

The kinetic energy of a flywheel (an electro-mechanical energy storage device using kinetic energy of rotation) is:

$$E = \frac{1}{2}I\omega^2 = 0.5 \cdot 10 \cdot \left(\frac{2\pi \times 1200}{60} \right) = 628.3 \text{ J}$$

1.2.1 ENERGY CONVERSION AND LOSSES

Strictly speaking, mass and energy cannot be lost; they only convert from one form to another. However, the term is used if the mass or energy goes in a direction that is not desirable. The term is also very often used without further clarification in energy systems in many different ways. It is very important for the understanding of energy conversion processes to have a full and clear understanding of energy and mass losses. Here are a few examples: (a) In one sense it is a leakage of mass. If, for example, a certain mass flow rate goes into a machine, and not all of it comes out where it should, there is a leakage. This could be due to an undesirable gap within the machine. (b) Another is a loss of energy such as heat transfer to the surroundings. (c) A third is a pressure loss, as what happens when there is a flow along a pipe. Energy can exist in various forms. Some of the energy forms are given here. Radiation energy from the sun contains energy, and also the radiation from a light or a fire. More solar energy is available when the radiation is more intense and when it is collected over a larger area. Light is the visible part of radiation. Chemical energy is the energy contained, for example in wood or oil. The same is true for all other material that can burn. The content of chemical energy is larger for larger heating values (calorific values) of material and, of course, the more material we have. Also animate energy (delivered by bodies of human beings and animals) is, in essence, chemical energy. Furthermore, batteries contain chemical energy. Potential energy is, for example, the energy of a water reservoir at a certain height. The water has the potential to fall, and therefore contains a certain amount of energy. More potential energy is available when there is more water and when it is at a higher height. Kinetic energy is the energy of movement, as in wind or in a water stream. The faster the stream flows and the more water it has, the more energy it can deliver. Similarly, more wind energy is available at higher wind speeds, and more of it can be tapped by bigger windmill rotors. Thermal energy or heat is the type indicated by temperature. The higher the temperature, the more energy is present in the form of heat. Also, a larger body contains more heat. Mechanical energy, or rotational energy, also called shaft power, is the energy of a rotating shaft. The available energy depends on the flywheel of the shaft, i.e., the power that makes the shaft rotate. Electrical energy is produced by a dynamo or an AC generator or released by a battery. Note that sometimes *energy form* means an energy source, or even a particular fuel (such as oil or coal).

The term *utilizing energy* always means converting energy from one form into another. For instance, in space heating, energy is used, by converting, for example, chemical energy of wood into heat. Or, in lift irrigation, a diesel engine converts oil chemical energy into mechanical energy for powering the shaft of a pump, which, in its turn, converts shaft power into potential energy of water (i.e., bringing the water to a higher height). "Generating" energy means converting energy from one form into another. For example, a diesel engine generates energy, meaning that the engine converts oil chemical energy into mechanical energy. Also, a wind turbine generates energy, which means it converts kinetic energy from wind into mechanical energy. And a solar photovoltaic cell generates energy by converting radiation energy into electricity. The generation of energy, in fact, deals with a source of energy, whereas the utilization of energy serves an end-use of energy. In between, the

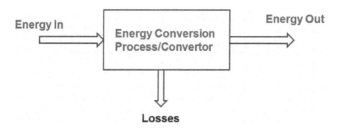

FIGURE 1.3 Diagram of energy conversion.

energy can flow through a number of conversion steps. The words *generation* and *utilization* are somewhat confusing because, in fact, no energy can be created or destroyed. All we can do is transform or convert energy from one form into another. In generating energy, we make energy available from a source, by converting it into another form. In utilizing energy, we also convert energy, often from some intermediate form into a useful form. In all conversions, we find that part of the energy is lost. This does not mean that it is destroyed, but rather that it is lost for our purposes, through dissipation in the form of heat or otherwise, as shown in Figure 1.3.

Energy conversions can take place from any one form of energy into almost any other form of energy. Some of these conversion processes have no practical value. Which conversion is desired depends on the application. For instance, for power generation, the potential energy from hydro resources is converted into mechanical energy, whereas, in water pumping for lift irrigation, the reverse process is performed. With photovoltaic cells, the solar radiation energy is converted directly into electricity, whereas with light bulbs the reverse process is taking place. A large number of energy conversion processes are taking place in nature or human-made systems, by means of various devices invented during our history. Such devices may be classified according to the type of construction used, according to the physical or chemical operating principles, or according to the forms of energy appearing before and after the action of the device. The energy flows through a number of forms, as well as conversion steps, between the source and the end use. The costs increase accordingly. We distinguish between primary, secondary, final, and useful energy. An example is an energy flow that is related to charcoal. Here, the primary energy form is wood. The wood is converted into charcoal in a charcoal kiln. Charcoal is the secondary form of energy, and it is transported to the consumer. What the consumer buys at the marketplace is charcoal, and this is called final energy. The consumer eventually converts the charcoal into heat for cooking. The heat is the useful energy. Another example of an energy flow is: primary energy in the form of a hydro resource, secondary energy in the form of electricity at the hydro power station, final energy in the form of electricity at a sawmill, and useful energy in the form of shaft power for sawing.

An energy quantity in a certain form is put into a device, for conversion into another form of energy. The output energy in the desired form is only a part of the *input energy*. The balance is the *energy losses* (usually in the form of heat), meaning that any converter has less than 100% efficiency. In some of these converters, intermediate forms of energy occur between the input energy and the output energy. For instance, with diesel engines, the intermediate form is thermal energy. Table 1.1 gives typical efficiency of some of the common types of energy converters. The efficiency of an energy conversion device is a quantitative expression of this balance between energy input and useful energy output, defined as their ratio. So the efficiency of an energy converter is defined as the quantity of energy in the desired form (*the output energy*) divided by the quantity of energy put in for conversion (*the input energy*), expressed as:

$$\eta = \frac{\text{Output Energy}}{\text{Input Energy}} = 1 - \frac{\text{Losses}}{\text{Input Energy}} \tag{1.15}$$

TABLE 1.1
Typical Efficiencies of Some Common Types of Energy Converters

Converter	Input Energy	Output Energy	Efficiency (%)
Gasoline engine	Chemical	Mechanical	25–35
Diesel engine	Chemical	Mechanical	30–45
Electric motor	Mechanical	Electrical	80–95
Steam turbine	Thermal	Mechanical	10–45
Gas turbine	Thermal	Mechanical	30–35
Hydro turbine	Mechanical	Mechanical	30–90
Generator	Mechanical	Electrical	80–95
Battery	Chemical	Electrical	80–90
Solar cell	Solar radiation	Electrical	10–25
Electric lamp	Electrical	Light	5–20
Water pump	Mechanical	Mechanical	50–60
Water heater	Electrical	Thermal	85–95
Home gas furnace	Chemical	Thermal	80–85

Example 1.4

An electric motor consumes 1 kW electric power and produces 0.9 kW mechanical power. What is its efficiency?

SOLUTION

Taking into consideration the relationship between energy and power, the motor efficiency can be expressed as:

$$\eta = \frac{\text{Power Out}}{\text{Power In}} = \frac{.90}{1.0} = 0.90 \text{ or } 90\%$$

In order to understand such processes and devices, a brief review of thermal energy and mechanics is needed and presented here. A pure substance is defined as a homogeneous collection of matter. Consider a fixed mass of a pure substance bounded by a closed, impenetrable, flexible surface. Such a mass is called a system, and needs to be defined carefully to ensure that the same particles are in the system at all times. All other matter that can interact with the system is called the surroundings. The combination of the system and the surroundings is termed the universe, used here not in a cosmological sense, but to include only the system and all matter that could interact with the system. Thermodynamics and energy conversion are concerned with changes in the system and in the system's interactions with the surroundings. Thermodynamic properties may be extensive and depend on the size of the material being considered. Examples are internal energy, enthalpy, and entropy. Intensive properties, on the other hand, do not depend on the size, examples being pressure and temperature. Extensive properties can be made specific, i.e., per unit mass. A simple compressible substance is defined as one in which only two independent properties are required to identify its state. In mixtures (such as moist air, for example), however, concentrations are also required to be specified. The mass contained within a system can exist in a variety of conditions called states. Qualitatively, the concept of state is familiar. For example, the system state of a gas might be described qualitatively by saying that the system is at a high temperature and a low pressure. Values of temperature and pressure are characteristics that identify a particular condition of the system. Thus a unique condition of the system is called a state.

A system is said to be in *thermodynamic equilibrium* if, over a period of time, no change in the state of the system is observed. It is a fundamental assumption that a state of thermodynamic equilibrium of a given system may be described by a few observable characteristics called thermo-dynamic properties, such as pressure, temperature, and volume. We now can define the following terminology and conventions:

1. A system is a component in a universe and is the subject of our thermodynamic analysis.
2. The surroundings and the system are in contact typically through heat, material, or charge transfer.
3. A boundary separates the system from the universe.
4. The universe is composed of the system and its surroundings.
5. Closed systems do not allow matter to enter or leave.
6. Open systems allow matter to enter or leave. Isolated systems are closed and also do not allow energy transport across the system boundaries.
7. Heat is energy in transit between a system and its surroundings or between two systems, driven by the temperature difference between the two. Heat energy has units of J (joules) and symbol Q; Q > 0 will denote heat owing into the system.
8. Intensive properties do not depend on the total amount of material present; examples include temperature T, pressure P, and molar quantities such as molar volume V (unless we specify by mass).
9. Extensive properties do depend on the quantity of material, such as volume V^t or mass m. Note the use of the t superscript.

1.2.2 Thermodynamics Essentials

Temperature is a characteristic of thermal energy of a system due to internal motion of molecules and atoms, being an indication of the thermal energy stored in a substance. In other words, we can identify hotness and coldness with the concept of temperature. The temperature of a substance may be expressed in either relative or absolute units. Two systems in physical contact are said to be in thermal equilibrium if both are at the same temperature. The two most common temperature scales are Celsius (°C) and Fahrenheit (°F). The Kelvin degree is a unit of temperature measurement; zero Kelvin (0 K) is absolute zero and is equal to −273.15°C. Increments of temperature in units of K and °C are equal. The modern temperature unit is defined based on the efficiency of ideal fluid working in a Carnot cycle (discussed later in this chapter), independent of the properties of any material. Temperature can be measured in a large number of ways by devices, with thermometers as the most common. Another way to observe changes in the state of a liquid or gaseous system is to connect a manometer to the system and to measure its *pressure*. It has been empirically observed that an equilibrium state of a system containing a single phase of a pure substance is defined by two thermodynamic properties. Thus, if the temperature and pressure of such a system is observed, we can identify when the system is in a particular thermodynamic state. Properties that are dependent on mass are known as extensive properties. For these properties that indicate quantity, a given property is the sum of the corresponding properties of the subsystems comprising the system. Examples are internal energy and volume. Thus, adding the internal energies and volumes of subsystems yields the internal energy and the volume of the system, respectively. In contrast, properties that may vary from point to point and that do not change with the mass of the system are called intensive proper-ties. Temperature and pressure are well-known examples. For instance, thermometers at different locations in a system may indicate differing temperatures. But if a system is in equilibrium, the tem-peratures of all its subsystems must be identical and equal to the temperature of the system. Thus, a system has a single, unique temperature only when it is at equilibrium. From basic mechanics, work, W, is defined as the energy provided by an entity that exerts a force, F, in moving one or more par-ticles through a distance, x. Thus work must be done by an external agent to decrease the volume, V, of a system of molecules. In a piston-cylinder arrangement, an infinitesimal system volume change due to the motion of the piston is related to the differential work through the force-distance product:

$$dW = F \cdot dx = p \cdot A \cdot dx = p \cdot dV \qquad (1.16)$$

or

$$dw = p \cdot dv \qquad (1.17)$$

where p is the system pressure, and A is the piston cross-sectional area. Note that in Equation (1.17), the lowercase letters w and v denote work and volume on a unit mass basis. All extensive properties, i.e., those state properties that are proportional to mass, are denoted by lowercase characters when on a unit mass basis. These are called *specific properties*. Thus, if V represents volume, then v denotes specific volume. Although work is not a property of state, it is dealt with in the same way. The units of different measurement systems are related to the conversion factors. Conversion factors are frequently required and are not explicitly included in many equations. When work decreases the volume of a system, the molecules of the system move closer together. Notice, that the state of the system may be changed by work done on the system. A thermodynamic system is a device or combination of devices that contains a certain quantity of matter. It is important to carefully define a system under consideration and its boundaries, such as open, closed, or isolated systems. A process is a physical or chemical change in the properties of matter or the conversion of energy from one form to another. In some processes, one property remains constant. A cycle is a series of thermodynamic processes in which the endpoint conditions or properties of the matter are identical to the initial conditions. It is known that all substances can hold a certain amount of heat; this property is their thermal capacity. When a liquid is heated, its temperature rises to the boiling point. This is the highest temperature that the liquid can reach at the measured pressure. The heat absorbed by the liquid in raising the temperature to the boiling point is called *sensible heat*. The heat required to convert the liquid to vapor at the same temperature and pressure is called *latent heat*. This is the change in enthalpy, defined in the next paragraphs during a state change (the heat absorbed or rejected at constant temperature at any pressure, or the difference in enthalpies of a pure condensable fluid between its dry saturated state and its saturated liquid state at the same pressure). A number of equations of state are in common use, such as the state equation for an ideal gas or an incompressible fluid approximation, given by:

$$p = \rho RT \qquad (1.18)$$

Here, R is the universal gas constant, equal to 8.314 kJ/kg K. For a system immersed in a hot fluid container, by virtue of a difference in temperature between the system and the surrounding fluid, energy passes from the fluid to the system, saying that *heat*, Q (measured in Btu or J), is transferred to the system. Heat transfer to or from the system, like work, can also change the state of the matter within the system. When the system and the surrounding fluid are at the same temperature, no heat is transferred. In this case the system and surroundings are said to be in *thermal equilibrium*. The term *adiabatic* is used to designate a system in which no heat crosses the system boundaries. A system is often approximated as an adiabatic system if it is well-insulated. Heat and work are not state properties but forms of energy that are transferred across system boundaries to or from the environment, referred to sometimes as the energy in transit or transfer. Energy conversion engineering is concerned with devices that use and create energy in transit. A property of a system that reflects the energy of the molecules of the system is called the internal energy; the U. *Law of Energy Conservation* states that energy can be neither created nor destroyed. Thus the internal energy of a system can change only when energy crosses a boundary of the system, i.e., when heat and/or work interact with the system. This is expressed in an equation known as the *First Law of Thermodynamics*, expressed in differential form as:

$$du = dq - dw \qquad (1.19)$$

Here, u is the internal energy per unit mass, a property of state, quantities q is the heat per unit of mass, and w is the work per unit of unit mass. The differentials indicate infinitesimal changes in quantity of each energy form. Here, we adopt the common sign convention of thermodynamics that both the heat entering the system and work done by the system are positive. We can convert energy from one form to another (e.g., mechanical-to-electrical), but the total energy is always constant. Total energy is not the same as usable energy, which leads to the concept of conversion efficiency and entropy. A special and important form of the First Law of Thermodynamics is obtained by integration of Equation 1.11 for a cyclic process. A system is said to be in a cyclic process, if after undergoing arbitrary change due to heat and work, returns to its initial state. The key points are: the integral of any state property differential is the difference of its limits, and the final state is the same as the initial state, meaning there is no change in internal energy of the system, as:

$$\oint du = u_f - u_i = 0$$

where the special integral sign indicates integration over a single cycle and subscripts i and f designate, respectively, initial and final states. As a consequence, the integration of Equation (1.18) for a cycle (cyclic processes) yields:

$$\oint dq = \oint dw \qquad (1.20)$$

This states that the integral of all transfers of heat into the system, taking into account the sign convention, is the integral of all work done by the system, the *net work* of the system. The integrals in Equation (1.18) are replaced by summations for a cyclic process that involves a finite number of heat and work terms. Many heat engines operate in cyclic processes, making it convenient to evaluate the net work of a cycle using Equation (1.18) with heat additions and losses rather than using work directly.

Example 1.5

During a cycle process, 60 kJ of heat flow into the system and 30 kJ are rejected from the system later in the cycle. What is the net work of this cycle?

SOLUTION

Applying Equation (1.12) for the case of finite number of heat and work terms yields

$$w = q_{in} - q_{out} = 60 - 30 = 30 \text{ kJ}$$

Another important form of the First Law of Thermodynamics is the integral of Equation 1.12 for an arbitrary process involving a system, expressed as:

$$q = u_f - u_i + w \qquad (1.21)$$

where q is the net heat transferred, w is the net work for the process, and u_f and u_i are the final and initial values of the internal energy. Equation (1.21), like Equation (1.18), shows that a system that is rigid ($w = 0$) and adiabatic ($q = 0$) has an unchanging internal energy. It also shows, like Equation (1.18), that for a cyclic process the heat transferred must equal the work done. If a system undergoes a process in which temperature and pressure gradients are always small, the process may be thought of as a sequence of near-equilibrium states. If each of the states can be restored in reverse sequence, the process is said to be internally reversible. If the environmental changes accompanying the process can also be reversed in sequence, the process is called externally reversible. Irreversibility occurs due to temperature, pressure, composition, and velocity gradients caused by heat transfer, solid and fluid friction, chemical reaction, and high rates of work applied to the system. An engineer's job frequently entails efforts to reduce irreversibility in machines and processes. Entropy and enthalpy are thermodynamic properties that, like internal energy, usually appear in the form of differences between initial and final values. The entropy change of a system, s (J/kg-K), is defined as the integral of the ratio of the system differential heat transfer to the absolute temperature for a reversible thermodynamic path, that is, a path consisting of a sequence of well-defined thermodynamic states. In differential form this is equivalent to:

$$ds = \frac{dq_{rev}}{T} \tag{1.22}$$

where the subscript *rev* denotes that the heat transfer must be evaluated along a reversible path made up of a sequence of neighboring thermodynamic states, meaning for such a path, the system may be returned to its condition before the process took place by traversing the states in the reverse order. The enthalpy, h, is a property of state defined in terms of other properties:

$$h = u + pv \tag{1.23}$$

where h, u, and v are the system specific enthalpy (J/kg), specific internal energy, and specific volume, respectively, and p is the pressure. Two other important forms of the First Law of Thermodynamics make use of these properties, expressed as:

$$Tds = du + pdv \tag{1.24}$$

and

$$Tds = dh - vdp \tag{1.25}$$

Equations 1.24 and 1.25 may be regarded as relating changes in entropy for reversible processes to changes in internal energy and volume in the former and to changes in enthalpy and pressure in the latter. All quantities in these equations are state properties implying that the entropy is also a thermodynamic property. Because entropy is a state property, the entropy change between two equilibrium states of a system is the same for all processes connecting them, reversible or irreversible.

Example 1.6

Calculate the entropy change of an infinite sink at 27°C temperature due to heat transfer into the sink of 1000 kJ.

SOLUTION

Because the sink temperature is constant, Equation (1.22), in finite difference form, shows that the entropy change of the sink is the heat transferred reversibly divided by the absolute temperature of the sink. This reversible process may be visualized as one in which heat is transferred from a source that is infinitesimally hotter than the system:

$$\Delta S_{rev} = \frac{\Delta q_{rev}}{T} = \frac{1000}{27 + 273.15} = 3.333 \text{ kJ/K}$$

Equation (1.24) may be used to determine the system entropy change. The Second Law of Thermodynamics is concerned with the entropy change of the universe, i.e., of both the system and the surroundings. Entropy being an extensive property, the entropy of a system is the sum of the entropy of its parts. Applying this to the universe, the entropy of the universe is the sum of the entropy of the system and its surroundings. The Second Law may be stated as: The entropy change of the universe is non-negative. The First Law of Thermodynamics deals with how the transfer of heat influences the system internal energy but says nothing about the nature of the heat transfer, i.e., whether the heat is transferred from hotter or colder surroundings. However, the experience tells us that the environment must be hotter to transfer heat to a cooler object, but the First Law is indifferent to the condition of the heat source. However, calculation of the entropy change for heat transfer from a cold body to a hot body yields a negative universe entropy change, violates the Second Law, and is therefore impossible. Thus the Second Law provides a way to distinguish between real and impossible processes. In general, apart from the material phase change, temperature of material increases as it absorbs heat, and the heat, Q, required to increase temperature by an amount ΔT is given by:

$$Q = mC\Delta T \tag{1.26}$$

Here, m is the mass (kg), and C is the specific heat, roughly independent of temperature. The specific heats or heat capacities at constant volume, c_V (kJ/kg·K) and at constant pressure, c_P (kJ/kg·K) respectively are given by:

$$c_V = \left(\frac{\partial u}{\partial T}\right)_V \tag{1.27}$$

and

$$c_P = \left(\frac{\partial h}{\partial T}\right)_P \tag{1.28}$$

As thermodynamic properties, the heat capacities are functions of two other thermodynamic properties. For solids and liquids, pressure changes have little effects on volume and internal energy, so to a good approximation $c_P = c_V$. The other simplifying feature of ideal-gas behavior is that, if assumed that the constant-pressure specific heat, c_P, and constant-volume specific heat, c_V, are constant, Equations 1.24 and 1.25, changes in specific internal energy and specific enthalpy can be calculated simply without referring to thermodynamic tables and graphs from the following expressions:

$$\Delta u = u_2 - u_1 = c_P (T_2 - T_1) \tag{1.29}$$

and

$$\Delta h = h_2 - h_1 = c_V (T_2 - T_1) \tag{1.30}$$

The following is another useful relation for ideal gases:

$$c_P - c_V = R \tag{1.31}$$

In above Equation (1.31), R and heat capacities need to be in consistent units. Another important gas property is the ration of the heat capacities, $\gamma = c_P/c_V$. This parameter is constant for gases at room temperatures.

In thermodynamics and thermal engineering, the thermal efficiency is a dimensionless parameter that is used to measure the performance of a device or system that uses thermal energy, such as an internal combustion engine, a steam turbine or a steam engine, a boiler, a furnace, or a refrigerator. Heat engines or thermal systems transform thermal energy, or heat, Q_{IN} into mechanical energy, or work, W_{OUT}, while some of the input heat energy is not converted into work, but is dissipated as waste heat Q_{OUT} into the environment:

$$Q_{IN} = W_{OUT} + Q_{OUT}$$

The thermal efficiency of a heat engine is the ratio of heat energy that is transformed into work to the input heat or thermal energy, defined as:

$$\eta = \frac{W_{OUT}}{Q_{IN}} = 1 - \frac{Q_{OUT}}{Q_{IN}} \tag{1.32}$$

The efficiency of even the best heat engines is low; usually lower than 50%, so the energy lost to the environment by heat engines is a major waste of energy resources, although modern cogeneration, combined cycle, and energy recycling schemes are beginning to use this heat for other purposes. Low-heat engine inefficiency is attributed to three causes. There is an overall theoretical limit to the efficiency of any heat engine due to temperature, the Carnot efficiency. Second, specific types of engines have lower limits on their efficiency due to the inherent engine cycle irreversibility used. Third, the behavior of real heat engines is non-ideal, i.e., mechanical friction and losses in the combustion process cause further efficiency losses. The Second Law of Thermodynamics puts a fundamental limit on the thermal efficiency of all heat engines. The limiting factors are the temperature at which the heat enters the engine, T_H, and the temperature of the environment into which the engine exhausts its waste heat, T_C, measured in Kelvin degrees. This limiting value is called the Carnot cycle efficiency because it is the efficiency of an unattainable, ideal, reversible engine cycle called the Carnot cycle. No device converting heat into mechanical energy, regardless of its construction, can exceed this efficiency. From Carnot's theorem, for any engine working between these two temperatures:

$$\eta \leq \eta_{Carnot} = 1 - \frac{T_C}{T_H} \tag{1.33}$$

Example 1.7

In a power plant, the steam from the boiler reaches the turbine at a temperature of 750°C. The spent steam leaves the turbine at 100°C. Calculate the maximum efficiency of the turbine. Compare it to the typical value listed in Table 1.1.

SOLUTION

From the above expression and noting that 750°C = 1023 K and 100°C = 373 K, the Carnot efficiency of the turbine is:

$$\eta_{max} = 1 - \frac{373}{1023} = 0.635 \text{ or } 63.5\%$$

This, as we expected, is larger than the typical efficiency of about 45%, shown in Table 1.1.

Most heat engines use a fossil fuel, or a product derived from it, such as natural gas, coal, or gasoline, to provide heat, which is then converted into mechanical work. So, in essence, these engines consist of two subsystems, as illustrated in Figure 1.4. We thus need to introduce the concept of

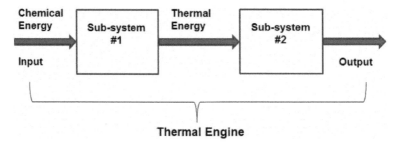

FIGURE 1.4   Energy conversion in a heat engine, consisting of two subsystems.

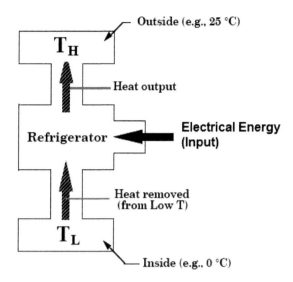

FIGURE 1.5 Schematic (thermodynamic) representation of a refrigerator.

system efficiency. A system means a well-defined space in which at least two energy conversions take place. It consists of two or more energy conversion devices. *The system efficiency is equal to the product of efficiencies of the individual devices (subsystems).*

In contrast to energy conversion devices, converting one energy form into another, the energy transfer devices just transfer the same energy form from one place to another. The interest in building energy systems is in the devices that transfer heat, e.g., refrigerators, air conditioners, and heat pumps. A refrigerator uses energy to maintain its temperature lower than that of the surroundings. The electric energy consumed by the refrigerator is used to reverse this process, to pump heat from inside (low-temperature reservoir or *source*) to the outside (high-temperature reservoir or *sink*), as in Figure 1.5. The air conditioner and the heat pump accomplish exactly the same task as a refrigerator, pumping heat *uphill*, from T_L to T_H.

The heat pump is an energy transfer device that may be very convenient for residential comfort in certain geographical areas. There are two heat exchangers instead of one (a condenser and an evaporator instead of just a condenser). A special liquid (Freon or antifreeze) is used as the working fluid because water would, of course, freeze in winter. In winter, its T_L is the outdoor air (say, at 20°F), and T_H is the indoor air (say, at 65°F). Electricity is used to increase the energy of this liquid, by compressing it, and the compressed liquid delivers this energy to the house by condensing in the internal heat exchanger. The external heat exchanger is necessary to bring it back into the gaseous state (by evaporation) so that it can be compressed again and the cycle repeated. In summer, the pump functions as an air conditioner, with T_L being the inside air and T_H the outside air. The interior heat exchanger is the evaporator and the exterior one is the condenser. Evaporation is a process that requires energy input. So the warm inside air flows past this Freon evaporator, it transfers to it some of its energy and thus becomes cooler. The efficiency of a heat transfer system, the coefficient of performance, or COP, is defined in the same way as the efficiency of an energy conversion device:

$$COP = \frac{\text{Useful Energy}}{\text{Energy Input}} \tag{1.34}$$

There is one important difference between energy conversion devices and energy transfer devices. In a conversion device, only a portion of the energy input is obtained as useful energy output, and the efficiency is necessarily a number between 0 and 1. In a transfer device, the useful energy output is the quantity of heat extracted from T_L, and this is not a portion of the energy input. In fact, the useful energy output can exceed the energy input, and this is why heat pumps can be extremely attractive for space heating purposes. The coefficient of performance (called also "energy efficiency ratio") can be larger than 1 and this does not violate the First Law of Thermodynamics; the larger it is, the more efficient the heat mover will be. From thermodynamic analysis, which again we do not need to go into, it is possible to define the *maximum (or ideal) coefficient of performance* (COP_{max}). The definition depends on whether the heat mover is used as a heater or as a cooler. If it is a heater, then the definition is:

$$COP_{max} = \frac{T_H}{T_H - T_L} \tag{1.35}$$

The temperature here is expressed in Kelvin. If the heat transfer device is a cooler or a refrigerator, then the definition is:

$$COP_{max} = \frac{T_L}{T_H - T_L} \tag{1.36}$$

1.2.3 Essentials of Fluid Mechanics

Given here is a very brief summary of the physical properties of fluids, needed to describe and understand energy conversion in industrial processes, such as building ventilation, heating, air-conditioning, or renewable energy. A fluid is either a liquid or a gas, and can flow. Energy can be either extracted from the flow, like in a turbine, or put into a flow, as in a pump. While the details of energy transfer from the fluid are complicated, being an individual subject in an engineering program, the overall input-output energy relationships are simple. The basic concepts of mass, linear momentum, angular momentum, and energy balance are also applicable for a flow, but the only difference is the way of looking at it. For a rigid body, one can identify a definite piece of material and follow it. For a flow, however, rather than following an identifiable body of material, as one would for rigid bodies, we consider a fixed region of space (called a control volume) through which the flow passes. Thus there are quantities such as *mass, linear momentum, angular momentum*, and *energy flow rates*, advecting past a given cross section of a pipe, with the word *rate* meant to signify that the quantity is per unit time. Once the flow rate is identified, the usual conservation or balance principles apply, which is why the rates are quantities of interest, referring per unit time, rather than in absolute quantities. The bulk physical properties of a fluid are stated in terms of fluid density (ρ) mass per unit volume, pressure (p, which is force per unit of area), and viscosity (the force per unit area due to internal friction arising from the relative motion between adjacent fluid layers, acting in tangential directions). A useful concept to visualize the fluid velocity field is the notion of the *streamline*, the lines parallel to the direction of motion in all points of the fluid. Any fluid mass element flows along a stream tube bounded by neighboring streamlines. In applications involving fluids, the forces within the fluid are often discussed more effectively in terms of fluid pressure, defined by:

$$\text{Fluid Pressure} = p \equiv \frac{Force}{Area} \tag{1.37}$$

The absolute pressure is any pressure for which the base for measurement is a complete vacuum, and is expressed in Pa (absolute). The gauge pressure is any pressure for which the base for measurement is atmospheric pressure expressed as Pa (gauge). Atmospheric pressure serves as a reference level for other types of pressure measurements, for example, gauge pressure. Absolute pressure is composed of the sum of the gauge pressure (positive or negative) and the atmospheric pressure. The pressure on a surface submerged at depth, h, in a fluid having density ρ (kg/m^3), is given by:

$$p = \rho g h \tag{1.38}$$

And, the force is expressed as:

$$F = \int_A p(h) \cdot dh$$

One of the fundamental laws of fluid mechanics is the *conservation of mass* (mass continuity). If the speed of a fluid and the stream tube cross-sectional area are, u and A, for the fluid confined in the stream tube, the mass flow per second is constant, expressed as:

$$\rho u A = Const. \tag{1.39}$$

In Equation (1.39) the left side represents the flow rate (in kg/s).

Example 1.8

What is the mass flow rate of water at a given section of a pipe of diameter 10 cm where the fluid velocity is 30 cm/s?

SOLUTION

Assuming the water density to be 1000 kg/m³, the mass flow rate is:

$$\dot{Q} = \rho \cdot u \cdot A = 1000 \cdot 0.3 \cdot \left(\frac{\pi}{4} 0.1^2 \right) = 2.355 \text{ kg/s}$$

In many practical applications, viscous forces are more negligible than forces due to gravity and pressure gradients over large areas of fluid flows. With this assumption, the equation of energy conservation in a fluid, the Bernoulli equation (theorem), for steady flow is of the form:

$$\frac{p}{\rho} + gz + \frac{1}{2}u^2 = Const. \tag{1.40}$$

For a stationary case, when $u = 0$ everywhere in the fluid, Equation (1.40) reduces to:

$$\frac{p}{\rho} + gz = Const. \tag{1.41}$$

Equation (1.41) is the equation of hydrostatic pressure, stating that the fluid at a given depth, z, is all at the same pressure, p. The significance of the Bernoulli equation is that it states that the pressure in a moving fluid decreases as the fluid speed increases.

Example 1.9

The atmospheric pressure on the surface of a lake is 10^5 N/m². Calculate the pressure at depth of 10 m, assuming that the water density is 10^3 kg/m³, and the water is stationary.

SOLUTION

From Equation (1.39), for acceleration due to the gravity 9.806 m/s², $z_1 = 0$, and $z_2 = 10$ m, then:

$$p_2 = p_{atm} - \rho g (z_2 - z_1) = 10^5 - (10^3)(9.806)(0 - 10) = 198060 \text{ N/m}^2$$

1.2.4 SI Units and Dimensional Analysis

The dimension of a variable is the kind of quantity that it is, having important significance in sciences and engineering fields. Fundamental dimensions are mass, length, time, temperature, and electric charge. It is important to distinguish between the quantity, for example the mass of an object m, and its dimension M, thus we can write $[m] = M$, to be read as: "the dimension of m is M." Two variables can have the same dimension, like, for example, the two sides, a and b, of a rectangle are both lengths, i.e., $[a] = L$, $[b] = L$, and so they can be added or subtracted; both a + b and a − b may have physical meanings. Variables that are of different dimensions cannot. It is meaningless for the length to be added to the area, or for a 10 km length to be added to a speed of 5 km/hr. On the other hand, two quantities of the same or different dimensions can be multiplied, and their product is that of the individual dimensions. Variables can also be nondimensional, e.g., the efficiency or coefficient of performance. All equations must be dimensionally homogeneous, i.e., all terms in the equation must have the same dimensions. One important reason for checking the equation homogeneity of dimensions is that inhomogeneity can indicate an error. However, a homogeneous equation can still be erroneous though not because of dimensionality. There is an important exception, quite common in engineering. Sometimes it seems that an empirically obtained equation is dimensionally inhomogeneous, but that may be because the dimensions are unknown or are suppressed. It is not enough to know a physical quantity dimension, in order to quantify, it also must be known how large it is. Thus every physical quantity is associated with a pure number, and corresponding units (except for dimensionless quantities). Though in principle we can use our own units, these units are only useful if they are understood by other people, though not necessarily by all. Fundamental units (listed in Table 1.2) are those from which all other units, so-called derived, are obtained.

   The quantification of units is arbitrary, as well as the choice of which units are fundamental and which are derived. For example, length can be a fundamental unit and area the derived, or vice versa. Dimensions and units should not be confused with each other; a variable can be described by different units that all have the same dimension. The fundamental units in Système International d'Unités (SI) are mass (kilogram, kg), length (meter, m), time (second, s), temperature difference (Kelvin, K), electrical current (ampere, A), luminous intensity (candela, cd), and moles (Mol). Other units can be derived from these using definitions or physical laws. Some units outside SI are also accepted for use with SI, such as minute (min), hour (hr.), day (d), degree of angle (°), minute of angle ('), second of angle ("), liter (L), and metric ton (t). To make comparisons of various quantities and to quantify the magnitude of physical quantities we need a good understanding of units, their definitions, and techniques of dimensional analysis. The International System (SI) of units is used throughout this book, but a comprehensive list of energy-related units, including the ones used in the United States, with conversions is also included. In engineering and science, dimensional analysis is the analysis of the relationships between different physical quantities by identifying

TABLE 1.2
SI Base Units

Base Quantity	Name	Symbol
Length	Meter	m
Mass	Kilogram	kg
Time	Second	s
Electric current	Ampere	A
Temperature	Kelvin	K
Amount of substance	Mole	Mol
Luminous intensity	Candela	cd

TABLE 1.3
Most Common Energy Units

Unit Name	Definition
Joule (J)	Work done by ace of 1 N acting through 1 m (also W-s)
Erg	Work done by 1 dyne force acting through 1 cm
Calorie (Cal)	Heat needed to raise the temperature of 1 g of water by 1°C
British thermal unit (BTU)	Heat needed to raise the temperature of 1 lb of water by 1°F
Kilowatt-hour (kWh)	Energy of 1 kW of power flowing for 1 hour
Quad	10^{15} BTU
Electron-Volt (eV)	Energy gained by an electron through 1 V potential difference
Foot-pound	Work done by 1 lb force acting through 1 ft
Megaton	Energy released when a million tons of TNT explodes

their fundamental dimensions (such as length, mass, time, and electric charge) and units of measure (such as miles vs. kilometers, or pounds vs. kilograms vs. grams) and tracking these dimensions as calculations or comparisons are performed. Converting from one dimensional unit to another is often somewhat complex.

Dimensional analysis, or more specifically the factor-label method, also known as the unit-factor method, is a widely used technique for performing such conversions using the rules of algebra. Any physically meaningful equation (and any inequality and in-equation) must have the same dimensions on the left and right sides. Checking this is a common application of performing dimensional analysis. Dimensional analysis is also routinely used as a check on the plausibility of derived equations and computations. It is generally used to categorize types of physical quantities and units based on their relationship to or dependence on other units. SI is founded on seven SI base units for seven base quantities assumed to be mutually independent, as given in Table 1.2. Other quantities, called derived quantities, are defined in terms of the seven base quantities via a system of quantity equations. The SI-derived units for these derived quantities are obtained from physics principles and equations and the seven SI base units. The fact that energy exists in many forms was one of the reasons that we have several units for this physical quantity. For example, for heat we have calories, British thermal units (BTUs), joules, ergs, and foot-pound for mechanical energy, kilowatt-hours (kWh) for electrical energy, and electron-Volts (eV) for nuclear and atomic energy. However, because all describe the same fundamental quantity, there are conversion relationships or factors relating them. Table 1.3 lists some of the most common energy units.

Example 1.10

Show that physical dimensions of the expression of hydrostatic pressure, $p = \rho gh$, are consistent with physical dimensions of pressure, fluid density, acceleration due to gravity, and height.

SOLUTION

Replacing the individual symbols in the equation of the hydrostatic pressure in terms of their fundamental physical units, we have:

$$p = \rho gh$$
$$[p]_{SI} = \left(kg \cdot m^{-3}\right)\left(ms^{-2}\right)(m) = \left(kg \cdot m \cdot s^{-2}\right)\left(m^{-2}\right) = N \cdot m^{-2}$$

The hydrostatic pressure, $p = \rho gh$, is consistent with pressure physical dimensions, force per unit of area.

It is clear then that physical quantities that have different dimensions cannot be converted from one to the other. In particular, for instance, weight, which is a force (normally gravitational) with which one body attracts another, cannot be converted to mass, though the two are frequently confused. Unfortunately in today's engineering and applied sciences areas, many power and energy units sound similar, so proper care must be taken when using them. The terms energy and power are quite often used informally as they are synonymous (e.g., electrical energy or power, wind energy, or wind power). It is important to have a clear distinction and understanding of these terms. The power unit is Watt (W). The imperial unit, still in use in the United States, is the British thermal unit (Btu, 1 Btu = 1055 J). However, the Btu is commonly used in thermal processes to designate thermal energy. For a system to produce or consume power, there must be forceful motion of some of its components. In practice, it is often convenient to express the energy in terms of power used for a specific period of time. For example, if the power of an electric motor is 1 KW, and the motor is running for 1 hour, the energy consumed is 1 kilowatt-hour (kWh). All the systems used for the performance of several tasks desired by the human society generate work and consume energy resources. Energy is also often measured simply in terms of fuel quantities used, such as tons of coal or oil. The motion of a body is a type of work. Heat represents another form of energy. Wind or running water is able to move the blades of a rotor. Similarly, sunlight can be converted into heat, being another form of energy. Energy transfer refers to movements of energy between systems that are closed to transfers of matter. The portion of the energy that is transferred by conservative forces over a distance is measured as the work the source system does on the receiving system. The portion of the energy that does not do work doing during the transfer is called heat. In other words, heat or thermal energy is a form of energy that is transferred from materials or systems at higher temperatures to ones at lower temperatures. It is usually produced by the combustion of fuels, which are regarded as energy sources.

1.3 SUSTAINABLE ENERGY AND ENERGY ECONOMICS

If the current level of consumption of fossil fuel continues, the entire fossil energy reserve sooner or later may be depleted, so the transition to renewable energy is inevitable. Currently, the utilization of renewable energy is still a small portion of total energy consumption. On the other hand, although currently the cost of solar or other renewable-energy-based electricity is higher than that from fossil fuels, its technology is constantly being improved and the cost is constantly being reduced. Inevitably, fossil fuels eventually will be replaced by solar energy. This is simply a geological fact: The total recoverable reserve of crude oil is finite. Energy conversion, the type that is concerned with converting solar radiant energy to electrical, chemical, or kinetic energy, is a key topic in this book, so we started with a review of the basics of energy, power, and radiant intensity unit definitions. Because of the limited reserve of fossil fuel and the cost, from the beginning of the industrial age, or even before, renewable energy resources have been explored and used. Although solar energy is by far the largest resource of renewable energy, other renewable energy resources, including hydropower, wind power, and shallow and deep geothermal energy, have been extensively utilized. Except for deep geothermal energy, all are derived from solar energy. When we examine the power production potential of wind, solar, and other renewable resources, we observe a number of important phenomena: (1) Renewable energy sources can introduce a high degree of dynamic (in time) variability in the power supply, resulting from wind variations and passing clouds that reduce the intensity of the solar radiation reaching the solar panels, (2) Temporal variations have multiple time scales, introducing higher-frequency fluctuations with time scales measured in seconds or minutes and diurnal variations that take place over the day, and (3) There is a strong incentive to include energy storage elements in the overall design to reduce power fluctuations and to allow for peak power delivery at demand peaks. To compensate for the variations introduced by solar and wind power sources, some type of an energy storage system is used to help match power production levels to demand, helping to compensate for the difference between production and demand, if any. The increasing use of renewable power sources complicates the energy picture and requires the integration of a number of energy generation, conversion, and storage technologies.

Among the renewable energy sources, solar energy, wind energy, and geothermal energy, as well as biofuels and biomass, are the most suitable to be integrated into buildings and industrial facilities, or to directly provide energy for industrial processes. The estimate of worldwide available wind power varies, with a conservative estimate that the total available wind power, 75 TW, is more than five times the world's total energy consumption. However, there are still technological and other issues on the full use of wind energy. The utilization of wind power has been widespread since medieval times. Windmills were used in the rural United States to power irrigation pumps and drive small electric generators used to charge batteries that provided electricity during the last century. A windmill or wind turbine converts the kinetic energy of moving air into mechanical motion, usually in the form of a rotating shaft. This mechanical motion can be used to drive a pump or to generate electric power. The energy content of the wind increases with the third power of the wind velocity, and wind power installations are economical in regions where winds of sufficient strength and regularity occur. Wind energy technology has progressed significantly over the last two decades. Over the many thousands of years of human history, until the Industrial Revolution when fossil fuels began to be used, the direct use of biomass was the main source of energy. Wood, straw, and animal wastes were used for space heating and cooking. Candle (made of whale fat) and vegetable oil were used for light. Currently there is a well-established industry to generate liquid fuel using biomass for transportation. By definition, geothermal energy is the extraction of energy stored in Earth. However, there are two distinct types of geothermal energy depending on its origin: shallow and deep geothermal energy. Shallow geothermal energy is the solar energy stored in Earth, the origin of which is described later in this book. The solar energy stored in Earth is universal and of very large quantity. The temperature is typically some 10°C off that of the surface. The major application of shallow geothermal energy is to enhance the efficiency of electrical heaters and coolers (air conditioner) by using a vapor compression heat pump or refrigerator. Deep geothermal energy is the heat stored in the core and mantel of Earth. The temperature could be hundreds of degrees Celsius. It can be used for generating electricity and large-scale space heating, as is discussed in a later book chapter.

1.3.1 Energy Economics and Cost Analysis

The design, construction, and implementation of any engineering system are ultimately decided by economic considerations of cost and economic analysis. Economic analysis of investments, including any engineering project, is one of the critical steps leading to a decision. Therefore, making informed decisions on the design and implementation of energy systems or projects requires comprehensive economic, life cycle, and life cost assessment and analysis. System environmental impacts, resource sustainability, tax credits, and incentives play key roles in economic analysis of any renewable energy development. Several methods are available for the evaluation of investments and project viability (Blank and Tarquin, 2012). In all engineering projects, including energy projects, it is necessary to justify the project development, implementation, and/or installation of new equipment. In weighing the various alternatives, the economic analysis problems can be classified as fixed input type, fixed output type, or situations where neither input nor output is fixed. Whatever the nature of the problem, the proper economic criteria are to optimize the benefit-cost ratio. Keep in mind that the lifetime of any energy project typically spans a period from 20 to 60 years. Therefore, it is important to compare savings and expenditures of various amounts of money properly over the project or development lifetime. In engineering economics, savings and expenditures of amounts of money during the project are usually called cash flows. To perform a sound economic analysis of a renewable energy system (RES) or a distributed generation (DG) project the most important economic parameters, data and project technical concepts must be well-defined and known. The important parameters and concepts that significantly affect the economic decision making on the project and development include:

1. The time value of money and interest rates including simple and compound interests;
2. Inflation rate and composite interest rate;
3. Taxes including sales, local, state, and federal tax charges;
4. Depreciation rate and salvage value;
5. Tax credits and incentives, local, state and federal.

In any project's financing, the two major costs are investment costs and operating costs. The investment cost relates to the fixed costs of materials and installation to deliver an energy system to the client or stakeholders. The client has to make this payment only once, and the system will last for decades. The price of both power and energy is measured in $/MWh, and because capacity is a flow like power and measured in MW, like power, it is priced like power, in $/MWh. Many find this confusing, but an examination of screening curves shows that this is traditional (as well as necessary). Because fixed costs are mainly the cost of capacity, they are measured in $/MWh and can be added to variable costs to find total cost in $/MWh. When generation cost data are presented, capacity cost is usually stated in $/kW. This is the cost of the flow of capacity produced by a generator over its lifetime, so the true (but unstated) units are $/kW-lifetime. Confusion over units causes too many different units to be used, and this requires unnecessary and sometimes impossible conversions. This chapter shows how to make almost all relevant economic calculations by expressing almost all prices and costs in dollars per megawatt-hour ($/MWh). Energy is measured in MWh, while power and capacity are measured in MW. All three are priced in $/MWh, as are fixed and variable costs. Other units with the same dimensions (money divided by energy) may be used, but this book will use only $/MWh. Power is the flow of energy and is measured in watts (W), kilowatts (kW), megawatts (MW), or gigawatts (GW). Energy is an accumulation of power over a period of time. For instance, a kilowatt flowing for 1 hour delivers a kilowatt-hour (kWh) of energy. The price of both energy and power is expressed in $/MWh.

Capacity is the potential to deliver power and is measured in megawatts. Like power, it is a flow concept. Power is the rate of flow of energy. This is true for any form of energy, not just electricity. If you wish to boil a cup of water you need a quantity of energy to get the job done, about 30 watt-hours. Any specific power level, say 1000 watts (kilowatt, or kW), may or may not make you a cup of tea depending on how long the power continues to flow. A typical microwave oven delivers power at a rate of about 1 kW (not 1 kW per hour). If it heats your water for 1 second, the water will receive power at the rate of 1000 watts, but it will gain very little energy and it will not make tea. Two minutes in the microwave will deliver the necessary energy, 1/30 of a kWh. Confusion arises because it is more common to have the time unit in the measurement of a flow than in the measurement of a quantity. Thus if you want to fill your gas tank, you buy a quantity of 15 gallons of gasoline, and that flows into your tank at the rate of 5 gallons per minute. But if you need a quantity of electric energy, that would be 30 watt-hours and it would be delivered at the rate of 1000 watts. Because a watt-hour is a unit of energy, it would make sense to speak of delivering 1000 watt-hours per hour, but that just boils down to a rate of 1000 Watts (1 kW) because a watt-hour per hour means watts times hours divided by hours, and the hours cancel out.

Example 1.11

Units—kilowatts, hours, and dollars—follow the normal laws of arithmetic. But it must be understood that a kWh means a (kW × h) and a $ per hour means a ($/h). Also note that "8760 hours per year" has the value of 1, because it equals (8760 h)/(1 year), and (8760 h) = (1 year). As an example:

$$\$100/\text{kWyr} = \frac{\$100}{\text{kW} \times \text{year}} \times \frac{1000 \text{ kW}}{1 \text{ MW}} \times \frac{1 \text{ year}}{8760} = \$11.42/\text{MW}$$

In any energy project, a payment of interest is expected for the invested capital, which depends on the economic factors and risks. The energy unit cost estimate without considering return is relatively simple. All costs over the generation system's lifetime are added (the total cost) and divided by the operating period in order to calculate the annual cost, and then by diving the annual cost by the annual generated energy, the cost per unit of energy is obtained. If an energy generation system produces annually, E_{anl}, expressed in kWh with a total annual cost C_{anl}, then the *specific energy cost*, SE_{Cost} is calculated by:

$$SE_{Cost} = \frac{C_{anl}}{E_{anl}} \ (\$/kWh) \tag{1.42}$$

The annual cost is calculated by dividing the total cost C_{total} by N, the number of the power plant or energy generation system operating years, the plant, or system lifespan. The total cost includes initial investment, INV_0, and all payments, P_k, for every operational year, k, for the entire lifespan of the system, N. For the power plant or generation system operating period of N years, the annual cost is calculated by:

$$C_{anl} = \frac{C_{total}}{N} = \frac{INV_0 + \sum_{k=1}^{N} P_k}{N} \tag{1.43}$$

The annual cost is also called the leveled electricity cost (LEC) in the case of power systems and the leveled heat cost (LHC) in the case of thermal or heating systems. Before discussing other issues, it is prudent to clarify the difference between *price* and *cost* of a product, often mistakenly used as synonyms. The product price is determined by the supply and demand for the product. The price of a product consists of its cost (production), the profit, taxes, installation, transportation, and maintenance costs, from which are subtracted incentives and tax credits (if any), with the resulting amount perhaps corrected by a scarcity factor.

Example 1.12

A 1.2 MW wind energy conversion system is generating an average of 2.85 millions of kWh per year. The investment cost was $1450/kW for the installation, and an average annual cost of 2.5% of the investment cost for annual operation costs for this wind turbine. Assuming the wind turbine lifespan of 20 years, calculate the specific energy cost.

SOLUTION

The total cost of the wind energy system is:

$$C_{total} = 1.2 \times 10^3 \cdot 1650 \cdot (1 + 20 \times 0.025) = \$2{,}610{,}000$$

The specific energy cost, computed with Equation (1.42) is:

$$SE_{Cost} = \frac{C_{total}}{20 \times 2.85 \times 10^6} = \frac{2.61 \times 10^4}{5.7 \times 10^7} = 0.045789 \ \$/kWh \approx 4.6 \ Cent/KWh$$

1.3.2 Basic Concepts, Definitions, and Approaches

Before proceeding to the discussion of the economic and cost analyses and methods, it is important to provide key definitions that are used in this field. These definitions have been adapted from the literature, and interested readers are directed to the reference and suggested readings of the chapter for additional information. The definitions of the most commonly used concepts and definitions in the fields of economics, cost analysis, and management, which are critical in the decision-making process for energy projects, are discussed here. The term *interest* can be defined as the money paid for the use of money. It is also referred to as the value or worth of money. Two important terms are *simple interest* and *compound interest.* Simple interest is always computed on the original principal (the borrowed money). Unlike simple interest, with compound interest, interest is added periodically to the original principal. The terms conversion or compounding of interest simply refer to the addition of interest to the principal. The interest (conversion) period in compound interest calculations is the time interval between successive conversions of the interest, while the interest period is the ratio of the stated annual rate to the number of interest periods in 1 year. The *present worth* is the current value of an amount of money due at a later time and the effect of the applied interest. *Average cost* represents the total of all fixed and variable costs calculated over a period of time, usually 1 year, divided by the total number of units produced, while the *average revenue* is the total revenue over a period of time, usually 1 year, divided by the total number of units produced. Average profit represents the difference between average revenue and average cost. *Fixed costs* are all costs not affected by the level of business activity or production level, e.g., rents, insurance, property taxes, administrative salaries, and the interest on borrowed capital. *Life cycle cost (LCC)* is the sum of all fixed and variable costs of a project from its initiation to its end of life. Life cycle costs also include planning costs, abandonment, disposal, or storage costs. *Marginal or incremental cost* is the cost associated with the production of one additional output unit (e.g., product or energy unit), while *marginal or incremental revenue* is the revenue resulting from the production of one additional output unit. The opportunity to use scarce resources, such as capital, to achieve monetary/financial advantage and the associated costs are the opportunity costs, e.g., an opportunity cost to building a new power plant for a company is not to build and invest its capital in 7%-interest-bearing securities. *Time horizon* is the time (expressed usually in years) from the project initiation to its end, including any disposal or storage of equipment and products. *Variable costs* are costs associated with the level of business activity, production, or output levels, such as fuel cost, materials cost, labor cost, distribution cost, and storage cost. The variable costs increase monotonically with the number of units produced. *Project term* is the planning horizon over which the project cash flow is assessed, usually divided by a specific number of years. Initial cost is the one-time expense occurring at the beginning of the project investment, such as purchasing major assets for an energy project. *Annuity* represents the annual increment of the project cash flows, as opposed to the one-time quantity, such as the initial cost. Annuities can be either positive (e.g., selling energy annual revenues) or negative (e.g., operating and maintenance [O&M] annual costs). *Salvage value*, usually very small compared to the initial cost, is the one-time positive cash flow at the project planning horizon end, consisting of the assets, equipment, buildings, or business sold at end of the project in actual condition. In order to gain a true perspective as to the economic value of renewable energy systems and projects, it is necessary to compare the system technologies to conventional energy technologies on the LCC basis. The LCC method allows the total system cost calculation during a specific time period, usually the system lifespan considering not only the initial investment but also the costs incurred during the useful system life or a specific period. The LCC is the *present value* life cycle cost of the initial investment cost and the long-term costs related to repair, operation, maintenance, transport to the site, and fuel used to run the system. *Present value* is the calculation of expenses that are realized in the future but applied in the present. An LCC analysis gives the total system cost, including all expenses incurred over the system lifetime. The main reasons for an LCC analysis are to compare different energy technologies, and to determine the most cost-effective system designs. For some renewable energy applications, there are no other options to such systems to produce electricity either because there is no power or the conventional energy methods are too expensive. For these applications, the initial cost of the system, the infrastructure to operate and maintain the system, and the price people pay for the energy are the main concerns. An LCC analysis allows the designer to study the effect of using different equipment and components with different characteristics, performances, and lifetimes. The common LCC relationship applicable for energy projects is:

$$LCC = INV0 + \sum OM_t + \sum FL_t + \sum LRC_t - \sum SV_{sys+parts} \qquad (1.44)$$

Here, *INV0* is the initial overall installation costs, consisting of the present value of the capital that will be used to pay for the equipment, system design, engineering, and installation (the initial cost incurred by the user), the $\sum OM_t$, represents the sum of all yearly O&M (operation and mainte- nance) costs, the present value of expenses due to operation and maintenance programs (O&M costs include the salary of the operator, site access, guarantees, and maintenance), the $\sum FL_t$ is the energy cost, sum of all yearly fuel costs, an expense that is the cost of fuel consumed by the conventional power or auxiliary equipment (the transport fuel cost to site must be included), the $\sum LRC_t$ is the sum of all yearly replacement costs, the present value of the cost of replacement parts anticipated over the life of the system, and the $\sum SV_{sys+parts}$, is salvage value, the net worth at the end of the final year, typically up to 10% for the energy equipment. Future costs must be discounted because of the time value of money, so the present worth is calculated for costs for each year. The RES lifespans are assumed to be in the range of 20 to 30 years. LCC analysis is the best way of making acquisition decisions. On the LCC analysis, many renewable energy systems are economically viable. Financial evaluations can be done on a yearly basis to obtain cash flow, breakeven point, and payback time, discussed later. Notice that social, environmental, and reliability factors are not included here, but if they are deemed important they can be included.

Example 1.13

A small hybrid power system (1kW wind turbine, PV array, and battery bank) was installed at a remote weather station, with the following costs: installed cost of $36,000, the loan for 10 years, with the total interest paid, minus the tax credit of $22,500, the total operation and maintenance of 2.5% of the initial cost, or $900 per year, and the total replacement cost of 3.5% of the invest- ment. If the system salvage value is estimated at 7.5% of the initial cost, and no fuel is used for the system operation, calculate the LCC for the investment period.

SOLUTION

Applying Equation (1.44), the LCC for the hybrid power system is:

$$LCC = 36000 + 10 \times 900 + 0 + 2700 - 1260 = \$46,440.00$$

Capital budgeting is defined as a process in which a business determines whether projects such as building a new power plant, developing a wind energy system, or investing in a long-term venture are worth pursuing. A *capital investment* is expenditure by an organization in equipment, land, or other assets that are used to carry out the objectives of the organization. Most of the time, a prospec- tive project's lifetime cash inflow and outflow are assessed in order to determine whether the returns generated meet a sufficient target benchmark. Capital budgeting, an essential managerial tool, is also known as *investment appraisal.* Project proposals that scale through the preliminary screening and evaluation phase are further subjected to rigorous financial appraisal to ascertain if they would add value to the organization. This stage is also referred to as *quantitative analysis, economic and financial appraisal, project evaluation,* or simply *project analysis.* The financial appraisal of the project may predict the expected future cash flows of the project, analyze the risk associated with those cash flows, develop alternative cash flow forecasts, examine the sensitivity of the results to possible changes in the predicted cash flows, subject the cash flows to simulation, and prepare alter- native estimates of the project's net present value For the clarification and good understanding of the economic evaluation and analysis of renewable energy sources, it is necessary to define basic terms and concepts used in common practice.

The *economic value* is the asset or equipment value expressed through money. Different experts or economic schools explain it differently. There are two basic approaches, the *subjective* and *objec- tive* understanding of the value. Subjective understanding of economic value is based on individual preferences of an individual, whereas objective understanding is the relationship between preferences (individual and collective) and the cost of meeting the needs. *Utility (use value)* represents the ability of an asset, equipment, installation, service, or product to meet specific needs. *Non-use value* or *passive-use value* is the utility of good for others (subjective economics). *Environmental (internal) value* is the result of the belief that nature has a positive value for the environment independently of human preferences and direct benefit to humankind. *Discounting* is a concept used to evaluate the present (the costs and benefits) higher than the future (costs and benefits), so there is decline in the value. Discounting relies on the premise that the value of money declines over time, therefore future values should be discounted relative to the present. Two important parameters relate to discounted cash flows: *Interest rate* is the investment percent return, or percent charged, on the amount of money borrowed at the beginning of the time horizon. Usually interest is compounded at the end of each year, and the 1-year unit is referred to as the *compounding period*. The *minimum attractive rate of return* (MARR) represents the minimum interest rate required for project returns to make the project financially attractive, as set by the business, government office, or other entities and organizations that are making decisions about the project and investment. Three major reasons for demanding the MARR on the foregoing present value of an amount of money invested are: inflation (reducing the future money value), the possible investment alternatives (even if there are incentives for energy projects, still the investors prefer the ones with high returns), and, most importantly, the risks associated with any investment. Notice that when the MARR equals the inflation rate, the real dollar return is zero, but there are no actual value losses. The *nominal discount rate* is a summary rate for investment of capital that includes inflation. The *real discount rate* is the net discount rate, a nominal rate minus the inflation rate.

In the decision-making process, a short list of ideas/alternatives is identified and selected, then a detailed economic evaluation for the projects in the short list of the project alternatives is performed and is critical for the project viability. This evaluation takes into account not only the economic aspects of the chosen alternatives and determines the profitability of the projects, but also other issues, which may not be quantified and do not affect materially the cash flow and the profitability of the project, the so-called intangible items, such as public good, national security, environmental issues, and so on. However, the project economic analysis treats the projects strictly as investments, and the final decisions are based on project profitability. The critical concept of the time value of money is based on the premise that one monetary unit today is worth more than the same monetary unit 1 year from now, the latter is worth more 2 years from now, and so on. The time value of monetary funds is intricately related to the concepts of capital return, stipulating that capital invested must yield more capital at the investment period end, the *interest (discount) rate*, r, the percentage of additional funds that must be earned for the lending of capital, and the current and expected future inflation, are increasing the cost of goods in the future. *Money time value* is an important parameter in all economic calculations. However, there are situations, such as short-lifespan investments, where the discount impact is quite limited, making it reasonable to use only the value monetary amount, ignoring the discounting adjustments. When capital investments, such as energy production or conservation investments, are appraised, there are inherent risks associated with any investment and all or part of the capital may be lost. *Investment risks* are one of the justifications for charging an interest rate and the expectation of higher return on invested capital. The higher the investment risk, the higher would be the expected return on the capital. Two concepts often used in economic analysis are the *simple payback* and the *capital recovery factor* (CRF). In simple payback, the value known as the *net present value* (NPV) is computed by summing all incoming and outgoing cash flows, such as initial costs, annuities, and salvage value amounts. If the NPV is positive, then the project is economically viable. In the case of a multiyear project with positive simple payback, the break-even point (BEP) is the year where the project total annuities equal the initial costs, which have been paid back at this point. CRF is a parameter that is used to measure the relationship between cash flows and investment costs, usually applied for short-term investments, up to 10 years. CRF is the ratio of the *annual capital cost* (ACC) to the NPV, estimated for a period of N years:

$$CRF = \frac{ACC}{NPV} \qquad (1.45a)$$

and

$$ACC = \text{Annuity} - \frac{NPV}{N} \qquad (1.45b)$$

Example 1.14

A utility is making an investment of $75 million to improve its transmission capacity and reduce its losses, over a period of 10 years with an annuity of $15 million. What is the CRF in this case?

SOLUTION

The investment net present value is the difference between the total 8-year period annuity and $75 million:

$$NPV = -75 + N \times Annuity = -75 + 10 \times 15 = 75 \text{ million}$$

By using Equation (1.45b) and (1.44a) the ACC and CRF are:

$$ACC = Annuity - \frac{NPV}{N} = 15 - \frac{75}{10} = 7.5 \text{ million}$$

and

$$CRF = \frac{7.5}{75} = 0.1 \text{ or } 10\%$$

This value is less than the maximum 15% CRF, as recommended by the U.S. Electric Power Research Institute (EPRI) for power and energy industries.

Factors affecting the evaluation of natural resources are the amount of expected future benefits from the use of resources, and the time factor. Time factor (discounting) corresponds to the idea that economic analysis is based on the fact that value falls over time. A positive discount rate expresses the rate of decline of economic indicators over time. Discounting is a normal part of the economic efficiency evaluation. Reasons for positive discount rates include preference for current benefits against future ones, and capital productivity (the expectation that the preference of investment instead of immediate consumption results in future higher consumption). In some cases, it is appropriate to use a zero discount rate. Assumptions of discounting use the facts that all incomes during the certain period of investment will be invested, and the future value of the evaluated good is decreasing (e.g., lower quality or utility), or its amount will rise. Rules for the investment process include: the marginal productivity of capital is higher than the marginal productivity of time (the income of last unit of input does not fall below the value of time preference), and the nominal discount rate is higher than the inflation rate. In general, when the electricity price produced from renewable energy sources is similar or less than the price produced from conventional energy sources, the economic considerations favor the development of more geothermal units, solar power plants, wind energy parks, etc. A combination of rising fossil fuel prices, incentives, and a favorable regulatory environment that provides tax credits, investment guaranties, and accelerated depreciation changes the economic and financial circumstances for alternative energy.

Requirements for relevant information and analysis of capital budgeting decisions taken by management have paved the way for a series of models to help organizations to amass the best of the allocated resources. The most common methods of capital budgeting techniques include: *the payback period, the net present value, the internal rate of return* (IRR), *and the real options approach*. The payback period method of financial appraisal is used to evaluate capital projects and to calculate the return per year from the start of the project until the accumulated returns are equal to the cost of the investment at which time the investment is said to have been paid back. The time taken to achieve this payback is referred to as the *payback period*. The payback decision rule states that acceptable projects must have less than some maximum payback period designated by management. By definition, the payback method takes into account project returns only up to the payback period. It is therefore important to use the payback method more as a measure of project liquidity rather than project profitability. The discounted payback period method was proposed as an improved measure of liquidity and project time risk over the conventional payback method, not a substitute for profitability measurement. It still ignores returns after the payback period, stating that the proper role for discounted payback period analysis is as a supplement to profitability measures, thus highlighting the supportive nature of the payback method, whether conventional or discounted payback period. The net present value is defined as the difference between the present value of the cost inflows and the present value of the cash outflows. In other words, a project's net present value, usually computed as of the time of the initial investment, is the present value of the project's cash flows from operations and disinvestment less the amount of the initial investment. NPV is used in capital budgeting to analyze the profitability of an investment or project and it is sensitive to the reliability of future cash flows that the investment or project will yield. For instance, the NPV compares the value of the dollar today to the value of that same dollar in the future taking into account inflation and returns. The internal rate of return is the discount rate often used in capital budgeting that makes the NPV of all cash flows from a certain project equal to zero. This in essence means that IRR is the rate of return that makes the sum of present value of future cash flows and the final market value of a project (or investment) equal its current market value. The higher a project's internal rate of return, the more desirable it is to undertake the project. Internal rate of return is the flip side of the NPV, where NPV is a discounted value of a stream of cash flows, generated from investment. IRR computes the break-even rate of return showing the discount rate. The real options approach applies financial options theory to real investments, such as manufacturing plants, line extensions, and research and development investments. This approach provides important insights about business and strategic investments, which are very vital given the rapid pace of economic change. A financial option gives the owner the right, but not the obligation, to buy or sell a security at a given price, e.g., companies that make strategic investments have the right but not the obligation to exploit these opportunities in the future.

1.3.3 CRITICAL PARAMETERS AND INDICATORS

A capital investment is an expenditure made by an organization in equipment, land, or other assets that are used to carry out the objectives of the organization. There are several that affect decisions between various investment alternatives, and knowledge and understanding of their significance is critical for a good business decision. Fundamental to finance over long spans of time of any project or the project lifetime is the time value of money. The present value (PV) is the worth of an asset, money, or cash flows in today's dollars when rate of return is specified. The *money future value* (FV) is the asset worth or cash flow, being evaluated in today's dollars. It should make sense that *fuel costs* (FC) or *fuel savings* (FS) have values follow as the annual (or other time interval) cash flows. A *discount rate* is sometimes referred to as a *hurdle rate, interest rate, cutoff rate, benchmark*, or the *cost of capital*. Many companies and organizations have a fixed discount rate for all projects. However, if a project has a higher level of risk, a higher discount rate commensurate with that risk must be used. When cash-flow-related consequences occur in a very short duration, the income and expenditures are added and the net cash balance can be calculated. When the time span is longer, the effect of interest on the investment needs to be calculated. An economy with low interest rates is encouraging money borrowing for investments and projects, whereas higher interest rates are encouraging money savings. Usually most RES and DG projects require higher initial investments, so interest rates are very important here. Therefore, if the money is borrowed for such projects the interest rates are good indicators of whether the projects are cost-effective. The capital borrowed to cover the partial or full initial cost of a RES or DG project or development, a fee, the *interest* (*I*) is charged for the *principal* (*P*), the borrowed money. Interest charges are expressed as a percentage of the total amount, the principal, *the interest (discount) rate (r)*, as:

$$r \text{ (interest/discount rate)} = \frac{P}{I} \tag{1.46}$$

When considering projects with a useful life of several years, the time value of money has to be taken into account. There are two ways to calculate the total interest charges over a project lifetime. In *simple interest charges*, the total interest fee, *I*, is paid at the loan end, *N*, and is proportional with the interest rate, *r*, and the lifetime, *N*, while both are expressed in the same unit (e.g., 1 year, 1 month), as:

$$I = N \cdot r \cdot P \tag{1.47}$$

The total amount paid, *TP*, due at the end of the loan period, includes the principal and the interest charge:

$$TP = P + I = P \cdot (1 + N \cdot r) \tag{1.48}$$

In the second method, the *compound interest charges*, the loan lifetime (period) is divided into smaller periods, the *interest period* (usually 1 year), and the interest fee is charged at the end of each interest period and accumulates from one interest period to the next. Therefore, the total payment at the end of an interest period is computed with Equation (1.48), *P* being replaced by total payment at the previous interest period end. If the principal is *P*, the total amount due at the loan period end is then expressed by:

$$TP = P \cdot (1 + r)^N \tag{1.49}$$

Equation (1.49) states that in the second method, the amount of loan payment increases exponentially with *N*, if the interest charges follow the law of compound interest. For most RES, DG, or energy efficiency and conservation projects, the interest rates are usually constant throughout the project lifetime; otherwise, a common practice in economic analysis is to use average interest rates.

Example 1.15

A wind energy developer decided to invest in a wind project to borrow a loan of $2,500,000. One bank offered a 10-year loan with 5% compound interest, while a second bank offered the same 10-year loan with a 6.5% simple fixed interest rate. Which is the more advantageous loan for the wind developer?

SOLUTION

If the 6.5% simple interest is paid over a 10-year period, the total amount, Equation (1.48), is:

$$TP_{fix-rate} = 2500000 \cdot (1 + 10 \cdot 0.065) = 4.125 \times 10^6 \text{ USD}$$

In the case of the 10-year loan with 5% compound interest rate, the total amount, Equation (1.49), is then:

$$TP = 2500000 \cdot (1+0.05)^{10} = 4.07223 \times 10^6 \text{ USD}$$

The loan with 5% compound interest is more advantageous.

The process of calculating the future value of money (future cash flow) in the present value is called *present worth*. The present worth, *PW*, is calculated from Equation (1.49), by replacing *TP* with *FV*, and solving for *PW* parameter:

$$PW = FV \cdot \left(\frac{1}{(1+i)^N} \right) \tag{1.50}$$

where *FV* is the money value or the cost expected at time, after *N* years (future value) and *i* is the interest (discount) rate. If *FV*, *N*, and the discount rate, *i*, are known, then the *PW* can be calculated.

Example 1.16

Assuming an interest rate of 7.5%, what is the present value of $100,000 that will be received 5 years from today?

SOLUTION

By using Equation (1.50) for the above values, the present value is:

$$PW = \frac{100000}{(1+0.075)^5} = 69{,}655.86 \text{ USD}$$

Inflation is occurring when the good and service costs increase from one period to the next; the interest rate defines the cost of money, the inflation rate, r_{ifl} expresses the cost increase of goods and services. Therefore, a future cost of a commodity or asset, *FC*, is higher than its present cost, *PC*. The inflation rate is usually considered constant over an energy project's life, similar to the interest rate. Notice also that if the cost of energy changes, especially if it increases (the so-called energy cost escalation), this important factor must be taken into consideration for any energy project (development) economic and cost analysis. The future commodity or asset cost, *FC*, estimated from the present commodity or asset cost, *PC*, and the future value (worth), *FV*, calculated from the present value (worth), *PW*, considering the combined effect of the interest rate, *r*, and the inflation rate, *i*, over the energy project (development and operation) period, *N*, is expressed by these relationships:

$$FC = PC(1+i) \tag{1.51}$$

And, respectively for the future value:

$$FV = PW \left(1 + \frac{r-i}{1+i} \right)^N \tag{1.52}$$

Example 1.17

A residential complex owner decided to invest $350,000 in a building-integrated PV system. The money is borrowed through a 10-year loan with an interest rate of 6.5%, and the inflation experienced by the economy is 2.75%. Determine the future value and the total cost of the investment.

SOLUTION

Applying Equations (1.51) and (1.48), the future value and the total investment cost are:

$$FV = 350000 \cdot \left(1 + \frac{0.065 - 0.0275}{1 + 0.0275}\right)^{10} = \$350,475.83$$

and

$$TP = 350000 \cdot (1 + 0.065)^{10} = \$656,998.11$$

An important parameter and indicator for the annual cash flow analysis is *future worth* (FV). Supposing that an amount A_{dep} is deposited at the end of every year for N years, and if r is the loan interest rate compounded annually, then the future worth amount FV is given by this relationship:

$$FV = A_{dep} \cdot \left[\frac{(1+r)^N - 1}{r}\right] \tag{1.53a}$$

Or for an annuity stream, and certain time horizon, the future value, *FVA,* of the annuity at the end of the Nth year is then given by:

$$FVA = A \cdot \left[\frac{(1+r)^N - 1}{r}\right] \tag{1.53b}$$

From an investment point of view, the translation of the future money to its present value, the discounted cash flow analysis, is important, and for N equal annual amount (annuity) A, and a discount rate, i, is calculated as:

$$V_P = A \cdot \left(\frac{(1+r)^N - 1}{r \cdot (1+r)^N}\right) \tag{1.54}$$

Equations (1.49) to (1.54) assume a constant (fixed) annuity value; a nonconstant (irregular) annuity case is considered below. If some parameters are known in these relationships, the others can be computed directly. In the case of nonconstant annuities, the present worth value, Equation (1.50), by treating each annuity as a single payment to be discounted from the future to the present, changes to:

$$PW = \sum_{k=1}^{N} \frac{A_k}{(1+r)^k} \tag{1.55}$$

Here A_k is the yearly predicted annuity in year, k, from 1 to N (the last project year). Predicted costs vary from year to year, they are predicted to the net present value in a similar way. If the capital cost is subtracted from the present value of the revenue, the *net present value* (NPV) is obtained, as:

$$NPV = V_P - C_{Capital} \tag{1.56}$$

From the above equation, the *rate of return* (ROR) can be computed. The rate of return, r_{rt}, is the discount rate that is making the net present value zero, and is found by solving numerically the equation:

$$0 = V_P - C_{capital} \implies C_{capital} = A \frac{1 - (1 + r_{rt})^{-N}}{r_{rt}} \tag{1.57}$$

This equation is usually solved numerically, e.g., Newton-Raphson iteration; however, an approximate value of the rate of return can be obtained by trial-and-error approach. The approximate method consists of finding two r_{rt} values, for which the NPV is slightly negative and slightly positive, and then by linear interpolation between these two values, the value of the rate of return is determined. It is important to keep in mind that a solution of this ROR may not exist. Once r_{rt} value is computed for any project alternative, the actual market discount rate or the minimum acceptable rate of return is compared to the found ROR value, and if ROR is larger, the project is cost-effective. From the determined discount rate (ROR), making the NPV zero, the *annual energy cost*, A_{cost}, can be computed by:

$$A_{cost} = \frac{C_{capital} \times r_{rt}}{1 - (1 + r_{rt})^{-N}} \qquad (1.58)$$

The cost of energy unit (specific energy), C_{eng}, considering the return rate computed by dividing A_{cost} by the annual generated energy, E_{anl}, is defined as:

$$C_{eng} = \frac{A_{cost}}{E_{anl}} \qquad (1.59)$$

When electric equipment, DG and RES installations and systems, such as PV systems, wind turbines, energy storage units, or transformers used in a wind farm are purchased, there are several associated component costs. These include the equipment capital cost, installation cost, maintenance cost, salvage value, and the annual returns. In addition, other factors related to income tax, depreciation, tax credit, property tax, and insurance must be taken into account in order to evaluate the project economic viability. The maintenance cost depends on the design and the installation location. Usually any electrical and energy equipment or installations are designed for certain useful life, depending on the equipment type, design life, equipment maintenance and operating conditions. Sometimes the equipment may have zero value at the useful life end, the so-called salvage cost. Cost comparisons between different energy sources and electricity generation systems are made by the levelized cost of energy (LCOE). Levelized costs represent the present value of building and operating a power plant or a renewable energy system over an assumed plant or system lifetime, expressed in real terms to remove the effect of inflation. The levelized cost of energy, the levelized cost of electricity, and/or the levelized energy cost are economic assessments of the average total cost to build and operate a power-generating system over its lifetime divided by the total power generated by the system during that lifetime. LCOE is often used as an alternative for the average price that the power-generating system is receiving in a market to break even over its lifetime. It is a first-order cost competitiveness economic assessment of an electricity-generating system that incorporates all costs over its lifetime accounting for the initial investment, O&M cost, fuel cost, and capital cost. LCOE is a metric used to assess the cost of electric generation and the total plant-level impact from technology design changes, which can be used to compare electricity generation costs. There are different methods to calculate LCOE; the ones included here are the most common found in the literature and used by practitioners. LCOE calculations are in the framework of the annual technology baseline studies that provide a summary of current and projected future cost and performance of primary electricity generation technology in the United States, including renewable energy technologies. For energy sources that require fuels, assumptions are also made about future fuel costs. The levelized construction and operations costs are then divided by the total energy obtained to allow direct comparisons across different energy sources. For renewables to be cost-competitive, their costs need to decrease to the wholesale electricity price, or the price at which fossil-fuel power plants sell electricity to the grid. This has already occurred for some renewable energy sources, such as hydropower, small solar-thermal installations, and biomass. Solar PV systems are still more expensive than other energy sources, but because PVs are often installed by individual consumers, the price of PV only needs to fall to the retail power price that consumers pay, which is greater than the wholesale price. Organizations use several different methods to determine which investment is best. There are many methods available for the evaluation of the economic efficiency of project options and alternatives, including the following: the net present value, payback period method, rate of return method, the benefit-cost analysis, and return on investment. Usually, the following assumptions are made: one of the parameters in the present worth analysis is assumed, which may be either the interest rate or the useful life; the salvage value is usually taken as zero; even if the interest rate is provided, it is likely to change in the future; and when comparing products from a superior technology versus the older technology, the latter is the attractive choice. In such cases, the economic advantages and the technical merits should be weighted together.

1.3.4   NATURAL RESOURCE EVALUATION AND METHODS FOR PAYBACK PERIOD ESTIMATES

For energy resources, it is important to have some estimate of the resource viability. Basic methods for natural and energy resources evaluation include: the comparative method (derived from the price of other similar goods); the cost method (according to the cost incurred in obtaining the subject); and the method of return (according to useful effects, which source provides). The resource evaluation consists primarily of the natural or energy source price, C_{NR}, computed from an application of the present value relationship:

$$C_{NR} = \sum_{t=1}^{N} \frac{An_t}{(1+r_t)^t} \tag{1.60}$$

Here, t is the time period, An_t is the expected value of the annuity for the period of time, r_t is expected value of interest (discount) rate for one period of time (coefficient), and N is the number of periods. Interest rate and the discount rate are considered to be and are quite often variable in time. The expected value of an annuity for a period of time is a function of several variables, such as the type and cost of production, input prices, taxes, interest rates, and inflation. For proper expected value estimates, these factors must be known and defined. The expected value of the interest rate for a period of time is also a function of several variables, such as the money time preference, risk, inflation and other characteristics, etc., that also must be defined in order to make an accurate estimate. Frequently used assumptions are: a constant of value of the expected annuity value, An, at the time (usually for long-term contracts) and a constant of value of the discount rate, r, at the time, as well as because in most of the applications, there is usually a very long time horizon, the infinite time series assumption, $N \rightarrow \infty$ is considered. Then a simpler relationship for natural or energy resource evaluation is used:

$$C_{NR(simple)} = \sum_{t=1}^{N} \frac{An}{(1+r)^t} = \frac{An}{r} \tag{1.61}$$

Example 1.18

A geothermal energy source has an estimated annual annuity of $180,000 and a discount rate of 7.5%. Compute the estimated source price.

SOLUTION

From Equation (1.61) the geothermal source price is estimated at:

$$C_{NR(simple)} = \frac{r}{i} = \frac{180000}{0.075} = \$2,400,000$$

The payback method is commonly used for the appraisal of capital investments and engineering projects despite its theoretical deficiencies. The payback period is the time required for the benefits of an investment to equal the investment costs. The payback method is often used when aspects such as project time risk and liquidity are the focus and where pure profit evaluation is the single criterion. In practice, the maximum acceptable payback period is often chosen as a fixed value, for instance for a certain number of years. For example, the payback period for the domestic consumer is up to 5 years, but often only 2 or 3 years. In some cases, the payback period limit value is chosen in relation to the project economic life, for example the payback period could be shorter than half the economic life of the investment. In many companies, the payback period is used as a measure of attractiveness of capital budgeting investments. Most often the payback method is used as a first screening device to sort out the obvious cases of profitable and unprofitable investments, leaving only the middle group to be scrutinized by means of more advanced and more time-consuming calculations based on discounted cash flows (DCF), such as the internal rate of return (IRR) and net present value (NPV) methods. However it should be noted that the payback method can be developed to handle cases with varying cash flows, although some of its simplicity is lost in the process. Because the decision situations in the evaluation of capital budgeting investment typically are uncertain concerning the time pattern and the duration of cash flows, the use of the simple and more robust payback method can be justified even if there will be time for more advanced analyses or methods. The payback period goes on decreasing for high-capacity systems used by commercial consumers if electricity is replaced. The payback period (PBP) method of capital budgeting calculates the time it takes to recover the initial investment cost. The two approaches are the short-cut (simple) payback method and the unequal cash flow method. Both base the calculations on the annual net cash flows (cash outflows minus cash inflows). As is true with NPV and IRR methods, each year may have different cash flow amounts. In the short-cut method, the amounts of annual operating cash flows expected from a potential capital asset acquisition are equal each year, and the short-cut calculation is used to determine the payback period. The simple payback period (SPP) is calculated as:

$$SPP = \frac{\text{Initial Investment (Cost)}}{\text{Net Benefits per Year}} \tag{1.62}$$

Example 1.19

To improve the power factor, a capacitor back is installed at a 100 kW wind turbine. The cost of the unit is $20,000 and the interest rate is 7.5%, the combined federal and state tax credits are 40%, and the incremental tax rate is 45%. The system-related costs are: the loss factor 0.4, maximum reactive power demand 80 kVAR, maximum demand cost of kW, kVAR, kVA is $3.80/kW/month, the cost of released transformer kVA is $12 per kVA/year, cost of reactive energy is equal to $0.0025/kVAR/h, the combined feeder and transformer resistance per phase is 0.040 Ω, the current before installation 55 A, and after installation 80 A. Assuming a power factor improving from 0.65 to 0.95 due to the power factor correction unit installation, calculate the simple payback period.

SOLUTION

Savings due to the reduction in the reactive power penalty per year is:

$$C_{PFP} = 3.80 \times \left(100 \frac{\sqrt{1-0.65^2}}{0.65} - 100 \frac{\sqrt{1-0.95^2}}{0.95} \right) \times 12 = \$3832.4$$

The savings due to the reduction in kVA demand transformer is:

$$C_{kVA} = \left(\frac{100}{.65} - \frac{100}{.95} \right) \times 12 = 582.9959 \approx \$583.0$$

Savings due to the transmission line loss reduction is:

$$C_{loss} = 3 \times 0.06 \times 0.040 \times \left(80^2 - 55^2\right) \times 8760 \times 0.4 \times 10^{-3} = \$85.1$$

Savings due to the reduction of the reactive energy cost is:

$$C_{Reactive} = 0.0025 \times \left(100 \frac{\sqrt{1-0.65^2}}{0.65} - 100 \frac{\sqrt{1-0.95^2}}{0.95}\right) \times 8760 = \$1840.6$$

The annual benefit is $6341.2 and the simple payback period is then:

$$SPP = \frac{20000}{6341.2} = 3.15 \text{ years}$$

The payback period indicates how long it takes to recover the investment used to acquire an asset or for an equipment installation and purchasing. The payback period is equal to the number of full years plus the final investment recovery year fraction. The simple payback period method ignores the money time value. One way to compare mutually exclusive economic aspects is to use the present time using the present worth method. The present worth of annual benefits is obtained by multiplying the benefits with the present worth factor defined for a discount (interest) rate, r and N years, as:

$$PVF = \frac{1}{(1+r)^N} \tag{1.63}$$

Then the cumulative value of the benefits, for AB, the benefits per year is given by:

$$B_{cum} = AB \left(\frac{(1+r)^N - 1}{r \cdot (1+r)^N}\right) \tag{1.64}$$

In the payback evaluation method, the period Y_{proj} (usually expressed in years) required recovering the cost of the project, the initial investment, $CF0$ is determined by the following equation, expressed in the interest (discount) rate, r and the average annual net savings, CF_k, as:

$$CF0 = \sum_{k=1}^{Y_{proj}} \frac{CF_k}{(1+r)^k} \tag{1.65}$$

If the payback period Y_{proj} (often called the *discount payback period*, *DPB*) is less than the project lifetime, N, then the project is economically viable. In a large majority of applications, the time value of money is neglected, and Y_{proj} is SPP, as discussed before, being the solution of the simplified Equation (1.65), the following equation:

$$CF0 = \sum_{k=1}^{Y_{proj}} CF_k \tag{1.66}$$

In cases where the annual net savings are constant, A_{SV}, the simple payback period is easily calculated as the ratio of initial investment and the annual saving:

$$Y_{proj}(SPP) = \frac{CF0}{A_{SV}} \tag{1.67}$$

The SPP values are shorter than the DPB values because the undiscounted net savings are larger than their discounted counterparts. Therefore, acceptable SPP values are usually significantly shorter than the project lifetimes. Usually, the investments and the economic returns are expressed on a time-varying basis. In energy projects the payback period calculations, using the present worth analysis, involve the present worth factor evaluation, annual benefits, and the present worth of the cumulative benefits. In order to perform the economic analysis, the income tax effects must be included, and the available tax credits, if any, for the project. Businesses can deduct a percentage of the new equipment costs as a tax credit, usually after the tax return. At the same time, the basis for depreciation remains the full cost of the equipment. The main components taking into account the tax effects are cash flows before taxes; depreciation; change in taxable income, equal to the cash flow before taxes, less the depreciation; income taxes (taxable income times the incremental income tax); and after-tax cash flows, equal to before-tax cash flows, minus the income tax. In general, the prices and cost of services change with time and hence the inflationary trend should be included in the payback period calculations. The inflation adjustments must be included in the payback period analysis. If the annual inflation rate is i, the actual cash is computed by cash flow before taxes multiplying with $(1+i)^N$, the depreciation and the income taxes are computed as before, while the after-tax cash flow is also adjusted by multiplying the inflation rate with $(1+i)^N$.

Example 1.20

In a DG system an old, less efficient (60%) boiler is about to be replaced by a more efficient (85%), having the same annual O&M costs as the older one, $150. The equipment cost is $20,000 and the discount rate is 5%, after adding the tax credit per year and its lifetime is 10 years. The boiler consumes 9000 gallons of fuel at $1.30 per gallon.

SOLUTION

The annual cash flow is computed from the net savings of using the more efficient equipment:

$$A = C_k \left(k = 1...10\right) = 9000 \times \left(1 - \frac{0.6}{0.85}\right) \times 1.30 = \$3573.53$$

Assuming a constant annual savings, $CF_1 = ... = CH_{10} = A_{SV}$, $d = 0.05$ and $N = 10$, and then solving numerically the Equation 1.65, the DPB is 4.68 years, shorter than the project lifetime, 10 years. Therefore, replacing the old equipment is cost-effective. The cumulative benefits equation is given here:

$$20000 = \sum_{k=1}^{Y} \frac{3573.53}{\left(1+0.05\right)^k} = 3573.53 \frac{1 - \left(1+1.05\right)^{Y+1}}{1 - 1 + 0.05}$$

When the dollar amount of operating cash flows is not expected to be the same amount each year, a longer interval is expected for the recovery, and is estimated by the so-called unequal cash flow method. In essence, this method begins with the acquisition cost, the amount to be recovered, and then subtracts the expected cash flows for each year until the point in time at which the cash is recovered. The calculation begins by subtracting the first-year cash inflows from the initial investment, and if the remaining cash flows to recover are greater than or equal to the cash flows of the second year, the cash flows expected for the second year are subtracted, and the process continues for each subsequent year until the remaining cash flows not yet recovered are less than the cash flows expected for the next year, and the point during the next year that the remaining cash is recovered. The payback period indicates how long it takes to recover the cash investment used to acquire the asset. It is expressed usually in years with two decimals, such as 4.25 years. Usually, shorter payback periods are more attractive than longer ones. If the cash is expected to be recovered in a time period shorter than the investment lifetime, it is tentatively deemed acceptable. However, other capital budgeting methods must always be used in conjunction with payback period method, because even when it appears to be an acceptable investment, it may not be acceptable under a method that considers the time value of money. If the shortcut method is used to calculate the SPP, the results may indicate a payback period greater than the useful asset life. It is not possible to have a useful life greater than the payback period because the *lifetime end* indicates the asset is no longer used in the production, so it no longer brings in economic resources. As such, when the numerical result using the shortcut method appears to have a payback period exceeding the useful life, *the interpretation is the investment is never recovered*. The payback period method has some faults that create limitations on its usage. First, it does not consider the total stream of cash flows. The second drawback of the payback period method is that it does not consider the time value of money in its calculations. The payback period method also ignores the timing of the cash flows and evaluates both investments as equal options.

Green and sustainable engineering, manufacturing, and design are emerging engineering practice fields, being also sustainable models for modern industries. Design, development, and production of products and processes that have reduced impacts on the environment and human health have become strategic goals of corporations and research institutions, in the United States and all over the world. Among reasons are better use of resources, increasing competition, regulations, and public awareness. Future technology will likely have very little to do with creating fossil-fuel-based products, processes, and services. On the other hand, in education, while knowledge, expertise, and skills in the areas of science and engineering fundamentals are required, knowledge and skills in areas such as energy systems, built environments, environmental and chemical engineering, electricity generation, and combustion processes will change significantly. Consequently, engineers need to be educated in energy efficiency, and green and sustainable design methods. It is so critical that engineering graduates are equipped with relevant knowledge and skills to effectively address such challenges in society. However, to make this educational transition, it is critical to address a number of key elements within the curriculum renewal process, such as industry needs, the demand for engineering graduates to be literate and competent in sustainable development, and the needs to meet changing student expectations.

1.4   CHAPTER SUMMARY

This chapter provides an overview of energy concepts; systems; brief reviews of thermal engineering, fluid mechanics, unit systems, energy units, power system structure and major subsystems; main power system components; and briefly reviews energy conversion, heat transfer processes, and fluid mechanics. In order to understand building energy systems, a basic understanding of thermodynamics and heat-transfer concepts and processes is needed. The electric power system is a complex technical system with the main function of delivering electricity between generation, consumption, and storage. A basic review of the concepts of work, energy, power, heat, heat and energy transfer, and conversion is also included in this chapter. Understanding these concepts is critical for the understanding of many other topics discussed in this book, including wind energy, thermal conversion, distributed generation, and heat and air-conditioning. A brief discussion of units, unit systems, dimensional analysis, and the most common units for power, work, energy, and energy conversion and transfer is also included in this chapter. Electricity for an industrial facility is provided by the local utility at a voltage determined by the size of the load and the topology of the utility. One-line or single-line diagrams are used as a simplified means of describing the topology of a power system. Although not truly a circuit diagram, one-lines are often used as a starting point for constructing circuit diagrams. A section briefly discussed the energy economics. Several examples and end-of-chapter problems and questions help the readers to fully understand the materials discussed here. A sustainable energy system involves key components of increasing use of renewable energy resources, increased energy efficiency and use of electricity in the transportation sector. As an effect of this, some key questions for the electric power system are integration of intermittent electricity generation (e.g., wind power, the connections of electrical vehicles to the electrical distribution system, can be different solutions for the energy storage). Electrical distribution networks, transmission lines, wiring devices, and equipment are essential building and industrial energy subsystems and components. Power engineers, design and construction professionals, and engineers, as well as people in charge of building operation and maintenance make judgments and calculations about power, lighting, heating and cooling requirements, space required for switchgear, transformers, vertical and horizontal power distribution, optimized location of major components and equipment, safety and construction, operation, and maintenance costs. This chapter provides a brief introduction to power transmission and distribution, introducing terms and concepts related to building electrical systems. The improvement of energy efficiency in industry is one of the main focuses of this book, so the understanding of energy and power efficiency concepts is critical in examining energy efficiency, energy conservation, energy management, and new technologies in the areas of electrical systems, distributed generation, as well as cogeneration in small-, medium-, and large-sized buildings and enterprises.

1.5 QUESTIONS AND PROBLEMS

1. What are the major components of a power system?
2. Define and describe the transmission and subtransmission subsections of a power system.
3. Is the quality of life better with electricity or without? Explain.
4. Describe how a specific energy-related process can affect the environment.
5. What are the ways in which water is currently used in electricity generation?
6. What are the mean electrical power production rates of the United States, European Union, and China? Report your results in terawatts (TW) and cite your sources of information.
7. What is the difference between energy and power? Define energy and power.
8. What is 1 HP in W and in kJ/s? Should 100 kWh be converted into W or J? Do the conversion.
9. What is the equivalent of 1 HP in: (a) BTU/hr., and (b) kW?
10. How much energy is needed to accelerate a shaft-mounted disk of mass moment of inertia of 10,000 kg·m^2 from 1000 rpm to 30,000 rpm?
11. What is the change in kinetic energy for a 1360 kg car that is slowing down from 100 km/hr. to 50 km/hr.?
12. What is the kinetic energy of a wheel of mass moment of inertia 100 kg·m^2 rotating at 500 rpm?
13. How much energy is generated in a year by a 135 MW geothermal power plant, if it is in operation about 88% of the time?
14. A cylinder with mass moment of inertia of 100 kg·m^2 is rotating freely at 50 rad/s about its axis of symmetry. What is the retarding torque that must be applied to reduce the velocity to zero in 1 min.?
15. The energy input into a light bulb is 40 W. The output is 950 lumens. Determine the efficiency of the light bulb. (Note: One lumen is equivalent to 0.001496 W.)
16. What is the torque developed in the shaft of an engine working at 1 HP and rotating at 6500 rpm?
17. What is efficiency and what is the Carnot efficiency?
18. A heat engine operates between a temperature of 20°C and 150°C. What is the efficiency of the engine if it works at 75% of the maximum possible Carnot efficiency?
19. Calculate the efficiency of a power plant if the efficiencies of the boiler, turbine, and generator are 88%, 40%, and 98%, respectively.
20. An automobile engine could operate between 2200°C (the combustion temperature of gasoline) and 20°C (ambient temperature). If it did, what would its maximum efficiency be?
21. The efficiency of a fossil-fuel-based power plant is limited by the Carnot efficiency of the thermal engine. Assuming that the expelled heat into the atmosphere has a temperature of 30°C and combustion temperature is 520°C, and an overall efficiency of the conversion to electricity of 85%, what is the maximum overall efficiency of the electricity generation?

22. A refrigerator operates between a temperature of 21°C and −8°C. What is the COP of the refrigerator if it works at 83% of the maximum possible Carnot COP?

23. Compare the heating efficiencies (maximum COP) of the same heat pump installed in Miami and in Buffalo. In Miami, because the climate is milder, assume that T_H is 70°F and that T_L is 40°F. In Buffalo, assume that T_H is the same, but that T_L (the outside temperature) is much lower, on average about, 15°F.

24. By how much does the efficiency of a turbine have to increase in order to raise the efficiency of a power plant from 35% to 40%? The efficiencies of the boiler and generator are 90% and 95%, respectively. In order to achieve the required efficiency increase, by how much does the inlet steam temperature have to increase? Assume that the efficiency is proportional to the temperature difference between hot and cold steam and that the cold steam is in both cases at 100°C.

25. The flow rate of water that enters and leaves a cooling chamber is 1 kg/s. What is the heat rate that must be extracted from the water to reduce its temperature by 20°C?

26. In the United States, the British thermal unit (Btu) is defined as the energy to raise the temperature of 1 pound water by 1 degree Fahrenheit. Show that, to a good approximation, 1 Btu equals 1 kJ.

27. Using a solar photovoltaic field of 1 square mile (2.59 km²) with efficiency of 18%, how many kilowatt-hours will this field generate annually at locations of average daily insolation (on flat ground) of 3 h (Alaska), and 6 h (Arizona)? An average household consumes 1000 kWh per month. How many households can this field support in the two states, respectively?

28. Calculate the future worth, FW, for a deposit of $60,000. The number of investment years is 6, and the interest rate is 4.5%.

29. If an amount of $7,500 annual deposit is made for 5 years, what is the future worth value of those deposits? The interest rate is 3.85%.

30. If a cash flow consists of an annuity of $175,000 over an 8-year period at 6% interest rate, calculate its future value at the end of this period.

31. An investor decided to invest $1.5 million, borrowed with 7.5% interest, in a geophysical district heat, having a lifespan of 20 years. Determine the cost of the money. Assuming the O&M cost per year of 8.5% of the initial investment, and 3.5% of the initial investment, in governmental tax credit per year, and electricity saving per year of 9.575 million of kWh at an average cost of 4 cent per kWh, calculate the annual revenue of the development.

REFERENCES AND FURTHER READINGS

1. F. Bueche, *Introduction to Physics for Scientists and Engineers*, McGraw-Hill, 1975.
2. E. L. Harder, *Fundamentals of Energy Production*, Wiley, New York, 1982.
3. T. D. Eastop and D. R. Croft, *Energy Efficiency for Engineers and Technologists*, Longman, Harlow, UK, 1990.
4. A. W. Culp, Jr., *Principles of Energy Conversion* (2nd ed.), McGraw-Hill, New York, NY, 1991.
5. D. T. Allen and D. R. Shonnard, *Green Engineering*, Prentice Hall, Upper Saddle River, NJ, 2001.
6. J. R. Cogdell, *Foundations of Electric Power*, Prentice Hall, Upper Saddle River, NJ, 1999.
7. S. J. Chapman, *Electrical Machinery and Power System Fundamentals* (2nd ed.), McGraw-Hill, 2002.
8. P. Schavemaker and L. van der Sluis, *Electrical Power System Essentials*, John Wiley and Sons, 2008.
9. M. E. El-Hawary, *Introduction to Power Systems*, Wiley-IEEE Press, 2008.
10. M. A. El-Sharkawi, *Electric Energy: An Introduction* (2nd ed.), CRC Press, 2009.
11. A. Von Mayer, *Electric Power Systems—A Conceptual Introduction*, Wiley-IEEE Press, 2006.
12. V. Quaschning, *Understanding Renewable Energy Systems*, Earthscan, 2006.
13. R. A. Ristinen and J. J. Kraushaar, *Energy and Environment*, Wiley, Hoboken, NJ, 2006.
14. J. Andrews and N. Jelley, *Energy Science, Principles, Technology and Impacts*, Oxford University Press, 2007.

15. E. L. McFarland, J. L. Hunt, and J. L. Campbell, *Energy, Physics and the Environment* (3rd ed.), Cengage Learning, 2007.

16. F. Kreith and D. Y. Goswami (eds.), *Handbook of Energy Efficiency and Renewable Energy*, CRC Press, Boca Raton, FL, 2007.

17. M. Kaltshmitt, W. Streicher, and A. Weise (eds.), *Renewable Energy—Technology, Economics and Environment*, Springer, 2007.

18. D. R. Patrick and S. W. Fardo, *Electrical Distribution Systems* (2nd ed.), CRC Press, 2008.

19. B. K. Hodge, *Alternative Energy Systems and Applications*, Wiley, 2010.

20. B. Everett and G. Boyle, *Energy Systems and Sustainability: Power for a Sustainable Future* (2nd ed.), Oxford University Press, 2012.

21. F. M. Vanek and L. D. Albright, *Energy Systems Engineering: Evaluation and Implementation* (2nd ed.), McGraw-Hill, 2012.

22. L. Blank and A. Tarquin, *Engineering Economy* (9th ed.), McGraw-Hill, New York, 2012.

23. International Energy Agency Report, *Key World Energy Statistics*, International Energy Agency, 20, 2013.

24. G. Petrecca, *Energy Conversion and Management: Principles and Applications*, Springer, NY, 2014.

25. R. A. Dunlap, *Sustainable Energy*, Cengage Learning, 2015.

26. V. Nelson and K. Starcher, *Introduction to Renewable Energy (Energy and the Environment)*, CRC Press, 2015.

27. R. Bansal (ed.), *Handbook of Distributed Generation—Electric Power Technologies, Economics, and Environmental Impacts*, Springer, 2017.

28. R. Belu, *Industrial Power Systems with Distributed and Embedded Generation*, The IET Press, 2018.

2 Review of Electric Circuits and Power System Basics

2.1 INTRODUCTION TO ELECTRIC CIRCUIT REVIEW

The study of electric circuits covers the fundamental phenomenon of electric charge carrier flows. The fundamental quantities of current and voltage are used to describe how rapidly charged particles move in a circuit and in what way they do so in the circuit. Current is sometimes referred to as the "through" quantity and voltage as the "across" quantity. In a physical context, current is the flow of electric charge through a component or apparatus, whereas voltage is the potential difference between two points in a circuit. Current flows from high potential to low potential. In particular, we define the current, $i(t)$, flowing in a component or apparatus as the amount of charge passing through that component or apparatus per unit of time. Denoting charge by $q(t)$, we may write current $i(t)$ as:

$$i(t) = \frac{dq(t)}{dt} \tag{2.1}$$

Two other important quantities frequently used to describe physical systems are *power* and *energy*. If a small quantity of electric charge Δq is displaced from a point A to a point B, then the change in its potential energy, or the work done, is equal to $V \cdot \Delta q$, where V is the voltage between A and B. The amount of work done per unit of time, the time rate, is called *power*, and is usually denoted by P, defined as:

$$P = \lim_{\Delta t \to 0} \frac{V \cdot \Delta q}{\Delta t} = V \frac{dq}{dt} \tag{2.2}$$

Current unit is the *ampere* (A), for the voltage is *volt* (V), for the energy is *joule* (J), and for the power is *watt* (W). Prefixes are often used to emphasize the significant figures when these magnitudes are too large or too small. Common engineering prefixes and their corresponding multipliers are given in Table 2.1.

An electric circuit is formed by interconnecting components having different electric properties. A collection of devices such as resistors and sources in which terminals are connected together by connecting wires is called an electric circuit. These wires converge in nodes, and the devices are called circuit branches. The general circuit problem is to find currents and voltages in the branches of the circuit when the intensities of the sources are known. Such problems are referred to as *circuit analysis*. It is therefore important, in electric circuit analysis, to know the involved component properties, as well as the way the components are connected to form the circuit. In this chapter, some ideal electric components and simple connection styles are introduced. Without resorting to advanced analysis techniques, we will attempt to solve simple problems involving circuits that contain a relatively small number of components connected in some relatively simple fashions. In particular, we will derive a set of useful formulae for analyzing circuits that involve such simple connection types as series, parallel, ladder, star, and delta. This chapter serves as a review of the basic properties of electric circuits. Electric network theory deals with two primitive quantities, which we will refer to as: potential (or voltage), and current. Current is the actual flow of charged carriers, while the difference in potential is the force that causes that flow. As we will see, potential is a single-valued function that may be uniquely defined over the nodes of a network. Current, on

TABLE 2.1

Prefixes of the Units

Prefix	Multiplier	Abbreviation
Peta	$\times 10^{15}$	P
Terra	$\times 10^{12}$	T
Giga	$\times 10^{9}$	G
Mega	$\times 10^{6}$	M
Kilo	$\times 10^{3}$	k
Hecto	$\times 10^{2}$	h
Deca	$\times 10$	da
Deci	$\times 10^{-1}$	d
Centi	$\times 10^{-2}$	c
Milli	$\times 10^{-3}$	m
Micro	$\times 10^{-6}$	μ
Nano	$\times 10^{-9}$	n
Pico	$\times 10^{-12}$	p
Femto	$\times 10^{-15}$	f

the other hand, flows through the branches of the network. Figure 2.1 shows the basic notion of a branch, in which a voltage is defined across the branch and a current is defined to flow through the branch. A network is a collection of such elements, connected together by wires.

Current direction indicates the direction of flow of positive charge, and voltage polarity indicates the relative potential between two points. Usually, + is assigned to a higher potential point and − to a lower potential point. However, during analysis, direction and polarity can be *arbitrarily* assigned on circuit diagrams. Actual direction and polarity will be governed by the sign of the value. When voltage is applied to a metal wire (as in Figure 2.1), the current $i(t)$ flowing through the wire is proportional to the voltage $v(t)$ across two points in the wire. This property is known as Ohm's law, expressed as:

$$v(t) = R \cdot i(t) \text{ or } i(t) = G \cdot v(t) \tag{2.3}$$

where R is called resistance, and G is called conductance. The resistance R and the conductance G of the same piece of wire is related by $R = 1/G$. Resistance is measured in Ohms (Ω) and conductance in Siemens (S). Any device that has this property is called a resistor. Physics shows that a resistance is proportional to the metal wire length, l, and inversely proportional to the cross-sectional area, A, that is,

$$R = \rho \frac{l}{A} \tag{2.4}$$

where the proportionality constant ρ is known as the resistivity of the metal (material). Two concepts are very useful in electric circuit analysis. An ideal voltage source provides a prescribed voltage across its terminals irrespective of the current flowing through it. The amount of current supplied by the source is determined by the circuit connected to it. An ideal current source provides

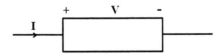

FIGURE 2.1 Current and voltage notation.

a prescribed current to any circuit connected to it. The voltage generated by the source is determined by the circuit connected to it. The definition of voltage as work per unit charge lends itself very conveniently to the introduction of power. Recall that power is defined as the work done per unit time. Thus, the power, P, either generated or dissipated by a circuit element can be represented by the product of voltage and current. It is important to realize that, just like voltage, power is a signed quantity, and that it is necessary to make a distinction between positive and negative power. The general expression for the power in an electric circuit, regardless of the voltage or current functional dependence of time, is:

$$p(t) = v(t) \cdot i(t) \tag{2.5}$$

Network topology is the interconnection of its elements. That, plus the constraints on voltage and current imposed by the elements themselves, determines the performance of the network, described by the distribution of voltages and currents throughout the network. Two important concepts must be described initially: the *loop* and *node*. A loop in the network is any closed path through two or more network elements. Any nontrivial network will have at least one such loop. A node is a point at which two or more elements are interconnected. The two fundamental laws of network theory are known as Kirchhoff's Voltage Law (KVL), and Kirchhoff's Current Law (KCL). These laws describe the topology of the network, and arise directly from the fundamental laws of electromagnetics.

- *Kirchhoff's voltage law* states that, around any loop of a network, the sum of all voltages, taken in the same direction, is zero:

$$\sum_{loop} v_k = 0 \tag{2.6}$$

- *Kirchhoff's current law* states that, at any network node, the sum of all currents is zero:

$$\sum_{node} i_k = 0 \tag{2.7}$$

Example 2.1

Apply both KVL and KCL to the circuit depicted below.

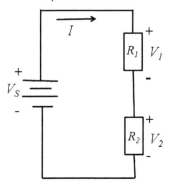

SOLUTION

Applying KVL we write:

$$V_S - V_1 - V_2 = 0$$
$$V_S = R_1 I + R_2 I$$

Applying KCL we obtain two equations, one at the top node, the other one at the node between the two resistors:

$$I - \frac{V_1}{R_1} = 0, \text{ and } \frac{V_1}{R_1} - \frac{V_2}{R_2} = 0$$

It is worth noting that KVL is a discrete version of Faraday's law, valid to the extent that no time-varying flux links the loop. KCL is just charge conservation, no charge accumulation at any node. Network elements affect voltages and currents in three ways: (1) voltage sources constrain the potential difference across their terminals to be of a fixed value (the source value); (2) current sources constrain the current through the branch to be of some fixed value; and (3) all other elements impose some sort of relationship, either linear or nonlinear, between voltage across and current through a branch. Voltage and current sources can be either independent or dependent. Independent sources have values that are, as the name implies, independent of other variables in a circuit. Dependent sources have values that depend on some other variable in a circuit. A common example of a dependent source is the equivalent current source used for modeling the collector junction in a transistor. Typically, this is modeled as a current-dependent current source, in which collector current is taken to be directly dependent on emitter current. Such dependent sources must be handled with some care, for certain tricks we will be discussing below do not work with them. For the present time, we will consider, in addition to voltage and current sources, only impedance elements, which impose a linear relationship between voltage and current.

2.1.1 LINEARITY AND SUPERPOSITION

An extraordinarily powerful notion of the network theory is *linearity*. This property has two essential elements, stated as follows:

- For any single input x yielding output y, the response to an input k·x is k·y for any value of k.
- If, in a multi-input network the input x_1 by itself yields output y_1 and a second input x_2 by itself yields y_2, then the combination of inputs x_1 and x_2 yields the output $y = y_1 + y_2$.

This is important to us at the present moment for two reasons:

1. It tells us that the solution to certain problems involving networks with multiple inputs is actually easier than we might expect: If a network is linear, we may solve for the output with each separate input, and then by adding the outputs, we can get the solution. This is called *superposition*.
2. It also tells us that, for networks that are linear, it is not necessary to actually consider the value of the inputs in calculating response. What is important is a system function, or a ratio of output to input.

Superposition is an important principle when dealing with linear networks, and can be used to make analysis easier. If a network has multiple independent sources, it is possible to find the response to each source separately, and then add up all of the responses to find the total response. A particularly important ramification of the property of linearity is expressed by equivalent circuits. If we are considering the response of a network at any given terminal pair, from the linearity properties, if the network is linear, the output at a single terminal pair (either voltage or current) is the sum of two components:

1. The response that would exist if the excitation at the terminal pair were zero, and
2. The response forced at the terminal pair by the exciting voltage or current.

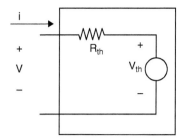

FIGURE 2.2 Thevenin equivalent network.

It may be expressed with either voltage or current as the response. These yield the Thevenin and Norton equivalent networks, which are exactly equivalent. At any terminal pair, a linear network may be expressed in terms of either Thevenin or Norton equivalents. The Thevenin equivalent circuit is shown in Figure 2.2, while the Norton equivalent circuit is shown in Figure 2.3. Thevenin and Norton equivalent circuits have the same impedance. Further, the equivalent sources are related by the simple relationship:

$$V_{Th} = R_{eq}I_N \tag{2.8}$$

Thevenin equivalent voltage, the source internal voltage to the Thevenin equivalent network, is the voltage of the circuit terminals that are opened. Similarly, the Norton equivalent current is the same minus the short-circuit current, when the circuit terminals are shortened.

2.1.2 DC vs. AC Circuits

The current of a direct current (DC) circuit, as shown in Figure 2.4, consisting of a battery and a pure resistive load, can be calculated by using Ohm's law as a ratio between the source voltage V_{DC} and the load resistance, R, as:

$$I_{DC} = \frac{V_{DC}}{R} \tag{2.9}$$

The power, P, provided by the voltage source is given by:

$$P_{DC} = V_{DC} \cdot I_{DC} = \frac{V_{DC}^2}{R} = RI_{DC}^2 \tag{2.10}$$

The DC voltage and current waveforms are shown in Figure 2.5. It is worth mentioning that these DC quantities are real numbers not complex numbers.

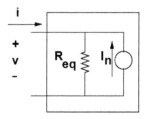

FIGURE 2.3 Norton equivalent network.

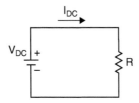

FIGURE 2.4   A simple DC circuit.

Example 2.2

Determine the minimum resistor size that can be connected to 1.5 V and 3.0 V batteries without exceeding the resistor's 0.5 W power rating.

SOLUTION

The power dissipated by the resistor is given by Equation (2.10). Because the maximum allowable power dissipation is 0.5 W, we can write:

$$P = \frac{V_{DC}^2}{R} < 0.5 \text{ W} => \begin{cases} \dfrac{1.5^2}{R} < 0.5, \text{ for the 1.5 V battery} \\[2mm] \dfrac{3.0^2}{R} < 0.5, \text{ for the 3.0 V battery} \end{cases}$$

Solving for the resistor minimum size, for each battery type, we have:

$$R_{1.5 \text{ V}} = \frac{1.5^2}{0.5} = 4.5 \ \Omega$$

$$R_{3.0 \text{ V}} = \frac{3.0^2}{0.5} = 18 \ \Omega$$

There is another category of circuits, the alternating current (AC) circuits. Electric power systems usually involve sinusoidally varying (or nearly so) voltages and currents. That is, voltage and current are functions of time that are nearly pure sine waves at fixed frequency. In North America, Canada, Brazil, and eastern Japan that frequency is 60 Hz. In most of the rest of the world it is 50 Hz. Normal power system operation is at this fixed frequency, which is why we study how systems operate in this mode. We will deal with transients later. Accordingly, this section opens with a review of complex numbers and with representation of voltage and current as complex amplitudes with complex exponential time dependence. The discussion proceeds, through impedance, to describe a pictorial representation of complex amplitudes, called phasors. Power is then defined, and, in sinusoidal steady state, reduced to complex form. Finally, flow of power through impedances and a

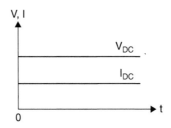

FIGURE 2.5   Voltage and current waveforms of the simple DC circuit.

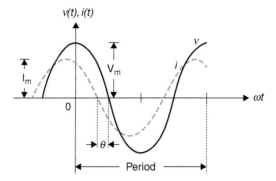

FIGURE 2.6 Typical voltage and current waveforms.

conservation law are discussed. This section deals with alternating voltages and currents and with associated energy flows. The focus is on sinusoidal steady-state conditions, in which virtually all quantities of interest may be represented by single, complex numbers. Because in power systems the sinusoidal voltages are generated, and consequently, most likely sinusoidal currents are flowed in the generation, transmission, and distribution systems, sinusoidal quantities are assumed throughout this material, unless otherwise specified. In general, a set of typical steady-state voltage and current waveforms of an AC circuit can be drawn as shown in Figure 2.6, and their mathematical expressions can be written as follows:

$$v(t) = V_m \cos\left(\omega t + \phi\right) \tag{2.11}$$

and

$$i(t) = I_m \cos\left(\omega t + \theta\right) \tag{2.12}$$

Here, V_m and I_m are the maximum (peak) values or the amplitudes of the voltage and current waveforms, ω is the angular frequency in radians/second, ϕ and θ are the voltage and current phase angle, respectively, with respect to the reference in degrees or in radians. Usually in power engineering the voltage is taken as reference. The period in Figure 2.6 can be 360° or 2π radians. In some cases, the period can be in time, for instance, 0.016667 second for 60 Hz. A sinusoidal function is specified by three parameters: amplitude, frequency, and phase. The amplitude gives the maximum value or height of the curve, as measured from the neutral position. (The total distance from crest to trough is thus twice the amplitude.) The frequency gives the number of complete oscillations per unit time. Alternatively, one can specify the rate of oscillation in terms of the inverse of frequency, the period. The period is simply the duration of one complete cycle. The phase indicates the starting point of the sinusoid. In other words, the phase angle specifies an angle by which the curve is ahead of or behind where it would be, had it started at time zero. The period (measured in seconds), frequency (measured in Hz or cycles/s), and the angular frequency (velocity) (measured in radians/s) are related to the following relationship:

$$T = \frac{1}{f} = \frac{2\pi}{\omega}$$

2.2 RESISTIVE, INDUCTIVE, AND CAPACITIVE CIRCUIT ELEMENTS

Consider a purely resistive circuit with a resistor connected to an AC generator, as shown in Figure 2.7. Applying Kirchhoff's voltage law yields the instantaneous voltage across the resistor:

$$v(t) - V_R(t) = v(t) - R \cdot i(t) = 0 \Rightarrow V_R(t) = R \cdot i(t) \tag{2.13}$$

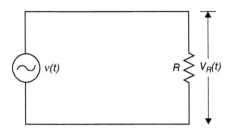

FIGURE 2.7   A purely inductive circuit.

The power dissipated by a resistor in AC is given by:

$$p(t) = Ri^2(t) \tag{2.14}$$

The discussed circuit elements have no memory, and, therefore, are characterized by instantaneous behavior. Algebraic (and for most such elements, linear too) expressions are used to calculate what these elements are doing. As it turns out, much of the circuitry we will be studying can be so characterized, with complex parameters. The behavior of inductors and capacitors is discussed briefly here.

Symbols for capacitive and inductive circuit elements are shown in Figure 2.8. They are characterized by the first-order derivative relationships between voltage and current:

$$i_C(t) = C \frac{dv_C(t)}{dt} \tag{2.15a}$$

And

$$v_L(t) = L \frac{di_L(t)}{dt} \tag{2.15b}$$

The expressions in Equation (2.15a), representing Ohm's law for a capacitor, while (2.15b) is the Ohm's law for an inductor, respectively. Both these equations are *nonlinear*, because time derivatives are involved. However, the expressions describing their behavior in networks become ordinary differential equations. If the $i(t)$ in Equation (2.13) is sinusoidal values, as one in Equation (2.11), then the voltage across a resistor is given by

$$v(t) = RI_m \sin(\omega t + \phi) = V_m \sin(\omega t + \phi) \quad \text{V} \tag{2.16}$$

Using the sinusoidal current and voltage in the expressions in Equations (2.15b), we get for inductor:

$$v(t) = L \frac{dI_m \sin(\omega t + \phi)}{dt} = \omega L I_m \cos(\omega t + \phi)$$
$$= \omega L I_m \sin(\omega t + \phi + \pi/2) = V_m \sin(\omega t + \phi + \pi/2) \quad \text{V} \tag{2.17}$$

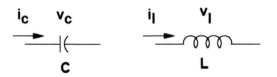

FIGURE 2.8   Symbols of capacitance and inductance.

In Equation (2.17) the voltage leads the current by 90°. For the capacitor voltage of Equation (2.15a), we get the following relationship:

$$i(t) = C\frac{dV_m \sin(\omega t + \theta)}{dt} = \omega C V_m \cos(\omega t + \theta)$$
$$= \omega C V_m \sin(\omega t + \theta + \pi/2) \quad \text{A}$$

(2.18)

Example 2.3

Given: A 40 Ω resistor in series with a 79.58 mH inductor. Find the impedance at 60 Hz.

SOLUTION

The inductive reactance and the circuit impedance are given by:

$$X_L = 2\pi f L = 2\pi 60 \times 79.58 \times 10^{-3} = 30 \ \Omega$$
$$Z = R + jX_L = 40 + j30 \ \Omega = 50 < 36.8° \ \Omega$$

Example 2.4

Given: A 40 Ω resistor in series with an 88.42 μF capacitor. Find the impedance at 60 Hz.

SOLUTION

The inductive reactance and the circuit impedance are given by:

$$X_C = \frac{1}{2\pi f C} = \frac{1}{2\pi 60 \times 88.42 \times 10^{-6}} = 30 \ \Omega$$
$$Z = R - jX_C = 40 - j30 \ \Omega = 50 < -36.8° \ \Omega$$

In this case the current leads the voltage by 90°. Equations (2.3) and (2.15a,b) can be written in a standard form using the impedance concept as:

$$v(t) = Zi(t)$$

(2.19)

The values of the impedance corresponding to the values of resistance, inductance, and capacitance are, respectively

$$Z = R + j0$$
$$Z = 0 + j\omega L$$
$$Z = 0 - \frac{j}{\omega C}$$

(2.20)

The impedances of a resistor in series with an inductor (R-L circuit), a resistor in series with a capacitor (R-C circuit), and a resistor in series with an inductor and a capacitor (R-L-C circuits) are expressed as:

$$Z_{R-L} = R + jX_L = R + j\omega L$$
$$Z_{R-C} = R + jX_C = R - j\omega C$$
$$Z = R + j(X_L - X_C)$$

(2.21)

The two principal types of electrical energy sources are voltage source and current source. Sources can be either independent or dependent upon some other quantities. An independent voltage source maintains a voltage (fixed or varying with time) that is not affected by any other quantity. Similarly, an independent current source maintains a current (fixed or time-varying) that is unaffected by any other quantity. Some voltage (current) sources have their voltage (current) values varying with some other variables. They are called dependent-voltage (current) sources or controlled-voltage (current) sources.

Example 2.5

Given $V_S(t) = 750 \cos(5000t + 30°)$ V, find the current $i(t)$ for the circuit below:

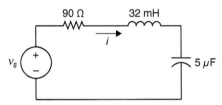

SOLUTION

The angular frequency of the source voltage is $\omega = 5000$ rad/s. The reactances for resistor, inductor and capacitor are, respectively:

$$Z_R = R + j0 = 90 \ \Omega$$

$$Z_L = 0 + j\omega L = j \times 5000 \times 32 \cdot 10^{-3} = j160 \ \Omega$$

$$Z_C = 0 - \frac{j}{\omega C} = -\frac{j}{5000 \times 5 \cdot 10^{-6}} = -j40 \ \Omega$$

And then the circuit impedance is:

$$Z = Z_R + j(Z_L - Z_C) = 90 + j(160 - 40) = 90 + j120 \ \Omega$$

2.3 PHASOR REPRESENTATION

In analyzing AC circuits or the real reactive power situation for a specific type of load, concepts such as *impedance, phasor,* and *complex power* are very useful. A phasor (a portmanteau of phase vector) is a complex number representing a sinusoidal function whose amplitude (A), angular frequency (ω), and initial phase (θ) are time-invariant. It is related to a more general concept, the analytic representation, which decomposes a sinusoid into the product of a complex constant and a factor that encapsulates the frequency and time dependence. These topics belong to an electrical circuit course. Before reviewing these concepts, it is useful to have a brief presentation of some important mathematical relationships. Euler's formula relates complex exponentials and trigonometric functions, through the expression:

$$e^{j\theta} = \cos\theta + j\sin\theta \tag{2.22}$$

Here $j = \sqrt{-1}$ is the complex unit. In electrical engineering, we use j not i, which traditionally is used for current. By adding and subtracting Euler's formula and its complex conjugate, we get two important relationships:

$$\cos\theta = \frac{1}{2}\left(e^{j\theta} + e^{-j\theta}\right) \text{ and } \sin\theta = \frac{1}{2j}\left(e^{j\theta} - e^{-j\theta}\right) \tag{2.23}$$

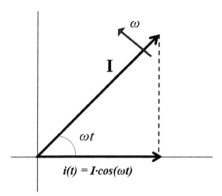

FIGURE 2.9 Rotating phasor *I* representing an AC current.

These relationships are called the "inverse Euler formulae," even called by some mathematicians, as *definitions* of cos (θ) and sin (θ), very important formulae that are extensively used in electrical and power engineering. Time-domain expressions in Equations (2.11) and (2.12) are expressed in phasor form, as:

$$\vec{V} = V_m \langle \theta_v \tag{2.24}$$

and

$$\vec{I} = I_m \langle \theta_i \tag{2.25}$$

However, we usually take the RMS values of voltage and current, rather than the amplitude or peak values as magnitudes of the phasors.

$$\vec{V} = V \langle \theta_v$$
$$\vec{I} = I \langle \theta_i \tag{2.26}$$

In other words, an AC current like one of Equation (2.12) is represented by a vector of length *I* rotating at an angular velocity (frequency) ω rad/sec as shown in Figure 2.9. Because the actual value of $i(t)$ depends on the phase angle of the rotating vector, the rotating vector is named *phasor*. A phasor represents a time-varying sinusoidal waveform by a fixed complex number. Depending on the sign of ($\theta_v - \theta_i$), we have a situation in which the current is either lagging ($\theta_v > \theta_i$) the voltage or leading ($\theta_v < \theta_i$) the voltage. The voltage and current phasors can have phase difference between their maximum values. However, the phasors and complex impedances are relevant only to sinusoidal sources. In the electrical power industry, because voltage is given by the generator or the utility, the power engineers always take the voltage as a reference phasor and then designate the current as leading or lagging the voltage.

Example 2.6

If the voltage and the current in an electric circuit are:

$$v(t) = \sqrt{2}\,(30)\cos(\omega t + 30°)$$
$$i(t) = \sqrt{2}\,(5)\cos(\omega t - 20°)$$

Write the phasor forms of $v(t)$ and $i(t)$, and then find the average power P into the network.

SOLUTION

$$\vec{V} = 30\langle 30° \ v, \ \ \vec{I} = 5\langle -20° \ A$$
$$\theta_V - \theta_i = 30° - (-20°) = 50°$$
$$P = 30 \times 5\cos(50°) = 96.5 \ W$$

Ohm's and Kirchhoff's laws and other network theorems are expressed straightforward in phasor notation, in the same manner as in DC, except that all numbers in AC are phasors (complex numbers), with magnitude and phase angle (see Figure 2.9). The nodal analysis, mesh analysis, superposition, and Thevenin and Norton equivalent source models are valid in phasor notation as they are in DC. For a general form of AC power delivered from a source to a load, we consider the voltage and current phasors:

$$\vec{V} = Z\vec{I} \tag{2.27}$$

Here $Z = R + jX$ is the circuit impedance. Kirchhoff's circuit laws work with phasors in complex form:

$$\sum_{k=1}^{N} \vec{I}_k = 0, \ \text{ and } \ \sum_{k=1}^{M} \vec{V}_k = 0 \tag{2.28}$$

The R-L-C circuit shown in Figure 2.10a is composed of a resistor, an inductor, and a capacitor connected in series. Its phasor diagram is shown in Figure 2.10b. The load impedance is expressed as:

$$\vec{Z} = \vec{R} + \vec{X}_L + \vec{X}_C$$
$$Z = R + j(X_L - X_C) = R + j\left(\omega L - \frac{1}{\omega C}\right) \tag{2.29}$$

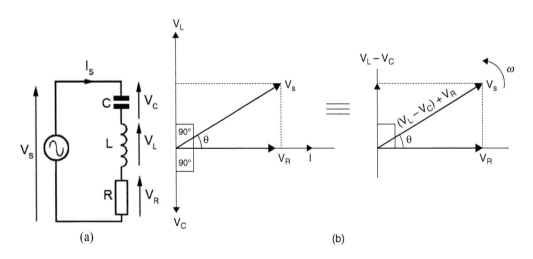

(a)   (b)

FIGURE 2.10 R-L-C circuit connection (a) and phasor diagram (b).

The magnitude of the impedance and the phase angle applied voltage and other voltages can be expressed as:

$$|Z| = \sqrt{R^2 + (X_L - X_C)^2}$$
$$\tan(\theta) = \frac{X_L - X_C}{R}$$

(2.30)

Depending of the magnitude of the inductive reactance X_L with respect to the capacitive reactance X_C, the phase angle can be positive, negative, or zero. When $X_L = X_C$, the impedance in the circuit in Figure 2.10a is equivalent to the resistance only, and the phase angle is zero. At the resonance, ωL equals $1/\omega C$, and:

$$\omega_0 = \frac{1}{\sqrt{LC}}, \text{ and } f_0 = \frac{1}{2\pi\sqrt{LC}}$$

(2.31)

Example 2.7

A 220 V adjustable frequency AC source is connected across R-L-C circuits, as one shown in Figure 2.10a. At 60 Hz, the resistance is 5 Ω, the inductive reactance is 4 Ω, and the capacitive reactance is 3 Ω. Calculate the following: (a) load impedance at 60 Hz, and (b) the resonant frequency of this circuit.

SOLUTION

a. $Z = 5 + j(-3) = 5 + j = \sqrt{5^2 + 1^2} \left\langle \tan^{-1}\left(\frac{1}{5}\right) = 5.10 \left\langle 78.7° \right. \Omega \right.$

b. $L = \dfrac{X_L}{2\pi f} = \dfrac{4}{2\pi 60} = 0.0006635$ H

$C = \dfrac{1}{2\pi f \times X_C} = \dfrac{1}{2\pi 60 \cdot 3} = 0.00088464$ F

and

$$f_0 = \frac{1}{2\pi\sqrt{LC}} = 207.85 \text{ Hz}$$

2.4   THREE-PHASE CIRCUITS

Almost all electric power generation and most of the power transmission in the world is in the form of three-phase. Three-phase power was first conceived by Nikola Tesla. In the early days of electric power generation, Tesla not only led the battle concerning whether the nation should be powered with low-voltage direct current or high-voltage alternating current, but he also proved that three-phase power was the most efficient way that electricity could be produced, transmitted, and consumed. A three-phase AC system consists of three-phase generators, transmission lines, and loads. An AC generator designed to develop a single sinusoidal voltage for each rotation of the shaft (rotor) is referred to as a *single-phase AC generator*. However, if the number of coils on the rotor is increased in a specified manner, the result is a *poly-phase AC generator*, which develops more than one AC phase voltage per rotation of the rotor. In general, three-phase systems are preferred over single-phase systems for the transmission of power for many reasons. First, thinner conductors can be used to transmit the same kVA at the same voltage, which reduces the amount of copper required (typically about 25% less). Second, the lighter lines are easier to install, and the supporting structures can be less massive and farther apart.

The generation, transmission, and distribution of electric power is accomplished by means of three-phase circuits. As we mentioned at the generating station, three sinusoidal voltages are generated having the same amplitude but displaced in phase by 120°, so-called a balanced source. If the generated voltages reach their peak values in the sequential order *abc*, the generator is said to have a positive phase sequence. If the phase order is *acb*, the generator is said to have a negative phase sequence. In a three-phase system, the instantaneous power delivered to the external loads is constant rather than pulsating as it is in a single-phase circuit. Also, three-phase motors, having constant torque, start and run much better than single-phase motors. This feature of three-phase power, coupled with the inherent efficiency of its transmission compared to single-phase (less wire for the same delivered power), accounts for its universal use. In general, three-phase systems are preferred over single-phase systems for the transmission of power for many reasons, such as:

1. Thinner conductors can be used to transmit the same kVA at the same voltage, which reduces the amount of copper required (typically about 25% less).
2. The lighter lines are easier to install, and the supporting structures are less massive and farther apart.
3. Three-phase equipment and motors run smoothly and have preferred running and starting characteristics compared to single-phase systems because of a more even flow of power to the transducer than can be delivered with a single-phase supply.
4. In general, the larger motors are three-phase because they are essentially self-starting and do not require a special design or additional starting circuitry.

Three-phase sinusoidal voltages and currents are generated with the same magnitude but are displaced in phase by 120°, by what is called a balanced source or generator, as shown in Figure 2.11. In a three-phase generator, three identical coils a, b, and c, are separated by an angle of 120° from each other, and the generator is turned by prime-mover, a system of three voltages, V_{an}, V_{bn}, and V_{cn} with the same magnitude and separated by 120° phase angles:

$$
\begin{aligned}
V_{an} &= V_m \sin(\omega t) = V_m \langle 0° \\
V_{bn} &= V_m \sin(\omega t - 120°) = V_m \langle -120° \\
V_{cn} &= V_m \sin(\omega t - 240°) = V_m \sin(\omega t + 120°) = V_m \langle -240°
\end{aligned}
\tag{2.32}
$$

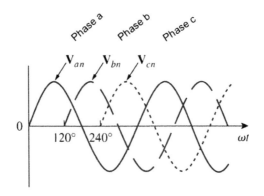

FIGURE 2.11   Three-phase voltages.

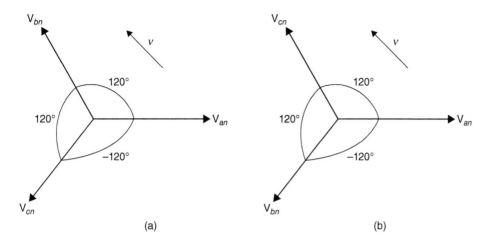

FIGURE 2.12 Positive (a) and negative (b) voltage sequences.

Here V_m is the peak value or the magnitude of the generated voltage. The sum of the three waveform voltages, by using trigonometric identities is

$$
\begin{aligned}
V &= V_{an} + V_{bn} + V_{cn} \\
&= V_m \left[\sin(\omega t) + \sin(\omega t - 120°) + \sin(\omega t - 240°) \right] = 0
\end{aligned}
$$
(2.33)

Three-phase systems may be labeled either by 1, 2, 3, or a, b, c, or sometimes, by using the three primary colors, red, yellow, and blue to represent them. The phase sequence is quite important for transmission, distribution, and use of electrical power. If the generated voltages reach their peak values in the sequential order *abc*, the generator is said to have a positive phase sequence, shown in Figure 2.12(a). If the phase order is *acb*, the generator is said to have a negative phase sequence, as shown in Figure 2.12(b).

The three single-phase voltages can be connected to form practical three-phase systems in two ways: (1) star or wye (Y) connections (circuits), or (2) delta (Δ) connections (circuits), as shown in Figure 2.13(a, b). In the Y-connection, one terminal of each generator coil is connected to a common point or neutral *n* and the other three terminals represent the three-phase supply. In a balanced three-phase system, knowledge of one of the phases provides the other two phases directly. However, this is not the case for an unbalanced supply. In a star-connected supply, it can be seen that the line current (current in the line) is equal to the phase current (current in a phase). However, the line voltage is not equal to the phase voltage. In a three-phase system, the instantaneous power delivered to the external loads is constant rather than pulsating as it is in a single-phase circuit. Three-phase

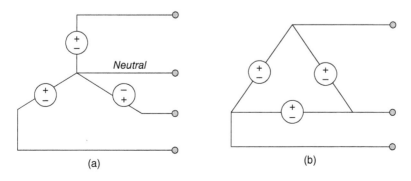

FIGURE 2.13 (a) Y-connected source; (b) Δ-connected source.

motors, having constant torque, start and run much better than single-phase motors. This three-phase power feature, coupled with the inherent higher efficiency of its transmission compared to single-phase (less wire for the same delivered power), accounts for its universal use. A power system has Y-connected generators and usually includes both Δ- and Y-connected loads. Generators are rarely Δ-connected, because if the voltages are not perfectly balanced, there is a net voltage, and consequently a circulating current, around the Δ. In a Y-connected phase, voltages are lower in the Y-connected generator, and thus less insulation is required. Figure 2.13 shows a Y-connected generator supplying balanced Y-connected loads through a three-phase line. In Y-connected circuits, the phase voltage is the voltage between any line (phase) and the neutral point, represented by V_{an}, V_{bn}, and V_{cn}, while the voltage between any two lines is called the line or line-to-line voltage, represented by V_{ab}, V_{bc}, and V_{ca}, respectively. For a balanced system, each phase voltage has the same magnitude, and we define:

$$|V_{an}| = |V_{bn}| = |V_{cn}| = V_P \qquad (2.34)$$

Here V_P denotes the effective magnitude of the phase voltage. We can show that

$$V_{ab} = V_{an} - V_{bn} = V_P\left(1 - 1\langle -120°\right) = \sqrt{3}V_P\,\langle 30° \qquad (2.35)$$

Similarly with relationships, we can obtain

$$V_{bc} = \sqrt{3}V_P\,\langle -90°$$
$$V_{ca} = \sqrt{3}V_P\,\langle 150° \qquad (2.36)$$

In a balanced three-phase y-connected voltage system, the line voltage V_L magnitude is related to the phase voltage magnitude through:

$$V_L = \sqrt{3}V_P \qquad (2.37)$$

Example 2.8

A three-phase generator is Y-connected, as shown in Figure 2.13a. The magnitude of each phase voltage is 220 V RMS. For *abc* phase sequence, write the three-phase voltage equations, and calculate the line voltage magnitude.

SOLUTION

The expressions of the phase voltages are:

$$V_{an} = 220\langle 0° \text{ V}$$
$$V_{bn} = 220\langle -120° \text{ V}$$
$$V_{cn} = 220\langle 120° \text{ V}$$

While the magnitude for the line voltage is:

$$V_{LL} = \sqrt{3} \times V_P = \sqrt{3} \times 220 = 380.6 \text{ V}$$

For a balanced system, the angles between the phases are 120° and the magnitudes are equal. Thus the line voltages would be 30° leading the nearest phase voltage. Calculation will easily show that the magnitude of the line voltage is $\sqrt{3}$ times the phase voltage. A current flowing out of line terminal, I_L, (the effective value of the line current) is the same as the phase current I_p (the effective value of the phase current) for the Y-connected circuits, thus:

$$I_L = I_P \tag{2.38}$$

In delta connection, as shown in Figure 2.13b, the line and the phase voltages have the same magnitude:

$$|V_L| = |V_P| \tag{2.39}$$

Similarly in the case of a delta connected supply, the current in the line is $\sqrt{3}$ times the current in the delta. In a manner similar to the Y-connected sources we can easily prove:

$$
\begin{aligned}
I_{ab} &= \sqrt{3}I_P \, \langle 0° \\
I_{bc} &= \sqrt{3}I_P \, \langle -90° \\
I_{ca} &= \sqrt{3}I_P \, \langle 150°
\end{aligned}
\tag{2.40}
$$

The balanced systems of three-phase currents are, as delta connection yields a corresponding set of balanced line currents, related as:

$$I_L = \sqrt{3}I_P \tag{2.41}$$

Where I_L denotes the magnitude of any of the three individual line currents.

2.4.1 BALANCED LOADS

A load on a three-phase supply usually consists of three impedances, either in wye (Y or star) or in delta (Δ) connection, as shown in Figure 2.14. A balanced load would have the impedances of the three phase equal in magnitude and in phase. Although the three phases would have the phase angles differing by 120° in a balanced supply, the current in each phase would also have phase angles differing by 120° with balanced currents. Thus if the current is lagging (or leading) the corresponding voltage by a particular angle in one phase, then it would lag (or lead) by the same angle in the other two phases as well (Figure 2.14a). For the same load, Y-connected impedance and the delta-connected impedance do not have the same value. However in both cases, each of the three phases has the same impedance as shown in Figures 2.14a,b. In a Y-connected loads, the line currents are taken from the supply are therefore equal to the phase currents of the loads. In Figure 2.15 the star point of the load is connected to the load star point, and the neutral current, in phasor notation is expressed as:

$$I_N = I_A + I_B + I_C \tag{2.42}$$

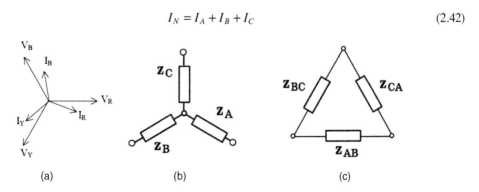

(a)   (b)   (c)

FIGURE 2.14 (a) Phasor diagram; (b) Y-connected load; (c) delta connection.

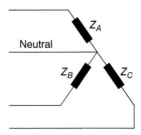

FIGURE 2.15   Star connected loads with neutral conductors.

With the neutral connected to the three-phase loads, the phase voltages across the corresponding phase of the load are:

$$V_A = Z_A I_A; \quad V_B = Z_B I_B; \quad V_C = Z_C I_C \tag{2.43}$$

For a balanced set of supply voltages and loads, the phasor sums of the load voltages and currents are zero, as well the neutral current. Because, for balanced loads, the neutral current is zero, the neutral current between the load and the source need not be made, and frequently is omitted. If the load impedances are not balanced, with the neutral conductor connected, a neutral current is flowing. However, if the neutral is not connected, then the star point neutral departs from the supply neutral (so-called floating neutral). The following equations then apply:

$$
\begin{aligned}
V_{AB} &= Z_A I_A - Z_B I_B \\
V_{BC} &= Z_B I_B - Z_C I_C \\
I_A + I_B + I_C &= 0
\end{aligned}
\tag{2.44}
$$

These equations are sufficient to solve for the load phase currents (equal to the source line currents) and hence the load voltages.

In the delta-connected loads, the load phase voltages are equal to the source line voltages for a balanced supply and loads. The relationships between supply line currents and the load phase currents in a phasor form are expressed as:

$$
\begin{aligned}
I_A &= I_{AB} - I_{CA} \\
I_B &= I_{BC} - I_{AB} \\
I_C &= I_{CA} - I_{BC}
\end{aligned}
\tag{2.45}
$$

A distinct advantage of a consistent set of notation adopted in a three-phase circuit analysis is the symmetry of the expressions, resulting in an additional way to check their consistency and correctitude. By using Ohm's law, the phase currents are given by:

$$I_{AB} = \frac{V_{AB}}{Z_{AB}}; \quad I_{BC} = \frac{V_{BC}}{Z_{BC}}; \quad I_{CA} = \frac{V_{CA}}{Z_{CA}} \tag{2.46}$$

Example 2.9

A Y-connected balanced three-phase load consists of three impedances of

$$Z_L = 44 \langle 30° \ \Omega$$

The loads are supplied with the balanced phase voltages:

$$V_{an} = 220\langle 0° \text{ V}$$
$$V_{bn} = 220\langle -120° \text{ V}$$
$$V_{cn} = 220\langle 120° \text{ V}$$

Calculate: (a) the phase currents; and (b) the line-to-line phasor voltages.

SOLUTION

a. The phase currents are computed as:

$$I_{an} = \frac{220\langle 0°}{44\langle 30°} = 5\langle -30° \text{ A}$$

$$I_{bn} = \frac{220\langle -120°}{44\langle 30°} = 5\langle -150° \text{ A}$$

$$I_{cn} = \frac{220\langle 120°}{44\langle 30°} = 5\langle 90° \text{ A}$$

b. Applying Equations (2.34) and (2.35), the line-to-line voltages are obtained as:

$$V_{ab} = V_{an} - V_{bn} = 220\langle 0° - 220\langle -120° = 220\sqrt{3}\langle 30° \text{ V}$$
$$V_{bc} = V_{bn} - V_{cn} = 220\langle -120° - 220\langle 120° = 220\sqrt{3}\langle -90° \text{ V}$$
$$V_{ca} = V_{cn} - V_{cn} = 220\langle 120° - 220\langle 0° = 220\sqrt{3}\langle 150° \text{ V}$$

Example 2.10

Repeat Example 2.9, in the case that the three impedances are Δ-connected.

SOLUTION

a. From the previous example, we have:

$$V_{ab} = 220\sqrt{3}\langle 30° \text{ V}$$
$$V_{bc} = 220\sqrt{3}\langle -90° \text{ V}$$
$$V_{ca} = 220\sqrt{3}\langle 150° \text{ V}$$

The currents in each of the load impedances are:

$$I_{ab} = \frac{220\sqrt{3}\langle 30°}{44\langle 30°} = 5\sqrt{3}\langle 0° \text{ A}$$

$$I_{bc} = \frac{220\sqrt{3}\langle -90°}{44\langle 30°} = 5\sqrt{3}\langle -120° \text{ A}$$

$$I_{ca} = \frac{220\sqrt{3}\langle 150°}{44\langle 30°} = 5\sqrt{3}\langle 120° \text{ A}$$

b. The line currents are computed using Equation (2.44), as:

$$I_a = I_{ab} - I_{ca} = 5\sqrt{3}\langle 0° - 5\sqrt{3}\langle -120° = 15\langle 30°\ \text{A}$$
$$I_b = I_{bc} - I_{ab} = 5\sqrt{3}\langle -120° - 5\sqrt{3}\langle 120° = 15\langle -90°\ \text{A}$$
$$I_c = I_{ca} - I_{bc} = 5\sqrt{3}\langle 120° - 5\sqrt{3}\langle -120° = 15\langle 150°\ \text{A}$$

2.4.2 Mixed Connection Circuits, Wye-Delta Transformation

The source and the load are not always connected in the same manner. For example, the load can be Δ-connected and the source Y-connected, or vice versa. In either case, attention must be given to the calculation of the line and phase quantities. The phase and line voltage and current relationships established in previous subsections apply here. We can infer this by examining the following example.

Example 2.11

Figure 2.16 shows a balanced Δ-connected load supplied by a balanced 120 V Y-connected source. The line current is $I_3 = 15\langle 75°$ A. Compute the load impedance.

SOLUTION

First we compute the voltage across the load:

$$\bar{V}_{12} = \sqrt{3}V_1\langle 30° = \sqrt{3}\left(120\langle 0°\right)\langle 30° = 208\langle 30°\ \text{V}$$

The load current is:

$$I_{12} = \frac{I_1}{\sqrt{3}\langle -30°} = \frac{I_3\langle -120°}{\sqrt{3}\langle -30°} = \frac{15\langle 75° - 120°}{\sqrt{3}\langle -30°} = 5.77\langle -15°\ \text{A}$$

The load impedance is computed by Ohm's law, as:

$$Z_\Delta = \frac{V_{12}}{I_{12}} = \frac{208\langle 30°}{5.77\langle -15°} = 36\langle 45°\ \Omega$$

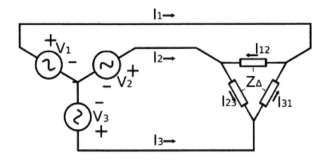

FIGURE 2.16 Wye-connected source supplying a delta-connected load.

2.4.3 POWER RELATIONSHIPS IN THREE-PHASE CIRCUITS

Assuming a three-phase is supplying a three-phase balanced load, as in Figure 2.14b with three-phase sinusoidal phase voltages:

$$v_a(t) = \sqrt{2}V_P \sin(\omega t)$$
$$v_b(t) = \sqrt{2}V_P \sin(\omega t - 120°)$$
$$v_c(t) = \sqrt{2}V_P \sin(\omega t + 120°)$$

(2.47)

With current flowing through the load given by

$$i_a(t) = \sqrt{2}I_P \sin(\omega t - \phi)$$
$$i_b(t) = \sqrt{2}I_P \sin(\omega t - 120° - \phi)$$
$$i_c(t) = \sqrt{2}I_P \sin(\omega t + 120° - \phi)$$

(2.48)

where ϕ is the phase angle between the voltage and current in each phase.

The instantaneous power supplied to one phase of the load (Equation [2.5]), as in Figure 2.17 is

$$p(t) = v(t) \cdot i(t)$$

Therefore, the instantaneous power in each of the three phases of the load is:

$$p_a(t) = v_a(t) \cdot i_a(t) = 2VI \sin(\omega t)\sin(\omega t - \theta)$$
$$p_b(t) = v_b(t) \cdot i_b(t) = 2VI \sin(\omega t - 120°)\sin(\omega t - 120° - \theta)$$
$$p_c(t) = v_c(t) \cdot i_c(t) = 2VI \sin(\omega t - 240°)\sin(\omega t - 240° - \theta)$$

(2.49)

The total instantaneous power flowing into the load is expressed as:

$$p_{3\phi}(t) = v_a(t)i_a(t) + v_b(t)i_b(t) + v_c(t)i_c(t) \text{ W}$$

(2.50)

By substituting expressions for phase voltages and currents, Equations (2.46) and (2.47), respectively, in Equation (2.48) and using additional trigonometric identity:

$$\cos(\alpha) + \cos(\alpha - 120°) + \cos(\alpha - 240°) = 0$$

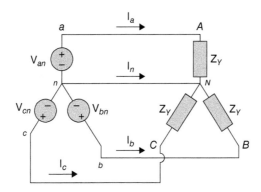

FIGURE 2.17   A Y-connected generator supplying a Y-connected load.

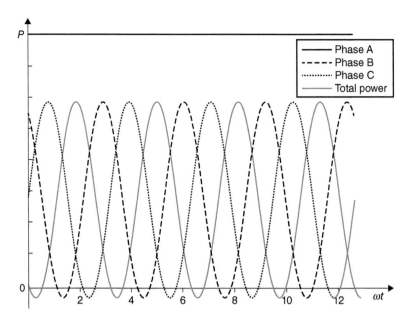

FIGURE 2.18 Power in a three-phase balanced system.

Equation (2.19) can be rewritten as the equation below:

$$p_{3\phi}(t) = 3|V||I|\cos(\phi) = 3P \text{ W} \tag{2.51}$$

Here: $|V| = \sqrt{2}V_P$ and $|I| = \sqrt{2}I_P$ are the peak magnitude (amplitude) of the phase voltage and current. Equation (2.51) represents a very important result. In other words: *In a balanced three-phase system the sum of the three individually pulsating phase powers adds to a constant, nonpulsating total active power of magnitude three times the real (active) power in each phase.* A graphical representation of this important consequence of three-phase balanced systems is shown in Figure 2.18. The single-phase power equations are applied to each phase of Y- or Δ-connected three-phase loads. The real, active, and apparent powers supplied to a balanced three-phase load are:

$$
\begin{aligned}
P &= 3V_\phi I_\phi \cos(\theta) = 3ZI_\phi^2 \cos(\theta) \\
Q &= 3V_\phi I_\phi \sin(\theta) = 3ZI_\phi^2 \sin(\theta) \\
S &= 3V_\phi I_\phi = 3ZI_\phi^2
\end{aligned}
\tag{2.52}
$$

The angle θ is again the angle between the voltage and the current in any of the load phase, and the power factor of the load is the cosine of this angle. We can express the powers of Equation (2.50) in terms of line quantities, regardless of the connection type (wye or delta), as:

$$
\begin{aligned}
P &= \sqrt{3}V_{LL}I_L \cos(\theta) \\
Q &= \sqrt{3}V_{LL}I_L \sin(\theta) \\
S &= \sqrt{3}V_{LL}I_L
\end{aligned}
\tag{2.53}
$$

We have to keep in mind the angle θ in Equations (2.53) is the angle between the *phase voltage* and the *phase current*, not the angle between the line-to-line voltage and the line current.

Example 2.12

The terminal line-to-line voltage of a three-phase generator equals 13.2 kV. It is symmetrically loaded and delivers an RMS current of 1.350 kA per phase at a phase angle of 24° lagging. Compute the power delivered by this generator.

SOLUTION

The RMS value of the phase voltage is

$$|V| = \frac{13.2}{\sqrt{3}} = 7.621 \text{ kV/phase}$$

The per-phase active (real) and reactive power are given by:

$$P = 7.621 \cdot 1.350 \cdot \cos(24°) = 9.399 \text{ MW/phase}$$
$$Q = 7.621 \cdot 1.350 \cdot \cos(24°) = 4.185 \text{ MVAR/phase}$$

The instantaneous powers in phases a, b, and c are pulsating and are given by:

$$p_a(t) = 9.399\left(1 - \cos(2\omega t)\right) - 4.185\sin(2\omega t)$$
$$p_b(t) = 9.399\left(1 - \cos(2\omega t - 120°)\right) - 4.185\sin(2\omega t - 120°) \qquad (2.54)$$
$$p_a(t) = 9.399\left(1 - \cos(2\omega t - 240°)\right) - 4.185\sin(2\omega t - 240°)$$

The total (constant) three-phase power is:

$$P_{3\phi} = 3 \times 9.399 = 28.197 \text{ MW}$$

The fact that three-phase *active (real) power* is constant tempts us to believe that the *reactive power* in a three-phase is zero (as in a DC circuit). However, the reactive power is very much present in *each phase* as shown in Equation (2.52). The reactive power per phase is 4.185 MVAR.

Example 2.13

A three-phase load draws 120 kW at a power factor of 0.85 lagging from a 440 V bus. In parallel with this load, a three-phase capacitor bank that is rated 50 kVAR is inserted. Find:

  a. The line current without the capacitor bank.
  b. The line current with the capacitor bank.
  c. The P.F. without the capacitor bank.
  d. The P.F. with the capacitor bank.

SOLUTION

  a. From the three-phase active power formula, the magnitude of the load current is:

$$I_{Load} = \frac{P}{\sqrt{3}V_L \times PF} = \frac{120 \times 10^3}{\sqrt{3}\,440(0.85)} = 185.25 \text{ A}$$
$$I_{Load} = 185.25 \langle -\cos^{-1}(0.85) = 185.25 \langle -31.8° \text{ A}$$

b. The line current of the capacitor bank (a pure reactive load) is:

$$I_{Cap} = \frac{50 \times 10^3}{\sqrt{3}\,440} = 65.6 \langle 90° \text{ A}$$

The line current is:

$$I_L = I_{Load} + I_{Cap} = 160.6 \langle -11.5° \text{ A}$$

c. The PF without capacitor bank is PF = 0.85
d. The PF with capacitor bank is

$$PF = \cos(11.5°) = 0.98$$

2.5 PER-UNIT SYSTEM

The per-unit (p.u.) value representation of electrical variables in power system and electric machine computation is a common and useful practice. In power system analysis, a per-unit system is the expression of system quantities as fractions of a defined base-unit quantity. An interconnected power system typically consists of many different voltage levels, given a system containing several transformers and/or rotating machines. The *per-unit system* simplifies the analysis of complex power systems by choosing a common set of base parameters in terms of which all system quantities are defined. The different voltage levels disappear and the overall system reduces to a set of impedances. Calculations are simplified because quantities expressed as per-unit do not change when they are referred from one side of a transformer to the other, a pronounced advantage in power system analysis where large numbers of transformers may be encountered. Moreover, similar types of apparatus have the impedances lying within a narrow numerical range when expressed as a per-unit fraction of the equipment rating, even if the unit size varies widely. Conversion of per-unit quantities to volts, ohms, or amperes requires knowledge of the base to which the per-unit quantities were referenced. The idea of the per-unit system is to absorb large differences in absolute values into base relationships. Representations of elements in the system with per-unit values become more uniform. The per-unit numerical value of any quality is the ratio of its value to a chosen base quantity of the same dimension. A per-unit quantity is a normalized quantity with respect to the chosen base value. There are several reasons for using a per-unit system:

- Similar apparatus (generators, transformers, lines) have similar per-unit impedances and losses expressed on their own rating, regardless of their absolute size. Per-unit data can be checked rapidly for gross errors. A per-unit value out of normal range is worth looking into for potential errors.
- The per-unit component values lie within a narrow range regardless of the equipment rating.
- Manufacturers usually specify the apparatus impedance in per-unit values.
- Use of the constant $\sqrt{3}$ is reduced in three-phase calculations.
- Per-unit quantities are the same on either side of a transformer, independent of voltage level.
- By normalizing quantities to a common base, both hand and automatic calculations are simplified.
- It improves numerical stability of automatic calculation methods.
- Per-unit data representation yields important information about relative magnitudes.
- It is ideal for computer simulations.

The definition of the per-unit value of a quantity is:

$$\text{p.u. value} = \frac{\text{Actual value}}{\text{Base (reference) value of the same dimension}} \tag{2.55}$$

The complete characterization of a per-unit system requires that all four base values be defined. Given the four base values, the per-unit quantities are defined as:

$$V_{p.u.} = \frac{V}{V_{base}}; \quad I_{p.u.} = \frac{I}{I_{base}}; \quad S_{p.u.} = \frac{S}{S_{base}}; \quad Z_{p.u.} = \frac{Z}{Z_{base}} \tag{2.56}$$

The per-unit system was developed to make manual analysis of power systems easier. Although power-system analysis is now done by computer, results are often expressed as per-unit values on a convenient system-wide base. The base value is always a real number and the per-unit value is dimensionless. Five quantities are involved in this calculation: the current, the voltage, the complex power, the impedance, and the phase angle. Phase angles are dimensionless, the other four quantities are completely described by the knowledge of only two of them. Usually the nominal line or equipment voltage is known as well as the apparent (complex) power, so these two quantities are often selected for base-value calculation. For example, considering a single-phase system, the expression of base current is:

$$I_{base} = \frac{S_{base(1-\phi)}}{V_{base(LN)}} \tag{2.57}$$

The expression of the base impedance is:

$$Z_{base} = \frac{V_{base(LN)}^2}{S_{base(1-\phi)}} = \frac{V_{base(LN)}}{I_{base}} \tag{2.58}$$

The magnitude of the base current in a three-phase system can be calculated as:

$$I_{base} = \frac{S_{base(3-\phi)}}{\sqrt{3}V_{base(LL)}} \tag{2.59}$$

The base impedance can be calculated as:

$$Z_{base} = \frac{V_{base(LL)}^2}{S_{base(3-\phi)}} = \frac{V_{base(LL)}}{\sqrt{3}I_{base}} \tag{2.60}$$

Per-unit quantities obey the circuit laws, thus:

$$S_{p.u.} = V_{p.u.}I_{p.u.}^*$$
$$V_{p.u.} = Z_{p.u.}I_{p.u.} \tag{2.61}$$

For a three-phase system, the phase impedance in per-unit is given by:

$$Z_{p.u.} = \frac{V_{p.u.}^2}{S_{L(p.u.)}^*} \tag{2.62}$$

Example 2.14

Assuming line voltage of 735 KV for a 120 MVA transmission line with the impedance:

$$Z = 4.50 + j75.30 \ \Omega$$

Calculate the per-unit transmission line resistance, reactance, and impedance.

SOLUTION

The line impedance is:

$$Z = \sqrt{4.5^2 + 75.3^2} \left\langle \tan^{-1}\left(\frac{75.3}{4.5}\right) = 75.4 \langle 86.6° \ \Omega \right.$$

The base impedance is:

$$Z_{base} = \frac{V_{base}^2}{S_{base}} = \frac{\left(735 \times 10^3\right)^2}{120 \times 10^6} = 4502 \ \Omega$$

The per-unit transmission line resistance, reactance, and impedance are computed as:

$$R_{p.u.} = \frac{4.5}{4502} = 9.996 \times 10^{-4} \ \text{p.u.}$$

$$X_{p.u.} = \frac{75.3}{4502} = 0.01673 \ \text{p.u.}$$

$$Z_{p.u.} = \frac{75.4}{4502} = 0.01675 \ \text{p.u.}$$

Usually if none is specified, the p.u. values given are the nameplate ratings. There are situations when the base for the system is different from the base for each particular generator or transformer; hence, it is important to be able to express the p.u. value in terms of different bases. The rule for the impedances is:

$$Z_{p.u.(new)} = Z_{p.u.(old)} \frac{S_{base(new)}}{S_{base(old)}} \cdot \frac{V_{base(new)}^2}{V_{base(old)}^2} \qquad (2.63)$$

Example 2.15

Convert the impedance value of Example 2.5 to the new base 240 MVA and 345 kV.

SOLUTION

We have

$$Z_{p.u.(old)} = 9.996 \times 10^{-4} + j0.01673$$

This is the value for the base for 120 MVA, and 735 kV base. With a new base 240 MVA and 345 kV, by using the impedance conversion relationship (Equation [3.30]):

$$Z_{p.u.(new)} = Z_{p.u.(old)}\left(\frac{240}{120}\right) \cdot \left(\frac{735}{345}\right)^2 = 9.0775 \cdot Z_{(p.u.)(old)}$$

and

$$Z_{p.u.(new)} = 0.0091 + j0.1519$$

Example 2.16

A three-phase generator rated at 350 MVA and 21 kV has a per-phase reactance of 0.35 p.u. on its own base. The generator is placed in a system where the bases are 100 MVA and 13.8 kV. Find the reactance of the generator in per-unit on the new base.

SOLUTION

Applying Equation (3.30), we have:

$$X_{p.u.(new)} = 0.35\left(\frac{100}{350}\right)\left(\frac{21}{13.8}\right)^2 = 0.232 \text{ p.u.}$$

2.6 CHAPTER SUMMARY

This chapter discusses and introduces the groundwork for the study of electrical energy systems. The fundamental DC and AC circuits and theorems, including phasors and three-phase circuits, power calculations and flows, and basic tools used to resolve them are introduced and reviewed in this chapter. A clear understanding of these concepts, principles, and calculation methods is critical in solving problems and design projects in building and industrial electrical and energy systems. Most power engineers and designers are working in power distribution and in the electrical and energy systems of buildings and industrial facilities. Very few of them work in power plants or in high-voltage transmission systems. Regardless of the working area, the engineers and technicians must have a good understanding of the fundamentals and calculation and design methods used. A major portion of the power systems and most of the electrical and energy systems of large buildings, and industrial and commercial facilities use three-phase AC circuits, networks, devices, and equipment. The conception, by Nikola Tesla, of a poly-phase AC system was one of the most important innovations in the history of engineering and applied sciences. The essential feature of a three-phase system is this: Although all currents are sinusoidal alternating type, if the system is balanced, the total instantaneous power of the system is constant. In the early days of power generation, Tesla not only led the battle of the power system using DC or AC, but he also proved that three-phase electric power was the most efficient way to generate, transfer, and use electricity. The per-unit system simplifies the analysis and calculations of complex power systems, power flows, and design by choosing a common set of base parameters in terms of which all system quantities are defined. The different voltage levels disappear and the overall system reduces to a set of impedances. A poly-phase electric generator or motor converts power from one form to another without fluctuations or pulsations, with constant energy stored in the electromagnetic field. Standard frequency used in North America is 60 Hz, while Europe and the rest of the world are using 50 Hz frequency. In summary, this chapter provides a brief but complete discussion and review of fundamental concepts, including power concepts, three-phase systems, phasors, per-unit system, energy units, and important theorems and calculation methods.

2.7 QUESTIONS AND PROBLEMS

1. What is the significance of apparent power?
2. What is the advantage of using phasors in AC electric circuits?
3. What is power factor?

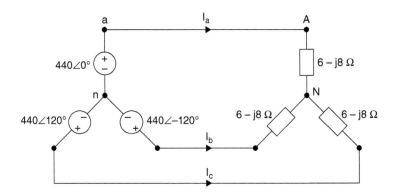

FIGURE P2.1

4. What are the advantages of the three-phase system compared with single-phase?
5. What are the advantages of using phasors?
6. What does reactive power do?
7. Why has the three-phase power system been adopted over the single-phase power system?
8. What is the real, reactive, and apparent power? In what units are they measured?
9. Why is it important to use apparent power near the unity power factor?
10. Why was AC adopted in power systems?
11. Consider a purely capacitive circuit (a capacitor connected to an AC source).
    a. How does the capacitive reactance change if the driving frequency is doubled? And, if it is halved?
    b. Are there any times when the capacitor is supplying power to the AC source?
12. A series RLC circuit with $C = 100 \ \mu F$, $L = 100 \ mH$, and $R = 40.0 \ \Omega$ is connected to a sinusoidal voltage $v(t) = 40.0\sin(\omega t)$, with $\omega = 200$ rad/s.
    a. What is the impedance of the circuit?
    b. Let the current at any instant in the circuit be $i(t) = I_0\sin(\omega t\text{-}\phi)$. Find I_0.
    c. What is the phase ϕ?
13. Obtain the line currents for the circuit in Figure P2.1
14. Two parallel resistors carry current of 3 A and 5 A. If the resistance of one is 90 Ω, what is the resistance of the other?
15. A DC generator transmits power over transmission lines having a resistance of 0.35 Ω to a 10 KW load. The load voltage is 135 V. What is the generator voltage? How much power is lost by the transmission line?
16. A single-phase source supplies a load of R = 30 Ω connected in parallel with capacitive reactance $X_C = 20 \ \Omega$. The voltage across the load is

$$v(t) = \sqrt{2}120\cos(\omega t - 45°) \ V$$

    a. Find the phasors of voltage, total current, instantaneous current, and complex power.
    b. Repeat (a) if instead of capacitive reactance we have an inductive reactance $X_L = 20 \ \Omega$.
17. An electric load consists of a 6 Ω resistance, a 6 Ω inductive reactance, and an 8 Ω capacitive reactance connected in series. The series impedance is connected across a voltage source of 120 V. Compute the following: (a) power factor of the load; (b) source current; parent, real, and reactive powers of the circuit.
18. A one-loop circuit has a source voltage of 120 V and a load impedance 10< + 84° Ω. Determine the current and the average power delivered to the load. What can you infer about the load nature?

19. A parallel circuit consisting of two impedances

$$Z_1 = 15\langle 45° \; \Omega$$
$$Z_2 = 30\langle -30° \; \Omega$$

Is connected across a voltage source: $\vec{V} = 120\langle 0° \; V$

Find: (a) the current in each circuit branch and the total current; (b) the complex power in each branch and the total complex power; (c) the overall power factor of the circuit; and (d) draw the power triangles for each branch circuit and the combined power triangle.

20. A 10 HP, 220 V single-phase electric motor operates at 90% efficiency while running at rated horsepower and has a power factor of 0.707 lagging. Find: (a) the current drawn by the motor; and (b) the real and reactive power absorbed by the motor.

21. A circuit element has a voltage of 270cos(314t) V, across its terminals and is drawing 30cos(314t-25°) A. Calculate the average power absorbed by this circuit element.

22. A transformer rated at 1200 kVA is operated at full load (i.e., carrying rated kVA) at a PF = 0.70 lagging. A capacitor bank is added to improve the overall power factor to become 0.9 leading. Find the capacitor bank reactive power required.

23. A parallel circuit consists of two impedances:

$$Z_1 = 30\langle 45° \; \Omega$$
$$Z_2 = 45\langle -30° \; \Omega$$

Find: (a) total impedance, the current in each branch and the total current; (b) the total complex power; and (c) the overall power factor of the circuit.

24. In a Y-Y circuit, the source is a balanced positive phase sequence, with the phase voltage: $V_{an} = 120 < 0° \; V$. It feeds a balanced load with the phase impedance: $Z_Y = 9 + j12 \; \Omega$ through a balanced line with line impedance: $Z_L = 0.5 + j0.5 \; \Omega$. Calculate the phase voltages and currents in the load.

25. A balanced star-connected load is fed from a 460 V, 60 Hz, three-phase supply. The resistance in each phase of the load is 30 Ω and the load draws a total power of 15 kW. Calculate (a) the line current drawn, (b) the load power factor, and (c) the load inductance.

26. A three-phase load is connected in star to a 460 V, 60 Hz supply. Each phase of the load consists of a coil having inductance 0.2 H and resistance 45 Ω. Calculate the line current.

27. If the load specified in previous problem is connected in delta, determine the values for phase and line currents.

28. A star-connected balanced generator has the phasor voltage $V_a = 120 < 0° \; V$. Determine the line-to-line voltage phasors and their phase angles with respect to V_a.

29. A 120 V RMS source is supplying three impedances, $10 + j20 \; \Omega$, $5 + j20 \; \Omega$, and $10 + j30 \; \Omega$, connected in parallel. Determine the real and reactive powers drawn by each of the impedances. Without any calculation, can you identify the impedance that is drawing the largest power?

30. If the applied voltage leads the current in a series RLC circuit, is the frequency above or below resonance?

31. Two parallel loads draw power from a 460 V voltage source. Load 1 draws 15 kW and 12 kVAR, while Load 2 draws 7.5 kW and 6 kVAR. Determine the combined power factor and the total apparent power drawn from the source.

32. How does the power factor in an RLC circuit change with resistance, R, inductance, L, and capacitance, C?

33. A series RC circuit with R = 4 × 10³ Ω, C = 0.40 μF is connected to an AC voltage source, $v(t) = 100 \sin(\omega t)$ V, with $\omega = 200$ rad/s.
    a. What is the RMS current in the circuit?
    b. What is the phase between the voltage and the current?
    c. Find the power dissipated in the circuit.
    d. Find the voltage drop both across the resistor and the capacitor.
34. A 220 V RMS source voltage is applied to 35 < 30° Ω load impedance. Calculate the average, peak, and RMS currents, and average power delivered to the load.
35. A three-phase, 460 V, 60 Hz, Y-connected generator delivers power to a Y-connected balanced load with 18 + j9 Ω in each load phase. Determine the line current and the power delivered to this load.
36. The loads in parallel draw power from a 480 V source. First load is 24 kW and 8 kVAR, the second one is 12 kW and 6 kVAR, and the third one is 18 kW and 6 kVAR. Determine the overall power factor, apparent, active, and reactive powers drawn by the three-load combination from the source. What is the capacitance of a capacitor bank needed to improve the overall power factor to 0.95?
37. A balanced Y-connected load has a phase impedance of 8 + j6 Ω connected to a three-phase supply of 460 V. Determine: (a) the phase voltage; (b) the phase and line currents; (c) the power factor at the load; and (d) the power consumed by the load.
38. Three loads are connected in parallel across a 12.47 kV power supply. One is resistive 63 kW load, the second one is an induction motor of 72 kW and 63 kVAR, and the last one is a capacitive load drawing 180 kW at 0.85 PF. Find the total apparent power, power factor, and power supply current.
39. The magnitude of each phase voltage of an unbalanced load is 220 V RMS. The load impedances are:

$$Z_A = 6 + j8 \ \Omega$$
$$Z_B = 4 + j6 \ \Omega$$
$$Z_A = 3 + j4 \ \Omega$$

    Calculate: (a) line currents; and (b) the neutral current.
40. The per-phase reactance, 10 kVA, 120 V, Y-connected synchronous generator is 12 Ω. Determine the per-unit reactance, considering the base values are 10 kVA and 120 V.
41. The per-phase load impedance of a three-phase delta-connected load is 4 + j6 Ω. If a 480 V, three-phase supply is connected to this load, find the magnitude of: (a) phase current; and (b) line current.
42. Calculate the RMS value, supply frequency and the phase shit in degrees for the AC voltage given by:

$$v(t) = 180 \sin(300t + 0.866) \ V$$

43. The magnitude of each phase voltage of an unbalanced load is 220 V RMS. The load impedances are:
    Calculate: (a) line currents; and (b) the neutral current.
44. The per-phase reactance, 10 kVA, 120 V, Y-connected synchronous generator is 12 Ω. Determine the per-unit reactance, considering the base values are 10 kVA and 120 V.
45. The per-phase load impedance of a three-phase delta-connected load is 4 + j6 Ω. If a 480 V, three-phase supply is connected to this load, find the magnitude of: (a) phase current; and (b) line current.

REFERENCES AND FURTHER READINGS

1. O. Elegerd, *Basic Electric Power Engineering*, Addison-Wesley, 1977.
2. A. Daniels, *Introduction to Electrical Machines*, The Macmillan Press, 1976.
3. H. M. Rustebakke, *Electric Utility Systems and Practices*, John Wiley & Sons, New York, 1983.
4. R. D. Schultz and R. A. Smith, *Introduction to Electric Power Engineering*, John Wiley & Sons, New York, 1988.
5. Y. Wallach, *Calculations and Programs for Power System Networks*, Prentice Hall, 1986.
6. A. C. Williamson, *Introduction to Electrical Energy Systems*, Longman Scientific & Technical, 1988.
7. J. R. Cogdell, *Foundations of Electric Power*, Prentice-Hall, Upper Saddle River, NJ, 1999.
8. N. Cohn, *Control of Generation and Power Flow on Interconnected Systems*, Wiley, New York, 1966.
9. D. J. Glover, *Power System Analysis and Design* (5th ed.), Cengage Learning, 2012.
10. F. A. Furfari, The evolution of power-line frequencies $133^{1}/_{3}$ to 25 Hz, *Industry Applications Magazine, IEEE*, Vol. 6(5), pp. 12–14, 2000.
11. E. L. Owen, The origins of 60-Hz as a power frequency, *Industry Applications Magazine, IEEE*, Vol. 3(6), pp. 8, 10, 12–14, 1997.
12. G. Neidhofer, 50-Hz frequency: How the standard emerged from a European jungle, *IEEE Power and Energy Magazine*, July/August, pp. 66–81, 2011.
13. S. J. Chapman, *Electrical Machinery and Power System Fundamentals* (2nd ed.), McGraw Hill, 2002.
14. G. Rizzoni, *Principles and Applications of Electrical Engineering*, McGraw-Hill, 2007.
15. P. Schavemaker and L. van der Sluis, *Electrical Power System Essentials*, John Wiley and Sons, 2008.
16. M. E. El-Hawary, *Introduction to Power Systems*, Wiley – IEEE Press, 2008.
17. M. A. El-Sharkawi, *Electric Energy: An Introduction* (2nd ed.), CRC Press, 2009.
18. J. L. Kirtley, *Electric Power Principles: Sources, Conversion, Distribution and Use*, Wiley, 2010.
19. C. K. Alexander and M. N. O. Sadiku, *Fundamentals of Electric Circuits*, McGraw-Hill, 2012.
20. F. M. Vanek, L. D. Albright, and L. T. Angenent, *Energy Systems Engineering*, McGraw-Hill, 2012.
21. N. Mohan, *Electric Power Systems – A First Course*, Wiley, 2012.
22. M. R. Patel, *Introduction to Electrical Power Systems and Power Electronics*, CRC Press, 2013.
23. J. E. Fleckenstein, *Three-Phase Electrical Power*, CRC Press, 2016.

3 Building Power Supply and Industrial Power Distribution

3.1 POWER SYSTEM STRUCTURE AND COMPONENTS

Electric utilities transfer power (or *energy*) from the power plant, in the most efficient and reliable ways. Electricity produced at generation stations is conveyed to the consumers through a complex network of transmission and distribution systems. In the United States, power companies provide electricity to medium or large buildings at 13,800 V (13.8 kV). For small commercial buildings or residential customers, power companies lower the voltage with transformers, located on power poles or mounted on the ground. From there, the electricity is fed through a meter and into the building. Entire distribution networks are complete systems, structured in the most suitable application architecture, according to local regulations, constraints related to the power supply, and to the type of loads. Power distribution equipment (panelboards, switchgears, circuit connections, etc.) are determined from building plans, structure, and the location and grouping of loads. The type of premises and allocation can influence their immunity to external disturbances. Electric power distribution, the section of the power delivery infrastructure, takes electricity from high-voltage (HV) transmission circuits and delivers it to customers. Primary distribution lines are medium-voltage (MV) circuits, usually from 600 V to 35 kV. At a distribution substation, a transformer steps down the incoming transmission-level voltage (35 kV to 230 kV or higher) to the primary distribution circuits, fanning out from the substation. Close to each end user, a power distribution transformer steps down to the primary-distribution voltages, and further to low-voltage secondary circuits (commonly 120/240 V or other utilization voltages). From the power distribution transformer, the secondary distribution circuits connect to the end users where the connection is made at the service entrance. The distribution system provides the infrastructure to deliver power from the substations to the loads and end users. Typically radial in nature, the primary distribution system includes feeders and connecting networks, the so-called laterals, with typical voltages of 34.5 kV, 14.4 kV, 13.8 kV, 13.2 kV, 12.5 kV, 12 kV, or lower voltages, such as 4.16 kV. The distribution voltages in a specific service territory are similar because it is easier and more cost effective to stock spare parts when the system voltages are consistent. The electric distribution system in a building, industrial facility, or an installation site transfers power from one or more power supply, in the most efficient way and at higher standards, to the individual loads and to all electrically operated devices. Modern industrial applications require the most technologically advanced systems and power quality. Industrial facilities are becoming more and more dependent on computer controls, automation, and power electronics in their processes, operation, and management. As a consequence, modern facilities require increased cleanliness and reliability in the electrical power supply. The main purpose of power supply design is to provide an optimum, secure, and high-quality power supply, equipment, feeder, branch circuits, protection, and control in the most economical way and with the right selection of electrical equipment. Because the service is a connection point to the utility, it provides the electric service location, the metering equipment, pertinent information regarding the electrical service, and often special requirements for designers. The service and the utility-facility connections must comply with all National Electric Code (NEC) or International Electrotechnical Commission (IEC) requirements and specifications. The building and facility electrical service consists of the conductors and cables connecting the building and facility electrical system to the local utility power source, the raceway system containing the electrical service entrance conductors and cables, metering equipment, the main service disconnect, and the main overcurrent protection devices. In practice, for many cases the electrical designer relies on information and assistance from specialists in related but separate fields. This applies in particular to the controls for heating and air-conditioning, which are designed by specialists in that field and not by the consultant or contractor employed for the general electrical system. Other services within a building include electrical equipment, electrical motors, heaters, air-conditioning, and controls. From the user's perspective, the electricity service in a building consists of switches, sockets, clock connectors, cooker control units, and similar outlets. Such fittings are collectively known as accessories (accessory to the wiring), which are the main substance of the installation from the designer and installer point of view. In the complete building electrical installation, the wiring and accessories are interdependent and can be fully understood without each other. This chapter discusses the major components of the power distribution system, and associated electrical equipment and controls, feeders, transformers, monitoring, control, and protection. The primary purpose of electricity distribution is to meet the customer's demands and requirements for energy after receiving the bulk electrical energy from power transmission and power subtransmission substations. The primary power distribution substations serve as load centers and the customer substation interfaces to the LV networks. Customer power distribution substations are referred to as power distribution centers, usually provided by the customers, accommodating several switchgears and power distribution transformers to enable LV connections to the customer's incoming switchboard or panelboards.

3.2   POWER DISTRIBUTION NETWORKS AND CONFIGURATIONS

Electric utilities transmit power from power plants most efficiently at very high voltage levels. In the United States, power companies provide electricity to medium or large buildings at 13.8 kV. For smaller buildings or residential customers, utilities lower the HV voltages with a system of transformers, located in substations, or in power distribution networks. From there, the electricity is fed through a meter and into the building. Power flow on a line is a function of voltage and current. Because the current itself is inherently bidirectional, power can typically readily flow in either direction. However, other operational constraints such as circuit breakers and control devices may not be able to accommodate reversal of power flow without replacement or modification. The electric power grid operates as a three-phase network down to the level of the service point to residential and small commercial loads. Feeders are usually three-phase overhead pole line or underground cables. As one gets closer to the loads (many of which are single-phase), three-phase or single-phase laterals provide spurs to the various customer connections. Depending on the needs of the customer, the voltage supplied can be as low as 120 V single-phase or 120/240 V single-phase, where the 240 V secondary of the distribution transformer has a center tap that also provides two 120 V single-phase circuits. Larger customers may utilize three-phase power, with 120/208 V or 277/480 V service. However, electric energy is commonly delivered to industrial users by three-phase networks at medium or high voltage. The values of medium and high voltage are defined by international standards; broadly, medium voltages range from 5 to 25 kV and high voltages range from 50 to 500 kV. Higher and lower values can be found according to local conditions. Individual industrial users have to reduce the levels indicated above to lower voltages inside factories. Although the transformer efficiency is quite high (generally not less than 95 %), it is worth reducing these losses as much as possible, because their quantity is related to the total electric energy consumed in the site. Transformer losses are no-load and load losses: no-load losses are almost exclusively iron losses (the squared function of voltage, occurs whenever the transformer is connected to a voltage source independently of the load value), whereas load losses are a squared function of current (or of apparent power load), being mainly copper losses, occurring only when a load is connected to the transformer.

The function of the electric power distribution system in a building, industrial facility, or an installation site is to transfer power from one or more supply, in the most efficient way and at higher standards to the individual loads and to all other electrically operated devices. Today's industrial applications require the most technologically advanced support systems to ensure state-of-the-art power quality needs. Electromechanical subsystems are being replaced by electronic logic. Harmonic interference, welding, variable-speed drives, and other in-plant noise have reliable mitigation procedures. Overhead distribution is used because it is generally less costly than underground. Where underground distribution is more cost effective, it should be used. When exceptions are considered, follow local requirements and practices. The building and facility electrical service consists of the conductors and cables connecting the building and facility electrical system to the local utility power source, the raceway system containing the electrical service entrance conductors and cables, metering equipment, the main service disconnect, and the main overcurrent protection devices. The purpose of the power distribution is to provide the optimum and the best quality power supply, protection, and control in a cost effective way. he service and the utility-facility connections must comply with all NEC, IEEE, or IEC code and standard requirements and specifications. Typical low-voltage values range from 260 V to 600 V with local variations often present. Transformers at site boundaries and related electrical equipment make up the electrical substations, through which all the purchased electric energy flows, and they are responsible for transformation losses, as are boiler plants with thermal energy. However, very large, industrial customers, for example, can include higher three-phase voltages, such as 2400 V, 4160 V, or greater. Depending on the voltage at which it acquires power, the customer is usually responsible for providing the transformers and other infrastructure to serve all of the lower-voltage requirements needed by its facility or site. A typical service transformer (ground mounted or pole mounted), supplying five to ten residences, converts the three-phase power at a distribution voltage of 13.8 kV (transformer primary), for example, to power at 120/240 V. Each of the 240 V, single-phase transformer secondaries has a center tap that provides two 120 V single-phase circuits. Therefore, an individual residence can be supplied with both 120 V and 240 V single-phase service. The higher voltage is necessary for appliances such as clothes dryers.

Functionally, distribution circuits are those that feed customers (this is how the term is used in this book, regardless of voltage or configuration). Some also think of distribution as anything that is radial or anything that is below 35 kV. The distribution infrastructure is extensive; after all, electricity has to be delivered to customers concentrated in cities, customers in the suburbs, and customers in very remote regions; few places in the industrialized world do not have electricity from a distribution system readily available. Distribution circuits are found along most secondary roads and streets. Urban construction is mainly underground; rural construction is mainly overhead. Suburban structures are a mix, with a good deal of new construction going underground. A mainly urban utility may have less than 50 ft. of distribution circuit for each customer. A rural utility can have over 300 ft. of primary circuit per customer. Several entities may own distribution systems: municipal governments, state agencies, federal agencies, rural cooperatives, or investor-owned utilities. In addition, large industrial facilities often need their own distribution systems. While there are some differences in approaches by each of these types of entities, the engineering issues are similar for all. Because of the extensive infrastructure, distribution systems are capital-intensive businesses. Cost-effective, simplified, and standardized designs are critical for good power distribution systems. Few components or installations are individually engineered; standardized equipment and standardized designs are the norm. Cookbook engineering methods are used for much of the power distribution planning, design, management, and operations.

To better understanding how an electrical power supply system in a building, plant, or facility is designed, the following design elements needed to be analyzed and considered. Load analysis, which includes the estimation of the power absorbed by the loads and their relevant positions; the definition of the position of power centers (switchboards, load centers, panelboards); path and calculation of the length of connection elements; and the definition of the total absorbed power, taking into account the utilization factors and demand factors, as defined in previous book chapters. After load analysis, the next phase consists of the dimensioning of transformers, feeders, load centers, and if the facility includes site-power generation, the size of the local generators. Codes and standards recommend a 15% to 30% margin should be considered for future expansion and changes. Dimensioning of the conductors, cables, and branch circuits is the next design phase. This phase consists of the estimate of the currents flowing through conductors, definition and selection of the conductor, cable type and insulation material, calculations of the cross section and the current carrying capacity, and the voltage drops at the load current under normal and transient (motor starting, inrush current, overload, etc.) operations. Another important step is the verification of the voltage drop limits at the final load; if the voltage drop is not within the limit, phase 3 must be modified. After the verification is completed, the short-circuit current calculations are conducted, which consist of the estimations of the maximum value at the busbar and minimum values at the end of the line, followed by the phase of the selection of protective circuit breakers (CBs), fuses, and other protective devices, having breaking capacity higher than the maximum prospective short-circuit current, rated current no lower than load current, and characteristics compatible with the type of protected loads (motors, capacitors, heaters, etc.). An important step of the verification of the protection of the conductors, cables, and branch circuits, consists of the verification against overload currents, the currents of the circuit breaker, fuses, or protection devices that are higher than the load currents but lower than the current capacity of conductors or cables. The protection verification against short-circuit, consisting of the specific load through current by the circuit breaker under short-circuit conditions, must be lower than the specific current that can be withstood by the cable or conductor, expressed with standard notations in terms of power and energy. In case of negative outcomes, all the above stages must be repeated from the third phase, sizing the conductors, cables, and branch circuits. The last step consists of the verification of the coordination with other equipment.

Modern power systems are made up of three major distinct components or subsystems: generation, transmission, and distribution, as shown in Figure 3.1. The generation subsystem includes generation stations, turbines, generators and operation, control, and protection equipment. The energy resources used to generate electricity in almost all power plants are fossil fuels (e.g., coal, natural gas, oil), nuclear, or hydropower. In recent decades, it has become an increasing trend to include renewable energy sources for electricity energy generation, especially in the distribution section. Nuclear power and hydroelectric power are nonpolluting energy sources, while the hydropower is also a renewable energy source. Currently in the United States, hydropower accounts for about 8%, and nuclear power for about 20% of the generated electricity. Generated electricity is transmitted by a complex network composed of transmission lines, transformers, monitoring, control, and protective equipment. Transformers are used to step up the voltage at power plants to very high values (up to 1200 kV), in order to reduce transmission losses, and control the size of transmission wires, and corral the overall system costs. Transmission lines carry power to the load centers, where the voltages are lowered at 35 kV or other levels at bulk power substations. At load centers, the transmission line voltages are reduced by step-down transformers to lower values (4.5 kV to 35 kV) for power distribution networks. Some large industrial customers are supplied from these substations. This setup is known as a transmission–subtransmission system or section. Distribution of power to commercial and residential

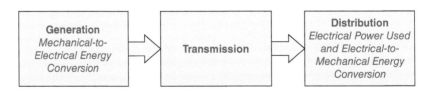

FIGURE 3.1   A simplified power system structure.

users takes place through a power distribution system consisting of substations where step-down transformers lower the voltages to a range of 2.4 kV to 69 kV. Power is carried by main feeders to specific areas where there are lateral feeders to step it down to end-user voltage levels. At customer sites, the voltage is further reduced to values, such as 120 V, 208 V, or 280/277 V, as required by the users. Power systems are extensively monitored, controlled, and protected. These complex transmission and distribution networks encompass larger areas or regions. Each power system has several levels of protection to minimize or avoid the effects of any damaged or improperly operating system component on the system's ability to provide safe reliable electricity to all customers. Any power system serves one important function and that is to supply the customers with electricity as economically and as reliably as possible. In summary the main functions of the subsystems of power systems are:

1. Generation subsystem—Generating and/or sources of electrical energy.
2. Transmission and subtransmission subsystems—Transporting electrical energy from its sources to load centers with high voltages (115 kV or higher) to reduce losses.
3. Distribution—Distributing electrical energy from substations to end users/customers.
4. Consumers, users, or utilization subsystem.

This basic structure of a power system (Figure 3.1) is complex network, in which the electromechanical energy-conversion systems, advanced control, and power electronics play key roles. An essential component of power systems is three-phase alternating current (AC) synchronous generators or alternators. The electric generator converts non-electrical energy provided by the prime movers (e.g., steam or hydro turbines), to electrical energy. The turbine function is to rotate electrical generators by converting the thermal energy of the steam or the kinetic energy of the water into rotating mechanical energy. In thermal power plants, fossil fuels or nuclear reactions are used to produce high-temperature steam, eventually passed through the turbine blades, causing the turbine to rotate. Typical hydroelectric plants include a dam, holding water upstream at high elevations with respect to the turbine. When electricity is needed, the water flows through the hydro turbine blades through penstocks rotating the generator. Because the generator is mounted on the turbine shaft, the generator rotates with the turbine generating electricity. At load level, the bulk of the energy is consumed through electrical motors, electric heaters, appliances, electronics, and computers.

3.2.1 Power Distribution Configurations

Electricity is delivered from the generation ends (power plants) to the loads' (consumers) transmission lines and transformers. The bulk of the electricity produced in power plants is transmitted to load centers over long-distance, high-voltage transmission lines, operating at very high voltages, 220 kV to 1200 kV. The lines that distribute the electrical power within an area are called medium-voltage distribution lines. There are several other categories, such as subtransmission and high-voltage distribution line or power distribution networks. The transmission lines are high-voltage conductors (wires) mounted on tall towers to prevent them from being in contact with humans, trees, animals, buildings, equipment, or ground. High-voltage towers are normally made of galvanized steel, 25 m to 45 m in height for strength and durability in harsh environments in which they may operate, similar to the ones shown in Figure 3.2. The higher the voltage of the wire, the higher is the tower. Because steel is an electrical-conductive material, high-voltage wires are not mounted directly on the towers. Instead, insulators, made of nonconductive materials mounted on the towers, are used to hold the conductors away from the tower structure. Insulators, in various shapes and designs, are strong enough to withstand the static and dynamic forces exerted by the conductors during windstorms, freezing rains, earthquakes, or animal impacts. Distribution circuits come in many different configurations and circuit lengths.

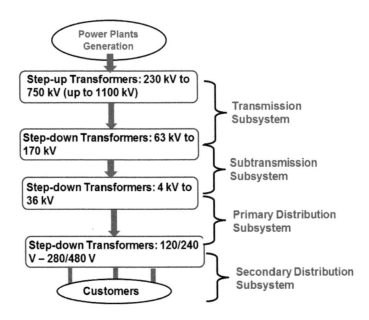

FIGURE 3.2   Electricity infrastructure from generation to power distribution.

Most share many common characteristics. Figure 3.2 shows a *typical* power distribution circuit, and Table 3.1 shows typical parameters of power distribution circuits. A *feeder* is one of the circuits out of the substation. The main feeder is the three-phase backbone of the circuit, which is often called the *mains* or *mainline*. The mainline is usually a medium-size conductor such as a 500- or 750-kcmil aluminum conductor. Utilities often design the main feeder for 400 A and often allow an emergency rating of 600 A. Branching from the mains are one or more *laterals*, which are also called taps, lateral taps, branches, or branch lines. These laterals may be single-phase, two-phase,

TABLE 3.1
Typical Distribution Circuit Parameters

Substation and Feeder Characteristics	Most Common Value	Other Common Values
Voltage	12.47 V	4.16, 4.8, 13.2, 13.8, 24.94, and 34.5 kV
Number of station transformers	2	1 to 6
Substation transformer size	21 MVA	5 to 60 MVA
Number of feeders per bus	4	1 to 8
Peak current	400 A	100 to 600 A
Peak load apparent power	7 MVA	1 to 15 MVA
Power factor	0.98 lagging	0.8 lagging to 0.95 leading
Number of customers	400	50 to 5000
Length of feeder mains	4 mi	2 to 15 mi
Length including laterals	8 mi	4 to 25 mi
Area covered	25 mi^2	0.5 to 500 mi^2
Mains wire size	500 kcmil	4/0 to 795 kcmil
Lateral tap wire size	1/0	#4 to 2/0
Lateral tap peak current	25 A	5 to 50 A
Distribution transformer size (1-ph)	25 kVA	10 to 150 kVA

or three-phase. The laterals normally have fuses to separate them from the mainline if they are faulted. The most common distribution primaries are four-wire, multigrounded systems: three-phase conductors plus a multigrounded neutral. Single-phase loads are served by transformers, being connected between one phase and the neutral. The neutral acts as a return conductor and as equipment safety ground (it is grounded at all equipment). A single-phase line has one phase conductor and the neutral, and a two-phase line has two phase and the neutral. Some distribution primaries are three-wire systems (with no neutral). On these, single-phase loads are connected phase to phase, and single-phase lines have two of the three phases.

An essential part of electricity systems, electrical distribution systems are the grid sections that distribute electricity to consumers. It is the portion of the power delivery infrastructure that takes electricity from highly meshed, high-voltage transmission networks and delivers it to the end users. In order to transfer electrical power from an AC or DC power supply to the place where it is used (loads), specific types of power distribution networks must be utilized. Complex power distribution systems are used to transfer electrical power from power plants to industries, homes, and commercial buildings. Power distribution systems are usually broken down into three components: *distribution substation*, *primary* and *secondary power distribution networks*. At a substation, the voltage is lowered as needed and the power is distributed to the customers, with a single substation supplying power to thousands of customers. Thus, the number of transmission lines in the power distribution systems is several times that of the transmission systems. Furthermore, most customers are connected to only one of the three phases in the power distribution system. Therefore, the power flow on each of the lines is different and the system is typically unbalanced. Primary distribution lines are medium-voltage networks, ranging from 600 V to 35 kV. At a distribution substation, transformers lower the incoming transmission-level voltage (35 to 230 kV) to several distribution primary circuits fanning out from the substation. Close to each end user, power distribution transformers step down the primary voltages to low-voltage secondary circuits at 120/240 V or other utilization voltages. Power distribution employs equipment such as transformers, monitoring units, metering, and protective devices in order to deliver safe and reliable power. Power distribution in the United States is in the form of three-phase, 60 Hz, AC current. The electric power is changed in many ways through the employment of electrical special circuitry and devices. Electricity distribution involves a system of interconnected transmission lines, originating at the power-generating stations located throughout the country. The ultimate purpose of these power transmission and distribution systems is to supply the electricity needed for industrial, residential, and commercial uses.

Power distribution infrastructure, a very extensive and complex, is designed to deliver power to various end-users or consumers with different requirements. Industrial facilities use almost 50% of all the electrical energy. Three-phase power is distributed directly to most large industrial and commercial facilities. Electrical substations use massive three-phase power transformers and associated equipment (circuit breakers, high-voltage conductors, and insulators) to distribute electricity. From these substations, electricity is distributed to industrial sites, residential, and commercial users. Facilities and buildings must provide maximum supply safety, consume few resources during construction and operation, and be flexible to adapt to any future power requirements. The intelligent integration of all building service installations optimum to be attained for safety, energy efficiency, flexibility, and environmental compatibility, offering maximum comfort. Buildings are responsible for about 40% of the total energy consumption within the United States and developed countries. In large buildings, the type of distribution depends on the building type, dimension, the length of supply cables, and the loads. The distribution system can be divided into: (a) the vertical supply system (rising mains), and (b) the horizontal supply (distribution at each floor level). Power distribution circuits and networks, as mentioned, are mainly as underground power distribution type in urban areas, while the rural power distribution is mainly overhead transmission type, and suburban networks are a mixture of the two.

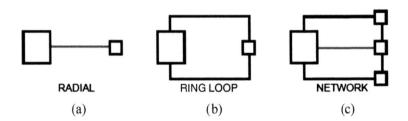

RADIAL   RING LOOP   NETWORK

(a)   (b)   (c)

FIGURE 3.3 Typical distribution feeder configurations: (a) Radial distribution system; (b) ring (loop) distribution system; (c) network distribution system.

The three general configurations of electrical power distribution are radial, ring (loop), and network systems (Figure 3.3a–c). Radial power distribution systems (Figure 3.3a) are the simplest type because electrical energy is coming from a single power source. A generating system supplies electricity from the substation to radial power lines that extend to various areas. Most distribution circuits are radial (both primary and secondary). Radial circuits have many advantages over networked circuits, including: easier voltage control, easier prediction and control of power flows, and lower cost. Although radial is the simplest power distribution system, it is also the least reliable in terms of continuous service because there is no power backup. If any power line opens, one or more loads are interrupted, leading to a high probability of power outages. This type of power distribution is usually used in remote areas where other distribution types are not economically feasible. Ring power distribution systems are used in heavily populated areas, and the power distribution lines encircle the service area (see Figure 3.3b). Electricity is delivered from one or usually more power sources into substations near the service area, and then the power is delivered through radial lines. In the event that a power line is opened, through alternative lines the electricity is still delivered to the loads. Ring systems provide a more continuous service than radial power distribution systems, but ring systems are more expensive due to additional power lines and greater circuit complexity. Network power distributions are a combination of radial and ring systems (Figure 3.3c). They usually result when one of the other two systems is expanded. Most US power distribution systems are network distribution type. The network distribution type is more complex but it provides reliable service to consumers, with each load being fed by two or more power circuits. Power flowing on a line is a function of voltage and current. Because the current itself is inherently bidirectional, power can typically readily flow in either direction. However, other operational constraints such as circuit breakers and control devices may not be able to accommodate power flow reversals, without replacement or modification. The electric power grid operates as a three-phase network down to the level of the service point to residential and small commercial loads. Feeders are usually three-phase overhead pole line or underground cables, and closer to the loads (many are single-phase), three- or single-phase laterals provide spurs to various customer connections. Three-phase electricity refers to voltage waveforms (and corresponding current) 120° out of phase with each other. This provides advantageous characteristics for rotating electric machines by inducing a smooth rotating magnetic field and more robust equipment. To ensure efficient operation, it is important to balance the phases so that they are approximately equal.

Power distribution feeder circuits consist of overhead and underground transmission line networks (circuits) in a mix of branching circuits (laterals) from the station to various customers. The main reason for the use of overhead distribution lines is because they are usually less expensive than underground to install, operate, and maintain. Where underground distribution is more cost effective, it should be used. When exceptions are considered, follow local requirements and practices. Each circuit is designed around various requirements such as the required peak load, voltage levels, distance to customers, and other local conditions such as: terrain configurations, street layout, visual and environmental regulations, or user requirements. The secondary voltage in North America consists of a split single-phase service that provides the customer with 240 V and 120 V; the customer is then connected to devices depending on their ratings. This is served from a three-phase distribution feeder normally connected in a Y configuration consisting of a neutral center conductor and a conductor for each phase. In most other parts of the world, the single-phase voltage of 220 V or 230 V is provided directly from a larger neighborhood power distribution transformer, providing also a secondary voltage circuit often serving hundreds of customers. The various branching laterals are operated in radial or in looped configurations, where two or more feeder parts are connected together usually through an open distribution switch. Power distribution networks are overhead or underground, highly redundant, and reliable complex networks. Overhead lines are mounted on concrete, wooden, or steel poles arranged to carry power distribution transformers or other needed equipment, along with the conductors. Underground distribution networks use conduits, cables, manholes, and necessary equipment installed beneath the street surface. The choice between the two systems depends on factors such as safety; initial, operation, and maintenance costs; flexibility; accessibility; appearance; life; fault probability; location and repairs; and interference with communication systems. For example, compared with overhead systems, underground distribution systems are more expensive, requiring higher investment, maintenance, and operation costs, and have lower fault probabilities, but they do make it more difficult to locate and repair a fault, lower cable capacities, and voltage drops. Each system has advantages and disadvantages; oftentimes the most important consideration is economic. However, non-economic factors are sometimes more important than budget considerations.

3.2.2 Voltage and Frequency Characteristics

Utility frequency or (power) line frequency is the frequency of the AC in an electric power grid transmitted from a power plant to an end user. In large parts of the world, this is 50 Hz, although in the Americas and parts of Asia it is 60 Hz. During the development of commercial electric power systems in the late nineteenth and early twentieth centuries, several different frequencies (and voltages) were used. Large investments in equipment at one frequency made standardization a slow process. However, as of the turn of the twenty-first century, places that now use the 50 Hz frequency tend to use 220–240 V, and those that now use 60 Hz tend to use 100–127 V. Both frequencies coexist today (Japan uses both) with no great technical reason to prefer one over the other and no apparent desire for complete worldwide standardization. Unless specified by the manufacturer to operate on both 50 Hz and 60 Hz, appliances may not operate efficiently or even safely if used on anything other than the intended frequency. Several factors influence the choice of frequency in an AC system. Lighting, electrical motors, transformers, generators, and transmission lines all have characteristics that depend on the power frequency. All of these factors interact and make selection of a power frequency a matter of considerable importance. The best frequency is a compromise between contradictory requirements. An accurate model for an electrical load is very important for a power system. Loads such as electrical motors, lighting, and heating show different characteristics with changes of voltage and frequency. In a power system, voltage and frequency at load bus always change due to disturbances, which cause fluctuations in load power. This load power variation is not the same for all loads, and depends on the characteristics of the load connected to the load bus. For the purpose of stability analysis of a system, it is essential to determine the effects of changes of loads due to voltage and frequency changes.

The primary reason for accurate frequency control is to allow the flow of AC power from multiple generators through the network to be controlled. Trends in system frequency are a measure of mismatch between demand and generation, being a critical parameter for load control in interconnected systems. System frequency varies as the load and/or generation change. Increasing the mechanical input power to a synchronous generator will not greatly affect the system frequency, but will produce more electric power from that unit. During a severe overload caused by tripping or failure of generators or transmission lines, the power system frequency will decline, due to an imbalance of load versus generation. Loss of an interconnection, while exporting power (relative to system total generation), will cause system frequency to rise. Automatic generation control is used to maintain scheduled frequency and interchange power flows. Control systems in power plants detect changes in the network-wide frequency and adjust mechanical power input to generators back to their target frequency. This counteracting usually takes a few tens of seconds due to the turbine-generator system inertia. Temporary frequency changes are an unavoidable consequence of changing demand. Unusual or rapidly changing frequency is often a sign that an electricity distribution network is operating near its capacity limits, sometimes being observed shortly before major outages. Frequency protective relays on the power system network sense the decline of frequency and automatically initiate load shedding or tripping of interconnection lines, to preserve the operation of at least part of the network. Small frequency deviations (i.e., 0.5 Hz on a 50 Hz or 60 Hz network) will result in automatic load shedding or other control actions to restore system frequency. Smaller power systems, not extensively interconnected with many generators and loads, will not maintain frequency with the same degree of accuracy. Where system frequency is not tightly regulated during heavy load periods, the system operators may allow system frequency to rise during periods of light load, to maintain a daily average frequency of acceptable accuracy. Portable generators, not connected to a utility system, need not tightly regulate their frequency, because typical loads are insensitive to small frequency deviations. Power system stability is classified as rotor angle stability and voltage stability. Voltage stability is in power systems that are heavily loaded, have disturbances, or have a shortage of reactive power. Nowadays, demand for electricity has dramatically increased and a modern power system becomes a complex network of transmission lines interconnecting the generating stations to the major load points in the overall power system in order to support the high demands of consumers.

System voltage is a term used to identify whether reference is being made to secondary or primary distribution systems. Residential, commercial, and small industrial loads are normally served with voltages under 600 V. Manufacturers have standardized the provision of insulated wire to have a maximum 600 V AC voltage rating for "secondary" services. For example, household wires such as extension cords have a 600 V insulation rating. Other than changing the plugs and sockets on either end, one could use this wire for higher AC voltages such as 240 V. The 34,000 V (34 kV) system's voltage is used differently among electric power companies. Some companies use 34.5 kV distribution system voltages to connect service transformers in order to provide secondary voltages to consumers, whereas other companies use 34.5 kV power lines between distribution substations and not for consumers. There are several common distribution system voltages between secondary and 34.5 kV used in the industry. For example, many power companies have standardized distribution at 12.5 kV while others use 25 kV. Some companies use 13.2 kV, 13.8 kV, 14.4 kV, 20 kV, and so on. Several areas still use 4.16 kV systems. These lower-voltage distribution systems are quickly being phased out due to their high losses and short-distance capabilities. The voltage category for distribution is usually high voltage (HV). Utilities often place HV warning signs on power poles and other associated electrical equipment.

3.2.3 Building Power Supply and Electricity Distribution

Depending on the needs of the customer, the voltage supplied can be as low as 120 V single-phase or 120/240 V single-phase, where the 240 V secondary of the distribution transformer has a center tap that also provides two 120 V single-phase circuits. Larger customers may utilize three-phase power, with 120/208 V or 277/480 V service. A common way to denote three-phase service is to list the line-to-neutral followed by the line-to-line voltage. Three-phase circuits are defined by the line-to-line voltage, with its line-to-neutral voltage multiplied by $\sqrt{3}$. Large industrial customers include usually higher three-phase voltages, 4.16 kV, or greater, usually providing the transformers and other infrastructure to serve all of their lower voltage requirements. For example, a typical service transformer converts the three-phase distribution voltage of 13.8 kV (transformer primary) to power at 120/240 V. Each of the 240 V, single-phase transformer secondaries has a center tap that provides two 120 V single-phase circuits, being able to supply both 120 V and 240 V, single phase services. Electricity for buildings is supplied from local utility distribution systems or can be generated in part or in full on-site, using conventional generators or renewable energy conversion systems, such as wind turbines, photovoltaic systems, or fuel cells. One appealing energy source is a combined heat, cooling, and power generation system, designed to produce heat, air-conditioning, cooling, and like by-product electricity, significantly increasing overall efficiency of building heat, ventilation, and air-conditioning systems. Several types of building power distribution systems are used, and the most appropriate for a specific building depends on the building size, destination, and load characteristics (power ratings, voltages, frequency, etc.). For example, the standard frequency used in the United States is 60 Hz, while Europe uses 50 Hz, or special frequency values for unusual or special equipment or applications.

Distributing and delivering electrical energy requires many types of specialized equipment. These include overhead transmission lines, underground cables, transformers, protection devices, circuit breakers, fuses, lightning arresters, power-factor correction capacitors, and power metering systems. The most common distribution primaries are four-wire, multi-grounded systems: three-phase conductors plus a multi-grounded neutral. Single-phase loads are served by transformers connected between one phase and the neutral. The neutral acts as a return conductor and as an equipment safety ground (it is grounded periodically and at all equipment). A single-phase line has one phase conductor and the neutral, and a two-phase line has two phases and the neutral. Some distribution primaries are three-wire systems (with no neutral). On these, single-phase loads are connected phase-to-phase, and single-phase lines have two of the three phases. Power distribution lines (feeders) are normally connected radially out of the substation. Radially means that only a single end of the distribution power line is connected to a source. Therefore, if the source end becomes opened (i.e., de-energized), the entire feeder is de-energized and all the consumers connected to that feeder are out of service. The transmission side of the substation normally has multiple transmission lines feeding the substation. Distribution feeders might have several disconnect switches located throughout the line. These disconnect switches allow for load transfer capability among the feeders, isolation of line sections for maintenance, and visual openings for safety purposes while working on the lines or other high-voltage equipment. Even though there might be several lines and open/closed disconnect switches connected throughout a distribution system, the distribution lines are still fed radially. Most three-phase power distribution feeders and transformer connections use the wye system because it offers more advantages than disadvantages. Although delta distribution systems do exist, much of the delta distribution has been converted to wye. The wye connection has one wire from each coil connected together to form the neutral. Most of the time, this neutral is grounded. Grounded implies that the three common wires are connected together and then connected to a ground rod, primary neutral, or ground grid, to provide a low-resistance connection to earth. Grounding gives earth an electrical reference; in the case of the wye connection, this neutral reference is zero volts.

3.2.4 Definitions, Loads, and End-User Parameters

Power distribution systems, or networks, are designed and installed to supply electricity to end users, so the loads and their characteristics are very important design and operation characteristics. Utilities supply a broad range of loads, from rural areas with load densities of 10 kVA/mi.^{2} to urban areas with load densities of 300 MVA/mi.^{2}, or even higher. A utility feeds houses with a 10 kVA to 20 kVA peak load on the circuit providing power to an industrial customer, having the peak load of 5 MVA. The feeder electrical load is the sum of all individual customer loads, while the electrical customer load is the sum of the load drawn by the customer appliances and equipment. Customer loads have common characteristics, and several parameters are used for characterization and description. Load levels vary throughout the day, usually peaking in the afternoon or early evening. The *peak power (load)* is the maximum power consumed by loads during a specific or given time interval and it is equal to the maximum actual power, generated by the power plant, neglecting the transmission line losses. *Average load* is the average power consumed by the loads during a certain time period, being equal to the actual power that is generated by the power plant during the same time period, neglecting transmission line losses. Several other parameters are used to quantify load characteristics at a given location on a circuit or feeder. The *load demand* is the load average over a specified time period, often 15, 20, or 30 min., used to characterize real (active) power, reactive power, total power, or current. Peak load demand over some period of time is the most common way utilities quantify a circuit's load. In substations, it is common to track the current demand. *Load factor* is the ratio of the average load over the peak load for a certain period of time. Basically, the load factor is the ratio of the load that equipment actually draws when it is in operation to the load it could draw (the full load). The *utilization (use) factor* is the ratio of the time that equipment is in use to the total time that it could be in use. If the period of time is a day, the load factor is a daily load factor and if the period of time is a month, the load factor is a monthly load factor, and similarly for the yearly load factor. The *load curve* shows the characteristics of the load over a certain period of time (day, month, or year). The curve is plotted by placing the ordinate (kW) with its proper time sequence as shown in the diagram of Figure 3.4. Peak load is normally the maximum demand but may be the instantaneous peak power, being a number between 0 and 1. A load factor close to 1.0 indicates that the load runs almost constantly. A low load factor indicates a widely varying load. From the utility point of view, it is desirable to have higher load factors. Load factor is usually found from the total energy used (kWh) as:

$$LF = \frac{E(kWh)}{d_{kW} \times hrs} \tag{3.1}$$

Here, LF is the load factor, $E(kWh)$ represents the energy use in kWh, d_{kW} is the peak load demand in kW, and hrs is the number of hours during that specific time period. Its value is always less than 1, because maximum demand is always more than the average demand. It is used for determining the overall cost per unit generated. The higher the load factor, the lesser is the cost per unit. Load factor is in fact the load that a piece of equipment actually draws over the load it could draw (full load). Load factor is a term that does not appear on utility bills, but it does affect electricity costs. Load factor indicates how efficiently the customer is using the peak demand. A high load factor means power usage is relatively constant. Low load factor shows that occasionally a high demand is set. To service that peak, capacity is sitting idle for long periods, thereby imposing higher costs on the system. Electrical rates are designed so that customers with

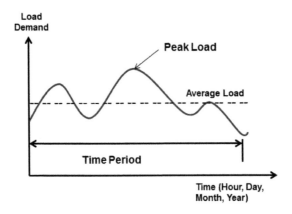

FIGURE 3.4 A typical load curve for a generic time period.

higher load factors are charged less overall per kWh. *Coincident factor* is the ratio of the peak demand of a whole system to the sum of the individual peak demands within that system. The peak demand of the whole system is referred to as the peak *diversified load* demand or as the peak *coincident* demand. The individual peak demands are the *noncoincident load* demands. The coincident factor is less than or equal to 1; usually, the coincident factor is much lower than 1 because each of the individual loads does not reach its peak at the same time (they are not coincident). *Diversity* is the relationship between the rated full loads of the equipment downstream of a connection point, and the rated load of the connection point. *Diversity factor* is the ratio of the sum of the individual peak load demands in a system to the peak load demand of the whole system, being greater than or equal to 1, the sum of individual maximum load demands is greater than the maximum facility demand, and is the reciprocal of the coincident factor. The diversity factor is also defined as the probability that particular equipment is coming on at the time of the facility peak load. Many designers prefer to use unity as the diversity factor in calculations for planning conservatism because of plant load growth uncertainties. Local experience can justify using a diversity factor larger than unity, and smaller service entrance conductors and transformers. The diversity factor for all other installations will be different, and would be based upon a local evaluation of the loads to be applied at different moments in time. The load is time-dependent as well as being dependent upon equipment characteristics (Figure 3.4). The diversity factor recognizes that the whole load does not equal the sum of its parts due to this time interdependence (i.e., diversity). *Responsibility factor* is the ratio of a load demand at the time of the system peak load to its peak load demand. A load with a responsibility factor of 1 is peaking at the same time as the overall system. The responsibility factor can be applied to individual customers, customer classes, or circuit sections. The loads of certain customer classes tend to vary in similar patterns. Commercial loads tend to run from 8 a.m. to 6 p.m., and industrial loads tend to run continuously, and as a class, have a higher load factor, while residential loads usually peak in the evening. Weather significantly changes loading levels, affecting both the load demand and peak demand. For example, on hot summer days, air-conditioning increases the demand and reduces the diversity among loads.

Example 3.1

An oversized motor, 20 HP drives a constant 15 HP load whenever it is ON, operating only 8 hours a day, and 50 weeks in a year. Estimate its load factor and the utilization factor.

SOLUTION

The motor is driving a constant load all the time when the load is ON, so the load factor is:

$$\text{Load Factor} = \frac{P_{Load}}{P_{Motor}} = \frac{15\ HP}{20\ HP} = 0.75 \text{ or } 75\%$$

The hours of operation would then be 2000 hr., and the motor use factor for a base of 8760 hr. per year would be 2000/8760 = 22.83%. With a base of 2000 hr. per year, the motor use factor would be 100%. The bottom line is that the use factor is applied to get the correct number of hours that the motor is in use.

Notice that any load is time-dependent as well as being dependent upon equipment characteristics. The diversity factor recognizes that the whole load does not equal the sum of its parts due to time interdependence (i.e., diversity). When the maximum load demand of a supply is assessed, it is not sufficient to simply add together the ratings of all electrical equipment that could be connected to that supply. If this is done, a figure somewhat higher than the true maximum demand will be produced. This is because it is unlikely that all the electrical equipment on a supply will be used simultaneously. The concept of being able to *derate* a potential maximum load to an actual maximum load demand is known as the application of a diversity factor. For example, 70% diversity means that the device in question operates at its nominal or maximum load level 70% of the time that it is connected and turned on. If everything (all electrical equipment) was running at full load at the same time, the diversity factor is equal to 1. *Notice that the greater the diversity factor, the lesser is the cost of generation of power.* Diversity factor in a power distribution network is the ratio of the sum of the peak demands of the individual customers to the peak demand of the network, being determined by the service type, i.e., residential, commercial, industrial, or combinations. Diversity factor is used for sizing the distribution feeders and transformers. Determining the maximum peak load and diversity factor is based on the process, by understanding what is ON or OFF at a given time for different buildings, to properly size the feeder. For typical buildings, the diversity factor is 1. To calculate the diversity factor, data records or estimates are needed. For example, from a 24-hour load graph, the maximum demand load for node is determined and then the feeder and transformer size. The feeder diversity factor is the sum of the maximum demands of the individual consumers divided by the maximum demand of the feeder. In the same way, the diversity factor on a substation, a transmission line, or a whole utility system is determined. Residential loads have the highest diversity factor. Industrial loads have lower diversity factors usually of 1.4, street lighting practically unity, and other loads vary between these limits. Diversity factors are used by utilities for distribution transformer sizing and load predictions. Demand factors are more conservative and are used by NEC for service and feeder sizing. Demand factors and diversity factors are used in design. For example, the sum of the connected loads supplied by a feeder is multiplied by the demand factor to determine the load for which the feeder must be sized. This load is termed the maximum demand of the feeder. The sum of the maximum demand loads for a number of subfeeders divided by the diversity factor for the subfeeders gives the maximum demand load to be supplied by the feeder from which the sub-feeders are derived. The utilization factor must be applied to each individual load, with particular attention to electric motors, which are very rarely operated at full load. In industrial installations, this factor may be estimated on an average at 0.75 for motors. For incandescent-lighting loads, the factor always equals 1, while for socket-outlet circuits, the factors depend entirely on the type of appliances being supplied from the sockets concerned.

Example 3.2

Four individual feeder-circuits have connected loads of 250 kVA, 200 kVA, 150 kVA, and 400 kVA and demand factors of 90%, 80%, 75%, and 85%, respectively. Assuming a diversity factor of 1.5, estimate the transformer size supplying these feeders.

SOLUTION

Calculating demand for feeder circuits:

Feeder 1: 250 kVA × 90% = 225 kVA
Feeder 2: 200 kVA × 80% = 160 kVA
Feeder 3: 150 kVA × 75% = 112.5 kVA
Feeder 4: 400 kVA × 85% = 340 kVA

The sum of the individual demands is equal to 837.5 kVA. If the main feeder-circuit were sized at unity diversity: kVA = 837.5 kVA ÷ 1.00 = 837.5 kVA. The main feeder-circuit would have to be supplied by an *850 kVA transformer.* However, using the diversity factor of 1.5, the kVA = 837.5 kVA ÷ 1.5 = 558 kVA for the main feeder. For diversity factor of 1.5, a 600 kVA transformer could be used.

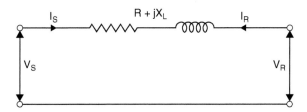

FIGURE 3.5 Cable impedance between source and load ends.

3.2.5 FEEDER VOLTAGE DROPS, ELECTRIC DISTRIBUTION LOSSES, AND POWER FACTOR CONTROL

Electricity is generated at large generation stations is transferred over high-voltage transmission lines to the utilization points. The employment of high-voltage lines is motived by increased line capacity and loss reduction, an important aspect of energy conservation. An electrical transmission line is modeled using series resistance, series inductance, shunt (parallel to ground) capacitance, and conductance, while series resistance and inductance are the most important and parallel elements often neglected in calculations. In power engineering, the term *wire* refers to one or more insulated conductors of small size (solid or stranded for flexibility), while the term *cable* refers to one or more insulated conductors of large size grouped in a common insulated jacket, quite often with a ground shield. A wire or combination of wires not insulated from one another is called a *conductor.* A stranded conductor consists of a group of wires, often twisted or braided together to minimize magnetic coupling. In the most common model used in power engineering, each cable phase conductor is represented by a resistance, R, in series with leakage reactance, X_L ($= \omega L$), where L is the leakage inductance between the current-carrying conductors, as shown in Figure 3.5. A cable design or calculation requires selecting the conductor size with required cable capacity (ampacity) at the operating temperature that meets the voltage limitations over the feeder length under the steady-state and inrush motor or equipment current. In such calculations, the per-phase current, I, power factor, resistance, and inductive reactance are required. The voltage drop of the transmission line in Figure 3.4 is then computed as:

$$V_{drop} = I \times (R \cdot \cos\theta + X_L \cdot \sin\theta) = I \times Z_{cable} \qquad (3.2)$$

Here θ is the phase angle (PF = $\cos\theta$), and all calculations can be performed in standard units or in percent or p.u. values, taking θ positive for lagging power factor and negative for leading power factor. Equation (3.1) is convenient for the definition of the effective cable impedance because its product with the cable current is simply giving the voltage drop magnitude or the cable voltage drop in amperes:

$$V_{drop/A} = Z_{eff(cable)} \, \text{V/A} \qquad (3.3)$$

Example 3.3

A 4.8 kV three-phase, line-to-line wye-connected feeder has a per-phase resistance of 50 mΩ and 0.33 mΩ inductive reactance per 1000 ft. Calculate the feeder voltage drops for: 0.85 lagging power factor, unit power factor, and 0.85 leading power factor at an apparent power of 1.2 MVA.

SOLUTION

The feeder phase voltage and the current at three-phase apparent power of 1.2 MVA are:

$$V_\phi = \frac{4800}{\sqrt{3}} = 2775.6 < 0° \text{ V}$$

$$I = \frac{1.2 \times 10^6}{\sqrt{3} \times 4800} = 144.5 \text{ A}$$

By using Equation (3.2) the voltage drops for 0.85 lagging power factor, unit power factor, and 0.85 leading power factor are:

$$V_{drop} = 114.5 \times \left(50 \times 10^{-3} \cdot 0.85 + 0.33 \times 10^{-3} \cdot 0.527\right) = 6.86 \text{ V}$$

$$V_{drop} = 114.5 \times \left(50 \times 10^{-3} \cdot 1.0 + 0.33 \times 10^{-3} \cdot 0.0\right) = 7.73 \text{ V}$$

$$V_{drop} = 114.5 \times \left(50 \times 10^{-3} \cdot 0.85 - 0.33 \times 10^{-3} \cdot 0.527\right) = 2.88 \text{ V}$$

The cable must be selected and manufactured to operate in the application-specific requirements and conditions. Notice that the power factor has quite a significant effect on the cable voltage drop.

Power distribution systems have automatic voltage regulation schemes and devices at one or usually at several power control stations, such as area substation or facility switchgears to maintain constant voltage regardless of load currents. Voltage drops in lines are in relation to the resistance and reactance of line, length, and the current drawn. For the same quantity of power handled by a transmission, the lower the voltage, the higher is the current drawn and the higher is the voltage drop. The current drawn is inversely proportional to the voltage for the same power handled, while the power loss in the line is proportional to the resistance and square of current (i.e., $P_{Loss} = I^2 R$). Higher-voltage transmission and distribution helps to minimize line voltage drop in the ratio of voltages, and the line power loss in the ratio of square of voltages. The feeder cables deliver power from the controlled *sending end* (power source) to the *receiving end* (the load), with the voltage gradually dropping form the initial maximum value to the minimum value at the load. For this reason, the power distribution system is sized not only for the required ampacity (capacity) but also to maintain required voltage levels during the steady-state operation and during the transients such as inrush current during motor starting. If the load current of a power factor *PF* (usually lagging the voltage) flow from the switchboard is controlled to maintain the constant bus voltage, the magnitude of the voltage drop between the ends, Equation (3.2) can be written for the cable impedance as:

$$V_{drop} = I \times \left(R \cdot PF + X_L \cdot \sqrt{1 - PF^2}\right) \tag{3.4}$$

Often, the cable manufacturers list the cable effective impedance at typical 0.85 lagging power factor; therefore, the catalog cable impedance values are:

$$Z_{cable} = R \times 0.85 + X_L \times \sqrt{1 - 0.85^2} = 0.85 \cdot R + 0.527 \cdot X_L \ \Omega/\text{phase} \tag{3.5}$$

However, if the power factor value is quite different from 0.85 lagging, Equation (3.5) must be adjusted accordingly. The cable size is usually selected to meet the capacity (ampacity) with up to 30% margin, limiting the solid-state voltage drop at 3% to 5% from the switchboard to the load. We have noticed that the feeder voltage drop depend on the cable impedance and also on the load power factor. Capacitors can improve the load power factor correcting also the feeder voltage drops, by reducing or eliminating the reactive terms in Equations (3.2), (3.4), and (3.5). The voltage boost estimation produced by the capacitors (bank) placed at the end of the feeder (load), starts with an estimate of the capacitor value, so the reactive power correction and values, Q, needed to improve power factor at the desired level is:

$$Q_{Cap} = Q_{Load} - Q_{Desired} \tag{3.6}$$

Expressing the reactive power, through the apparent power and the usual relations between power factor, the apparent power, reactive power, and active (real) power for a feeder supplying a load, the load and the desired reactive power are expressed as:

$$Q_{Load} = \sqrt{\left(\frac{P_{Load}}{PF_{Load}}\right)^2 - P_{Load}^2}$$

$$Q_{Desired} = \sqrt{\left(\frac{P_{Load}}{PF_{Desired}}\right)^2 - P_{Load}^2}$$

(3.7)

From the above relationships, the needed capacitor (bank) reactive power to improve the power factor at the desired level is given by:

$$Q_{Cap} = P_{Load} \times \left[\sqrt{\frac{1}{PF_{Load}^2} - 1} - \sqrt{\frac{1}{PF_{Desired}^2} - 1}\right]$$

(3.8)

The pre-phase capacitor (bank) value is then computed by using the well-known relationship:

$$C = \frac{Q_{Cap}}{2\pi f \cdot V_{LL}^2}$$

(3.9)

Here f is the electrical supply frequency and V_{LL} is the feeder line-to-line voltage. In addition to improving the power factor, the capacitor (bank) reduces the feeder voltage drop and the feeder cable losses by producing a voltage rise, due to the leading capacitor current. To understand the voltage rise calculation, the circuit of Figure 3.6 is considered. The rated capacitor (bank) current is given by:

$$I_C = \frac{Q_{Cap}}{\sqrt{3} \times V_{LL}}$$

(3.10)

The voltage rise due to the capacitor (bank) is equal to:

$$V_{Rise} = I_C \cdot X_L = X_L \frac{Q_{Cap}}{\sqrt{3} \times V_{LL}}$$

(3.11)

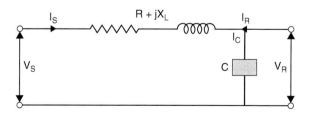

FIGURE 3.6 Voltage rise calculation circuit diagram.

Example 3.4

An industrial facility has a power demand of 1000 kW at a power factor of 0.75 lagging. Determine the capacitor bank reactive power ratings required to improve the power factor to 0.85 and 0.95, respectively. What is the voltage rise in each case, if the feeder inductive reactance is 0.015 Ω/phase and the line-to-line voltage is 460 V?

SOLUTION

Direct application of Equation (3.8) leads to:

$$Q_{Cap-0.85} = 1000 \times \left[\sqrt{\frac{1}{0.75^2} - 1} - \sqrt{\frac{1}{0.85^2} - 1} \right] = 262.2 \text{ kVAR}$$

$$Q_{Cap-0.95} = 1000 \times \left[\sqrt{\frac{1}{0.75^2} - 1} - \sqrt{\frac{1}{0.95^2} - 1} \right] = 553.2 \text{ kVAR}$$

The voltage boosts in the case and for the feeder characteristics (Equation [3.10]) are:

$$V_{Rise-0.85} = 0.015 \times \frac{262.2 \times 10^3}{\sqrt{3} \times 460} = 4.94 \text{ V}$$

$$V_{Rise-0.95} = 0.015 \times \frac{553.2 \times 10^3}{\sqrt{3} \times 460} = 10.4 \text{ V}$$

The voltage rise (boost) due to the capacitor can be expressed in per-phase values as:

$$V_{Rise-phase} = X_L \frac{Q_{Cap-phase}}{V_\phi} \text{ V/phase and } \%V = \frac{V_{Rise-phase}}{V_\phi} \times 100 \qquad (3.12)$$

Notice that in Equation (3.12) the voltage drop is zero for leading power factor that gives the $R/X_L = tan(\theta)$; therefore, the receiving-end voltage is constant. Typical R/X_L ratios for power distribution cables are in the range 0.2 to 0.3, with an average value of 0.25 and a power factor 0.97 leading, giving a flat receiving-end voltage value, regardless of the load current. Even if only a small part (3% or less) of the total electric power transferred through the transmission lines in a power distribution network is lost through joule effect and additional losses (related to the square of the line current), it is still important to estimate such losses. Cable manufacturers usually provide the AC resistance and the series leakage reactance, including the skin and the proximity effects, and the conduit/raceway type. Notice that the difference between the DC and high-frequency inductance is often neglected in power engineering studies. Electric distribution losses at any site are mostly due to the joule effect, which depends on the current squared and on the line resistance. However, most electric losses occur at end users, and are grouped as: electrical machinery and drives, electrical heating (e.g. furnaces, ovens, boilers, induction heating equipment, resistors, microwave and lighting equipment), and other load types such as electrochemical equipment, monitoring, control and communication systems. The basic relationship is expressed as follows:

$$P = n \times R \times I_{line}^2 \text{ (W)} \qquad (3.13)$$

Here n is the number of phase conductors, R (Ω) is the phase conductor resistance, and I_{line} the RMS (effective) line current (A). If the current is flowing through a series combination of N_S conductors the total resistance is given by:

$$R_{total} = \sum_{k=1}^{N_S} \frac{\rho_k \times l_k}{A_k} \text{ } (\Omega) \qquad (3.14)$$

In the case of N_P conductors connected in parallel sharing the total current, the lower the resistance the higher is the current flowing in each conductor. The total resistance of the parallel configuration of N_P conductors is expressed as:

$$R_{total} = \frac{1}{\displaystyle\sum_{k=1}^{N_P} \frac{A_k}{\rho_k \times l_k}} \ (\Omega) \tag{3.15}$$

In the above equations, ρ_k is the conductor material resistivity (Ωm), A_k is the cross-sectional conductor area (m^2), and l_k is the length of k conductor (m). The DC resistance, as we discussed in a previous chapter, is also dependent on temperature and must be adjusted for AC system operation. These changes need also to be reflected in voltage drop or cable loss calculations. The AC resistance of the cable conductors is the DC resistance corrected for skin and proximity effects:

$$R_{AC} = R_{DC} \cdot \left(1 + F_{skin} + F_{proximity}\right) \tag{3.16}$$

Here R_{DC} is the cable DC resistance, while F_{skin} and $F_{proximity}$ are the skin effect and proximity effect correction factors. The skin effect depends on the electrical frequency, the size of the conductor, the amount of current flowing, and the diameter of the conductor, through a rather complex relationship, having net results in an increase in the cable/conductor effective resistance. The proximity effect factor varies depending on the conductor or cable geometry. The proximity effect also increases the effective resistance and is associated with the magnetic fields of the cable conductors, which are close together. The proximity effect decreases with increases in spacing between cables. A few empirical relationships are used to calculate both these factors, or they are provided by the cable manufacturers. Notice that skin and proximity effects may be ignored with small conductors carrying low currents, but become increasingly significant with larger conductors. It is often desirable for technical and economic reasons to design conductors to minimize both these effects. AC resistances are important for calculation of current-carrying capacity. Values for standard designs of distribution and transmission cables are included in the tables in any cable design handbook. The series inductive reactance of a cable, an important parameter in voltage drop calculations, can be approximated by the following relationship:

$$X_L = 2\pi f \cdot \left[K + 0.2\ln\left(\frac{2 \cdot s}{D_{cond}}\right) \right] \times 10^{-3} \ (\Omega / km) \tag{3.17}$$

Where X_L is the conductor inductive reactance (Ω/km), f is the electrical supply frequency (Hz), s is the axial spacing between conductors (mm), D_{cond} is the diameter of the conductor, or for shaped conductors, the diameter of an equivalent circular conductor of equal cross-sectional area and degree of compaction (mm), and K is a constant factor pertaining to conductor formation and geometry. The most common values of the constant K are 0.0500 for solid conductors, 0.0778 for a three-wire conductor, and 0.0664 for a seven-wire conductor.

Example 3.5

Calculate the power losses per km of length, in a six-wire poly-phase system having the resistance per unit of length and phase of 0.018 Ω/km, carrying 2000 A.

SOLUTION

From Equation (3.13) the power loss per length is:

$$P = 6 \times 0.018 \times (2000)^2 = 0.432 \ (MW/km)$$

3.3 BUILDING POWER DISTRIBUTION AND ELECTRICAL SYSTEMS

The function of the electric power distribution system in a building, facility, or an installation site is to receive power at one or several supply points and to deliver it to the loads, lighting, motors, and all other electrically operated devices and equipment. The importance of building power distribution to the facility operation makes it imperative that the best system be designed and installed. In order to design the best distribution system, the system information concerning the loads and knowledge of the types of distribution systems is needed. Various categories of buildings have specific design challenges, but certain basic principles are common to all. Such principles, if followed, provide a soundly executed design. The specific use of commercial and industrial buildings, rather than the nature of the overall area development of which these buildings and facilities are part, determines their electrical system design category, structure, and type. While industrial plants are primarily machine- and production-oriented, commercial, residential, and institutional buildings are primarily people- and public-oriented types. The fundamental objective of commercial building design is to provide a safe, comfortable, energy-efficient, and attractive environment for living, working, and enjoyment. The electrical design must satisfy these criteria if it is to be successful. Modern commercial buildings, because of their increasing size and complexity, have become more and more dependent upon adequate and reliable electric systems. A power distribution system includes all parts of an electrical system between the power source and the customer service entrance. The power source can be either a local generating station or a HV transmission line feeding a substation that is reducing the high voltage to a voltage level suitable for local power distribution. Depending on the type of system, the distribution system consists of a combination of substations, distribution transformers, power distribution lines, secondary circuits, secondary service drops, and safety and switching equipment. A distribution transformer is used to transfer electrical energy from a primary distribution circuit to a secondary distribution circuit. Distribution transformers are installed in the vicinity of each customer to reduce the voltage of the distribution circuit to a usable voltage, usually 120 V or 240 V. The distribution system may be underground, overhead, or a combination of the two. Regardless of the type of installation or arrangement, power transformers must be protected by fused cutouts or circuit breakers, and lightning arresters should be installed between the high-voltage line and the fused cutouts. Three general types of single-phase power distribution transformers are in use today. The conventional type requires a lightning arrester and fuse cutout on the primary-phase conductor feeding the transformer. The self-protected type has a built-in lightning protector, and the completely self-protected types have the lightning arrester and overload devices connected to the transformer and no separate protective devices. A service drop is the combined conductors used to provide an electrical connection between a secondary power distribution circuit and facility. There are different ways of installing the service drop, as specified by the NEC codes, standards, or local utility codes. In a building, the service distribution board feeds all lighting systems, electric stairs, water pumps, and all building electromechanical equipment. Based on the load list(s) and understanding the demands, a stable network to supply the required power is designed. During the design, feeding type mentioned in the load list and voltage level(s) are taken into account to properly feed the loads. Normal loads are fed from the common busbar and the essential loads such as emergency lighting are fed from the emergency busbar. While single-line diagrams are prepared, load-balance studies are conducted simultaneously to calculate the required power supply and to size the transformer and busbars. In addition, active power, reactive power, and power factor are calculated for each bus and the entire system. The calculation aims to obtain the total active and reactive powers considering load factor from downstream (loads) to upstream (generator end or feeder). Load-flow studies are carried out in order to calculate all bus voltages, branch power factors, currents, and power flows throughout the plant electrical system.

Electric energy, delivered to sites by means of electric grids at high- and/or medium-voltage power lines, is distributed to the end users through medium- and low-voltage power distribution networks, which deliver electricity from the distribution substations to the service-entrance equipment located at residential, commercial, and industrial facilities. Depending on the power demand of loads and on the network layout, transformer substations, whose main task is to step-down the supply voltage downstream of the utility-delivering node, are located at a site or area boundary, or distributed around that area. The choice among different power distribution systems, such as radial or loop-feeder systems, is based on technical and economic evaluations that do not generally consider energy-saving targets, because very-low-energy losses are involved. Most US distribution systems operate usually at primary voltages from 12.5 kV to 24.9 kV, while some operate at 34.5 kV or low-voltage distribution such as 4 kV. However, most of the low-voltage (LV) distribution systems are being phased out. Distribution transformers convert the primary voltage to secondary consumer voltages to a low-voltage secondary circuit, usually 120/240 V, and the secondary distribution circuits connect to the end users at the service entrance. Electrical code defines service as, "The conductors and equipment for delivering electric energy from the serving utility to the wiring system of the premises served," indicating that the service is where a utility company provides electricity to a building or a structure, most buildings being served directly by the utility company, and the service includes a revenue meter. The requirements of NEC Section 130.5(A) refer to this service and to this meter. However, not all buildings or structures are connected directly to utilities and not all services have revenue-measuring meters. For example, a college campus can purchase bulk power from an electric utility company, in order to save energy costs. Usually, the revenue meters are located where the electric utility company connects the customers to the power distribution system. From this point of view, the customer owns the electrical system and becomes the serving utility for the NEC, IEEE, or IEC code purposes. Codes also recommend that the customer's service conductors shall be continuous from the point of electricity delivery to the metering equipment. Conductors and cable installed in raceways or conduits comply with NEC Article 300.3 for safety and supply security reasons. The rating of the service-entrance equipment must satisfy the general requirements stated by the NEC and local building codes. We have to keep in mind that considerable efforts are required in order to maintain the power supply within the requirements of various consumers, such as proper voltage, availability of the power on demand or supply reliability, and security.

The power distribution systems, in large building and facilities can be divided in two main categories: (a) the vertical supply system (rising mains), and (b) the horizontal supply (power distribution unit at each floor level). In most cases a high-voltage supply and transformer substation is required. Normally high-voltage (HV) switchgear and substation transformers are installed at the ground floor (or basement). However, often there are appliances with large power demand installed on the top floors (converters and motors for lifts, air-conditioning equipment, and electric kitchens). As it is desirable to bring the high-voltage supply as close as possible to the load centers, transformers are installed at the top floor, or if required, additional ones are installed on one of the intermediate floors. In such cases, transformers with noninflammable insulation and cooling are used. The arrangement of the rising mains depends on the size and shape of the building and suitable size of shafts for installing cables and bus ducts are needed.

3.3.1 Branch Circuits and Feeders

Several definitions are essential to understanding branch circuits and feeders, their characteristics, purpose, and designations, from the power company terminals to the main service disconnect. Feeders are conductors and cables that originate at the main depower distribution or main disconnect device and terminate at another distribution center, panelboard, or load center, while subfeeders are the conductors or cables originating at power distribution centers other than the main power distribution center and extending to panelboards, load centers, and disconnect switches that supply branch circuits. A panelboard can be a single panel or multiple panels containing switches, fuses, and circuit breakers for switching, controlling, and protecting circuits. Branch circuits represent the section of the wiring system extending past the final overcurrent device. These circuits usually originate at a panel and transfer power to load devices. Any circuit that extends beyond the final overcurrent protective device is called a branch circuit, including the circuits servicing single motors (individual) and circuits serving several lights or lighting systems and receptacles (multiwire). As specified by codes and standards, branch circuits are usually low current (30 A or less), but in some instances they also supply higher currents. A basic branch circuit is made up of conductors or cables extending from the final overcurrent protective device to the load. Some branch circuits originate at safety switches (disconnects), but most originate at a panelboard. Based on their designation, the branch circuits are classified as: *individual branch circuit*, a branch circuit that supplies a single load; *multi-outlet branch circuit*, a branch circuit with multiple loads; *general-purpose branch circuit*, a multi-outlet branch circuit that supplies multiple outlets for appliances and lighting; *appliance branch circuit,* a branch circuit that supplies a single appliance load; or *multiwire (conductor) branch circuit,* a branch circuit with two or more ungrounded conductors and one grounded conductor, designed to supply specific loads. A branch circuit is sized for the supplied load. Sizing the circuit for additional future loads is good engineering practice. The rating of a branch circuit depends on the rating of the overcurrent device protecting the circuit. Branch circuits serving only one device can have any rating, while a circuit supplying several loads is limited to ratings of 15 A, 20 A, 30 A, 40 A, or 50 A. Branch-circuit voltage limits are contained in Section 210.6 of the NEC code, with similar provisions in the IEC standards. These limits are based on the equipment or loads that are supplied by the circuit. In residences and hotel rooms, circuits supplying lighting fixtures and small receptacle loads cannot exceed 120 V. Circuits that are 120 V and less may be used to supply lamp-holders, auxiliary equipment of electric-discharge lamps, receptacles, and permanently wired equipment. Branch circuits exceeding 120 V but less than 277 V may supply lamp-holders, ballasts for fluorescent lighting, ballasts for electric-discharge lighting, plug-connected appliances, and hard-wired appliances. Incandescent lamps operating over 150 V are permitted in commercial construction. Circuits exceeding 277 V and less than 600 V can supply mercury-vapor and fluorescent lighting. NEC Section 210.21(B) specifies that receptacle ratings are permitted on multi-outlet branch circuits, having a rating equal to or greater to the overcurrent device rating. NEC Sections 210.52 through 210.63 specify the requirements for receptacle locations and ratings. Section 210.70 lists the requirements and specifications for lighting outlets. NEC Section 210.70 specifies the lighting outlet locations for houses and other facilities. Ground-fault circuit interruption (GFCI), as specified by Section 210.8, is required on 125 V, 15 A, or 20 A receptacle outlets in certain locations. This NEC section specifies GFCI locations, as well as the exceptions that apply. In the event that two energized conductors come into contact with each other or a different electrical potential surface, an electric arc is initiated at the contact point, causing a fault current. This fault current may not be of sufficient magnitude to trip the circuit breaker or to blow the fuse, requiring a special electronic circuit breaker, arc fault circuit interrupter to be used. NEC Section 210.12(B) requires arc fault circuit interruption on all 15 A and 20 A, 125 V supplying outlets in dwelling unit bedrooms. NEC code also places load limitations on any branch circuit with continuous loads (loads with duration longer than 3 hours, such as lighting loads). Continuous loads must not exceed 80% of the circuit rating allotted for them. If the overcurrent protective device is listed for continuous operation at 100% of its rating, the 80% factor is not used. Branch-circuit loads are classified into five categories: lighting loads, receptacle loads, equipment loads, heating and cooling loads, and electrical motor loads.

    The amperage rating of branch-circuit conductors must be greater than the maximum load the circuit is supplying. For multiple-load branch circuits, the conductor ampacity must correspond to the rating of the overcurrent protective device. However, for branch circuits that are supplying hardwired devices, such as electric heaters, air-conditioning units, and water heaters, the fuse or circuit breaker is usually rated at the next higher rating. The conductor is acceptable if its rating is at least that of the load current, even if the overcurrent protective device rating is higher. After the required receptacles, outlets, lighting outlets, branch circuit for appliances, equipment, and motor supply circuits are calculated and located, it is necessary to determine the minimum number of the feeders and the branch circuits needed to supply them. The actual number of branch circuits usually exceeds the minimum specified by the codes and standards. Once the receptacle and lighting outlets are located, the number of branch circuits is estimated and the branch circuits must be designed. The total number of the required branch circuits includes the ones supplying receptacle and lighting outlets for general lighting, small appliances, the laundry and bathroom branch circuits, dedicated equipment loads, and outdoor and unfinished areas receptacle outlets branch circuits. Usually the actual number of branch circuit exceeds the code or standard minimum number. NEC Section 210.11(A) requires that the minimum number of branch circuits to supply the general lighting load, before the application of any demand factor is determined by:

$$N_{Min\text{-}BrCirts} = \frac{\text{General Lighting Load (VA)}}{\text{Maximum Load per Branch Circuit (VA)}} \qquad (3.18)$$

Here $N_{Min\text{-}BrCirts}$ is the minimum number of branch circuits as required by the NEC, while the maximum branch-circuit load is based on the branch-circuit rating. Usually, for residential applications, 15 A and 20 A branch circuits are used to supply general lighting, while 20 A circuits are used in commercial and industrial facilities. A good design practice is to limit the maximum load per branch circuit to 80% of the circuit rating. An alternative approach is to determine the permitted maximum area (m² or sq. ft.) to serve a branch circuit, and then by dividing the total by this value, the minimum number of branch circuits supplying the general lighting as required by the code is found.

Example 3.6

What is the minimum number of branch circuits supplying the general lighting load for the following occupancies, a 9600 sq. ft. office (at 20 A rating) and a 200 m² residence (at 15 A)? The NEC specifies a unit load per sq. ft. of office space is 3.5 VA plus 1 VA is required for general receptacles, or 33 VA for a residence load per unit area in m².

SOLUTION

For the office space, the branch-circuit rating is 120 V and 20 A, while for the house (residential space) the branch-circuit rating is 120 V and 15 A. Therefore, the maximum load permitted for the first circuit is 1920 VA (120·0.8·20), while for the second circuit is 1440 VA (120·15·0.8). By applying Equation (3.18), the minimum number of branch circuits for general lighting in each case is:

$$N_{office} = \frac{4.5 \text{ (VA/sq. ft)} \times 9600}{1920 \text{ VA/circuit}} = \frac{43200}{1920} = 22.5$$

$$N_{residence} = \frac{33 \text{ (VA/m}^2) \times 200}{1440 \text{ VA/circuit}} = \frac{6600}{1440} = 4.58$$

Therefore, a minimum 23 branch circuits to supply the general lighting load are required for the office space and 5 branch circuits for the residence space.

NEC Section 210.11(C) states the requirements, provisions, and specifications for branch circuits installed in residential specific areas, such as kitchen, laundry, bathroom, outdoor, and for dedicated loads and equipment. As a general recommendation, these branch circuits need to supply only receptacles in their designated area, while lighting outlets may not be connected to the space branch circuits. NEC Section 210.11(C) requires that the laundry or bathroom use 125 V and 20 A branch-circuit ratings, and only receptacles in that area be connected. NEC or IEC do not limit the number of receptacles or lighting outlets permitted on 15 A, 20 A, and 125 V, or IEC rating branch circuits; the designers must determine this based on their experience and engineering practice. However, where specific equipment ratings are known, these values can be used to properly design the branch circuits. Where such values are not available, a load of 180 VA can be assigned to each general-purpose receptacle, while good design practice limits the maximum number of receptacles. Certain loads in industrial and small commercial facilities are connected to multiwire branch circuits. A multiwire circuit consists of two or more ungrounded conductors and one grounded conductor, having the same voltage between each ungrounded and grounded conductor. The loads in multiwire branch circuits can be connected between two ungrounded conductors (line-to-line) or between an ungrounded and grounded conductor (line-to-neutral). Notice that for a fully balanced system, the neutral current is zero.

The conductors between the service equipment and the branch-circuit overcurrent devices are called feeders. NEC Article 215—Feeders provides information regarding the safe, adequate sizing and installation of feeders. This article also applies to subfeeders, which provide power to branch-circuit panels but originate at power distribution centers rather than at service equipment. Feeder loading is dependent on the total system power requirements. When all connected loads operate simultaneously, the feeder must be of sufficient ampacity to meet that load demand. If only 75% of connected loads are operating at the same time, then the feeder is sized larger than the service conductors. Prior to installation, there are factors that must be considered to ensure the feeder size, type, and overcurrent protection is correct for that application. The feeder can be copper or aluminum; the environment (damp, hot, corrosive) around the feeder must be taken into consideration; the feeder can be run in conduits, cable trays, or other systems; and single- or multiconductor cable can be used. Feeders can be paralleled or individual, while voltage drop becomes a consideration in long-run feeders. Conductor sizing includes several factors: conduit fill, ambient temperature, and connected load demand, and a neutral wire may not be necessary with the feeder, while the continuous loads affect the feeder size. Various overcurrent protective devices are also used with each feeder. Besides the previous adjusting factors, the two important design factors for calculating loads in electrical systems are demand factor and diversity factors. The demand factor (always less than one) is the ratio of the maximum demand of an electrical system, or part of a system, to the total connected load on the system, or part of the system under consideration. The diversity factor is the ratio of the sum of the individual maximum demands of the various subdivisions of an electrical system, or part of a system, to the maximum demand of the whole system, or part of the system, under consideration. Diversity factor is usually higher than one. For example, these terms, when used in an electrical design, should be applied as follows. The total connected loads supplied by a feeder-circuit can be multiplied by the demand factor to determine the load used to size the system components. The sum of the maximum demand loads for two or more feeders is divided by the diversity factor for the feeders to derive the maximum demand load.

Example 3.7

Consider three individual feeder-circuits with connected loads of 200 kVA, 300 kVA, and 400 kVA and demand factors of 85%, 75%, and 80% respectively. If the diversity factor is 1.5 for all feeders, calculate the feeder load demands.

SOLUTION

The load demand for each feeder circuit is:

$$200 \times 0.85 = 170 \text{ kVA}$$
$$300 \times 0.75 = 225 \text{ kVA}$$
$$400 \times 0.80 = 320 \text{ kVA}$$

The sum of the individual load demands is equal to 715 kVA. If the main feeder circuit is sized at unity diversity: kVA = 715 kVA ÷ 1.00 = 715 kVA, so the main feeder circuit would have to be supplied by a 720 kVA transformer. However, using the diversity factor of 1.5, the kVA = 715 kVA ÷ 1.5 = 476.7 kVA for the main feeder. For a diversity factor of 1.5, a 500 kVA transformer could be used. Notice that although feeder-circuit conductors should have an ampacity sufficient to carry the load, the ampacity of the feeder need not always be equal to the total of all loads on all branch circuits connected to it. The demand factor permits a feeder circuit ampacity to be less than 100% of the sum of all branch-circuit loads connected to the feeder. Section 220.3(A) of the NEC states that the load on a service or feeder is the sum of all of the branch loads subject to their demand factors as permitted by the code rules.

3.3.2 Power Distribution in Small Buildings

Small commercial or residential buildings have a very simple power distribution system. The local utility owns the transformer, which is sited often on a pad outside the building or is attached to a utility pole. The transformer reduces the voltage from 13.8 kV down to 120/240 V or 120/208 V and then passes the electricity to a meter, which is owned by the utility and keeps a record of power consumption. After leaving the meter, the power is transmitted into the building at which point all wiring, panels, electric devices, outlets, control, and protection devices are the property of the building owner. Cables (wires) transfer the electricity from the meter to a panelboard, which is usually located in the building basement or garage. In small commercial buildings, the panelboard may be located in a special utility closet. The panelboard has a main service breaker and a series of circuit breakers, which control the flow of power to various circuits in the building. Each branch circuit serves a device, equipment (some appliances and building equipment, pumps, elevators, etc., require heavy loads), or a number of devices such as convenience outlets or lights. The electrical service in a building consists of conductors and accessories connecting the building or facility to the utility power supply, the raceways containing electric service-entrance conductors and cables, metering and monitoring infrastructure and equipment, main service disconnect, and main overcurrent protection devices, usually a main circuit breaker. Fuses are not usually employed for such purpose and designation. Electrical service is the connection point with the local utility or to the grid in the case of large buildings or large commercial industrial facilities. Building electrical service must meet all requirements and provisions of the NEC code, similar national codes, or IEC standards, as well as those of the local utility. The service entrance includes all the wires, cables, devices, and fittings that transfer the electricity from the utility power transformer to consumers. The type of equipment used for an electrical service entrance of a building may include high-current conductors and insulators, disconnect switches, protective equipment for each load circuit that is connected to the main power system, and the meters needed to measure power, voltage, current, or frequency (Figure 3.7).

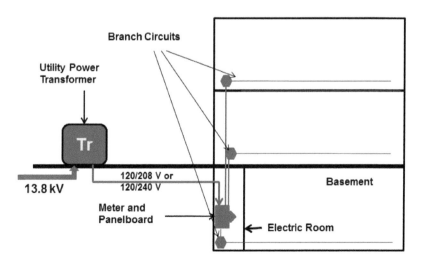

FIGURE 3.7 Small-building typical electric service configuration and components.

The service components protect, meter, monitor, and distribute electrical energy to the feeders and branch circuits. Conductors brought to the building and run overhead, coming from a utility pole are called the service drop, or routed underground, from either a pole or a transformer pad, are called service laterals. In either case, these service conductors are connected to the service-entrance conductors at the building. A single building is in general supplied by a single electrical service, as specified in NEC Section 230.2, meaning that only one set of service conductors between the utility and the building exists. A structure can have only one service, one service drop, or one service lateral. While this is the basic rule, there are practical exceptions, which are specified by the codes or the local utility. However, if the building size and conditions require additional services to supply fire pumps, emergency and backup power systems, or a parallel power supply, is included, as specified in NEC Section 230.2(A). If the building size requires, in multi-occupancy buildings or when the capacity requirements (over 2000 A and 600 V or less) dictate, multiple or additional services are permitted as stated in NEC Section 230.2. Other exceptions requiring additional service include: generators, driven by thermal engines or by wind turbines, which can provide additional services; the electric power supplied by the solar photovoltaic sources; or when different voltages and phases may be required within the same structure for special uses. As general guidelines, the service conductors must be kept as short as possible in order to minimize the voltage drops, service conductors must enter the building as close as possible to the service panel, and the service disconnect must be at or very near the building point of entry. NEC Section 230.2(D) permits additional services when different service characteristics are required, for example a building supplied by a 120/240 V, single-phase, three-wire service and a 240 V, three-phase, four-wire service. It is quite unlikely that the utility is providing both services, and usually a transformer is installed to provide the needed services.

3.3.3 POWER DISTRIBUTION IN LARGE BUILDINGS

In large buildings, the type of distribution depends on the building type, dimension, the length of supply cables, and the loads. In large buildings and facilities, usually the HV switchgear and substation transformers are installed at the ground floor or basement. However, if there are appliances with large power demands installed on the upper floors, e.g. power converters, elevator electrical motors, heating and air-conditioning equipment, monitoring and security equipment switchgears and transformers can installed at higher floors. As common sense rule, it is desirable to bring the high-voltage supply closer to the loads. In such cases, when transformers are installed at the top and intermediate floors, transformers with noninflammable insulation and cooling are used. Any arrangement of the building electric distribution is determined by the building size, shape, building equipment and designation. However, regardless the building electrical system structure and configuration, suitable means for installing cables, and bus ducts must be provided. Building vertical power distribution supply systems are implemented in several ways, some of which are:

1. The vertical supply system (rising mains); and
2. The horizontal supply (distribution at each floor level).

In most cases, a high-voltage supply and transformer substation is required.

Large buildings have a much higher electrical load than small buildings; therefore, the electrical equipment must be larger and more robust. Large-building owners often purchase electricity at high voltages (in the US, 13.8 kV) because it comes at a cheaper rate. In this case, the building owners provide and maintain their own transformer(s), which lowers the voltage to a more usable level (in the US, 480/277 V). This transformer can be mounted on a pad outside the building or in a transformer room inside the building. The electricity is then transmitted to switchgear. The switchgear role is to distribute electricity safely and efficiently to the various electrical closets throughout the building. The equipment has numerous safety features including circuit breakers, which allow power to be disrupted downstream. This may occur due to a fault or problem, but it can also be done intentionally to allow technicians to work on specific branches of the power system. It should be noted that very large buildings or buildings with complex electrical systems may have multiple transformers, which may feed multiple pieces of switchgear. We are keeping this section simple by sharing only the basic concepts. The electricity will leave the switchgear and travel along a primary feeder or bus. The bus or feeder is a heavy-gauge conductor that is capable of carrying high-amperage current throughout a building safely and efficiently. The bus or feeder is tapped as needed and a conductor is run to an electric closet, which serves a zone or floor of a building. In large buildings, the type of distribution depends on the building type, dimension, the length of supply cables, and the loads.

3.3.4 Emergency Generators and Uninterruptible Power Supply Systems

The term uninterruptible power supply (UPS) has been applied to both an actual uninterruptible power system and to specific battery (energy-storage unit) equipment incorporated into the system. Historically, the specific equipment (UPSs) were developed in parallel with IT equipment and computers to provide a reliable, secure, higher-power, quality, and continuous electric service (supply) that the utility could not provide. The concept, as illustrated in Figure 3.8, consists of battery bank(s) and/or other energy storage device(s), DC-AC power converter(s) or inverter(s), DC-DC power converter(s) or voltage regulator(s), AC-DC converters or rectifier(s), control unit, switches, bypass circuit(s), optional gen-set, and eventual secondary energy source connection. The rectifier, DC-DC converter, and power management form the so-called energy storage power management unit. Emergency generators and backup uninterruptible power supply systems are used to provide critical loads with power supply in the case of main failures (e.g., operating theaters and intensive care units in hospitals, first-response units, communication and data centers). Emergency generators are usually driven by diesel engines, and connected to the load in the following way: when the generator is of the same size as the power supply transformer, and when the generator is of a smaller size as compared with the power supply transformer. The basic version

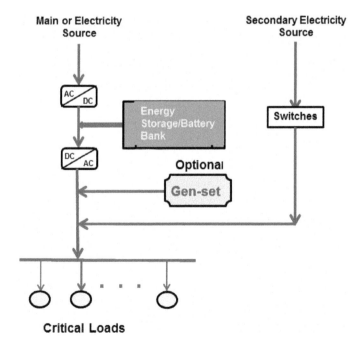

FIGURE 3.8 Typical uninterruptible power supply configuration.

of the UPS consists of a rectifier, inverter, DC-DC converter, and batteries. During normal operation, the inverter supplies the critical load and ensures that the output voltage is stable and precisely controlled. The inverter receives its power from mains via the rectifier, which, at all times, float charges the batteries. In the event of an interruption in the power supply, the batteries take over the task of supplying power to the inverter. When the main's power has been restored, the inverter is disconnected and the rectifier, via DC-DC converter, resumes the power input to the battery, automatically recharging the battery banks.

Commercially available or specific-application UPS equipment and systems are categorized based on power ratings, voltage type, application, manufacturer, configurations, topologies, structure, and components. In order of their power ratings, UPS systems are divided in low-power (3 kVA or lower), medium-power range (3 kVA to 10 kVA), high-power range (10 kVA up to 120 kVA), and very-high-power range, above 120 kVA, often about 1000 kVA. Low-power type are usually AC single-phase modules (120 V or 240 V), while the ones in higher power ranges are only three-phase systems (460 V or 480 V). Modern UPS systems employ, besides the battery bank(s), a broad category of energy-storage devices, such as flywheels, super-capacitors of fuel cell stack(s) used in modern UPS systems and equipment. The UPS requirements for the energy-storage unit are determined by the time duration of the energy (power) delivery and by the energy (power) level. The time duration requirements are further categorized into: short-term (5 to 30 min.), medium-term (systems able to deliver up to 1000 kW for 5 min. or lower power for longer periods, usually until the gen-sets start), and long-term units, able to deliver power for about 8 hours. Specific UPS requirements are determined by the application and the local conditions. However, still one of the most-used UPS systems is battery inverter equipment, available in power ranges from 100 VA to 1000 kVA (and even higher) at all commercial input and output voltage and phase configurations and topologies. In modern systems, there is a clear tendency to use a diverse portfolio of energy-storage devices, e.g., flooded battery systems, fuel cells, flywheels.

3.4   INDUSTRIAL POWER DISTRIBUTION

Depending on the needs of the customers, the voltage supplied can be as low as 120 V single-phase or 120/240 V single-phase, where the 240 V secondary of the distribution transformer has a center tap, providing, if needed, also two 120 V single-phase circuits. However, larger commercial customers usually utilize three-phase power, with the most common 208/120 V or 480/277 V services. However, very large commercial and industrial facilities can include higher three-phase services, such as 2.4 kV, 4.16 kV, 13.8 kV or often higher voltage levels. In such cases, the customers are responsible for providing power transformers and the needed electric infrastructure to serve all of the voltage requirements and needs. In residential areas, the service transformers, either ground mounted or pole mounted, convert the three-phase voltages, e.g. 13.8 kV (primary power distribution transformer) to power at a voltage level of 120/240 V, usually single-phase or the needed voltage levels. The secondary of such 240 V, single-phase power transformers has a center tap, in order to provide two 120 V single-phase circuits. Any individual residence can be supplied with both 120 V and 240 V single-phase service, if higher voltages are necessary for appliances and equipment, e.g. water pumps. Electricity for buildings can be supplied from local utility or can be generated, in part or entirely on site, using conventional energy generators or alternative energy conversion systems (e.g., wind turbines, photovoltaic systems, fuel cells). One appealing energy system is a combined heat, cooling, and power generation (CCHP) system designed for heat, air-conditioning, cooling, and like by-product the electricity, significantly increasing building energy system overall efficiency. Several types of building power distribution systems are used, and the most appropriate for a specific building depends on the building size, destination, and load characteristics (power ratings, voltages, frequency, etc.). Requirements for electrical power distribution systems apply to all nonresidential and residential buildings. The intention is to save energy and to allow future systems for power use, expansion, monitoring, and control to be added when expected changes in the marketplace occur.

The enclosure of any equipment serves to keep out dirt, dust, moisture, and intruders, and to protect equipment and personnel. This is a separate matter from protection against explosions, where a piece of electrical equipment must be mounted outdoors and be protected against the weather where there is no risk of explosion, or it may be indoors in a particularly dusty but non-flammable atmosphere. An internationally agreed-upon system has been developed to designate the degree of protection afforded by any enclosure. It consists of the letters IP, which stands for International Protection, followed by two digits, indicating the degree of protection. The first digit, 0 to 6, describes the protection against ingress of solids, while the second digit, 0 to 8, describes protection against ingress of liquids. In both cases, the higher numbers indicate greater degrees of protection. Panelboards are metal cabinets, enclosing the main disconnect switch and the branch-circuit protective equipment. Distribution panelboards are located between the power feed lines within a building and the branch circuits connected to it. MV switchgears are used any time when a facility is directly connected to the MV section of power transmission and distribution. MV metal-clad switchgears have grounded metal barriers separating compartments and structures within the assembly. LV panelboards and switchboards typically do not have such types of protection. Metal-enclosed LV switchgear is used in industrial and commercial buildings, as a power distribution control center to house the circuit breakers, bus-bars, and terminal connections, which are part of the power distribution system. NEMA standard 250-2003 defines criteria for electrical enclosures based on their ability to withstand external environments. Metal-enclosed switchgear is governed by IEEE standard C37.20.3, while metal-clad switchgears are covered by multiple standards and codes. Metal-clad switchgear meets or exceeds the requirements for metal-enclosed switchgear, but not vice versa. Usually, a combination of switchgear and distribution transformers is placed in adjacent metal enclosures. This combination is referred to as a load-center unit substation because it is the central control for several loads. The rating of these load centers is usually 15 kV or lower for the high-voltage section and 600 V or less for the low-voltage section, but higher values such as 35 kV are available. Load centers provide flexibility in the electrical power distribution design of industrial plants and commercial buildings.

3.4.1 Panelboard and Switchboard Calculations and Ratings

The panelboard and switchboard functions are to supply and distribute power to the branch circuits and feeders, through overcurrent protection devices. The panelboard structure must allow the settings of proper voltages to the supplied branch circuits and feeders. In addition to circuit protection, power distribution systems must have equipment that can be used to connect or disconnect the entire system or parts of the system. Safety switches are used only to turn a circuit off or on; however, fuses are often mounted in the same enclosure with the safety switch. Single-pole, double-pole, and three-pole circuit breakers are available to be installed into the most common panelboard types. In order to understand how a panelboard distributes the power and voltage, its internal structure and the electrical diagram must be available. Panelboards are classified as power panelboards or as lighting and appliance branch-circuit panelboards. A panelboard that has less than 10% of its protection devices supplying lighting and appliance branch circuits is considered a power panelboard, as defined by NEC Section 408.14(A). Otherwise, a panelboard with 10% or more of its protection devices supplying lighting and appliance branch circuits is defined as a lighting and appliance branch-circuit panelboard. NEC defines a lighting and appliance branch circuit as one rated at 30 A or less, for example a 20 A and 277 V branch circuit supplying lighting in commercial buildings.

Many panelboards are provided with necessary overcurrent protection by use as a main circuit breaker, or by the feeder overcurrent protection device. Power panelboards, which do not have system neutral to the panel, supply branch-circuit loads with only ungrounded conductors not requiring overcurrent protection. However, good design practice recommends providing protection.

Usually, two separate panels are installed for systems supplied by 240/120 V, three-phase, four-wire systems, one panel supplying 120/240 V, single-phase, three-wire loads, and the other supplying 240 V, three-phase, three-wire loads. Panelboards supplied by three-phase, three-wire systems, with voltage ratings of 240 V and 480 V supply only three-phase loads, such as motors, insulation, step-down transformers, and electric heaters. In order to keep a record of the branch circuits, each panelboard has a panel schedule, designed to allow easy tabulation of the loads on the individual branch circuits. This load data is used to determine load balance among ungrounded conductors. Loads on the branch circuits may be expressed in amperes (A) or volt-amperes (VA). The panel schedule also contains information about neutral and ground bus, very useful in applications where the neutral must be separated from the ground bus. The neutral bus is usually rated to carry 100% of the panelboard-rated current; however, there are panels with neutrals rated to carry 200% of the panel-rated current.

Example 3.8

Determine the loading, expressed in A, on the each of the phase (ungrounded) conductors for the specified loads: (a) 120 V, single-phase, and 1800 VA, (b) 277 V, single-phase, and 2800 VA, and (c) 480 V, three-phase, and 36,000 VA.

SOLUTION

The current (loading) in each case is:

$$I = \frac{1800}{120} = 15.0 \text{ A}$$

$$I = \frac{2800}{277} = 10.1 \text{ A}$$

$$I = \frac{36000}{\sqrt{3} \times 480} = 43.35 \text{ A}$$

In order to ensure the safety of the electrical installation, proper and required clearances must be ensured and maintained around panelboards, switchboards, and switchgears. They are needed to provide adequate working space around electrical equipment and devices and safety protection space in the event of equipment failure. NEC Section 110.26(A) specifies a working space of at least 30 in., unless the equipment is wider than 30 in., in which case the required working space in the front of the equipment must be at least the width of the equipment. It is also required that hinged doors or panels can open at least 90° without any obstruction, while the height of working space must be at least 6.5 ft or equipment height, whichever is larger. The required depth is a function of the nominal voltage to the ground and the installation equipment condition. In addition to the electrical equipment required working space, Section 110.26(C) specifies the clearance requirements and specifications for access to the working spaces.

3.4.2 SWITCHGEARS, LOAD, AND MOTOR CENTERS

Power distribution systems used in large commercial and industrial and large commercial applications are complex, being distributed through switchgear, switchboards, transformers, and panelboards. Power distributed throughout a commercial or industrial application is used for a variety of applications, such as heating, cooling, lighting, and motor-driven machinery. The terms *switchgear* (SWGR) and *motor control center* (MCC) are used in general to describe various combinations of special enclosures, busbars, circuit breakers, power contactors, power fuses, protective relays, controls, indicators, and monitoring devices. The standards used in Europe often refer to IEC60050 for definitions of general terms. Particular IEC standards tend to give additional definitions that relate to the equipment being described, e.g., IEC60439 and IEC60947 for low-voltage equipment;

IEC60056, IEC60298, and IEC60694 for high-voltage equipment. In general, switchgear may be more closely associated with switchboards that contain circuit breaker or contactor cubicles for power distribution to other switchboards and motor control centers, and which receive their power from generators or incoming lines or cables. Motor control centers tend to contain outgoing cubicles specifically for supplying and controlling power to motors. However, motor control centers may contain outgoing cubicles for interconnection to other switchboards or motor control centers, and circuit breakers for their incomers and busbar sectioning. Switchboards may be a combination of switchgear and motor control centers. Switchgear tends to be operated infrequently, whereas motor control centers operate frequently as required by the process that uses the motor. Apart from the incomers and busbar section circuit breakers, the motor control centers are designed with contactors and fuses (or some types of molded-case circuit breakers in low-voltage equipment) that will interrupt fault currents within a fraction of a cycle of AC current. Circuit breakers need several cycles of fault current to flow before interruption is complete. Consequently, the components within a circuit breaker must withstand higher forces and heat produced when several complete cycles of fault current flow. Switchgear is available up to 400 kV or even higher, whereas motor control centers are only designed for voltages up to approximately 15 kV because this is the normal limit for high-voltage motors.

The SWGR and MCC equipment are invariably housed in a building or enclosed module, or at least effectively protected against bad weather and aggressive environmental conditions. The construction is therefore of the metal-clad type, in which all the live parts are housed in a mild-steel sheet metal enclosure. The enclosure is subdivided so that personnel may work safely on some compartments without danger or risk of electric shock. Various degrees of personnel and ingress protection are commonly available. The degree of protection is defined in various international standards, e.g., NEMA and NEC in United States, IEC in United Kingdom and Europe. For use inside buildings where manual operation and interference is infrequent and where the atmosphere is cool, dry, and clean, an enclosure of the IEC60529 type IP40, 41, or 42 or NEMA type 1 or 2 is usually adequate. Both types include main electrical components, such as main busbars, earthing busbar, incoming and busbar section circuit breakers, outgoing switching devices, contactors or circuit breakers, fuses for MCC or SWGR outgoing circuits, safety interlocking devices, electrical protective relays and devices for all power circuits, control, monitoring, and indication devices, communication or network interfacing system, and main connections and terminal compartments. The requirements for control and indication vary considerably depending upon the type of circuit, e.g., incoming, busbar section or outgoing circuit, whether the equipment is a switchboard or a motor control center, high or low voltage, process duty, the need for remote indication and control, and owner preferences. Most switchboards and motor control centers are fitted with a variety of electrical and mechanical safety interlocking devices. Their purposes are to protect against hazards.

Motor control centers receive this power through complex distribution systems, which include power distribution lines and related equipment. Transformers used with three-phase power require three interconnected coils in both the primary and the secondary. These transformers can be connected in either a wye or a delta configuration. The type of transformer and the actual voltage depend on the requirements and capability of the power company and customer needs. Unlike other types of power distribution equipment, which are used with a variety of load types, motor control centers primarily control the distribution of power to electric motors. Usually in industrial and large commercial facilities and structures, the electricity is delivered to loads from designated systems, known as load centers, containing the necessary equipment to monitor, protect, and control the loads. Motor control centers are centralized hubs containing motor control units sharing a common power bus. Used in low-voltage (230 V to 600 V) and medium-voltage (2.3 kV to 15 kV) three-phase applications, MCCs traditionally house startup and drive units. Auxiliary equipment is often found in almost all load centers, such as control switches, indicator lights, and metering equipment are sometimes installed in smaller metallic enclosures whose capabilities to withstand the external environment are determined by NEMA Standard 250. Today's units go far beyond these basic functions. It is not uncommon for modern "smart" centers to include programmable controllers and metering equipment for complex control schemes and safety features, in addition to the overload relays and contactors. The main equipment used to isolate a circuit is some type of switchgear. Switchgears provide a means to stop the flow of electricity to the entire power distribution system or just a portion of it. Switchgear can be found in the substation, at the junction of feeders, in between two circuits, or anywhere the engineer decided it was necessary to provide a means to isolate the circuit. The main purpose of switchgear is to isolate the circuit. Switchgear incorporates switches, circuit breakers, disconnects, and fuses used to route power and in the case of a fault, isolate parts of an electric circuit. In general, switchgear has three basic functions: (1) protection and safety for equipment and workers, (2) electrical isolation to permit work and testing, and (3) local or remote circuit switching. Developments in switchgear design have led to the introduction of network support for monitoring and control as well as advanced diagnostic capabilities for the purposes of monitoring usage, loading, and a host of other operational parameters. Among the advantages of modern switchgear are: (a) control and monitoring over a network, (b) remote control of the switchgear over a network, by removing the risk of arc flash to the operator, (c) individual analysis and power monitoring, involving the power monitoring and analysis of the main power and individual circuit breakers, and (d) advanced technology, the latest technology in circuit breakers and switching. There are different types of load centers, their selection based primarily on the electrical requirements of the loads and the installation environment. Several types of load centers are housed in metal enclosures to protect enclosed equipment, nearby objects, and personnel in the event of equipment malfunction. Load centers supplying large motors and/or other industrial loads are referred to as switchgear, while smaller load centers specialized to supply small to midsize motors will be called motor control centers (MCCs). Auxiliary equipment such as control switches, indicator lights, monitoring, and metering equipment is sometimes installed in smaller metallic enclosures whose capabilities to withstand the external environment are determined by the specifications of the NEMA 250 standard. Switchgears are specialized load centers designed to supply large loads, such as large electric motors or smaller load centers. Switchgears come in two varieties: metal enclosed and metal-clad. Metal-clad switchgear is the more robust design, requiring shutters between the bus and the equipment front, compartmentalization of live parts, and insulation of the bus and primary components.

All switching and interrupting devices must be drawing-out mounted, allowing removal from the switchgear without unwiring. Metal-enclosed switchgear does not need to meet these criteria and often has lower interrupting ratings, lower breaker duty cycle, and may use fused or non-fused switches instead of circuit breakers. Metal-enclosed switchgear is governed by IEEE standard C37.20.3, while metal-clad switchgear is covered by multiple standards. Metal-clad switchgear meets or exceeds the requirements for metal-enclosed switchgear, but not vice versa. A switchgear lineup is made up of multiple sections or cubicles joined side-by-side. The front of the switchgear is hinged, and opening the doors exposes the circuit breakers. When a breaker is removed from metal-clad switchgear, an insulated barrier separating the cubicle from the energized bus-work running the length of the switchgear is visible. This is an important safety feature, because metal-clad switchgear commonly is maintained while the main bus is energized. An MCC is a load center customized to serve small to midsize motors. Circuit breakers are replaced with combination motor starters. Because the starters are much smaller than switchgear breakers, stacking more than two units per horizontal section and reducing the width of the horizontal sections from 36″ to 20″ can save considerable space. MCCs are centralized hubs containing motor control units sharing a common power bus. Used in low-voltage (230 V to 600 V) and medium-voltage (2.3 kV to 15 kV) three-phase applications, MCCs traditionally house startup and drive units. However, modern MCC units go far beyond these basic functions. Modern MCC units often include programmable and smart controllers and smart metering equipment for advanced and adaptive control schemes and improved safety capabilities, in addition to the usually overload relays, switches and contactors. MCCs are simply physical groupings of combination starters in one assembly. A combination starter is a single enclosure containing the motor starter, fuses or circuit breaker, and a disconnecting power device.

Other devices associated with the motor, such as pushbuttons and indicator lights, may also be included. Switchgear incorporates switches, circuit breakers, disconnects, and fuses used to route power and in the case of a fault, to isolate electric circuit sections. Switchgear has three basic functions: (1) protection and safety for equipment and workers; (2) electrical isolation to permit work and testing; and (3) local or remote circuit switching. Developments in switchgear design have led to the introduction of network support for monitoring and control as well as advanced diagnostic capabilities for the purposes of monitoring usage, loading, and a host of other operational parameters.

3.4.3 Load Center, Switchgear, and Motor Control Center Ratings

The standards controlling the design and testing of metal-clad switchgear were developed by the American National Standards Institute (ANSI) in conjunction with the IEEE. IEEE standard C37.20.2 stipulates metal-clad switchgear-rating criteria. Low-voltage switchgear (600 V class and below) is governed by IEEE standard C37.20.1, while IEEE standards C37.04, C37.06, and C37.09 specify medium-voltage circuit breaker rating and testing criteria. Several important ratings are given to the main buswork in metal-clad switchgear. Both copper and aluminum bus carry these ratings. A continuous-current rating is assigned to limit the temperature rise of the bus bars to a value that will not compromise the bus insulation. The busbars typically are insulated with an epoxy-type material that can withstand fairly high operating temperatures. Insulators, either porcelain or polymeric composite, support the bus bars. The insulators also have a maximum operating temperature. A main (horizontal) bus similar to that used in switchgear is implemented in MCCs. Because the loads supplied by an MCC are smaller than the loads fed from switchgear, the available bus ratings tend to be lower in an MCC. A system of vertical buswork must be incorporated into the design to provide power to each unit in each horizontal section. Commonly available bus ratings range from 600 A to 2500 A for the horizontal bus, and 300 A to 1200 A for the vertical bus, although some manufacturers may offer other ratings. Interrupting ratings of 65 kA and 100 kA are common. Although MCCs are usually implemented at low voltages (240 or 480 V), medium-voltage MCCs are also available. It is important to determine the continuous-current rating of switchgear bus by temperature rise. While some equipment specifications cite a maximum allowable current density to determine the bus rating, this practice is not allowable. Facilities and structures requiring several panelboards located through the building or facility are using main power distribution panelboards (MDPs), serving as service-entrance equipment for larger services and supply feeders for other panelboards. The common MDP ratings are in the range 600 A to 5 kA. MDP schedules provide pictorial information of the major MDP components, ratings, distributions, and subsections.

3.5 CHAPTER SUMMARY

The basic layout, structure and operations of the power distribution infrastructure have remained quite the same over the last half of the twentieth century. However, in the last three decades, the equipment and power distribution structure have undergone steady improvements: transformers are more efficient, cables are much less expensive and easier to use, and metering infrastructure, control, monitoring and protection equipment and devices are better and computerized. Utilities operate more distribution circuits at higher voltages and use more underground circuits. But the concepts are much the same: alternating current, three-phase systems, radial circuits, fused laterals, overcurrent relays, and so on. Advances in computer technology have opened up possibilities for more automation and more effective protection, monitoring infrastructure, operation, and control. In industrial environments, electricity is provided to loads from load centers, switchgear, and motor control centers. These load centers have hinged doors for easy internal access for maintenance and operation purposes. Switchgear can be metal enclosed or metal-clad for equipment protection. A bus system runs inside the load center in order to provide power to each unit, section, or cubicle. Switchgear with two power sources (double-headed switchgear) requires a specific source transfer method to switch between sources, such as fast transfer, slow transfer, or parallel transfer. Power distribution inside buildings, structures, and industrial facilities is accomplished through electric services, panelboards, and switchboards. Electric services, feeders, branch circuits, and panelboards are critical and essential components of power distribution inside buildings, structures, and industrial facilities. These elements are important aspects of designing the electrical systems of buildings and facilities. The information required for feeder and branch-circuit design is often presented as cabling diagrams. Feeders are conductors and cables that originate at the main depower distribution or main disconnect device and terminate at another distribution center, panelboard, or load center, while subfeeders are the conductors or cables originating at the power distribution centers other than the main power distribution center and extending to panelboards, load centers, and disconnect switches that supply branch circuits. A cable tray system is a unit or assembly of units or sections and fittings forming a rigid structural system used to securely fasten and/or support cables, conductors, and raceways, being part of a facility's structural system. The electrical considerations must be included in any design of cable tray systems. Several standards and guidelines exist for the design of cable tray systems, as NEC sections, NEMA, and IEC specifications are designed to address various aspects of cable tray systems. Most manufacturers provide cable trays in a variety of materials, designs, and shapes. In industrial facilities, the electricity is provided to the loads from the so-called load centers, with switchgear supplying power to large electrical motors, equipment, and smaller MCCs, and motor control centers supplying electricity to midsize motors and loads. Switchgear with two power sources (double-headed switchgear) requires a method of source transfer to switch between sources. Several important ratings apply to both switchgear and MCCs. A continuous-current rating limits the temperature rise of the load center components, particularly the insulation. A short-circuit current rating determines both the interrupting capability of the circuit breakers and fuses and the mechanical strength of the bracing and support systems that must resist the severe mechanical forces exerted by the intense magnetic fields present during a fault. The basic impulse level, or BIL rating, determines the electrical strength of the insulation. Among other topics covered, this chapter gives a comprehensive presentation of the wiring and electrical protection systems in commercial and industrial building, fuses, circuit breakers, instrument transformers and protective relays, grounding and ground-fault protection, feeder design and branch circuits for lighting, equipment, and electrical motors. A special notice for engineers and contractors, the above demand factors are the most widely used on a regular basis because of their uniqueness to electrical design. With the application of demand factors, smaller components can be utilized in the electrical system and greater savings can be passed on to the consumer. Due to the high cost of wiring, designers need to utilize these techniques more than ever before to reduce costs. Cables are contained in some type of raceway, typically conduit, duct, or cable tray.

3.6   QUESTIONS AND PROBLEMS

1. What is the purpose of power distribution systems?
2. Discuss the three classifications (configurations) of power distribution systems.
3. List the advantages and disadvantages of overhead and underground distribution systems.
4. What is the purpose of power distribution substations?
5. Discuss the advantages and disadvantages of overhead and underground power systems.
6. List the advantages and disadvantages of three-phase delta (Δ) transmission lines.
7. List the advantages and disadvantages of three-phase star (Y) transmission lines.
8. List the voltage levels used in the various stages of transmission and distribution in electric power systems.
9. Define the load factor, the diversity factor, and the utilization factors.
10. How is the utilization factor used in applications?
11. What is a load curve? A coincidence factor?

12. A substation has three outgoing feeders: feeder 1 has maximum demand 10 MW at 10 a.m., feeder 2 has maximum demand 12 MW at 7 p.m., and feeder 3 has maximum demand 15 MW at 9 p.m., while the maximum demand of all three feeders is 33 MW at 8 p.m. What is the substation diversity factor?

13. Four individual feeder-circuits have connected loads of 225 kVA, 210 kVA, 160 kVA, and 360 kVA and demand factors of 90%, 85%, 82%, and 88%, respectively. By assuming a diversity factor of 1.35, estimate the transformer size supplying these feeders.

14. Briefly describe the importance and the use of diversity and demand factors.

15. What are the effects of high primary voltages on distribution systems?

16. Explain what is meant by service drop and service lateral.

17. Describe the typical elements of a low-voltage service entrance and the elements of a low-voltage underground service entrance.

18. What devices can be used to protect electric motors from overloads? What is used to protect from short-circuits?

19. What are the applications of single-phase electrical power? Of three-phase electrical power?

20. Why is voltage drop important in electrical power distribution systems?

21. How can a capacitor improve not only the power factor, but also the voltage regulation (i.e., by reducing the voltage drop)?

22. Three 25 kW loads are connected in a power distribution system, through a three-phase 600 V. The load power factors are 0.9 lagging. Compute the load currents.

23. List the NEMA motor control center's enclosure types.

24. What are the voltage levels of a motor control center?

25. How many voltages can be set from a three-phase, wye-connected feeder transformer?

26. List the man devices used in power distribution protection, control, and operation.

27. List the major components of a service entrance.

28. Describe the purpose and location of the main service-entrance components.

29. In your own words, explain the importance of the short-circuit capabilities of a panelboard.

30. Although a smart motor control center can reduce operating costs, increase system efficiency, and provide more system information, it is typically more expensive to install and commission. Is this true or false?

31. A three-phase 13.8 kV cable to power 24,000 HP induction motor has 95% efficiency and 0.93 PF lagging. Determine the voltage drop if the switchboard is 200 ft., and the cable has a resistance of 0.1 Ω and 0.085 Ω per phase and per 1000 ft.

32. Describe the typical elements of a medium-voltage service entrance and the elements of a medium-voltage underground service entrance.

33. A utility decided to increases the primary power distribution voltage level from 12.47 kV to 34.50 kV. Estimate the voltage drop reduction, and the increases into the power and covered area.

34. A three-phase, 250 HP, 60 Hz, 480 V induction motor has an efficiency of 93% and is operating at a power factor of 0.86 lagging. If the motor cable reactance is 0.0165 Ω/phase, compute the capacitance needed to improve the motor power factor to 0.97. What is the voltage rise (boost) at the motor terminals?

35. A 450 kVAR, 480 V, three-phase capacitor bank is installed at the main switchgear of 1.25 kA service. The service transformer is rated at 1500 kVA, 4.16 kV, 480/277 V and has impedance referred to its low-voltage side of 0.003 + j0.0135 Ω/phase. Compute the percentage voltage boost due to the capacitor bank installation.

36. A three-phase, 180 HP, 460 V, 60 Hz induction motor is started direct from 480 V supply. The combined source and cable impedance is 0.01 + j0.028 Ω/phase. If the motor starting power factor is 0.28 lagging, determine the voltage drop of the line voltage.

37. A 480/277 V, wye-connected, four-wire service is supplied by a 500 kVA service transformer. The transformer impedance is $0.0065 + j0.0285$ Ω/phase. If a 250 kVAR, 480 V capacitor bank is installed at the distribution switchgear, what is the voltage rise?

38. An industrial facility has a power demand of 1500 kW at a power factor of 0.70 lagging. Determine the capacitor bank reactive power ratings required to improve the power factor to 0.85, 0.90, 0.95 and unit power factor, respectively. What is the voltage rise in each case, if the feeder inductive reactance is 0.015 Ω/phase and the line-to-line voltage is 460 V?

REFERENCES AND FURTHER READINGS

1. R. D. Schultz and R. A. Smith, *Introduction to Electric Power Engineering*, John Wiley & Sons, New York, 1988.
2. K. Chen, *Industrial Power Distribution and Illuminating Systems*, M. Dekker, New York, 1990.
3. J. R. Cogdell, *Foundations of Electric Power*, Prentice Hall, Upper Saddle River, NJ, 1999.
4. L. L. Grisby (ed.), *Electrical Power Engineering*, CRC Press, Boca Raton, FL, 2001.
5. T. R. Bosela, *Electrical Systems Design*, Prentice Hall, 2002, ISBN-13: 978-0139754753.
6. R. C. Dorf (ed.), *The Electrical Engineering Handbook*, CRC Press, Boca Raton, FL, 1993.
7. G. F. Moore, *Electric Cables Handbook* (3rd ed.), Blackwell Science, 1997.
8. A. S. Pabla, *Electric Power Distribution*, McGraw-Hill, 2005.
9. P. Schavemaker and L. van der Sluis, *Electrical Power System Essentials*, John Wiley and Sons, 2008.
10. M. E. El-Hawary, *Introduction to Electrical Power Systems*, Wiley – IEEE Press, 2008.
11. A. Daniels, *Introduction to Electrical Machines*, The Macmillan Press, 1976.
12. A. C. Williamson, *Introduction to Electrical Energy Systems*, Longman Scientific & Technical, 1988.
13. S. J. Chapman, *Electrical Machinery and Power System Fundamentals* (2nd ed.), McGraw-Hill, 2002.
14. H. L. Willis, *Power Distribution Planning Reference Book* (2nd ed.), CRC Press, 2004.
15. M. A. El-Sharkawi, *Electric Energy: An Introduction* (2nd ed.), CRC Press, 2009.
16. J. L. Kirtley, *Electric Power Principles: Sources, Conversion, Distribution and Use*, Wiley, 2010.
17. F. M. Vanek, L. D. Albright, and L. T. Angenent, *Energy Systems Engineering*, McGraw-Hill, 2012.
18. N. Mohan, *Electric Power Systems – A First Course*, Wiley, 2012.
19. S. W. Blume, *Electric Power System Basics for the Nonelectrical Professional* (2nd ed.), Wiley and Sons, 2016.
20. S. L. Herman, *Electrical Transformers and Rotating Machines*, Cengage Learning, 2017.
21. W. T. Grondzik, et al., *Mechanical and Electrical Equipment for Buildings* (7th ed.), Wiley, 2010.
22. R. Fehr, *Industrial Power Distribution* (2nd ed.), Wiley, 2015, ISBN: 978-1-119-06334-6.
23. W. A. Thue, *Electrical Power Cable Engineering*, Marcel Dekker, New York, 1999.
24. EPRI 1000419, *Engineering Guide for Integration of Distributed Generation and Storage into Power Distribution Systems*, Electric Power Research Institute, Palo Alto, CA, 2000.
25. EPRI 1016097, *Distribution System Losses Evaluation*, Electric Power Research Institute, Palo Alto, CA, 2008.
26. EPRI 1023518, *Green Circuits: Distribution Efficiency Case Studies*, Electric Power Research Institute, Palo Alto, CA, 2011.
27. EPRI 1024101, *Understanding the Grid Impacts of Plug-In Electric Vehicles (PEV): Phase 1 Study – Distribution Impact Case Studies*, Electric Power Research Institute, Palo Alto, CA, 2012.
28. EPRI TR-109178, *Distribution Cost Structure – Methodology and Generic Data*, Electric Power Research Institute, Palo Alto, CA, 1998.
29. EPRI TR-111683, *Distribution Systems Redesign*, Electric Power Research Institute, Palo Alto, CA, 1998.
30. G. Petrecca, *Energy Conversion and Management – Principles and Applications*, Springer, 2014.
31. F. P. Hartwell, J. F. McPartland, and B. J. McPartland, *McGraw-Hill's National Electrical Code 2017 Handbook* (29th ed.), McGraw-Hill, New York, 2017.
32. A. Short (ed.), *Handbook of Power Distribution*, CRC Press, Boca Raton, FL, 2014.
33. D. R. Patrick and S. W. Fardo, *Electrical Distribution Systems* (2nd ed.), CRC Press, Boca Raton, FL, 1999.
34. J. E. Fleckenstein, *Three-phase Electrical Power*, CRC Press, 2016.

4 Electrical System Design, Load Calculation, and Wiring

4.1 BUILDING ENERGY ANALYSIS AND ELECTRICAL DESIGN PROCEDURES

Electrical distribution networks, transmission lines, electrical service, wiring devices, protection, and equipment are essential building subsystems and components. Power engineers are concerned with every step and aspect in the process of electricity generation, transmission, distribution, and utilization. Providing adequate electricity amounts and utilizing electricity efficiently are essential for the growth and development of a country. Past development of power distribution often resulted in higher system losses and poor-quality power service, so efficient and effective power distribution networks, and building or industrial electric systems have become important issues. Optimizing power distribution in order to reduce capital costs and power losses, and improving power quality are critical issues in power system operation and management, resulting in substantial energy savings. On the other hand, electric loads vary with time and place, customer types, and power production. The distribution system must respond to the customers' load demand at any time. Therefore, modern electricity distribution utilities need accurate load data for pricing and tariff planning, distribution network planning and operation, power generation planning, load management, customer service and billing, and finally to provide information to customers and public authorities. Power supply system exploitation for industry, mining, transportation, agricultural, commercial, or residential use is characterized often with significant overestimates in installed power capacity, leading to power use overestimation and lower efficiency, leading to increased supply cable cross sections, wires, conductors, circuit breakers, fuses, or wiring devices, which leads to problems ranging from power losses to power protection malfunctions. This is the result of objective factors, such as: reduced generation and power consumption and proactively overestimated parameter values and electrical loads. All electricity supply system elements (e.g., transformers, power lines, wires, switches, protection devices) are selected based on electrical load estimates (Williamson, 1988; Cogdell, 1999). Small errors in load calculations may easily lead to incorrect selection of system components, which can lead to design, operation, and management issues. It is not a coincidence that special attention is given to load calculations in any electric design. In order to operate efficiently and provide accurate information to customers, the public, and authorities, modern power distribution systems need precise load estimates for planning, design, operation, and management of electrical systems. However, the issue of electrical load calculation is far from solved, because significant and nontolerable discrepancies between calculated and actual loads still exist. Some reasons for this include both imperfect calculation methods and incorrect use of regulatory factors.

Building electrical system design is an important part of the overall building design process. With very few exceptions, all modern mechanical systems, air condition, heating, ventilation, and appliances are electrically powered. In addition, a building's communication, monitoring, control, and security systems demand proper electrical power to operate, which adds further complexity to the building electrical system. The design of a building's electrical system and its component selection are strongly influenced by the mechanical, heating, security, control, monitoring, ventilation, air-conditioning, and control systems. The complexity of these multiple modern systems suggests that the standards and electrical codes in use for decades should be updated. However, modern building electrical systems require less space, when compared with mechanical and heating counterparts. Most electrical operating devices are normally exposed in building occupied spaces of the building. Therefore, their location, configuration, appearance, and design must be closely coordinated with the building architecture and interior design. Unfortunately, quite often these electrical operating or wiring devices (switches, outlets, receptacles, lighting, controls, and alarms) are installed without regard to the location, size, shape, or color. Building electrical system design involves analyzing building requirements, electrical load estimates, electrical system selection, and component selection; preparing electrical plans, diagrams, and specifications; and coordinating with other building design groups, and with overall building design processes. Electrical system design is an integral and important part of the overall building design and analysis. Most of the mechanical, air-conditioning, thermal, heating, building equipment, and appliances are electricity based or use electric power in order to operate. Design, planning, and selection of building electrical systems are strongly influenced by the selection and characteristics of these systems and equipment.

This chapter concentrates on the most common and practical methods of electric load estimates, and specifications and provisions of electric codes, standards, and utilities, mainly for building electric design and analysis purposes. It presents overall issues, the methods and procedures for load estimate and calculation, conductor and cable selection and sizing, building electrical system design, operation, upgrading, expansion, and restructuring. Special attention is given to the standards, codes, and utility regulations. Energy conservation, efficiency, and new technical developments and trends are also presented. Wiring devices, power ratings, and selection and sizing procedures are also discussed. These are devices used to control the power flow and to allow the operation and control of other devices and building equipment. The concepts of unit load and demand load, as well as practical examples are presented and discussed in this chapter. The overall chapter objectives are to enable readers to (1) accurately calculate loads for the electrical system of a building so that one can read or prepare an electrical plan for the building; (2) as part of the construction plans, understand and take the electrical plan for building permit applications; and (3) with a full understanding of the electrical plan, better estimate the cost of a construction project.

4.1.1 Electrical Design Procedure and Building Energy Analysis

Designing a building electrical system involves the following important steps: analyze and understand building energy needs and requirements, determine and estimate electrical loads at highest degree of accuracy, select appropriate electrical systems and equipment, coordinate with other deign decisions, and prepare the detailed electrical plans and specifications. In practice, these steps may be bypassed or performed in a sequence that is more appropriate with a specific project or application, which may depend on the local utility policy and procedures, etc. The first step in any electrical system design consist of the accurate analysis of building or facility energy needs and requirements as part of the overall architectural program. Major factors affecting building electrical systems are, among others, as follows: Occupation factors involving building occupancy types, present and future number of occupants, installed and future equipment and appliances, types, etc. Architectural factors, such as building size, structure, number of floors, floor plans, building footprint, and elevation have strong effects on the electric system design and structure. Important design factors refer to illumination criteria, building environment such as air-conditioning and type of heating systems (e.g., central or unitary types), type of lighting systems, other mechanical and thermal systems (e.g., hot water, sewage disposal), fire protection and monitoring systems, etc. Building equipment, such as vertical transportation systems, food preparation and recreational facilities, computing and processing equipment and systems, and production equipment requiring electric power are also important factors taken into account in the electric system design process. Designers also must take into account building auxiliary systems, fire and security protection systems, monitoring equipment, and television and communication systems. For hospitals, banks, hotels, and industrial facilities, auxiliary and special equipment, devices, and systems must be included with the electric system design and analysis. However, for most building types, the auxiliary equipment power requirements are generally minimal. Last but not least, the cost and the availability of funds are also important design factors.

Accurate estimates of demand load on each branch circuit, feeder, or service is critical and necessary to determine the ratings of these circuits, as well as to set the number of branch circuits and feeders. Article 220 of the National Electrical Code (NEC) discusses the rules and procedures for minimum estimated demand loads for branch circuits, feeders, and services. The basic rule is that each branch circuit, feeder, or service must have enough rating to supply the demand load needing to be served. A circuit demand load must be equal to the minimum estimated load as determined by the applying the NEC or the actual calculated load, whichever is greater. System design is based on the greater of these two estimates. Designers must follow the information from the NEC, Institute of Electrical and Electronics Engineers (IEEE), and International Electrotechnical Commission (IEC) standards and technical guidelines to determine the minimum estimated loads; however, designers also must apply common sense, judgment, and technical experience when determining the load to be served. The quantity and location of receptacles, wiring devices, and lighting fixtures depend on the nature of occupancy and use. For example, a general office area usually requires only functional lighting, providing a certain illumination level, whereas a restaurant or a theater may have decorative fixtures necessary to create specific ambiance. Future power and energy needs and requirements change e more frequently than any other factors or components. The general trend and tendency was that building power and energy requirements keep increasing from year to year. The question is, how much capacity should be included into electric system design, and for what duration. This trend toward increased energy use with less conservation has started to reverse, with the advent of more efficient and less-energy-consuming building equipment and systems, smart and intelligent energy management, green design methods, and the use of the natural environment for lighting, heating, or ventilation needs.

4.1.2 Branch Circuits and Feeders

Power, or more accurately stated, *electrical energy* is transferred in any electrical system from the service equipment and network to lighting systems, machines, equipment, and outlets. Regardless of the wiring methods and systems, power carrying conductors and cables are divided into feeders and branch circuit conductors. Service conductors are conductors that extend from the power company or utility terminals to the main disconnects; feeders originate at the main disconnect device or main distribution and terminate at another distribution center, panel-board, or load center. Subfeeders are conductors originating at a distribution center other than the main distribution and terminating to panelboard, load center, and disconnect switches to supply branch circuits. Panelboards, in single or multiple configurations, are power equipment, containing switches, indicators, fuses, and circuit breakers, intended for power switching and controlling and for circuit and equipment protection. Branch circuits are the section of wiring systems extending beyond the final overcurrent device, usually originating from a panel and transferring power to loads. Branch circuits include circuits servicing single motors (individual circuits) and circuits servicing multiple lighting systems and receptacles (multiwire); they usually operate at lower currents, 30 A or less, but can also supply higher currents. A branch circuit consists of conductors originating at safety switches (disconnects), or more often at a panelboard. The most common types of branch circuits include: *individual branch circuits* that supply a single load, *multi-outlet branch circuits* that supply multiple loads, *general-purpose branch circuits* a multi-outlet type that supplies multiple outlets for lighting and appliances, *appliance branch circuits* that supply a single appliance load, and *multiwire branch circuits* with two or more ungrounded conductors and one grounded conductor. Any branch circuit is sized for the supplied load, while sizing the circuit for future loads is good design practice. Its rating depends on the rating of the overcurrent device protecting the circuit. Branch circuits serving a single load can have any rating, while the ones serving multiple loads have limited ratings of 15, 20, 30, 40 or 50 A. NEC Section 210.6 specifies, in agreement to the supplied circuits, the branch circuit voltage limits. The feeders, the conductors between the service equipment and the branch circuit overcurrent devices, provide the limits regarding safe and adequate sizing and installations of these conductors. Feeder loading depends on the overall system power requirements, assuming that all connected loads are operated as required and the feeder has sufficient capacity (ampacity) to meet the load demand. Prior to the installation, several factors must be taken into account to ensure the proper feeder size and type, and adequate overcurrent protection for that application. Feeder materials are copper or aluminum, and can be single or multiconductor cable. The feeder environment (e.g., damp, hot, corrosive) must be considered, as well as the voltage drop for determining the feeder length. Feeder can run in conduit, cable trays, or other systems, and can be paralleled or individual, while the neutral may not be necessary. Conductor sizing includes several factors: conduit fill, ambient temperature, connected load demand, demand and derating factors. Various overcurrent protection devices are used with feeders.

Generally, loads are classified as *continuous* or *noncontinuous*. A continuous load is any load that is on for 3 hours or more (NEC Article 100). Examples of continuous loads are lighting, computer terminals, copy and printing machines, HVAC equipment, and other devices and equipment found in offices. Most industrial facility equipment and devices, as well as industrial processes, are also considered to be continuous. Noncontinuous loads are general-purpose receptacle outlets, residential lighting outlets, and so on. If there is doubt whether to classify a load as continuous or noncontinuous, design practice is to consider it as continuous. Overcurrent protection for continuous loads must be sized at no less than 125% of the load. The increase in the fuse or breaker size is not because the current from a load increases when a load operates for more than 3 hours. Continuous loads must not exceed 80% of the circuit rating allotted to them. The problem with continuous loads is that heat is generated in both conductors and in terminals where conductors are connected. Heat is produced any time current flows through conductors and through any circuit terminals. If the conductors are sized correctly and the loads do not run continuously, the heat produced is insignificant. However, the longer the current flows through a conductor or through circuit terminals, the more the conductor or terminals are heated. The NEC has found that loads operated for 3 hours or more can cause significant heat to build up, which adversely affects conductors, terminations, and overcurrent devices. Increasing the rating of overcurrent devices and the ampacity of conductors for continuous loads by 125% compensates for the heat buildup at the overcurrent device terminals. Circuit breakers are thermal-magnetic devices; increased heat at the breaker terminals can cause a breaker to trip at a current below the breaker's rating. A larger breaker will compensate for the heat buildup of continuous loads by having a higher thermal trip point. A circuit breaker thermal setting is to protect the circuit from overloads, not short circuits. If an overcurrent protective device is listed for continuous operation at 100% of its rating, the 80% correction factor is not applied. Branch circuit loads are classified into the following use categories: lighting, receptacle, equipment, heating and cooling, and motor loads. Branch circuit conductor current ratings must be greater than the maximum load supplied by the circuit. In the event of multiple loads, the conductor ampacity must correspond to the overcurrent protective device. However, for circuits supplying hard-wired devices (e.g., electric heaters, air-conditioning equipment, water heaters), the fuse or circuit breaker can be rated at the next higher rating.

4.2 LOAD ESTIMATES AND CALCULATIONS

Analysis of a building's energy requirements and usage is critical in the building's electrical system design, restructuring, upgrading, and/or expansion, or in planning for energy and electricity conservation and efficient use. An important step of this process consists of the most accurate identification of the current and, if possible, future, energy requirements and use. Major factors affecting building electrical systems include: present, anticipated, and future building occupancy; electrical appliances and equipment; cost; economic and architectural factors; building environment; air condition, heating and ventilation types; illumination criteria; lighting system type; mechanical systems; equipment; building transportation, monitoring, control, security, and recreational systems; computing, production, and processing equipment; and the electricity supply availability.

The electric system analysis also includes the most accurate evaluation of building auxiliary systems, building management, communication, and specialty equipment (e.g., equipment used in hospitals, hotels, industrial facilities, schools, banks), and the most accurate possible future energy and electricity demand estimates and usage changes. As a rule of thumb, a minimum 25% spare electricity capacity should be included at building main or distribution centers. Building electricity usage has increased significantly in recent decades, especially in office buildings, due to the increased use of computers, communication equipment, printers, copy machines, and control and monitoring systems. However, new trends toward energy conservation, higher equipment efficiency, new lighting systems, high efficient electrical motors, sustainable design, and energy management have significantly reduced building electricity use and needs.

Accurate estimates of the demand load on a particular branch circuit, feeder, or service are necessary to determine the required rate of these circuits. Actual quantities of branch circuits and feeders, conductors, wires, and cable size and selection are also based on the determination of the estimated demand load and electricity needs. The required rules for determining the minimum estimated demand load for branch circuits, feeders, and services are discussed in NEC Article 200, and similar guidelines in the IEC (in French: Commission électrotechnique internationale [CEI]) standards. *The basic rule is that the branch circuit, feeder, or service must have a rating sufficient to supply the demand load to be served.* The load demand on an electric circuit is equal to the minimum estimated load as determined by applying the *NEC minimum estimates or the actual load, whichever is greater.* Electrical circuit load capacity represents the total amount of electrical power that a home, hospital, school, industrial, or commercial facility is actually using. In order to determine the size of electrical service needed in a residential, commercial, or industrial facility, the most accurate load estimate possible is required. For example, older homes have 60 A of electrical service, connected to a fused panelboard, whereas newer homes have 100 A or 200 A electrical services connected to a breaker panelboard. In order to calculate power requirements, explicit knowledge is needed about existing equipment, electrical appliances, lighting, facility mechanical, heating, air-conditioning, monitoring, and processing equipment, as well as how much power the equipment uses during the operation periods. As technology continues to advance, it seems we add more and more electrical loads to our homes. However, the general trends are that newer equipment, devices, and appliances require less power and are more efficient in terms of energy use than older ones. Although codes and standards, such as NEC or IEC, provide necessary information to determine and estimate minimum loads, the designers and engineers must exercise common sense and judgment when estimating the load to be served, by taking into account not only present power requirements but also future changes.

4.2.1 CONVENIENCE POWER, CONNECTED, AND DEMAND LOADS

Electrical power and energy needs and demands in buildings, residential, industrial, and commercial, especially in office buildings, increased dramatically during the last part of the twentieth century, due to the extended use of computing, information technology (IT), telecommunication, and printing equipment, systems, and devices. On the other hand, recent trends and applications of sustainability principles, including energy conservation and management, and sustainable designs, such as new and more efficient equipment, motors, and devices, have reversed energy use tendencies. Electrical system design is an integral and important part of the overall building design and analysis. Most of the mechanical, air-conditioning, thermal, heating, and building equipment and appliances are electricity based or are using electric power in order to operate. Design, planning, and selection of building electrical systems are strongly influenced by the selection and characteristics of these systems and equipment. Electrical power for mechanical equipment and systems varies widely with different types, and with building designation, location, and architectural design. Mechanical systems and equipment include HVAC, plumbing, and fire protection. Such equipment and systems require large amounts of power in order to operate, being more efficient and economically designed

TABLE 4.1

Typical Power Ratings of Some Common Household Appliances

Appliance or Equipment	Power Range (VA)	Appliance or Equipment	Power Range (VA)
Window air conditioner (115 V, 0.5 ton)	500–700	Oven (115 V/230 V)	2000–5000
Window air conditioner (115 V, 1.0 ton)	1200–1400	Microwave oven (115 V)	300–1000
Window air conditioner (230 V, 2.0 ton)	2400–2800	Television set (115 V)	200–1000
Refrigerator (115 V)	300–1000	Desktop PC	80–200
Freezer (115 V)	300–800	Printer	100–200
Washing machine (115 V)	800–1200	Laptop	20–75
Dryer (115 V)	3000–5000	Scanner	10–20
Water heater (115 V/230 V)	3000–6000	Coffee maker	800–1400
Dishwasher (115 V)	1200–1500	Ceiling fan	20–100
Water filter and cooler (115 V)	75–100	Vacuum cleaner	200–700
		Toaster	800–1800

at higher voltages, such as 208 V, 240 V, and 480 V, and three-phase. However, such types of residential equipment are normally designed for 120 V, 240 V, and single-phase power.

Building equipment includes elevators, escalators, and other vertical transportation systems; food service equipment; and recreational, household, and miscellaneous equipment and systems. Power requirements and voltage levels for such equipment and systems vary widely in capacity and operating characteristics. Building auxiliary equipment and systems include those deigned for life safety, communication, monitoring, surveillance, and security, which do not require large amounts of power to operate and are usually powered at single-phase 120 V or 240 V. Convenience loads refer to plug-in equipment and systems, such as personal computers, laptops, printers, laboratory instruments or equipment, portable lights, audio and video equipment, and some household appliances. The recommendations are to reserve about 180 VA per duplex outlet for convenience power, plus any large and/or special equipment for sizing the service. Typical power ratings for some common household appliances are given in Table 4.1, while the recommended power ratings for various interior spaces and buildings are included in Table 4.2. The electrical load requirements for commercial, industrial, and residential installations and systems result in a great deal of diversity and complexity in usage, meaning that some types of equipment or electrical loads are in use for extended periods, and others are used only occasionally or for short periods of time. In addition, there are often electrical loads on the same service or feeder that are not in service simultaneously by their very nature, such as heating and air-conditioning. For this reason, demand factors are applied when calculating service and feeder loads. Different sets of demand factors apply for different electrical loads, and even for different types of commercial, industrial, or residential buildings. Although most of the service and feeder commercial load calculation requirements are found in NEC Article 220, other rules affecting these loads are scattered throughout the NEC. For example,

TABLE 4.2

Typical Power Specifications for Convenience Power (Portable and Fixed Loads)

Space Type	Power Allowance (W/ft.^{2})	Power Allowance (W/m^2)
Office	1–4	11–44
Classroom (no PCs)	2–3	22–33
Classroom (with PCs)	3–5	33–55
Meeting room	3–5	33–55

NEC Chapter 3 focuses on wiring methods, while other articles provide more in-depth requirements for particular equipment or applications (e.g., specific requirements for motor circuits [NEC Article 430]). The loads of an individual customer or a group of customers are constantly changing. Every time a light bulb or an electrical appliance is switched on or off, the load seen by the distribution feeder changes. The building electrical load calculations consist of several components of the electrical power plan, which is part of the building construction plans. In order to describe the changing load, the following terms and parameters used to characterize the load are defined in the next paragraphs. *Load* is the power consumed in a closed circuit. The *average load* is the average power consumed in the circuit system during a specific time period. The *maximum load* is the maximum value of the consumed loads during the same time period under all conditions. The *demand load* is the required power to satisfy adequate device or circuit operation. The load at the receiving terminal averaged over a specific period of time, represents the load demand. Loads are expressed in kW, kVA, or even in A, only.

Factors used for load calculations and analysis use the following equations or ratios. In Chapter 3, these factors and parameters were briefly discussed from a power system point of view. This discussion focuses on load estimates and analysis perspectives. *Demand factor* (DF) is the ratio of the maximum demand load to the total load connected. When the power input is connected to a circuit system, whether the switch is closed or open, the maximum demand load should always be equal or smaller than the connected load because of a system loss and other technical reasons. *Coincident factor* is the maximum system demand divided by the sum of individual maximum demand, while diversity factor is its reciprocal, i.e., the sum of individual maximum demands divided by the maximum system demand. The power planning is more effective for a greater DF because the value of the nominator is fixed but the value of the denominator or the maximum system demand changes with different designs. Load factor (LF) is the ratio of the average load to the maximum load. The maximum load should occur under the worst-case load event. In this context, the average load is always less than the maximum load. It would likely be more economical to develop a power plan if these two values were closer. This shall depend on the qualities of the devices and equipment in the building. *Maximum demand* represents the greatest of all demands that occur during a specific time, and must include demand interval, period, and units in the definition. The average demand of a load curve in kW equals, and is given by:

$$P = \frac{W}{\Delta T} \tag{4.1}$$

Here W is the electrical energy consumption in period of ΔT is in hours, and is the time interval in hours (=24 hours for the daily load curve, and 8760 hours for the annual load curve). *LF* is an indication how well the utility's facilities are being utilized. From a utility standpoint, optimal *LF* would be 1.0 (system consumption approaches the maximum). As reflected on the electricity bill, the bigger *LF* value the better for consumers or electricity users. *LF* is defined as:

$$LF = \frac{\text{Average Demand}}{\text{Maximum Demand}} = \frac{P_{Average}}{P_{Maximum}} \tag{4.2}$$

Demand factor is the ratio of the maximum demand of a system to the building or facility total connected load (maximum demand when all are used), being usually less than 1. It gives the fractional amount of some quantity being used relative to the maximum amount that could be used by the same system. The lower the demand factor, the less system capacity required to serve the connected load. It is computed as:

$$DF = \frac{\text{Maximum Load Demand}}{\text{Maximum Possible Load Demand}} \tag{4.3}$$

Example 4.1

A large industrial facility might have a connected load of 20 MW, but if only 75% of its electrical equipment is operating, what is the demand factor?

SOLUTION

The demand factor would be only 75% or 0.75 of the maximum load power.

Coincident factor is the ratio of the peak demand of a whole system to the sum of the individual peak demands within that system. The peak demand of the whole system is referred to as the peak diversified demand, or the peak coincident demand. The individual peak demands are the noncoincident demands. The coincident factor is less than or equal to 1. Normally, the coincident factor is much less than 1 because each individual load does not hit its peak at the same time (i.e., they are not coincident). Diversified demand represents the sum of demands imposed by a group of loads over a particular period. It must include load demand interval, period, and the units. Diversity factor represents the ratio of the sum of the individual peak demands in a system to the peak demand of the whole system. The diversity factor is greater than or equal to 1 and is the reciprocal of the coincident factor. Diversity factor (also known as simultaneity factor [ks]) is also expressed as the ratio of the maximum noncoincident load demand to the maximum diversified demand (the inverse of ks). For a quantitative measure of the inherent diversity of individual load peaks, a diversity factor is defined as:

$$\text{Div F} = \frac{\sum_{i=1}^{N} P_{max,i}}{P_{max,T}} \tag{4.4}$$

The greater the diversity factor, the lesser is the cost of generation of power. The term *simultaneity* is used by some engineers, designers, and technicians, being the inverse of the diversity factor:

$$ks = \frac{1}{\text{Div F}}$$

Being thus greater than 1. The residential load has the highest diversity factor. Industrial loads have low diversity factors, usually of 1.4, streetlights practically unity, and other loads vary between these limits (see Table 4.3, adapted from IEC standards). The *responsibility factor* is

TABLE 4.3
Diversity Factors (Adapted from IEC Standard)

System Component	Residential	Commercial	General Power	Large Industrial
Between individual users	2.00	1.46	1.45	—
Between transformers	1.30	1.30	1.35	1.05
Between feeders	1.15	1.15	1.15	1.05
Between substations	1.10	1.10	1.10	1.10
From users to transformers	2.00	1.46	1.44	—
From users to feeder	2.60	1.90	1.95	1.15
From users to substation	3.00	2.18	2.24	1.32
From users to generating station	3.29	2.40	2.40	1.45

the ratio of a load's demand at the time of the system peak to its peak demand. A load with a responsibility factor of 1 has a peak at the same time as the overall system. The responsibility factor can be applied to individual customers, customer classes, or circuit sections. The loads of certain customer classes tend to vary in similar patterns. Commercial loads are usually highest from 8 a.m. to 6 p.m. Residential loads peak in the evening, with a second peak (usually smaller than the evening one) earlier in the morning. Weather significantly changes loading levels. For example, on hot summer days, the air-conditioning increases the demand and reduces the diversity among loads. Loads are separated into: (a) *fixed loads*, which are permanently wired loads, and (b) *convenience loads*, the plug-in ones. In general, each load has a nameplate specifying its current rating (A), real (active) power (W), or apparent power (VA). These power ratings are used to estimate the total system and/or building loads and power requirements. The *connected load* of a system or building represents the sum of all connected electrical loads, regardless of the way that they are used, expressed by a simple relationship between connected load (CL) of various load groups, lighting, equipment, and convenience loads and the total (gross) connected load of a power system or a building, as:

$$TCL = \sum_{k=1}^{N} CL_k \tag{4.5}$$

Demand load (DL) of an electrical power system or of a facility is the net load that is likely used at the same time by each load group. The demand load is always smaller than the connected load, except when all connected loads are used at the same time, at which time the demand load is equal to the connected load. Straight and simple relationships between the connected load (CL), demand load (DL), demand factor (DF), and the total (gross) demand load (TDL) are given by the following two relationships, respectively:

$$DL = CL \times DF \tag{4.6}$$

and

$$TDL = \sum_{k \geq 1} DL_k \tag{4.7}$$

Diversity factor (sometimes called diversity coefficient, to avoid confusion with demand factor) can be defined in terms of the parameters introduced in Equations (4.6) and (4.7). It counts on the demand diversity of different load groups, being a time-consuming parameter to calculate. As a rule of thumb, as recommended in the literature, a value equal to 1.0 is considered for systems that do not have load diversification, and a value of 1.2 is for cases of large systems and/or ones with diversified loads. Most buildings have values about 1.4. In the case, when the diversification is greater than suggested by the diversification factor, a complete analysis of the behavior of the loads at each hour, or even at every minute, is needed, under all operating conditions. If the diversity coefficient is neglected during the design stage, the system may be oversized and have exorbitant costs. Net demand load (NDL), total demand load, and diversity factor (coefficient) are related by the following relationship:

$$NDL = \frac{TDL}{Div\ F} \tag{4.8}$$

Example 4.2

A building electrical system is calculated to have the following load groups: lighting 150 kW and diversity factor 0.9, receptacles 105 kW and 0.2, building equipment 200 kW and 0.6, mechanical equipment 240 kW and 0.85, and auxiliary equipment 50 kW and 0.2. Calculate the connected load, if the diversity factor is 1.15 the net load demand of the system.

SOLUTION

Total connected load and total demand load are calculated as:

$$TCL = 150 + 105 + 200 + 240 + 50 = 745 \text{ kW}$$

$$TDL = 0.9 \times 150 + 0.2 \times 105 + 0.6 \times 200 + 0.85 \times 240 + 0.2 \times 50 = 490 \text{ kW}$$

$$NDL = \frac{TDL}{Div \ F} = \frac{490}{1.15} = 426.1 \text{ kW}$$

The recommendation is to use 430 kW.

4.2.2 LIGHTING LOAD ESTIMATE METHODS

Lighting accounts for a significant part of the overall electrical load for most buildings or facilities. In a broad sense, lighting loads are categorized in: general lighting, show-window, track, sign, outline, and other lighting. General lighting is the overhead lighting within a building and facility, with an adequate intensity for any type of work performed in that area. Lighting fixtures are generally designed for 120 V and single-phase power service, but may be designed for 208 V, 240 V, and 277 V single-phase power systems. Estimating the general lighting load is based on either a load per area method or the actual full load of the fixtures used, whichever is greater. A typical building or facility usually has several different area types, such as storage, office, hallways, and cafeterias; each area type must be considered separately. Continuous improvements and advances in lighting technology and systems have increased the electricity to light conversion efficiency, reducing the needs for electrical power for building lighting. Back in the 1980s, it was common to require 3–5 VA/ft.^{2} for office areas. In the 1990s this decreased to only 2 VA/ft.^{2}, and now the requirement is down to about 1 VA/ft^2. For planning electrical power for building lighting, the NEC has prescriptive requirements tabulated for various space types. In general, these exceed the actual needs and requirements of the ASHRAE/IES Standard 90-2007. The general lighting load applied to a particular occupancy is based on specific unit load per unit of area (square foot or square meter) expressed in volt-ampere per unit area. The total minimum estimated load is found by multiplying by the total occupancy area. The unit loads for various occupancies are given in Table 4.4 (NEC Table 220.3[A]). If the type occupancy is not listed in NEC Table 220.3(A), the load is based on the actual connected equipment. In addition, only habitable areas of dwelling are used in the square-footage determination. Unfinished basements, garages, or porches are not considered habitable and are not usually included in the square-footage determination. If the load is continuous, the computed load is multiplied by 1.25 to determine the circuit requirements. However, the general lighting load is not required if the load of each fixture is determined separately with the same provision if it is considered continuous load.

Example 4.3

Determine the lighting load for an office area of 6300 square feet using actual lights and the information in Table 4.1 (NEC Table 220.3[A]). The office contains 30 fluorescent lights having a ballast of 280 V and 0.8 A.

TABLE 4.4
General Lighting Loads (Table 220.3[A] of the NEC)

Occupancy Type	Unit Load (VA/m²)	Unit Load (VA/ft.²)
Atrium	11	1
Auditorium	11	1
Bank	39	3.5
Barber shop	33	3
Church	11	1
Club	22	2
Courtroom	22	2
Dwelling unit	33	3
Garage (commercial)	6	0.5
Hospital	22	2
Hotel and motel	22	2
Industrial building	22	2
Lodge room	17	1.5
Museum	22	2
Office building	39	3.5
Restaurant	22	2
School	33	3
Store	33	3
Warehouse	6	0.25

SOLUTION

By using actual office lighting, the estimated lighting load is:

$$\text{VA Lighting Load} = 30 \times 280 \times 0.8 = 6720 \text{ VA}$$

By using Table 220.3(A) of the NEC the estimate load is:

$$\text{VA Lighting Load (NEC)} = 6300 \times 3.5 = 22050 \text{ VA}$$

Either the actual load estimate is 6.720 kVA, the NEC estimate of the load demand is: 22.05 kVA.

If the type of occupancy is not listed in NEC Table 220.3(A), the load is determined by using the power requirements of the actual connected loads. In addition, only the habitable area of buildings (dwelling units) are included in the square-footage determination; other spaces, such as unfinished basement, garages, and porches, are not included in such estimates. A dwelling unit is a single unit that provides complete and independent living facilities, according to the NEC definition (NEC Article 100). NEC Table 220.3 is applied for lighting load estimates; however, the load from general-purpose receptacle outlets in habitable areas (residential buildings, hotels, motels, and dwelling units) can be included with the general lighting load estimates, being taken into account in the unit lighting load of this table. For office buildings, banks, designed school rooms, and data centers an additional 1 VA per square foot should be added if the number of actual receptacles is not known, due to the fact that in such areas the final furniture layout and the plug-in equipment are not known, and extra provision for the load must be included.

Example 4.4

Compute the estimated lighting load of an office area of 250 m², containing 30 fluorescent lighting fixtures, each having a ballast of 277 V and 0.7 A.

SOLUTION

From Table 4.4 the total lighting load is:

$$S_{lighting} = 250 \times 39 = 9750 \text{ VA}$$

The lighting load of the connected equipment (fluorescent lighting fixtures) is calculated from actual power requirements:

$$S_{fixture} = 30 \times 277 \times 0.7 = 5817 \text{ VA}$$

Actual connected load, based on the connected equipment, is only 5817 VA, the estimated demand load is 9750 VA based on the NEC minimum requirements, and so the number of the branch circuits for this area. The greater of the load estimates applies.

Show-window lighting is any window designed for display of goods and advertising materials, whether it is entirely open at the rear, fully or partially enclosed, or has or not a platform raised higher than the street level. Show-window lighting load is not considered part of the general lighting load. Section 220.43(A) and 220.14(G) of the NEC requires that 200 VA per linear foot of the show window or the maximum volt-ampere rating of the equipment, whichever is greater, be used to estimate this type of lighting load. The NEC also requires, in addition to the show-window lighting load, at least one receptacle of 180 VA for every 12 feet of show-window space measured horizontally. Chain-supported fixtures in a show window are permitted to be externally wired, with another externally wired fixture allowed.

Example 4.5

Compute the lighting load for two show windows of a store, 36 feet and 48 feet long, respectively.

SOLUTION

Total show-window length is: 36′ + 48′ = 84′. The show-window lighting load is:

$$\text{Show-window Load} = 84' \times 200 = 16800 \text{ VA}$$

This lighting is a continuous load and the circuit load requirement is: $16,800 \times 1.25 = 21,000$ VA. In addition, receptacles are required at every 12 ft. of window and a total of 7 (3 and 4, respectively) receptacles are required, and the receptacle load is: $7 \times 180 = 1260$ VA. The total load is then:

$$\text{Total Load} = 16800 + 1260 = 18060 \text{ VA}$$

Sign and outline lighting loads, discussed in NEC Article 600, which requires that a structure have a least one circuit exclusively used to supply such lighting, are considered continuous loads.

This circuit must be designed for a minimum load of 1200 VA. Track lighting, often used in commercial buildings for accent lighting, is discussed in NEC Article 410 (part XV). It is a manufactured assembly designed to support and energize luminaires (lighting fixtures) that are easily repositioned on the track. Note that the track length can be altered by the addition or reduction of track sections. The minimum track lighting load requirement is to assign 150 VA for every 2 feet of track length. Any other and additional lighting loads are computed separately from the general lighting, and added to the total lighting load. Such lighting loads include security, parking area, sidewalk, roadside, and stadium lighting, and are calculated using the actual loads of equipment and devices.

Example 4.6

A furniture store has a total area of 120′ × 250′. The store uses 100′ × 60′ for storage, uses 10′ × 20′ as a small office, and uses the remainder of the building as showroom. There are also a total of 48′ show windows, a 40′ track, and two outdoor signs. What is the total lighting load?

SOLUTION

The storage, office, and the showroom areas, and the corresponding lighting loads are:

Storage area: $100 \times 60 = 6000$ sq. ft
The storage lighting load: $6000 \times 0.25 = 1500$ VA

Office area: $10 \times 20 = 200$ sq. ft
The storage lighting load: $200 \times 3.5 = 700$ VA

Store total area: $250 \times 120 = 30000$ sq. ft
Showroom area: $30000 - 6000 - 200 = 23,800$ sq. ft
The showroom lighting load: $23800 \times 3 = 71,400$ VA

General lighting load: $1500 + 700 + 71400 = 73,600$ VA
Show window lighting load: $200 \times 48 = 9,600$ VA
Sign lighting load: $1200 \times 2 = 2400$ VA
Track lighting load: $0.5 \times 40 \times 150 = 3000$ VA

The sign lighting load is estimated at a minimum 1200 VA, so a total of 2400 VA.

Total lighting load: $73600 + 9600 + 2400 + 3000 = 88,600$ VA
Circuit requirement (continuous load): $98600 \times 1.25 = 110,750$ VA

Lighting loads have a high degree of diversity, especially in commercial and industrial facilities, or hospitals. For example, it is very unlikely that every light in a hospital or a department store will be operated at the same time. Lighting derating factors are specified in Section 220.42 of the NEC, allowing the power reduction (derating) of the feed, panelboard, or service. However, there are some areas in specific structures in which the derating factors are not recommended. Some special areas may require lighting all the time (e.g., operating rooms in hospitals, emergency

TABLE 4.5
Derated Factors for Lighting Loads

Occupancy Type	Load Portion (VA)	Demand Factor (%)
Dwelling unit	0–3000	100
	3001–120,000	35
	Over 120,000	25
Hotel and motel	0–20,000	50
	20,001–100,000	40
	Over 120,000	30
Hospital	0–50,000	40
	Over 50,000	20
Warehouse	0–12,500	100
	Over 12,500	50
All others	Total VA	100

departments, intensive care units, data centers, stairways). Table 4.5, an adapted version of NEC Table 220.42, gives derating factors for some structures or facilities. Notice that these derating factors are not applied to the branch circuit conductor or branch circuit overcurrent protection device calculations.

4.2.3 DEDICATED AND GENERAL-PURPOSE RECEPTACLE LOAD ESTIMATES

Most receptacles installed in commercial facilities and apartment buildings are not supplying continuous loads. It can be quite difficult to predict the loads that they are supplying, unless the receptacle is dedicated (assigned to a specific purpose). The NEC does not require a specific or minimum number of outlets for commercial or apartment buildings. Usually, many receptacles are required and installed. If a receptacle is the load of a branch circuit, its current rating must be equal to or higher than that of the branch circuit. When there are multiple receptacles on a branch circuit, the receptacle ratings vary with the current rating, as shown in Table 4.6. A load of 180 VA is assigned to each receptacle, whether it is single type, duplex, or triplex. If the receptacle is dedicated to a specific load, the actual load is used. If the dedicated load is continuous, then a 125% correction factor is applied. When a receptacle is one of the loads supplied by an individual branch circuit, the receptacle current (ampere) rating must be equal to or greater than that of the branch circuit. The NEC has in such cases specific requirements and recommendations, and the specific receptacle ratings are given Table 4.6 (adapted from NEC Table 210.21[B][3]). If a receptacle is dedicated for a specific device or equipment, the actual load is used, and in the case of continuous loads then a 125% overrate is applied. To calculate the allowable number of receptacles on a branch circuit, multiply the voltage and current (the circuit power) and divide the result by 180 VA, then round off the lower integer.

TABLE 4.6
Receptacle Rating as Determined by the Circuit Rating

Circuit Rating (A)	Receptacle Rating (A)
15	Up to 15
20	15 or 20
30	30
40	40 or 50
50	50

Example 4.7

How many receptacles can be placed on a 120 V, 15 A circuit? How many on a 120 V, 20 A circuit?

SOLUTION

The maximum circuit power in each of the two cases is:

$$P_{15\,A} = 120 \times 15 = 1800 \text{ VA}$$
$$P_{20\,A} = 120 \times 20 = 2400 \text{ VA}$$

The number of receptacles in each case is:

$$\text{Nr. receptacles} = \frac{1800 \text{ VA}}{180 \text{ VA}} = 10$$
$$\text{Nr. receptacles} = \frac{2400 \text{ VA}}{180 \text{ VA}} = 13.3$$

A 120 V, 15 A branch circuit can supply 10 receptacles, while a 120 V, 20 A branch circuit can supply 13 receptacles. (We rounded off to the lower value.)

The receptacle load can be included in the general lighting load by adding a value of 1 VA/ft.^{2} or 11 VA/m^2 to the general lighting unit loads of Table 4.4 (NEC Table 220.12). However, the recommendation is that this approach should be used only when the number of receptacles is unknown.

Example 4.8

Calculate the receptacle load for a 200′ × 150′ department store, and number of 15 A circuits needed to supply the load at 120 V power supply. The number of receptacles is unknown.

SOLUTION

For unknown a load of 1 VA/ft^2 is used, so:

$$Area = 200 \times 150 = 30000 \text{ sq ft.}$$
$$\text{Receptacle load} = 1 \times 30000 = 30000 \text{ VA}$$

Maximum single circuit load: 120 × 15 = 1800 VA

$$\text{Number of Circuits} = \frac{30000}{1800} = 16.67$$

Maximum number of circuits is 17.

Multi-outlet systems are often installed in repair shops, lighting display areas, electronics departments, research and educational laboratories, data and computing centers, small industrial facilities, and other locations where multiple outlets are needed. These multi-outlet assemblies require 180 VA for each 5 ft. of length. NEC allows systems in laboratories, repair shops, and stores to be derated in accordance with Table 4.5 (NEC Table 220.42), if the load exceeds 10 kVA. The diversity and loading intermittency and inconstancy allows the total receptacle load to be derated, as specified in NEC Section 220.44. If the load exceeds 10 kVA, the first 10 kVA are counted at 100%, while the additional load is derated (reduced) at 50%. However, this procedure may not be used if the NEC requires that specific appliances or equipment cannot be derated (refer to NEC Sections 22.12 and 220.444, and Table 220.44).

NEC and IEC standards require that certain branch circuits, the dedicated branch circuits in dwelling units and apartment buildings, be designated for receptacle outlets only, prohibiting the connection of lighting outlets to them. Included into the dedicated branch circuits are a minimum of two 20 A branch circuits to supply small appliance loads in kitchen areas as well as the laundry equipment of the dwelling units. NEC Section 220.16 limits the computed loads on each of this type of branch circuit to 1500 VA, which is in addition to the unit VA load computed from the general lighting. In addition, NEC Section 220.3(B).9 requires that a minimum 180 VA be assigned to general-use receptacles, not located in the habitable areas or connected to dedicated branch circuits. Notice that the 180 VA applies to a device located on a single yoke or mounting strap. However, the estimated load can always be greater than the minimum NEC load determination. Dwelling units have special requirements for load calculations. Most of the actual load calculation requirements are in Article 220; others are scattered throughout the Code and come into play when making certain calculations. In addition to the branch circuits required for dedicated appliances and those needed to serve the general lighting and receptacle loads, a dwelling unit must have the following branch circuits: (a) a minimum of two 20 A, 120 V small-appliance branch circuits for receptacles in the kitchen, dining room, breakfast room, pantry, or similar dining areas, as specified in Section 220.11(C)(1). These circuits must not be used to serve other outlets, such as lighting outlets or receptacles from other areas, as described in the 210.52(B)(2) Ex. These circuits are included in the feeder/service calculation at 1500 VA for each circuit (220.52[A]). (b) One 20 A, 120 V branch circuit for the laundry receptacle(s). It can't serve any other outlet(s), such as lighting, and can serve only receptacle outlets in the laundry area, in agreement with 210.52(F) and 210.11(C)(2). The feeder or service load calculation must include 1500 VA for the 20 A laundry receptacle circuit, as described in 220.52(B).

Example 4.9

Determine receptacle load of a 120′ by 200′ hardware store that has one duplex receptacle at every 12′ of wall around the store area, a total of 24 floor receptacles, and 24′ and 30′ show windows.

SOLUTION

The number of receptacles placed on each wall is:

$$\text{Receptacles per 120' wall} = \frac{120}{12} = 10$$

$$\text{Receptacles per 200' wall} = \frac{200}{12} = 16.67$$

There are 10 receptacles for each of the 120′ walls and 17 receptacles for each of the 200′ walls, a total of:

$$\text{Wall Recepacles} = 2 \times 10 + 2 \times 17 = 54$$

One receptacle is needed for every 12′ of show window, so the 24′ show window requires 2 receptacles and the 30′ show window requires 3 receptacles, for a total of 5. The total number of store receptacles is then:

$$\text{Total Recepacles} = 54 + 5 + 24 = 83$$

The total receptacle load is:

$$\text{Total Receptacle Load} = 83 \times 180 = 14{,}940 \text{ VA}$$

For feeder sizing, the first 10 kVA must count at 100%, and the excess of 4940 VA is 50% derated, so the feeder load is:

$$\text{Feeder Load} = 10000 + 0.5 \times 4940 = 12{,}470 \text{ VA or } 12.47 \text{ kVA}$$

4.2.4 EQUIPMENT, AUXILIARY, INDUSTRIAL, AND MOTOR LOAD CALCULATIONS

Equipment, such as appliances, water heaters, washers, dryers, cooking equipment, and some laboratory, industrial, and commercial equipment, usually used for short periods of time is considered as noncontinuous loads and is included in the branch circuit loads. Such equipment can be hard-wired, but quite often it is the cord-and-plug type connected to a receptacle. Branch circuits for appliances and equipment must have ratings equal to or higher than the appliance or equipment rating. The current rating is marked on the appliance by the manufacturer or is found on the equipment nameplate. If the appliance or equipment has an electrical motor, the current rating of the branch circuit conductors must be 125% of the motor current rating. Motor and motor load calculations are some of the most challenging calculations performed when dealing with the NEC, IEC, and other codes and standards. While they are challenging, they are easier to understand if we compute each section separately rather than as a whole. For the purposes of this textbook, we will concentrate on the more common calculations pertaining to motors rather than try to understand the entire article philosophy and technical details. If you have one available, you should also follow along with your NEC book. In NEC Article 430, there are 13 sections that discuss all of the code requirements for motors. Specifications from this article are summarized in Table 4.7 and Table 4.8. These sections can easily be used with the addition of tables and formulas

TABLE 4.7
Full-Load Current (A) for Single-Phase AC Motors (NEC Table 430.248)

Horsepower	115 V	200 V	208 V	230 V
1/6	4.4	2.5	2.4	2.2
1/4	5.8	3.3	3.2	2.9
1/3	7.2	4.1	4.0	3.6
1/2	9.8	5.6	5.4	4.9
3/4	13.8	7.9	7.6	6.9
1.0	16.0	9.2	8.8	8.0
1.5	20.0	11.5	11.0	10.0
2.0	24.0	13.8	13.2	12.0
3.0	34.0	19.6	18.7	17.0
5.0	56.0	32.2	30.8	28.0
7.5	80.0	46.0	44.0	40.0
10.0	100.0	57.5	55.0	50.0

Note: Listed voltages are the rated motor voltages, and the listed currents are permitted for system voltage ranges of 110 to 120 V, and 220 to 240 V. FLA values are for motors running at usual speeds and motors with normal torque characteristics.

TABLE 4.8

Full-Load Current (A) for Three-Phase Squirrel-Cage Induction Motors (NEC Table 430.250)

Horsepower	115 V	200 V	208 V	230 V	460 V	575 V	2300 V
1/2	4.4	2.5	2.4	2.2	1.1	0.9	—
1/3	6.4	3.7	3.5	3.2	1.6	1.3	—
1.0	8.4	4.8	4.6	4.2	2.1	1.7	—
1.5	9.8	6.9	6.6	6.0	3.0	2.4	—
2.0	12.0	7.8	7.5	6.8	3.4	2.7	—
3.0	13.6	11.0	10.6	9.6	4.8	3.9	—
5.0	—	17.5	16.7	15.2	7.6	6.1	—
7.5	—	25.3	24.2	22.0	11.0	9.0	—
10	—	32.2	30.8	28.0	14.0	11.0	—
15	—	48.3	46.2	42.0	21.0	17.0	—
20	—	62.1	59.4	54.0	27.0	22.0	—
25	—	78.2	74.8	68.0	34.0	27.0	—
30	—	92.0	88.0	80.0	40.0	32.0	—
40	—	120.0	114.0	104.0	52.0	41.0	—
50	—	150.0	143.0	130.0	65.0	52.0	—
60	—	177.0	169.0	154.0	77.0	62.0	16.0
75	—	221.0	211.0	192.0	96.0	77.0	20.0
100	—	285.0	273.0	248.0	124.0	99.0	26.0
125	—	359.0	343.0	312.0	156.0	125.0	31.0
150	—	414.0	396.0	360.0	180.0	144.0	37.0
200	—	552.0	528.0	480.0	240.0	192.0	49.0
250	—	—	—	—	302.0	242.0	60.0
300	—	—	—	—	361.0	289.0	72.0
350	—	—	—	—	414.0	336.0	83.0
400	—	—	—	—	477.0	382.0	95.0
450	—	—	—	—	515.0	412.0	103.0
500	—	—	—	—	590.0	472.0	118.0

Note: Listed voltages are the rated motor voltages, and the listed currents are the ones permitted for system voltage ranges of 110 to 120, 220 to 240, 440 to 480, and 550 to 660 V. FLA values are for motors running at usual speeds for belted motors and motors with normal torque characteristics.

For 80% and 90% power factors, the values must be multiplied by 1.25 and 1.1, respectively.

for sizing overcurrent protection, conductors, and motor full load currents. Most calculations dealing with a motor circuit are based off of the full load amperage (FLA) of the motor or motors connected. Therefore, it is important to start with the correct amperage. To determine this, the NEC provides several tables that are to be used to find the motor amperage. The code reference that explains this follows in this chapter subsection.

The highest rated or smallest rated motor is determined in compliance with NEC Sections 430.24, 430.53(B), and 430.53(C). The highest rated or smallest rated motor shall be on the rated full-load current as selected from Table 430.247, Table 430.248, Table 430.249, and Table 430.250. There are six NEC tables displaying motor full load currents; however, the two used most often are Table 430.248 and Table 430.250. Electric motor loads are calculated based on the FLA, with 25% added to allow for overload operating conditions. For load estimating purposes, the FLA for various motors is found in these NEC tables. In addition, for the feeders or branch circuits supplying multiple electric motor loads, 25% of the load of the largest motor in

TABLE 4.9
Full-load Current (A) for Three-Phase Synchronous Motors, Running at Unity Power Factor (NEC Table 430.250)

Horsepower	230 V	460 V	575 V	2300 V
25	53.0	26.0	21.0	—
30	63.0	32.0	26.0	—
40	83.0	41.0	33.0	—
50	104.0	52.0	42.0	—
60	123.0	61.0	49.0	12.0
75	155.0	78.0	62.0	15.0
100	202.0	101.0	81.0	20.0
125	253.0	126.0	101.0	25.0
150	302.0	151.0	121.0	30.0
200	400.0	201.0	161.0	40.0

Note: Listed voltages are the rated motor voltages, and the listed currents are permitted for system voltage ranges of 110 to 120, 220 to 240, 440 to 480, and 550 to 660 V. FLA values are for motors running at usual speeds for belted motors and motors with normal torque characteristics.

the group must be added to the total motor load estimate. Tables 4.7, 4.8, and 4.9, adapted from NEC Tables 430.248 and 430.250, provide the FLA values for most common single-phase and three-phase AC motors.

Example 4.10

Determine the estimated load for the following single-phase motors, supplied by a dedicated branch circuit: (a) 1/3 HP at 115 V, (b) 3 HP at 230 V, and (c) 7.5 HP at 208 V.

SOLUTION

From Table 4.7 (NEC Table 430.248), we can find the full load currents and the estimated load are calculated as:

    a. FLA = 7.2 A, and the estimated load is: S = 7.2 A × 115 V = 828 VA
    b. FLA = 17.0 A, and the estimated load is: S = 17.0 A × 230 V = 3910 VA
    c. FLA = 44.0 A, and the estimated load is: S = 44.0 A × 208 V = 9152 VA

Example 4.11

Determine the estimated load for the following three-phase motors, supplied by a dedicated branch circuit: (a) 30 HP at 460 V, induction motor; (b) 20 HP at 200 V; (c) 200 HP, 2800 V synchronous motor; and (d) 10 HP at 208 V, induction motor.

SOLUTION

From Tables 4.6 and 4.7 (NEC Table 430.50), we can find the full load currents and the estimated load are calculated as:

    a. FLA = 40.0 A, and the estimated load is: S = 40.0 A × 460 V = 18,400 VA
    b. FLA = 62.1 A, and the estimated load is: S = 62.1 A × 200 V = 12,420 VA
    c. FLA = 400.0 A, and the estimated load is: S = 400.0 A × 230 V = 92,000 VA
    d. FLA = 30.8 A, and the estimated load is: S = 30.8 A × 208 V = 6406.4 VA

TABLE 4.10
Feeder Demand Factor for Multiple Pieces
of Commercial Kitchen Equipment

Equipment Units	Demand Factor (%)
1–2	100
3	90
4	80
5	70
6+	65

4.2.5 Heating, Cooling, Electric Cooking, and Laundry Equipment

Section 220.56 of the NEC discusses commercial cooking equipment related load calculations. Total feeder or panel load is simple: The nameplate ratings of such types of appliances, along with derating factors (listed in Table 4.10), apply in the case of three or more cooking equipment units. However, the branch circuit loads are not derated using these values. Ovens, fryers, grills, food heaters, large vat blending machines, large mixers, booster heaters, conveyors, and tray assemblies are considered kitchen equipment and may be derated in agreement with NEC Table 220.56. Auxiliary equipment, such as exhaust fans, space heaters, and air-conditioning units, is not included in this category and cannot be derated. Notice that derating factors are applied to kitchen equipment that has either thermostatic control or intermittent use. Electric cooking equipment usually consists of electric ranges, counter-mounted cooktops, or well-mounted electric ovens. The counter-mounted cooktops and wall-mounted electric ovens are separated units and are permitted to be supplied from the branch circuit; however, it is preferable to supply from separate circuits. Freestanding electric ranges have both a cooktop and an oven built in a single unit, and therefore are serviced from a separate branch circuit, determined based on the estimated load. The ratings for branch circuits supplying counter-mounted cooktops or wall-mounted electric ovens are based on the estimated load of the individual units. For purposes of load estimation, electric ranges are considered individually. A counter-mounted cooktop and up to two wall-mounted electric ovens supplied from the same branch circuit may be considered as a single unit for load estimation; the equivalent rating is the sum of the individual ratings. The demand load for household electric cooking appliances is taken from Table 4.11, adapted from NEC Table 220.19. NEC recommends for equally rated ranges rated over 12 kW but no more than 27 kW, an adjustment of 5% for each additional kW of rating (including major fractions) is added to the maximum demand in Column C, while for unequally rated ranges, an average value is computed by adding all range ratings to obtain the total connected load (by using 12 kW for any range rated less than 12 kW) and dividing by the range numbers. Then the maximum demand (Column C) is increased by adding 5% for each kW or major fraction exceeding 12 kW. For power ratings between 1.75 up to 8.75 kW, it is permissible to add nameplate ratings of all household cooking appliances in this rating range and multiply the sum by the demand factors specified in Columns A and B. It is also permissible, according to the NEC, to calculate the branch circuit load for one range using Table 220.19, using nameplate ratings, adding them for more than one and considering the total rating.

Example 4.12

Determine the estimated load for a branch circuit supplying: (a) both a 4.5 kW counter-mounted cooktop and a 6 kW wall-mounted oven; and (b) both a 7.5 kW counter-mounted cooktop and an 8 kW wall-mounted oven.

TABLE 4.11
Household Cooking Appliances Over 1.75 kW, Demand Loads (NEC Table 220.19)

Number of Appliances	A. Demand Load (%) Less than 3.5 kW	B. Demand Load (%) 3.5 kW to 8.75 kW	C. Maximum Load (kW) Not Over 12 kW
1	80	80	8
2	75	65	11
3	70	55	14
4	66	50	17
5	62	45	20
6	59	43	21
7	56	40	23
8	53	36	23
9	51	35	24
10	49	34	25
11	47	32	26
12	45	32	27
13	43	32	28
14	41	32	29
15	40	32	30
16	39	28	31
17	38	28	32
18	37	28	33
19	36	28	34
20	35	28	35
21	34	26	36
22	33	26	37
23	32	26	38
24	31	26	39
25	30	26	40
26–30	30	24	15kW + 1kW
31–40	30	22	For each range
41–50	30	20	25 kW + 0.75 kW
51–60	30	18	For each range
61 and over	30	16	

SOLUTION

(a) It is permitted to consider both cooking appliances as a single unit with the combined rating of 4.5 + 6 = 10.5 kW. From Column C of Table 4.11 treating as a single range of 10.5 kW, the estimated load demand is 8 kW. (b) Again, it is permitted to consider as a single range (unit) for the purpose of determining the load demand. The combined rating is 7.5 + 8 = 15.5 kW, and the excess over 12 kW is 3.5 kW, meaning an increase of 17.5% added to the 8 kW. The estimated load demand is 8 × 1.175 = 9.4 kW for the units.

Example 4.13

Determine the estimated demand load for the following household equipment: (a) a 15 kW range, (b) an 8 kW oven, (c) a 20 kW range, (d) a 3.5 kW electric dryer, and (e) a 7.5 kW electric dryer.

SOLUTION

(a) The rating is 3 kW in excess of 12 kW; therefore, the demand load is

$$Demand\ load = 8\ kW + 8\ kW \times 0.15(15\%) = 9.2\ kW$$

(b) The branch circuit demand load in this case is the unit rated power, or 8 KW.
(c) For a 20 kW range, there is an excess of 8 kW of 12 kW; therefore, the demand load is:

$$Demand\ load = 8\ kW + 8\ kW \times 0.40\ (40\%) = 11.2\ kW$$

(d) A minimum of 5 kW applies even if the actual power rating is 3.5 kW.
(e) The actual dryer power rating applies; because it is larger than 5 KW, the demand load is 7.5 kW.

According to NEC Section 220.18, the required demand load for household clothes dryers is equal to 5000 VA or nameplate, whichever is larger, and the branch circuit must have a rating sufficient to supply the dryer load. Demand factors from more than one dryer connected to a feeder or service were shown in Table 4.12, an adaptation of NEC Table 220.54.

Example 4.14

Determine the estimated load for two household dryers: (a) one rated at 4.5 kW and (b) another rated at 8 kW.

SOLUTION

(a) The minimum load demand of 5000 VA applies, even if the nameplate rating is 4.5 kW.
(b) The nameplate rating for the second dryer is 8 kW. According to the NEC, this applies, being larger than 5000 VA.

The NEC requires that regardless of the building or facility type—commercial, industrial, or residential—the heating loads are computed at the rated value of the nameplate unit, while the supplying branch circuit requires specific considerations. Article 424 of the NEC covers fixed electric space

---

TABLE 4.12
Demand Factors for Household Electric Clothes Dryers (NEC Table 220.54)

Number of Dryers	Demand Factor (%)
1–4	100
5	85
6	75
7	65
8	60
9	50
10	47
11	
12–23	47% minus 1% for each dryer exceeding 11
24–42	35% minus 0.5% for each dryer exceeding 23
> 42	25%

heating equipment, including central heating, boilers, heater cables, and unit heaters (baseboard, panel, and duct heaters). The NEC also requires disconnects for heater and motor controllers and supplementary overcurrent protection for any fixed electric space heating units. The nameplate rating is used to determine the load, and if the current operates more than 3 hours (continuous load), its rating is increased by a factor of 1.25. Fixed electric space heating is considered a continuous load. Similarly, air-conditioning system ratings are determined from the nameplate values. The nameplate rated current is used to size the branch circuit supplying the unit. However, the recommendation is to use the larger of the branch circuit current rating or the full-load current rating to size the branch circuit. If a unit has two or more motors, the circuit rating is computed at 125% of the largest motor plus the sum of the other motors, while for a single motor unit disconnect, we must use 115% of the full-load current.

4.2.6 Load and Correction Factors Estimate Applications

Receptacle outlets (if known) and fixed multi-outlet assemblies (if any) are entered into the calculation and if the load is great enough, a demand factor is applied. As with dwelling-unit load calculations, general lighting is computed using outside dimensions, as discussed in the previous chapter section. Other items, such as a sign outlet (where required) and show window(s) (if present) are part of the nondwelling load calculation. Continuous loads are very important analysis and design computation parameters. All continuous load ratings must be *increased by 25% for load estimate purposes*. While kitchen equipment is not included in every type of calculation, be aware that kitchen equipment is not limited to restaurants. For instance, kitchen equipment could be a portion of a load calculation for a school. All other loads are referring to any load not included otherwise in the calculation. The load calculation form has little room for listing individual items, such as motors, equipment, and so on. Depending upon the size of the occupancy, the calculation could contain hundreds, if not thousands, of individual items. Utilization factor, another load computation and characterization parameter, is the time that equipment is in use divided by the total time that it could be in use. In normal operating conditions the power consumption of a load is sometimes less than that indicated as its nominal power rating, a fairly common occurrence that justifies the application of a utilization factor (ku) in the estimation of realistic values. For example, a motor may be used for only 8 hours a day, 50 weeks a year. The hours of operation would then be 2000 hours, and the motor utilization factor for a base of 8760 hours per year would be 2000/8760 = 22.83%. With a base of 2000 hours per year, the motor utilization factor would be 100%. The bottom line is that the use factor is applied to get the correct number of hours that the motor is in use. In an industrial installation this factor is estimated at an average at 0.75 for motors. For incandescent lighting loads, the factor always equals 1, while for socket-outlet circuits, the factors depend on the type of appliances being supplied from the sockets concerned. IEC recommendations for estimating the diversified peak demand of residential buildings consisting of multi-dwelling units are:

1. Illumination: 50% of total connected load.
2. Small appliance circuits: 100% of rated load for maximum outlet wattage in the circuit plus 40% of the total connected loads of other outlets in the circuit.
3. Fixed appliance circuits and fixed electric ranges: 100% of rated load of largest equipment plus 50% for rated load for the first equipment following the largest one plus 33% for the second equipment following the largest load plus 20% of total connected load of other equipment.
4. Electric water heaters: 100% of rated load of largest equipment plus 100% for rated load for the first equipment following the largest one plus 25% of total connected load of other equipment.
5. Air-conditioning units: 100% of total connected load in all cases.

This way of calculating the DF is per standard and does NOT depend on social category or environment.

Example 4.15

Compute the diversity factor for the connected loads of a high load diversity apartment of an area of 200 m². Loads are given in the table below.

Load Type	Specifications	Unit Load	Total Load
L_1: General	Lighting and general-use receptacles	30 VA/m² at 200 m²	6000 VA
L_2: Small-appliance circuits	Vacuum cleaner	1000 W	4400 VA
	Refrigerator	500 W	
	Small oven	900 W	
	Kitchen appliances	2000 W	
L_3: Fixed-appliance circuits	Washing machine	2500 W	10,000 VA
	Dishwasher	2500 W	
	Fixed appliance	2000 W	
	Water heater	2000 W	
	Ironing	1000 W	
L_4: Electric cooker	Electric cooker	1500 W	1500 VA
L_5: HVAC	2 Units @ 4500 VA	4500 VA	9000 VA

SOLUTION

The load values are given in the above table. Applying IEC recommendations for the connected loads of the high load density apartment building gives:

$$L_1 = 0.5 \times 6000 = 3000 \text{ VA}$$
$$L_2 = 1000 + 0.4(500 + 900 + 2000) = 2360 \text{ VA}$$
$$L_3 = 2500 + 0.5 \times 2500 + 0.33 \times (2000 + 2000 + 1000) = 5400 \text{ VA}$$
$$L_4 = 1500 \text{ VA}$$
$$L_5 = 9000 \text{ VA}$$

$$P_{max} = 3000 + 2360 + 5400 + 1500 + 9000 = 22160 \text{ W}$$
$$\Sigma \text{ Connected Load Demand} = 30900 \text{ VA}$$
$$Div \ F = \frac{22160}{30900} = 0.72$$

This is a very common value for the diversity factor in residential buildings. Notice that the more equipment that is included, the lower the DF is.

To understand the concept, if we have a 15 A circuit, the safe operating amperage would be no greater than 12 A. The total wattage would be 1800 W, meaning the safe wattage usage would be 1440 W. If you had an 1100 W hair dryer plugged into this circuit, notice that just one device uses almost the entire desired load capabilities. For a 20 A circuit, the safe operating amperage would be no greater than 16 A. If the total wattage is 2400 W, the safe wattage usage is 1920 W. In this instance, a hair dryer, radio, and electric razor running can be connected on the same circuit, but not much else. This area should include additional bathroom circuits to cover lighting, exhaust fans, and heat lamps for drying. On a 30 A circuit, the safe operating amperage would be no greater than 24 A. The total wattage is 3600 W, meaning the safe wattage usage would be 2880 W. Similar information comes with central air conditioners, electric dryers, electric ranges, and electric ovens.

NEC Table 220.3(A) applies to general lighting loads. The load from general-purpose receptacle outlets in habitable areas of dwelling units and motels is permitted to be included with the general lighting load for the purpose of load estimate. For office buildings and banks, an additional load of 1 VA per square foot shall be included if the actual number of general-purpose receptacles is unknown. The designer may not know at the time of design what the final layout is, so this extra provision for load must be included. NEC Section 220.3(B) requires a minimum of 180 VA be assigned to other general-use receptacle outlets that are not located in habitable areas or connected to a dedicated branch circuit. The estimated load is usually greater than the required NEC minimum of 180 VA. We must use common sense and judgment when estimating loads on circuits where large loads might be connected. In addition to lighting and receptacle loads, other loads may be presented in the occupancy. These loads may consist of electric heating, electric hot water systems, electrical motors and other loads, and/or other receptacle loads. For non-electrical motor operated appliances and equipment, the apparent power ratings are calculated by simply multiplying the actual voltage and current ratings of the device. Motor loads are calculated using the full-load current rating with the addition of 25% to allow slight overload conditions on the motor. NEC specifications can be used for load estimating purposes of the electric motors (the full load current). In addition, for feeders or branch circuits supplying multiple motor loads, 25% of the load of the largest motor in the group must be added to the total motor load.

4.3 CONDUCTORS AND CABLES

Two important parameters need to be determined in order to estimate the size of the electrical system that will power any building environment: the total critical loads and the total noncritical loads. In general, the electrical supply must be large enough to support the sum of these two numbers, plus the related building additional electrical loads and services. The steady-state power consumption of the loads within a building establishes the power consumption for purposes of determining electricity management, electrical costs, design or restructuring, and upgrading. However, the electrical service and the backup generator power sources, if any, that provide power to the building cannot be sized to the steady-state values. They must be sized to the peak power consumption of the loads, plus any derating or oversizing margins required by code or standard engineering practice. Once the total electrical capacity is estimated from the process described above, two critical determinations can be made: (1) an estimate of the electrical service needed to supply the building or facility, and (2) the size of any standby generator capacity, needed to achieve the desired availability. Component parts of an electric circuit or service and their protection are determined such that all normal and abnormal operating conditions are satisfied and electricity is delivered at the highest level of quality and reliability. The cabling and its protection at each level must satisfy several conditions at the same time, in order to ensure a safe and reliable installation, such as: carry the permanent full load current and normal short-time overcurrents; prevent voltage drops that result in an inferior performance of certain loads; protect the cabling and busbars for all levels of overcurrent, up to and including short-circuit currents; and ensure personnel protection against indirect contact hazards. The methods for calculation of the conductor cross-sectional areas are described later in this chapter. Apart from this method, some national standards may prescribe a minimum cross-sectional area to be observed for reasons of mechanical endurance. In particular, some loads may require that the cable supplying them be oversized, and that the protection of the circuit be likewise modified.

4.3.1 CONDUCTOR TYPES AND SIZES

The term *wire* generally refers to one or more insulated conductors (solid or stranded for flexibility), whereas *cable* refers to one or more insulated conductor of larger size grouped in a common insulated jacket. However, in common understanding, a wire is metal drawn or rolled to long lengths, normally understood to be a solid wire. Wires may or may not be insulated. A conductor is one or

TABLE 4.13

Nominal or Minimum Properties of Conductor Wire Materials (Adapted from IEC Standard)

Cable Type/Property	Int. Annealed Copper Stranded	Hard-Drawn Copper Wire	Standard 1350-H19 Aluminum Wire	Galvanized Steel Core Wire	Aluminum Clad Steel
Resistivity (20°C Ω in.²/1000 ft.)	0.008145	0.008397	0.01331	0.101819	0.04007
Thermal resistivity coefficient per °C	0.00393	0.00381	0.00404	0.00327	0.00360
Density at 20°C lb./in.³	0.3212	0.3212	0.0977	0.2811	0.2381
Linear expansion coefficient 10⁻⁶ per °C	16.9	16.9	23.0	11.5	13.0

more wires suitable for carrying electric current. Often the term *wire* is used to mean conductor. Conductors used in electrical power distribution and building electrical systems are made of copper or aluminum (see Table 4.13 for material characteristics). The conductors, usually copper or aluminum, may be solid or stranded, depending on the size and the required flexibility. They are packaged in several ways to form electric power cables. The most common cable construction type for low-voltage (600 V or less) power is the single-conductor cable covered with single-layer insulation, or by an outer nylon jacket. They are typically installed in conduit or other suitable raceway systems. Insulation materials are usually extruded around the electrical conductors to provide insulation between the conductors and for the cable. Several variations exist with respect to the number and size of conductors in cable assemblies. The thickness of the insulating material is generally determined by the voltage rating of the cable. Common voltage classes for cables are 600 V, 2 kV, 5 kV, 15 kV, 25 kV, and 35 kV. Various insulation thicknesses are permitted within a given class. Each conductor size is measured in American wire gauge (AWG) or in metric gauge (SI units) by the conductor cross-section area in square millimeters. The AWG numbers are log-inverse measures of the conductor diameter, *d* in inches, expressed by this relationship:

$$AWG = 20\log\left(\frac{0.325}{d(inch)}\right) \tag{4.9}$$

A decrease in gauge number by 1 increases the conductor diameter by 1.225 and the conductor area by 1.26. The diameter doubles at every sixth gauge down, while the area doubles at every three gauge down. A multiconductor cable is an assembly of two or more conductors (often four conductors), having an outer jacket and insulation for easy installation or replacement. Utilities use aluminum for almost all new overhead installations. However, the aluminum alloys are used only in larger conductors, provided that it is an approved alloy. Copper has very low resistivity and is widely used as a power conductor, although use as an overhead conductor has become rare because copper is heavier and more expensive than aluminum. It has significantly lower resistance than the aluminum ones, so a copper conductor has equivalent ampacity (resistance) of an aluminum conductor that is two AWG sizes larger. Copper has very good resistance to corrosion. It melts at 1083°C, starts to anneal at about 100°C, and anneals most rapidly between 200°C and 325°C, depending on the presence of impurities and amount of hardening. Different sizes of conductors are specified with gauge numbers or area in circular mils. Smaller wires are normally referred to using the AWG system. The gauge is a numbering scheme that progresses geometrically. A number 36 solid wire has a defined diameter of 0.005 in. (0.0127 cm), and the largest size, a number 0000 (referred to as 4/0 and pronounced "four-ought") solid wire has a 0.46-in.

(1.17-cm) diameter. The larger gauge sizes in sequence of increasing conductor size are: 4, 3, 2, 1, and 0 (1/0), 00 (2/0), 000 (3/0), 0000 (4/0). Going to the next bigger size (smaller gauge number) increases the diameter by 1.1229. Some other useful rules are: (a) an increase of three gage sizes doubles the area and weight and halves the dc resistance, and (b) an increase of six gauge sizes doubles the diameter.

The smallest wire size used in power distribution networks and building electrical systems is #14 AWG, consisting typically of a single conductor having an outside diameter of 0.0641 inches, or 64.1 mils (one mil is equal to 1/1000 of an inch). The largest AWG designation is #4/0 AWG, having a diameter of 0.072 inches or 72 mils. Each individual strand of a seven-strand #4/0 AWG conductor has a diameter of 173.9 mils. Larger conductors are specified in circular mils of cross-sectional area. One circular mil is the area of a circle with a diameter of one mil (one mil is one thousandth of an inch). Conductor sizes are often given in kcmil, thousands of circular mils. In the past, the abbreviation MCM was used, which means thousands of circular mils (M is thousands, not mega, in this case). By definition, a solid 1000 kcmil wire has a diameter of 1 inch. The diameter of a solid wire in mils is related to the area in circular mils by $d = \sqrt{A}$.

Outside of America, most conductors are specified in mm². Some useful conversion relationships are:

$$1 \text{ kcmil} = 1000 \text{ cmil} = 785.4 \times 10^{-6} \text{ in}^2 = 0.5067 \text{ mm}^2 \qquad (4.10)$$

The IEEE standard uses kcmil to specify conductor sizes, which increase with the conductor cross-sectional area. The European and international standards measure conductors in metric gauge equal to square millimeters of the net conductor area (see the tables in the Appendixes at the end of this book for AWG, IEEE, and metric gauges). Stranded conductors have better flexibility than regular ones. Typically a two-layer arrangement has seven wires, while a three-layer arrangement has 19 wires. A four-layer arrangement has 37 wires. The cross-sectional area of a stranded conductor is the metal cross-sectional area, having a larger diameter than a solid conductor of the same area. The ACSR conductor area is defined by the conductor aluminum area. Utilities in heavy tree areas often use covered conductors, with a thin insulation, which is not rated for full conductor line-to-ground voltage, being thick enough to reduce the chance of flashover if a tree branch falls between conductors. Covered conductors are also called tree wire or weatherproof wire. Tree wire also helps with animal faults and allows utilities to use armless or candlestick designs or other tight configurations. Tree wire is available with a variety of covering types. The insulation materials polyethylene, XLPE, and EPR are most common. Insulation thicknesses typically range from 30 to 150 mils (1 mil = 0.001 in. = 0.00254 cm). From a design and operating viewpoint, covered conductors must be treated as bare conductors according to the National Electrical Safety Code (NESC) or IEEE C2-2000, with the only difference that tighter conductor spacing is allowed. Some of the conductor characteristics of aluminum and copper conductors are given in Table 4.13.

Multicore and multiconductor cables comprise more than one conductor, which may eventually include bare conductors. The term *three-core cable* is used to designate the cable making up the phases of a three-phase system. A single-core cable comprises a single insulated conductor. The term *single-core cable* is used to designate a cable making up one of the phases of a three-phase system. A wiring system is an assembly made up of one or more electric conductors and the devices ensuring their fixation and, if necessary, their mechanical protection. Conductor sizes are specified in terms of AWG or thousand circular mils (kcmil). The AWG designation for conductors ranges from #14 up through #4/10, while the kcmil designation is used from sizes from 250 up to 2000 kcmil. For power distribution circuits including feeders and branch circuits, wire sizes in the range of #14 AWG to 500 kcmil are the most commonly specified. Wire sizes larger than 500 kcmil are frequently avoided due to difficulty with installation into raceways and conduits. The electrical resistance of an electrical conductor is a measure of the difficulty to pass an electric current through that conductor. The inverse quantity is electrical conductance, and is the ease with which an electric current passes. Electrical resistance shares some conceptual parallels with the notion of mechanical friction. The SI unit of electrical resistance is the ohm (Ω), while electrical conductance is measured in siemens (S). The resistivity and conductivity are proportionality constants, and therefore depend only on the material the wire is made of, not the geometry of the wire. Resistivity is a measure of the material's ability to oppose electric current. The conductor resistance's effect on the current-carrying capability is given by:

$$R = \frac{\rho l}{A} = \frac{l}{\sigma A} \qquad (4.11)$$

Here, $\sigma = 1/\rho$ is the material conductivity, while ρ is the conductor material resistivity, in $\Omega \cdot m$ or $\Omega \cdot cmil/ft$. The resistivity values at 25°C are 2.65×10^{-8} $\Omega \cdot m$, or 17.291 $\Omega \cdot cmil/ft$. for hard-drawn aluminum, and 1.724×10^{-8} $\Omega \cdot m$, or 10.571 $\Omega \cdot cmil/ft$. for hard-drawn copper, respectively. A practical relationship used in the United States is expressing the conductor resistance by:

$$R = K \frac{\text{Length (feet)}}{Area \text{ (cicular mils)}}, \Omega \qquad (4.12)$$

For DC and low-frequency currents at 20°C, K is equal to 10.372 for copper and 18.046 for aluminum, both of electric grades. Conductor resistance changes with temperature, expressed by the following relationship:

$$R_{t2} = R_{t1} \frac{M - t_2}{M - t_1} \qquad (4.13)$$

Where R_{t2} is the resistance at temperature t_2 given in °C, R_{t1} = resistance at temperature t_1 given in °C, and M is a temperature coefficient for the given material, 228.1 for aluminum, and 241.5 for annealed hard-drawn copper. For a wide range of temperatures, resistance rises almost linearly with temperature for both aluminum and copper. The resistivity of a material constant is temperature dependent, so the element resistance is expressed as:

$$R = R_{25}\left[1 + \alpha\left(T - 25°\right)\right] \qquad (4.14)$$

Where, R is the resistance at the new temperature, R_{25} is the resistance of the conductor at 25°C, T is the new temperature in degrees Celsius, and α is the temperature coefficient of the resistivity, 0.00385 for copper, and 0.00395 for aluminum, respectively.

Example 4.16

Determine the resistance of a conductor 100 m long, with a diameter 0.5 cm at 25°C, made from (a) hard-drawn copper, and (b) hard-drawn aluminum. Determine the resistances of these conductors at 45°C.

SOLUTION

The area of both conductors is:

$$A = \frac{\pi D^2}{4} = \frac{3.14 \times \left(0.5 \times 10^{-2}\right)^2}{4} = 0.0019625 \text{ m}^2$$

From Equation (4.11), the resistances are:

$$R_{Cu} = \frac{1.724 \times 10^{-8} \cdot 100}{0.0019625} = 0.000879 \ \Omega$$

$$R_{Al} = \frac{2.65 \times 10^{-8} \cdot 100}{0.0019625} = 0.001351 \ \Omega$$

By using Equation (4.12), we can estimate the resistances at 45°C, as:

$$R_{Cu(45)} = 0.000879 \cdot \left[1 + 0.00385 \cdot (45 - 25)\right] = 0.000947 \ \Omega$$

$$R_{Al(45)} = 0.0013510 \cdot \left[1 + 0.00395 \cdot (45 - 25)\right] = 0.001458 \ \Omega$$

We can also linearly interpolate using resistances provided at two different temperatures as:

$$R(t_c) = R(t_1) \frac{R(t_2) - R(t_1)}{t_2 - t_1} (t_c - t_1) \tag{4.15}$$

Where $R(t_c)$ is the conductor resistance at temperature t_c, $R(t_2)$ is the resistance at the higher temperature, t_2, and $R(t_1)$ is the resistance at the lower temperature, t_1. With alternating current, skin effects raise the resistance of a conductor relative to its dc resistance. A conductor offers a greater resistance to a flow of alternating current than it does to direct current. This increased resistance is generally expressed as the AC/DC resistance ratio. At 60 Hz, the resistance of a conductor is very close to its DC resistance except for very large conductors, and such effects are ignored in this textbook. Skin effects are much more important for high-frequency analysis such as switching surges and power-line carrier problems, playing a larger role in larger conductors. For most distribution power-frequency applications, we can ignore skin effects (and they are included in AC resistance tables). Power distribution and building electrical systems use a large variety of sizes and types of conductors. Several electrical, mechanical, and economic characteristics affect conductor selection, such as:

- Ampacity, the peak current-carrying conductor capability limits the current, and the power capability.
- Economics, often a conductor operates well below its ampacity rating, and the cost of the extra aluminum or copper pays for itself with lower I²R losses. The conductor runs cooler, also leaving room for expansion.
- Mechanical strength on lines with long span lengths plays an important role in conductor size and type. Stronger conductors such as ACSR are often used; ice and wind loadings also must be taken into account.
- Corrosion, while not usually a problem, sometimes can limit certain conductors in certain applications.

4.3.2 CABLE IMPEDANCE CALCULATIONS

This section provides details on the calculation of cable impedances—DC resistance, AC resistance, and inductive reactance. The DC and AC resistance of cable conductors can be calculated based on IEC 60287-1 Clause 2.1. The DC cable conductor resistance is calculated by using:

$$R_{DC} = \frac{1.02 \times 10^6 \rho_{20}}{A} \left[1 + \alpha_{20} \left(\theta - 20°\right)\right] \tag{4.16}$$

Here, A is the conductor cross-sectional area (mm^2), R_{DC} is the DC resistance at the conductor operating temperature θ (Ω/m), ρ_{20} is the conductor resistivity, at 20°C (Ω.m). For copper conductors, $\rho_{20} = 1.7241 \times 10^{-8}$, while for the aluminum conductors, $\rho_{20} = 2.8264 \times 10^{-8}$, α is the temperature coefficient of the conductor material per K at 20°C (for copper conductors, $\alpha = 3.93 \times 10^{-3}$, while for aluminum conductors, $\alpha = 4.03 \times 10^{-3}$), and θ is the conductor operating temperature (°C).

The AC resistance of cable conductors is the DC resistance corrected for skin and proximity effects. *Skin effect* is the AC current tendency to become distributed within a conductor such that the current density is largest near the surface of the conductor, and decreases with conductor depths. The electric current flows mainly at the "skin" of the conductor, between the outer surface and a level called the *skin depth*. The skin effect causes the effective resistance of the conductor to increase at higher frequencies where the skin depth is smaller, thus reducing the effective cross-section of the conductor. The skin effect is due to opposing eddy currents induced by the changing magnetic field resulting from the alternating current. At 60 Hz in copper, the skin depth is about 8.5 mm. In a conductor carrying alternating current, if currents are flowing through one or more other nearby conductors, such as within a closely wound coil of wire, the distribution of current within the first conductor will be constrained to smaller regions. The resulting current crowding is called the *proximity effect*. This crowding gives an increase in the effective resistance of the circuit, which increases with frequency. The AC conductor resistance is given by:

$$R_{AC} = R_{DC}\left(1 + y_{skf} + y_{prf}\right) \tag{4.17}$$

Here R_{AC} is the AC resistance at the conductor operating temperature θ (Ω/m), R_{DC} is the DC resistance at the conductor operating temperature θ (Ω/m), y_{skf} is the skin effect factor, and y_{prf} is the proximity effect factor. Skin effect and proximity effect factors, for a current frequency f(Hz), diameter D$_{con}$ (mm), and the distance between conductor axes, x (mm) are calculated with the following relationships:

$$y_{skf} = \frac{x_s^4}{192 + 0.8x_s^4} \tag{4.18}$$

And, for 2C and 2 × 1C cables:

$$y_{prf} = \frac{x_p^4}{192 + 0.8x_p^4}\left(\frac{D_{con}}{s}\right)^2 \times 2.9 \tag{4.19a}$$

While for 3C and 3 × 1C cables:

$$y_{prf} = \frac{x_p^4}{192 + 0.8x_p^4}\left(\frac{D_{con}}{s}\right)^2\left[0.312\left(\frac{D_{con}}{s}\right)^2 + \frac{1.18}{\dfrac{x_p^4}{192 + 0.8x_p^4} + 0.27}\right] \tag{4.19b}$$

Here:

$$x_s^4 = \left(\frac{8\pi f}{R_{DC}}k_s \times 10^{-7}\right)^2 \text{, and } x_p^4 = \left(\frac{8\pi f}{R_{DC}}k_p \times 10^{-7}\right)^2 \tag{4.20}$$

TABLE 4.14

Skin Effect and Proximity Effect Factors

Type of Conductor	Dried and Impregnated?	k_s	k_p
Copper			
Round, stranded	Yes	1	0.8
Round, stranded	No	1	1
Round, segmental	—	0.435	0.37
Sector-shaped	Yes	1	0.8
Sector-shaped	No	1	1
Aluminum			
Round, stranded	Either	1	1
Round, 4 segment	Either	0.28	0.37
Round, 5 segment	Either	0.19	0.37
Round, 6 segment	Either	0.12	0.37

The factors k_s and k_p used in Equation (4.20) are given in Table 4.14. The above relationships for skin and proximity effects are accurate provided that k_s and k_p are less than or equal to 2.8. For shaped conductors, the proximity effect factor is two-thirds the above values, and with D_{con} equal to the diameter of an equivalent circular conductor of equal cross-sectional area and degree of compaction (mm), and:

$$s = D_{con} + t$$

where t is the thickness of the insulation between conductors (mm).

The cable series inductive reactance X_c (Ω/km) can be approximated by the following equation:

$$X_C = 2\pi f \left[K + 0.2 \ln \left(\frac{2s}{D_{con}} \right) \right] \times 10^{-3} \qquad (4.21)$$

Here s is the axial spacing between conductors (mm), and D_{con} is the diameter of the conductor. For shaped conductors, the diameter of an equivalent circular conductor of equal cross-sectional area and degree of compaction (mm), K is a constant factor pertaining to conductor formation, with typical values given in Table 4.15. For 3C and $3 \times 1C$ cables, the axial spacing parameter depends on the geometry of the conductors, as shown in Figure 4.1.

TABLE 4.15

Typical K Values

No. of Wire Strands in Conductor	K
3	0.0778
7	0.0642
19	0.0554
37	0.0528
> 60	0.0514
1 (solid)	0.0500

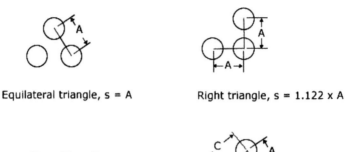

Equilateral triangle, s = A   Right triangle, s = 1.122 x A

Flat, s = 1.26 x A   Unequal, s = (A x B x C)$^{1/3}$

FIGURE 4.1   Typical three-wire cable configurations.

4.3.3   CABLE INSULATION AND CONDUCTOR AMPACITY

The ampacity is the maximum designed current of a conductor, and is given in amperes. Each conductor in a cable is sized to meet the required current-carrying capacity (ampacity) and the voltage drop limitation under normal and inrush (abnormal) conditions. The cable ampacity is limited by the continuous operating temperature limit, which is set by the cable insulation type. The cable temperature is typically in the range 60°C to 75°C for safety, depending on the application. A conductor may have several ampacities, depending on its application and the assumptions used. It also is determined by the weather conditions for overhead cables. The ampacity is dependent mainly on the ambient temperature, which is varying, usually from 40°C to 65°C, depending on the cable routing (e.g., outdoor, indoor, heating system and boiler rooms) and on the grouping in raceways, due to the mutual heating. Both higher ambient temperature and the raceway grouping require derating of the conductor ampacity from the rated values as given in the manufacturer specifications. When higher current flows, higher conductor and insulation temperatures result as the conductor power losses increase. One limiting factor of the conductor ampacity is the current need to bring the conductor to a certain temperature. Internal generated heat due to power losses must be dissipated in order for the temperature to reach equilibrium condition. Moreover, different types of insulation designations reflect their ability to withstand various temperatures, such as: 60°C, 75°C, or 90°C. The conductor and insulation ability to dissipate heat is affected by the ambient temperature and the proximity of other current-carrying conductors. Higher ambient temperatures reduce the conductor heat dissipation capability, while the current-carrying conductors in close proximity to one another generate heat and raise the surface temperature of the conductors. Ampacities of three-phase cables with insulation temperature ratings in 30°C (86°F) ambient air are listed in a table of this book's appendix section. All electrical equipment and devices, not only cable, are affected by temperature increases that are limiting their power ratings. Industry-standard ambient temperature is 40°C, and up to 65°C, special areas, such as boiler rooms or near heating systems. The maximum permissible operating temperature depends on the equipment insulation class, here the cable insulation type. The temperature rise is determined by the cooling medium, equipment geometry, and any cooling system that is dissipating the internal equipment power losses (heat). For electrical equipment usually operating up to 100°C, the temperature rises above ambient temperature can be estimated by this empirical relationship:

$$\Delta T\left(°C\right) = k_{tmp} \cdot P_{Loss}(W) \tag{4.22}$$

Here, k_{tmp} is an equipment-specific constant, and P_{Loss} is the power losses that need to be dissipated, in watts. The power loss by a cable carrying a specific current is proportional to that current square. A large variety of insulation materials and types are used in cables for the required temperature ratings in specific applications (see Table 4.18). Higher temperature ratings allow higher currents, or higher conductor ampacities. The industry and the standards have assigned letter designations to indicate temperature and moisture resistance, for operations of the cable insulations in wet and water environments. At higher ambient air temperature, the cable ampere rating must be reduced and vice versa.

Example 4.17

A Teflon-insulated cable has nominal capacity of 180 A in standard 40°C ambient air temperature. Determine its ampacity if the cable is installed in an industrial process room where the ambient air temperature is 75°C.

SOLUTION

The Teflon insulation (Table 4.16) has maximum temperature of 200°C, a permitted temperature rise of 160°C for standard operation. The new temperature rise is 125°C, and considering the cable power losses proportional to the cable current, the ampacity can be estimated by:

$$I_{new} = I_{old}\left(\frac{\Delta T_{new}}{\Delta T_{old}}\right)^{0.8\times2} = 180\left(\frac{125}{160}\right)^{1.6} \approx 121.3 \text{ A}$$

The cable can be used only to about 121 A in this room, a derating factor of about 66%.

Some frequently consulted tables of the NEC, which are used to determine the conductor ampacity, are 310.16 and 310.17. These are included in Appendix B of this book. These tables list the conductor ampacity for copper and aluminum conductors, for an ambient temperature of 30°C. Table 310.16 applies where there are no more than three current-carrying conductors in a raceway or conduit, and Table 310.17 refers to a single conductor in free air. The temperature rating and insulation type are shown at the table head, and the conductor size designation is in the left-side column. The conductor ampacity for no more than three current-carrying conductors, in a conduit or raceway or for a single current-carrying conductor in free air, are read directly for Table 310.16 and Table 310.17,

---

TABLE 4.16
Cable and Wire Insulations and Operating Temperatures

Cable and Wire Insulation (Letter-Identified Material)	Operating Temperatures (°C)	Operating Temperatures (°F)
Thermoplastic (T)	−40 to 60	−40 to 140
Rubber (R)	−40 to 75	−40 to 167
Vinyl	−20 to 80	−4 to 176
Polyethylene (X)	−60 to 80	−76 to 176
Neoprene	−30 to 90	−22 to 194
Polypropylene	−20 to 105	−4 to 221
Propylene (PF)	−40 to 150	40 to 302
Teflon	−70 to 200	−94 to 392
Silicon rubber (S)	−70 to 200	−94 to 392

respectively. However, the selection of proper conductor size requires the value of load current supplied by the conductor. For applications involving no more than three current-carrying conductors in a raceway and a 30°C ambient temperature, the conductor ampacity must be equal to or greater than the load current.

Example 4.18

Determine the ampacity of the following conductors: (a) #8 TW copper, (b) 600 kcmil THW copper, and (c) 300 kcmil RHH aluminum. There are no more than three conductors in the raceway.

SOLUTION

(a) For #8 TW copper, the ampacity is 40 A. (b) For 600 kcmil THW copper, the ampacity is 460 A. (c) For 300 kcmil RHH aluminum, the ampacity is 255 A.

4.3.4 CABLE CORRECTION FACTORS

The power transmission capacity of an insulated cable system is the product of the operating voltage and the maximum current that can be transmitted. Power transmission systems operate at fixed voltage levels so that the delivery capability of a cable system at a given voltage is dictated by the current-carrying capacity of the conductors. The delivery capability is defined as the "ampacity" of the cable system. The operating voltage determines the dielectric insulation requirements of a cable, while the conductor size is dictated by the ampacity rating. These two independent parameters (insulation and conductor size) of the cable system are interrelated by thermal considerations; a bigger conductor size (less I^2R losses) results in higher ampacity, while increases in insulation material (lower heat dissipation) result in lower ampacity. The parameters of great influence in determining ampacity are the cable size, insulation characteristics, thermal resistivity of the soil, depth of burial for underground cables, and the horizontal spacing between the circuits. However, a given conductor can have several ampacities, depending on its application and the assumptions used. Sun, wind, and ambient temperature change a conductor's ampacity. A conductor's temperature depends on the thermal balance of heat inputs and losses. Current driven through a conductor's resistance creates heat (I^2R). The sun is another source of heat into the conductor. Heat escapes from the conductor through radiation and from convection. Considering the balance of inputs and outputs, the ampacity of a conductor is expressed as:

$$I = f(q_c, q_r, q_s, R_{ac}) \tag{4.23}$$

Where, q_c is the convected heat loss, W/ft., q_r is the radiated heat loss, W/ft., q_s is the solar heat gain, W/ft, and R_{ac} is the nominal AC resistance (at low frequencies is the same as the conductor DC resistance) at operating temperature t, W/ft. The cables are designed to operate within the NEC-prescribed operating characteristics. Ampacity is the current that a conductor can carry continuously, without exceeding its temperature rating. If the conductors get too hot they will burn up and short out, as the conductor heats up the current-carrying capacity goes down. If we overload the capacity of the conductors, they will heat up and short out. Methodology for calculation of ampacities is described in the NEC articles and specifications. Conductor ampacity is presented in the tables, along with factors that are applicable for different laying formations. An alternative approach to the NEC approach is the use of equations for determining cable current-carrying rating. Some of the main factors impacting ampacity are: (1) the allowable conductor temperature, the ampacity increases significantly with higher allowed temperatures; (2) ambient temperature, the ampacity increases about 1% for each 1°C decrease in ambient temperature; and

(3) for aerial conductors, the wind speed, even a small wind helps cool conductors significantly. With no wind, ampacities are significantly lower than with a 2 ft/s crosswind. For electrical wiring systems, especially those built into walls having heating elements, it is necessary to reduce current-carrying capacities by applying the reduction factors specified in codes and standards. This supposes that the temperature distribution inside the heated walls in contact with the electrical wiring system is known.

A conductor's temperature depends on the thermal balance of heat inputs and losses. Current driven through a conductor's resistance creates heat (I^2R losses). Heat escapes from the conductor through radiation and from convection. The ability of an underground cable conductor to conduct current depends on a number of factors. The most important factors of the utmost concern to the designers of electrical transmission and distribution systems or building electrical installations are the following: weather conditions, thermal details of the surrounding medium, ambient temperature, heat generated by adjacent conductors, and heat generated by the conductor due to its own losses. Once the load has been sized and confirmed, the cable system must be designed in a way to transfer the required power from the generation to the end user. The total number of overhead or underground cable circuits; their size; the layout methods; and crossing with other utilities such as roads, telecommunication, gas, or water networks are of crucial importance when designing cable systems. In addition, overhead, aerial, or underground cable circuits must be sized adequately to carry the required load without overheating. Heat is released from the conductor as it transmits electrical current. The cable type, its construction details, and installation method determine how many elements of heat generation exist. These elements can be Joule losses (I^2R losses), sheath losses, etc. Heat that is generated in these elements is transmitted through a series of thermal resistances to the surrounding environment. The foregoing adjustment factors apply where all current-carrying conductors carry current continuously. Where load diversity is involved, such as may be the case in numerous industrial applications, for more than nine conductors in a raceway or cable, Table B310-11 (NEC) provides factors with less severe reduction in ampacities than the values shown above. Conductor sizes and types influence the amount of current a conductor can carry where the conductor is installed in close proximity to other current-carrying conductors. For practical reasons the numbers given for the adjustment factors are not exact. However, they serve well to ensure minimum levels of safety that can be achieved by design, installation, and verification.

Example 4.19

Suppose that a cable has an ambient temperature derating factor of $k_{amb} = 0.94$, and a grouping derating factor of $k_g = 0.85$. If the base current rating is $I_b = 45$ A, what is the installed current rating?

SOLUTION

The overall derating factor is:

$$k_{d(total)} = k_{amb} \times k_g = 0.94 \times 0.85 = 0.799$$

and

$$I_c = k_{d(total)} \cdot I_b = 0.799 \times 45 \approx 36.0 \text{ A}$$

As discussed in Section 4.3.3 and in above paragraphs, conductor ampacities are determined based on ambient temperature of 30°C. If the ambient temperature is higher than 30°C, a temperature correction factor must be applied to the table-listed ampacity to determine the derated conductor ampacity. The correction factors are found at the bottom of NEC Table 310.16 and Table 310.17. By adapting Equation 4.22, the derated conductor ampacity is calculated as:

$$\text{Derated Ampacity} = \text{Table-listed Ampacity} \times k_{Amb} \tag{4.24}$$

Here k_{Amb} is the ambient temperature correction factor, as found in the above mentioned tables.

Current-carrying conductors in raceways and conduits are dissipating heat, and the more such conductors are in the proximity of each other, the more heat is disbursed. NEC Table 310.16 was developed on the assumption of no more than three current-carrying conductors in the raceway, which is reasonable for three-phase balanced circuits, with the load current flowing into the ungrounded conductors. If there are more than three current-carrying conductors in the raceway, a fill or grouping adjustment factor, listed in Table 4.15 must apply to reduce the table-listed conductor ampacity, as:

$$\text{Derated Ampacity} = \text{Table-listed Ampacity} \times k_{grp} \tag{4.25}$$

Here k_{grp} is the grouping adjustment factor, as found in Table 4.17. The number of current-carrying conductors in a raceway or conduit consists of all ungrounded (phase) conductors, plus any neutral conductor that is considered to carry current under normal operating conditions. For example, the neutral conductor of a three-phase, four-wire, wye-connected circuit supplying linear loads, is carrying only the unbalanced load current in the circuit and is not included in the counting. However, if this system is supplying nonlinear loads, such as motor drivers, fluorescent lamps, rectifiers, etc., which may generate harmonics that add up in the neutral, the neutral current may have enough magnitude to be counted as a current-carrying conductor. Also, any equipment-grounding conductor occupying the same raceway as ungrounded conductors is not counted as a current-carrying conductor. When derating is required for both ambient temperature and raceway fill, a total derating factor, the product of the temperature correction factor and adjustment factor, is applied. The derated conductor ampacity is now given by:

$$\text{Derated Ampacity} = \text{Table-listed Ampacity} \times CF_{Total}$$
$$CF_{Totla} = k_{Amb} \times k_{grp} \tag{4.26}$$

TABLE 4.17

Bundle Correction Factor (Adapted from Table 310-15[b][2][a]: Adjustment Factors for More Than Three Current-Carrying Conductors in a Raceway or Cable [Applies also to single conductors or multiconductor cables in free air, stacked, or bundled more than 24 in. (0.61 m)])

Number of Percent of Current-Carrying Conductors	Values in Tables 310-16 or 316-17 (%)
1–3	100
4–6	80
7–9	60
10–20	50
21–30	45
31–40	40
Above 40	35

Example 4.20

Determine the derated ampacity of a 300 kcmil THW copper conductor used in a three-phase, four-wire circuit supplying an induction motor drive, operating at 42°C.

SOLUTION

Because of nonlinear load, the grounded conductor is counted as a current-carrying conductor, and the raceway fill adjustment factor is 80%. From NEC Table 310.16 the ambient temperature correction factor for this type of conductor is 0.82, while the table-listed ampacity is 285 A. The derated ampacity is given by the Equation (4.25).

$$\text{Derated Ampacity} = 285 \times 0.8 \times 0.82 = 197 \text{ A}$$

4.3.5 VOLTAGE DROP CALCULATION

Voltage drop is an important subject that unfortunately gets little attention in the NEC because it is perceived to be more important to equipment performance than to safety. Some manufacturers specify the minimum voltage to be supplied to their equipment. This becomes a requirement under NEC 110.3(B). In addition, special precautions must be taken to provide adequate voltage during startup of certain equipment utilizing a large motor. Completely ignored is energy efficiency, which is a major consideration today. The NEC states in two Informational Notes that a maximum voltage drop of 3% for branch circuit or feeder conductors, and 5% for branch circuit and feeder conductors together, will provide reasonable efficiency of operation for general-use circuits. For sensitive electronic equipment operating within its scope, NEC 647.4(D) requires that the voltage drop on any branch circuit shall not exceed 1.5%, and that the combined voltage drops of branch circuit and feeder conductors shall not exceed 2.5%. Correct voltage is critical for optimal load operation. Low operating voltage causes some equipment to draw higher-than-normal load current. For constant wattage loads, load current increases to make up the difference from low voltage to maintain the power output. Consequently, low voltage can cause motors and certain equipment and components to run hotter than normal, and can cause components to fail prematurely. In addition, significant voltage drop across phase and neutral conductors can result in incorrect operation of computers and other sensitive electronic equipment. Manufacturers of air-conditioning and refrigeration equipment, fire pumps, submersible pumps, and many other types of motor-driven equipment specify acceptable ranges of operating voltages. During startup, motors typically draw several times their rated operating currents. Sizing conductors based only on operating currents may not provide sufficient voltage to allow motors to start. Sensitive electronic equipment is also particularly susceptible to mal-operation or failure if voltage drop is excessive. In electrical wiring circuits, voltage drops also occur from the distribution board to the different sub circuit and final sub circuits, but for sub circuits and final sub circuits, the value of voltage drop should be half that of allowable voltage drops (i.e., 2.75 V of 5.5 V in the above case). Normally, voltage drop in tables is described in ampere per meter (A/m). For example, what would be the voltage drop in a 1 meter cable carrying 1 A current? There are two methods to calculate the voltage drop in a cable. In SI, voltage drop is described by ampere per meter. In English, the voltage drop is computed per 100 feet cable length.

Example 4.21

If the supply voltage is 220 V, then the value of allowable voltage drop should be:

SOLUTION

$$\textit{Allowable Voltage Drop} = 220 \times \frac{2.5}{100} = 5.5 \text{ V}$$

Current flowing through a conductor with a finite resistance causes a voltage drop over the length of the conductor. Voltage drop can be calculated using a variation of the familiar Ohm's law (E = IR), as follows:

$$VD = kRi(t) \qquad (4.27)$$

Where *VD* is the voltage drop across the length of the conductor in volts, *k* is a constant depending upon whether the system is single-phase or three-phase, *i(t)* is the load current flowing through the conductor in amperes, and *R* is the direct-current resistance of uncoated copper conductors at 75°C (167°F). R is calculated by multiplying ohms per 1000 feet (the value given in Table B.2-I of the Appendix adapted from NEC Chapter 9, Table 8) by the one-way circuit length in thousands of feet. For single-phase loads, the constant *k* is 2. For three-phase loads, the constant *k* is the square root of 3, or 1.732. For simplicity, the examples in this chapter use direct current resistance *R* values, as shown in NEC Table 8, Chapter 9. However, if the geometrical dimensions and material of the conductor are known, Equation (4.11) can be used. For alternating current systems that are found in practice, the values are slightly different, and for loads that are not purely resistive, impedance value *Z* should be used, taking power factor into consideration. This is also covered in NEC Chapter 9. In AC circuits, the cable voltage drop, with the parameter of Figure 3.5, is calculated by using Equations (3.4) and (3.5) by:

$$VD_{AC} = I \times Z_{cable} = I \times (R \cdot \cos\theta + X_L \cdot \sin\theta) \qquad (4.28)$$

The cable resistance and inductance are often provided by manufacturers or can be found from tables in reference books or by using appropriate relationships. The conductor resistance can be estimated by one of the previous resistance equations, and similar for the cable leakage (inductance) reactance. For example, for three-phase cable, consisting of three insulated conductors, having a diameter *D*, and the bare conductor diameter *d*, the preface leakage reactance per phase is estimated by:

$$X_{1-phase} = 52.9 \frac{f}{60} \log_{10}\left(\frac{d}{D}\right), \text{ in } m\Omega/\text{ft/phase} \qquad (4.29)$$

4.3.6 CABLE CONSTRUCTION

The basic characteristics of a cable's physical construction include: conductor materials, shapes, types, surface coating, insulation types, number of conductors, installation conditions and arrangements (underground or aboveground, cable bunching and spacing), ambient temperature, soil characteristics, depth of laying, and so on. Conductor materials normally used are copper or aluminum, while the conductor shape can be circular, rectangular, or square. The most common conductor types are stranded or solid types. Conductor surface coating can be plain (no coating), tinned, silver, or nickel, while the most-used insulation types are PVC, XLPE, or EPR. The number of conductor cores includes single-core or multicore (e.g., 2, 3, or 4 conductors) types. Installation conditions refer to how the cable will be installed, which includes above the ground or underground cables. Installation arrangements refer, for example, for underground cables, if they are directly buried or buried in the conduit, while for the aboveground cables, if it is installed on cable tray, conduit, or ladder, against a wall, in air, etc. Ambient or soil temperature of the installation site is important for cable sizing, ampacity calculations, and sizing the protective devices. Cable bunching is the number of cables that are bunched together, without affecting the cable performances and characteristics. Cable spacing refers to whether cables are installed touching or are spaced. Soil thermal resistivity is an important factor for the calculation of the characteristics of underground cables.

Cable selection is based on the current rating estimates. Current flowing through a cable generates heat through the resistive losses in the conductors, dielectric losses through the insulation, and resistive losses from current flowing through any cable screens/shields and armoring. The component parts that make up the cable (e.g., conductors, insulation, bedding, sheath, armor) must be capable of withstanding the temperature rise and heat emanating from the cable. The current-carrying capacity of a cable is the maximum current that can flow continuously through a cable without damaging the cable's insulation and other components (e.g., bedding, sheath). It is sometimes also referred to as the continuous current rating or ampacity of a cable. Cables with larger conductor cross-sectional areas (i.e., more copper or aluminum) have lower resistive losses and are able to dissipate the heat better than smaller cables. Therefore, a 16 mm^2 cable will have a higher current-carrying capacity than a 4 mm^2 cable. International standards and manufacturers of cables will quote base current ratings of different types of cables in tables such as the one shown on the right. Each of these tables pertains to a specific type of cable construction (e.g., copper conductor, PVC insulated, voltage grade) and a base set of installation conditions (e.g., ambient temperature, installation method). It is important to note that the current ratings are valid for only the quoted types of cables and base installation conditions. In the absence of any guidance, the following reference-based current ratings may be used. In general, there are several ways to construct a cable and no one standard to which all vendors are adhering. Most cables tend to have common characteristics, and from an electrical system design point of view, cables are divided into the following main categories: (a) low-voltage power and control cables pertain to electrical cables, (b) low-voltage instrumentation cables pertain to cables for use in instrument applications, and (c) medium- or high-voltage cables pertain to cables used for electric power transmission at medium and high voltage.

Conductor consists usually of stranded copper (Cu) or aluminum (Al). Copper is denser and heavier, and makes a better conductor than aluminum. Electrically equivalent aluminum conductors have a cross-sectional area approximately 1.6 times larger than copper, but are half the weight (which may save on material cost) and lower support cost, also. *Annealing* is the process of gradually heating and cooling the conductor material to make it more malleable and less brittle. *Coating*, or the surface coating (e.g., tin, nickel, silver, lead alloy) of copper conductors, is common to prevent the insulation from attacking or adhering to the copper conductor and prevents copper deterioration at higher temperatures. Tin coatings were used in the past to protect against corrosion from rubber insulation, which contained traces of the sulfur used in the vulcanizing process. *Conductor screen* consists of a semiconducting tape to maintain a uniform electric field and minimize electrostatic stresses (for MV/HV power cables). *Insulation* is a critical characteristic of cables used in electrical systems. Commonly used materials are thermoplastic (e.g., PVC) or thermosetting (e.g., EPR, XLPE). Mineral insulation is sometimes used, but constructing such types of cables is entirely different from constructing normal plastic/rubber insulated cables. Typically used are a thermosetting (e.g., EPR, XLPE) or paper/lead insulation for cables under 22 kV. Paper-based insulation in combination with oil- or gas-filled cables is generally used for higher voltages. *Plastics* are one of the more commonly used types of insulating materials for electrical conductors. Plastics have good insulating, flexibility, and moisture-resistant qualities. Although there are many types of plastic insulating materials, thermoplastic is one of the most common. With the use of thermoplastic, the conductor temperature can be higher than with some other types of insulating materials without damage to the insulating quality of the material. Plastic insulation is normally used for low- or medium-range voltage. The designators used with thermoplastics are much like those used with rubber insulators. The following letters are used when dealing with NEC-type designators for thermoplastics: T, for thermoplastic, H, for heat-resistant, W, for moisture-resistant, A, for asbestos, N, for outer nylon jacket, and M, for oil-resistant.

Paper has little insulation value alone. However, when impregnated with a high grade of mineral oil, it serves as a satisfactory insulation for extremely high-voltage cables. The oil has a high dielectric strength, and tends to prevent breakdown of the paper insulation. The paper must be thoroughly saturated with the oil. The thin paper tape is wrapped in many layers around the conductors, and then soaked with oil. *Enamel,* the wire used on the coils of meters, relays, small transformers, motor windings, and so forth, is called magnet wire. This wire is insulated with an enamel coating. The enamel is a synthetic compound of cellulose acetate (wood pulp and magnesium). In the manufacturing process, the bare wire is passed through a solution of hot enamel and then cooled. This process is repeated until the wire acquires from six to ten coatings. Thickness for thickness, enamel has higher dielectric strength than rubber. It is not practical for large wires because of the expense and because the insulation is readily fractured when large wires are bent. *Mineral-insulated (MI) cable* was developed to meet the needs of a noncombustible, high heat-resistant and water-resistant cable. MI cable has from one to seven electrical conductors. These conductors are insulated in a highly compressed mineral, normally magnesium oxide, and sealed in a liquid-tight, gas-tight metallic tube, normally made of seamless copper. *Silk and cotton* are used in certain types of circuits (e.g., communications circuits), where a large number of conductors are needed. Because the insulation in this type of cable is not subjected to high voltage, the use of thin layers of silk and cotton is satisfactory. Silk and cotton insulation keeps the size of the cable small enough to be handled easily. The silk and cotton threads are wrapped around the individual conductors in reverse directions. The covering is then impregnated with a special wax compound for extra protection and strength.

Insulation screen represents a semiconducting material that has a similar function as the conductor screen (i.e., control of the electric field for MV/HV power cables). *Conductor sheath* represents a conductive sheath/shield, typically of copper tape or sometimes lead alloy, used as a shield to keep electromagnetic radiation in, and also to provide a path for fault and leakage currents (sheaths are earthed at one cable end). Lead sheaths are heavier and potentially more difficult to terminate than copper tape, but generally provide better earth fault capacity. *Filler* refers to the interstices of the insulated conductor bundle sometimes being filled, usually with a soft polymer material. *Bedding* or *inner sheath*, typically a thermoplastic (e.g., PVC) or thermosetting (e.g., CSP) compound, is included to keep the bundle together and to provide a bedding for the cable armor. *Individual screen*, typically for instrument cables, is an individual screen occasionally applied over each insulated conductor bundle for shielding against noise/radiation and interference from other conductor bundles. Screens are usually a metallic (copper, aluminum) or semimetallic (PETP/Al) tape or braid. These are typically used in instrument cables, but not in power cables.

Drain wires (instrument cables) are those that have an individual screen that has an associated drain wire, designed to assist in the termination of the screen, and are typically used in instrument cables, not in power cables. *Overall screens* (instrument cables) can be applied over all the insulated conductor bundles for shielding against noise/radiation, interference from other cables, and surge/lightning protection. Screens are usually a metallic (copper, aluminum) or semimetallic tape or braid. These are typically used in instrument cables, but not in power cables. *Armor* is the cable part designed for mechanical protection of the conductor bundle. Steel wire armour or braid is typically used. Tinning or galvanizing is used for rust prevention. Phosphor bronze or tinned copper braid is also used when steel armor is not allowed. The two categories of cable armor are: (a) steel wire armor, typically used in multicore cables (magnetic), and (b) aluminum wire armor, typically used in single-core cables (nonmagnetic). When an electric current passes through a cable, it produces a magnetic field (the higher the voltage the stronger the field), which induces an electric current in steel armor (eddy currents), which can overheat the AC systems. The nonmagnetic aluminum armor prevents this from happening. An *outer sheath* is applied over the armor for overall mechanical, weather, chemical, and electrical protection. Materials typically used are a thermoplastic (e.g., PVC) or a thermosetting (e.g., CSP) compound, and often the same material is used as the bedding. Outer sheath is normally color-coded to differentiate between LV, HV, and instrumentation cables. Manufacturer's markings and length markings are also printed on the outer sheath. *Termite protection* is used for underground cables, consisting of a nylon jacket that can be applied for termite protection, although sometimes a phosphor bronze tape can be used.

As discussed, a multiconductor cable is an assembly of two or more conductors having an outer jacket, for holding the conductors and easy installation. Common examples are the service entrance cable and nonmetallic sheathed cable. Type AC armored cable consists of insulated conductors contained in a flexible metal raceway. There are several variations with respect to conductor size and numbers. For example, an armored cable assembly is available with two, three, or four insulated conductors; solid or stranded; and in sizes from #14 AWG to #2 AWG. NEC Article 320 permits this type of cable to be installed in dry areas, embedded in masonry or plaster, or running to the masonry hollows. Its outer armor may be used for equipment grounding. A metal-clad (MC) cable consists of three or four individual insulated conductors with one or more grounding conductors grouped together and enclosed in an outer sheath. Variations include cables containing power and control conductors in the same sheath and cables used for #8 AWG or larger for easy installation. Fillers are used between conductors to keep the conductors in place within the cable. Article 320 permits MC cables to be installed indoors or outdoors, in wet (not submerged) or dry locations, as cable open runs (not exceeding 6 feet), in cable tray, in raceway or conduit, or directly buried where listed for direct burial.

The nonmetallic (NC), sheathed cable is the most used for residential branch circuits, being available with two or three insulated conductors, and with or without a bare copper ground, jacketed with an outer polyvinylchloride sheath. It is usually rated at 90°C with a nylon jacket, having common sizes from #14 AWG to #2 AWG. NEC Article 334 restricts NM cables to be used in any structure more than three floors above grade, for service entrance cable, in commercial garages, or embedded in poured concrete. Underground feeded (UF) cable has a similar construction to the NM cable, being used in residential applications to supply lamp posts, pumps, or other similar outdoor equipment, as discussed in Article 340. Variations of NM cables are permitted to be used in dry, damp, moist, or corrosive locations, or consist of assembly, containing power, signaling, and other communication conductors in the same sheath. Service entrance cables, usually identified as SE, USE, SEU, or SER generally consist of two or three insulated conductors wrapped by a concentrically neutral bare, jacketed with a PVC outer covering. NEC Article 338 describes such cables, while Article 230 discusses service entrance requirements. Regarding cable insulation, the two major types of insulation used for buildings are thermoplastic and thermoset. The thermoplastic type begins to melt at temperatures higher than rated; common examples are polyethylene and polyvinylchloride. Thermoset insulations, such as rubber, do not begin to melt at temperatures reasonably higher than rated; however, they do deteriorate more quickly at these higher temperatures. As we mentioned, heat produced by the conductor currents is transmitted to the insulation, so the conductor and the insulation temperatures are basically the same. Temperature ratings typically assigned to conductor insulations are: 60°C, 75°C, or 90°C. Operation above these rated temperatures causes premature insulation degradation. Table 4.18 summarizes the most common cable designations used in building electrical systems.

TABLE 4.18
Cable Insulation Types and Designations

Identifier	Designation
THW-2	Thermoplastic insulation (PVC), heat resistant (90°C), wet locations
THHN	Thermoplastic insulation (PVC), high heat resistant (90°C), dry locations, nylon jacket
XHHW-2	Cross-linked polyethylene insulation (X), high heat resistant (90°C), wet and dry locations
RHH	Rubber insulation, high heat resistant (90°C), dry locations
RHW-2	Rubber insulation, heat resistant (90°C), wet locations
NM-B	Nonmetallic sheathed cable, ampacity limited at 60°C, thermoplastic, overall PVC insulation, nylon jacketed
SEU	Service entrance cable, unarmored, usually XHHW type, rated at 90°C for dry locations, and 75°C for wet locations

4.4   WIRING DEVICES

This section describes the various types of wiring devices used in power distribution to control the flow of electrical current, such as switches, receptacles, and disconnects. These devices allow the operation of other devices and equipment or are used to control and isolate specific equipment. Various types of wiring devices used in power distribution networks include switches, conductors, cables, receptacles, and disconnects. They are used to control the flow of electrical current and electricity and to allow for desired operation of other devices and equipment. The types and ratings of wiring devices are discussed in this section. Disconnect switches can be used to control and isolate electric equipment. A wiring (connection) diagram is a diagram that shows the connection of an installation or its component devices or parts. It shows, as closely as possible, the actual location of each component in a circuit, including the control circuit and the power circuit. A line (ladder) diagram is a diagram that shows the logic of an electrical circuit or system using standard symbols, helping in operation, maintenance, or restructuring of the actual electric circuits in a building or industrial facility. A line diagram is used to show the relationship between circuits and their components but not the actual location of the components. Line diagrams provide a fast, easy understanding of the connections and use of components. Selecting the correct cable for the application is imperative to ensure a satisfactory life of conductors and insulation subjected to the thermal effects of carrying current for prolonged periods of time in normal service. Choosing the minimum size cross-sectional area of conductors is essential to meet the requirements for protection against electric shock, protection against thermal effects, overcurrent protection, voltage drop, and limiting temperatures for terminals of equipment to which the conductors are connected. A properly engineered and installed cable tray wiring system provides some highly desirable safety features that are not obtainable with a conduit wiring system.

4.4.1   Switches

An electrical switch is a device for making or breaking an electrical circuit. The definition suggests the ultimate in simplicity, that a switch need be no more than the bare ends of two wires that can be touched to make a circuit or separated to break a circuit. There are many different types of switches: toggle, rotary, pushbutton, rocker, pull-chain, slide, magnetic, mercury, timer, voice-activated, touch-sensitive, and many others. Each switch type has specific characteristics and features suitable for specific applications. The switches used in most residential and commercial applications, for controlling light fixtures, and occasionally to control small motor loads, are toggle switches. Most of these switches have two positions, and many of them are marked with the ON and OFF positions. Many toggle switches are self-contained with no externally visible switch contacts, with the exception of disconnect switches, while the switch arrangement is specified by the number of poles (the number of inputs to the switch) and by the number of throws (the number of outputs affected by the switch operation). Single-pole single-throw (SPST) is the most common switch arrangement for toggle switches, with a single conductor entering and leaving the switch. The switch has two states: ON or OFF. SPST switches are commonly used to control the switching operation of light fixtures from a single location. The double-pole single-throw (DPST) has two separate inputs that are switched to two separate outputs by its operation, so two separate outputs can be switched ON or OFF from two separate inputs, usually switching two ungrounded (hot) conductors in a circuit. Also known as a *three-way switch*, the single-pole double-throw (SPDT) switching arrangement is designed to switch a single input to two outputs. The double-pole double-throw (DPDT) has two distinct inputs connected to two distinct outputs. Figure 4.2 shows the diagrams of these toggle switches. The most common application of toggle switches is to connect a load to a source. Switches must be connected to an ungrounded conductor in order to comply with NEC requirements.

Three major terms designate a switch's functions: pole, throw, and break. *Pole* refers to the number of circuits that can be controlled by the switch. In the example below, the single-pole switch is

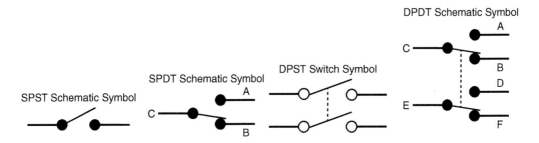

FIGURE 4.2 Toggle switches, contacts, and arrangements.

capable of interrupting the current in a single circuit. A double-pole switch, on the other hand, is capable of simultaneously interrupting the current in two separate circuits. The term *throw* indicates the number of conductors or paths the switch can control. In the example below, the movable contact member of the single-throw switch completes a circuit to only one conductor. However, a circuit to a double-throw switch permits its movable contact element to alternately complete two different paths. The term *position* refers to the number of stops the switch actuator will make when moved from one extreme position to the opposite position. For example, an on-none-off is a two-position switch and an on-off-on is a three-position switch. The term *break* is self-explanatory. It refers to the breaking or opening of a circuit. For example, single-break means that the contacts are separated at only one place. A double-break switch has two pairs of contacts that open the circuit at two places. A double-break switch provides greater volume of contact material, permitting greater heat dissipation and thereby longer switch life. The double-break switch also has twice the voltage breaking capacity, a desirable feature for DC circuit applications. Despite the great number of switches, these devices have a common denominator in basic components: the operator that initiates switch operation; the contacts, low-resistance metal that makes or breaks the electrical circuit; and the switch mechanism that is linked to the operator and opens and closes the contacts. The next type of switch is the double-pole single-throw (DPST). DPST switches are used when there are two live lines to switch but they can only turn on or off (single throw). These switches are not used much and are usually found in 240 V applications. A single-pole double-throw switch with a common terminal is the middle terminal in the SPDT knife switch; if you are using a household switch, it would be the brass-colored terminal (the other two would be silver-colored). This circuit clearly demonstrates what happens when the SPDT switch is moved back and forth. Light A goes on and B goes off, light B goes on and A goes off, and so forth. You can see that this popular switch has many practical applications: the transmit/receive button on a two-way radio, the high/low beam switch for car headlights, the pulse/tone dialing switch on your telephone, and so on. If you are using the SPDT knife switch, you have a "center off" position, which an ordinary wall switch does not have. In that case, you will need to add an SPST switch for shutting off this circuit. (In electronics work, many SPDT switches have a middle position in which the electricity is turned off to both circuits. It is an SPDT "center off" switch. Also, some electronic SPDT switches have a "center on" position. The best example of this type of switch is the pickup selector on an electric guitar, which can select the rhythm, treble, or both pickups for three varieties of sounds.)

4.4.2 RECEPTACLES

Receptacles are wiring devices designed to allow portable appliances and other systems, the so-called cord-and-plug connected equipment to be connected to a power supply. The majority of receptacles in commercial and residential structures are not supplying continuous loads, so it is quite difficult to predict the load size supplied by a receptacle unless it is dedicated (assigned to a specific purpose). Receptacle requirements are covered in NEC Section 406. Notice that the code does not require a minimum number of outlets for commercial buildings. However, many receptacles are usually required. Load connections to the power sources are done through a variety of receptacle configurations. The load ungrounded (hot) conductor is connected to the shorter receptacle slot. The cord plug that is connected must have the same configuration as the receptacle. Connections to the receptacles are usually made by screw terminal connections on the receptacle sides, in order to allow for feed-through receptacle wiring. Typical receptacle current ratings are 15 A and 20 A. These types of receptacles are grounded, and the requirement is not to use them if the circuit is not actually grounded. The equipment-grounding conductor terminal, green in color, is located near the receptacle top or bottom, while ungrounded conductor terminals are bronze in color and the grounded conductor terminals are silver in color. The National Electrical Manufacturers Association (NEMA) has established a standard slot-and-prong configuration for various ampere- and voltage-rated receptacles. To prevent accidental plug disconnection from the receptacle, in locking-type receptacles and plugs, after the plug is inserted in the receptacle, it is then slightly twisted to lock into place. Most receptacles are non-locking-type. Ground fault circuit interrupted (GFCI) receptacles are designed to minimize electrical shock hazards. A typical rating for a GFCI receptacle is 20 A feed-through (allowing standard receptacles to be used downstream) and 15 A through its own terminals. The main component of a GFCI receptacle is the current sensor that is monitoring the current through the ungrounded and grounded conductors. Under normal operation conditions, the currents through the ungrounded and grounded conductors are equal in magnitude, and the net sensor current is zero. When a fault occurs between ungrounded conductor and the ground, the load current returns to the source through the equipment ground, bypassing the ground fault sensor, the current through the sensor is unbalanced, resulting in opening of the GFCI contacts, de-energizing the load. The GFCI receptacles respond only to the load faults and short-circuit faults involving ground, and not to short-circuit between the ungrounded and grounded circuit conductors. A typical GFCI is tripping at a current of 5 mA in approximately 25 ms time range. Once the GFCI is tripped due to a fault or testing, the device must be reset. A GFCI receptacle can be used as a replacement for an ungrounded receptacle outlet.

Another special type of receptacle is the isolated ground receptacle, identified by an orange triangle on the faceplate designed to power sensitive electronic equipment, computers, or data processing systems by eliminating the ground loops, formed by the interconnections of equipment grounding conductors. In this way, the electromagnetic interference generated in these current loops is minimized. The receptacle mounting strap, metal boxes, and metal faceplate (if used) are grounded through a raceway system, by using a separate bare or insulated ground. Hospital-grade receptacles are heavy-duty receptacles used in areas where they may be subject to impact and sudden cord removal or insertion. This receptacle is identified by a green dot on the faceplate. Its internal contact action usually provides for multiple levels of wiping action between the plug prongs and contacts, improving the receptacle–plug conductivity and extending its service lifetime. Tamper-resistant receptacles, designed to prevent the insertion of small objects into the receptacle, usually have some form of sliding barrier incorporated into their design. They are usually installed in pediatric care areas, kindergartens, or schools. To protect sensitive electronic equipment against transient overvoltages, transient voltage surge suppressor receptacles are used. Transient overvoltages result from lightning strikes to power lines or switching surges. These receptacles come in a variety of forms and structures, often incorporating metal oxide varistors (MOV) to absorb the surge transient energy and divert it to the ground. Some of them can have light indicators to show when the suppression feature is functional, others have sound alarms that make noise when the surge suppression is no longer functioning. They are available in standard duplex, isolated ground, hospital-grade, and isolated ground hospital-grade receptacle types. In residential facilities, receptacles that are dedicated for stationary appliances where the receptacle is inaccessible, or for those located 2 m above the floor or finished grade are not required to be tamper-resistant.

Receptacles are wired to permit parallel load connections, so the rated voltage is applied across each load connected. The ungrounded (hot) conductor is connected to the terminals on receptacle side, while the grounded (neutral) conductor is connected to the other set of terminals. The grounding conductor is connected to the green grounding conductor. Duplex receptacles are often split-wired in order to allow for a load division between two circuits supplying the receptacles. The receptacles in this case may be supplied by two circuits from two pole-breaker circuits to ensure that both ungrounded conductors are de-energized in the event of short circuit on either of the two supplying circuits. The voltage between the two hot terminals of a split-wire duplex receptacle is 240 V for a single-phase three-wire system. The grounding conductor is pigtailed to each receptacle grounding terminal, and the neutral conductor is pigtailed to each receptacle terminal. If one receptacle is removed, the pigtail is disconnected from the receptacle, with continuity of the grounded conductor maintained to the other receptacle in the duplex. As previously stated, a load of 180 VA is assigned to each general-purpose receptacle, single, duplex, or triplex. The multi-outlet receptacle assemblies, such as plug strips, which are common in commercial applications, require that for each 5 ft. (or fraction thereof) of multi-outlet receptacle assembly, use 180 VA in your feeder/service calculations. This is assuming that it's unlikely for the appliances plugged into this assembly to operate simultaneously, as stipulated in Article 220.14(H). If we are expecting several appliances to operate simultaneously from the same multi-outlet receptacle assembly, consider each foot (or fraction of a foot) as 180 VA for feeder/service calculations. A multi-outlet receptacle assembly is not generally considered a continuous load.

4.4.3 DISCONNECT SWITCHES

Disconnect switches are used to connect or disconnect a load from the power supply. Compared with conventional switches, they are heavy-duty devices, having higher ratings than toggle switches. The most common applications are in motor circuits. They are usually enclosed in a metal housing, with an operating handle located on the right side of the enclosure. The disconnect switch contacts are usually of a knife-blade type, with both movable and stationary contact assembly. A switch contact mechanism is designed for quick make-and-break operation in order to minimize the arcing. Disconnect switches can be unfused or fused. The fuse-type disconnects have a fuse-holder in addition to the switch contacts, allowing the switch to accommodate appropriate rated fuses for overcurrent and short-circuit protection of controlled equipment. A rejection kit is also available from some manufacturers, to permit installation of current-limiting fuses only, to limit let-through fault current in short-circuit events. Physical construction of the last one differs from that of standard non-current-limiting fuses, making it impossible to install a non-current-limiting fuse in a fuse holder with the rejection feature. Disconnect switches are typically rated at 240 V for general-duty or 600 V for heavy-duty applications, with provisions that voltage ratings must equal or exceed the line-to-line system voltage. Common disconnect switches current ratings are 30 A, 60 A, 100 A, 200 A, 400 A, 600 A, 800 A, 1200 A, 1600 A, 2000 A, 4000 A, and 6000 A. A disconnect installed fuse rating must be equal or less to its rating. Notice that the physical fuse dimensions deem it impossible to insert it into a fuse holder with a rating lower than that of the fuse. In addition to the current ratings, disconnects also have horsepower ratings. The disconnect horsepower rating must exceed the horsepower rating of any motor connected to the load side of the switch.

The disconnect enclosure types, as specified in NEMA standards are: NEMA 1, NEMA 3R, NEMA 4, NEMA 4X, NEMA 12, and NEMA 13, with the first two being the most common. NEMA 1 is designated for indoor use, while the NEMA 3R is for general outdoor use, both providing protection against contact with internal energized parts, and against the falling dirt inside. In addition, NEMA 3R is waterproof under normal rainfall conditions, but not watertight, so it is unsuitable in applications where water spray may result. NEMA 4 and 4X protect not only against the physical contact with energized parts, but also falling dirt, dust, and wash-down with noncorrosive elements, while NEMA 4X also protects against corrosion-prone gasketed doors. Both may be used for indoor and outdoor applications. NEMA 12 and 13 enclosures protect against contact with energized parts, falling dirt, dust, and oil seepage, while NEMA 13 also has oil-spray-proof capabilities. Both are designed for indoor applications. The basic disconnect contact arrangements include: three-pole single-throw (3PST), four-pole single-throw (4PST), three-pole double-throw (3PDT), and four-pole double-throw (4PDT). The first two configurations can be used to connect three or four distinct outputs to three or four distinct inputs, while the last two can switch either three or four distinct inputs to three or four distinct outputs. For example, a 4PST switch can be used to disconnect three ungrounded circuit conductors and the neutral (grounded) conductor simultaneously. A 3PDT switch can be used to switch three hot conductors of a three-phase system simultaneously, while a 4PDT could be used to switch all three ungrounded conductors and neutral of a three-phase, four-wire system simultaneously. The most common applications are three-phase loads, while 3PDT and 4PDT switches are used to switch hot and neutral conductors of a three-phase system. 3PDT and 4PDT switches are often used in transfer switches applied to a system with ground fault protection. This type of switching arrangement is often used in conjunction with an emergency generator to switch from utility supply to the emergency supply. An application not usually found in the 4PST disconnect is a two-phase, four-wire system.

4.5   SUMMARY OF THE LOAD COMPUTATION PROCEDURE AND CABLE SIZING

For proper sizing of an electrical cable, the cable load bearing is very important to ensure that the cable can: operate continuously under full load without being damaged, withstand the worst short-circuit currents flowing through the cable, provide the load with a suitable voltage (and avoid excessive voltage drops), and optionally can ensure operation of protective devices during an earth fault. This calculation can be done individually for each power cable that needs to be sized, or alternatively, it can be used to produce cable sizing waterfall charts for groups of cables with similar characteristics (e.g., cables installed on ladder feeding induction motors). Cable sizing methods more or less follow the same basic five-step process, involving: (1) gather data about the cable, its installation conditions, the load that it will carry, etc.; (2) determine the minimum cable size based on continuous current-carrying capacity and the minimum cable size based on voltage drop considerations; (3) determine the minimum cable size based on short-circuit temperature rise; (4) determine the minimum cable size based on earth fault loop impedance; and (5) select the cable based on the highest of the sizes calculated in steps 2, 3, 4, and 5. The first step is to collate the relevant information that is required to perform the sizing calculation. Typically, you will need to obtain the following information about the load details. The characteristics of the load that the cable will supply include: load type (motor or feeder); three-phase, single-phase, or DC system/source voltage; the full-load current (A), actual or calculate this if the load is defined in terms of power (kW); full-load power factor; locked rotor or load-starting current (A); starting power factor; and distance/length of cable run from source to load. (This length should be as close as possible to the actual route of the cable and include enough contingency for vertical drops/rises and termination of the cable tails.)

The component parts that make up the cable (e.g., conductors, insulation, bedding, sheath, armor) must be capable of withstanding the temperature rise and heat emanating from the cable. The current-carrying capacity of a cable is the maximum current that can flow continuously through a cable without damaging the cable's insulation and other components (e.g., bedding, sheath). It is sometimes also referred to as the continuous current rating or ampacity of a cable. Cables with larger conductor cross-sectional areas (i.e., more copper or aluminum) have lower resistive losses and are able to dissipate the heat better than smaller cables. Therefore a 16 mm^2 cable has a higher current-carrying capacity than a 4 mm^2 cable. International standards and cable manufacturers quote the base current ratings of different cable types, in tabular format. Each of these tables pertains to a specific type of cable construction (e.g., copper conductor, PVC insulated, voltage grade) and a base set of installation conditions (e.g., ambient temperature, installation method). It is important to note that the current ratings are only valid for the quoted cable types and base installation conditions.

In the absence of any guidance, the following reference-based current ratings from standards or codes may be used. When the proposed installation conditions differ from the base conditions, derating (or correction) factors can be applied to the base current ratings to obtain the actual installed current ratings. International standards and cable manufacturers will provide derating factors for a range of installation conditions, for example ambient/soil temperature, grouping or bunching of cables, soil thermal resistivity, and so on. The installed current rating is calculated by multiplying the base current rating with each of the derating factors.

The main steps needed in the load service computation procedure are the following: First, determine the general lighting load based on the total square footage and space designation (NEC Table 220.12), or on the actual lighting fixtures if known, and select whichever is greater, as the NEC specifies. If the lighting is continuous, as is the case for most commercial structures, adjust the estimated value with 125%. For feeder and service loads, use NEC Table 220.42 and apply the lighting demand factor for that specific building type. Compute the heating and air-conditioning and discard the smaller value, as specify in NEC Section 220.60. Compute the load of the receptacle outlets, as specified in NEC Section 220.14, as for general-purpose ones, and 200 VA per show-window foot, 600 VA per heavy-duty lamp holders, and for others (Article 220.14). Apply demand factors as specified, add sign lighting load (1200 VA minimum), motor loads (if any), and finally, size the service and conductors.

Example 4.22

A three-story office structure (100′ × 60′) is supplied with 120/208 V service. Loads consist of 85 kVA air-conditioning, 105 kVA heating, 270 duplex receptacles, 30 linear feet of show window, 32 exterior lighting fixtures (at 175 VA each), 4 blower motors (at 3.5 A), and kitchen equipment of a total of 6.3 kW. Copper THW conductors are used. What is the conductor size?

SOLUTION

The total load is computed by types. For the motors the current is 3.5 A, and the line-to-line voltage is 1.73 × 208 = 360 V, three-phase service.

1. General lighting load: 3 × 100 × 60 × 3.5 = 63,000 VA
2. Show-window load: 30 × 200 = 6000 VA
3. Exterior lighting load: 32 × 175 = 5600 VA
4. Heating/air-conditioning load: 105,000 VA (keeping the largest of coincidental loads)
5. Receptacle load: 270 × 180 = 48,600 VA
6. Motor load: 4 × 3.5 × 360 = 5040 VA
7. Overall kitchen load: 6300 VA

Total load (sum of the individual ones): 239,540 VA. The total current is then:

$$I = \frac{P}{V_{LL}} = \frac{239540}{360} = 665.3 \approx 665 \text{ A}$$

From NEC Table 310.16, a 1750 kcmil THW cable is suitable for this application.

4.6   CHAPTER SUMMARY

Accurate demand load estimates on a particular branch circuit, feeder, or service are critical and necessary to determine the required ratings of these circuits. The actual and required number of branch circuits or feeders is also based on the estimated demand load determination. This chapter provides a comprehensive overview of load calculation and estimation methods, steps required for accurate load estimates for commercial, industrial, and residential facilities and buildings, cable sizing, and types and construction. The calculation of electrical loads to be served in a specific building, concepts of unit loads, load demand, load correction, and derating factors are also presented in detail. Several examples for load calculation to determine service entrance requirements for residential, commercial, and industrial facilities and occupancies are also included. The specification of the type, proper size, and insulation type for conductors supplying electrical loads is an important part of electrical system design. The cable ampacity is limited by the insulation type, ambient temperature, and operation conditions. The most important factor in conductor selection is that the conductor must be sized to safely carry the load currents, which are accurately calculated if the demand load is accurately estimated. Likewise, the overcurrent protection devices must also be rated to carry the load currents. Important sections of this chapter focus on cable and conductor construction, types, insulation characteristics, materials and types, cable voltage drop, ampacity estimates, derating and correction factors, and their applications in cable calculations, wiring devices, their characteristics, and applications. After completing this chapter, readers are able to estimate and compute demand load; apply demand factors; determine demand load for motor, equipment, and appliances; understand methods to calculate cable and conductor sizing and capacity, voltage drop calculations; and service entrance operation, parameters, and characteristics of wiring devices and their applications, and to develop an understanding and appreciation of the importance of codes and standards.

4.7 QUESTIONS AND PROBLEMS

1. Define the diversity factor and the load demand factor.
2. Define the net load demand and the total connected load.
3. Explain the importance of derating factors in load determination.
4. List the main factors that limit the cable ampacity.
5. What color insulation is used on equipment grounding conductor?
6. Define a single-conductor and a multiconductor cable.
7. Define continuous and noncontinuous loads.
8. How is the rating of a branch circuit estimated?
9. What are load types of a branch circuit?
10. Describe the most common types of receptacles.
11. What are the general types of branch circuit loads?
12. Describe the construction of (a) type AC cable, (b) type MC cable, and (c) type NM cable. For each type of cable, describe typical applications.
13. A cable with silicon rubber insulation has nominal ampacity of 300 A in 30°C ambient air. What is its ampacity if it is installed in a room with 60°C temperature?
14. What types of receptacles are used to connect sensitive electronic equipment?
15. What are the major disconnect applications?
16. What type of conductor extends beyond the final overcurrent protection device?
17. List the special receptacle types and their applications.
18. Explain the operation of a GFCI receptacle.
19. Calculate the minimum estimated lighting load for: (a) a school building with a total floor area of 36,000 ft.², being illuminated with 150 fluorescent light fixtures, each having a ballast input of 277 V, 0.8 A; and (b) an office area that is 125 ft. by 180 ft., illuminated with 120 fluorescent light fixtures, each having a ballast input of 277 V, 0.8 A.
20. An electric water heater is rated 120 V, 3.6 kW. Determine the estimated demand load and the heater-rated current.
21. Determine the branch circuit estimated load supplying both: (a) a 5 kW counter-mounted cooktop unit and a 6 kW wall-mounted oven; and (b) a 6.5 kW counter-mounted cooktop unit and a 9 kW wall-mounted oven.

22. Determine the demand load (VA) for the following motors: (a) single-phase 10 HP, 230 V induction motor; (b) single-phase 1.5 HP, 115 V induction motor; and (c) three-phase 100 HP, 460 V induction motor.

23. A hardware store has a total area of $100' \times 200'$, uses for storage an area of $100' \times 50'$, a small office of $15' \times 20'$, and the remainder of the building is used as showroom. There are also a total of $36'$ show windows, a $50'$ track lighting, and three outdoor signs. What is the total lighting load?

24. Determine the load on a 120 V, 20 A branch circuit that has eight general-use duplex circuits connected.

25. Determine the minimum estimated lighting load for a school that has a total floor area of 28,000 ft.^{2}, illuminated by 135 fluorescent light fixtures, each having a ballast of 277 V, 0.85 A.

26. Compute the receptacle load for a grocery store having 210 duplex and 90 single receptacles.

27. Compute the general lighting load of a large 3000 ft.^{2} hallway of a hotel, assuming continuous load.

28. What is the general lighting load for a 25-room motel, if each room is $15' \times 18'$?

29. Repeat Problem 22 for the following three-phase motors: (a) 30 HP, 208 V, induction motor; (b) 75 HP, 575 V, induction motor; and (c) 40 HP, 230 V synchronous motor.

30. Determine the resistance of a conductor of 200 ft. length at 25°C, made of: (a) copper; and (b) aluminum.

31. Determine the resistance of the cable of Problem 30 at 60°C.

32. A high school is supplied with 208/120 V, single-phase power. Calculate the total load, assuming continuous load. The school has the following dimensions and loads:
    a. 36,000 ft.^{2} classroom space
    b. 7500 ft.^{2} auditorium
    c. 5500 ft.^{2} cafeteria
    d. 12.5 kW outside lighting
    e. 250 duplex receptacles (120 V)
    f. Kitchen equipment: 15 kW oven, 12 kW range, 4.5 kW fryer, 10 kW water heater, and 3.6 kW dishwasher

33. A residential building has a living space of 3200 square feet. There are six small-appliance branch circuits supplying the kitchen and dining areas and two laundry circuits. There are a total of 24 duplex receptacles for miscellaneous use in the residence's unfinished areas. Determine the estimated load demand.

34. A 24-hour restaurant has 105 duplex receptacles and 45 single receptacles. Calculate the total load for feeder calculations.

35. Sketch the schematic wire diagrams, showing the connection of the ungrounded, grounded, and equipment grounding conductors and wiring devices for: (a) single lighting outlet controlled by two switches; (b) two lighting outlets controlled by two switches; and (c) two duplex receptacles with a receptacle of each duplex hot and the other receptacle outlet on each duplex switched simultaneously from one switching location.

36. Specify all situations in which a grounded conductor must be included among the current-carrying conductor for ampacity adjustment purposes.

37. A small retail store ($90' \times 50'$) is supplied with a single-phase 120/240 V service. Loads consist of: 45 kVA air-conditioning, 55 kVA heating equipment, 1 HP, 240 V ventilating unit, 90 duplex receptacles, 30 linear feet of show window, 1200 VA exterior lighting sign, and auxiliary equipment of a total of 4.2 kW. Copper THW conductors are used. What is the conductor size?

38. What is the ampacity of #1 AWG THW copper conductor in a raceway containing four current-carrying conductors, in a 41°C, dry location?

REFERENCES AND FURTHER READINGS

1. D. Maxwell and R. Van der Vorst, Developing sustainable products and services, *J. Clean. Prod.* Vol. 11, pp. 883–895, 2003.
2. D. T. Allen and D. R. Shonnard, *Green Engineering*, Prentice Hall, Upper Saddle River, NJ, 2003.
3. A. R. Daniels, *Introduction to Electrical Machines*, The Macmillan Press, 1976.
4. Y. Wallach, *Calculations and Programs for Power System Networks*, Prentice Hall, 1986.
5. A. C. Williamson, *Introduction to Electrical Energy Systems*, Longman Scientific & Technical, 1988.
6. V. Del Toro, *Basic Electric Machines*, Prentice Hall, Englewood Cliffs, NJ, 1985.
7. R. D. Schultz and R. A. Smith, *Introduction to Electric Power Engineering*, John Wiley & Sons, New York, 1988.
8. J. R. Cogdell, *Foundations of Electric Power*, Prentice Hall, Upper Saddle River, NJ, 1999.
9. S. J. Chapman, *Electrical Machinery and Power System Fundamentals* (2nd ed.), McGraw-Hill, 2002.
10. M. A. Pai, *Power Circuits and Electro-Mechanics*, Stipes Publishing L.L.C., Champaign, IL, 2003.
11. T. R. Bosela, *Electrical Systems Design*, Prentice Hall, 2002, ISBN-13: 9780139754753.
12. R. C. Dorf (ed.), *The Electrical Engineering Handbook*, CRC Press, Boca Raton, FL, 1993.
13. G. F. Moore, *Electric Cables Handbook* (3rd ed.), Blackwell Science, 1997.
14. W. A. Thue, *Electrical Power Cable Engineering*, Marcel Dekker, New York, 1999.
15. L. L. Grisby (ed.), *Electrical Power Engineering*, CRC Press, Boca Raton, FL, 2001.
16. F. P. Hartwell, J. F. McPartland, and B. J. McPartland, *McGraw-Hill's National Electrical Code 2017 Handbook* (29th ed.), McGraw-Hill, New York, 2017.
17. T. A. Short (ed.), *Handbook of Power Distribution*, CRC Press, Boca Raton, FL, 2014.
18. D. R. Patrick and S. W. Fardo, *Electrical Distribution Systems* (2nd ed.), CRC Press, Boca Raton, FL, 1999.
19. P. Schavemaker and L. van der Sluis, *Electrical Power System Essentials*, John Wiley and Sons, 2008.
20. M. E. El-Hawary, *Introduction to Power Systems*, Wiley – IEEE Press, 2008.
21. Y. Goswami and F. Keith (eds.), *Energy Conversion*, CRC Press, 2008.
22. M. A. El-Sharkawi, *Electric Energy: An Introduction* (2nd ed.), CRC Press, 2009.
23. L. James Kirtley, *Electric Power Principles: Sources, Conversion, Distribution and Use*, Wiley, 2010.
24. N. Mohan, *Electric Power Systems – A First Course*, Wiley, 2012.
25. F. M. Vanek, L. D. Albright, and L. T. Angenent, *Energy Systems Engineering*, McGraw-Hill, 2012.
26. M. R. Patel, *Introduction to Electrical Power Systems and Power Electronics*, CRC Press, 2013.
27. R. Fehr, *Industrial Power Distribution* (2nd ed.), Wiley, 2015.
28. J. E. Fleckenstein, *Three-Phase Electrical Power*, CRC Press, 2016.

5 Circuit Protection, Grounding, and Service

5.1 INTRODUCTION

Utility-scale generated and distributed electricity is the cheapest, cleanest, most convenient and safest form of energy used in our modern society. An *electric installation* is a combination of electrical equipment, devices, and circuits to perform a specific task or for a specific purpose having coordinated characteristics. In dealing with electrical installations, it is necessary to ensure safety of personnel and protection of equipment installation from electric faults or malfunctions. Misapplication, misuse, or accidents can unleash the frightening powers of electrical energy, invariably resulting in damages to both life and property through fault currents, electrical shocks, and fire hazards. The uncontrolled flow of electrical energy is generally the result of such misapplication, misuse, or accident and manifests itself in the form of electrical fault currents either in intended or unintended electrical circuits or paths. Electrical fault currents fall into three main categories: *overload currents*, *short-circuit currents*, and *earth fault currents*. Electrical power systems operate at various voltage levels from 120 V to 750 kV or even higher, making difficult the system protection job. The electrical apparatus and equipment used may be enclosed (e.g., generators and motors) or placed in open space (e.g., transmission lines or distribution transformers). All equipment and apparatus may undergo abnormalities, faults, and malfunctions for various reasons. For example, insulation can be degraded by pollution, weather, or misoperation of other components, leading to breakdowns or short-circuits. It is necessary to avoid these abnormal operations for equipment protection and for personnel safety. System protection is the art and science of detecting problems and malfunctions with system components and isolating these components while protecting the rest of the electric network and personnel. Protection systems and methods can be classified into equipment/apparatus protection and system protection. Equipment and apparatus protection deals with detection of a fault in the equipment or apparatus and consequent protection, which is further classified into transmission line and feeder protection, transformer protection, generator protection, motor protection, and busbar protection. Monitoring of system behavior, taking corrective measures to maintain synchronous operation, and protecting the power, operation, and protection system apparatus and equipment from harmful operating states is referred as *system protection*. System protection deals also with detection of system proximity to unstable operating regions and consequent control actions to restore stable operating points and prevent damage to equipment and circuits. System stability loss can lead to partial or complete system blackouts. Devices, such as under-frequency relays, out-of-step protection, islanding systems, rate of change of frequency relays, reverse power flow relays, and voltage surge relays, are used for system protection. Control actions associated with system protection may be classified into preventive or emergency control actions. Designing electrical equipment and circuits from a safety perspective is a critical design issue. Power system problems include short-circuits, abnormal conditions, overloads, and equipment and system component failures. The main objectives and the purpose of system protection include public protection, system stability improvement, damage minimization, and overload protection.

Changes in electric circuits can cause conditions that are potentially dangerous to the circuit itself or to people living or working near the circuits, requiring electric circuit protection. The conditions that require circuit protection are direct short-circuits, excessive currents, and excessive heat. Any wrong or misapplication, misuse, malfunction, or accident can release uncontrolled electrical energy, which almost invariably results in equipment and circuit damages, which can threaten lives and property through electrical shock and fire hazards. All of the conditions mentioned are potentially dangerous and require the use of circuit-protection devices and equipment. Circuit-protection devices are used to stop current flow or open the circuit. To do this, a circuit-protection device must *always* be connected in series with the circuit it is protecting. For example, when a circuit is overloaded, the insulation materials are damaged or even vaporized, and the insulation becomes brittle. Even brittle insulation has good electrical insulating properties, but the conductor movement can crack the insulation, and a fault can result. Electric faults occur in two ways: a ground fault (i.e., between a conductor and an enclosure) or a short-circuit fault (i.e., between two conductors). In summary, electrical fault currents fall into three main categories: overload currents, short-circuit currents, and earth fault currents. The newsworthy nature of the often-spectacular or tragic results of electrical accidents in the form of fire or electrocution has resulted in a proliferation of technical articles and papers pertaining to the protection against short-circuit currents and earth fault or shock hazard currents. On the other hand, while much has been written about the components and technologies for protection against overload current, the specification requirements for overload current protection remain clouded in confusion and contradiction.

The common types of faults in electric systems and equipment are: (a) *short-circuit faults* (phase-to-neutral and phase-to-phase faults), resulting in large current flows, which may damage wires, insulators, switches, and devices, due to overheating; and (b) *insulation failure* (fault between the phase conductor and non-current-carrying metallic parts of equipment), resulting in high voltages that may appear on equipment frames and may be dangerous to a person coming into contact with them. Faults are often unexpected events, causing malfunctions and interruptions to power systems. Production downtimes and high repair costs are often the consequences. Such effects can be minimized by good protection design of the individual devices, equipment and with coordinated protection of device groups. Therefore, in all electrical circuits, the equipment associated with the electrical wiring systems must be protected to: (a) prevent damage by fires or electro-shocks, (b) maintain supply and service continuity, (c) disconnect the faulty component from the remainder of the system, (d) prevent damages to wiring and equipment, and (e) minimize system losses and interruptions. Each protection function must be set to ensure the best possible power system operation in all conditions. The best settings are the result of complete analysis based on the detailed characteristics of various elements in the installation, usually carried out by specialized software tools that analyze power or industrial systems during faults, and provide the settings for each protection function. One common fault is overcurrent, caused by equipment overloads, short-circuits, or ground faults. An overload occurs when equipment is subjected to current above its rated capacity and excessive heat is produced. A short-circuit occurs when there is a direct unintended connection between conductors. Short-circuits can generate temperatures thousands of degrees above designated ratings. A ground fault occurs when electrical current flows from a conductor to an uninsulated equipment part that is not designed to conduct electricity. There are many overcurrent protection devices, and their uses must meet code and standard requirements to ensure that the equipment and conductors within a system are protected against abnormal currents that can produce destructive temperatures above specified rating and design limits.

Increasing demands for high-quality, efficient electric service in buildings and industrial facilities have led to increasingly complex power systems. At the same time, the requirements for safety and availability are increasing because equipment, device, component, or entire-system failure can result in significant costs. A well-designed safety plan for individual circuits, devices, or entire systems makes a significant contribution to higher operational reliability. This also includes the selection of a robust power supply and suitable protection systems that safely protect against short-circuit and overload currents. It is advantageous to provide protection for each device or equipment, so only those affected by abnormal currents shut off. Protective equipment must possess the following features: (a) operation certainty and reliability under fault conditions, (b) non-operation under normal conditions, (c) rapid operation response and discrimination, (d) simplicity and lower costs (initial and maintenance), and (e) easy adjustment and testing. Some common protection methods include: equipment earthing or grounding, fuses or circuit breakers, and use of earth leakage and residual-current circuit breakers. The protection scheme function is to ensure the maximum continuity of power supply, which is achieved by determining the fault location and disconnecting the minimal circuitry and equipment needed to clear it. When a fault occurs, several relays may detect the fault; however, only those directly associated with the faulty equipment are required to operate. A protection device protects the circuit, components, connectors, and wires from damages caused by the excessive current flow (e.g., short-circuits or overcurrents). Selection of the most suitable protection devices, coming in different shapes, current ratings, or types, must be carefully made. Relays, which are discriminating by time, protect not only the associated equipment, but they also can act as backup protection for other relays. The major disadvantage of such a protection scheme is that there is a time delay in the fault removal, tending to increase the damage to the faulty equipment and the possibility of damage to healthy equipment, by carrying the fault current for a longer time. In contrast to relays, unit protection protects only the equipment with which it is associated. The basic principle of unit protection is that if the current entering and leaving the unit is the same, then it is healthy; if there is a difference then the unit is faulty. Relays, discriminating by magnitude, can be used only where there is a large change in the fault current (e.g., the transformer primary and secondary windings). Discrimination by time is the basis for many simple protection devices, where time delays are inversely proportional to current level. These schemes are applied to low-voltage (LV) systems as an integral part of the circuit breaker, where it is uneconomical to provide protection relays. System grounding is designed to isolate faulted sections of the electrical system by selective relaying of ground faults. Sensitivity and time delays of protective circuitry can be adjusted, so that a fault in an area causes a local breaker to sense the malfunction and quickly remove power from only the affected section. The relaying system must be arranged so that even at the lowest level of the power-distribution chain, sufficient fault current flows to enable the protective circuitry to sense the fault and take remedial actions.

5.2 SYSTEM, CIRCUIT, EQUIPMENT, AND DEVICE PROTECTION

To understand system and equipment protection methods and approaches, knowledge of the power system operation and behavior is critical. Power system behavior is described in terms of differential and algebraic equations, used to describe the behavior of generators, transmission lines, motors, transformers, feeders, etc. The depth of details depends upon the time scale of investigation. The dynamics involved in switching, lighting, load rejection, etc., have high-frequency components that die down quickly. In analysis of such dynamics, differential equations associated with inductances and capacitances of transmission lines are employed. Such analysis is restricted to a few cycles. At a larger time scale (order of seconds), the response of the electromechanical elements is perceived. These transients are typically excited by faults that disturb the system equilibrium by upsetting the system generator-load balance. As a consequence of a fault, the electrical power output reduces instantaneously while the mechanical input does not change instantaneously. The resulting imbalance in power (and torque) excites the electromechanical transients, which are essentially slow because of the inertia of the mechanical elements. Detection and removal of fault is the task of the protection system (equipment and apparatus protection). Post-fault, the system may or may not return to an equilibrium position. Transient stability studies are required to determine the post-fault system stability. In practice, out-of-step relaying, under-frequency load shedding, islanding, etc., are measures used to enhance system stability and prevent blackouts. However, the distinction between system protection and control is a fine one.

Over current protection or short-circuit protection is very important on any electrical power system, and the distribution system is no exception. The four major components of any protection system are *instrument transformers*, *protective relays*, *circuit breakers and fuses*, and *power control circuits*. Circuit breakers and reclosers, expulsion fuses, and current-limiting fuses interrupt the fault current, a vital function in the power network operation. Current and voltage transformers supply input signals to protective relays. Protective relays provide a wide range of protection functions, including but not limited to short-circuit protection. When tripped by protective relays, the circuit breakers interrupt the fault current to isolate the affected zone from the rest of the power system. In high-voltage applications, the differential and directional comparison schemes, as well as the under-reaching distance and overcurrent elements, provide instantaneous protection against short-circuits. Short-circuit protection is the equipment selection and placement, selection of the settings, and coordination of devices to efficiently isolate and clear faults with minimal impacts on customers and the power system. Protection units monitor the electrical status of power system components and de-energize them, when they are the site of a serious disturbance such as short-circuits or insulation faults. The choice of a protection device is not the result of an isolated study, but rather one of the most important steps in the design of the power system. Based on an analysis of the behavior of electrical equipment (e.g., motors, transformers) during faults and the phenomena produced, this guide is intended to facilitate your choice of the most suitable protective devices. Among their multiple purposes, protection devices: contribute to protecting people against electrical hazards; avoid damage to equipment (a three-phase short-circuit on medium-voltage busbars can melt up to 50 kg of copper in 1 second and the temperature at the center of the arc can exceed 10,000°C); limit thermal, dielectric, and mechanical stress on equipment; maintain stability and service continuity and quality in power systems; and protect adjacent installations (e.g., by reducing induced voltage in adjacent circuits or equipment). In order to attain these objectives, a protection system must be fast and reliable and ensure discrimination. However, protection has its limits—faults must first occur before the protection system can react. Protection systems and devices therefore cannot prevent disturbances, only limit their effects and duration. Furthermore, the choice of a protection system is often a technical and economic compromise between the availability and safety of the electrical power supply. The basic idea of an overcurrent protective device is to make a weak link in the circuit. In the case of a fuse, the fuse is destroyed before another part of the system is destroyed. In the case of a circuit breaker, a set of contacts opens the circuit. Unlike a fuse, a circuit breaker can be reused by reclosing the contacts. Fuses and circuit breakers are designed to protect equipment and facilities, and in so doing, they also provide considerable protection against shock in most situations. However, the only electrical protective device whose sole purpose is to protect people is a ground-fault circuit interrupter (GFCI).

Electrical protective devices include fuses, circuit breakers, and GFCIs; they are critically important to electrical safety. *Overcurrent devices* should be installed where required. They are sized and selected to interrupt current flow when it exceeds the capacity of the conductor. Proper selection takes into account not only the conductor capacity, but also the rating of the power supply and potential short-circuits. An overcurrent protection device is making a weak link in the circuit, e.g., a fuse is destroyed before another system component is destroyed, or in the case of a circuit breaker the contacts opened preventing current flow. For example, a branch-circuit protection device is intended to be the weakest link in the branch circuit. Rating protection devices are provided by the manufacturers, and protection devices are selected in agreement with their ratings. Short-circuits can produce enough thermal and electromagnetic forces to destroy any protective device or even the whole equipment. When selecting a protective device, it is very important to consider the available short-circuit amperage (SCA), which is the potential amperage at any system site. SCAs are measured at the equipment terminals, the utility transformer, and the distribution panel. The highest values are at the power transformer. The material conductivity, its size and length reduce the SCA down the line from the transformer. The correct size overcurrent protection device is chosen when the system SCA values are accurately estimated. Fuses and circuit breakers are assigned an amperage-interrupting capacity, indicating the SCA values sustained before tripping. Unless otherwise designated, fuses and circuit breakers are rated as specified by standards and codes. Protection devices are rated to manage both the normal maximum load and the potential short-circuit currents at any system section. Equipment controls should have a short-circuit rating that enables them to absorb current while the protective device clears the faulty circuit. If the rating is lower, a fast-clearing fuse with a lower rating than the controller should be used. In summary, the protection device's function is to prompt removal from the service of faulty electrical system components, devices, or equipment, protecting the rest of the power system, other devices, and equipment from damage and electrical instabilities. Every item of electrical equipment must have some form of electrical protection, disconnecting the equipment or section of the equipment from the electrical power in the event of its becoming faulty or overloaded.

5.2.1 Fault Types and Currents: Terminology and Definitions

Overload currents and *short-circuit currents* are unexpected, often causing malfunctions and interruptions to the system or equipment ongoing operations, with production downtimes and high repair costs. Fault effects are minimized by device protection and through appropriate coordination with the protection device groups. System areas not in the affected circuit(s) can continue to operate without interruption, whenever the overall process allows, ensuring higher system availability. The different nominal load currents illustrate the usefulness of separate protection for individual circuits. Suitable device circuit breakers are available for every nominal current level. *Overload currents* occur if terminal devices unexpectedly require higher current than the rated current. Such situations may arise, for example, due to a blocked drive or motor load malfunction. Temporary starting currents from machines are also considered overload currents. The occurrence of such currents can be calculated, but nonetheless varies depending upon the load status at the moment it starts. When selecting suitable fuses or circuit breakers for such circuits, these conditions should be taken into account. A safe shutdown should be carried out in a matter of seconds to lower-minute ranges. Damage to the conductor insulation and contact between conductors can cause *short-circuits*. A *short-circuit* is the accidental or intentional connection, by a relatively low resistance or impedance, of two or more points in a circuit that are normally at different voltages. Typical protective devices for short-circuits include fuses and circuit breakers with various tripping mechanisms. Short-circuit currents should be shut down safely and reliably in the millisecond range.

Selecting the right protective devices for protecting circuits and loads ensures the safe, optimized operation of electrical systems, even in the event of a fault. *Tripping characteristics* provide essential information for determining the protective device's suitability for a particular application, indicating the operating range of current-limiting protective devices in a current vs. time characteristic curve. The width or tolerance of the operating range depends on the type of protective device. The *internal resistance* of a protective device also has an influence on the characteristic curve. It is specified either as a resistance value or as a voltage drop over the internal resistance. In principle, you should try to achieve a very low internal resistance. The power dissipation in the circuit breaker then drops and it is better suited for use in circuits with low nominal voltage. However, in comparison, the tripping characteristic shifts slightly to the right. This results in a slightly delayed tripping time. *Overload* represents the operating conditions in an electrically undamaged circuit that cause an overcurrent. *Residual current* (I_Δ) represents the vectorial sum of the currents flowing in the main circuit of the circuit breaker. A *rated operational voltage* of a piece of equipment is the voltage which, combined with a rated operational current, determines the application of the equipment to which the relevant tests and the utilization categories are referred. *Rated insulation voltage* (U_{IS}) of equipment is the value of voltage to which dielectric tests voltage and creepage distances are referred. In no case shall the maximum value of the rated operational voltage exceed that of the rated insulation voltage. *Rated impulse withstand voltage* (U_{imp}) represents the peak value of an impulse voltage of prescribed form and polarity, which the equipment is capable of withstanding without failure under specified conditions of test and to which the values of the clearances are referred. *Rated frequency* is the supply frequency for which equipment is designed and to which the other characteristic values correspond. *Rated uninterrupted current* (I_{RUC}) is the circuit-breaker-rated uninterrupted current value that the circuit breaker can carry during uninterrupted service. *Rated residual operating current* (I_{RROC}) is the RMS value of a sinusoidal residual operating current assigned to the circuit breaker by the manufacturer, at which the circuit breaker operates under specified conditions. To a fuse or circuit breaker, *ground-fault current* is sensed just as any other current. If the ground-fault current is high enough, the fuse or circuit breaker responds before the ground-fault relay, depending on the ground-fault relay setting, overcurrent device characteristics, speed of response of the overcurrent device, and ground-fault current magnitude. Therefore, when analyzing ground-fault protection, it is necessary to study the characteristics of the ground-fault relay and overcurrent protective device in combination.

A circuit protection device operates by opening and interrupting current to the circuit. The *rating* of a circuit breaker or other protection device denotes its capabilities under specified conditions of use and behavior. The *capabilities* of a circuit breaker or a protection device are proved by conducting type tests as per the recommendations of the standards and codes. The opening of a protection device shows that something is wrong in the circuit and should be corrected before the current is restored. When a problem exists and the protection device opens, the device should isolate the faulty circuit from the other unaffected circuits, and should respond in time to protect unaffected components in the faulty circuit. The protection device should NOT open during normal circuit operation. Protection against temperature is termed *overcurrent protection*. Overcurrents are caused by equipment overloads, by short-circuits, or by ground faults. An overload occurs when equipment is subjected to current above its rated capacity and excessive heat is produced. A short-circuit occurs when there is a direct but unintended connection between line-to-line or line-to-neutral conductors. Short-circuits can generate temperatures thousands of degrees above designated ratings. A ground fault occurs when electrical current flows from a conductor to uninsulated metal that is not designed to conduct electricity. These uninsulated currents can be lethal. Coordination is the act of isolating a faulted circuit from the remainder of the electrical system, thereby eliminating unnecessary power outages. However, the term *coordination* is sometimes interpreted to mean a *degree of coordination* where more than one protective device is allowed to open under a given short-circuit condition. Therefore, the term *selective coordination* or *selectivity* means positive coordination over the entire range of possible fault currents, assuring that the faulted circuit is cleared and that other parts of the system are not affected. Circuit breakers are selected based on the nominal voltage, nominal current, and, if required, the starting current of a terminal device. In addition, the shutdown behavior of the circuit breaker must correspond to the expected error situations. There are differences in the error situations that involve short-circuits and those involving overload conditions.

5.2.2 Fuses

A fuse is the simplest form of overcurrent protective device, but it can be used only once before it must be replaced. It derives its name from the Latin word *fusus*, meaning to melt. Fuses have been used almost from the beginning of the use of electricity. A fuse breaks the circuit under conditions such as heating by a short-circuit, sustained inrush current, or overcurrent conditions. A fuse provides protection from an overload by opening only once, and then it must be replaced. Fuses are current-sensitive devices designed to serve as an intentional weak link in an electrical circuit. Their function is to provide protection of individual components, complete circuits, or equipment, by reliably melting under current overload conditions. A fuse consists of a *conducting element* enclosed in a glass, ceramic, or other nonconductive tube and connected by ferrules at each end of the tube. The ferrules fit into slots at each end to complete a split in an electric circuit. Excess current flowing through the fuse melts the conducting element and interrupts current. The fuse heart is the wire that is heated to its melting point by the excessive current. The circuit current flow decreases to zero as the wire melts open. Fuses are an inexpensive, dependable means of protecting equipment and wiring, offering relatively high current-interrupting capacity at acceptable cost. Power fuses must have a means to extinguish the arc to prevent the arc restriking, which can lead to undesirable and even catastrophic effects. Disadvantages to using a fuse are that they are destroyed in the protection process, and they are also somewhat imprecise due to the effects of ambient temperature changes and

FIGURE 5.1   Typical fuses.

repeated exposure to minor surges or overloads (i.e., pre-damage). Pre-damage can cause the fuse to blow when the current level is lower than the rated fuse-operation current. Fuses are rated by the current they can carry before heat melts the element. Fuses are manufactured in many shapes and sizes, as shown in Figure 5.1. While such large fuse varieties may be confusing, there are basically only two main types: *plug-type fuses* and *cartridge fuses*. Both types use either a single wire or a ribbon as the fuse (melting) element. The fuse condition is determined by visual inspection, providing information about fuse status and condition, but made safer by testing. Plug-type fuses can be screwed into a socket mounted on a control panel or electrical distribution center. The fuse link is enclosed in an insulated housing of porcelain or glass. The construction is arranged so the fuse link is visible through a glass or mica window. The plug-type fuses are primarily used in low-voltage and low-current circuits, with an operating range up to 150 V and from 0.5 A to 30 A. This type of fuse is found in older circuit-protection devices, being almost replaced by circuit breakers. A cartridge fuse operates exactly like a plug-type fuse. In a cartridge fuse, the fuse link is enclosed in a tube of insulating material with metal ferrules at each end (for contact with the fuse holder). Common insulating materials are glass, Bakelite, or a fiber tube filled with insulating powder. Cartridge fuses are available in a variety of physical sizes and are used in many different circuit applications. They can be rated at voltages up to 10,000 V and have current ratings from 1/500 (0.002) A to 800 A. Cartridge fuses may also be used to protect against excessive heat and open at temperatures from 165°F to 410°F (74°C to 210°C).

A fuse is ideal for protection against short-circuits. Short-circuits produce enough amperage to vaporize a fuse element and break connection in one cycle of a 60 Hz system. Fuses are more commonly used in devices connected to a system than within the system circuit. A fuse breaks the circuit under conditions such as heating by short-circuits, sustained inrush currents, or overcurrent conditions. The design fuse characteristics refer to how rapidly the fuse responds to various current overloads. Fuses are classified into three general categories, *fast-acting fuse*, which has virtually no delay; *medium-blowing fuse*, which protects devices subjected to moderate to high inrush currents; and *slow-blowing fuse*, which can withstand higher inrush currents. Current ratings and characteristics are both needed to define a fuse because fuses with the same current rating can be represented by considerably different time-current curves. Slow-blowing fuses have additional thermal inertia designed in so that they can withstand inrush or transient currents. Some factors involved in the selection of a fuse include operating current, voltage type (AC or DC), physical size limitations, agency approvals, as well as overload current (in breaking capacity) and length of the time in which the fuse must be open. Fuses are rated by current, voltage, and time-delay characteristics to aid in the proper fuse selection. To select the proper fuse, the fuse ratings must be known, as well as the normal circuit operating current. To protect a circuit from overloads (excessive current), a fuse rated at 125% of the normal circuit current is selected. In other words, if a circuit has a normal current of 10 A, a 12.5 A fuse provides overload protection. To protect against direct short-circuits only, a fuse rated at 150% of the normal circuit current is selected. For example, in the case of 10 A of circuit current, a 15 A fuse protects against direct short-circuits, but it is not adequate protection against overloads. The fuse voltage rating is *not* an indication of the voltage the fuse is designed to withstand while carrying current. The voltage rating indicates the fuse's ability to quickly extinguish the arc after the fuse element melts as well as the maximum voltage the open fuse blocks. In other words, once the fuse has opened, any voltage less than the fuse voltage rating will not be able to *jump the gap of the fuse*. Because of the way the voltage rating is used, it is a maximum RMS voltage value.

A great many electrical and electronic circuits require protection. In some of these circuits, it is important to protect against temporary or transient current increases, while other protected devices are very sensitive to current and cannot withstand any current increase. In such cases, a fuse must open very quickly if the current increases. Other circuits and equipment have large currents for shorter periods and a normal (smaller) current most of the time. An electric motor, for instance, draws a large current when starting, but its normal operating current is much smaller. A fuse used to protect a motor must be able to allow such large temporary current for short periods, but it must operate if such large current continues. Fuses are an inexpensive, dependable means of protecting equipment and wiring, offering relatively high current-interrupting capacity for their cost. The fuse design characteristics refer to how rapidly the fuse responds to various current overloads. Fuses are time-delay rated to indicate the relationship between the current through the fuse and the time it takes for the fuse to open. The three time delay ratings are *delay*, *standard*, and *fast*. A delay or slow-blowing fuse has a built-in delay that is activated when the current through the fuse is greater than the current rating of the fuse. This fuse allows temporary increases in current (surge) without opening. Some delay fuses can have two elements, allowing very long time delays. If the overcurrent condition continues, a delay fuse opens, but it takes longer to open compared to a standard or a fast fuse. *Delay fuses* are used for circuits with high surge or starting currents, such as motors, solenoids, and transformers. *Standard fuses* have no built-in time delay; they are not designed to be very fast acting. Standard fuses are sometimes used to protect against direct short-circuits only, being wired in series with a delay fuse to provide faster direct-short protection. A standard fuse can be used in any circuit where surge currents are not expected and a very fast fuse is not needed. A standard fuse opens faster than a delay fuse, but slower than a fast-rated fuse. Standard fuses can be used for lighting circuits or electrical power circuits. *Fast fuses* are designed to open very quickly when the current through the fuse exceeds the current rating of the fuse. Fast fuses are used to protect devices that are very sensitive to increased currents, e.g., delicate instruments or semiconductor devices. Notice that in each of the fuses, the time required to open the fuse decreases as the rated current increases. Fuses have identifications printed on them. The printing on the fuse identifies the physical size, the fuse type, and the fuse ratings. Unlike traditional fuses, polymeric fuses provide resettable circuit protection by limiting fault current. Polymeric fuses are made from a conductive plastic (a nonconductive crystalline polymer and a highly conductive carbon black) formed into thin sheets, with electrodes attached to either side. The electrodes ensure even distribution of power through the device, and provide a surface for leads to be attached or for custom mounting. The phenomenon that allows conductive plastic materials to be used for resettable overcurrent protection devices is their very large nonlinear positive temperature coefficient when heated, so that resistance increases with temperature.

Fuse parameters, characteristics, and application concepts must be well-understood in order to properly select a fuse for a given application. Ambient temperature refers to the temperature of the air immediately surrounding the fuse and is not to be confused with *room temperature*. The fuse ambient temperature is appreciably higher in many cases, because it is enclosed (as in a panel-mount fuse holder) or mounted near other heat-producing components, such as resistors and transformers. Breaking capacity, also known as interrupting rating or short-circuit rating, is the maximum approved current at which the fuse can safely break at rated voltage. *Current rating* is the fuse nominal amperage value, established by the manufacturer as a value of current which the fuse can carry, based on controlled sets of test conditions. For 25°C ambient temperatures, standards and codes recommend that fuses be operated at no more than 75% of the nominal current rating established using the controlled test conditions. Fuse resistance is usually an insignificant part of the total circuit resistance. Because the resistance of fractional amperage fuses can be several ohms, it must be considered when using them in low-voltage circuits. The interrupting rating is the maximum approved current that the fuse can safely interrupt at rated voltage. During a fault or short-circuit condition, a fuse may receive an instantaneous overload current many times greater than its normal operating current. Safe operation requires that the fuse remain intact (no explosion or body rupture) and clear the circuit. The voltage rating, as marked on a fuse, indicates that the fuse can be relied upon to safely interrupt its rated short-circuit current in a circuit where the voltage is equal to, or less than, its rated voltage. This system of voltage rating is covered by NEC regulations and is a requirement of Underwriters Laboratories as a protection against fire risks. Laboratory tests are conducted on each fuse design to determine the amount of energy required to melt the fusing element. This energy is described as nominal melting I^2t, expressed in A^2s.

The fuse time characteristic is a graph showing the dependence upon current of the time before arcing starts (the pre-arcing time), and the fuse total time consisting of the pre-arcing time plus the arcing time. The *conventional time*, the *conventional fusing current,* and the *conventional no-fusing current* are usually shown on the fuse time-current characteristics. These values are defined by standards. All fuses must operate within conventional time when carrying the conventional fusing current, and they must not operate within the conventional time when carrying conventional nonfusing current. The I^2t rating is related to the energy amount let through by the fuse element when it clears an electrical fault. This term is used in short-circuit conditions and the values are used to perform coordination studies in electrical networks. I^2t parameters are provided in manufacturer data sheets for each fuse family. For coordination of the fuse operation with upstream or downstream devices, both melting I^2t and clearing I^2t are specified. The melting I^2t is proportional to the energy required to begin melting the fuse element. The clearing I^2t is proportional to the total energy let through by the fuse when clearing a fault. The energy is mainly dependent on current, time, and the available fault level and system voltage. Because the fuse I^2t rating is proportional to the energy, it is a measure of the thermal damage from the heat and magnetic forces that will be produced by a fault. The breaking capacity is the maximum current that can safely be interrupted by a fuse, and it must be higher than the prospective short-circuit current. Fuses for small, low-voltage, residential wiring systems are commonly rated, in North American practice, to interrupt 10 kA. Fuses for commercial or industrial power networks have higher interrupting ratings, with some low-voltage current-limiting high-interrupting fuses rated for 300 kA. Fuses for HV equipment, up to 115 kV, are rated by the total apparent power (usually in MVA) of the fault level on the circuit. Low-voltage high-rupture-capacity (HRC) fuses are used in the areas of main distribution panelboards in LV networks where there is a high probability of short-circuit currents. They are generally larger than screw-type fuses, and have ferrule cap or blade contacts, being rated to interrupt current of 120 kA, in industrial installations and in power grids, e.g., in transformer stations, main distribution boards, or in building junction boxes and as meter fuses. In some countries, because of the high fault current available where such fuses are used, local regulations permit only trained personnel to change them. Some types of HRC fuses include special handling features. The fuse voltage rating must be equal to, or greater than, the open-circuit voltage. For example, a glass tube fuse rated at 32 V is not reliable to interrupt current from a voltage source of 120 V or 230 V. If a 32 V fuse attempts to interrupt the 120 V or 230 V source, an arc may result. Plasma inside the glass tube may continue to conduct current until the current diminishes to the point where the plasma becomes a nonconducting gas. Rated voltage should be higher than the maximum voltage source to be disconnected. Connecting fuses in series does not increase the rated voltage of the combination, nor of any one fuse. MV fuses rated for a few thousand volts are never used on low-voltage circuits, because of their cost and because they cannot properly clear the circuit when operating at very low voltages. Ambient temperature will change a fuse's operational parameters. A fuse rated for 1 A at 25°C may conduct up to 10% or 20% more current at −40°C and may open at 80% of its rated value at 100°C. Operating values vary with each fuse family and are provided in the manufacturer data sheets.

5.2.3 Circuit Breakers

While a fuse protects a circuit, it is destroyed in the process of opening the circuit, and once the problem is corrected, a new fuse must be placed in the circuit. A circuit protection device that can be used more than once solves the problem of fuse replacement. Such a device is safer, more reliable, and tamper-proof, being resettable and reusable without replacing any parts. A circuit breaker, designed for a broad range of current applications, interrupts the circuit during short-circuits or inrush currents. It is called a *circuit breaker* because it breaks (opens) the circuit. An advantage this product offers is its repeatability. Circuit breakers may be used several times. The three types of circuit breakers are *thermal, magnetic,* and *electronic circuit breakers.* With simple operation through the heating effect of current, thermal circuit breakers offer one of the most reliable and cost-effective forms of protection device, being well-suited for the protection of a broad range of components and systems, from motors and transformers, through printed circuit boards, to LV power distribution to a broad range of equipment and devices. Thermal circuit breakers utilize a bimetallic strip electrically in series with the circuit. The excess heat generated by the current during an overload deforms the bimetallic strip and trips the breaker. Thermal breakers have the reusability advantage over fuses, after tripping. They can be used as the main ON/OFF switch for the equipment being protected. Magnetic circuit breakers provide highly precise, reliable, and cost-effective solutions to most design problems. They have the advantages of thermal breakers without their disadvantages. Magnetic circuit breakers are temperature-stable, being not appreciably affected by temperature changes. These circuit breakers are designed to offer reliable automatic circuit protection, power switching, and circuit control in a versatile, compact, cost-effective package. Magnetic circuit breakers offer a more precise trip time for delicate circuits than thermal circuit breakers. Usually, wherever precise and reliable circuit protection is required, a magnetic circuit breaker is the choice.

A *circuit breaker* is a circuit protection device that, like a fuse, stops the current in the circuit if there is a direct short, excessive current, or excessive heat. Circuit breakers can also be used as circuit control devices. By manually opening and closing the contacts of a circuit breaker, you can switch the power on and off. Circuit breakers are tripped with external relays. The relays provide the brains to control the circuit breaker, so a breaker can be coordinated with other devices. The relays also perform reclosing functions. Circuit breakers are available in a large variety of sizes and types. Circuit breakers have five main components: the frame, the operating mechanism, the arc extinguishers and contacts, the terminal connectors, and the trip elements. A circuit breaker is an automatically operated electrical switch designed to protect an electrical circuit from damage caused by excess current, typically resulting from an overload or short-circuit. Its basic function is to interrupt current flow after a fault is detected. Unlike a fuse, which operates once and then must be replaced, a circuit breaker can be reset (either manually or automatically) to resume normal operation. The *rated voltage* of a circuit breaker corresponds to the higher system voltage for which the circuit breaker is intended. The rated voltage is expressed in kV (RMS) and refers to phase-to-phase voltage for three-phase circuits. Circuit breakers in power systems are historically rated as constant-MVA devices. The standard power system circuit breakers are rated at 60 Hz or 50 Hz, the common electric frequencies. The *rated continuous current* (the nominal circuit breaker current rating) of a circuit breaker is the RMS current that the circuit breaker is able to carry continuously at rated voltage and frequency. The *rated short-circuit current* is the highest value of the RMS short-circuit current. The symmetrical interrupting current, at any operating voltage is given by:

$$I_{Sym-Inter} = I_{SC}\left(\frac{V_{Max}}{V_{Operating}}\right) \tag{5.1}$$

A symmetrical short-circuit rating (SSCR) is specified at the maximum rated voltage (for more ratings information, see the appropriate IEEE, IEC, or ANSI standards designed for protection). Below the maximum rated voltage (to a specified minimum value), the circuit breaker has more interrupting capability. The minimum value where the circuit breaker is a constant-MVA device is specified by the constant K_{CB}, as expressed in by the circuit breaker's symmetrical interrupting capability (SIC):

$$SIC = \begin{cases} \dfrac{V_R}{V_{Op}} \cdot I_R, \text{ for } \dfrac{V_R}{K_{CB}} < V_{Op} \le V_R \\[4mm] K_{CB} \cdot I_R, \text{ for } V_{Op} \le \dfrac{V_R}{K_{CB}} \end{cases} \tag{5.2}$$

Here I_R, is the rated symmetrical RMS short-circuit current operating at the maximum voltage level, V_R is the maximum RMS line-to-line rated voltage, V_{Op} is the operating RMS line-to-line voltage, and K_{CB} is the voltage range factor, the ratio of the maximum rated voltage to the lower limit in which the circuit breaker is a constant-MVA device. Newer circuit breakers are rated as constant-current devices ($K_{CB} = 1$).

Example 5.1

A 15 kV class circuit breaker, so V_R equal to 15 kV, and I_R equal to 18 kA, is used on a 12.47 kV secondary power-distribution system where the maximum voltage is assumed to be 13.1 kV (105%). For an ANSI-rated 500 MVA CB class, K_{CB} equal to 1.3, estimate the CB symmetrical interrupting capability.

SOLUTION

The CB operating voltage 12.47 kV is higher than 15 kV/1.3 and from Equation (5.2), the symmetrical interrupting capability is:

$$SIC = \frac{15}{13.1} \times 18 = 20.6 \text{ kA}$$

The CB *rated insulation level* refers to the power frequency withstand voltage and the impulse voltages withstand values that characterize the CB insulation. A circuit breaker has three main ratings: the breaking capacity, the making capacity, and the short-time capacity. Breaking capacity is the RMS current that a circuit breaker is able to break at given recovery voltage, under specific conditions. Recovery voltage is RMS voltage at power system frequency, appearing across the CB contacts after final arc extinction. The breaking capacity is always stated as the fault current RMS value at the instant moment of contact separation. It is expressed in MVA units. Making capacity is the peak value of current during the first current wave after the closure of the circuit breaker. The making capacity depends upon the ability to withstand and close successfully against the effects of electromagnetic forces, which are proportional to the square of maximum instantaneous current on closing, expressed usually in terms of kA. Short-time rating is the period for which the circuit breaker is able to carry fault current while remaining closed. The short-time rating of a circuit breaker depends upon its ability to withstand electromagnetic force effects and temperature rise. In order to minimize the power interruption time, after a breaker trips, a circuit breaker is closed after a short time delay. Circuit breakers have a specified limit of 3 seconds when the ratio of symmetrical breaking current to the rated current does not exceed 40. However, if this ratio is higher than 40, then the specified limit is equal to 1 second. The circuit breaker closing (latching) capability defines its ability to withstand the magnetic developed forces occurring during the first short-circuit cycle, being expressed in RMS and/or peak currents, as:

$$I_{\text{Close-Latch}}(RMS) = 1.6 \times \text{Max-Symmetrical Fault RMS Current} \qquad (5.3a)$$

and

$$I_{\text{Close-Latch}}(Peak) = 2.7 \times \text{Max-Symmetrical Fault RMS Current} \qquad (5.3b)$$

The symmetrical breaking current is the RMS value of the AC at the separation time, while the asymmetrical breaking current is the total current comprising the AC and DC components at the moment of the contact separation, expressed as:

$$I_{Sym} = \frac{I_{AC}}{\sqrt{2}} \qquad (5.4a)$$

and

$$I_{Asym} = \sqrt{\left(\frac{I_{AC}}{\sqrt{2}}\right)^2 + I_{DC}^2} \qquad (5.4b)$$

Example 5.2

A 25 kV class circuit breaker has specifications: three-phase, 60 Hz frequency, 1.2 kA, and 2000 MVA. What is the latching RMS current?

SOLUTION

The symmetrical breaking current is the RMS phase line current:

$$I_{Sym} = \left(\frac{2200 \text{ MVA}}{\sqrt{3} \times 25 \text{ kV}}\right)\frac{1}{\sqrt{2}} = 35.822 \text{ kA}$$

The latching (closing) RMS current, from Equation (5.3a) is equal to $1.6 \cdot 35.822 = 57.32$ kA.

Circuit breaker sizes range from small devices that protect low-current circuits or individual household appliances, up to large switchgear designed to protect high-voltage circuits feeding an entire city. The circuit breaker operating mechanism provides a means of opening and closing the breaker contacts (turning ON and OFF the circuit). The toggle mechanism is usually the quick-make, quick-break type, which means the contacts snap open or closed quickly, regardless of how fast the handle is moved. In addition to indicating whether the breaker is ON or OFF, the operating mechanism handle indicates when the breaker has automatically tripped by moving to a position between ON and OFF. The *arc extinguisher* confines, divides, and extinguishes the arc drawn between contacts each time the circuit breaker interrupts current. It is actually a series of contacts that open gradually, dividing the arc and making it easier to confine and extinguish. Arc extinguishers are used in circuit breakers that control a large amount of power, such as those found in power distribution panels. Low-power circuit breakers (such as those found in lighting panels) may not have arc extinguishers. *Terminal connectors* are used to connect the circuit breaker to the power source and the load. They are electrically connected to the contacts of the circuit breaker and provide the means of connecting the circuit breaker to the circuit. The *trip element* is the part of the circuit breaker that senses the overload condition and causes the circuit breaker to trip. Some circuit breakers use solid-state trip units and solid-state circuitry. A thermal trip element circuit breaker uses a bimetallic element that is heated by the load current. The bimetallic element is made from strips of two different metals bonded together, expanding at different rates as they are heated, causing the bimetallic element to bend and open the circuit. The amount of time it takes for the bimetallic element to bend and trip the circuit breaker depends on the heat rates. A large overload heats the element quickly. A small overload requires a longer time to trip the circuit breaker. A magnetic trip element circuit breaker uses an electromagnet in series with the circuit load. With normal current, the electromagnet has not enough attraction to trip the contact bar, and the contacts remain closed. The strength of the magnetic field increases as current through the coil increases; if it becomes large enough to trip the contact bar, the contacts are opened, and the current stops. The current needed to trip the circuit breaker depends on the size of the gap between the trip bar and the magnetic element. On some circuit breakers, this gap (and therefore the trip current) is adjustable. The thermal trip element circuit breaker, like a delay fuse, protects a circuit against a small overload that continues for a long time. The larger the overload, the faster the circuit breaker will trip. The thermal element also protects the circuit against temperature increases. A magnetic circuit breaker trips instantly when the preset current is present. In some applications, both types of protection are desired. Rather than use two separate circuit breakers, a single trip element combining thermal and magnetic trip elements is used. In the thermal-magnetic trip element circuit breaker, a magnetic element (electromagnet) is connected in series with the circuit load, and a bimetallic element is heated by the load current. With normal circuit current, the bimetallic element does not bend, and the magnetic element does not attract the trip bar. If the temperature or current increases over a sustained period of time, the bimetallic element bends, pushing the trip bar and releasing the latch.

Molded-case dual-element circuit breakers, usually used for small-size LV circuits, are made of two tripping elements and contacts enclosed in a molded plastic enclosure (case). The trip is initiated by the thermal element with a time inverse with the square of the current or by magnetic element tripping almost instantaneously. There are several variations of this type of circuit breaker. The continuous-current ratings of molded-case CBs are 15, 30, 40, 125, 175, 200, 250, 300, 400, 500, 700, 1000, and 1200 A. *Air circuit breakers* and *air-blast circuit breakers* are suitable for MV networks, being usually used in indoor applications, for fire protection reasons. The air is an insulation and cooling medium between the conductors, making such CBs larger in size. In the air-blast versions, the arc and contacts during current interruption are blown away quickly by an air blast or a magnetic field. In such CB types, their main objective is to create a setting where the contact gap withstands the system recovery voltage in order to prevent the reestablishment of arcing after the current is zero. *Cross-blast* and *axial-blast circuit breakers* also employ moving and fixed contacts, and the blast eliminates or quenches the produced arc. In *vacuum circuit breakers* or *pressurized gas-filled circuit breakers*, the vacuum or pressurized gas (SF_6 or nitrogen) is the insulation medium, and such CB types are more compact than previous types. Such CBs are used indoors or outdoors in high-voltage, above 35 kV, and high-power distribution systems, especially where the space is critical. *Oil-filled circuit breakers*, one of the oldest types of circuit breaker, are widely used for outdoor applications for high-power distribution networks, land-based grids, or large industrial facilities with voltage ranges in hundreds of kilovolts. Oil is one of the best insulation and cooling media for arc extension. There are two types, the bulk-oil circuit breaker and low-oil circuit breaker. Bulk-oil circuit breakers consume a large amount of oil in their tanks. On the other hand, low-oil circuit breakers only use less oil compared to the bulk-oil circuit breaker. These are not recommended for indoor applications due to fire hazards in the case of oil tank damages following a severe fault. All such CBs use mechanical contacts that arc and wear when opened to interrupt fault currents.

Power electronic circuit breakers with no mechanical contacts or moving parts are developed and used in high-power applications, having extended uses in modern power systems. Oil circuit breakers are very cheap and very reliable in operation, without requirements for any special devices for controlling the arc caused by moving contacts. An arc is formed between the separated contacts when the current carried by the contacts in the oil is separated. A hybrid DC breaker system consists of the combination main breaker with a bypass system. The bypass for this hybrid DC breaker consists of the series connection of an ultra-fast disconnection and load commutation switch. In this DC breaker, the important semiconductor part is separated into a few sections for full voltage and current-breaking capability that consists of individual arrester banks. In addition, the disconnecting CB will interrupt residual current at the same time it also islanded the fault condition within the DC grid after the fault clearance. This function is to cover the arrester banks within the main semiconductor breaker to overheat. During a routine condition of the hybrid DC breaker, the current only flows past the bypass that consists of the series connection of load commutation switch and fast disconnection. When a DC fault occurs, the fast disconnection immediately opens and the load commutation switch also will fast-function to direct the current flows towards the main breaker. The main DC breaker will play its important function to block the current from flowing when the mechanical switch in the fast disconnection opens. Solid-state CBs are usually used in DC systems of high voltage supply. The advantages of this DC breaker are it has a fast response of switching speed, efficient current interruption, and it is capable of withstanding high-voltage conditions. Other advantages are the lack of mechanical components and moving parts, making the solid-state topology more responsive and effective, which results in less turn-off time.

5.2.4 Protection Relays

Formally, a relay is a logical element that processes the inputs (usually voltages and currents) from the system, apparatus, or circuits and issues a trip decision if a fault within the relay jurisdiction is detected. Protective relays monitor the currents and/or voltages of the power system to detect problems with the power system. When two protection devices are required to discriminate the chosen settings depends on how closely the devices can be guaranteed to conform to their characteristic curves. A conceptual diagram of relay is shown in Figure 5.2a. A relay is a device that makes a measurement or receives a signal, causing its operation, having effects on the operation of other equipment or devices. A protection relay is a device that responds to abnormal conditions in an electrical power system to operate a circuit breaker to disconnect the faulty section of the system with the minimum interruption of supply. In Figure 5.2b, the relay Rel_1 is used to protect the transmission line under fault Flt_1. An identical system is connected at the other end of the transmission line relay Rel_3 to the open circuit from the other ends as well. To monitor the health of the apparatus, the relay senses current through a current transformer (CT) and voltage through a voltage or potential transformer (V/PT). Instrument transformers (i.e., current and voltage transformers) insulate secondary circuits from primary (power) circuits and provide quantities that are proportional to those in primary. These quantities are used for metering and relaying circuits. Current and voltage transformers are parts of the protection system and must be carefully matched with the protective relay. Measuring current transformers for metering are required to accurately perform their function over a normal range of

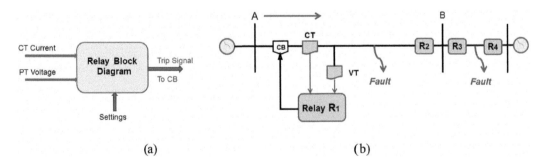

(a)   (b)

FIGURE 5.2   (a) Conceptual relay diagram; (b) typical circuit configuration with relays.

load currents, whereas protective current transformers are required to provide accurate secondary current to provide satisfactory protection over a wide range of fault current from a fraction of full load to many times full load. Therefore, separate types of current transformers are used for measuring and protection. In cases of voltage transformers, the same transformers can serve both purposes. Protective relays in power systems are connected in the secondary circuits of the instrument transformers. In current transformers, primary current is not controlled by condition of the secondary circuit. Hence, primary current is dominant in the operation of current transformers. The relay element analyzes these inputs and decides whether there is an abnormal condition or a fault and whether it is within relay jurisdiction. The jurisdiction of relay Rel_1 is restricted to bus B where the transmission line terminates. If the fault is in its jurisdiction, the relay sends a tripping signal to the circuit breaker, opening the circuit. In power transmission and distribution subsystems, the currents and voltages to the relays are supplied via instrument transformers. A *current transformer* is a device that transforms the current on the power system from large primary values to safe secondary values. The secondary current is proportional (as per the ratio) to the primary current. A *potential (voltage) transformer* transforms the voltage on the power system from primary values to safe secondary values of the relay. Several types of relays are in use to control distribution circuit breakers. Distribution circuits are almost always protected by overcurrent relays that use inverse time-overcurrent characteristics, meaning that the relay operates faster with increased current. Historically, there are three different relay types: electromechanical relays, solid-state relays, and digital (numerical) relays.

An *electromechanical relay* operates on the electromechanical energy conversion principle, being first generation of relay. Let us consider a simple example of an overcurrent relay, which issues a trip signal if current in the apparatus or system is above a reference value. By proper geometrical placement of current-carrying conductor in the magnetic field, Lorentz force, $\vec{F} = I\vec{B} \times \vec{l}$ is produced in the operating coil. This force creates the operating torque and if the magnetic field, B, is constant (for example by a permanent magnet), then the produced instantaneous torque is proportional to the instantaneous value of the current. Because the instantaneous current is sinusoidal, the instantaneous torque is also sinusoidal, which has a zero average value. Thus, no net deflection of operating coil is perceived. On the other hand, if the B is also made proportional to the instantaneous value of the current, then the instantaneous torque will be proportional to square of the instantaneous current (non-negative quantity). The average torque will be proportional to square of the RMS current. Movement of the relay contact caused by the operating torque may be restrained by a spring in the overcurrent relay. If the spring has an elastic constant k, then the deflection is proportional to the operating torque (in this case proportional to I_{RMS}^2). When the deflection exceeds a preset value, the relay contacts close and a trip decision is issued. Electromechanical relays are known for their ruggedness and immunity to electromagnetic interference.

With the advent of transistors, operational amplifiers, solid-state devices, etc., *solid-state relays* were developed and employed. They realize their functionality through various operations, such as comparators, while providing more flexibility and consuming less power than their electromechanical counterparts. A major advantage of solid-state relays is their ability to provide self-checking, i.e., relays can monitor their own health and raise a flag or alarm if their own component fails. Some advantages of solid-state relays are low-burden, improved dynamic performance characteristics, high seismic withstand capacity, and reduced panel space. Relay burden refers to the amount of apparent power (VA) consumed by the relay. The higher is this value, the more is the corresponding load on the current and voltage sensors (i.e., CT and VT transformers) that energize such relays. Higher loading of the sensors leads to deterioration in their performance. A performance of CT or VT is gauged by the quality of the replication of the corresponding primary waveform signal. Higher burden leads to the problem of CT saturation and inaccuracies in measurements. Thus, it is desirable to keep CT/VT burdens as low as possible. These relays have now been superseded by microprocessor-based relays or digital relays. A digital relay involves analog-to-digital conversion of analog voltage and currents obtained from secondary CTs and VTs. If a fault is diagnosed, a trip decision is issued. Digital relays provide maximum flexibility in defining relaying logic. The hardware of the numerical relay can be made scalable, i.e., the maximum number of voltage and current input signals can be scaled up easily. A generic hardware board can provide multiple functionalities. Changing the relaying functionality is achieved by simply changing the relaying program or software. Also, various relaying functionalities can be multiplexed in a single relay. It has all the advantages of solid-state relays, such as self-checking, etc. Enabled with communication capabilities, it is an *intelligent electronic device*, which can perform both control and protection functionality. Also, a relay that can communicate can be made adaptive, i.e., it can adjust to changing apparatus or system conditions. For example, a differential protection scheme can adapt to transformer tap changes. An overcurrent relay can adapt to different loading conditions. It is not exaggeration to say that numerical relays are both the present and the future of CB technology. A relay is said to be dependable if it trips only when it is expected to trip, either when the fault is in its primary jurisdiction or when it is relied upon to provide backup protection. However, false tripping of relays or tripping for a fault that is either not within its jurisdiction or within its purview is compromising and affects the system operation. The power system may get unnecessarily stressed or, worse, there can be loss of service. *Dependability* is the degree of certainty that the relay operates correctly, being defined by this relationship:

$$Dependability~(\%) = \frac{Number~of~Correct~Trips}{Number~of~Desired~Trips} \times 100 \tag{5.5}$$

Dependability can be improved by increasing the relaying system *sensitivity*. For example, consider the case of overcurrent protection. The protective system must have the ability to detect the smallest possible fault current. The smaller the detected current, the more sensitive it is. One way to improve sensitivity is to determine fault characteristic signature. It is unique to the fault type and it does not occur in normal operations. For example, earth faults involve zero sequence current, providing a very sensitive method to detect earth faults. Faults in a three-phase system can be single line-to-ground, double line-to-ground, and line-to-line or three-phase types. Charles L. Fortescue, in 1918, came up with the idea that any unbalanced power system has six degrees of freedom, whereas a balanced system has only two degrees of freedom. Hence an unbalanced system having six degrees of freedom can be synthesized by three sets of a balanced system each with two degrees of freedom. For a three-phase system with phase sequence *a-b-c*, the three sets of balanced phasors are called *positive, negative,* and *zero* sequences. The zero-sequence phasors are the same set of balanced phasors, as in the positive or negative sequences, but with 0° phase difference. Again there are two degrees of freedom in placing the zero-sequence phasors as the positive or negative sequences. Application of zero sequence does not create any rotation to the rotor of an induction machine. This is because the net *mmf* induced in the air gap is zero. An unbalanced set of phasors can be synthesized by a linear combination (superposition of the positive-, negative-, and zero-sequence phasors). Positive- and negative-sequence phasors are used to represent any balanced set of phasors (each of equal magnitude and phase difference of 120°) but in which the order of *b* and *c* phase are reversed, in the negative sequence with respect to the positive-sequence phasor. Power system operation during any of these faults can be analyzed using sequence components. Once this signature is detected, the abnormality is correctly classified and hence appropriate action is initialized and performed. On the other hand, *security* is a property used to characterize false tripping on the relays. A relay is said to be secure if it does not trip when it is not expected to trip. *Security index*, the degree of certainty that the relay will not operate incorrectly, is defined by:

$$Security~Index~(\%) = \frac{Number~of~Correct~Trips}{Total~Number~of~Trips} \times 100 \tag{5.6}$$

False trips do not simply create nuisance. They can compromise system security. For example, tripping of a tie-line in a two-area system can result in load-generation imbalance in each area, which can be dangerous. Even when multiple paths for power flow are available, under the peak load conditions, overloads or congestion in the power system may occur. Dependability and security are contrasting system requirements. Usually, there are biases with setting towards dependability, which may cause some nuisance tripping, which, in the worst case, can trigger a partial or complete system shutdown (blackout). Security of the relaying system can be improved by improving selectivity of the relaying system. Similar to sensitivity, *selectivity* implies the reliability to discriminate. A relay should not confuse some peculiarities of an equipment or apparatus with a fault. For example, a transformer when energized can draw up to 20 times rated current, the inrush current, which can confuse both overcurrent and transformer differential protection. Typically, inrush currents are characterized by large second harmonic content. This discriminant is used to inhibit relay operation during inrush, by improving the selectivity in transformer protection. Also, a relay should be smart enough, not just to identify a fault, but also to be able to decide whether or not a fault is in its jurisdiction. For example, a relay for a feeder should be able to discriminate a fault on its own feeder from faults on adjacent feeders. This implies that it should detect first existence of fault in its vicinity in the system and then make a decision whether it is in its jurisdiction. Directional overcurrent relays were introduced to improve selectivity of overcurrent relay. This jurisdiction of a relay is also called the *protection zone*. Typically, protection zones are classified into primary and backup zones. In detecting a fault and isolating the faulty element, the protective system must be very selective. Ideally, the protective system should zero in on the faulty element and isolate only it, thus causing minimal disruption to the system. Selectivity is usually provided by using time discrimination and by applying differential protection principles. With overcurrent and distance relays, such boundaries are not properly demarcated. This is a very important consideration in operation of power systems. A relaying system has to be reliable. *Reliability* is achieved by redundancy, i.e., duplicating the relaying system. Obviously, redundancy is costly. Another way to improve reliability is to ask an existing relay's protecting equipment to be the backup protection for another. Both approaches are used simultaneously in practice. However, it is important to realize that backup protection must be provided for safe operation of relaying systems. Redundancy in protection also depends upon the criticality of the power equipment. For example, a 400 kV transmission line will have independent (duplicated) protection using the same or a different philosophy; on the other hand, a power distribution system will not have such local backup, which may not be needed. A quantitative measure for reliability is defined as:

$$Reliability\ (\%) = \frac{Number\ of\ Correct\ Trips}{Number\ of\ Desired\ Trips + Number\ of\ Incorrect\ Trips} \times 100 \qquad (5.7)$$

Example 5.3

The performance of an overcurrent relay was monitored over a period of 3 years. It was found that the relay operated 40 times, out of which 36 were correct trips. If the relay failed to issue a trip decision on 6 occasions, compute the relay dependability, security index, and reliability.

SOLUTION

The number of correct trips is 36, while the number of desired trips is 36 + 6 = 42. Applying Equations (5.5), (5.6), and (5.7), the relay dependability, security index, and reliability are:

$$Dependability\ (\%) = \frac{36}{42} \times 100 \approx 85.7\%$$

$$Security\ Index\ (\%) = \frac{36}{40} \times 100 = 90\%$$

and

$$Reliability\ (\%) = \frac{36}{42 + 4} \times 100 \approx 78.3\%$$

A *relay protection zone* is a region defined by relay jurisdiction. It is shown by demarcating the boundary. This demarcation for differential protection is quite crisp and is defined by the current transformer location. On the other hand, such boundaries for overcurrent and distance relays are not very crisp. It is essential that primary zones of protection should always overlap to ensure that no position of the system ever remains unprotected. This overlap also accounts for faults in circuit breakers. To provide this overlap, additional CTs are required. To maximize safety, and minimize equipment damage and system instability, a fault should be cleared as quickly as possible. This implies that relay should quickly arrive at a decision and circuit breaker operation should be fast enough. Typically, a fast circuit breaker operates in about two cycles. A reasonable time estimate for ascertaining presence of fault is one cycle. This implies approximately a three-cycle fault-clearing time for primary protection. On the other hand, if a five-cycle circuit breaker is used, the fault-clearing time increases to six cycles. So long as short-circuit faults exist in a transmission system, the electrical output of the generator remains below the mechanical input.

5.2.5 GROUND-FAULT CIRCUIT INTERRUPTER

Ground-fault protection is equipment protection from the effects of ground faults. A GFCI is a fast-acting circuit breaker that senses small imbalances in an electrical circuit caused by the electrical current leaking to ground. If this imbalance occurs, the GFCI shuts off the electricity within a fraction of a second. Specific ground-fault equipment protection requirements can be found in NEC sections 215.10, 230.95, 240.13, and 517.17. The GFCI continually matches the amount of current going to an electrical device against the amount of current returning from the device along the electrical circuit path. Whenever the amount *going* differs from the amount *returning* by approximately 5 mA or a threshold, the GFCI interrupts the electric power by closing the circuit within as little as 1/40 of a second. GFCIs must always be used in wet environments—both indoors (e.g., around sinks or wet floors) and outside. When a cord connector is wet, current leakage can occur to the grounding conductor and shock anyone who touches the connector, if the person provides a path to ground. Ground-fault (GF) relays or sensors are used to sense low-magnitude ground faults. When the ground-fault current magnitude and time reach the GF relay pickup setting, the control scheme signals the circuit-disconnect to open. GF relays can only offer protection for equipment from the effects of low-magnitude ground faults. Equipment protection against the effects of higher-magnitude ground faults depends on the response speed of the conventional overcurrent protective devices, fuses, or circuit breakers. A GFCI is not an overcurrent device. A GFCI is used to open a circuit if the current flowing to the load does not return by the prescribed route. In a simple 120 V circuit, we usually think of the current flowing through the black (ungrounded) wire to the load and returning to the source through the white (grounded) wire. If it does not return through the grounded wire, then it must have gone somewhere else, usually to ground. The GFCI is designed to limit electric shock to a current- and time-duration value below that which can produce serious injury. Ground-fault protection does not effectively protect the people, being designed for equipment; it will not prevent shock; prevent ground faults; protect from three-phase, phase-phase, or phase-neutral faults; protect from high-level ground faults; or guarantee a selectively coordinated system. In fact, coordination may be compromised. Ground-fault relays are not simple and the ultimate reliability depends on the reliability of each element such as the solid-state sensor, monitor, control wiring, control power source, shunt trip, and circuit disconnecting means. If one element is incorrectly wired, inoperative, miscalibrated, or damaged, the low-level ground-fault protection may be negated. If the system neutral is incorrectly or accidentally grounded on the load side of the sensor, a ground fault can have a return path over the neutral and never trip the relay. *Ground-fault protection of the equipment (GFPE)* systems are intended for equipment protection, by disconnecting all ungrounded conductors of a circuit at current levels less than that of a supply circuit overcurrent protective device. This device is designed typically to trip in the current ranges from mA or higher current ranges, and therefore is not used for personnel protection. The reliability of the ground-fault relays depends on the reliability of its elements, and incorrectly wired, miscalibrated, or damaged element, the effectiveness of the low level ground fault protection is compromised. A system neutral is incorrectly or accidentally grounded on the sensor load side, provides a return path for a ground fault over the neutral, preventing the relay's trips. Notice that GFCIs need to be used when work is performed in wet or damp locations, outdoors, or when receptacles are being used near sources of water.

5.3 PROTECTION SCHEMES AND SELECTING PROTECTION DEVICES

The main objectives of power system equipment protection are: (1) to limit the extent and duration of service interruption whenever equipment failure, human error, or adverse natural events occur on any portion of the system, (2) to minimize damages to system components involved in the failure, and (3) to prevent human injuries. Protection schemes are specialized equipment that monitor the power system, detect faults or abnormal conditions, and then initiate corrective actions. The main electrical system faults are short-circuits and overloads. Short-circuits may be caused in many ways, including failure of insulation due to excessive heat or moisture, mechanical damage to electrical distribution equipment, and failure of utilization equipment as a result of overloading or other abuse. Short-circuits may occur between two-phase conductors, between all phases of a poly-phase system, or between one or more phase conductors and ground. The short-circuit may be solid (bolted) or welded, where the short-circuit is permanent and has relatively low impedance. The main types of faults in a power system are: short-circuit faults (three-phase, phase-to-phase, phase-to-ground, phase-to-phase-to-ground), open-circuit faults, complex faults (intercircuit, broken conductor, cross-country), and interturn faults in windings. Series unbalances, such as a broken conductor or a blown fuse, are uncommon, except in the lower-voltage system in which fuses are used. Fault occurrence can be quite variable, depending on the type of power system (e.g., overhead vs. underground lines) and the local natural or weather conditions. The five basic protection features, as discussed before, underlying foundations of all system, equipment, or circuit protection are:

1. Reliability: assurance that the protection will perform correctly.
2. Selectivity: maximum continuity of service with minimum system disconnection.
3. Speed of operation: minimum fault duration and consequent equipment damage and system instability.
4. Simplicity: minimum protective equipment and associated circuitry to achieve the protection objectives.
5. Economics: maximum and the most efficient protection at minimal total costs.

Reliability has two aspects, dependability and security. Dependability is defined as *the degree of certainty that a relay or relay system* operates correctly, as defined by the IEEE C 37.2 standard. Security is related to the certainty degree that a relay or relay system is not operating incorrectly, as defined by the IEEE C 37.2 standard. In other words, dependability indicates the ability of the protection system to perform correctly when required, whereas security is its ability to avoid unnecessary operation during normal operations or faults and problems outside the designated operation zone. Thus, the protection must be secure (e.g., not operate on tolerable transients), yet dependable (e.g., operate on intolerable transients and permanent faults). Dependability is easy to ascertain by testing the protection system to assure that it is operating as intended when the operating thresholds are exceeded. Security is more difficult to ascertain, because there are large varieties of transients or tolerable abnormal conditions that might upset the protective system, and predetermination of all these possibilities is difficult if not impossible. Relays have an assigned area known as the primary protection zone, but they may properly operate in response to conditions outside this zone. In these instances, they provide backup protection for the area outside their primary zone. This is designated as the backup or overreached zone. Selectivity (also known as relay coordination) is the process of applying and setting the protective relays that overreach other relays such that they operate as fast as possible within their primary zone, but have delayed operation in their backup zone. This is necessary to permit the primary relays assigned to this backup or overreached area time to operate. Obviously, it is desirable that the protection isolates a trouble zone as rapidly as possible. However, this is not possible all the time. Modern high-speed circuit breakers operate in the range of 17 to 50 ms (one to three cycles at 60 Hz); others operate at less than 83 ms (five cycles at 60 Hz). Relay speed is especially important when the protected facility exists in a stability-sensitive area of the power system network. A protective relay system should be kept as simple and straightforward as possible while still accomplishing its intended goals. However, the increasing use of solid-state and digital technologies in protective relaying provides many convenient possibilities for increased sophistication. It is also important to obtain the maximum protection for the minimum cost, as cost is always a major factor.

Fault currents can be considered with a good degree as sinusoidal waveforms with the power system frequency, but not symmetric about the time (horizontal) axis, due to the presence of the DC (offset) current, occurring when the fault starts at a time other than at a zero-current crossing. The DC component is higher and persist longer for higher inductance-to-resistance (X/R) ratios of the power system at the fault point. The DC component can be substantial and can double the symmetrical component magnitude at the fault point. The symmetric fault current and especially the asymmetry factor, which is essential, must be determined and are critical in sizing the protection device. The fault currents of interest in protection are first-cycle symmetrical current, the maximum symmetrical current experienced by the system, used with a multiplier to account for the asymmetry factor due to the DC offset, being used to determine the short-circuit withstand requirements for the system components and to determine the actual current that a circuit breaker must interrupt. A critical factor to determine the decay level is the X/R ratio at the fault point, setting the decay and the tripping breaker time. The X/R ratios are determined by the existing transformers, cables, transmission lines, and the distance of generators to the fault location. Notice that the larger is the X/R ratio, the longer is the fault current decay. Notice that, systems or circuits with higher X/R ratios are difficult, putting additional and significant stress on circuit breakers because of the higher asymmetrical peaks for the same symmetrical RMS values of the fault currents. When the current-carrying contacts open, the energy stored in the system leakage inductance keeps flowing until all the stored energy is diverted and/or dissipated, while the contacts continue arcing until the inductor-stored energy is depleted or the current naturally comes to zero. On the other hand, if the circuit breaker interrupting rating is not adequate, but the continuous-current rating is valid, some options may be less expensive than the use of larger circuit breaker frame size to achieve adequate interrupting ratings. Fused LV circuit breakers are rated based on the first-cycle symmetrical currents as specified by the IEEE C37 standards. The asymmetry permitted in the symmetrical current ratings of the fused CB corresponds to an X/R ratio of 4.9, if the X/R at the fault point exceeds 4.9 a correction factor (*MF*), as specified by the IEEE C37.13 standard, must be used to adjust the short-circuit current, as expressed by:

$$MF_{LV-CB} = \frac{\sqrt{2 \cdot \exp\left(-\frac{2\pi}{X/R}\right) + 1}}{1.25}, \quad \text{for X/R} > 4.9 \qquad (5.8)$$

Circuit breakers have several ratings, such as: rated continuous current (carried into allowable temperature rise), rated line voltage, symmetrical short-circuit current (in MVA) or fault current that a circuit breaker can interrupt without thermal damage, tripping (interruption) time, switching overvoltage capability without prestriking, lightning impulse voltage capability, and the first symmetrical peak current withstand capability without mechanical damage. These rating factors or parameters are critical for circuit breaker selection. However, the key selection factors are:

1. The continuous-current rating must be higher than the maximum load current that it carries;
2. The current-interruption capability must be greater than the maximum fault current; and
3. Circuit breaker voltage rating must be higher than the line operating voltage.

Similar to molded-case circuit breakers, the ratings are based on the first-cycle symmetrical current (IEEE C37 standards), for an asymmetry degree, corresponding to an X/R ratio of 6.6. If the X/R ratio at the fault location is above 6.6, the short-circuit current must be derated by a correction factor, accounting for the slower fault current decay rate due to the higher X/R ratio, as specified by the IEEE C37.13 standard, expressed by this equation:

$$MF_{Molded-case-CB} = \frac{\sqrt{2}\left(\exp\left(-\frac{\pi}{X/R}\right)+1\right)}{2.29}, \quad \text{for X/R} > 6.6 \tag{5.9}$$

Example 5.4

If the first-cycle symmetrical current of a bus is 52.5 kA and the bus X/R ratio is 12.8, calculate the correction factor and the adjusted first-cycle symmetrical fault current for the additional asymmetry due to the higher X/R ratio.

SOLUTION

From Equation (5.9) the multiplication factor in this case is:

$$MF = \frac{\sqrt{2}\left(\exp\left(-\frac{\pi}{12.8}\right)+1\right)}{2.29} \simeq 1.095$$

And the adjusted first-cycle symmetrical fault current is then:

$$I_{Adj-SC-Sym(Bus)} = 52.5 \times 1.095 \simeq 57.5 \text{ kA}$$

The fuses or groups of fuses should have the following capabilities for protection: required continuous-current ratings, switching inrush and surge current ratings, and rated fuse voltage ratings. Power fuses are tested in circuits having an X/R ratio of 15. If at the fault point the X/R ratio is higher than 15, a similar derating factor, as for circuit breakers, must be used to adjust the calculated short-circuit current for the slower decay rate due to the higher X/R ratios. The correction factor is calculated by:

$$MF_{MV-Fuse} = \frac{\sqrt{2\cdot\left(2\cdot\exp\left(-\frac{2\pi}{X/R}\right)+1\right)}}{1.52}, \quad \text{for X/R} > 15 \tag{5.10}$$

Selection of current-limiting and LV fuses is based on similar criteria as discussed above. Current-limiting fuses are not protective against overloads and low-magnitude fault currents, but they do operate in very short time, less than a quarter of a cycle, protecting downstream devices and equipment against very-high-fault currents. Current-limiting fuses are tested, as specified by the standards, in circuits having an X/R ratio of 10. If the X/R ratio is higher than 10, the correction factor is:

$$MF_{MV-Fuse} = \frac{\sqrt{2 \cdot \left(2 \cdot \exp\left(-\frac{2\pi}{X/R}\right) + 1\right)}}{1.44}, \quad \text{for X/R} > 10 \tag{5.11}$$

Similar to LV circuit breakers, the LV fuses are tested in circuits having an X/R ratio of 4.9. If the X/R at the fault point is higher than 4.9, the correction factor, applied to adjust the first-cycle fault current for the slower decay rate caused by the higher X/R ratio, as from standards, is expressed by:

$$MF_{LV-Fuse} = \frac{\sqrt{2 \cdot \exp\left(-\frac{2\pi}{X/R}\right) + 1}}{1.25}, \quad \text{for X/R} > 4.9 \tag{5.12}$$

The fuse and circuit breakers clear the faults only if the overcurrent level is higher than 200% of the circuit protection device current rating, leaving basically a protection gap between 115%, the overload level allowed in most equipment and circuits for up to 2 hours, and 200% overcurrent level. This gap is bridged by overcurrent protection devices, selected by using overcurrent ratings.

Example 5.5

If the first-cycle symmetrical fault current of a low-voltage power distribution bus is 52.85 kA having an X/R ratio of 8.95, estimate the adjusted first-cycle symmetrical fault current for an LV fuse.

SOLUTION

The multiplying factor, for LV fuse, for an X/R exceeding 4.9 is:

$$MF_{LV-Fuse} = \frac{\sqrt{2 \cdot \exp\left(-\frac{2\pi}{9.80}\right) + 1}}{1.25} \approx 1.147$$

The first-cycle symmetrical fault current is then:

$$I_{Sym-Fault-Adj} = 1.147 \times 52.85 = 60.64 \text{ kA}$$

Fuses and circuit breakers clear the faults usually if the overcurrent level is greater than 200% of their own rated currents, leaving a protection gap between 115% overload level, allowed by standards and codes for most equipment and devices for a time horizon of 1 hour. Such gaps are bridged by the overload protection circuitry and devices. As a rule of thumb, the nominal current ratings of fuses operating in high ambient temperature must be derated by 0.5% per °C above 25°C, for time-delay fuses, and by 0.2% per °C above 25°C for standard general-purpose fuses. For branch circuits a design of 30% increase in the amperage is applied to the estimated load current.

Example 5.6

Determine the general-purpose fuse current rating for a single-phase 120 V, 2.4 kW load, having an efficiency of 0.93, and a power factor of 0.89. The load is operating at 50°C.

SOLUTION

The load current and the branch design current are estimated as:

$$I_{Load} = \frac{2850}{120 \times 0.93 \times 0.89} = 28.69 \text{ A}$$

and, respectively

$$I_{Load-Branch} = 1.3 \times I_{Load} = 37.30 \text{ A}$$

The derating temperature correction is

$$DT = (50 - 25) \times 0.002 = 0.05$$

and the fuse rating is:

$$\text{Fuse Rating} = \frac{I_{Load-Branch}}{0.95} = 39.3 \text{ A}$$

So, if a 40 A fuse is chosen, there is a 30% margin in the current.

5.3.1 OVERCURRENT, DIRECTIONAL, AND DISTANCE PROTECTION

The approach to overcurrent protection is based on the intuition that faults, typically short-circuits, lead to currents much above the load (rated) current. Overcurrent relaying and fuse protection use the principle that when the current exceeds a predetermined value (threshold), it indicates presence of a fault (short-circuit). Such protection schemes are useful in radial power distribution systems with a single source. It is quite simple to implement such a protection scheme. As shown in Figure 5.3a, the

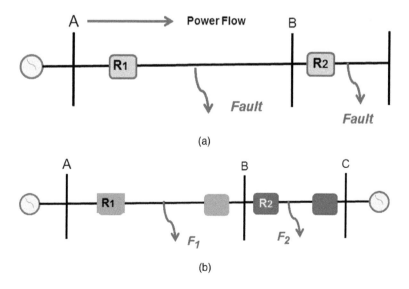

FIGURE 5.3   (a) Radial power distribution; (b) radial power distribution with generation at each ends.

fault current is fed from only one end of the feeder. For this network, to the relay R_1, both the downstream faults F_1 and F_2 are visible. That is, I_{F1} as well as I_{F2} pass through the current transformer of R_1, while to the relay R_2, fault F_1, an upstream fault is not seen, only F_2 is seen. This is because no component of I_{F1} passes through CT of R_2. Thus, selectivity is achieved naturally. Relaying decisions are based solely on the magnitude of fault current. Such protection schemes are nondirectional. In contrast, there can be situations where for the purpose of the selectivity, the phase-angle information (always relative to a reference phasor) may be required. Figure 5.3b shows such a case for a radial system with sources at both ends. Consequently, fault is fed from both ends of the feeder. To interrupt the fault current, relays at both ends of the feeder are required. In this case, from the magnitude of the current seen by the relay R_2, it is not possible to distinguish whether the fault is in section AB or BC. Because faults in section AB are not in its jurisdiction, it should not trip. To obtain selectivity, a directional overcurrent relay is required. It uses both magnitude of current and phase-angle information for decision making. It is commonly implemented in subtransmission networks where ring mains are used.

Consider a simple radial distribution system that is fed from a single power source. The apparent impedance at the sending (generator) end is estimated by the ratio of the line voltage to the line current (V/I). For an unloaded system, the current is zero and the apparent impedance seen by the relay is infinite. If the system is loaded, the apparent impedance reduces to a finite value, $Z_{Load} + Z_{Line}$, the sum of the load impedance, Z_{Load}, and the transmission line impedance, Z_{Line}. In the presence of a fault, located at a per-unit distance, x, on the transmission line, which is the line section to the fault location, the impedance that is seen by the protection relay drops to a value $x \cdot Z_{Line}$, as shown in Figure 5.4.

The basic principle of the distance protection relay is that the apparent impedance seen by the relay, defined as the ratio of the line phase voltage to the line current of a transmission line (Z_{App}) is reduced drastically in the presence of a line fault. Typically, distance relays protect transmission lines from power system faults by using a method of step distance protection. They use the line impedance as the basis to form zones of protection, and each zone is calculated by a predetermined percentage of the line impendence. The impedance setting establishes the relay impedance characteristic, which is graphically displayed as circles or quadrilaterals in the R-jX plane. A distance relay is capable of detecting faults, indicated by a drop in the impedance of the line, when the observed impedance is inside the relay's defined impedance characteristic. The relay calculates the impedance by using the measured voltage and current phasors at each processing interval. A distance relay has the ability to detect a fault within a preset distance along a transmission line or power cable from its location. Every power line has a line characteristic resistance and reactance per kilometer or per mile related to its design and construction, so its total impedance is a function of its length. A distance relay therefore looks at current and voltage and compares these two quantities on the basis of Ohm's law. Because the impedance of a transmission line is proportional to its length, for distance measurement it is appropriate to use a relay capable of measuring the impedance of a line up to a predetermined point (the reach point). A distance relay is designed to operate only for faults occurring between the relay location and the predetermined (reach) point, thus discriminating for faults that may occur in different line sections. The basic principle of distance protection involves the division of voltage at

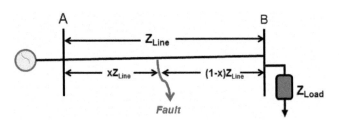

FIGURE 5.4   Fault location and the distance from the relay on a transmission line.

the relaying point by the measured current. The calculated apparent impedance is compared with the reach-point impedance. If the measured impedance is less than the reach-point impedance, it is assumed that a fault exists on the line between the relay and the reach point. If the relay operating boundary is plotted on an R-jX diagram, its impedance characteristic is a circle with its center at the origin of the coordinates and its radius is setting (the reach point) in ohms. The relay operates for all values less than its setting (i.e., for all points within the circle). This type of relay, however, is nondirectional. It can operate for faults behind the relaying point. It does not account for the phase angle between voltage and current. It is also sensitive to power swings and load encroachment due to the large impedance circle. The limitation of the impedance characteristic can be improved by a technique known as the self-polarization approach. Additional voltages are fed into the comparator in order to compare the relative phase angles of voltage and current, so providing a directional feature. This has the effect of moving the circle such that the circumference of the circle passes through the origin. The mho element restrains until Z plots inside the circle. A distance relay compares the apparent line impedance (this ratio) with the positive sequence impedance ($Z_{Positive-Seq}$) of the transmission line. If the fraction $Z_{App}/Z_{Positive-Seq}$ is less than one, there is a fault presence on the line. This ratio also indicates the distance of the fault location from the relay. Basically, during steady-state or normal conditions, the load impedance is large enough that it remains far enough from the relay characteristic value. Because the impedance is a complex quantity, the distance protection is inherently a directional one. The first quadrant is the forward direction, i.e., impedance of the transmission line to be protected lies in this quadrant. However, if only magnitude information is used, nondirectional impedance relay results. Diagrams in Figure 5.5a and in Figure 5.5b show the characteristic of an *impedance relay* and *mho relay* both belonging to this class of overcurrent distance protection relays. The *impedance relay* trips if the impedance magnitude is within the circular region. Because the circle spans all the quadrants, such a protection scheme is nondirectional. In contrast, the *mho relay*, which covers primarily the first quadrant, is directional in its nature. Thus, the tripping relationship for the impedance relay can be written as follows:

$$Z_{App} = \frac{|V_R|}{|I_R|} \leq |Z_{Relay-Set}| \tag{5.13}$$

In this case there is a trip, while elsewhere the relay is restraining to trip. While impedance relay has only one design parameter, $Z_{Relay-Set}$, the mho relays have two design parameters, Z and the torque angle, T.

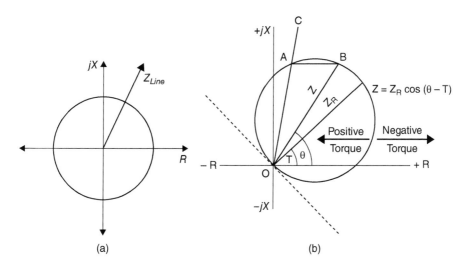

(a)   (b)

FIGURE 5.5   Diagrams of (a) an impedance relay, and (b) a mho relay.

The mho circle is composed of its impedance maximum reach, maximum torque angle, and relay characteristic angle. The mho circle maximum reach is set by the impedance reach Z_R of the protective zone. These impedance reaches vary depending on the zone of protection. Each impedance value determines the diameter of the mho circle. There are several relay parameters of interest, as discussed in this paragraph. The angle of maximum torque of a distance relay using the mho characteristic is the angle at which it has the maximum reach. For microprocessor relays, the MTA is the same as the positive sequence line impedance angle. The relay characteristic angle (RCA) of a mho circle is 90°. For purposes of calculating the maximum relay loadability, the RCA is the angle whose vertices are made between the load impedance vector and the difference between the line impedance and load impedance vectors. The trip relationship for the mho relays is given by Equation (5.14), and if apparent impedance is within the mho circle, then the relay trips; elsewhere the relay refrains from tripping.

$$|Z_{App}| \leq |Z| \cdot \cos(\theta - T) \tag{5.14}$$

Example 5.7

Calculate the value of Z (circle radius) for a mho relay with torque angle of 75° that has to give 100% protection to a 50 km long 110 kV transmission line with impedance per km, 0.8 Ω/km, and a phase angle of 80°.

SOLUTION

The transmission line impedance is $Z_{Line} = 0.8 \cdot 50 = 40$ Ω, and then:

$$Z = \frac{40}{\cos(80° - 75°)} \approx 40.2 \text{ Ω}$$

Differential protection is based on the fact that any fault within electrical equipment would cause the current entering the equipment to differ from the current leaving it. Thus, by comparing the two currents either in magnitude, in phase, or in both, we can determine a fault and issue a trip decision if the difference exceeds a predetermined set value. Figure 5.6 shows a short transmission line in which shunt parameters are neglected. Then under no-fault condition, the phasor sum of line currents is zero, $I_R + I_S = 0$. Thus, we can say that differential current under no-fault condition is zero. However in case of a fault in the transmission line, the sum of the currents (the differential current) is no longer zero, as expressed by:

$$I_S + I_R + I_F \neq 0 \tag{5.15}$$

This principle of checking the differential current on the line is known as a differential protection scheme. In case of transmission lines, implementation of differential protection requires a communication channel to transmit current values to the other end. It can be used for short feeders,

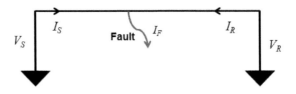

FIGURE 5.6 Differential protection on a short transmission line.

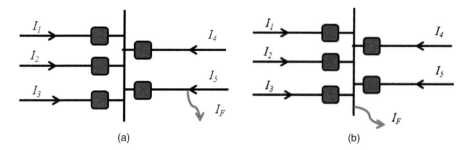

FIGURE 5.7   An illustration of the basic idea of differential busbar protection: (a) external fault to the busbar; (b) internal busbar fault.

and a specific implementation is known as pilot wire protection. Differential protection tends to be extremely accurate. Its zone is clearly demarcated by the CTs that provide the boundary. Differential protection can be used for tapped lines (multiterminal lines) for boundary conditions, transformers, busbars, etc. Differential protection for detecting faults is an attractive option when both ends of the equipment are physically located near each other, e.g., on a transformer, a generator, or a busbar. For example, differential protection is the ideal solution for busbar protection (Figure 5.7). If the fault is external to the busbar, the algebraic sum of the currents entering the busbar is zero (). On the other hand, if a fault is on the bus (internal fault), this sum is not zero ($\Sigma I_k = 0$). Thus, differential protection can be used to protect a busbar ($\Sigma I_k + I_F \neq 0$).

Modern distance relays offer quadrilateral characteristics, whose resistive and reactive reach can be set independently. They therefore provide better resistive coverage than any mho-type characteristic for short lines. This is especially true for earth fault impedance measurement, where the arc resistances and fault resistance to earth contribute to the highest values of fault resistance. Polygonal impedance characteristics are highly flexible in terms of fault impedance coverage for both phase and earth faults. For this reason, most digital (numerical) relays offer these types of features.

5.4   EQUIPMENT GROUNDING

Safety, system protection, and performance are the three main reasons to earth or ground a system. However, not all electronic and electric equipment and installations need to be connected to earth to work safely; some examples of exempt equipment are satellites, drones, and aircrafts. A ground fault occurs when electricity leaks from a current-carrying conductor to the enclosure of a piece of equipment. System grounding or earthing involves the ground connection of power services and separately derived systems. The two different ways to protect against a ground fault are grounding, and double insulating equipment such as appliances, machines, and power tools (referred to collectively as simply "equipment" throughout the remainder of this section). An added measure of protection can be afforded by using GFCIs. Grounding works by limiting potential gradients, limiting energy, or controlling overvoltages. The circuit that helps ground equipment is known as a grounding circuit, designed to protect people and property by providing an alternate path for current to flow from the enclosure back to the service entrance if a ground fault occurs. A grounding circuit is not normally energized. A grounding circuit must be designed and installed properly to perform the following functions: (1) prevent people or other objects from being energized (people safety), and (2) provide a path for large currents to flow back to their source, thus blowing the fuse or tripping the circuit breaker. Wrong or poorly designed grounding configurations, oriented to satisfy the special power and performance requirements of loading equipment, can compromise safety rules and create dangerous situations for personnel and equipment. Personnel safety, equipment safety, and performance grounding issues must be analyzed together. Grounding must be taken into account wherever electrical current flows. It can never be stressed too strongly: Proper grounding and bonding *must* be correctly applied if the system, the equipment, and the people that come in contact with them are to be protected. If a grounding circuit is improperly designed or installed, or even perhaps damaged, then a sufficiently large amount of current will not be able to flow back to the source, which could result in an electrocution. The grounding circuit must be continuous so that it is not interrupted from the equipment back to the electrical service.

Effective grounding means that the path to ground: (1) is permanent and continuous, (2) has ample current-carrying capacity to conduct safely any currents liable to be imposed on it, and (3) has impedance sufficiently low to limit the potential above ground and to facilitate the operation of the overcurrent devices in the circuit. *Effective bonding* means that the electrical continuity of the grounding circuit is assured by proper connections between service raceways, service cable armor, all service equipment enclosures containing service entrance conductors, and any conduit or armor that forms part of the grounding conductor to the service raceway. *Effective grounding has no function unless and until there is electrical leakage from a current-carrying conductor to its enclosure.* When such a *ground fault* occurs, the equipment grounding conductor *prevents voltages* between the electrical enclosure and other enclosures or surroundings, and it *provides a path* for large amounts of fault or overload current to flow back to the service entrance, thus blowing the fuse or tripping the circuit breaker. Metal parts of equipment enclosures, racks, raceways and equipment grounding conductors susceptible of being energized by electrical currents (due to circuit faults or lightning) must be effectively grounded for reasons of personnel safety, fire-hazard reduction, equipment protection, and equipment performance. Grounding these metallic objects facilitates the operation of overcurrent protective devices during ground faults and permits return current from electromagnetic interference filters and surge-protective devices, connecting line to ground or line to chassis, to flow in proper fashion. All metallic conduits and raceways in areas containing electronic load equipment have to be carefully bonded to form an electrically continuous conductor. All mechanical equipment in electronic equipment areas should be carefully bonded for electrical safety and for noise current control. Such equipment should be grounded or bonded to local building steel using direct or higher-frequency grounding and bounding means. Some pieces of equipment include double layers of insulating material or reinforced insulation to protect live parts from energizing their enclosure. These products are not grounded and should never be grounded.

Grounding limits the potential gradients between conducting materials in a given area. For example, during a ground fault, a phase conductor comes into contact with a machine frame. Current flows through the equipment and the potential of the equipment becomes elevated above the ground potential by an amount equal to the voltage on the conductor. If a person touches the machine while simultaneously in contact with the ground (earth), the potential of the person's body is elevated. The maximum potential when touching a machine frame is equal to the voltage drop along the grounding conductors. The grounding system must provide a low-resistance path for fault current to return to the source. Ground conductors should have low resistances to allow the conductors to carry the maximum expected fault current without excessive voltage drop. Grounding limits the energy that leads to premature failures, reduced component life, and nuisance trips. Most sources of transient overvoltages can be reduced or possibly eliminated by grounding. During a surge event, the grounding conductor becomes ionized, making the conductor capable of carrying tremendous amounts of current. High-energy faults can vaporize breakers, switchgear, and phase conductors. Protective enclosures may be blown apart with explosive force. Controlling the maximum and allowable current significantly reduces the fire danger and minimizes equipment damage. An overvoltage condition may occur by accidental contact of equipment with a higher voltage system or from transients due to lightning strikes, intermittent ground faults, and switching surges. The maximum ratings may be temporarily exceeded in cable insulation, transformer windings, and relay contactors. Components of the electrical system are successively overstressed and weakened by repeated exposures. Overvoltages lead to premature failures and reduced component life. There are two basic requirements for grounding power services and separately derived systems or sources (transformer, generators, etc.). The first requirement is to bond the neutral or secondary grounded circuit conductor to the equipment grounding terminal or bus. For power service entrances, the incoming neutral conductor is connected to the equipment ground bus in the switchboard by means of the main bonding jumper. For separately derived sources, the neutral must be bonded to the equipment grounding terminal or bus. The second requirement is that the equipment grounding terminal or bus must be connected to the nearest effective grounded electrode by means of the grounding electrode conductor. If no effective grounded electrode or building steel is available, then a separately derived source should be connected to the service entrance grounding point via a grounding electrode conductor installed in the most direct and shortest path practicable. OSHA provides requirements for grounding that assume grounding electrodes can be directly inserted at locations proximate to power usage; however, in some applications, cables carrying grounding conductors to exterior grounding beds must be used. Proper grounding requires connecting all of the enclosures (equipment housings, boxes, conduit, etc.) together, and back to the service-entrance enclosure. This is accomplished by means of the green wire in the cord (portable equipment), and the conduit system or a bare wire in the fixed wiring of the building. When a ground fault occurs, as in a defective tool, *the grounding conductor must carry enough current to immediately trip the circuit breaker or blow the fuse.* This means that the ground-fault path must have low impedance. The only low-impedance path is the green wire (in portable cord) and the metallic conduit system, or an additional bare wire if conduit is not used. In most cases, insulation and grounding are used to prevent injury from electrical wiring systems or equipment. However, there are instances when these methods do not provide the degree of protection required. The basic resistance-grounded system consists of a resistor inserted between the power-system neutral point and the ground. Concerns when selecting the grounding resistor are resistance, time rating, insulation, and available connection to the power-system neutral. Ground current can be limited at a level less than the restricted maximum for high-resistance grounding. The smallest value chosen has two concerns: ground-fault relaying and charging current. For maximum safety ground, protective circuitry should sense ground current at a fraction of the current limit. Reliable relay operation with electromechanical devices can be a problem if the maximum current is less than 15 A to 20 A. The limitation is that the ground-fault current should always be greater than the system-charging current (the current required to charge system capacitance) when the system is energized. When very low ground-relay settings are used, charging current may itself cause tripping.

5.4.1 LIGHTNING AND SWITCHING PROTECTION

Often, transmission line outages result from lightning strikes that hit overhead transmission lines. Certain weather conditions can cause an abundance of negative charges to gather on the bottom of clouds, while positive charges accumulate on buildings, trees, or any objects that project above the ground. When the potential between the cloud and ground becomes great enough, a streamer of negative charges moves erratically toward the earth. At the same time, a short leader of positive charges may move a short distance up into the air. When the two charges meet, the downward moving streamer completes the grounding path as the positive charges instantaneously move back up the path to the clouds. The resulting flash is lightning. A lightning strike happens very quickly and contains a great deal of electrical energy. A typical lightning strike can be up to 1000 MV, 300 kA, and can generate heat to the surrounding air to an excess of 50,000°F. A discharge between the clouds and the ground occurring near a power cable or communication cable generates a magnetic field due to its surge current. When the magnetic waves propagated within the field reach the cable, a lightning surge is induced. When lightning strikes a building or a lightning rod, high current flows to the ground and the ground potential rises. In addition, if a building is in a charged state due to the thunderstorm's electrostatic induction, an atmospheric discharge dissipating the electric charge at the bottom of the cloud causes the electric charge on the building to flow toward the ground. This also leads the ground potential to rise and the lightning surge to be directed to the cable from the ground. The effects of induced surge on a connected device can be more severe when the connected cable is longer and the device is located closer to the place where the actual lightning strikes. Lightning discharges normally produce large overvoltage surges, which may last for a fraction of a second and are extremely harmful. The line outages can be reduced to an acceptable level by protection schemes such as installation of earth wires and earthing of the towers. The effect of lightning surges depends upon the cable location and its environment. Surge protectors installed in those paths absorb and eliminate the high-voltage impulse energy and protect the electric instruments from damage. Lightning overvoltages can be classified as follows: (a) induced overvoltages that occur when lightning strikes reach the ground near the line; (b) overvoltages due to shielding failures that occur when lightning strikes reach the phase conductors; and (c) overvoltages by back flashovers that occur when lightning strike reaches the tower or the shield wire. The most commonly used devices for protection against lightning surges include shielding by earth wires; usually, the transmission lines are equipped with earth wires to shield against lightning discharges, and the earth wires are placed above the line conductor at such a position that the lightning strikes are intercepted by them. In addition to this, earthing of towers is also essential. Surge voltage generated between a cable and the ground might reach several tens of thousands of volt; however, lightning occurs at a voltage typically 5000 volts between the cable and the ground, while it induces several hundred volts between lines. Lightning surges cause a very high potential (voltage) difference between two conductors and the ground, and a discharge occurs between some part of an electronic circuit and those electrically connected to the ground, such as metal housings.

There are five main parts of any lightning protection system: air terminals, main conductors, secondary conductors, lightning arresters, and ground rods. The required size for all of the parts is determined by NFPA 780. NFPA 780 divides structures into 12 categories. Lightning arresters are an alternative to the use of earth wire for protection of conductors against direct lightning strikes. Using lightning arrestors in parallel with insulator strings can prevent dangerous surges from entering the building wiring when lightning strikes the power lines. Usually these are installed at electrical entrances and on antennas. The use of lightning arresters is more economical than other measures. Proper grounds are also critical to ensure the lightning is dissipated without damage. Types of grounding and number of grounds depend on the soil type and the size of the building. As recommended by standards, a lightning protection system must provide a low resistant circuit to the earth. All of the air terminals should be connected in a common grid, while multiple grounding rods must be used for correct and proper protection. Main conductors connect the air terminals to the ground, and they must have a minimum of two wires, and must be sized correctly as required by codes. ZnO varistors are usually used as lightning arresters because their resistance varies with applied voltage, i.e., their resistance is a nonlinear inverse function of applied voltage. At normal voltage, its resistance is high, but when high-voltage surges such as lightning strokes appear across the varistor, its resistance decreases to a very low value and the energy is dissipated in it, giving protection against lightning. Surge arresters, installed near the ends of longer conductors, are designed to divert electric transient, generated by lightning strikes, switch faults in a high-voltage system to the ground without damaging any electrical equipment. Surge arresters (*surge protection devices* or *transient voltage surge suppressors*) are designed to protect against electrical transients only. Although the transients are smaller than the original electrical surge, they still can carry enough energy to cause arcing between different circuit pathways. These transients are usually initiated at a point between the two conductors; often surge arresters are installed at each conductor end. It is required that each conductor provide a pathway to earth to safely divert the transients from the protected component or equipment. The one exception is in HV distribution systems, in which the induced voltage is not high enough to cause damage at the electric generation end of the lines, so no surge arrester is installed. However, an arrester installation at the service entrance to a building is a key measure to protecting downstream circuits and equipment. The effectiveness of a surge arrester depends on its ability to limit the transient rate-of-rise overvoltages and to reduce system characteristic impedance, by discharging energy associated with a transient overvoltage, limiting and interrupting current that follows transient current through the arrester, and returning to an insulating state without interrupting the power supply to the loads.

5.5 RACEWAYS AND CONDUITS

Cable installations make up a large portion of the initial distribution system investment, contribute to a lesser extent to the annual maintenance and operating costs, and affect system reliability. Power, control, and instrumentation cables and conductors are usually installed in raceways such as conduit or cable tray to protect the cables from physical damage and for easy maintenance and replacement. Conduits are metallic or nonmetallic tubing, circular in shape, designed to protect electrical cables. Metallic conduits can be of galvanized steel or nonmagnetic (aluminum). Electrical conduits are used to enclose, support, and protect electrical conductors for power, control, and communication. Rigid metallic conduits (RMC) have a circular cross-section, trade size in the range 0.5" to 6", the heaviest wall thickness of all metallic conduits, and a standard length of 10 ft. NEC Article 344 specifies the requirements for RMC installation and use. RMC conduits have the outer finish galvanized, hot-dipped for corrosion protection, and can be used in all environments and applications. Intermediate metallic conduit (IMC), a rigid steel type has a circular cross-section, trade size of 0.5" to 4", and is lighter in weight than RMC. NEC Article 358 specifies requirements for electrical metallic tubing (EMT) as a circular conduit, made of steel or aluminum, trade size of 0.5" to 4". EMT conduits are used in indoor applications and in noncorrosive environments. Nonmetallic conduits, made of materials such as fiber-based composites or polyvinyl chloride (PVC), are often encased in concrete to form a duct bank. Cable tray is a prefabricated structure often resembling a ladder. Aluminum is the most common material used to fabricate cable trays, but steel (raw, galvanized, or stainless), fiber-reinforced plastic, and vinyl-coated steel trays are also used.

Electric conduits are usually in a circular cross-section and may be rigid or flexible, and often are covered with outer PVC or other plastic materials for corrosion resistance and/or weatherproofing. Therefore, underground cables and their accompanying protective and operating devices should be selected in accordance with criteria set forth in the following paragraphs. The joint specifications of the Insulated Cable Engineers Association and National Electrical Manufacturers Association (ICEA-NEMA) and the specifications of the Association of Edison Illuminating Companies (AEIC) are used as covered by NFGS-16301. ICEA-NEMA specifications cover medium-voltage cables that are manufactured as stock items. AEIC specifications describe medium-voltage cable that is not a stock item (35 kV rating) or where the footage installed is large enough to make a special run. Single-conductor cables are usually used in power distribution, making the installed costs less than that of multiconductor cables. The use of multiconductor cables is recommended where justified by considerations such as installation in cable trays, twisted to provide lower inductance for 400 Hz distribution systems, and for high-altitude electromagnetic pulse (HEMP) hardened systems. In harsh industrial environments, lids are often installed on cable trays to keep contaminants away from the cables. If lids are used, the natural cooling effect of the cables is reduced greatly, so the cables in a covered tray must be derated. A full range of accessories to join tray sections and support them from walls or ceilings is available, such as couplings to form three- and four-way intersections, supports, or boxes. The NEC code states which kinds of cables can be installed in the same raceway, and which cables must be separated. In a conduit system, the cables are drawn into tubing called conduit. The conduit can be steel or plastic. The different sizes of conduit are identified by their inner diameter, and in the case of electrical conduit the nominal inner diameter is always the same as the outside diameter of the tube. Heavy-gauge conduit is normally joined together by screwed fittings; there is a standard electrical thread which is different from other threads of the same nominal diameter. Conduit is thick enough for the cross-sectional area of the metal to provide a good ground (earth) low-impedance path. The conduit can be used as the earth continuity conductor and no separate cable or wire need be used for this purpose. Because metallic raceways can carry ground currents, grounding and bonding of the raceways are critical. It is essential that the conduit, with all its fittings and screwed joints, should form a continuous conducting path of low impedance and the safety of the installation depends on good electrical contact at all the joints. Even though it may be decided not to use steel conduit as the circuit protective conductor, in preference for a separate protective conductor, usually copper, the conduit must be erected properly with tight joints. The final connection to machines and mechanical equipment such as pumps, boilers, fans, fan heaters, workshop equipment, and so on is usually made in flexible conduit. The fixed wiring terminates in a box either in the wall near the equipment to be connected or on the surface of the wall, and from this box a short length of flexible conduit is taken to the equipment. Solid conduit from this to the machine could involve a large number of bends in a short distance, which would be difficult to make and impossible to pull cable through. Flexible conduit can take up a gentle curve, serves to isolate the fixed wiring from any mechanical vibrations on the connected machine, and allows for belt tension adjustment of the motor. There are several types of flexible conduit. However, flexible conduit cannot be used as a protective conductor. A conduit system must be completely installed before any cables are pulled into it.

Conduit systems are intended to be easily rewired, which means that at 20 or 30 years after the building has been erected, it should still be possible to pull all the cables out of the conduit and pull new replacement cables into each conduit. If this is possible, then quite regardless of what happens when the building is first constructed, the layout of the conduit must be such that cables can be drawn into it when it is complete and finished. The original reason for wanting to have electrical systems that could be recabled during the life of the building was that old cable types deteriorated in about 20 years or so to the point at which they should be removed. However, PVC cable appears to last indefinitely, so all modern installations that use this cable should not need rewiring. New cables do have to be run where there were no cables previously and the original conduit has at best to be added to and at worst abandoned altogether. Re-wireability is then no help, and, in fact, the need for a re-wireable system is not as great as is often supposed. Notice, there is always the possibility that a cable may become damaged during the construction of a building, and it is obviously an advantage if it can be replaced without difficulty after the building has been finished. If the conduit is installed so that the system is re-wireable, repairs are always possible. To achieve re-wireability, draw-in boxes must be accessible from the surface. In other words, their covers must be flush with the finished surface. When conduit and duct systems are designed, proper conduit size is selected. Conduits are sized properly so the cables can be pulled without damaging the cable jacket or insulation and adequate cooling exists when the cables are energized. The NEC limits conduit fill; the sum of the cross-sectional cable areas in a conduit cannot exceed the jam ratio of the inner conduit cross-sectional area, as shown in Table 5.1, as specified in the NEC.

In addition to the percent fill, a quantity called the jam ratio must also be determined. The jam ratio predicts the likelihood that the cables are jamming by lining up along the conduit inner

TABLE 5.1

Conduit or Tubing Percent for Conductors

Number of Conductors	Maximum Percent Fill (%)
1	53
2	31
3 or more	40

diameter. Jamming occurs as the conductors are installed or pulled around the bends, and become twisted inside the conduit or tubing. Jam ratio (JR) is defined as the ratio of the conduit inner diameter (ID_{Cond}) over the cable outer diameter (OD_{Cable}). If the jamming ratio has a value in the range 2.4 to 3.2 there is a good probability of cable jamming during the installation, and large conduits are recommended.

$$JR = \frac{ID_{Cond}}{OD_{Cable}} \tag{5.16}$$

Example 5.8

Determine the jamming ratio for a 3 in. conduit and 500 kcmil XHHW-2 copper conductors.

SOLUTION

The cross-section area of the conductors is 2.7936 sq. in., with a conductor outer diameter of 0.943 in. and inner conduit diameter is 3.09 in. (see Appendix B, Table B.2-II). The jam ratio is:

$$JR = \frac{ID_{Cond}}{OD_{Cable}} = \frac{3.09}{0.943} = 3.28$$

The jam ratio lies outside the jamming range in this case. A minimum bending radius must be determined according to the cables being pulled.

Shielded cables contain a metallic layer just beneath the jacket to distribute evenly the electric field gradient throughout the insulation. This shield is sometimes provided in the form of a thin copper tape, or tape shield, which is grounded at the cable terminations. Typically, in three-phase circuits, each cable has enough copper strands in its concentric neutral to make up one-third of the system neutral. Conduits subjected to large temperature variations, usually in outdoor and some industrial applications, can expand or contract depending on the temperature fluctuations. The amount of the conduit expansion or contraction is computed by using the thermal expansion relationship:

$$\Delta L_{cond} = L_{run} \times TEC_{Cond} \times \Delta T \tag{5.17}$$

Here, L_{run} is the run length (m or ft.), TEC_{Cond} is the coefficient of thermal expansion for the conduit material (m/°C or ft./°F), and ΔT is the change in temperature, in Celsius or Fahrenheit degrees, respectively.

Pull and junction boxes are used to provide conduit termination endings at specific locations and to enable cables and conductors to be pulled into a conduit system. Cable pulling calculations are needed to ensure that maximum permitted tension on the cable is not exceeded and to prevent the damage of the cable insulation or the cable. Pulling tension is the tensile force that must be applied to the cable to overcome friction as the cable is pulled through the raceway. Sidewall pressure is the crushing force applied to the cable by the conduit in the radial direction as the cable is pulled around a bend. Exceeding the maximum allowable pulling tension is not usually a problem for large power cables because that maximum value is usually quite large. Control and instrument cables, however, tend to have lower maximum allowable pulling tensions. When more than one cable is pulled in a conduit, a weight correction factor (w_c) is needed to account for the additional friction forces that exist between the cables or conductors. The most common case is to install three cables of the same size in a conduit for a three-phase power circuit. The weight correction factor depends on whether the cables are arranged in a triangular or cradled configuration. The

TABLE 5.2
Representative Dynamic Friction Coefficients

	Conduit Type			
Cable Insulation	**Metallic (Steel or Aluminum)**	**PVC**	**Fiber Conduit**	**Asbestos Cement**
Polyvinyl chloride (PVC)	0.40	0.35	0.50	0.50
High molecular weight polyethylene (HMW PE)	0.35	0.35	0.50	0.50
Cross-linked polyethylene (XLPE)	0.35	0.35	0.50	0.50
Hypalon (CSPE)	0.50	0.50	0.70	0.60
Chlorinated polyethylene (PE-C)	0.50	0.50	0.70	0.60
Nylon	0.40	0.35	0.50	0.50

tension required to pull a cable or group of cables through a straight section of horizontal conduit and an inclined section, up or down, respectively, is expressed as:

$$T_{OUT} = T_{IN} + w_c \mu_f L W \tag{5.18a}$$

and

$$T_{OUT} = T_{IN} \pm L W \left(\sin \theta \pm w_c \mu_f \cos \theta \right) \tag{5.18b}$$

where T_{OUT} is the pulling tension in pounds at the duct or conduit output end, T_{IN} is the conduit or duct input tension, w_c is the weight correction factor (dimensionless), μ_f is the coefficient of dynamic friction (dimensionless), L is the length of straight section of conduit in feet, W is the weight of cable in pounds per foot, and θ is the incline angle of the conduit. Table 5.2 gives the representative dynamic friction coefficients for the most common types of insulation and conduit. The coefficient of friction ranges from more than 0.5 for dry cable to less than 0.2 for well-lubricated cable. Suitable lubricants include wire soaps, waxes, and synthetic polymer compounds. Notice that if the coefficient of friction cannot be reduced sufficiently, reducing the length or changing the pull direction is strongly recommended. Installing pull boxes in the raceway can reduce the length of the pull. In Equation (5.4b), the plus sign is for cases of pulling up a straight section of conduit, while the minus sign is for pulling down a straight conduit section. The tension required to pull a cable through a horizontal bend is given by:

$$T_{OUT} = T_{IN} \exp\left(w_c \mu_f \theta \right) \tag{5.19}$$

where T_{OUT} is the tension out of the bend in pounds, T_{IN} is the tension coming into the bend in pounds, and θ is the bend angle in radians. A minimum bending radius must be determined according to the cables being pulled. As IEEE 576-2000 specifies, when pulling an eye attached to copper conductors, the maximum pulling tension should not exceed 0.008 times the circular-mil area (C_m), and maximum pulling tension should not exceed 0.006 times the circular-mil area (C_m), when pulling an eye attached to aluminum conductors, expressed as:

$$T_m = k \times n \times C_m \tag{5.20}$$

Here, k is equal to 0.008 for copper and 0.006 for aluminum, when the maximum tension, T_m is expressed in lbf, k is 0.036 for copper and 0.027 for aluminum, n is the number of conductors, and C_m is the circular mil area of each conductor. Maximum limitation for this calculation is 22,240 N

(2268 kgf) (5000 lbf) for single-conductor (1/C) cables and 44,480 N (4536 kgf) (10,000 lbf) for three multiconductor cables. This limitation is due to unequal distribution of tension forces when pulling multiple conductors. When pulling cable through a vertical bend, the tension is calculated as for a horizontal bend, then the cable weight in the vertical section is either added (if the cable is pulled uphill) or subtracted (if the cable is pulled downhill) from the required tension. When pulling cable downhill, a negative tension can be calculated. The coefficient of dynamic friction depends on the cable insulation type and conduit. Note the significant effect on tension that small changes in μ_f (friction coefficient) can cause, especially in conduit bends where this friction coefficient is in the exponent. Inaccurate friction coefficients lead to poor correlation of tension calculations with actual tensions. Unfortunately, it is in multibend pulls, where the tension and sidewall pressure are of most concern, that the use of an inaccurate coefficient of friction produces the greatest error. For wiring applications in which three cables are pulled, the weight correction factor is typically in the range 1.15 to 1.35, while a value of 1.40 is recommended for situations involving four or more cables. Weight correction factors for cradled and triangular configurations are calculated using Equations (5.18a) and (5.18b), respectively.

$$w_c = 1 + \frac{4}{3}\left(\frac{OD_{Cable}}{ID_{Cond} - OD_{Cable}}\right) \quad \text{(Cradled)} \qquad (5.21a)$$

and

$$w_c = \frac{1}{\sqrt{1 - \left(\frac{OD_{Cable}}{ID_{Cond} - OD_{Cable}}\right)^2}} \quad \text{(Triangular)} \qquad (5.21b)$$

Example 5.9

Determine the maximum pulling tension for a 250 kcmil, three-conductor copper cable.

SOLUTION

The maximum pulling tension, Equation (5.20), is given by:

$$T_m = k \times n \times C_m = 0.008 \times 3 \times 250,000 = 6000 \text{ lbf}$$

When cable grip is used over non lead-jacketed cable, the pulling tension should not exceed 1000 lb. or 1000 lb. per grip (when used with multiconductor cables) and the tension calculated in Equation (5.20). If the coefficient of friction is unknown, a 0.5 friction coefficient is recommended. Sometimes physical constraints require that the cable be pulled in one specific direction. If no such constraints exist, it is good engineering practice to attempt to pull the cable in the direction requiring the minimum pulling tension. The maximum permissible pulling length, L_m, for one straight cable section is given by:

$$L_{max} = \frac{T_{max}}{w_c \mu_f W} \qquad (5.22)$$

Here T_{max} is the maximum tension in kgf (lbf), W is the linear weight of the cable(s) kg/m (lb./ft.), and μ_f is the coefficient of dynamic friction. Sidewall pressure in a raceway or conduit is caused by the tension in the cable acting horizontally and the weight of the cable acting vertically.

Sidewall pressure is the crushing force applied to the cable by the conduit in the radial direction as the cable is pulled around a bend, being a function of pulling tension and bending radius of the conduit, and differs according to the arrangement of the cables in the conduit. Sidewall pressure is typically the controlling factor in raceway design for large power cable. Sufficiently large conduit bending radii must be used, and pulling tension may need to be limited to values well below the maximum allowable pulling tension for the cable to keep the sidewall pressure below the maximum allowed by the cable. Generally, the tension of a cable immediately as it leaves a bend must not be greater than 300 times the bend radius (in feet), and the maximum sidewall pressure must not exceed 300 lb./ft. Shown below are formulas to calculate the maximum allowable tension at a bend and the actual sidewall pressure. The maximum allowable pulling tension at bend, T_{bm}, is the limit with which the calculated pulling tension, T_b, should be compared. If T_b, as computed by using Equation (5.20) is greater than T_{bm}, the possibility of redesign or rerouting should be considered. The formulas to calculate sidewall pressure for various cable arrangements when the conduit bend radius is r are given in Equations (5.23a), (5.23b), and (5.23c).

$$P_{SW} = \frac{T_{bm}}{r} \text{ (one single-conductor cable)} \tag{5.23a}$$

$$P_{SW} = \frac{3w_c - 2}{3}\left(\frac{T_{bm}}{r}\right) \text{ (three single-conductor cable)} \tag{5.23b}$$

and

$$P_{SW} = \frac{w_c}{2}\left(\frac{T_{bm}}{r}\right)\left(\begin{array}{l} \text{three single-conductor cable triangular} \\ \text{or one three-conductor cable} \end{array}\right) \tag{5.23c}$$

As was stated before, a weight correction factor of 1.4 is usually used if four or more cables are installed in a conduit. Under these conditions, the sidewall pressure, given by Equation (5.24a) may be estimated as:

$$P_{SW} \approx 0.75\left(\frac{T_{bm(out)}}{r}\right) \tag{5.24}$$

where P_{SW} is the sidewall pressure, in N/m (lbf/ft.) of radius, T_{bm} is the maximum tension (leaving the bend), N (lbf), and r is the inside radius of conduit in m (ft.). One of the critical limitations to be considered in the installation of electrical cables is sidewall pressure. The sidewall pressure is the force exerted on the insulation and sheath of the cable at a bend point when the cable is under tension, and is normally the limiting factor in an installation where cable bends are involved. When installing single-conductor cables, or multiconductor cables in duct or conduit, the sidewall pressure acting on the cable at a bend is the ratio of the pulling tension out of the bend to the radius of the bend, Equations (5.23a), (5.23b), and (5.23c). The normal maximum sidewall pressure per meter (foot) of radius is given in Table 5.3. However, in order to minimize cable damage because of excessive sidewall pressure, it is recommended to check the cable manufacturer recommendations for each cable type. Although the normal maximum allowable sidewall pressure is as stated in Table 5.3, specific installation procedures and specific cable type or construction may cause this maximum to be increased or decreased. The weight correction factor for one cable (single or one multiconductor cable) is taken as 1.0. During installation of the cable, it is recommended that the radius of bends be 1.5 times that of the minimum bending radius (MBR) for the final training. The

TABLE 5.3
Recommended Maximum Sidewall Pressure

Cable Type	Maximum Sidewall Pressure (lbf/ft.)	Maximum Sidewall Pressure (N/m)
600 V nonshielded multiconductor control	500	7300
600 V and 1 kV single conductor (size 8 and smaller)	300	4400
600 V and 1 kV single conductor (size 6 and larger)	500	7300
5 kV to 15 kV power cable	500	7300
25 kV and 35 kV power cable	300	4400
Interlocked armored cable (all voltage classes)	300	4400
Instrumentation cable (single pair)	300	4400
Instrumentation cable (multipair)	500	7300

MBRs for the common wires and cables under tension are given in Table 5.4. Notice that the minimum bending radius factor is applied to the overall cable diameter unless otherwise stated. These limits do not apply to conduit bends, sheaves, or other curved surfaces around which the cable may be pulled under tension while being installed. Larger radii bends are required for such conditions. In all cases, the minimum radius specified refers to the cable inner surface and not to the axis of the cable. The minimum radii values to which such cables may be bent while being pulled into an installation, being under tension, is determined by the formula given below. This value will greatly depend on the tension the cable is experiencing as it exits the bend in question. For instance, the greater the exiting tension, the greater the minimum bending radius is for the cable. In the case of dynamic condition, the minimum bending ratio, function of sidewall pressure, and maximum out tension is expressed as:

$$MBR = \frac{T_{bm}}{P_{SW}} \times 12 \tag{5.25}$$

TABLE 5.4
Minimum Bending Radius of Single and Multiconductor Nonmetallic Portable Cable as a Multiple of Cable Diameter

Cable Type	Minimum Bending Radius (MBR)
0 kV to 5 kV	6
Over 5 kV	8
Control cable (7 conductors and over)	20
Single-conductor cable (metallic tape shielded cable, concentric neutral wire shielded cable, lead sheath cable, metallic fine wire shield)	12
Multiconductor cable (metallic tape shielded cable, concentric neutral wire shielded cable, lead sheath cable, metallic fine wire shield)	7

Example 5.10

Determine the tension in the following cables, and the sidewall pressure where applicable for four 600 kcmil THWN-2 aluminum conductors, having a weight of 0.698 lb/ft., in a 6 in. (154.8 mm inner diameter) in a steel (metallic) conduit, friction coefficient 0.35. Assume a tension of 150 lb. at the input to each of these conduits, a weight correction factor of 1.35 for all calculations. The three conduits used in this application are: (a) a 200 ft. straight horizontal conduit, (b) a 50 ft. pull up inclined at 45°, and (c) a bend of 90° having radius of 36 in. Calculate also the jamming ratio and fill factor for the first conduit.

SOLUTION

(a) Tension for this conduit is given by Equation (5.18a), and being a horizontal straight, there is no sidewall pressure.

$$T_{OUT} = T_{IN} + w_c \mu_f LW = 150 + 1.35 \times 0.35 \times 200 \times 0.689 = 215.96 \text{ lbf} \approx 216 \text{ lbf}$$

The diameter of a 600 kcmil, THWN-2 conductor is 1.051 in. and 0.8676 sq. in. area, while the inner diameter of rigid metallic conduit of 6 in. is exactly 6.093 in., as found in Table B.2-II of Appendix B. The jamming ratio is computed by using Equation (5.16):

$$JR = \frac{ID_{Cond}}{OD_{Cable}} = \frac{6.093}{1.051} = 5.80$$

This value is well out of the jamming range, a 5 in. conduit (5.073 in.) may be used, the jamming ratio in this case is 4.82, still well outside the jamming range. To estimate the percent fill, the combined four-conductor area is divided by the conduit area, 29.158 sq. in. (as found in the same Table B2):

$$\%Fill = \frac{4 \times 0.8676}{29.158} = 0.12 \text{ or } 12\%$$

The value is less than the required 40% fill by the codes.

(b) For the second conduit, the pull tensions are computed by using Equation (5.18b) as:

$$T_{OUT} = 150 + 50 \cdot 0.698 \left(\sin 45° + 1.35 \cdot 0.35 \cos 45° \right) = 185.8647 \approx 185.9 \text{ lbf}$$

(c) The swept angle converter in radians is 1.5707 rad and the tension is given by Equation (519):

$$T_{OUT} = T_{IN} \exp \left(w_c \mu_f \theta \right) = 150 \times \exp \left(1.35 \cdot 0.35 \cdot 1.5707 \right) = 315.0688 \approx 315.1 \text{ lbf}$$

The sidewall pressure in this case, by Equation (5.23b), is:

$$P_{SW} = \frac{3 \times 1.35 - 2}{3} \left(\frac{315.1}{3} \right) = 0.683 \times 105.033 = 71.77 \text{ lb/ft.}$$

This value for sidewall pressure is acceptable for this application.

5.6   SAFETY AND ELECTRICITY HAZARDS

The primary hazards associated with electricity and the use of electricity include direct and indirect shocks, burns, arc blasts, explosions, and fires. However, the most common types of injury are electric shock and burns, with burn injuries arising from either current passing through the body or from the effects of arcing and flashovers. Electric shocks occur when a person closes an electric circuit. Currents, even at 6 mA level, can be fatal due to heart fibrillation and respiratory paralysis, and tissues burn at higher current levels (~5 A). In addition to these direct forms of electrical injury, the following secondary types of injury can occur: bum injuries and the adverse effects of smoke or fume inhalation from fire of electrical origin, the effects of an explosion that has an electrical source of ignition, and physical injuries arising from the reaction to electric shock. An arc flash represents a dangerous condition associated with the release of energy caused by an electrical arc. Arc flash hazards are dangerous conditions deriving from the release of energy due to phase-to-phase or phase-to-ground faults. Arc flash analysis aims to define the procedures necessary to minimize the dangerous effects of arc flash on personnel. Arc rating is the maximum resistance of a determined material to the incident energy, while the flash protection boundary is the distance from exposed conductive parts, within which a person could receive a second-degree burn in case of an arc fault. Arc faults are the short-circuit currents resulting from the conductors at different voltages making less than solid contact. This results in a relatively high-resistance connection compared to a bolted fault. Bolted faults are the short-circuit currents resulting from conductors at different voltages becoming solidly connected together. Arcing current, I_a, represents current flowing through the electric arc plasma, also called arc fault current or arc current. Normalized incident energy En is the amount of energy measured on a surface, at 24" (610 mm) from the source, generated during an electrical arc event of 0.2 s. Incident energy, E, is the amount of energy measured on a surface, a certain distance from the source, generated during an electrical arc event. Clearing time is the total time between the beginning of the overcurrent and the final opening of the circuit at rated voltage by an overcurrent protective device. Limited approach boundary is the electrical shock protection boundary to be crossed by qualified personnel only, not to be crossed by unqualified personnel unless escorted by a qualified person. Restricted approach boundary is the electrical shock protection boundary to be crossed by qualified personnel only, which, due to its proximity to a shock hazard, requires the use of shock protection techniques and equipment when crossed. Prohibited approach boundary is the shock protection boundary to be crossed by qualified personnel only, which, when crossed by a body part or object, requires the same protection as if direct contact is made with a live part. Codes and standards, such as IEEE 1584-2002, "Guide for Performing Arc Flash Calculations," provide guidance for the calculation of incident energy and arc flash protection boundaries. The first step in conducting arc flash calculations is to estimate the arcing short-circuit current. The magnitude of an arcing fault is less than a bolted fault due to the arc impedance. The calculation is based on using the bolted three-phase short-circuit current obtained from a standard short-circuit study. This value, along with other variables, e.g. arc gap distance, is input into the equation. Normal system analysis determines the fault current available at various points throughout the electrical system. For incident energy, it is first necessary to calculate free-air short-circuit arcing currents. Here it is noted that these arcing currents are significantly less than the available bolted-fault short-circuit currents because the arc provides significant circuit impedance. The IEEE 1584 equation for determining arcing currents, for system voltages lower than or up to 1000 V, is expressed by:

$$Log(I_a) = K + 0.662 \cdot \log(I_{bf}) + 0.0966 \cdot V + 0.000526 \cdot G$$
$$+ 0.5588 \cdot V(\log I_{bf}) - 0.00304 \cdot G(\log I_{bf}) \tag{5.26a}$$

And for voltages above 1000 V, the less complicated relationship is given by:

$$Log(I_a) = 0.00402 + 0.983 \cdot \log(I_{bf}) \tag{5.26b}$$

The incident energy, determined from the arcing current, is given by:

$$log(E_n) = K_1 + K_2 + 1.081 \cdot \log(I_a) + 0.0011 \cdot G \tag{5.27}$$

TABLE 5.5

Factors for Equipment and Voltage Classes (Adapted from IEEE 1584)

System Voltage (kV)	Equipment Type	Conductor Gap (mm)	x-Factor Distance
0.208 to 1.0	Open-air installation	10–40	2.000
	Switchgear	32	1.473
	Motor control center	25	1.641
	Cable	13	2.000
1.0 to 5.0	Open-air installation	13–102	2.000
	Switchgear	102	0.973
	Cable	13	2.000
5.0 to 15.0	Open-air installation	13–153	2.000
	Switchgear	153	0.973
	Cable	13	2.000

where I_a is arcing current (kA), K is a constant equal to -0.153 for open configurations and -0.097 for box configurations, I_{bf} is bolted fault current for three-phase faults (symmetrical RMS) (kA), V is system voltage (kV), G is the gap between conductors (in mm), K_1 is -0.792 for open configurations and -0.555 for enclosed, K_2 is 0 for ungrounded and high-resistance, and -0.113 for grounded systems. Finally, the incident energy is converted to normalized energy:

$$E = C_f \cdot E_n (t / 0.2)(610^x / D^x) \qquad (5.28)$$

Here, E is incident density energy in Cal/cm^2, C_f is a calculation factor, equal to 1.0 for voltages larger than 1 kV, and 1.5 for voltages up to 1 kV, E_n is incident energy normalized for time and distance, t is arcing time (seconds), D is distance from arc to person (mm), typically to 455 mm (18") for MCC, and x is distance exponent from IEEE table (based upon equipment type-conductor gap), 1.641 for MCC. The arc flash boundary (AFB), known also as the flash hazard boundary in IEEE 1584, is defined as an "approach limit at a distance from exposed live parts within which a person could receive a second-degree burn if an electrical arc flash were to occur." This boundary is typically calculated as the distance where the incident energy falls off to 1.2 Cal/cm. The arc flash boundary (D_B) is given by:

$$D_B = \left[4.184 \cdot C_f \cdot E_n (t / 0.2)(610^x / E_B) \right]^{1/x} \qquad (5.29)$$

In summary, the first step in conducting arc flash calculations is to estimate the arcing short-circuit current. The magnitude of an arcing fault will be less than a bolted fault due to the arc impedance.

The calculation is based on using the bolted three-phase short-circuit current obtained from a standard short-circuit study. This value along with other variables such as arc gap distance is input into the above equations. The common parameters, specified form the standards, are listed in Table 5.5.

Example 5.11

A short-circuit study indicates the available bolted three-phase fault current at a panelboard is 30 kA. The 480 V power distribution system is considered to be solidly grounded and the arc flash is in a box representative of an enclosure. Estimate the normalized incident energy.

SOLUTION

The voltage system is less than 1000 V, Equation (5.26a) is used here, for I_{bf} of 30 kA, and K = −0.097 (for an arc in a box), and I_a is calculated as:

$$log(I_a) = -0.097 + 0.662 \cdot log(30) + 0.0966 \cdot (0.480) + 0.000526 \cdot (25)$$
$$+0.5588 \cdot (0.48) \cdot log(30) - 0.00304 \cdot (25) \cdot log(30) = 1.22431$$

Then the arcing current is

$$I_a = 10^{1.224321} = 16.761 \, kA$$

In this case, the energy is normalized for arcing time of 0.2 seconds and working distance of 24 inches (610 mm), and must be adjusted for other than 0.2 s and working distance of 24 inches (610 mm), K_1 is equal to −0.555, for arcs in a box, and K_2 is equal to −0.113 for grounded systems, and is calculated by using Equation (5.28), as:

$$log(E_n) = 1.0811 \cdot log(16.761) + 0.0011 \times (25 \, mm) - 0.555 - 0.113 = 0.68310$$

And, then the normalized energy density, for distance 610 mm and C_f, is equal to 1.5 (voltages less than 1.0 kV) is then, 0.96910 Cal/cm² or 4.05473 J/cm².

Electric shock occurs when the human body becomes part of a path through which electrons can flow. The resulting effect on the body can be either *direct* or *indirect*. Injury or death can occur whenever electric current flows through the human body. Even currents of less than 30 mA can result in death. Electric current passing through the body, particularly alternating current at power frequencies of 50 Hz and 60 Hz, may disrupt and affect the nervous system, causing muscular reaction and the painful sensation of electric shock. The consequences can be deadly from cardiac arrest, ventricular fibrillation, or from respiratory arrest, while physiological effects are largely determined by the magnitude and frequency of the current, its duration, and the path it takes through the body. An authoritative guide on the topic is published in IEC 60479. The magnitude of the current is the applied voltage divided by the impedance of the body. The overall circuit impedance comprises the body of the casualty and the other components in the shock circuit, including that of the power source and the interconnecting cables. For this reason, the voltage applied to the body, which is commonly known as the touch voltage, is often lower than the source voltage. Although the electric current through the human body may be well below the values required to cause noticeable injury, human reaction can result in falls from ladders or scaffolds, or movement into operating machinery. Burns can result when a person touches electrical wiring or equipment that is improperly used or maintained. Arc blasts occur from high currents arcing through air. This abnormal current flow (arc blast) is initiated by contact between two energized points. This contact can be caused by persons who have an accident while working on energized components, or equipment failure due to fatigue or abuse. Temperatures about 35,000°F have been recorded in arc-blast research. The primary hazards associated with an arc blast are thermal radiation and pressure waves. In most cases, the radiated thermal energy is only part of the total arc available energy. Various factors, including skin color, area of skin exposed, and type of clothing, have an effect on the degree of injury. Proper clothing, work distances, and overcurrent protection can improve the chances of curable burns. A high-energy arcing fault can produce a considerable pressure wave. In addition, such a pressure wave can cause serious ear damage and memory losses due to mild concussions. In some cases, the pressure wave may propel the victim away from the arc blast, reducing the exposure to thermal energy. However, such rapid movement could also cause serious physical injury. The pressure wave can propel relatively large objects over a considerable distance. The high-energy arc also causes many of the copper and aluminum components in electrical equipment to become molten, while the droplets of molten metal can be propelled large distances by the pressure wave. Electrical systems that are poorly designed, or that have certain fault conditions, may overheat due to excess current flowing to such an extent that adjacent flammable materials may be ignited. Fires may be started by arcs and sparks evolved from short-circuit faults, most frequently resulting from a breakdown in insulation. Hot spots in circuits can develop, when poorly made connections have sufficiently high resistance to cause localized heating, which may lead to fires and other dangerous situations. If standard electrical equipment is installed in hazardous areas, where a flammable or explosive atmosphere exists, the arcs, sparks, electrostatic discharges, or hot surfaces created during normal operation or under fault conditions may have enough energy for them to act as an ignition source. For this reason, correct and proper measures are critical for personnel and equipment safety.

5.7 CHAPTER SUMMARY

Power distribution inside buildings, structures, and industrial facilities is accomplished through the electric services, panelboards, and switchboards. Electric services, feeders, branch circuits, and panelboards are critical and essential components of power distribution inside buildings, structures, and industrial facilities. Feeders are conductors and cables that originate at the main power distribution or main disconnect device and terminate at another distribution center, panelboard, or load center, while subfeeders are the conductors or cables originating at power distribution centers other than the main power distribution center and extending to panelboards, load centers, and disconnect switches that supply branch circuits. A cable tray system is a unit or assembly of units or sections and fittings forming a rigid structural system used to securely fasten and/or support cables, conductors, and raceways, being part of a facility's structural system. Electrical considerations must be included in any design of cable tray systems. Several standards and guidelines exist for the design of cable tray systems, such as NEC sections, NEMA, and IEC specifications. Most manufacturers provide cable trays in a variety of materials, designs, and shapes. The main function or purpose for a circuit protection device is to limit excessive current from flowing into a circuit, network, or equipment that could result in permanent damages, malfunction to other circuit components or equipment, or may affect personnel safety. Among other topics, this chapter gives a comprehensive presentation of the wiring and electrical protection systems in commercial and industrial buildings, fuses, circuit breakers, instrument transformers and protective relays, grounding and ground-fault protection, feeder design, and branch circuits for lighting equipment and electrical motors. A special notice for the engineers and contractors, the above demand factors are the most widely used on a regular basis because of their uniqueness to electrical design. With the application of demand factors, smaller components can be utilized in the electrical system and greater savings can be passed on to the consumer. Due to the high cost of wiring, designers need to utilize these techniques more than ever before to reduce costs. Cables are contained in some type of raceway, typically conduit, duct, or cable tray. A short-circuit current rating determines both the interrupting capability of the circuit breakers and fuses and the mechanical strength of the bracing and support systems that must resist the severe mechanical forces exerted by the intense magnetic fields present during a fault. The basic impulse level, or rating, determines the electrical strength of the insulation. Electrical shock occurs when a person becomes part of an electrical circuit. When electricity passes through the human body, the results can range from death to a slight, uncomfortable stinging sensation, depending upon the amount of electricity that passes through the body, the path that the electricity takes, and the amount of time that the electricity flows. The severity of injury from electric shock is directly related to the current flow path through the body. Understanding arc flash hazards is a critical element in order to reduce the risk of electrical accidents and personal injuries. Sudden releases of energy due to an uncontrolled electric arc, which are the product of short-circuit current and arc duration, can generate arc shock and arc flash hazards. Arc flash hazards must be addressed to provide safe working procedures around energized equipment. NFPA Standard 70E and IEEE Standard 1584 are the arc flash hazard and effects guides for calculations needed to determine the incident energy during an arc flash event.

5.8   QUESTIONS AND PROBLEMS

1. Why are circuit protection devices necessary? What are the three conditions that require circuit protection?
2. What are the two types of protection?
3. Why is system protection required?
4. How are circuit protection devices connected to the electric circuit they are intended to protect, and why are they connected in this way?
5. Briefly describe various generation of relays.
6. What are the common types of circuit protection devices?
7. List the three types of overcurrent conditions.
8. Which overcurrent condition usually has the smallest magnitude of current?
9. What does the current rating of a fuse indicate?
10. What does the voltage rating of a fuse indicate?
11. What are the advantages and disadvantages of fuses?
12. List the major advantages of circuit breakers over fuses.
13. Briefly describe the common system protection relays in use in industry today.
14. What are the five main components of a circuit breaker?
15. What are the three types of circuit breaker trip elements? How does each type of trip element react to an overload?
16. How is reliability achieved in a protective system? Briefly describe dependability.
17. Distinguish between dependability and security of a relay. Briefly describe them.
18. Briefly explain why different protection devices are designed and tested for different X/R ratios.
19. Explain the operation and the purpose of an arc-fault interruption device.
20. Define the following terms: (a) % dependability; (b) % security; and (c) % reliability.
21. What are the most commonly used devices for protection against lightning surges?
22. Is a surge protector necessary for a power supply line? Explain.
23. Why are the short-circuit capabilities of a panelboard important?
24. The performance of an overcurrent relay was monitored over a period of 1 year. It was found that the relay operated 14 times, out of which 12 were correct trips. If the relay failed to issue trip decision on 3 occasions, compute dependability, security, and reliability of the relay.
25. A 460 V switchgear has a peak first-cycle symmetrical fault current of 14,800 A and an X/R ratio of 11.2. What is the minimum symmetrical rating of a fused LV circuit breaker installed in this switchgear?
26. The X/R ratio of a certain bus, having a first-cycle symmetrical fault current 28.95 kA, is 28.5. Calculate the adjusted first-cycle symmetrical fault current for the fuse.
27. A 12.8 kV bus has a 24.5 kA first-cycle symmetrical fault current. What is the highest X/R ratio that a current-limiting fuse rated at 15.5 kA can be installed on this bus?
28. Select a current-limiting fuse for a single-phase 400 kVAR, 13.8 kV capacitor bank.
29. What is the multiplication factor to adjust the first-cycle symmetrical fault current of a fused LV circuit breaker of a bus, having an X/R ratio of 10.2 and a first-cycle symmetrical current of 51.85 kA?
30. The first-cycle symmetrical fault current of feeder is 18.506 kA, and its X/R ratio is 13.5. What is the adjusted first-cycle symmetrical component for a current-limiting fuse used for this feeder?
31. Briefly discuss the basic principle of distance protection.
32. How is a differential protection scheme used in transmission line? In busbar?
33. Determine the length change of 300 ft. PVC conduit, having a thermal expansion coefficient, if the temperature range is from 10°F to 90°F, and the PVC thermal expansion coefficient for PVC is $4.05 \cdot 10^{-4}$ in/ft/°F.

34. How many cables, each with an outside diameter of 0.78 in., can be pulled in a 3 in. rigid steel conduit?

35. Determine the number of 20 A branch circuits required to supply general lighting load in a residence of 2400 sq. ft.

36. What is the minimum number of branch circuits supplying the general lighting load for the following occupancies, a 3600 sq. ft. house (at 15 A rating) and a 1500 m² office space (at 20 A)?

37. What is minimum bending permitted on a 2 in. rigid conduit?

38. The minimum number of 20 A and 120 V branch circuits required to serve the general lighting load for a 3000 square foot, one-family dwelling unit is _____ circuits.

39. NEC Section 220-3(b)(9) requires that receptacles in other than dwelling units be computed at 180 VA each. The maximum number of receptacles that can be installed on a 20 V and 120 V branch circuit in a store is _____. The maximum number of receptacles that can be installed on a 15 A and 120 V branch circuit in a school is _____.

40. List the most common types of conduits.

41. What are the main reasons why cables are installed into conduits?

42. What is the purpose of calculating the jam ratio when pulling cables into conduits?

43. Determine which of the following service disconnects requires ground-fault protection: (a) 700 A, 480Y/277 V; (b) 2000 A, 480Y/277 V; and (c) 2000A, 208Y/120 V.

44. What is the size of rigid steel conduit that can be used to contain three cables, each with an outside diameter of 1.65 in.?

45. Determine the loading, expressed in A, on each of the phase (ungrounded) conductors for the specified loads: (a) 120 V, single-phase, 2400 VA; (b) 277 V, single-phase, 4800 VA; (c) 240 V, three-phase, 7500 VA; and (d) 480 V, three-phase, 27,000 VA.

46. Determine the tension in the following cables, and the sidewall pressure where applicable for three 750 kcmil THWN-2 copper conductors in a 5 in. (128.2 mm inner diameter) PVC conduit, friction coefficient 0.35. Assume a tension of 200 lb. at the input to the conduit, a weight correction factor of 1.4 for all calculations. The three conduits used in this application are: (a) a 300 ft. straight horizontal conduit, (b) a 60 ft. pull up inclined at 45°, and (c) a bend of 90° having radius of 36 in.

47. Consider a raceway layout, in horizontal plane consisting of 300 ft. linear run, a 90° bend with 1.5 ft. radius, followed by a 150 ft. linear run. A four-conductor 500 kcmil copper power cable is used. The cable has an outer diameter 1.2 in., weighs 1.83 pound per foot. Assuming a friction coefficient of 0.25, design the raceway.

48. Consider the conduit raceway layout shown in the figure below. All bends lie in the horizontal plane. Three single-conductor 500 kcmil tape-shielded 5 kV power cables need to be pulled through the conduit. The cable has a 1.093 in. outside diameter and weighs 1.83 lb/ft. A coefficient of friction of 0.25 is anticipated and weight correction factor of 1.4. Determine the optimum conduit for this application.

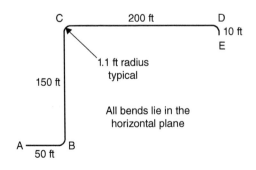

REFERENCES AND FURTHER READINGS

1. T. R. Bosela, *Electrical Systems Design*, Prentice Hall, 2002, ISBN-13: 978-0139754753.
2. ANSI Standard 242 – IEEE Recommended Practice for Protection and Coordination of Industrial and Commercial Power Systems, 1986.
3. ANSI Z535.4-1998, Product Safety Signs and Labels, 1998.
4. R. D. Schultz and R. A. Smith, *Introduction to Electric Power Engineering*, John Wiley & Sons, New York, 1988.
5. D. R. Patrick and S. W. Fardo, *Electric Distribution Systems*, Prentice Hall, 1999.
6. K. Chen, *Industrial Power Distribution and Illuminating Systems*, Marcel Dekker, 1990.
7. H. Ungrand, W. Winkler, and A. Wiesxniewski, *Protection Techniques in Electrical Energy Systems*, Marcel Dekker, 1995.
8. P. M. Anderson, *Power System Protection, IEEE Press Power Engineering Series, Part I–VI*, McGraw-Hill, 1999.
9. IEEE Standard, for Low-voltage AC Power Circuit Breakers used in Enclosures, IEEE Standard C37.13, 1990.
10. IEEE Application Guide for AC High-Voltage Circuit Breakers Rated on a Symmetrical Current Basis, IEEE Standard C37.010, 1999.
11. J. J. Burke, *Power Distribution Engineering*, Marcel Dekker, NY, 1994.
12. J. R. Cogdell, *Foundations of Electric Power*, Prentice Hall, Upper Saddle River, NJ, 1999.
13. L. L. Grisby (ed.), *Electrical Power Engineering*, CRC Press, Boca Raton, FL, 2001.
14. R. C. Dorf (ed.), *The Electrical Engineering Handbook*, CRC Press, Boca Raton, FL, 1993.
15. G. F. Moore, *Electric Cables Handbook* (3rd ed.), Blackwell Science, 1997.
16. A. S. Pabla, *Electric Power Distribution*, McGraw-Hill, 2005.
17. IEEE Guide for Performing Arc Flash Hazard Calculations, IEEE Standard 1584, 2002.
18. NFPA 70: *National Electric Code (NEC) Handbook*, National Fire Protection Association, Quincy, MA, 2014.
19. P. Schavemaker and L. van der Sluis, *Electrical Power System Essentials*, John Wiley and Sons, 2008.
20. M. E. El-Hawary, *Introduction to Electrical Power Systems*, Willey-IEEE Press, 2008.
21. R. D. Schultz and R. A. Smith, *Introduction to Electric Power Engineering*, John Wiley & Sons, New York, 1988.
22. A. C. Williamson, *Introduction to Electrical Energy Systems*, Longman Scientific & Technical, 1988.
23. R. Fehr, *Industrial Power Distribution* (2nd ed.), Wiley, 2015, ISBN: 978-1-119-06334-6.
24. W. A. Thue, *Electrical Power Cable Engineering*, Marcel Dekker, New York, 1999.
25. NFPA 70E-2000, Standard for Electrical Safety Requirements for Employee Workplaces, 2002.
26. NFPA-70E, Standard for Electrical Safety in the Workplace, National Fire Protection Association, 2015.
27. S. W. Blume, *Electric Power System Basics for the Nonelectrical Professional* (2nd ed.), Wiley and Sons, 2016.
28. S. L. Herman, *Electrical Transformers and Rotating Machines*, Cengage Learning, 2017.
29. W. T. Grondzik, et al., *Mechanical and Electrical Equipment for Buildings* (7th ed.), Wiley, 2010.
30. IEEE Standard-1584, Guide for Performing Arc Flash Hazard Calculations, 2008.
31. A. Selva, *Protective Relay Principles*, CRC Press, Boca Raton, FL, 2009.
32. J. C. Das, *Arc Flash Hazard Analysis and Mitigation*, Wiley-IEEE Press, 2012.
33. F. P. Hartwell, J. F. McPartland, and B. J. McPartland, *McGraw-Hill's National Electrical Code 2017 Handbook* (29th ed.), McGraw-Hill, 2017.
34. T. A. Short (ed.), *Handbook of Power Distribution*, CRC Press, 2014.
35. D. R. Patrick and S. W. Fardo, *Electrical Distribution Systems* (2nd ed.), CRC Press, 1999.
36. J. E. Fleckenstein, *Three-Phase Electrical Power*, CRC Press, 2016.

6 Lighting Fundamentals, Lighting Equipment, and Systems

6.1 INTRODUCTION AND HISTORICAL NOTES

Illuminating spaces, both interior and exterior, is a fundamental engineering function and requirement. In order to perform this function properly, knowledge of lights, lighting systems, and equipment characteristics and behavior, as well as a complete understanding of the use of space is necessary. Various activities that take place within the space have different illumination requirements and specifications. The Illuminating Engineering Society of North America (IESNA) and other regulatory bodies have established criteria for lighting applications, some of which are included and discussed in this chapter. Light is not only needed in visual task areas, but also for perception in rooms or designated spaces. Rooms and designated spaces should be illuminated in the proper manner and in agreement with codes and standards. An extensive range of light sources and luminaires are available for this task, to provide improved and adequate lighting. Due to technical progress, the scope of lighting technology has expanded, which has in turn led to the development of more specialized and efficient lighting systems and equipment. Lighting principles, concepts, parameters, design process, methods, equipment and systems, characteristics, and performances are discussed here. The factors involved in determining illumination requirements are discussed in relation to lighting levels for various tasks and the possible use of daylight. Lighting system design considerations related to luminaires are addressed in detail. The purpose of a luminaire is twofold: to hold, protect, and connect a lamp(s) to the electrical system; and to photometrically control the light output. The needs for trade-offs and qualitative decisions when selecting luminaires for a particular space are emphasized. Standards, codes, and specifications for lighting systems are also included and discussed. Examples of lighting calculations and applications are also included. These examples demonstrate how modern and advanced lighting technologies can be integrated, eventually with daylight, to ensure very efficient and high-quality lighting applications and environments. Readers will gain an understanding of the fundamentals of lighting system design, particularly for indoor areas, the various lighting technologies available for commercial and industrial applications, their advantages and disadvantages, and how they work. However, this chapter does not intend to compete with the existing comprehensive range of lighting engineering textbooks and manuals, or to be added to the limited number of beautifully illustrated volumes containing finished projects. On the other hand, an understanding of lighting fundamentals, lighting equipment, codes, and standards is essential for specifiers and decision makers who are evaluating lighting upgrades, improvement, and/or changes.

Light is that part of the electromagnetic spectrum that is perceived by our eyes, which makes things visible. It is defined as electromagnetic radiation or energy transmitted through space or any medium in the form of electromagnetic waves, being defined as visually evaluated radiant energy. Light is that part of the electromagnetic spectrum visible by the human eye, in agreement with the illuminating engineering definition. All electromagnetic radiation is similar. The physical difference between radio waves, infrared, visible light, ultraviolet, and x-rays is their wavelength. A spectral color is a specific wavelength light, exhibiting deep chromatic saturation. Hue is the attribute of color perception denoted by what we call red, orange, yellow, green, blue, and violet. Lighting in conformity with relevant standards is decisive for ensuring that a visual task can be identified and the related activities can be carried out. Consideration of traditional lighting quality characteristics has a major impact on visual task performances. Illuminating spaces, both interior and exterior, is a fundamental engineering function. To perform this function properly, a working knowledge of the characteristics and behavior of light, and a thorough understanding of the use of the space are necessary. Various activities that will take place within the space will have different illumination requirements. Light represents a form of electromagnetic radiation generated from physical processes from the sun or converted from other energy forms in natural or human-made processes and systems. Lighting represents the natural or converted light energy utilization to provide a desired or required visual environment for working and living. For the most part of history, from the origin of humans to most of the eighteenth century, there were basically two sources of light available. The first one is daylight, the natural light originating from the sun, to whose properties the eye has adapted over millions of years. The second source of light was the flame, discovered as an artificial light source during the Stone Age era, with its development of cultural techniques and tools. Lighting conditions remained almost the same for a considerable period of time. Lighting was limited to daylight and flame, and it was for this very reason that humans have continued to perfect the application of these two light sources for tens of thousands of years. In the case of daylight, this meant adapting architecture to the natural lighting requirements. Entire buildings and individual rooms were aligned to the incidence of the rays of the sun. Light not only serves to render spatial bodies three-dimensional, it is also an excellent means for our perception control on a psychological level. Similar processes also took place in the realm of artificial lighting, a development that was clearly driven by the inadequate luminous power provided by the available light sources. The story began when the flame, the source of light, was separated from fire, the source of warmth. Likely, burning branches were removed from the fire and used for a specific purpose. It soon became obvious that it was an advantage to select pieces of wood that combust and emit light particularly well, and the branch was replaced by especially resinous pinewood. The next step involved not only relying on wood's natural features, but also applying flammable materials to torches to produce more light artificially. The development of the oil lamp and the candle meant that humans then had compact, relatively safe light sources to use, and economical fuel selection; eventually the torch holder was reduced to a wick as a means of transport for wax or oil. The oil lamp, which was actually developed in prehistoric times, represented the highest form of lighting engineering progress for a very long time. The lamp itself, later to be joined by the candlestick, continued to be developed over the course of history.

In contrast to the oil lamp and gas lighting, which both started as weak light sources and were developed to become ever more efficient, the electric lamp embarked on its journey in its brightest forms. From the beginning of the nineteenth century, it was a known fact that by creating a voltage between two carbon electrodes, an extremely bright arc is produced. Incandescent gas light was doomed to go the way of most lighting discoveries that were fated to be overtaken by new light sources just as they are nearing perfection. This also applies to the candle, which only received an optimized wick in 1824 to prevent it from smoking too much. Similarly, the Argand lamp was developed as part of gas lighting development, using incandescent mantles, which in turn had to compete with newly developed electric light systems. However, the earlier electric lights required continuous manual adjustment, making it difficult for this new light source to gain acceptance. Also, arc lamps first had to be operated on batteries, a costly endeavor at that time. Following the arc lamp and the incandescent lamp, discharge lamps took their place in electric lighting. Again physical findings were available long before the lamp was put to any practical use. The three main functions of lighting are: (1) ensure the safety of people, (2) facilitate the performance of visual tasks, and (3) help create an appropriate visual environment (appearance and character). In approximately the mid-century, self-adjusting lamps were developed, thereby eliminating the hassle of manual adjustment.

At the earlier stages of its use, the incandescent lamp failed to establish itself as a new light source for technical reasons, much the same as the arc lamp. Only a few materials have a melting point high enough to create incandescence before melting; these higher resistances require thin filaments, which were difficult to produce, broke easily, and burnt up quickly in the air. The initial experiments with platinum wires or carbon filaments did not produce more than a minimum service life. The lifetime was extended when filament predominantly made of carbon or graphite was prevented from burning up by surrounding it with a glass bulb, either vacuum or inert-gas-filled. This breakthrough was made by Thomas Edison in 1879, which succeeded in developing industrial mass products from the experimental systems. This product corresponded in many ways to today's incandescent lamp, right down to the construction of the screw cap. The filament was the only element that needed improvements. Following the arc and the incandescent lamps, discharge lamps took their place as the third electric lamp form. Again physical findings were available long before the lamp was put to any practical use. For over 100 years after scientific research into new light sources began, all the standard electric lamps that we know today had been created in their basic forms. However, to this point, the only sufficient light available was during daylight hours. From then on, artificial light changed dramatically human life. It was no longer a temporary daylight replacement, but a form of lighting ranking with natural light. Illuminance levels similar to those of daylight are technically now produced in interior living and working spaces, or in exterior areas, such as street lighting, public spaces, or for the floodlighting of buildings.

6.2 LIGHTING IN ENGINEERING, ARCHITECTURE, AND INDUSTRIAL PROCESSES

In new lighting design, inadequate light sources had been the main problem, while lighting specialists also face the challenges of controlling excessive amounts of light, managing efficiency and cost, determining how much light is required in which situations, and choosing the optimum lighting systems. Task lighting in particular was examined in detail to establish to what degree illuminance and the kind of lighting influenced productivity. The result of these perceptual physiological investigations was a comprehensive work of reference that contained the illuminance levels required for certain visual tasks, plus minimum color rendering qualities and glare-limitation requirements. Although these standards were set predominantly as an aid for lighting planning for workplaces, they soon became a guideline for lighting in general, and even today these standards are used in lighting design. The perception of an object is more than a mere visual task, and, in addition to a physiological process, vision is also a psychological process, which usually was not considered in lighting design. Quantitative lighting design intends to provide uniform ambient lighting that meets the requirements of the most difficult visual task to be performed in the given space, while at the same time adhering to the standards with regard to glare limitation and color distortion. How we see architecture, for instance, under a given light, whether its structure is clearly legible and its aesthetic quality has been enhanced by lighting, goes beyond the realm of a set of rules. Lighting engineers still tended to practice a quantitative lighting philosophy. Architects began to develop new concepts for lighting design, with daylight as a defining agent. The significance of light and shadow and the way light can structure a building is something with which every architect is familiar. With the development of more efficient artificial light sources, the knowledge that has been gained of daylight technology was used in artificial lighting design. Increasing demand for higher-quality lighting design was accompanied by high-quality lighting equipment demands. Differentiated lighting required specialized luminaires, designed to cope with specific lighting tasks. Different luminaires are required to achieve uniform light over specific areas, for accentuating individual objects, or for the variable lighting required in halls or exhibition spaces. Technical progress in lighting applications led to productive correlations: industry had to meet new designer demands, and further developments in lamp technologies and luminaire design were promoted to suit particular applications required by designers. New lighting developments allow spatial differentiation and more flexible lighting, followed by new developments that offer time-related differentiation in lighting control systems. With the use of advanced control systems, it possible to plan lighting installations that not only offer one fixed application, but are able to define a range of light scenes. Lighting control systems are a logical consequence of spatial differentiation, allowing a lighting installation to be utilized to the fullest, a seamless transition between individual scenes, which is simply not feasible via manual switching.

General lighting is the main source of illumination in a living or working space. Most of our information received about the world around us is through our eyes. Light is not only an essential prerequisite and the medium by which humans are able to see, but its intensity, properties, and distribution throughout a space create specific conditions influencing our perception. Lighting design is, in fact, visual environment planning. Good lighting design aims to create perceptual conditions to allow us to work effectively and to orient safely while promoting a sense of well-being in a particular environment and at the same time enhancing that environment in an aesthetic sense. The physical qualities of a lighting situation can be calculated and measured. Ultimately, the way in which lighting affects the user of the space (i.e., the user's subjective perception) determines whether or not a lighting concept is successful. Lighting design must not be restricted to the creation of technical concepts only. Human perception must be a key consideration in the lighting design process. For example, the uniform, base level of lighting can easily become the focus of energy reduction, as the light levels from other fixtures can be lowered, by using LED and metal halide lamps, for example. Recommended light levels for general lighting are 30–50 foot-candles, providing the area with overall illumination, more specifically for orientation, general tasks, and to control contrast ratios. Diffused general lighting ensures a sense of well-being, and makes employees feel comfortable. A simple way to achieve this is by arranging recessed fixtures using reflectors, baffles, and lens trims. Lighting is one of the best and easiest ways to improve a building, living, and office environment. The challenge is that office lighting plans must be cohesive and effectively illuminate different space types, coexisting under one roof: the reception area, open office space, and private offices of varying sizes. There also are energy codes to follow, concerns about energy costs and efficiency of the lighting system, as well as the need to incorporate flexibility for easy adjustments as the company grows and lighting needs change. Scopes of lighting design are:

- Create an environment enhancing the personnel's well-being and productivity.
- Create flexible lighting settings enabling personnel to perform tasks and operations comfortably, effectively, and safely.
- Integrate and balance ambient, task, accent, and decorative lighting into any facility areas, allowing a comfortable transition from space to space.
- Design lighting systems for long-term employee comfort, higher lighting quality, with a proper energy-use balance, savings and conservation, and future lighting system upgrades.
- Integrate, monitor, and control daylight to improve performance, while reducing energy use.
- Address energy efficiency and conservation, while complying with energy codes and standards.

Designing a facility lighting plan involves more than calculations, lamps, equipment, and luminaire selection. Lighting solutions affect facility ambiance, as well as employee psychological well-being, interests, and enthusiasm, which can enhance positive attitudes and productivity, so efforts must be made to design and create stimulating workplaces. Employees need to perform tasks comfortably and effectively in workplace environments where they spend about one-third of their lives. Lighting is one of the best and easiest ways to improve work and living environments. Lighting design and plans must be cohesive and effectively illuminate spaces coexisting in the same structure, such as: reception area, open office space, home spaces, and private offices of varying sizes, at desired or required lighting levels, while representing and reinforcing the corporate or organization image. There are also energy codes to follow, concerns about energy costs and lighting system efficiency, and the need to incorporate flexibility for easy adjustments as the company and lighting needs change. Light has a triple effect on human behavior:

1. Light is used for visual functions, with the area of illumination must conform to standards, codes, and regulations and be glare-free and convenient;
2. Light systems create biological effects, stimulating or relaxing; and
3. Light is used for emotional perceptions, enhancing architecture, and creating scenes and effects.

6.3 LIGHTING THEORY AND ILLUMINATION CALCULATION METHODS

Light is an electromagnetic radiation form, emitted by natural or artificial sources. Human eyes are sensitive to only a very narrow band of electromagnetic spectrum, the visible spectrum. The visible spectrum covers a narrow band of wavelength from approximately 380 nm to 770 nm. Wavelengths shorter or longer than these do not stimulate the human eye receptors. Electromagnetic radiation is transmitted to the free space at a speed of 3×10^8 m/s (186,282 mi/s). Light and other electromagnetic radiation travel at lower speeds in other media, such as air or water. All waves differ in wavelengths and frequencies, expressed by:

$$c = \lambda \times f \qquad\qquad (6.1)$$

Where, c is the electromagnetic radiation (light) speed in m/s, f is the frequency in Hz, and λ is the wavelength in m. Electromagnetic radiation can be converted from other energy forms, via chemical, thermal, or electrical processes, the last one being the most efficient. Electrical energy is also part of the electromagnetic spectrum, at a range of lower frequencies. Light generated inside of buildings is nearly all converted from electricity. With the advent of modern technologies, the efficiency of electrical energy to lighting has advanced significantly, and modern light sources are electrical light sources. Each light source is characterized by a spectral power distribution or spectrum. The spectral power distribution curve (spectrum) of a light source shows the radiant power that is emitted by the source at each wavelength, over the electromagnetic spectrum (primarily in the visible region). With color temperature and color rendering index ratings, defined in the next sections of this chapter, the spectrum can provide a complete picture of the color composition of a lamp's light output. Incandescent lamps and natural light produce a smooth, continuous spectrum, while other types of lamps may produce spectra with discrete lines or bands. For example, fluorescent lamps produce spectra with a continuous curve and superimposed discrete bands. The continuous spectrum results from the halo-phosphor and rare earth phosphor coating, while the discrete band or line spectrum results from the mercury discharge.

6.3.1 Basic Parameters Used in Lighting Physics

In lighting technology a number of technical terms and units are used to describe the properties of light sources and the effects that are produced. Radiometry is the study of optical radiation (e.g., light, ultraviolet radiation, and infrared radiation). Photometry, on the other hand, is concerned with human visual response to light. Radiometry deals with the total energy content of the radiation, while photometry examines only the radiation that humans sense. The common unit in radiometry is the watt (W), which measures radiant flux (power), while the most common unit in photometry is the lumen (lm), which measures luminous flux. For monochromatic light of 555 nm, 1 W is equal to 683 lumens. For light at other wavelengths, the conversion between watts and lumens is different, as the human eye responds differently to each wavelength. Similarly, the radiant intensity is measured in watts/steradian (W/sr), while luminous intensity is measured in candelas (cd, or lm/sr). Luminous flux describes the total amount of light emitted by a light source. This luminous flux could be measured or expressed in watts, too. This does not, however, describe the light source optical effects adequately, because the varying eye spectral sensitivity is not taken into account. To include the spectral sensitivity of the eye, the luminous flux is measured in lumen. Radiant flux of 1 W emitted at the peak of the spectral sensitivity (in the photonic range at 555 nm) produces a luminous flux of 683 lm. The radiant flux produced is frequency-dependent.

Luminous efficacy describes the luminous flux of a lamp in relation to its power consumption and is therefore expressed in lumen per watt (lm/W). The maximum value theoretically attainable when the total radiant power is transformed into visible light is 683 lm/W. The luminous efficacy varies from one light source to another, but always remains well below this optimum value. The light quantity, or luminous energy, is a product of the luminous flux emitted multiplied by time, usually expressed in klm·h.

Luminous flux, Φ, is used to describe the quantity of light emitted by a light source. Its unit is lumen (lm). The luminous efficiency is the ratio of the luminous flux to the electrical power consumed (lm/W). It represents a measure of a light source's efficiency, directly related to the light energy. Light energy degrades very quickly into heat and cannot be stored. In order to maintain a certain light level in any space, electricity must be supplied continuously. Light energy, W_{Light} measured in lm·s (or lm·hr) is defined by:

$$W_{Light} = \int \Phi \cdot dt \tag{6.2}$$

In physics, power is the rate of energy change, while in lighting, light power is the luminous flux emitted by a light source in unit of time, given by:

$$\Phi = F \text{ (Luminous Power)} = \frac{dW_{Light}}{dt} \text{ (lm)} \tag{6.3}$$

In the case of constant light power, the light energy is expressed as:

$$W_{Light} = F \times t$$

The *luminous intensity* describes the quantity of light that is radiated in a particular direction. An ideal point-source lamp radiates luminous flux uniformly into the space in all directions; its luminous intensity is the same in all directions. In practice, however, luminous flux is not distributed uniformly. This results partly from the design of the light source, and partly on the way the light is intentionally directed. It makes sense, therefore, to have a way of presenting the spatial distribution of luminous flux, i.e., the luminous intensity distribution of the light source.

The unit for measuring luminous intensity is candela (cd). The candela is the primary basic unit in lighting technology from which all others are derived. The candela was originally defined by the luminous intensity of a standardized candle. In other words, the intensity of light is defined as the light flux density in a given direction, or the ability of a light source to produce illumination (or illuminance, defined bellow) in a given direction. Candela (cd) is defined as 1 lumen per steradian of solid angle, a useful measurement for directive lighting elements such as reflectors. The light source luminous intensity distribution throughout a space produces a three-dimensional graph. A section through this graph results in a luminous intensity distribution curve (LDC), describing the luminous intensity on one plane, usually given in a polar coordinate system as the function of the beam angle. To allow comparison between different light sources to be made, light distribution curves are based on a 1000 lm output. In the case of symmetrical luminaires, one light distribution curve is sufficient to describe them, while axially symmetrical luminaires require two curves, which are usually depicted in one diagram. The polar coordinate diagram is not sufficiently accurate for narrow-beam luminaires, e.g., stage projectors. In this case, it is typical to provide a Cartesian coordinate system. The luminous intensity is expressed as:

$$I = \frac{d\Phi}{d\Omega} \tag{6.4}$$

Here, Ω is the solid angle (measured steradian, or sr) into which luminous flux is emitted, given by:

$$\Omega = \frac{dA}{r^2} \text{ (here, r is the radius of an imginary sphere)}$$

Radiant intensity, I, is the amount of power radiated per unit solid angle, measured in W/sr. Luminous intensity is the amount of visible power per unit solid angle, measured in candelas (cd, or lm/sr). Luminous intensity (I_v) is the fundamental SI quantity for photometry. The candela is the fundamental unit from which all other photometric units are derived.

   Illuminance describes the quantity of luminous flux falling on a surface and its unit is Lux (lx). The light quantity is expressed, in engineering calculations, in terms of average number of lumens per unit of area. In the British system, the quantity of light is expressed in foot-candles (fc), with 1 foot-candle being equal to 1 lumen per square foot, while in the SI system, the light quantity is measured in lux (lx), where 1 lux is equal to 1 lumen per square meter. Note that the conversion relation between foot-candle and lux is: 1 foot-candle = 10.764 lux.

Example 6.1

Determine the illuminance of a 12 ft. by 12 ft. room, illuminated by a 6480 lumen light source.

SOLUTION

$$\text{Illuminance} = \frac{6480 \text{ lumens}}{12 \times 12 \text{ square ft}} = 45 \text{ fc.}$$

In physical term the illuminance is expressed by the inverse square law, stating that it decreases by the square of the distance. Relevant standards specify the required illuminance. Illuminance is the means of evaluating the luminous flux density, indicating the luminous flux from a light source falling on a given area. Illuminance need not necessarily be related to a real surface. It can be measured at any point within a space. Illuminance is determined from the light source luminous intensity. Illuminance decreases with the squared distance from the light source. Luminance is the only basic lighting parameter perceived by the eye. It specifies the brightness of a surface and is essentially dependent on its reflectance.

$$E = \frac{d\Phi}{dA} = I \frac{d\Omega}{r^2 \cdot d\Omega} = \frac{I}{r^2} \tag{6.5}$$

where A is the area hit by the luminous flux, measured in m². This is also known as the *inverse square law*, stating that the illuminance on a surface is directly proportional to the intensity, I, and inversely proportional to the square of the source-surface distance, the surface being normal to the direction of the light source. When the light-receiving surface is not perpendicular to the light beams, the luminous flux will cover a larger area, and Equation (6.5) is modified as:

$$E = \frac{I \times \cos(\theta)}{r^2} \tag{6.6}$$

Here, θ is the angle between luminous flux direction and the surface normal.

Example 6.2

A lighting fixture has a luminous intensity of 8500 cd. Calculate the illuminance on a work desk 2.5 m below the source.

SOLUTION

From Equation (6.6) the illuminance is:

$$E = \frac{8500 \cdot \cos(0°)}{2.5^2} = 1360 \text{ lx}$$

If the source is aimed at an angle toward a target on a vertical surface, the reduction in illuminance at the target is equal to the sine of the angle of incidence or tilt, θ, and is given by:

$$E = \frac{I \times \sin(\theta)}{r^2} \qquad (6.7)$$

where again I is the intensity of the source (in candelas) in the direction of the light ray, θ is the angle of tilt between nadir and the direction of the target, and r is the distance from the source to the target.

Example 6.3

A lamp has an intensity of 18,000 cd. Calculate the illuminance if the lamp is tilted at 30° from nadir to cast light on a vertical surface 6 ft. away.

SOLUTION

From Equation (6.7) the illuminance is:

$$E = \frac{I \times \sin(\theta)}{r^2} = \frac{18000 \times \sin(30°)}{(6)^2} = 250 \text{ fc}$$

When the surface receiving the luminous lux is not perpendicular to the light source, the flux will cover a larger area, in relation to the cosine of the angle, and for a distance, h, between the light source and the surface, at the normal to the surface, so Equation (6.5) became:

$$E = \frac{I \cdot \cos^3(\theta)}{h^2} \qquad (6.8)$$

Example 6.4

If a lamp with intensity of 6000 cd is focused at a painting on a wall 6 ft. from the light at an angle of 45°, calculate the illuminance.

SOLUTION

From Equation (6.8):

$$E = \frac{I \cdot \cos^3(\theta)}{h^2} = \frac{6000 \times (0.707)^3}{6^2} = 58.9 \text{ f c or } 10.76 \times 58.9 = 633.8 \text{ lx}$$

In summary and following the practical notation, as a surface that is illuminated by a light source moves away from the light source, the surface appears dimmer. In fact, it becomes dimmer much faster than it moves away from the source. The inverse square law, which quantifies this effect, as shown in Figure 6.1, relates illuminance (E_v) and intensity (I_v) as follows:

$$E_v = \frac{I_v}{d^2} \qquad (6.9)$$

where d is the distance from the light source. For example, if the illuminance on a surface is 40 lux (lm/m^2) at a distance of 0.5 meters from the light source, the illuminance decreases to

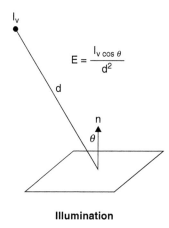

Illumination

FIGURE 6.1 Illumination diagram.

10 lux at a distance of 1 meter. Note that, the inverse square law can only be used in cases where the light source approximates a point source. If the light reaches the surface under an angle of incidence, θ, other than 90°, the illuminance is due to the perpendicular component and is expressed as:

$$E_v = \frac{I_v}{d^2}\cos\theta \qquad (6.10)$$

Example 6.5

A point source generates 3000 cd in the direction of interest at an angle of 36°. Determine the illuminance at a point 13.5 ft. from the source.

SOLUTION

The illuminance is given by:

$$E_v = \frac{I_v}{d^2}\cos\theta = \frac{3000}{(13.5)^2}\cos 36° = 13.3 \text{ fc}$$

Notice that the inverse square method yields only a rough idea of what is perceived. Its main use is for comparison, as when establishing the illuminance ratio between an object and its surrounding. The inverse square method does not account for any inter-reflections within the space. Even more significantly, perceived brightness depends on the surface reflectance and the observer position. For the so-called Lambertian light sources, a useful guideline to use for illuminance measurements is the *five times rule*: the distance from the measurement point to the light source needs to be greater than five times the largest source dimension, for an accurate measurement. However, this rule does not work for a directional light source. In the case of a constant luminous flux, F, the illuminance is computed by:

$$E = \frac{F}{A} \qquad (6.11)$$

Example 6.6

A room 5 m × 8 m is illuminated by 12 lighting fixtures, each of 3500 lumens. If 75% of the light energy can be used at the desk level, what is the average desk illuminance?

SOLUTION

From Equation (6.11), the illuminance is:

$$E = \frac{12 \times 0.75 \times 3500}{5 \times 8} = 787.5 \text{ lx}$$

When the light reaches the surface at incidence angles other than 90°, then the illuminance is given by the normal component of the light ray reaching the surface. For a light ray reaching the surface at an angle, θ, Equation (6.11) becomes:

$$E = E_0 = \frac{F}{A}\cos(\theta) \tag{6.12}$$

Example 6.7

A lamp produces 3000 lumens in the direction of interest. If the angle of incidence with respect to the vertical is 30°, what is the average desk illuminance for the case, the geometry, and 100% light energy used, as in Example 6.4?

SOLUTION

From Equation (6.12), the illuminance is:

$$E = \frac{F}{A}\cos(\theta) = \frac{3500}{40}\cos(30°) = 75.8 \text{ lx}$$

The terms *illumination level* and *illumination* are still used in the lighting industry, though to a lesser extent than in the past. These terms are more distinguishable than *luminance*, which means "brightness" to a layperson. ***Exposure*** is described as the product of the illuminance with exposure time. Exposure is an important issue, regarding the light exposure calculation on exhibits in museums. A Lambertian surface reflects or emits equal (isotropic) fluxes in every direction. For example, an evenly illuminated diffuse flat surface such as a piece of paper is approximately Lambertian, the reflected light being the same in all directions from which the paper surface is seen. However, it does not have isotropic intensity, because the intensity varies according to the cosine law. Figure 6.2 shows

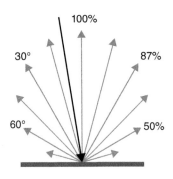

FIGURE 6.2   A Lambertian surface.

a Lambertian reflection from a surface. The reflection follows the cosine law, so the reflected energy in a particular direction (the intensity) is proportional to the cosine of the reflected angle. Remember that luminance is intensity per unit area. Because both intensity and apparent area follow the cosine law, they remain in proportion to each other as the viewing angle changes. Therefore, luminance remains constant while luminous intensity does not. To compare illuminance and luminance on a Lambertian surface, consider the following example: a surface with a luminance of 1 lm/m²/sr radiates a total of πA lumens, where A is the area of the surface, into a hemisphere (which is 2π steradians). The illuminance of the surface is equal to the total luminous flux divided by the total area πlx/m². In other words, if you were to illuminate a perfectly diffuse reflecting surface with 3.1416 lm/m², its luminance would be 1 lm/m²/sr.

Luminance, L, is the only basic lighting parameter that is perceived by the eye. Luminance is apparent brightness, how bright an object appears to the human eye. Whereas illuminance indicates the amount of luminous flux falling on a given surface, luminance describes the brightness of an illuminated or luminous surface. Luminance is defined as the ratio of luminous intensity of a surface (cd) to the projected area of this surface (m²). In the case of illumination, the light can be reflected by the surface or transmitted through the surface. In the case of diffuse reflecting (matte) and diffuse transmitting (opaque) materials, luminance can be calculated from the illuminance and the reflectance or transmittance. Luminance is the basis for describing perceived brightness, and the actual brightness is still influenced by the eye adaptation state, the surrounding contrast ratios, and the information content of the perceived surface. So when we look at the world what we see is varying luminance patterns, specifying the surface brightness, being essentially dependent on its reflectance (finish and color). In other words, it is directly related to the perceived brightness of a real or imaginary surface, the visual appearance of a surface by the illuminance of a surface. Brightness is a surface-subjective evaluation, while illuminance is an objective characteristic of surface. In fact, we do not see the surface illuminance, but rather the surface brightness or the surface luminance differences (contrast), measured in cd/m², and defined as the luminous intensity, I, of a surface in a given direction, θ, per unit of projected area, A_θ, from that direction:

$$L = \frac{dI_\theta}{dA_\theta} \qquad (6.13)$$

Exitance, M, measured in lm/m² or lm/ft.², represents the total luminous flux leaving a surface regardless of the direction. It is the ability of a surface to emit light expressed as the luminous flux per unit area at a specified point on the surface M. The luminous exitance or luminous emittance of a radiator is the total flux emitted in all directions from a unit area of the radiator. For a sphere with radius R, the exitance is expressed as:

$$M = \frac{d\Phi}{dA} = \frac{1}{R^2}\int L\cos(\theta)\cdot dA \qquad (6.14)$$

If the illuminance, E, (in lx or fc) is known, the exitance in the case of surface reflectance and in the case of surface transmittance may be computed as:

$$M = \rho \times E$$
$$M = \tau \times E \qquad (6.15)$$

Here, ρ and τ are the per-unit reflectance and transmittance of a surface, respectively.

Example 6.8

The illuminance of a surface is measured to be 750 lx, and surface reflectance is 0.75. What is the approximate exitance?

SOLUTION

From Equation (6.11), the exitance is:

$$M = 0.75 \times 750 = 562.50 \text{ lm/m}^2$$

Contrast, C, is not a physical property, but rather our visual perception and the response to the light energy emitted and transmitted from object surfaces. For the human eye to see details, there must be either a difference in color (wavelength), in luminous intensity (brightness), or both. Luminance is either generated by a light source or may be reflected or transmitted from an object. For example, to differentiate two objects of the same color situated side-by-side, one must be more reflective than the other. This is called *contrast*, and is defined as:

$$C = \frac{|L_o - L_b|}{L_b} \tag{6.16}$$

Here, the contrast, C, is unitless, and L_o and L_b are the object and the background luminances, in candela per unit area.

Example 6.9

If the illuminance of an object is 2000 cd/m² and the illuminance of the background is 100 cd/m², what is the luminance contrast?

SOLUTION

From Equation (6.12), the luminance contrast is:

$$C = \frac{|2000 - 100|}{100} = 19$$

6.3.2   THE VISIBLE SPECTRUM AND COLOR

The visible spectrum is the portion of the electromagnetic spectrum that is visible to the human eye. Electromagnetic radiation in this range of wavelengths is called visible light or simply light. A typical human eye responds only to the wavelengths from about 390 to 700 nm. In terms of frequency, this corresponds to a band of 430 to 770 THz. The spectrum does not, however, contain all the colors that the human eyes and brain can distinguish. Unsaturated colors such as pink, or purple variations such as magenta, are absent, for example, because they can be made only by a mix of multiple wavelengths. Colors containing only one wavelength are also called pure colors or spectral colors. Visible wavelengths pass through the *optical window*, the electromagnetic spectrum region allowing wavelengths to pass largely unattenuated through the earth's atmosphere. An example of this phenomenon is that clean air scatters blue light more than red wavelengths, and so the midday sky appears blue. The optical window is also referred to as the *visible window*, because it overlaps the human visible response spectrum. The near infrared (IR) window lies just out of the human vision, as well as the medium wavelength IR window, and the far infrared (FIR) window, although other animals may experience them. The human eye is our primary source of information about the outside world. It is first and foremost an optical system, often described as a camera, creating images on the retina, where the pattern of luminances is translated into nervous impulses. The retina poses light-sensitive receptors that allow high-resolution visual images. One of the most remarkable eye properties is its ability to easily adapt to lighting conditions, from moonlight or sunlight, although there is a difference by a factor of 10^5 in the illuminance. Eyes are capable of performing an extremely wide range of tasks, e.g., perceiving a faintly glowing star in the night sky, producing only an illuminance of about 10 lux on the eye, by its capacity of regulating incident light in a 1-to-16 ratio. Adaptation is performed to a large extent by the retina. Although vision is therefore possible over an extremely wide range of luminances, there are strict limits with regard to contrast perception in each individual lighting situation, lying in the fact that the eye cannot cover the entire possible luminances' range instantaneously; rather, it adapts to cover a narrow range in which differentiated perception is possible. Objects possessing high luminance for a particular adaptation level cause glare, appearing as extremely bright, whereas low-luminance objects often appear as very dark. The eye is able to adjust to new luminance conditions, by simply selecting a different restricted range. This adaptation process needs time, however. Adaptation from dark to light situations occurs quite rapidly, whereas the adaptation from light to darkness requires a longer time. The fact that luminance contrast is processed by the eye within a certain range, needing time to adapt to new light levels or brightness, has impacts on lighting design, or when adjusting lighting levels in adjacent spaces.

Objects possessing luminance too high for a particular adaptation level can cause glare, appearing as extremely bright, whereas low-luminance objects appear as very dark. Attempts to describe visual perception effectively must take into account the criteria by which the perceived information is affected. Any particular information is related to the observer's current activity (e.g., movement-related or any other activity) and work, requiring visual information. The specific information received depends on the activity type, with most of the required information from the need to feel safe, to be able to evaluate danger, or to comprehend the environment structure, which applies both to orientation, route, destination, and knowledge about environment qualities and peculiarities. This information or lack of it determines the way a person feels and his or her behavior, which may lead to tension and unrest in unknown or potentially dangerous situations, but relaxation and tranquility in a familiar and safe environment. The information about the world around us allows us to adapt our behavior to specific situations, including knowledge of weather conditions and the time of day as well as information relating to other activities occurring in the given environment. Areas that promise significant information be it in their own right, or through accentuation with the aid of light, are perceived first, attracting our attention. The information content of a given object is responsible for its being selected as an object of perception. Moreover, the information content also has an influence on the way in which an object is perceived and evaluated. The human eye utilizes the part of the spectrum of electromagnetic waves available to gather information about the world around us, perceiving the light amount and distribution that is radiated or reflected from objects to gain information about their characteristics, perceiving the light color to acquire additional information about these objects. We see in two ways: the color differences and luminance contrasts (brightness) transmit information from the eye to brain, where we process the information and transmit the information back to the eye. For eyes to see, there must be color differences and/or luminance contrasts, while high luminance contrast tends to improve visual activities. However, high luminance contrast for longer durations may cause discomfort and must be taken into account in lighting applications and design. If the light source or object luminance is high enough to interfere with our vision, excessive contrast or luminance becomes distracting and annoying, and brightness has a negative side, glare. Extreme glare cripples vision by reducing or destroying the ability to see accurately. Glare is often misunderstood as *too much light*, while in fact, the light coming from the wrong direction results in an extreme luminance within the normal field of view. An essential feature of good lighting is the extent to which glare is limited. Glare strong enough to cause physiological discomfort is called *discomfort glare*; when it affects the ability to see, it is called *disability glare*. The two aspects to glare are the objective depreciation of visual performance, and the subjective disturbance felt by individuals through excessive luminance levels or stark contrasts in luminance levels within the field of vision. *Direct glare* originates from a light source, and indirect (reflected) glare reflects from a surface. Direct glare can be avoided by changing the light source position, while indirect glare is minimized by replacing the reflecting surface with a matte (non-glaring) or low-reflectance (dark) surface. The evaluation of luminance and luminance contrasts that may lead to unwanted glare is predominantly dependent on the type of environment and the requirement that the lighting aims to fulfill. In the case of direct glare, there is the quantitative method of evaluating luminance-limiting curves. Reflected glare is excessive uncontrolled luminance reflected from objects or surfaces in the field of view, which includes the reflected luminance from interior surfaces and/or the luminance of the lighting system. Specular surfaces have reflecting properties similar to those of a mirror. The luminance reflected is the mirrored image of the light source and/or of another lighted surface within the reflected field of view. Reflected glare can be evaluated only according to qualitative criteria, however.

In the case of objective depreciation in visual performance, the term *physiological glare* is applied. In this case, the light from a glare source superposes the luminance pattern of the visual task, thereby reducing visibility. The reason for this superimposition of the luminous intensities of visual task and glare source may be the direct overlay of the images on the retina. Superimposition of scattered or disturbing light, which arises through the dispersion of the light from the glare source within the eye, is often enough to reduce visual performance. The degree of light scattering depends primarily on the opacity of the inner eye. The latter increases with age and is the reason why older people are considerably more sensitive to glare. The most extreme case of physiological glare is disability glare. This arises when luminance levels higher than 10^4 cd/m^2 are evident in the field of vision, e.g., when we look directly at artificial light sources or at the sun. Disability glare does not depend upon the luminance contrast in the environment. Suitable glare limitation can be achieved by the correct choice of luminaires. Especially developed reflectors can guarantee that luminaires positioned above the critical angle do not produce any unacceptable luminances. Installing luminaires that emit only minimal direct light downward can also substantially limit reflected glare.

6.3.3 Color Specifications and Characteristics

Different wavelengths in the visible spectrum (the band from 380 to 780 nm) produce different color sensations. In order to quantify color, the spectrum or wavelength composition of light must be known. A spectral power distribution (SPD), defined as the radiant power at each wavelength or band of wavelengths in the visible region, is typically used to characterize light. Depending on how light is generated by the source, the light SPD can vary from continuous across the visible spectrum to discrete across the spectrum to a narrow band at a particular wavelength. The common visible spectrum colors are blue, cyan (blue-green), green, yellow (green-red), red, and violet (magenta or red-purple). White light is a mixture of all colors in the visible spectrum, with a perfect white (an idealization) being a mixture of all colors of equal energy level. Daylight at noon is the closest match to the perfect white. Identical colors are produced not only by identical SPDs but also by many different SPDs that produce the same visual response. Physically different SPDs appearing to have the same color are called metamers. Wavelengths shorter than violet, in the band between 200 and 380 nm, are called ultraviolet, while the region with wavelengths higher than 780 nm is called infrared. Of all colors, red, green, and blue are dominant, being the basis for forming all other colors, the *primary colors*, while the other colors are the *secondary colors*. Secondary colors are the result of mixing primary colors. Mixing all three primary lights at equal energy levels produces white light. Light is reflected from opaque or translucent materials; we see most objects through the reflected light. Colors are a matter of visual perception's subjective interpretation. The color of a light or an object can be described by the following characteristics: (a) the basic colors and the mixture of these colors (the hue), (b) the color shade (the value), and (c) the intensity or degree of color saturation (the chroma). There several systems for color specification, with Munsell color system and Commission Internationale de l'Eclairage (CIE) the most commonly used. Brightness is referred to in this case as the reflecting coefficient of an object's color, hue is the color tone, and the term *saturation* refers to the degree of color strength from pure color to the noncolored grey scale.

The color of illuminated objects is the result of the spectral composition of the light falling on a body and its ability to absorb or transmit certain components of this light and only reflect or absorb the remaining frequency ranges. In addition to the resulting, objective calculable or measurable color stimulus, the eye color adaptation also plays a role in the actual perception of things. The eye is able to gradually adapt to the predominant luminous color, similar in its adaptation to luminance levels, meaning that in the case of a lighting situation that comprises different luminous colors, virtually constant perception of the scale of object colors is guaranteed. The deviation degree is referred to as the light source color rendering. Color rendering is defined as the degree of change that occurs in the color effect of objects through lighting with a specific light source in contrast to the lighting with a comparative reference light source, being a comparison of the similarity of color effects under two lighting types. Because the eye can adapt to different color temperatures, color rendering is determined in relation to luminous color. In order to determine the color rendering of a light source, the color effects of a scale of eight object colors are calculated under the lighting types to be evaluated, and under comparison standard lighting, and then the two are compared. The quality of the color rendering established is expressed as a color rendering index, which can be related to both general color rendering, and the rendering of individual colors. The maximum index of 100 represents optimum color rendering, and lower values correspondingly less adequate color rendering.

Human eyes are sensitive to the electromagnetic spectrum visible range, while the colors are a matter of visual perception, subjective to each person. Color is a perception caused by the unbalanced mixture of light wavelengths. An object can be described in terms of color, but terms such as plain, deep, or bright are inadequate to describe the color precisely. The light color can accurately be described by the following three terms. *Hue* is the basic color, such as red, yellow, green, or blue and the mixture of them. *Value* represents the color shade, such as light or dark blue. It has more significance in painting or architecture rather than in lighting. *Chroma, purity,* or *saturation* is the color saturation degree or intensity, meaning if a color is dull or vivid, these terms indicate the freedom of the color from white or gray. *Tints, tones,* or *shades* are colors diluted with white, grey, and black. In lighting, this represents the dilution of the spectrum saturated light with white lights. Colors are specified using the above-mentioned systems. A colored lamp or lighting system has most of its energy concentrated around a specific wavelength, while a colored object reflects most incident light in a narrow wavelength band. Correlated color temperature, measured in Kelvin units (K), describes the light appearance generated by a hot object (a heated object produces light), such as an incandescent filament, generated light being correlated to the blackbody curve. At lower temperatures, reddish light is generated, and as the temperature increases the light appears to shift from red to reddish-yellow to yellowish-white to white to bluish-white at high temperatures. The light source color can be quantified using absolute temperature (K) to describe the light emitted by a blackbody radiator. For blackbody temperature of 800 K, red light is emitted, yellow corresponds to 3000 K, white to 5000 K, pale blue to 8000 K, and bright blue to 60,000 K. Lamp color temperature rating is a convenient way to describe the lamp's color characteristics. However, this has nothing with the lamp's operating temperature or lamp color rendering. The illuminated object color is the result of the spectral composition of the light falling on a body and its ability to absorb or transmit certain light components, while only reflecting or absorbing the remaining frequency ranges. In addition to the resulting, objective measurable color stimulus, the eye color adaptation also plays a role in the actual perception. The eye is able to gradually adapt to the predominant luminous color, in a similar way that it adapts to a luminance level, meaning that in the case of lighting situation, comprising different luminous colors virtually constant perception of the scale of object colors is guaranteed.

The degree of deviation is referred to as the color rendering of the light source. Color rendering is defined as the degree of change that occurs in the color effect of objects through lighting with a specific light source in contrast to the lighting with a comparative reference light source. It is therefore a comparison of the similarity of color effects under two types of lighting. Because the eye can adapt to different color temperatures, color rendering must be determined in relation to luminous color. One single light source cannot therefore serve as the reference source—the comparison standard is rather a comparable light source with a continuous spectrum, a thermal radiator of comparable color temperature, or daylight. Other useful color measures can be derived from colorimetry. The most commonly used are color rendering index (CRI) and correlated color temperature (CCT). Color rendering index is a measure of how colors of surfaces will appear when illuminated by a light source. Light that has an even spectral distribution across the visible spectrum, such as daylight or incandescent light, has a high CRI (the maximum is 100). Light that has gaps in its spectral distribution has a lower CRI. To determine the light source color rendering, the color effects of a scale of eight object colors are calculated under lighting types to be evaluated as well as under comparison standard lighting, and the two then are compared. CRI is one of the ways to quantify the color rendering quality of a light source, being a measure of the shifts when standard color samples are illuminated by the light source, compared to the reference (standard) light source. The established color rendering quality, expressed as color rendering index, relates to both general color rendering (Ra) and the rendering of individual colors. The maximum index of 100 represents optimum color rendering, and lower values correspondingly less adequate color rendering. The color rendering quality is an important criterion when choosing light sources. The degree of color fidelity represents, therefore, which illuminated objects are rendered in comparison to reference lighting.

6.3.4 LIGHT CONTROL AND BASIC CONCEPTS IN OPTICS

Light travels in clean air or a similar transparent medium without bending or notable loss until it is intercepted by another medium, which is reflecting, absorbing, transmitting, refracting, diffusing, or polarizing the light. These material characteristics are used as methods to control light and to achieve better and more appropriate lighting. When light encounters a surface, it can be either reflected away from the surface or refracted through the surface to the material beneath. Once in the material, the light can be transmitted, absorbed, or diffused (or any combination) by the material. Note that these properties usually apply not only to light, but to all forms of electromagnetic radiation. However, we are limiting this discussion to light. The three general types of reflection are: *specular*, *spread*, and *diffuse*. A specular reflection, such as what you see in a mirror or a polished surface, occurs when light is reflected away from the surface at the same angle as the incoming light's angle. A spread reflection occurs when an uneven surface reflects light at more than one angle, but the reflected angles are all more or less the same as the incident angle. A diffuse reflection, called also *Lambertian scattering* or *diffusion*, occurs when a rough or matte surface reflects the light at many different angles. Specular reflections demonstrate the law of reflection, which states that the angle between the incident ray and a line that is normal (perpendicular) to the surface is equal to the angle between the reflected ray and the normal, as shown in Figure 6.3a. The angle between an incident ray and the normal is called the incident angle, denoted by the symbol θ. The angle between a reflected ray and the normal is called the reflected angle, denoted by the symbol θ'.

When light travels from one material to another (such as from air to glass), it refracts, bends, and changes velocity. Refraction depends on two factors: the incident angle (θ) and the refractive index of the material, denoted by the letter n. The index of refraction for a particular material is the ratio of the speed of light in a vacuum to the speed of light in that material:

$$n = \frac{\text{light speed in vaccum}}{\text{light speed in material}} = \frac{c}{v} \tag{6.17}$$

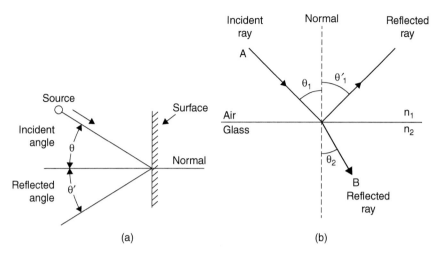

FIGURE 6.3 Diagrams of (a) law of reflection, and (b) refraction and Snell's law.

The speed of light in air is almost identical to the speed of light in a vacuum, so the index of refraction for air is considered to be 1 ($n_{air} = 1.000293$). The index of refraction for almost all other substances is greater than 1, because the speed of light is lower as it passes through them. As shown in Figure 6.3b, Snell's law of refraction shows the relationship between the incident angle and the refractive index:

$$n_1 \cdot \sin(\theta_1) = n_2 \cdot \sin(\theta_2)$$
(6.18)

Where n_1 is the refractive index of medium 1, n_2 is the refractive index of medium 2, θ_1 is the incident angle of the light ray (with respect to the normal), θ'_1 is the reflected angle (with respect to the normal), and θ_2 is the refracted angle (with respect to the normal). Using Snell's law (for $\sin 0° = 0$) means that light with a normal incident angle does not bend at a boundary. Snell's law also shows that light traveling from a medium with a low index to one with a high index ($n_1 < n_2$) bends toward the normal, while light traveling from a medium with a high index to one with a low index ($n_1 > n_2$) bends away from the normal.

Example 6.10

Determine the angle of refraction, for a light ray entering a piece of crown glass ($n = 1.52$) from the air ($n = 1$) at an incident angle of 45°.

SOLUTION

From Equation (6.14), the refraction angle is:

$$1 \times \sin(45°) = 1.52 \times \sin(\theta_2) => \theta_2 = 28°$$

A transparent substance transmits almost all light, reflecting only a small fraction of light at each of its two surfaces. This reflection occurs whenever light travels through a change in the refractive index. At normal incidence (incident angle = 0°), Fresnel's law of reflection quantifies the effect:

$$r_\lambda = \left(\frac{n_2 - n_1}{n_2 + n_1} \right)^2$$
(6.19)

where r_λ is the reflection loss, n_1 is the refractive index of medium 1, and n_2 is the refractive index of medium 2. For example, when light strikes a material that has a refractive index of 1.5 (such as glass) at a normal incident angle, each of the two material boundaries with air reflects less than 5% of the incident light. As the angle of incidence increases, so does the amount of reflected light. As Snell's law shows for light traveling from a material with a higher index of refraction to one with a lower index of refraction (such as light moving through a piece of glass toward air), the refracted light bends away from the normal, leading to the total internal reflection. If a light ray is incident on the interface at an angle greater than the critical angle, it is totally reflected into the same medium from which it came. If a beam of light's angle of incidence increases far from normal, it reaches an angle (the critical angle, θ_c) at which the light is refracted along the boundary between the materials instead of being reflected or passing through the boundary. The phenomenon of total internal reflection is exploited when designing light propagation in fibers by trapping the light in the fiber through successive internal reflections along the fiber or in lighting control systems. At even higher angles of incidence, all the light is reflected back into the medium, which allows fiber optics to transport light along their length with little or no loss except for absorption. Critical angle calculation (θ_c) based on the Snell's law, and for n_r and n_i the refractive indexes of the medium with lower refractive index and the incidence medium, respectively is then:

$$\theta_c = \sin^{-1}\left(\frac{n_r}{n_i}\right)$$

(6.20)

Example 6.11

Find the critical angle θ_c in the core, at the core-cladding interface of fiber optics having a core index of 1.53 and a cladding index of 1.39.

SOLUTION

From Equation (6.20) the critical angle is:

$$\theta_c = \sin^{-1}\left(\frac{n_r}{n_i}\right) = \sin^{-1}\left(\frac{1.39}{1.53}\right) = 65.3°$$

The index of refraction depends on the wavelength of the incident light. Materials typically have higher refraction indexes for shorter wavelengths, so blue light bends more than red light. This phenomenon is called *dispersion*. White light passing through the nonparallel faces of a prism is spreading into its spectral components, revealing the dispersion effects. *Transmission* represents the light passing through a material layer or an object. Absorption, reflection, refraction, and diffusion (explained in the next sections) affect light transmission. Instead of completely transmitting light, objects and materials can absorb part or the entire incident light, usually converting it into heat. Materials absorb usually some wavelengths, while transmitting others, the so-called selective absorption. Lambert's law of absorption states that equal thicknesses of homogenous materials absorb the same fraction of light, as expressed by this relationship:

$$I = I_0 \exp(-\alpha \cdot x)$$

(6.21)

where I is the intensity of transmitted light, I_0 is the intensity of incident light entering the material (excluding surface reflection), α is the absorption coefficient in inverse length units, and x is the sample thickness (measured in the same unit as α). Beer's law further breaks down the absorption coefficient α into two variables: β, the absorption per unit concentration coefficient, and c, the material concentration. Beer's law states that equal amounts of absorbing material absorb equal fractions of light. As with Lambert's law, each wavelength should be considered separately for Beer's law. The two laws are combined in a single equation, including the material thickness and the concentration, in the Beer-Lambert law:

$$I = I_0 \exp(-\beta \cdot c \cdot x) \tag{6.22}$$

Here, β is the absorption per concentration coefficient (inverse length per inverse grams or moles per liter), c is the concentration of the absorbing material, and x is the path length. When light strikes a perfectly smooth surface, the reflection is specular, and when striking a rough surface, the light is reflected and transmitted in several directions, which is *diffusion* and *scattering*. The amount of diffuse transmission or reflection, occurring when light moves through one material to strike another material, depends on two factors: the difference in refractive index between the materials, and the size and shape of the particles in the diffusing material compared to the light wavelength. For example, the molecules in air are of the right size to scatter light with shorter wavelengths, giving us blue sky. A transmissive filter is a material that absorbs some wavelengths and transmits others, while a reflective filter absorbs some wavelengths and reflects others. For example, a red filter absorbs all but the longest visible light wavelengths, a reflective red filter reflects the longest wavelengths, and a transmissive red filter transmits the longest wavelengths. The light amount absorbed by a filter depends on its thickness.

6.4 LIGHTING EQUIPMENT AND SYSTEMS

Light may originate in many ways, from solar energy, daylight, chemical reactions, combustion, or electricity conversion. Apart from all other light sources, daylight is plentiful and free of charge, with its only drawbacks being variability (fluctuating), available only during the day, and sometimes being too bright for visual comfort. However, when daylight is properly controlled, it is the most economical of all light sources. The electric lamps used for providing illumination are divided into three classes: incandescent, discharge, and solid-state lamps. Incandescent lamps produce light by heating a filament until it glows. Discharge lamps produce light by ionizing a gas through electric discharge inside. Solid-state lamps use electroluminescence effects to convert electricity directly to light. In addition to manufactured light sources, the daylight, the sunlight received on the earth directly from the sun or scattered and reflected by the atmosphere, or reflected by the moon, provides illumination. Daylight's prime characteristics are variability, magnitude, spectral content, and distribution, which is a function of the meteorological conditions, at different times of the day and year, and at different latitudes. Earth's surface illuminances due to daylight cover a large range, from 150,000 lx on a sunny summer day to 1000 lx on a heavily overcast winter day. The daylight spectral composition varies with the atmosphere and the path length through it.

6.4.1 LIGHT SOURCES AND SYSTEMS

Light, the basis for all vision, is an element of our lives that we take for granted. We are so familiar with brightness, darkness, and the spectrum of visible colors that another form of perception in a different frequency range and with different color sensitivity is difficult for us to imagine. Visible light is in fact just a small part of an essentially broader spectrum of electromagnetic waves, which range from cosmic rays to radio waves. The different kinds of electric light sources are categorized according to the means of their light production. The lighting industry makes millions of electric light sources, called lamps. Those used for providing illumination can be divided into three general classes: incandescent, discharge, and solid-state lamps. Incandescent lamps produce light by heating a filament until it glows. Discharge lamps produce light by ionizing a gas through electric discharge inside the lamp. Solid-state lamps use a phenomenon called electroluminescence to convert

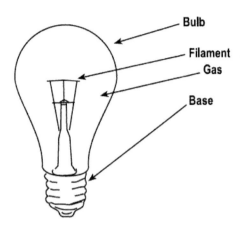

FIGURE 6.4 A typical incandescent lamp.

electrical energy directly to light. In the case of technical lamps the main distinction is between thermal radiators and discharge lamps. Discharge lamps are further subdivided into high-pressure and low-pressure lamps. Current developments show a marked trend towards the development of compact light sources such as low-voltage halogen lamps, compact fluorescent lamps, metal-halide lamps, fiber optic sources, and LED systems.

6.4.1.1   Incandescent Lamps

The incandescent lamp is a thermal radiator most commonly with a bulb shape, as shown in Figure 6.4. The filament wire begins to glow when it is heated to a sufficiently high temperature by an electric current. With increases in temperature, the spectrum of the radiated light shifts towards the shorter wavelength range, and the red heat of the filament shifts to the warm white light of the incandescent lamp. Although simple in operation, a substantial number of practical problems are associated with the construction of an incandescent lamp. There are only a few conducting materials that have a sufficiently high melting point and at the same time a sufficiently low evaporation rate below melting point that render them suitable for filament wire use. Depending on lamp type and wattage, the temperature of the filament can reach up to 3000 K, in the case of halogen lamps over 3000 K. Maximum radiation at these temperatures still lies in the infrared range, with the result that in comparison to the visible spectrum there is a high degree of thermal radiation and very little UV radiation. However, the lack of a suitable material for the filament means that it is not possible to increase the temperature further, which would increase the luminous efficacy and produce a cool white luminous color.

Nowadays for the most part, only tungsten is used for the manufacture of filament wires, because it only melts at a temperature of 3653 K and has a low evaporation rate. The tungsten is made into fine wires, wounded to make single- or double-coiled filaments, and the filament is located inside a soft glass bulb, which is relatively large in order to keep light loss, due to deposits of evaporated tungsten (blackening), to a minimum. To prevent the filament from oxidizing, the outer envelope is evacuated for low wattages and filled with nitrogen or a nitrogen-based inert gas mixture for higher wattages. The thermal insulation properties of the gas used to fill the bulb increases the temperature of the wire filament, but at the same time reduces the evaporation rate of the tungsten, which in turn leads to increased luminous efficacy and a longer lamp life. The inert gases predominantly used are argon and krypton. Krypton permits a higher operating temperature, and greater luminous efficacy, but due to its high costs, krypton is used only in special applications. A characteristic feature of these lamps is their low color temperature, so their light is warm in comparison to daylight. The continuous color spectrum of the incandescent lamp provides excellent color rendition. As a point source with a high luminance, sparkling effects can be produced on shiny surfaces and the light can be easily controlled using optical equipment.

Incandescent lamps can be easily dimmed. No additional control gear is required for their operation, and the lamps can be operated in any burning position. Despite of these advantages, there are a number of disadvantages: low luminous efficacy, for example, and a relatively short lamp life, while the lamp life relates significantly to the operating voltage. Special incandescent lamps are available with a dichroic coating inside the bulb that reflects the infrared component back to the wire filament, which increases the luminous efficacy by up to 40%. General-purpose service lamps (A lamps) are available in a variety of shapes and sizes. The glass bulbs are of clear, matte, or opal type. Special forms are available for critical applications (e.g., rooms subject to the danger of explosion, or lamps exposed to mechanical loads), as well as a wide range of special models available for decorative purposes. A second basic model is the reflector lamp (R lamp). The bulbs of these lamps are also from soft glass, and in contrast with the A lamps, which radiate light in all directions, the R lamps control the light via their form and a partly silvered area inside the lamp. Another type of incandescent lamp is the parabolic reflector (PAR) lamp. The PAR lamp is made of pressed glass to provide a higher resistance to changes in temperature and a more exact form; the parabolic reflector produces a well-defined beam spread. In the case of cool-beam lamps, a subgroup of PAR lamps, a dichroic, i.e., selectively reflective coating, is applied. Dichroic reflectors reflect visible light, but allow a large part of the IR radiation to pass the reflector. The thermal load on illuminated objects can therefore be reduced by half. Incandescent lamps are strongly affected by input voltage. For example, reducing input voltage from the normal 110 V to 104.5 V (95%) can double the life of a standard incandescent lamp, while increasing voltage to just 115.5 V (105% of normal) can halve its life. In summary, the main advantages of incandescent lamps include: inexpensive, easy to use, small and does not need auxiliary equipment, easy to dim by changing the voltage, excellent color rendering properties, directly work at power supplies with fixed voltage, no toxic components, and instant switching. Disadvantages of incandescent lamps include: short lamp life (about 1000 hours), low luminous efficacy, high heat generation, lamp life and other characteristics are strongly dependent on the supply voltage, and the total costs are high due to high operation costs. Traditional incandescent lamps are progressively replaced with more efficient light sources.

6.4.1.2 Halogen Lamps

Unlike incandescent lamps, halogen lamps use a halogen gas fill (typically iodine or bromine), to produce what is called a "halogen cycle" inside the lamp. Halogen lamps are available in a wide range of models and shapes (from small capsules to linear double-ended lamps), with or without reflectors. There are reflectors designed to redirect forward only the visible light, allowing infrared radiation to escape from the lamp back. There are halogen lamps available for main voltages or low voltages (6–24 V), the latter needing a step-down transformer. Low-voltage lamps have better luminous efficacy and longer lamp life than high-voltage lamps, but the transformer implicates energy losses in itself. The latest progress in halogen lamps has been reached by introducing selective-IR mirror coatings in the bulb. The infrared coating redirects infrared radiation back to the filament. This increases the luminous efficacy by 40%–60% compared to other designs, and lamp life extends to 4000 hours. In the halogen cycle, halogen gas combines with the tungsten that evaporates from the lamp filament, and eventually redeposits the tungsten on the filament instead of allowing it to accumulate on the bulb wall as it does in standard incandescent lamps. However, it is not so much the melting point of the tungsten (which, at 3653 K, is still relatively far from about 2800 K, the incandescent lamp operating temperature), hindering the construction of more efficient incandescent lamps, but rather the increase in the filament evaporation rate accompanying the temperature increase. This initially leads to lower performance due to the blackening of the surrounding glass bulb until finally the filament burns through. The increase in luminous efficiency is at a cost of shorter lamp life. One technical way of preventing the blackening of the glass is adding halogens to the gas mixture inside the lamp. The evaporated tungsten combines with the halogen to form a metal halide, which takes on the form of a gas at the temperature in the outer section of the lamp and can therefore leave no deposits on the glass bulb. The metal halide is split into tungsten and halogen once again at the considerably hotter filament and the tungsten is then returned to the coil. The temperature of the outer glass envelope has to be over 250°C to allow the halogen cycle to take place. To achieve this, a quartz glass compact bulb is fitted tightly over the filament, which results in an increase in temperature and in gas pressure, which in turn reduces the evaporation rate of the tungsten. Compared with conventional incandescents, halogen lamps give a whiter light, a result of its higher operating temperature of 3000 to 3300 K, and its luminous color is still in the warm white range. The tungsten-halogen lamp has several differences from incandescent lamps:

1. The lamps have a longer life (2000–3500 hours).
2. The bulb wall remains cleaner, because the evaporated tungsten is constantly redeposited on the filament by the halogen cycle, allowing it to maintain its lumen output throughout its life.
3. The higher operating temperature of the filament improves luminous efficacy.
4. The lamp produces a "whiter" or "cooler" light, which has a higher correlated color temperature (CCT) than standard incandescent lamps.
5. The bulbs are more compact, offering opportunities for better optical control.

Halogen lamps are sometimes called "quartz" lamps because their higher temperature requires quartz envelopes instead of the softer glass used for other incandescent lamps. In summary, their man advantages include: small size, directional light (narrow beams for some types), low-voltage options, easy to dim, instant switching, full light output, and excellent color rendering properties, while the disadvantages of tungsten halogen lamps are: low luminous efficacy, surface temperature is high, and lamp life and other characteristics are strongly dependent on the supply voltage. The recommended design practice is to use halogen lamps in applications requiring instant switch on, instant full light, excellent color rendering, easy dimming, frequent switching or short on-period, directional light, and compact light source size.

6.4.1.3 Light-Emitting Diode

Light-emitting diode (LED) sources are made from semiconductor material that emits light when energized. An LED is an electronic semiconductor component that emits light when a current flows through it. The wavelength of the light depends on the semiconductor material and its doping. The spectrum of LEDs offers a major benefit: only light (radiation in the visible range) and no ultraviolet or infrared radiation emitted. LEDs provide instant on/off capability and can be dimmed. LEDs are small and durable, usually providing much longer lamp life than other sources. The plastic encapsulate and the lead frame occupy most of the volume. The light-generating chip is quite small (a cub with one side equal to 0.25 mm). Light is generated inside the chip, a solid crystal material, when current flows across the junctions of different materials. The composition of the materials determines the wavelength and therefore the color of light. Major LED features include: long service life (e.g., 50,000 hours at 70% luminous flux), emitted light is only in the visible range—no UV or infrared radiation, compact size, high luminous efficiency (lm/W), good to excellent color rendering index (Ra), luminous flux and service life highly temperature-sensitive, no or little environmentally harmful materials (e.g., mercury), resistant to vibrations and impact, saturated color, about 100% luminous flux after switching on, no ignition, boosting or cooling time, high-precision digital dimming via pulse-width modulation, and no shifting of color locations during dimming. There are basically three major types of LED:

1. Standard through-hole LED, often used as indicator light sources, although with low light output. Due to their shorter service life, higher failure probability, and UV sensitivity, these are not used in lighting.

2. Surface-mounted device LED, in which an LED is reflow-soldered to the surface of a printed circuit board (using a reflow oven). Basically, it consists of an LED chip protected by silicon coating mounted in or on housing or a ceramic plate with contacts.
3. Chip-on-board LED, where the LED chip is mounted directly on the printed circuit board. This allows a dense arrangement of chips close to each other.

Heat management is one of the main concerns for LED luminaires. Proper heat management will allow for long LED life at or above 50,000 hours and is necessary to maintain proper light output. The efficacies for LEDs luminaires are as good as fluorescent and HID sources and continue to be improved. The more efficacious LEDs look whiter with typically high CCT (correlated color temperature) in the range of 5000 to 8000 K. Warmer color temperatures are available but their efficacy is reduced. LEDs can generate red, yellow, green, blue, or white light, with a life up to 100,000 hours, being widely used in traffic signals and for decorative purposes. White-light LEDs are a recent advance and may have a great potential market for some general lighting applications. However, a critical issue when comparing LED luminaires of different suppliers is the indication of luminous flux levels. Catalogs provide details regarding the luminous flux and efficiency of individual LEDs at a chip junction temperature of 25°C, details regarding the luminous flux levels of the LED boards used, or details regarding the luminous flux levels of luminaires and luminaire efficiencies, including the ballast power losses and any potential loss of efficiency through lighting optics, such as lenses, reflectors, or mixing chambers. Future lighting systems are expected to have intelligent features. In this regard, LED-based lighting systems have an important advantage due to their easy controllability. Intelligent features combined with the inherent high energy-saving potential of LEDs will be an unbeatable combination in a wide range of applications. Advantages of LEDs are: small size (heat sink can be large), physically very robust, long lifetime expectancy (with proper thermal management), switching has no effect on life, very short rise time, contains no mercury or other potentially dangerous materials, excellent low-ambient-temperature operation, and high luminous efficacy (LEDs are developing fast and their range of luminous efficacies is wide), new luminaire design possibilities, possibility to change colors, and no optical heat on radiation. Major disadvantages of LEDs are: expense (although prices keep declining), low luminous flux/package, often low CRI, risk of glare due to high output with small lamp size, need for thermal management, and lack of standardization.

6.4.1.4 Discharge Lamps

Discharge lamps produce light by passing an electric current through a gas that emits light when ionized by the current. In contrast to incandescent lamps, light from discharge lamps is not produced by heating a filament, but by exciting gases or metal vapors, by applying voltage between two electrodes located in a discharge tube filled with inert gases or metal vapors. Through the voltage, current is produced between the two electrodes. On their way through the discharge tube, the electrons collide with gas atoms, which are in turn excited to radiate light, when the electrons are travelling at a sufficiently high speed. For every type of gas there is a certain wavelength combination, and the radiation, i.e., light, is produced from one or a few narrow-frequency ranges. An auxiliary device, the ballast supplies voltage to the lamp electrodes, which are coated with a mixture of alkaline earth oxides to enhance electron emission. Two categories of discharge lamps are currently used: *high-intensity discharge* and *fluorescent lamps*. It soon becomes evident that discharge lamps have properties different from incandescent lamps. Whereas incandescent lamps have a continuous spectrum dependent on the temperature of the filament, discharge lamps produce a narrow-band spectrum, typical for the respective gases or metal vapors. The spectral lines can occur in all regions of the spectrum, from infrared through the visible region to ultraviolet. The number and distribution of the spectral lines results in light of different colors. These can be determined by the choice of gas or metal vapor in the discharge tube, and as a result white light of various color temperatures can be produced. Moreover, it is possible to exceed the given limit for thermal radiators of 3650 K and produce daylight-quality light of higher color temperatures. Another method for the effective production of luminous colors is through the application of fluorescent coatings on the interior surfaces of the discharge tube. Ultraviolet radiation in particular, which occurs during certain gas discharge processes, is transformed into visible light by means of these fluorescent substances, through which specific luminous colors can be produced by the appropriate selection and mixing of the fluorescent material. The quality of the discharge lamp can also be influenced by changing the pressure inside the discharge tube. The spectral lines spread out as the pressure increases, approaching continuous spectral distribution. This results in enhanced color rendering and luminous efficacy.

To ignite a discharge lamp there must be sufficient electron current in the discharge tube. As the gas that is to be excited is not ionized before ignition, these electrons must be made available via a special starting device. Once the discharge lamp has been ignited, there is an avalanche-like ionization of the excited gases, which in turn leads to a continuously increasing operating current, which would increase and destroy the lamp in a relatively short time. To prevent this from happening, the operating current must be controlled by means of ballast. Additional equipment is necessary for both the ignition and operation of discharge lamps. In some cases, this equipment is integrated into the lamp; but it is normally installed separate from the lamp, in the luminaire. Discharge lamps can be divided into two main groups depending on the operating pressure. Each of these groups has different properties. One group comprises low-pressure discharge lamps. These lamps contain inert gases or a mixture of inert gas and metal vapor at a pressure well below 1 bar. Due to the low pressure in the discharge tube, there is hardly any interaction between the gas molecules. The result is a pure line spectrum. The luminous efficacy of low-pressure discharge lamps is mainly dependent on lamp volume. To attain adequate luminous power, the lamps must have large discharge tubes. High-pressure discharge lamps, on the other hand, are operated at a pressure well above 1 bar. Due to the high pressure and the resulting high temperatures, there is a great deal of interaction in the discharge gas. Light is no longer radiated in narrow spectral lines but in broader frequency ranges. The radiation shifts with increasing pressure into the long-wave region of the spectrum.

6.4.1.5   High-Intensity Discharge (HID)

High-intensity discharge sources include mercury vapor, metal halide, and high-pressure sodium (HPS) lamps. Light is produced in HID and low-pressure sodium (LPS) sources through a gaseous arc discharge using a variety of elements. Each HID lamp consists of an arc tube that contains certain elements or mixtures of elements which, when an arc is created between the electrodes at each end, gasify and generate visible radiation. The major advantages of HID sources are their high efficacy in lumens per watt, long lamp life, and point-source characteristic for good light control. Disadvantages include the need for a ballast to regulate lamp current and voltage as well as a starting aid for HPS and some MH and the delay in restriking after a momentary power interruption.

6.4.1.6   High-Pressure Sodium (HPS)

In the 1970s, as increasing energy costs placed more emphasis on the efficiency of lighting, high-pressure sodium lamps (developed in the 1960s) gained widespread usage. With efficacies ranging from 80 to 140 lm/W, these lamps provide about seven times as much light per watt as incandescent and about twice as much as some mercury or fluorescent. The efficacy of this source is not its only advantage. An HPS lamp also offers the longest life (over 24,000 hours) and the best lumen maintenance characteristics of all HID sources. The major objection to the use of HPS is its yellowish color and low color rendition. It is ideal mainly for some warehouse and outdoor applications. The most common application today is in street and road lighting. Their characteristics include high luminous efficacy (80–100 lm/W), lamp life of 12,000 h (16,000 h), and the CCT is 2000 K. Improvement of the CRI is possible by pulse operation or elevated pressure, but this reduces the luminous efficacy. Color improved high-pressure sodium lamps have CRI of about 65 and white high-pressure sodium lamps of more than 80. Their CCT is 2200 and 2700, respectively. Among advantages of high-pressure sodium lamps are: very good luminous efficacy, long lamp life (12,000 or 16,000 h), high luminous flux from one unit for street and area lighting. Disadvantages include: low CCT (about 2200 K), low CRI (about 20, or color improved 65, white 80), and long starting and restarting time (2–5 min.).

6.4.1.7 Metal Halide

Metal-halide lamps are similar in construction to mercury lamps with the addition of various other metallic elements in the arc tube. The major benefits of this change are an increase in efficacy to 60 to 100 lumens per watt and an improvement in color rendition to the degree that this source is suitable for commercial areas. Light control of a metal-halide lamp is also more precise than that of a deluxe mercury lamp because light emanates from the small arc tube, not the total outer bulb of the coated lamp. Pulse-start metal halide lamps have several advantages over standard (probe-start) metal halide: higher efficacy (110 lumens per watt), longer life, and better lumen maintenance. A disadvantage of the metal-halide lamp is its shorter life (7,500 to 20,000 hours) as compared to induction, LEDs, and high-pressure sodium. Starting time of the metal-halide lamp is approximately 4–7 minutes depending on ambient temperatures. Restriking after a voltage dip has extinguished the lamp, however, can take substantially longer, depending on the time required for the lamp to cool. Among advantages of metal-halide lamps are: good luminous efficacy, alternatives with good color rendering available, and different color temperatures available. Major disadvantages include: expense, long starting and restarting time (2 to 5 min.), differences in CCT between individual lamps, and changes of CCT during burning hours. These differences are much reduced with new ceramic metal-halide lamps.

6.4.1.8 Low-Pressure Sodium

Low-pressure sodium (LPS) offers the highest initial efficacy of all lamps on the market today, ranging from 100 to 180 lumens per watt. However, because all of the LPS output is in the yellow portion of the visible spectrum, it produces extremely poor and unattractive color rendition. Control of this source is more difficult than with HID sources because of the large size of the arc tube. The average life of low-pressure sodium lamps is 18,000 hours. While lumen maintenance through life is good with LPS, there is an offsetting increase in lamp watts, reducing the lamp efficacy with use.

6.4.1.9 Fluorescent Lamps

Fluorescent lighting accounts for two-thirds of all electric light in the United States. Fluorescent lamps are the most commonly used commercial light source in North America. In fact, fluorescent lamps illuminate 71% of the commercial space in the United States. Their popularity can be attributed to their relatively high efficacy, diffuse light distribution characteristics, and long operating life. Fluorescent lamp construction consists of a glass tube with the following features: filled with argon or argon-krypton gas and a small amount of mercury; coated on the inside with phosphors; and equipped with an electrode at both ends, as shown in Figure 6.5. The fluorescent

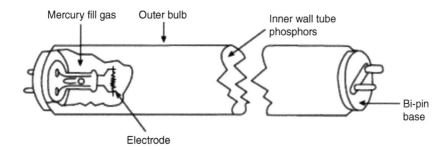

FIGURE 6.5 Construction of a linear fluorescent lamp.

lamp is a low-pressure discharge lamp using mercury vapor. It has an elongated discharge tube with an electrode at each end. The gas used to fill the tube comprises inert gas, which ignites easily and controls the discharge, plus a small amount of mercury, the vapor of which produces ultraviolet radiation when excited. The inner surface of the discharge tube is coated with a fluorescent substance that transforms the ultraviolet radiation produced by the lamp into visible light by means of fluorescence. The fluorescent lamp is a gas discharge source that contains mercury vapor at low pressure, with a small amount of inert gas for starting. Once an arc is established, the mercury vapor emits ultraviolet radiation. Fluorescent powders (phosphors) coating the inner walls of the glass bulb respond to this ultraviolet radiation by emitting wavelengths in the visible spectrum. Ballasts, which are required by both fluorescent and HID lamps, provide the necessary circuit conditions (voltage, current, and wave form) to start and operate the lamps. Two general types of ballasts are available for fluorescent lamps: magnetic and electronic. Electronic ballasts are often more expensive, but lighter and quieter, and they eliminate the lamp flicker associated with magnetic ballasts. Fluorescent lamps are described in terms of the diameter of the lamp tube, with the diameter given in eighths of an inch.

Full-size fluorescent lamps are available in several shapes, such as straight, U-shaped, and circular configurations, with diameters from 1" to 2.5". The most common lamp type is the 4 foot (F40), 1.5" diameter (T12), straight fluorescent lamp. More efficient fluorescent lamps are now available in smaller diameters, including the T10 (1.25") and T8 (1"). Fluorescent lamps are available in color temperatures ranging from warm (2700 K) "incandescent-like" colors to very cool (6500 K) "daylight" colors. Cool white (4100 K) is the most common fluorescent lamp color. Neutral white (3500 K) is becoming popular for office and retail lighting. Improvements in the fluorescent lamp phosphor coating have improved color rendering and made fluorescent lamps acceptable in applications previously dominated by incandescent lamps. Linear fluorescent lamps range in length from 6 inches to 8 feet, and in diameter from 2/8 inch (T2) to 2-1/8 inches (T17), with power ranges from 14 W to 215 W. Figure 6.6 shows the construction of a linear fluorescent lamp. Compact fluorescent lamps (CFLs) produce light in the same manner as linear fluorescent lamps. Their tube diameter is usually 5/8 inch (T5) or smaller. CFL power ranges from 5 to 55 watts. Their advantages include: inexpensive, good luminous efficacy, long lamp life (10,000 to 16,000 hours), and large variety of CCT and CRI. The main disadvantages of fluorescent lamps are: ambient temperature affects the switch-on and light output, need of auxiliary ballast and starter or electronic ballast, light output depreciates with age, contains mercury, and short burning cycles shorten lamp life. Fluorescent lamps are ideal for general lighting in most working places (shops, hospitals, open spaces, etc.), but also in some residential applications. The choice of the lamp is always related to the application. Always consider the correlated color temperature and the color rendering index.

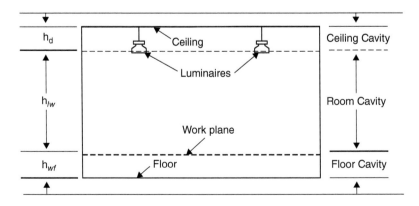

FIGURE 6.6 Zonal cavity method parameters.

6.4.2   Lamp Efficiencies, Control, and Electrical Requirements

Electric lighting consumes about 19% of the world's total electricity use, so any improvement in energy-efficient lighting has significant impacts on energy conservation and savings. Every change in technology, in customers' consumption behavior, even in lifestyle, has influences on global energy consumption and, indirectly, on the environment. Therefore, lighting energy saving and the methods of achieving this goal should be considered at different company and corporation levels, town, city, state, country, region, and by international organizations, too. Artificial lighting is based on systems: lamps, ballasts, starters, luminaires, and controls, and each component must be considered while figuring energy management, saving, use, and conservation. IESNA defines lamp efficacy as "the quotient of the total luminous flux emitted divided by the total lamp power input." It is expressed in lumens per watt (lm/W). For fluorescent and HID lamps, you must also include both the ballast wattage and any reduction in lumen output associated with the lamp-ballast combination to determine the system efficacy. Table 6.1 compares efficacies of some common lamp types. To summarize, energy savings and efficiency and economies are dependent on: improvement of lighting technologies, better use of available cost-effective and energy-efficient lighting technologies, optimum and good lighting design (identify needs, avoid misuses, proper interaction of technologies, automatic controls, daylight integration), proper building design, through daylight integration and architecture, knowledge dissemination to final users and operators (designers, sellers, decision makers), reduction of resources by recycling and proper disposal, size reduction, by using less aluminum, mercury, and finally the life cycle cost assessment.

Auxiliary equipment for lighting consists of two major categories: transformers and ballasts. All discharge-type lighting systems (all lighting technologies discussed in this chapter except incandescent and LED) require a ballast to supply electricity to the lamp. This auxiliary equipment usually consumes a small amount of electrical power, adding to the total amount of lighting system wattage. Low-voltage light sources require the use of a transformer to step down the standard building service of 120 V or 220 V to 6 V, 12 V, or 24 V. Transformers are placed either within (integral to) the luminaire or in a remote location. The smaller size of low-voltage light sources allows for the design of smaller luminaires. In the case of recessed luminaires, the transformer is hidden above the ceiling and out of view. Surface- or pendant-mounted luminaires usually have their transformers enclosed within the housing; however, the luminaire volume increases. Where ceiling conditions permit, surface- and pendant-mounted luminaires can be designed with the transformer recessed in the ceiling and out of view. Track-mounted luminaires usually contain their transformers. It is also possible to provide low-voltage service to a length of track, locating the transformer in the ceiling

TABLE 6.1
Efficacies of Common Light Sources (Adapted from IESNA Lighting Handbook)

Light Source	Power (W)	Efficiency (lumen/watt)	Color Rendering Index	Typical Life (hours)
Standard incandescent	100	18	Excellent	1000
Linear tungsten-halogen	300	20	Excellent	2000–4000
Fluorescent lamp (T-5)	28	50	Good	5000
Compact fluorescent lamp	100	60	Very good	8000–10,000
Metal-halide (low-voltage)	100	80		
Metal-halide (high-voltage)	400	90		
High-pressure mercury lamp	1000	50	Fair	5000
Xenon short-arc lamp	1000	30		
High-pressure sodium	70	90	Fair	6000–12,000
Low-pressure sodium	180	180	Poor	6000–12,000

or in an ancillary space. The high amperage of low-voltage lamps strictly limits the number of track luminaires per transformer. If remote transformers are used to maintain the compactness of the lighting element, the increased distance between the source and its transformer requires larger wire sizes to prevent a voltage drop from occurring over the longer wiring run. In modern lighting systems, electromagnetic conventional transformers are replaced with power electronics converters, which are cheaper, more compact, smaller in size, and more efficient. Rectangular-shape transformers are relatively large and heavy. However, properly sized for the lamp load, they have a long life expectancy. They can cause a noise problem by producing an audible 60 Hz hum. Toroidal magnetic transformers are quieter, but they also hum when controlled by electronic dimmers. The hum grows with the number of luminaires in a room, and the luminaires, if improperly designed, will resonate with their transformers. Lamps, with the exception of incandescent lamps and LEDs, require ballasts to operate properly and safely. Every discharge source has negative resistance characteristics. If the arc discharge is placed directly across a nonregulated voltage supply, it will draw large currents almost instantly and the lamp is quickly destroyed. Therefore, a current-limiting device called a ballast is inserted between the discharge lamp and the power supply to limit the electric current through the arc discharge. Besides limiting the current flow, the ballast also provides the correct voltage to start the arc discharge, by adjusting the available line voltage to that required by the lamp. Most of the old types of ballasts are not interchangeable, being designed to provide the proper operating characteristics for only one kind of lamp. However, modern electronic ballasts are designed to operate more than one connected load. Lamp wattage is controlled by the ballast, not by the lamp. Unlike incandescent lamps, the rated wattage of a discharge lamp is the wattage at which it is designed to operate, not the wattage at which it operates. Therefore, in order to reduce discharge, system energy use must change not only the lamp's wattage, but also the ballast must be changed. Notice that the lamp wattage is controlled by the ballast, not by the lamp. A 100 W HPS lamp operated by 400 W ballast operates at the ballast power, 400 W, to the detriment of the lamp's performance, which may lead to premature ballast failure.

Ballasts are devices, either of electromagnetic or electronic type, designed for: (a) providing sufficient ignition voltage to start the lamp, (b) acting as a constant current source after the lamp starts, and (c) acting as a constant power source when the lamp is operating. These functions can be implemented using an iron-core reactor (inductor) for current limitation in combination with a capacitor (starter) to provide ignition voltage, or by a variety of solid-state circuits. Solid-state ballasts are used on all systems today; however, core-and-coil ballasts were used for many years, so some may be still in service. During the ignition phase, the ballast provides sufficient voltage across the lamp electrodes to initiate and maintain discharge, and must also provide sufficient current at discharge voltage to force a transition from glow to arc. These voltages and currents are specific to the lamp type, so ballasts must be correctly matched to the lamps they are supplying. During the warm-up phase, the resistance of the lamp continuously increases, and the ballast must provide a constant current to the lamp, which linearly increases power to the lamp. In the operating phase, the lamp resistance takes on a value close to the arc impedance. This low resistance requires current limitation to prevent damage to the lamp. As the voltage supplied to the lighting system changes, the lamp reacts differently depending on the rate of change. If the change in voltage occurs over many seconds, a corresponding change in lamp current will try to occur, and a constant current supply is needed to keep the lamp operating properly. Sudden changes in voltage could cause extinction of the arc or a sudden increase in current. The ballast constant current characteristic allows proper and safe lamp operation for most voltage transients. If the magnitude or duration of the transient exceeds the capabilities of the ballast, the lamp will shut down, and will have to cool before restrike of the arc is possible. High-pressure sodium lamps experience a rise in lamp voltage over their lifetime. This voltage rise is high, often as much as 170% of the lamp voltage experienced when the lamp is new. Therefore, the ballast must keep the lamp power within an acceptable power range over the life of the lamp. The ballast *power factor* shows how effective the ballast converts the supplied power by the electrical distribution system to the power delivered by the ballast to the lamp. Perfect phase relationship would result in a power factor of 100%. The power factor of an inductive circuit is lagging, while the one of a capacitive circuit is leading. When discharge lamps are operated in conjunction with simple inductive ballasts, the overall power factor is 50% to 60%. With a capacitor, the leading current drawn by the capacitor compensates for the lagging current in the remainder of the circuit, improving the power factor. Ballasts are classified according to one of the following three categories: (1) high power factor: 90% or greater; (2) power-factor corrected: 80% to 89%; and (3) low (normal) power factor: 79% or less. High-power-factor ballasts use the lowest level of current for the specific amount of power needed, reducing wiring costs by permitting more luminaires on branch circuits. Low-power-factor ballasts use higher levels of current, about twice the line current needed by high-power-factor ballasts, so fewer luminaires are connected per branch circuit, increasing wiring costs. Power factor is not an indication of the lamp-ballast system's ability to produce light—the ballast's measurements pertain only to the ballast's ability to use the power that is supplied. The initial lumen and mean-lumen ratings published by lamp manufacturers are based on the operation of the rated lamps by *reference ballasts*. In practice, when a lamp is operated by commercially available ballasts, it provides fewer lumens than the rated value. Because of the electrical resistance created by the passage of a current through the core-and-coil of electromagnetic ballast, some power is converted to heat; the lost power, ballast loss, is not used to produce light from the lamp. The disparity between light provided by the reference ballast and the commercially available ballast is called the ballast factor, defined as the ratio of light output produced by lamps operated by commercially available ballasts to that which is theoretically supplied by lamps powered by laboratory-reference ballasts. The ballast efficacy factor is a ratio of the ballast factor to the input watts of the ballast. This measurement is used to compare the efficiency of various lamp-ballast systems. Ballast efficacy factors are meaningful only for comparing different ballasts when operating the same quantity and kind of lamp.

6.4.3 COMMON LAMP LUMINANCES AND LUMINAIRES

Different light sources generate a wide range of luminances. The direction of light is based on three fundamental principles: reflection, refraction, and diffraction, as discussed in previous sections of this chapter. These principles are applied to define the photometric properties of luminaires in terms of lighting patterns. Table 6.2 shows the approximate luminances of common light sources.

TABLE 6.2
Luminances of Common Light Sources
(Adapted from the IESNA Lighting Handbook)

Light Source	Approximate Average Luminance (cd/m²)
Sun	1.6×10^9
Moon	6×10^6
Clear sky	8×10^3
Overcast	2×10^3
60 W incandescent lamp	1.2×10^5
Tungsten-halogen lamp	1.3×10^7
T-5 fluorescent lamp	2×10^4
T-8 fluorescent lamp	1×10^4
High-pressure mercury lamp	2×10^8

Luminaires are devices that produce, distribute, control, filter, and/or transform the light emitted from one or more lamps. A luminaire is a device that distributes filters or transforms the light emitted from one or more lamps. The luminaire includes all the parts necessary for fixing and protecting the lamps, except the lamps themselves. In some cases, luminaires also include the necessary circuit auxiliaries, together with the means for connecting them to the electric supply. A wide variety of luminaire designs are available to meet virtually any lighting application. In general, considerations when selecting a luminaire for a particular application include construction and installation codes, standards, physical and environmental conditions, electrical and mechanical requirements, thermal characteristics, economics, and most important, safety. The principles used in optical luminaires are reflection, absorption, transmission, and refraction. The luminaire includes all the parts necessary for fixing and protecting the lamps, except the lamps themselves. In some cases, luminaires also include the necessary circuit auxiliaries, together with the means for connecting them to the electric supply. Most luminaires are fitted with reflectors, refractors, and/or diffusers, in order to control the distribution of light. Photometric data, including plots of candela distribution and isoilluminance, are available from the manufacturer. The performance of any luminaire system depends on how well its components work together. For fluorescent lamp-ballast systems, light output, input power, and efficacy are sensitive to the ambient temperature. When the ambient temperature around the lamp is significantly above or below 25°C (77°F), the system performances can change significantly. Luminaires are usually classified both by applications, used by the lighting manufacturers to present their products, and by their photometric characteristics. Applications include residential, commercial, and industrial with subclassifications by lighting technology, mounting method, and luminaire construction types. Photometric classifications are developed by professional organizations, such as the International Commission on Illumination (CIE), NEMA, and IESNA. The CIE classification system is based on upward-directed light to downward-directed light ratio, being applied to indoor luminaires. Categories include direct lighting (90% to 100% downward light), semidirect lighting (60% to 90% downward light), general diffuse lighting (approximately equal upward and downward components), semi-indirect lighting (with 60% to 90% upward light), and indirect lighting (90% to 100% upward light).

Outdoor luminaires are usually described by cutoff characteristics. Three methods are used in photometric reports: physical cutoff, optical cutoff, and shielding angle. Physical cutoff is the angle measured from the downward-directed vertical axis, or nadir, to the point where the lamp is fully occluded. Optical cutoff measures the angle from the nadir to the point where reflection of the lamp in the luminaire's reflector is fully occluded. The shielding angle is the angle measured from the horizontal at which the lamp is just visible. The NEMA classification system is based on the distribution of luminous flux, used primarily for athletic lighting and flood lighting, considering the spread of the light beam in degrees and the projection distance in feet, separated in seven beam types, as given in Table 6.3 and is Based on the shape of the area illuminated, the IESNA

TABLE 6.3

NEMA Beam Types

Beam Type	Beam Spread (degrees)	Projection Distance (ft.)
1	10–18	>240
2	18–29	200–240
3	29–46	175–200
4	46–70	145–175
5	70–100	105–145
6	100–130	80–105
7	>130	<80

TABLE 6.4
IESNA Cutoff Classification

Classification	Intensity Distribution Description
Full cutoff	At 90° and greater above nadir, 0 candela intensity, not exceeding 10% of the maximum candela intensity at 80° above nadir at all lateral angles around luminaire.
Cutoff	At 90° above nadir, candela intensity, not exceeding 2.5% of the maximum candela intensity, and not exceeding 10% at 80° above nadir, and at all lateral angles around luminaire.
Semi-cutoff	At 90° above nadir, candela intensity, not exceeding 5% of the maximum candela intensity, and not exceeding 20% at 80° above nadir, and at all lateral angles around luminaire.
Noncutoff	No candela limitations in the zone above maximum candela intensity.

classification system applies to outdoor luminaires commonly used in lighting for roadways, parking lots, and street lighting. Six intensity-distribution types are defined by IESNA, separated into four cutoff classifications, as shown in Table 6.4.

The luminaire efficiency is the percentage of lamp lumens produced that actually exits the fixture. The use of louvers can improve visual comfort; however, they do reduce the lumen output of the fixture, so efficiency is reduced. In general, the most efficient fixtures have the poorest visual comfort (e.g., bare strip industrial fixtures), while fixtures providing the highest visual comfort level are the least efficient. When selecting luminaires, a lighting designer must determine the best compromise between efficiency and visual comfort. In recent years, manufacturers began to offer fixtures with excellent visual comfort and efficiency. These so-called super fixtures combine state-of-the-art lens or louver designs to provide the best of both worlds. Surface deterioration and accumulated dirt in older, poorly maintained fixtures can also reduce luminaire efficiency. Each luminaire consists of several components, designed to work together to produce and direct light in the best way possible for a specific application. Reflectors are designed to redirect the light emitted from a lamp in order to achieve a desired distribution of light intensity outside of the luminaire. In most incandescent spot and flood lights, highly specular reflectors are usually built into the lamps. An energy-efficient upgrade option is to install custom-designed reflectors to enhance the light control and the fixture efficiency, allowing partial delamping. Retrofit reflectors are useful for upgrading the efficiency of older, deteriorated luminaire surfaces. A variety of reflector materials are available: highly reflective white paint, silver film laminate, and two grades of anodized aluminum sheet (standard or enhanced reflectivity). Silver film laminate is generally considered to have the highest reflectance, but is considered less durable. Proper design and installation of reflectors can have more effect on performance than the reflector materials. In combination with delamping, the use of reflectors may result in reduced light output and may redistribute the light, which may not be acceptable for a specific space or application. To ensure acceptable performance from reflectors, trial installation, calculations, and measurements of "before" and "after" light levels are required.

Most indoor commercial fluorescent fixtures use either a lens or a louver to prevent direct viewing of the lamps. Light that is emitted in the *glare zone* (angles above 45° from the fixture vertical axis) can cause visual discomfort and reflections, reducing contrast on work surfaces or computer screens. By using lenses and louvers, these problems can be controlled. Lenses made from clear ultraviolet-stabilized acrylic plastic deliver the most light output and uniformity of all shielding media, tending to provide less glare control than louvered fixtures. Clear lens types include prismatic, batwing, linear batwing, and polarized lenses. Lenses are usually cheaper than louvers. White translucent diffusers are less efficient than clear lenses, resulting in relatively low visual comfort probability. New low-glare lens materials are available for retrofit and provide higher visual comfort (over 80) and higher efficiency. Louvers provide superior glare control and higher visual comfort compared with lens diffusers. Common louver application is to eliminate the fixture glare reflected on computer screens. Deep-cell parabolic louvers (with 5"–7" cell apertures and depths of 2"–4") provide a good balance between visual comfort and luminaire efficiency. Although small-cell parabolic louvers provide the highest visual comfort, they tend to reduce luminaire efficiency to about 35% to 45%. For retrofit applications, both deep-cell and small-cell louvers are available for use with existing fixtures. However, the deep-cell louver retrofit adds 2"–4" to the overall depth of a troffer, and the available plenum depth must be checked before specifying the deep-cell retrofit.

Distribution is one of the primary functions of a luminaire, meaning to direct the light to where it is needed. The light distribution produced by luminaires is characterized by the Illuminating Engineering Society as follows:

1. Direct, 90% to 100% of the light is directed downward for maximum use.
2. Indirect, 90% to 100% of the light is directed to the ceilings or upper walls, being reflected to the room.
3. Semidirect, 60% to 90% of the light is directed downward with the rest directed upward.
4. General diffuse (direct-indirect), in which equal light portions are directed upward and downward.

Highlighting, beam-projection distance, and focusing ability characterize luminaires. The lighting distribution that is characteristic of a given luminaire is described using the candela distribution, intensity distribution curve, or candlepower curve, provided by the luminaire and lamp manufacturers. It represents the amount of luminous intensity generated in each direction by a light source in a plane through the center of the source, giving a picture of the total light pattern produced by a source. Luminous intensity distribution curves are available from luminaire manufacturers and are often found on the back of the manufacturer's product data sheet. A polar graph is used to represent the distributional intensity of a luminaire, and a rectilinear or Cartesian graph to represent the distributional intensity of a directional lamp. The candlepower distribution is represented by a curve on a polar graph showing the relative luminous intensity 360° around the fixture, looking at a cross-section of the fixture. In the candlepower polar graph, the luminaire or the lamp is located at the center of the radiating lines. The radiating lines represent specific degrees of angular rotation from the 0° axis of the luminaire (nadir). The concentric circles represent graduating intensity, with values entered along the vertical scale. For luminaires with symmetrical light distributions, a single curve fully describes the luminaire light intensity distribution. Often only one side of the polar graph is shown, because the other side is a symmetrical (mirror) image. A luminaire with an asymmetrical distribution, such as a linear fluorescent downlight, requires curves in a number of planes to adequately represent its distribution, with one curve parallel to the luminaire and another perpendicular to the luminaire, and either a third plane at 45° or other three planes at 22½° intervals. Reflector lamp sources or luminaires with directional distributions and abrupt cutoffs, having abrupt light intensity changes within a small angular area, have values that are difficult to read on a polar graph. Consequently, a Cartesian graph is substituted to portray the candlepower distribution or the data is tabulated. On this graph, the horizontal scale represents degrees from the beam axis and the vertical scale represents the light intensity. In general, for nonsymmetrical lighting distributions, the candlepower data, representing various vertical through the fixture, are presented in tabular format, as shown in Table 6.5. From the fixture top, the plane intersecting the fixture long dimension is designated 0°, while the direction across the fixture is designated 90°. When required, the data must be interpolated for intermediate angles.

When selecting luminaires for a lighting application, the proposed luminaire *and its source* must be precisely those shown in the manufacturer's photometric test data. It is inaccurate to extrapolate from one source or reflector finish to another unless the photometric report includes multipliers for various tested sources and reflector finishes. The candlepower distribution information is useful because it shows how much light is emitted in each direction and the relative proportions

TABLE 6.5
Candlepower Data

Angle	0° Plane	45° Plane	90° Plane
0	3670	3670	3670
10	3600	3630	3680
20	3410	3520	3600
30	3080	3270	3410
40	2480	2750	2800
50	1600	1700	1850
60	1010	800	1000
70	500	370	480
80	230	290	240
90	0	0	0

of down-lighting and up-lighting. The cutoff angle is the angle, measured from straight down, where the fixture begins to shield the light source and no direct light from the source is visible. The **shielding angle** is the angle, measured from horizontal, through which the fixture provides shielding to prevent direct viewing of the light source. The shielding and cutoff angles add up to 90°. The candlepower distribution provides the light intensity at the point source and at various points away from the fixture. The candlepower distribution curve and data are specified for fixture type, number of lamps, ballast, and lamp lumen output.

Example 6.12

Using the data from Table 6.4 and appropriate interpolation, determine the candlepower along the fixture, at an incidence angle of 30° and 45°, and the candlepower in a perpendicular plane and at an incidence angle of 30°.

SOLUTION

From table data, at 30°, I = 3080 cd. For 45°, by using linear interpolation I = 2100 cd, while for perpendicular plane at 30°, I = 3410 cd.

6.5   INDOOR AND OUTDOOR LIGHTING DESIGN

Human dependency on light is vital for existence and for all activities. Light is a natural phenomenon, taken often for granted. In fact, life involves day-night cycles beginning with sunrise and ending with sunset. Activities of prehistoric humans were limited only to day time. The introduction of artificial light has enabled extended activity periods employed in a planned, optimized manner, while minimizing the resources. The primary function of lighting in workspaces is to support work and to enhance personnel performance. Vision is the most important sense, accounting for 80% of human information acquisition. Information may be acquired through sunlight (direct) or moonlight (reflected), or by using artificial light. It is well-accepted that the lighting is good when our eyes can clearly and pleasantly perceive the things around us, often achieved by employing multiple light sources. However, the light sources must be economic and energy-efficient. All light sources today employ electrical energy. The lighting system in an office affects the ambiance, creating the company's impression to employees, clients, or customers, and having profound effects on feelings of well-being and productivity of staff. Lighting design must be a vital consideration to the successful operation of any business. A designer must consider a variety of characteristics when developing a lighting plan, including lamp life, system efficiency, lumen maintenance, color rendering and appearance, daylight integration and control, light distribution, system cost, control, and flexibility. The prime objectives behind a lighting system design are as follows:

1. **Safety and comfort of occupants.** The nature of a task or process performed in a space will dictate the illuminance level that must be provided by the lighting system (lx or lm/m²). Tasks involving high degrees of visual acuity will require higher lighting levels.
2. **Minimization of energy consumption.** Energy consumption reduction involves the development of the most energy-efficient lighting systems suitable for the task, which can be achieved by selecting high-efficiency equipment and making use of available daylight.
3. **Color rendering or the creation of a specific atmosphere.** The color characteristics of a lighting scheme will affect tasks performed when the lighting system is on. For example, tasks that require accurate color representation need spectral characteristics of daylight. Alternatively, creating a "warm atmosphere" in a restaurant requires the selection of lights from the red end of the spectrum.

Achieving the required illuminance level does not necessarily ensure good lighting quality. The quality as well as the illuminance quantity is important in producing a comfortable, productive, aesthetically pleasing lighted environment. The lighting system quality includes aspects of lighting such as proper color, good uniformity, proper room surface luminances, adequate brightness control, and minimal glare. The lighting system can affect impressions of visual clarity, spaciousness, and pleasantness, when they occur in spaces that are uniformly lighted with emphasis on higher luminances on room surfaces. A lighting system design has several important steps, such as:

1. Identification of the requirements for the lighting system (e.g., illuminance levels, color requirements, available space).
2. Selection of equipment, lamps, and luminaires. Lighting systems consist of numerous components, the two most important of which are *lamps*, which influence the lighting level, color characteristics, and efficiency of the lighting system, and *luminaires*, which affect the efficiency with which the light is distributed and so affect lighting efficiency and uniformity.
3. Design of the lighting system to achieve a reasonably uniform distribution of light on a particular plane (usually horizontal), avoidance of glare, with a minimum expenditure of energy.
4. Include the optimum system control for the designed lighting system to make maximum daylight use, through selection of appropriate switching mechanisms and daylight responsive controls.

A designer must consider a variety of key characteristics when developing the lighting plan, including lamp life, system efficiency, lumen maintenance, color rendering and appearance, daylight integration and control, light distribution, points of interest, cost, system control, and flexibility. Lighting requirements are primarily dictated by the function of a space or the tasks being performed within it, being usually specified by the required lighting level and the color rendering requirements. Selection of components follows from the identification of systems requirements. The luminaires are normally chosen first, particular types of luminaire for specific tasks. These also come with different types of reflectors, lens, etc., for different applications. Lamps are selected based on those that are compatible (lamp type, dimensions, frequency of operation, efficiency, etc.)

with the selected luminaire and that have the appropriate color rendering index. A lighting designer has four major objectives:

1. Provide the visibility required based on the task to be performed and the economic objectives.
2. Furnish high-quality lighting by providing a uniform illuminance level, where required, and by minimizing the negative effects of direct and reflected glare.
3. Choose luminaires aesthetically complimentary to the installation with mechanical, electrical, and maintenance characteristics designed to minimize operational expense.
4. Choose sustainable products that minimize energy usage while achieving the visibility, quality, and aesthetic objectives.

6.5.1 Factors Affecting the Selection of Light Sources and Equipment

The most common measure of light output (or luminous flux) is the lumen. Light sources and light fixture outputs are labeled with an output rating in lumens. For example, a 40 watt fluorescent lamp may have a rating of 3050 lumens. As lamps and fixtures age and become dirty, their lumen output decreases, i.e., lumen depreciation occurs. Most lamp ratings are based on initial lumens (when the lamp is new). Light intensity measured on a plane at a specific location is the illuminance, measured in foot-candles, which are work-plane lumens per square foot. It can be measured using a light meter located on the work surface where tasks are performed. Using simple arithmetic and manufacturer's photometric data, illuminance for a defined space can be estimated. Lux is the SI unit for illuminance, measured in lumens per square meter, and to convert foot-candles to lux, we multiply foot-candles by 10.76. Another light measurement, the luminance, sometimes called brightness, measures the light *leaving* a surface in a particular direction, considering the illuminance on the surface and the reflectance of the surface. The eye does not see illuminance, but luminance. Therefore, the amount of light delivered into the space and the surface reflectances in the space affect our ability to see. IESNA has developed a procedure for determining the appropriate average light level for a particular space, which is used extensively by designers and engineers. It recommends a target light level by considering the following factors: the performed task(s) (contrast, size, etc.), occupant ages, and the importance of speed and accuracy. Then, the appropriate type and quantity of lamps and light fixtures is selected based on the following: fixture efficiency, lamp lumen output, surrounding surface reflectances, the effects of light losses from lamp lumen depreciation and dirt accumulation, room size and shape, and daylight availability.

When designing a new or upgraded lighting system, one must be careful to avoid over-lighting a space. In the past, spaces were designed for as much as 200 fc in places where 50 fc may not only be adequate, but superior. This was partly due to the misconception that the more light in a space, the higher the quality. Not only does over-lighting waste energy, but it can also reduce lighting quality. Light levels are specified by the IESNA codes. Within a listed range of illuminance, three factors dictate the proper level: age of the occupant(s), speed and accuracy requirements, and the background contrast. For example, to light a space with computers, the overhead light fixtures should provide up to 30 fc of ambient lighting. The task lights should provide the additional foot-candles needed to achieve a total illuminance of up to 50 fc for reading and writing. For illuminance recommendations for specific visual tasks, refer to the IES Lighting Handbook or to the IES Recommended Practice No. 24. Improvements in lighting quality can yield high dividends for businesses. Worker productivity gains may come from providing corrected light levels with reduced glare. Although the cost of energy for lighting is substantial, it is small compared with the cost of labor. Therefore, these gains in productivity may be even more valuable than the energy savings associated with new lighting technologies. In retail spaces, attractive and comfortable lighting designs can attract clientele and enhance sales. Three quality issues should be addressed in the design processes: glare, uniformity of illuminance, and color rendition. The glare or luminance ratio, not exceeding the suggested light levels, is reduced by using lighting equipment designed to reduce glare. A louver or lens is commonly used to block direct viewing of a light source. Indirect lighting, or up-lighting, can create a low-glare environment by uniformly lighting the ceiling. Also, proper fixture placement can reduce reflected glare on work surfaces or computer screens. Standard data now provided with luminaire specifications include tables of its visual comfort probability (VCP) ratings for various room geometries. The VCP index provides an indication of the percentage of people in a given space that find a fixture glare acceptable. A minimum VCP of 70 is recommended for commercial interiors, while luminaires with VCPs exceeding 80 are recommended in computer areas. The uniformity of illuminance is a quality issue that addresses how evenly light spreads over a task area. Although a room's average illuminance may be appropriate, two factors may compromise uniformity: improper fixture placement based on the luminaire's spacing criteria (ratio of maximum recommended fixture spacing distance to mounting height above task height), and fixtures that are retrofit with reflectors that narrow the light distribution. Nonuniform illuminance causes several problems: inadequate light levels in some areas, visual discomfort when tasks require frequent shifting of view from underlit to overlit areas, bright spots, and patches of light on floors and walls that cause distraction and generate a low-quality appearance.

Electric light sources have three characteristics: efficiency, color temperature, and color rendering index. Some lamp types are more efficient in converting energy into visible light than others. The efficacy of a lamp refers to the number of lumens leaving the lamp compared to the number of watts required by the lamp (and ballast). It is expressed in lumens per watt. Sources with higher efficacy require less electrical energy to light a space. Another characteristic of a light source is the color temperature, which is a measurement of *warmth* or *coolness* provided by the lamp. Warmer sources in lower illuminance areas, such as dining areas and living rooms, and cooler sources in higher illuminance areas, such as grocery stores, are preferred. Color temperature refers to the color of a blackbody radiator at a given absolute temperature, expressed in Kelvins. A blackbody radiator changes color as its temperature increases (first to red, then to orange, yellow, and finally bluish white at the highest temperature). A warm color light source actually has a lower color temperature. For example, a cool-white fluorescent lamp appears bluish in color with a color temperature of around 4100 K. A warmer fluorescent lamp appears more yellowish with a color temperature around 3000 K. The CRI is a relative scale (ranging from 0–100), indicating how perceived colors match actual colors. We have to remember that the higher the color rendering index, the less color shift or distortion occurs. The CRI number does not indicate which colors will shift or by how much; it is rather an indication of the average shift of eight standard colors. Two different light sources may have identical CRI values, but colors may appear quite different under these two sources.

Selection of the right lamp or lighting equipment depends on what is required of the lighting and where. Incandescent lamps were very popular for private domestic use for many years. However, due to their poor efficiency and short service life, they are now being replaced by more environmentally compatible alternatives of higher quality such as LED lamps. Discharge lamps are the perfect choice for professional applications thanks to their efficient operating mode. LED light sources are taking over in application areas, due to higher luminous efficiency and longer service life. They can legitimately be regarded as the light source of the future. Thus part of the expertise is to find the most suitable lamp for a specific lighting task. The performance characteristics of lamps are defined by the following concepts:

1. The electric power is the power consumed by a light source. The system power takes into account the power consumption of the control gear as well as that of the light source.
2. Luminous flux defines the total quantity of light emitted from a light source. The unit used is the lumen [lm]. The ratio of luminous flux to the required electric power gives the luminous efficiency [lm/W]. The system luminous efficiency also takes the ballasts' losses into account. Luminous efficiency describes the efficiency of a light source and is now one of the most important performance characteristics of all.

3. The average service life is usually quoted, which is the time after half of the lamps are statistically still serviceable, in other words half of the lamps failed. This test is subject to standardized operating conditions. Lamp manufacturers display this failure rate by curves, and they are shown as maintenance factors (LSF). Special service-life data apply to some light sources such as LEDs.
4. Drop in luminous flux, meaning that the initial luminous flux of a new lamp decreases over its time of operation (lumen maintenance), due to the aging of its chemical and physical components. Lamp manufacturers display this drop in luminous flux by curves, shown as maintenance factors (LLWF).
5. The color code is a three-digit number (e.g., 840), describing the lighting quality of a white light source. The first digit denotes color rendering, the second and third digits color temperature (light color). The light color describes the color impression made by a white light source as relatively warm (ww) or relatively cool (nw, intermediate; tw, cool). It is affected by the spectrum red and blue components.
6. The spectral components of the light determine how well various object colors can be reproduced. The higher the color rendering index (Ra or CRI), or the lower the color rendering group number, the better the color rendering in comparison with the optimum reference light. The maximum color rendering index value is 100. Values in excess of 80 are considered to be very good. Eight test color samples (R1 to R8) are used for the general color rendering index, and there are another six more vivid high-saturation colors (R9 to R14). The color rendering index is calculated for a light source relative to a "known" reference light source. Color fields can only convey an impression of the original reflection patterns.
7. Discharge lamps in particular need between 30 seconds and several minutes to warm up and output the full luminous flux, while high-pressure discharge lamps need to cool down for several minutes before they can be started again.
8. Incandescent and halogen incandescent lamps and almost all fluorescent and compact fluorescent lamps can be dimmed as required today. Most metal-halide lamps continue to be incompatible with dimming, which may have uncontrolled effects on lighting quality and lamp service life. The new series of special models for indoor and outdoor applications constitute an exception. The output of sodium vapor lamps and high-pressure mercury lamps can be restricted in stages. LED light sources can be switched and dimmed as required.
9. Manufacturers specify the permitted operating positions for their lamps. For some metal-halide lamps, only certain operating positions are allowed so as to avoid unstable operating states. Compact fluorescent lamps may usually be used in any operating position; however, important properties such as the luminous flux vs. temperature curve may vary with the position.

6.5.2 LIGHTING DESIGN PROJECT STRUCTURE AND CRITERIA

Achieving lighting energy savings is considered one of the fundamental energy efficiency measures with numerous opportunities and supporting benefits. Lighting design is done in the framework of guidance, specifications, and recommendations, rather than fixed design rules. In general, there is more than a single optimum solution for a lighting problem. Quite often there is a set of requirements to which priorities need to be set before a satisfactory compromise can be made. General guidance involves: lighting requirements, design process decisions, and selected calculation procedures. Lighting projects and design executed properly and comprehensively can be easily justified for a number of reasons including:

1. Energy savings, often a 25% internal rate of return or even better.
2. Emission reductions, direct correlation between energy and emission reduction.
3. Maintenance cost savings by replacing inefficient and/or old systems.
4. Increasing light levels for occupant or employee comfort, improved safety, and productivity.
5. Improved CRI to enhance comfort or productivity.

The objective of a "quality" lighting design is to provide a safe and productive environment—whether for business or pleasure. A good lighting design must and should provide the proper amount of light in every room; to be built and constructed within budget, code, guideline, and other constraints; environmentally responsible; respond to the architecture and interior design requirements and specifications; produce good color, while achieving the desired moods of each space; and be able to control the lights. Last but not least, it must look good. The lighting objectives can be considered in three broad categories: (1) safety and health, (2) performance, and (3) appearance and comfort. This is accomplished by a redesign or upgrade to ensure that the appropriate quality and quantity of light is provided for the users of a space, at the lowest operating and maintenance cost. A quality lighting design addresses more than "first cost" issues. Proper evaluation of the data, planning, and execution are essential for successful implementation. Building systems are interrelated. For example, removing 10 kW of lighting energy from a commercial building will have a significant impact on the heating, ventilation, and air-conditioning system. Cooling costs will be reduced, but replacement heating may be required. It is necessary for the lighting designer to have a clear understanding of all the building systems and how they interrelate. The methodology used to evaluate the energy savings for a lighting project, either for a retrofit or a comparison for new projects, is critical to the success of installing a complete energy-efficient solution. Too often, the simple payback method is used, which undervalues the financial benefit to the organization.

Light level, or more correctly, illuminance level, is easily measured using an illuminance meter. Illuminance is the light energy striking a surface. The IESNA regularly publishes tables of recommended illuminance levels for all possible tasks. It is also important to realize that the illuminance level has no relevance to the lighting quality, being entirely possible to have the recommended illuminance in a space but with light sources producing enough glares to impede work. This accounts for many of the complaints of either too much or not enough light. There are a number of methods for determining whether a lighting installation is efficient. One method is for the lighting designer to check with the current version of the ASHRAE/IESNA 90.1 lighting standard. This document, which is revised regularly, provides a recommendation for lighting power density. It is usually possible for a capable lighting designer to achieve better results than the ASHRAE/IESNA 90.1 recommendations.

6.5.3 INDOOR LIGHTING DESIGN METHODS

It is essential that the decision about the method of lighting is made at an early stage in the design of the building, and the architect should consult the lighting engineer and others concerned during the conceptual stage. The first step is to establish the general requirements for the artificial lighting in terms of the main visual tasks to be carried out in the building. The illuminance levels are usually the ones required for designated spaces and activities, made available by the several authorities publishing the general accepted values for the lighting design use, the desired values for specific applications. Table 6.6 summarizes the recommended and minimal illuminance levels for common space and activities. The next step is to determine the lighting requirements in terms of revealing the form of the building and helping to create the right character of the interior lighting. The architect and the lighting engineer should be able to consider more detailed aspects of the lighting design under the following headings:

1. The extent to which artificial lighting is used alone, or to supplement the day lighting.
2. The illuminances required for lighting specific visual tasks.
3. The required luminance throughout the interior.
4. The evaluation of discomfort glare in terms of the whole visual environment.
5. The directional lighting characteristics required to give the desired modeling effects and to reveal form and texture.
6. The main features of the building interior color schemes in terms of type, chrome, and color rendering.

TABLE 6.6
Recommended and Minimal Illuminance Levels

Application	Illuminance (lx)
Emergency lighting	0.2
Suburban street lighting	5
Dwelling	50–150
Corridor	100
Rough task with large-detail storeroom	200
General office, retail shop	400
Drawing office	600
Prolonged task with small detail	900

Illumination calculations involved in lighting design are based on the principle of luminous flux transfer from the light source to the designated surfaces (areas). In order to design a luminaire layout that best meets the illuminance and uniformity workplace requirements, two types of information are generally needed: average illuminance level and illuminance level at a given point. Calculation of illuminance at specific points is often done to help the designer to evaluate the lighting uniformity, especially when using luminaires, and the spacing recommendations are not supplied, or where task lighting levels must be checked against ambient. The most important quantities involved in these calculations are the lighting power density (illuminance), and to a lesser degree, the luminance (brightness) and the contrast (luminance ratio of the surfaces). The most-used methods to calculate the illuminance are:

1. *Lumens method* that is based on the definition of illuminance (light power density), E on a surface, $E = F/A$ (Equation [6.11]), expressed in foot-candles, if the area is expressed in square feet, and in lux if the area is expressed in square meters.
2. *Point method* using the fact that the square law of illuminance, Equations (6.9) and (6.10), with illuminance expressed in foot-candles or in lux, depending on the units used for the distance.

When these illuminance definitions (equations) are applied in actual lighting design and analysis, they are modified to include correction factors because not all the luminous flux is coming from the light source(s) to the surface. Furthermore, the light flux degrades over the lifetime of the light source(s) and the depreciation must be accounted for. Depending on the receiving surface and the light source(s), either one or both methods may be applicable. If the light source is small or very small in connection with the surface of interest, then the point method can yield more accurate results, while for large light sources, such as the fluorescent lamps or diffused ones, such as luminaires with defusing lens, the lumen method is the choice. Most computer-based lighting calculation packages are based on the point method and ray-tracing algorithms. The point method is also used frequently in outdoor lighting calculations and design.

The lumen method or zonal cavity method is a widely used approach to the systematic design of electric lighting, especially for indoor lighting design. It is the most-used hand calculation method to estimate the average illuminance levels for indoor areas, unless the light distribution is radically asymmetric. It is an accurate hand method for indoor applications because it takes into consideration the effect that interreflectance has on the level of illuminance. Although it takes into account several variables, the basic premise that foot-candles are equal to luminous flux over an area is not violated. The basis of the zonal cavity method is that a room is divided into three spaces or cavities. The space between the ceiling and the fixtures, if they are suspended, is defined as the *ceiling cavity*; the space between the work plane and the floor, the *floor cavity*; and the space between the fixtures and the work plane, the *room cavity*. Once the concept of these cavities is understood, it is possible to calculate numerical relationships called *cavity ratios*, which are used to determine the effective reflectance of the ceiling and floor cavities and then to find the coefficient of utilization. The method depends essentially on the accuracy of the utilization factor (e.g., the estimation of the ratio of the lumens that are received on the working plane to the total output of the lamps in the room). The aim of the lumen method is to give a reasonably even spread of light over the horizontal working plane. How this spread of light is achieved depends upon the way the light is distributed from the fittings, not only in relation to fittings but related to the height at which the fittings are mounted over the working plane. The ratio mounting height to the spacing of the fittings will vary with the choice of fitting: the greater the concentration of light distribution from the fitting, the closer must be the spacing relative to the mounting height. There are four basic steps in any calculation of illuminance level: (1) determine cavity ratios; (2) determine effective cavity reflectances; (3) elect coefficient of utilization; and (4) compute average illuminance level. In the case of rectangular surfaces, the cavity ratios may be computed using the following relationships:

$$CCR = \frac{5h_{cl}(L+W)}{L \times W} \tag{6.23a}$$

$$RCR = \frac{5h_{lw}(L+W)}{L \times W} \tag{6.23b}$$

$$CCR = \frac{5h_{wf}(L+W)}{L \times W} \tag{6.23c}$$

Here *CCR* is the ceiling cavity ratio, *RCR* is the room cavity ratio, *FCR* is the floor cavity ratio, *CR* is the cavity ratio, h_{cl} is the distance from luminaire to ceiling, h_{lw} is the distance from luminaire to work plane, h_{wf} is the distance from work plane to floor, *L* is the room length, and *W* is the room width. All distances, length, and width are in feet or meters (as shown in Figure 6.6). Effective cavity reflectances are then determined for the ceiling cavity and for the floor cavity, using the values in Tables B.4-I through Table B.4-III (Appendix B) under the applicable combination of cavity ratio and actual reflectance of ceiling, walls, and floor. The computed reflectance values are ρ_{cc} (effective ceiling cavity reflectance) and ρ_{fc} (effective floor cavity reflectance). If the luminaire is recessed or surface-mounted, or if the floor is the work plane, the CCR or FCR is 0 and then the actual reflectance of the ceiling or floor will also be the effective reflectance. With the values of ρ_{cc}, ρ_{fc}, and ρ_w (wall reflectance), and the calculated RCR, the coefficient of utilization is found in the luminaire coefficient of utilization (CU) table (Table B.4-IV, Appendix B), linear interpolations may be used for exact cavity ratios and reflectance combinations. The coefficient of utilization found is for a 20% effective floor cavity reflectance. Thus, it is necessary to correct for the determined ρ_{fc}. This is done by multiplying the previously determined CU by the factor from Table B.4-IV, as:

$$CU_{final} = CU(20\% \text{ floor}) \times \text{Multiplier (Actual } \rho_{fc}) \tag{6.24}$$

If it is other than 10% or 30%, then it is required to interpolate or to extrapolate and multiply by this factor. Next, computation of the illuminance level is performed with the standard lumen method, as discussed in a later paragraph, using Equation (6.22).

An alternate relationship for calculating any cavity ratio (*CR*) is given by:

$$CR = 2.5h_c \times \frac{P_{cav}}{A_{cav}} = 2.5h_c \times PAR_{cav} \tag{6.25}$$

where p_{cav} is the cavity perimeter (ft. or m), A_{cav} is the area of the cavity base (ft.^{2} or m^2), and PAR_{cav} represents the ratio of the perimeter to the floor area (ft^{-1} or m^{-1}), determined by the room geometry, and expressed by:

$$PAR_{cav} = \begin{cases} \dfrac{2(L+W)}{L \times W}, & \text{for recangular rooms} \\ \dfrac{4}{D}, & \text{for circular rooms} \\ \dfrac{3.27}{D}, & \text{for semi-circular rooms} \end{cases} \quad (6.26)$$

Here, L and W, as previously defined, and D is the diameter of the circular room (m or ft.).

Example 6.13

A room is 25 ft. by 32 ft., and h_c is equal with 7.5 ft. Find the room cavity ratio.

SOLUTION

From Equation (6.20) the ratio of the perimeter to the floor area, for a rectangular room is:

$$PAR_{cav} = \frac{2 \times (25 + 32)}{25 \times 32} = 0.143 \text{ ft}^{-1}$$

and from Equation (6.20), the cavity ratio for this room is:

$$CR = 2.5 \times 7.5 \times 0.143 = 2.67$$

Once the cavity ratios are determined, the next step consists of finding the effective cavity reflectances for ceiling and floor cavities. The ability of a surface to reflect incident light is given by its luminance factor. Most values of reflectances are given in Tables B.4-I through Table B.4-III (see Appendix B for details) under the applicable combination of cavity ratio and actual reflectance of ceiling, walls, and floor. Samples of the luminance factor, determined by the reflectance values, are included in Table 6.7. Effective reflectance values found are ρ_{cc} (effective ceiling cavity reflectance) and ρ_{fc} (effective floor cavity reflectance). Note that if the luminaire is recessed or surface-mounted, or if the floor is the work plane, the CCR or FCR are 0 and then the actual reflectance of the ceiling or floor is the effective reflectance. The recommendations are to use the actual ceiling reflectance value for when fixtures are surface-mounted or recessed, and to use the actual floor reflectance if the floor is the work plane.

TABLE 6.7
Luminance Factors for Painted Surfaces

Surface	Typical Color	Luminance Factor Range (%)
Ceiling	White, cream	70%–80%
Ceiling	Sky blue	50%–60%
Ceiling	Light brown	20%–30%
Wall	Light stone	50%–60%
Wall	Dark grey	20%–30%
Wall	Black	10
Floor	—	10

With these values of ρ_{cc}, ρ_{fc}, and ρ_w (wall reflectance), and knowing the room cavity ratio (RCR) previously calculated, find the coefficient of utilization in the luminaire coefficient of utilization (CU) table. Note that because the table is linear, linear interpolations can be made for exact cavity ratios and reflectance combinations. The coefficient or factor of utilization (CU) found is for a 20% effective floor cavity reflectance. Thus, it is necessary to correct for the previously determined ρ_{fc}. This is done by multiplying the previously determined CU by the factor from Table B.4-IV (see Appendix B). CU final is equal to the CU, at 20% floor time the multiplier for actual ρ_{fc}. If it is other than 10% or 30%, interpolate or extrapolate and multiply by this factor, to get the utilization coefficient. Computation of the illuminance level is performed using the standard lumen method relationship used for lighting design:

$$E = \frac{F \times N \times CU \times LLF}{A} \qquad (6.27)$$

where, E is the average horizontal illumination at the working plane in lx or fc, F is the rated lamp ***lumens*** (as published by each lamp manufacturer), N is the number of lamps (number of luminaires times number of lamps per luminaire), CU is the utilization factor or coefficient, LLF is the light loss factor, and A is the area of the working plane (ft.^{2} or m^2). CU is determined as discussed above, based on the room size, configuration, surface reflectances, and the performance characteristics of each of the luminaires. The total light loss factor (LLF) consists of three basic factors: lamp lumen depreciation (LLD), luminaire dirt depreciation (LDD). and ballast factor (BF). If the initial levels are to be found, a multiplier of 1 is used. Light loss factors, along with the total lamp lumen output, vary with manufacturer and type of lamp or luminaire and are determined from the manufacturer's published data. Ballast factor is defined as the ratio between the published lamp lumens and the lumens delivered by the lamp on the ballast used. Typical HID ballast factors vary between 0.9 and 0.95. Occasionally, other light loss factors may be applicable. Some of these are luminaire ambient temperature, voltage factor, and room surface dirt depreciation. When the initial illuminance level required (specified in codes or standards) is known and the number of fixtures (luminaires), N_{lm} needed to obtain the desired or required illuminance level, a variation of the standard lumen formula may be used:

$$N_{lm} = \frac{E_{desired/required} \, (\text{fc or lx}) \times A \left(\text{ft}^2 \text{ or m}^2 \right)}{\# \dfrac{lamp}{fixture} \times F \times CU \times LLF} \qquad (6.28)$$

Example 6.14

A university amphitheater is 72′ long and 40′ wide with a 15′ ceiling height. Reflectances are: ceiling 80%, walls 30%, floor 10%. Four-lamp, Prisma-wrap type per each luminaire are used, having 3000 lm per lamp, on 6′ stems and the work plane is 2′ above the floor. Find the illuminance level if there are 24 luminaires in the room.

SOLUTION

From Equations (6.19a), (6.19b), and (6.19c), the CCR, RCR, and FCR are calculated as:

$$CCR = \frac{5h_{cl} \, (L+W)}{L \times W} = \frac{5 \times 4 \times (72+40)}{72 \times 40} = 0.78$$

$$RCR = \frac{5h_{lw} \, (L+W)}{L \times W} = \frac{5 \times (15-4-2)(72+40)}{72 \times 40} = 1.75$$

$$FCR = \frac{5h_{wf} \, (L+W)}{L \times W} = \frac{5 \times 2(72+40)}{72 \times 40} = 0.39$$

In Tables B.4-I, B.4-II, and B.4-III (Appendix B), for the above ceiling and floor cavities, the effective reflectance is determined as 65, while ρ_{fc} for the floor cavity is 10%. The effective cavity reflectances for the ceiling and floor cavities are 65% and 10%, respectively. Using the computed RCR value, 1.75 and the effective reflectance values, the coefficient of utilization is calculated (Appendix Table B.4-IV) by interpolation as 0.63. However, this CU is for an effective reflectance of 20% while the actual effective reflectance of the floor ρ_{fc} is 10%. To correct for this, locate the appropriate multiplier in Table B6 for the RCR = 1.75, which is 0.952. The final CU is computed: $CU_{final} = 0.63 \times 0.952 = 0.60$. Assuming LLF = 1 in Equation (6.22), the initial (desired) illumination level is:

$$E = \frac{24 \times 4 \times 3000 \times 0.60 \times 1}{72 \times 40} = 60 \text{ lm/sq.ft}$$

Notice that once the CU coefficient is found, we can determine the required total lumens (F_{tot}), number of luminaires (N), and area per luminaire (A_{lm}), by using the relationships given bellow. An approximation, but quite accurate one, for LLF is the product of LLD and LDD, only.

$$F_{tot} = \frac{E \cdot A}{CU \cdot LLF} \tag{6.29a}$$

$$N = \frac{F_{tot}}{F} \tag{6.29b}$$

$$A_{lm} = \frac{A}{N} \tag{6.29c}$$

The notations here are the ones defined above and with the specified units. The LDD factor is determined from the luminaire category, provided by manufacturers or found in the tables of the codes and standards, and it depends on the lamp and the replacement schedule (see Table 6.7). The coefficient of utilization is basically the ratio of the lumens received on the working plane to the total flux output of lamps.

Example 6.15

For a room of 20′ by 30′, the adjusted utilization factor was found 0.47, the LLD is 0.84, and LDD is 0.85. Find the total illuminance level, number of luminaires, and the area per luminaire, if the desired horizontal average illuminance is 50 lm/ft.^{2}. Each luminaire has two lumps of 3150 lm.

SOLUTION

From Equation (6.24a) the total desired lumens for this room are:

$$F_{tot} = \frac{E \cdot A}{CU \cdot LLF} = \frac{50 \cdot (30 \times 20)}{0.47 \times 0.84 \times 0.85} = 89397 \text{ lm}$$

The total number of luminaires is given, by Equation (6.24b) as:

$$N = \frac{F_{tot}}{F} = \frac{89397}{2 \times 3150} \simeq 14.2$$

The closest integer is 14, which is the selected number of luminaires. Area per luminaire is computed by using Equation (6.24c), as:

$$A_{lm} = \frac{A}{N} = \frac{20 \times 30}{14} = 42.86 \simeq 43 \text{ ft}^2$$

The aim of the lumen method is to give a reasonably even spread of light over the horizontal working plane. How this spread of light is achieved depends upon the way the light is distributed from the fittings, not only in relation to fittings but related to the height at which the fittings are mounted over the working plane. The ratio of the mounting height to the spacing of the fittings will vary with the choice of fitting: the greater the concentration of light distribution from the fitting, the closer must be the spacing relative to the mounting height. A slightly modified version of Equation (6.21) is given below, in which the LLF is replaced by the maintenance factor or coefficient (M). M represents the ratio taking into account the light lost due to average expectation of dirtiness of light fittings and room surfaces. The maintenance factor is often determined individually, and takes into account the installation reduction in luminous flux caused by soiling and aging of lamps, luminaires, and room surfaces. For normal conditions a factor of 0.8 may be used, for air-conditioned rooms a factor of 0.9 may be used, while for an industrial atmosphere where cleaning is difficult, a factor as low as 0.5 may sometimes be used. In order to have accurate M estimates, the maintenance schedule (the cleaning and maintenance intervals for the lamps and installation) must also be documented. With previous notation, with units specified above for each of the parameters involved, the E, the average horizontal illumination at working plane is expressed as:

$$E = \frac{F \times N \times CU \times M}{A} \tag{6.30}$$

Another design factor that is usually considered in lighting design is the spacing-to-height ratio (SHR) is the center-to-center (S) distance between adjacent luminaires to their mounting height (H) above the working plane. Manufacturer's catalogs can be consulted to determine maximum SHR, the ratio of the fitting spacing (S) to the work plane to fitting distance or height (H) (e.g., for a luminaire with trough reflector is about 1.65 and an enclosed diffuser about 1.4). We have to keep in mind, that there is a very large range of room dimensions; however, the behavior of light in rooms is a function not of the room dimensions but of the room index (RI), which is the ratio of the area of the horizontal surfaces to that of the vertical surfaces in the room. For the lumen method of design, the vertical surfaces are measured from the working plane to the center of the fitting. This is expressed by the equation:

$$RI = \frac{A}{H(L+W)} = \frac{L \times W}{H(L+W)} \tag{6.31}$$

Here, H is the room height (ft. or m). Equation (6.24) is for rectangular rooms, only. However, adaptations for other room geometries are easily made.

Example 6.16

A general office measuring 15 m × 9 m × 3 m high is to be illuminated to a design level of 400 lux using 85 W fluorescent fittings. The fittings are to be flush with the ceiling and the working plane is to be 0.85 m above the floor. Design the lighting system for the office when the installed flux is 8000 lumens per fitting. Assume that the maintenance factor is 0.8, and the height of fitting above the working plane, H, is 1.65. For the room index of this space of 3.4, calculated by using Equation (6.24), utilization factor from the tables is found to be 0.56.

SOLUTION

From Equation (6.23), the number of luminaires is:

$$N = \frac{E \times A}{F \times CU \times M} = \frac{400 \times (15 \times 9)}{8000 \times 0.56 \times 0.80} \approx 15$$

In terms of illumination, 15 fittings would provide about 398 lx and would probably be satisfactory. In terms of spacing arrangement, however, 16 fittings are required, which would provide the following illumination level:

$$E = \frac{F \times N \times CU \times M}{A} = \frac{8000 \times 16 \times 0.56 \times 0.8}{15 \times 9} \approx 425 \text{ lx}$$

The mounting height for fittings is: $3 - 0.85 = 2.15$ m.

6.5.4   LUMINAIRE TYPES, SELECTION, DESIGN, AND OPERATION

A *luminaire* is designed to provide physical support, electrical connections, and light control for the electric lamp(s). It is commonly known as the lighting fixture, being a complete unit that contains one or several lamps, structural supports, accessories, light control devices, and electrical components and parts. Ideally, the luminaires direct the light to where it is needed while shielding the lamp from the eyes at normal angles of view. Luminaires are composed of several components and parts that provide these functions: the housing, the electrical connection, control and protection, the light-controlling element, and the glare-controlling element. Depending on the design requirements and optical control desired, some of these functions may be combined. Various luminaire types are available, such as: stationary luminaires, louvered luminaires, integral luminaires, movable luminaires, secondary reflector luminaires, and decorative luminaires (where outward appearance is more important than their light). The electrical connection and physical support for the light source are provided by the luminaire housing. Often its electrical auxiliary equipment, when required, is also incorporated. The luminaires are classified according to their construction, physical configurations, and topologies, photometric distribution, type of light source(s), control, and electrical characteristics. Luminaires are classified by their features, such as:

1. Lamp type (incandescent, tungsten halogen, FL, CFL, HID, etc.);
2. Application (general lighting, down-light, wall-washer, accent light, spotlight, etc.);
3. Function (technical, decorative or effect luminaires);
4. Protection class (e.g., ingress protection IP-code);
5. Installation (suspended, recessed, surface-mounted, free-standing, wall-mounted, etc.); and
6. Type of construction (open, closed, with reflectors and/or refractors, high-specular louvers, secondary optics, projectors, etc.).

Luminaire housings are divided into five categories based on how they are supported: recessed, semi-recessed, surface-mounted, pendant-mounted, and track-mounted. Light structures are systems comprising modular elements that take integrated luminaires. Movable luminaires, e.g., spotlights, can be mounted and operated on light structures. Combinations of up-lighting and down-light systems are luminaires that provide a controlled downward distribution and a wider upward distribution. They have the advantage of providing efficient direct light onto the task, while providing an upward component to increase the background luminance. This reduces glare and makes the space appear larger. Luminaires are available in different combinations of upward and downward components. If the upward component is too high, the problems with a full up-light installation remain. Surface-mounted luminaires come in two basis families, bright sides and dark sides. Bright-sided luminaires have a prismatic lens or opal panel that diffuses the lamp image and directs the light. Diffuse opal fittings tend to have the same luminance on the sides and the bottom and, although they reduce the lamp luminance by increasing the area of the source, they do little to control the direction of the light and should only be considered as a decorative fitting. Prismatic lens panels control the luminance of the fitting by directing the light away from the angles of view and into the useful angles. Standard flat lens panels are designed to direct the light perpendicular to the panel. This improves the efficiency of the installation, as it directs the light onto the work surface. The side panels should direct light away from the angle of view by directing some of the light onto the ceiling and some towards the work surface. This reduces the luminance of the luminaire, when viewed from the side, and increases the luminance of the background. With this style of fitting it is only an increase in the immediate vicinity of the luminaire but it graduates the edge of the bright patch. Some of the less expensive surface-mounted fittings simply bend the same panel as the base onto the sides. This then directs light horizontally from the fitting. Therefore, they do not provide adequate glare control for a work environment. Well-designed surface-mounted fittings, on light ceilings, provide high efficiency and a good visual environment; however, aesthetic, building construction, and cost issues make recessed luminaires more popular. Surface-mounted fittings with dark sides should only be used in special circumstances as the construction of the luminaire prevents light from getting to the ceiling. As a result they have the highest contrast between the fitting and the background. This makes glare control difficult. Recessed luminaires come in two basic forms: fittings with prismatic panels to control the light and fittings that use reflectors to shield the lamp and control the distribution. The latter are commonly referred to as low-brightness fittings. Reflector or low-brightness fittings come in a wide range of configurations. The reflector is usually divided into cells to cut off the view of the lamp from the end. The luminaires work on the basis that the depth of the lamp in the fitting and the cross-blades shields the direct view of the lamp, while the reflectors are contoured to direct the light into specific angles of the distribution.

The lighting system illuminance reduces as soon it starts the service, when it is switched ON. There are several reasons for this: lump output reduction and lamp failure with the operation hours, dirt on luminaires and on room surface, or improper equipment maintenance. A lighting system needs to be serviced by replacing the defective lamps; maintaining equipment; and cleaning the lamps, luminaires, and rooms. Lamp replacement may be done individually when a lamp has failed (spot replacement) or all the lamps in an installation may be replaced periodically (group replacement), a cheaper option. Light loss factors (LLFs) are scalar multipliers that account for the degradation. LLFs may be either recoverable due to the maintenance of the lighting system and room or non-recoverable being a fixed value. Light loss factors include dirt accumulation on luminaires and room surfaces, lamp depreciation, ballast and electric system factors, and thermal application effects. Three factors must be considered in its determination: (a) the type of luminaire, (b) atmospheric conditions, and (c) maintenance period interval. Room surface maintenance factor (RSMF) is the proportion of the illuminance provided by a lighting installation in a room after a set time compared with what occurred when the room was clean. Luminaires, as mentioned before, accumulate dust and dirt during their operation time period, and different luminaire types collect dust and dirt at different rates. Luminaire degradation depends also on the room conditions, maintenance, air-conditioning, ventilation, and cleanliness. There are six types designed by CIE/IEC, A (bare lamp), B (ventilated open-top reflector, self-cleaning), C (unventilated closed-top reflector), D (enclosed), E (dust-proof), and F (indirect up-lighter type). The luminaire type, cleaning intervals, and environment in which it is operated (e.g., clean, normal, or dirty) are used to determine LMFs. Maintenance factor (MF) is the proportion of the initial light output from a luminaire after a set time to the initial light output from a lamp after a set time, taking into account all factors and losses:

$$MF = LLMF \times LSF \times LMF \times RSMF \tag{6.32}$$

TABLE 6.8
Typical Relamping Interval

Lamp Type	Lamp Life (hr.)
High-pressure sodium (HPS)	15,000
HPSDL High-pressure sodium (improved color)	6000
High-pressure mercury	9000
Metal-halide (MH) lamp	6000
Fluorescent tubular lamp	9000
LED	50,000

Lamp lumen maintenance factor (LLMF) accounts for the reduction in the lumen output of the lamp, a characteristic to all lamps, regardless of the type. However, it depends on the lamp type, being provided by manufacturer through graphs or tables of the lamp survival (initial lumens) versus rated life. These data are based on test procedures in standard conditions. Any test aspects of the lamp design, such as higher ambient temperature, vibration, switching cycle, or operating altitude, may affect the lamp life and/or output. Lamp performance under such conditions needs to be checked with the manufacturer. Life survival factor (LSF) accounts for how many lamps are still working after an operation interval. Table 6.8 gives relamping intervals for some typical lamps. If the lamps are replaced when failed, the LSF can be taken as 1.00. LSF is obtained from the manufacturer data, being usually expressed as *percent lamp survival*.

An important aspect in lighting design and the estimations of the maintenance factor values is *workplace cleanliness categories*, which include: *clean*, *normal*, and *dirty* categories used in the design of the interior environments. These categories are needed for determining the *LMF* and room surface maintenance factor (*RSMF*). RSMF is determined by the interior environment, room cleaning interval, and luminaire type. A luminaire that gives 90% or more light downward is categorized as *direct*, one that give 90% or more upward is considered as *indirect*, all others are classified as *direct-indirect* type. Table 6.9 shows RSMF data, adapted from standards for typical room indexes in the range 1.5 to 5.

TABLE 6.9
RSMF Values for Room Index 1.5 to 5

Cleaning Interval (yr.)				Environment	Type
0.5	1.0	2.0	3.0		
0.98	0.98	0.96	0.96	Clean	Direct
0.97	0.96	0.95	0.95	Normal	
0.96	0.95	0.94	0.94	Dirty	
0.95	0.92	0.89	0.86	Clean	Direct-Indirect
0.90	0.88	0.85	0.82	Normal	
0.86	0.85	0.81	0.78	Dirty	
0.92	0.88	0.84	0.78	Clean	Indirect
0.87	0.82	0.77	0.72	Normal	
0.83	0.77	0.70	0.64	Dirty	

Example 6.17

Find the maintenance factor for an office, with air-conditioning with the room index 3.0 using (direct) louver luminaires, and the lamps LLMF is 0.95. Assuming a cleaning at every 2 years, 10-hour, 50-day, 50-week operation, spot replacement is made after 4 years.

SOLUTION

Operating hours are: $50 \times 5 \times 10 = 2500$ hours per year. The spot lamp replacement as recommended by codes is this case is 1.00. The LMF value considering a clean environment, luminaire maintenance A, air-conditioning from Table B.8-II (Appendix B) is 0.96, the RSMF from Table 6.9, normal environment, direct luminaire is 0.95. The MF factor, Equation (6.27) is then:

$$MF = LLMF \times LSF \times LMF \times RSMF = 0.95 \times 1.00 \times 0.96 \times 0.95 = 0.87$$

As discussed before, the number of light fittings or luminaires is computed from the total lumens needed at the working plane, Equation (6.25) and the illumination provided by each fitting, the so-called lighting design lumens (LDL), which is determined by the fitting lamp type and number, as expressed by:

$$E = \frac{n \times F \times N \times CU \times M}{A} = \frac{LDL \times N \times CU \times M}{A} \tag{6.33}$$

Example 6.18

A 21 by 12.5 m and 3 m high design office room has white ceiling and walls and the working plane is at 0.85 m. If the two-lamp luminaires with an output of 5000 lumens are selected, and their SHR is 1.75 how many luminaires are needed? Assume a maintenance factor of 0.90 and a coefficient of utilization of 0.78.

SOLUTION

The required illuminance (Table 6.5) per unit of area is 600 lm/m², then the number of fittings (luminaires) is estimated from Equation (6.29) is:

$$N = \frac{600 \times (21 \times 12.5)}{0.79 \times 0.90 \times 5000} = 44.87 \text{ fittings}$$

The fitting number is raised to 45. The spacing between rows is calculated from the height of the luminaire above the working plane, H = 3 – 0.85 = 2.15 m, and the SHR:

$$S = H \times SHR = 2.15 \times 1.75 \approx 3.76 \text{ m}$$

The rows are assumed to be along the long side of the room in three 15-luminaire rows, spaced at 3.76 m of each other.

6.5.5 OUTDOOR LIGHTING DESIGN

One of the main lighting design applications involves the lighting of open outdoor areas, such as parking lots, sidewalks, building entrance areas, and loading docks. In such applications, lighting is necessary for vehicular and pedestrian safety, protection against crime, and user convenience. The outdoor lighting design objectives need to conform to those specified in the codes,

TABLE 6.10

Recommended Illuminance Values for Parking Lots and Loading Docks

Perimeter	Basic Use (fc)	Enhanced Security (fc)
Minimum horizontal illuminance	0.2	0.5
Uniformity ratio, maximum-to-minimum	20:1	15:1
Minimum vertical illuminance	0.1	0.25

standards, and technical manuals, as well as to the local and federal regulations and requirements and other requesting organizations. The aim for road and public space lighting schemes can include any or all of the following:

1. Facilitation of safe movement of vehicles and people.
2. Discouragement of illegal acts.
3. Contributing to the prestige and amenity of an area through increased aesthetic appeal.
4. Minimum light spill and glare.
5. Cost and efficient use of energy.

This lighting design type is based on maintaining specific illuminance levels and keeping the ratio of maximum illuminance to minimum illuminance below a predetermined value. The values for maintained illuminance and *uniformity ratio*, the ratio of maximum illuminance to minimum illuminance levels, are listed in the *IESNA Lighting Handbook* for various applications, as shown in Table 6.10. Horizontal illuminance is defined as the quantity of luminous flux falling on a horizontal plane. Similarly, vertical illuminance is the amount of luminous flux striking a vertical plane. These are minimum values whereas the uniformity ratio is a maximum value. Actual requirements for a particular application may surpass these values. Parking lot, sidewalk, and loading dock lighting are typically designed based on illuminance criteria. The IESNA recommends a minimum horizontal illuminance level of 0.2 ft-cd and a minimum vertical illuminance level of 0.1 ft-cd for low-activity open parking lots and loading docks to help assure safety and to deter crimes. A uniformity ratio, which compares the area with highest illuminance to the area with lowest illuminance, of 20:1 or lower is also specified for parking lots. A suitable lighting technology must be selected to start the design process. A luminaire style and lamp type must also be selected for the design to proceed. For outdoor lighting, it is reasonable to treat the light source as a point source, considering the luminaire and area of interest dimensions. In this case the inverse square, as shown in Figure 6.7, can be applied to calculate the illuminance for a point P as:

$$E = \frac{I(\theta,\phi)}{d^2}\cos(\theta)$$
(6.34)

or

$$E = \frac{I(\theta,\phi)}{h^2}\cos^3(\theta)$$
(6.35)

Where E is the illuminance (lx), d is the distance from the source to the point (m), θ is the angle of the light from the normal, $I(\theta,\phi)$ is the intensity of the source in the direction of θ *(cd)* as specified by manufacturer, h is the perpendicular distance from the source to the plane (m), and ϕ is the lateral angle. The values are given in tabular format, as the one found in Appendix B Table B.9 for parking

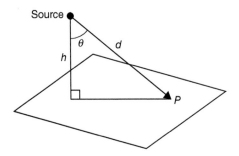

FIGURE 6.7   The illumination plan and geometry.

areas and other outdoor applications. In a tridimensional space, with x, y, the horizontal coordinates, and z, the vertical coordinate, then d and the lateral and vertical angles are calculated as:

$$d = \sqrt{x^2 + y^2 + z^2} \tag{6.36}$$

$$\theta = \tan^{-1}\left(\frac{\sqrt{x^2 + y^2}}{z}\right) \tag{6.37}$$

and

$$\phi = \tan^{-1}\left(\frac{y}{x}\right) \tag{6.38}$$

A variation of Equation (6.23), used for roadway lighting and sometimes for other outdoor applications, calculates how far apart the fixtures must be spaced to produce the necessary average illuminance, as given in Equation (6.32). A utilization curve shows the percent of light that falls onto an area having a designated width and an infinite length. This width is expressed on the utilization curve in terms of a ratio of the width of the area to the luminaire mounting height. Separate CUs are given for the area to the street side and area to the house side of the fixture and may be used to find illumination on the roadway or sidewalk areas or added to find the total light on the street in the case of median mounted luminaires. A relationship for street lamp separation or spacing, taking into account the lamp characteristics, CU and LLF values, and the desired foot-candle level is:

$$Spacing = \frac{I_{lamp} \times CU \times LLF}{\text{Avg MTD fc} \times W_{road}} \tag{6.39}$$

Here, I_{lamp} is the lamp intensity, W_{road} is the road width, and Avg MTD is the desired foot-candle level as required in codes and/or guides.

Example 6.19

A roadway 24 ft. wide is lightened to maintain illumination level of 1.0 fc. HPS lamps (35,000 cd), mounted on 30 ft. poles that are set back 30 ft. from the road, are used. Assuming a CU = 0.23, and LLF = 0.82, determine the light spacing.

SOLUTION

Spacing, by using Equation (6.33) is then calculated as:

$$Spacing = \frac{35000 \times 0.23 \times 0.82}{1.0 \times 24} = 275 \text{ ft}$$

6.6   CHAPTER SUMMARY

Artificial illumination for both functional and decorative purposes is a major consumer of primary energy, and developed civilizations have become used to very high illumination standards with consequently high electricity consumption. However, lighting represents the utilization of the natural or artificial lighting energy to provide the desired visual environment of working and living. Lighting is an essential service in all the industries and for residential sectors. Artificial lighting is provided to supplement daylight on a temporary or permanent basis, for both functional and decorative purposes, being a major consumer of primary energy. The power consumption by industrial lighting varies between 2% to 10% of the total power depending on the type of industry. Innovation and continuous improvement in the field of lighting have given rise to tremendous energy-saving opportunities in this area. Lighting is an area that provides a major scope to achieve energy efficiency at the design stage, by incorporation of modern energy-efficient lamps, luminaires, and gears, apart from good operational practices. Lighting is, obviously, a visual subject and it may be treated as such for design purposes. It is not necessarily helpful to study lighting only as applied mathematics. Designs are an artistic and engineering combination of architecture, interior design, decoration, illumination functionality, economical use of electrical energy, maintainability, safety, environmental health, controllability, prestige, and the overall and specific requirements of the user. Illumination intensity, illuminance, measured in lux on the working plane, is determined by the size of detail to be discerned, the contrast of the detail with its background, the accuracy and speed with which the task must be performed, the age of the worker, the type of space within which the task is to be performed and the length of time continuously spent on the task. Luminous intensity, measured in cd, is one of the seven basic SI measurement units. The quantity of light emitted by a source is the luminous flux, and is measured in lumens, where 1 lumen equals 1 candela-steradian. The total luminous flux incident on a surface per unit area is called illuminance, and is measured in foot-candles, where 1 foot-candle equals 1 lumen per square foot. Luminous efficacy quantifies the conversion efficiency from electricity to light. Various lighting technologies and systems have been developed and evolved over time, including incandescent, low-pressure discharge types such as fluorescent and low-pressure sodium, and HID types such as mercury vapor, high-pressure sodium, and metal-halide. LEDs produce photons when electrons change energy states while propagating through semiconductor material. Luminaires are fixtures designed to enclose lamps and provide specific dispersion patterns for the light. Luminaires are complete systems that contain the lamp(s), the housing, the light emitter(s), and the lighting and electrical controls. Technical luminaires are optimized for a certain function (e.g., a special luminous intensity distribution according to the task, prevention of glare), whereas decorative luminaires are designed with more focus on aesthetical aspects. Different lamp technologies require different luminaire construction principles and features. Choosing luminaires that efficiently provide appropriate luminance patterns for the application is an important part of energy-efficient lighting design. Lighting systems are usually designed using illumination criteria determined by IESNA, with criteria based on the anticipated use of the space being illuminated. This chapter also discusses the methods, tools, and analysis needed to design and to produce lighting applications, using modern and advanced light sources, luminaires, and control techniques. An overall presentation and discussion of lighting design methods and processes, including lighting quality issues, guidelines, and recommendations for advanced lighting design are presented in detail. Several examples, questions, and problems, as well as further readings and essential references are included here.

6.7   QUESTIONS AND PROBLEMS

1. What is the visible spectrum range?
2. What is the relationship between wavelength and frequency for EM waves?
3. What are the three primary colors of light?
4. Which waves from the EM spectrum does the sun emit?

5. What is illuminance? What is its measure unit?
6. Define glare, direct and reflected glare, discomfort, and disability glare.
7. What are the basic characteristics of color?
8. Define the luminous intensity.
9. What is the standard unit of luminous intensity?
10. What is the standard procedure to measure luminosity?
11. What is the wavelength range of the visible spectrum?
12. What is the refracted angle inside a glass (n = 1.52) if the light enters from air at 45° from the normal axis?
13. What are the types of commonly used lamps?
14. What do the following terms mean?
    a. Luminosity
    b. Light level
    c. Lumens
    d. Lux
    e. Lumen per watt
    f. Illumination intensity
15. What are the main characteristics for selecting a lighting system?
16. What are the primary benefits of an LED lighting system?
17. What do the following terms mean?
    a. Illuminance
    b. Luminous efficacy
    c. Luminaire
    d. Control gear
    e. Color rendering index
18. What is a lumen? (*Select answer from below.*)
    a. Unit of lighting not presently in use.
    b. 1000 candela/m².
    c. SI unit of light output or received.
    d. A directional measurement of light.
    e. Lighting power of a source.
19. What is the color temperature of fluorescent lamps?
20. What is meant by lamp efficiency?
21. Define lamp glare.
22. Describe how an incandescent lamp, a fluorescent lamp, and LED systems produce light.
23. What is the function of ballasts in a lighting system?
24. Define the terms used in lighting system design.
25. The three primary colors are?
26. What does luminaire mean?
27. List the luminaire major components.
28. What is the purpose of lighting system ballasts?
29. List some of the applications of point method.
30. What are the most important criteria in developing an outdoor lighting design strategy?
31. Determine the illuminance of a 12 ft. by 9 ft. office area if a total 6300 lumens are directed from a light source.
32. A point light source has 2800 cd intensity in the direction of interest. Determine the illuminance at a distance of 9 ft. and 6 m.
33. If 900 lm fall on 1.8 m² desk surface from a lamp, and a light meter measures that 360 lm are reflected, find the luminance on this surface.

34. Compare the energy efficiency and color rendering of different lamp types, stating suitable applications for each.

35. If a point source has lighting intensity of 2800 cd in the direction of interest, determine the illuminance at a point 15 ft from the source, and the angle of incidence with respect to the vertical is 36°.

36. By using data from Table 6.4, determine: (a) the candlepower along the fixture at an incidence angle of 35°; and (b) the candlepower in a vertical plane, at an incidence angle of 45°.

37. Calculate the room index for an office 24 m × 12 m in plane, 3.2 m high, where the working plane is 0.85 m above floor level.

38. Find the utilization factor for a bare fluorescent tube light fitting having two 58 W, 1500 mm lamps in a room 5 m by 3.5 m in plane and 2.5 m high. The working plane is 0.85 m above floor level. Walls and ceiling are light stone and white respectively.

39. On what factor does the arrangement of luminaires depend?

40. Ten incandescent lamps of 500 W and 10,800 lm are used for an area of 60 m². If the utilization coefficient and light loss factor are 0.65 and 0.80, respectively, calculate the illuminance and the lamp efficiency.

41. Determine the room cavity ratio for a space 60 ft. long, 30 ft. wide, and 15 ft. high. If the room described has a work surface 2.5 ft. above the floor, and the luminaire are suspended 2.5 ft. below, calculate the floor cavity ratio and the ceiling cavity ratio.

42. For the room in Problem 41, determine the 20,000 lm luminaires are needed to maintain an illuminance of 50 fc. Assume a ceiling reflectivity of 80%, wall reflectivity of 50%, a floor reflectivity of 20%, and a light loss factor of 0.85.

43. If an illuminance of 40 foot-candles must be maintained at the work surface of the room described in Problem 23, how many 20,000 lm luminaires are needed assuming a ceiling reflectivity of 77%, a wall reflectivity of 50%, a floor reflectivity of 20%, and a light loss factor of 0.88?

44. An office 12 m long by 8 m long requires an illumination level of 400 lux on the working plane. It is proposed to use 80 W fluorescent light fittings having a rated output of 7350 lumen each. Assuming a utilization factor of 0.6 and a maintenance factor of 0.85, calculate the number of light fittings required.

45. Determine the room cavity ratio for a space 90 ft. long, 50 ft. wide and 13.5 ft. high. If the room has a work surface of 2.85 ft. above the floor, and the luminaires are suspended 2 ft. below the ceiling, calculate the ceiling cavity ratio and the floor cavity ratio.

46. Find CU of a luminaire installed in a 15 × 20 m room with 3.5 m ceiling height, 0.75 work plane, and luminaires mounted 0.65 m below the ceiling. The reflectance coefficients are: $\rho_{cc} = 0.80$, $\rho_w = 0.60$, and $\rho_{fc} = 0.20$. Assuming LLF = 0.85, lamp intensity of 3200 lm, and two lamps per fixture, how many fixtures are required to maintain 540 lx?

47. Determine the number of luminaires and their space distribution across the room to provide the lighting for an industrial workshop, having dimensions 150' by 120', requiring 60 fc, if each luminaire has a rated output power of 5500 lm.

48. An office is to be illuminated for 12 h per day, for 5 days per week for 50 weeks per year. The floor is 24 m long and 12.5 m wide. An overall illumination of 450 lx is to be maintained over the whole floor. The total light loss factor for the installation is 72%. The designers have the choice of using 100 W tungsten filament lamps, which have an efficacy of 12 lm/W and need replacing every 3000 h, or 65 W fluorescent lamps, which have an initial output of 5400 lm and are expected to provide 12,000 h of service. The room layout requires an even number of lamps. Electricity costs 8 cents/kWh. The tungsten lamps cost $2 each while the fluorescent tubes cost $16.50 each. Compare the total costs of each lighting system and make a recommendation as to which is preferable, stating your reasons.

49. A supermarket of dimensions 24 m by 15 m and 4 m high has a white ceiling and mainly dark walls. The working plane is 1 m above floor level. Bare fluorescent tube light fittings with two 58 W, 1500 mm lamps are to be used, of 5100 lighting design lumens, to provide 400 lx. Their normal spacing-to-height ratio is 1.75 and total power consumption is 140 W. Calculate the number of luminaires needed, the electrical loading per square meter of floor area. Draw the layout of the luminaires.

REFERENCES AND FURTHER READINGS

1. R. Wolfson and M. J. Pasachoff, *Physics for Engineers and Scientists* (3rd ed.), Addison-Wesley, 1999.
2. R. R. Janis and W. K. Y. Tao, *Mechanical and Electrical Systems in Buildings*, Pearson, 2014.
3. T. R. Bosela, *Electrical Systems Design*, Prentice Hall, 2002, ISBN-13: 9780139754753.
4. G. Wyszeski and W. S. Stiles, *Color Science: Concepts and Methods, Quantitative Data and Formulae* (2nd ed.), John Wiley & Sons, 1982.
5. J. B. Murdoch, *Illumination Engineering*, Macmillan, NY, 1985.
6. R. McCluney, *Introduction to Radiometry and Photometry*, Artech House, 1994.
7. *IESNA Lighting Handbook* (9th ed.), Illuminating Engineering Society of North America, 2000. ISBN: 978-0-879-95150-4.
8. R. N. Helms, *Illumination Engineering for Energy Efficient Luminous Environments*, Prentice Hall, 1980.
9. W. B. Elmer, *The Optical Design of Reflectors*, John Wiley & Sons, 1980.
10. R. N. Helms and M. C. Beicher, *Lighting for Energy Efficient Luminous Environments*, Prentice Hall, 1991.
11. M. Schiler, *Simplified Design of Building Lighting*, John Wiley & Sons, 1992.
12. J. L. Lindsey, *Applied Illumination Engineering*, Fairmont Press, 1996.
13. A. Žukauskas, M., Shur, and R. Gaska, *Introduction to Solid-State Lighting*, Wiley, NY, 2002.
14. C. Cuttle, *Lighting by Design*, Oxford, 2003, ISBN: 0 7506 5130X.
15. P. R. Boyce, *Human Factors in Lighting* (2nd ed.), Taylor & Francis, London, UK, 2003.
16. R. Fehr, *Industrial Power Distribution* (2nd ed.), Wiley, 2015, ISBN: 978-1-119-06334-6.
17. F. J. Trost and I. Choudhury, *Mechanical and Electrical Systems in Buildings*, Pearson Prentice Hall, 2004.
18. H. W. Beaty, *Handbook of Electrical Power Calculations*, McGraw-Hill, 2001.
19. G. Gordon, *Interior Lighting for Designers* (4th ed.), John Wiley & Sons, 2003.
20. F. Kreith and D. Y. Goswami (eds.), *Energy Management and Conservation Handbook*, CRC Press, 2008.
21. A. Sumper and A. Baggini, *Electrical Energy Efficiency – Technology and Applications*, Wiley, 2012.
22. G. Petrecca, *Energy Conversion and Management – Principles and Applications*, Springer, 2014.
23. R. R. Janis and W. K. Y. Tao, *Mechanical and Electrical Systems in Buildings* (5th ed.), Pearson, 2014.
24. R. Fehr, *Industrial Power Distribution* (2nd ed.), Wiley, 2015, ISBN: 978-1-119-06334-6.
25. G. Gordon, *Interior Lighting for Designers* (5th ed.), Wiley and Sons, 2015.

7 Transformers and Electrical Motors

7.1 TRANSFORMERS AND ELECTROMECHANICAL ENERGY CONVERSION

The chapter objectives are to introduce the fundamental concepts of electromechanical energy conversion, leading to an understanding of the operation of various electromechanical energy converters, such as transformers and various electric motors. Electromechanical energy conversion is a subject that should be of particular interest to both the electrical and non-electrical engineers; it forms an important contact point between engineering disciplines. The emphasis of this chapter is on explaining the properties and characteristics of each type of electrical machine, with its advantages and disadvantages with regard to other types, and on classifying these machines in terms of their performances and characteristics and preferred field of applications. At the end of this chapter, the reader should understand the principles of operation of transformers, DC and AC motors and generators, nameplate data of electric machines, voltage-current of an electric generator, and torque-speed characteristics of an electric motor. Finally, the reader learns, understands, and specifies the requirements given for an application of an electric machine.

7.1.1 Transformers in Electrical Systems

Transformers are commonly used in applications that require the conversion of AC voltage from one voltage level to another. They are an essential part of power systems, having the role of converting electrical energy at one voltage to some other voltage. In order to effectively transmit power over long distances without prohibitive line losses, the voltage from the generator (a maximum output voltage of approximately 25–30 kV) must be increased to a significantly higher level (from approximately 150 kV up to 750 kV). Transformers must also be utilized on the distribution end of the line to step down the voltage (in stages) to the voltage levels required by the consumer. Transformers have a very wide range of applications outside the power area. Transformers are essential components in the design of DC power supplies. They can provide DC isolation between two parts of a circuit. Transformers can be used for impedance matching between sources and loads or sources and transmission lines. They can also be used to physically insulate one circuit from another for safety. The transformer is a valuable apparatus in electrical power systems, for it enables us to utilize different voltage levels across the system for the most economical value. The two broad categories of transformers are: electronic transformers, which operate at very low power levels, and power transformers, which process thousands of watts of power. The basic principle of operation of both types of transformers is the same. Power transformers are used in power generation, transmission, and distribution systems to raise or lower the level of voltage to the desired levels. In power systems, transmission lines are typically operated at voltages that are substantially higher than either generation or utilization voltages, for two reasons:

- Transmission line losses are proportional to the square of the current; higher voltages generally produce lower transmission line losses.
- Transmission line capacity to carry electrical power is roughly proportional to the square of the voltage, so higher voltages allow intensive use of the transmission line right-of-way: to enable transmission lines to carry high power levels, high transmission voltages are required. This is true for both underground cables and overhead transmission lines.

A power distribution transformer or service transformer is a transformer that provides the final voltage transformation in the electric power distribution system, stepping down the voltage used in the distribution lines to the level used by the customer. An electric power distribution system is the final stage in the delivery of electric power; it carries electricity from the transmission system to individual consumers. Distribution substations connect to the transmission system and lower the transmission voltage to medium voltage (MV) ranging from 2 kV to 35 kV with the use of transformers. Primary distribution lines carry this MV power to distribution transformers located near the user premises. Distribution transformers lower the voltage to the equipment utilization levels, typically feeding several customers through secondary distribution lines at this voltage. Commercial and residential customers are connected to the secondary distribution lines through service drops. Customers demanding a much larger amount of power may be connected directly to the primary distribution level or the subtransmission level. If mounted on a utility pole, these are called pole-mount transformers. If the distribution lines are located at ground level or underground, distribution transformers are mounted on concrete pads and locked in steel cases, thus known as pad-mount transformers. Distribution transformers normally have ratings less than 200 kVA, although some national standards specify units up to 5000 kVA as power distribution transformers. Because distribution transformers are energized for 24 hours a day (even when they don't carry any load), reducing iron losses has an important role in their design. As they usually don't operate at full load, they are designed to have maximum efficiency at lower loads.

7.1.2 Electromechanical Energy Conversion Systems

Electric machines are a means of converting energy. Motors take electrical energy and produce mechanical energy. Electric motors are used to power hundreds of devices we use in everyday life. Motors come in various sizes. Huge motors that can take loads of thousands of horsepower (HP) are typically used in the industry. Some examples of large motor applications include elevators, electric trains, hoists, and heavy metal rolling mills. Examples of small motor applications include motors used in automobiles, robots, hand power tools, and food blenders. Micromachines are electric machines with parts the size of red blood cells, and find many applications in medicine, instrumentation, or control. Electric motors are broadly classified into two different categories: DC (direct current) and AC (alternating current). Within these categories are several types, each offering unique abilities and characteristics that make each of them suitable for a specific application. In most cases, regardless of type, electric motors consist of a stator (stationary field) and a rotor (the rotating field or armature) and operate through the interaction of magnetic flux and electric current to produce rotational speed and torque. DC motors are distinguished by their ability to operate from direct current. There are different kinds of DC and AC motors, but they all work on similar principles. In this chapter, we will study their basic principle of operation and their characteristics. It's important to understand motor characteristics so we can choose the right one for our application requirement. An electromechanical energy conversion device is essentially a medium of transfer between an input side and an output side. Three electrical machines (DC, induction, and synchronous) are used extensively for electromechanical energy conversion. Electromechanical energy conversion occurs when there is a change in magnetic flux linking a coil, associated with mechanical motion. Generators and motors are primarily rotating machines. The rotating machines are called motors when they consume electrical energy, and are referred to as generators when they produce electrical energy, as shown in Figure 7.1. In practical applications, while DC machines are almost always single-phase, AC machines can be single-phase or three-phase types.

7.2 TRANSFORMER THEORY, CONSTRUCTION, AND DESIGN

Fundamentally, a transformer consists of two or more windings that are magnetically coupled using a ferromagnetic core. For a two-winding transformer, the winding connected to the AC supply is typically referred to as the *primary* while the winding connected to the load is referred to as

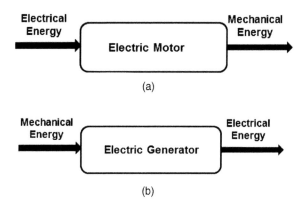

(a)

(b)

FIGURE 7.1   Electromechanical energy conversion: (a) motors, and (b) generators.

the *secondary*. A time-varying current passing through the primary coil produces a time-varying magnetic flux density within the core. According to Faraday's law, the time-changing flux passing through the secondary induces a voltage in the secondary terminals. In order to understand how a transformer operates, we will examine two inductors that are placed in close proximity to one another. The concepts of such magnetic-coupled circuits will be extended to the development of transformers. After understanding the relationships between voltages and currents, we will look at some practical considerations regarding the use of transformers. The purpose of a distribution transformer is to reduce the primary voltage of the electric distribution system to the utilization voltage serving the customer. Distribution transformers are static devices constructed with two or more windings used to transfer AC power from one circuit to another at the same frequency but with different values of voltage and current. Magnetic fields are created due to movement of electrical charge, and are present around permanent magnets and wires carrying current (electromagnet). In permanent magnets, spinning electrons produce a net external field. If a current-carrying wire is wound in the form of a coil of many turns, the net magnetic field is stronger than that of a single wire. This electromagnet field is further intensified if this coil is wound on an iron core. In many applications, the strength of magnetic fields needs to vary. Electromagnets are very commonly used in such applications. The magnetic field is represented by *lines of flux*, helping us to visualize the magnetic field of any magnet even though they represent only an invisible phenomena. Magnetic field forms an essential link between transfer of energy from mechanical to electrical form and vice versa, forming the basis for the operation of transformers, generators, and motors. If we examine the cross-sectional area of the magnet shown in Figure 7.2 and assume that the flux is uniformly distributed over the area, the magnetic flux density is defined as the magnetic flux per cross-sectional area:

$$B = \frac{\phi}{A} \tag{7.1}$$

where A is the cross-sectional area, and B is the magnetic flux density.

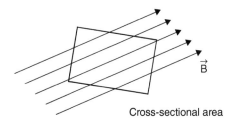

Cross-sectional area

FIGURE 7.2   Magnetic flux density.

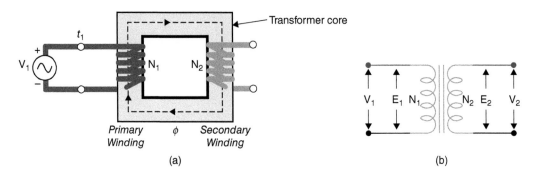

FIGURE 7.3   (a) Ideal transformer diagram; and (b) symbol for the ideal transformer.

Some characteristics of magnetic flux are, the magnetic flux passing through a surface bounded by a coil is said to link the coil (see Figure 7.3). The flux passing through a coil is the product of number of turns, N, and lux passing a single turn, ϕ, and the product is called the magnetic flux leakage of the coil, λ, expressed as:

$$\lambda = N\phi \text{ weber-turns}$$

Faraday discovered, the electromagnetic induction (Faraday's law), which states that whenever a conductor is moved through a magnetic field, or whenever the magnetic field near is changed, current flows in the conductor. Voltage induced in a single loop, due to a time-dependent magnetic flux ϕ is:

$$e(t) = \frac{d\phi(t)}{dt}$$

For a coil with N number of turns, the total induced voltage can be calculated by adding the voltage induced in all the turns:

$$e(t) = N\frac{d\phi(t)}{dt} = \frac{d\lambda}{dt} \text{ V} \tag{7.2}$$

Usually power transformers are built as three-phase units and may have multiple and complicated winding patterns. But three-phase transformers are at the core, understandable as three single-phase units with only a single-phase electric power, so our discussion starts with single-phase transformers (Figure 7.3). Transformers are fairly complex electromagnetic systems, subject to involved analysis, but for the purpose of learning about electric power systems, it will be sufficient to start with the model of an ideal transformer. A transformer contains two or more windings linked by a mutual magnetic field flowing through the transformer core. The simplest one consists of two windings on a core (Figure 7.3a). Figure 7.3 shows an ideal transformer and its symbol. We shall refer to the windings as HV (high-voltage) and LV (low-voltage). The designations *primary* and *secondary* are also common. Often additional windings (*tertiaries*, etc.) are added. In short, the action of a transformer is a particular case of the principle of mutual inductance, and a transformer consists essentially of two windings, the primary and secondary on a common magnetic core. A transformer is either core-type or shell-type construction. It can be seen from this figure that in core-type construction the primary and the secondary windings are wound as a pair of concentric coils on each limb, whereas for shell-type construction the primary and secondary windings form interleaved layers on a single limb. In all cases, the core will be of laminated construction in order to reduce iron losses to a minimum.

An ideal transformer is one with negligible winding resistances and reactances, no exciting losses, infinite magnetic permeability of the core, and all magnetic flux remaining into the transformer core. It consists of two conducting coils wound on a common core, made of high-grade iron. There is no electrical connection between the coils, being connected to each other through magnetic flux. The coil on the input side is called the primary winding (coil) and that on the output side the secondary. The essence of transformer action requires only the existence of time-varying mutual flux linking two windings. Such action can occur for two windings coupled through air. However, coupling between the windings can be made much more effective through the use of a core of iron or other ferromagnetic material because most of the flux will be confined to a definite, high-permeability path linking the windings. Such a transformer is commonly called an iron-core transformer. Most transformers are of this type. The following discussion is concerned almost wholly with iron-core transformers. When an AC voltage is applied to the primary winding, time-varying current flows in the primary winding and causes an AC magnetic flux to appear in the transformer core. The primary is connected to AC voltage sources, resulting in the flow of an alternating magnetic flux whose magnitude depends on the voltage and the number of turns of primary winding. The alternating flux links the secondary winding and induces a voltage in it with a value that depends on the number of turns of the secondary windings. The transformer operation is subject to Faraday's and Ampere's laws. If the primary voltage is $v_1(t)$, the core flux $\phi(t)$ is established such that the counter-emf $e(t)$ equals the impressed voltage (neglecting winding resistance), as:

$$v_1(t) = e_1(t) = N_1 \frac{d\phi(t)}{dt} \tag{7.3}$$

Here N_1 is the number of turns of the primary winding. The emf $e_2(t)$ is induced in the secondary by the alternating core magnetic flux $\phi(t)$, expressed as:

$$v_2(t) = e_2(t) = N_2 \frac{d\phi(t)}{dt} \tag{7.4}$$

Taking the ratio of Equations (7.3) and (7.4), we obtain:

$$\frac{v_1}{v_2} = \frac{N_1}{N_2} = a \tag{7.5}$$

Here a is the transformer turns ratio. If $a > 1$ the transformer is step-down, if $a < 1$, the transformer is step-up, while if $a = 1$, the transformer is a so-called impedance transformer, used to separate electric tow circuits. A load connected across the secondary terminals (Figure 7.4) will result in current i_2. This current will cause the change in the mmf in the amount $N_2 i_2$. Ohm's law for magnetic

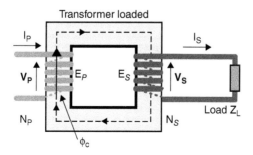

FIGURE 7.4 Loaded ideal transformer.

circuits must be satisfied. The only way in which this can be achieved is for the primary current to arise, such as:

$$N_1 i_1 - N_2 i_2 = \Re\phi$$

Here $\Re$ is the core magnetic reluctance. The reluctance or magnetic resistance for a magnetic core is simply calculated as:

$$\Re = \frac{l_c}{\mu A} = \frac{l_c}{\mu_r \mu_0 A} \qquad (7.6)$$

Here, l_c is the mean (average) core length (m), A is the core cross-sectional area (m^2), μ, and μ_r are the magnetic permeability, relative magnetic permeability of the core respectively, and $\mu_0 = 4\pi x 10^{-7} H/m$ is the magnetic permeability of the free space. For an ideal transformer we assume infinite permeability or $\Re = 0$. Neglecting losses (ideal transformer), the instantaneous power is equal on both transformer sides (power conservations):

$$v_1 i_1 = v_2 i_2$$
$$V_P I_P = V_S I_S$$

Combining the above relationship with Equation (7.5), we get transformer relationship of the primary and secondary currents:

$$\frac{i_1}{i_2} = \frac{N_2}{N_1} = \frac{1}{a}$$
$$\frac{I_P}{I_S} = \frac{1}{a} \qquad (7.7)$$

If all variables are sinusoidal, this equation applies also to the voltage and current phasors:

$$\frac{V_P}{V_S} = \frac{I_S}{I_P} = a \qquad (7.8)$$

Example 7.1

A 220/20 V transformer has 50 turns on its low-voltage side. Calculate:

a. The turns ratio a, when it is used as a step-down transformer.
b. The number of turns on its high side.
c. The turns ratio a, when it is used as a step-up transformer.

SOLUTION

a. For a transformer used as a step-down one, the turns ratio is

$$a_{SD} = \frac{220}{20} = 11$$

b. The number of turns in the high-voltage side is then:

$$N_P = aN_S = 11 \cdot 50 = 550 \text{ turns}$$

c. The turns ratio when the transformer is used as step-up is:

$$a_{SU} = \frac{1}{a_{SD}} = \frac{1}{11} = 0.091$$

Depending on the ratio of turns in the primary and secondary winding, the RMS secondary voltage can be greater or less than the RMS primary voltage. One can see from Equations (7.5) and (7.6) that almost any desired voltage ratio, or ratio of transformation, can be obtained by adjusting the number of turns of the transformer windings. Transformer action requires a magnetic flux to link two windings (coils). This will be obtained more effectively if an iron (or iron-based) core is used because the iron core confines the flux to a definite path linking both windings. However, a magnetic material such as iron undergoes losses of energy due to the application of alternating voltage in the B-H loop. These losses are composed of two parts. The first is called the *eddy-current loss*, and the second is the *hysteresis loss*. Eddy-current loss is basically an I^2R loss due to induced current in magnetic materials of the core due to the alternating magnetic flux linking the windings. To reduce these losses, the magnetic core is usually made by a stack of thin iron-alloy laminations. For analyzing an ideal transformer, we make the following assumptions. The resistances of the windings can be neglected, while the reluctance of the core is negligible. All the magnetic flux is linked by all the turns of the coil and there is no leakage of flux. We can write the equations for sinusoidal voltage in this ideal transformer as follows. The primary winding of turns N_1 is supplied by a sinusoidal voltage:

$$v_1(t) = V_{1m} \cos(\omega t) \tag{7.9}$$

Maximum value of voltage and the RMS value are related through:

$$V_1 = \frac{V_{1m}}{\sqrt{2}} = 0.707 V_{1m} \tag{7.10}$$

In the case of sinusoidal excitation with supply frequency f Hz, the RMS value of the primary emf, from Equation (7.3) is given by:

$$V_1 = 4.44 f N_1 \Phi_m \tag{7.11}$$

Here Φ_m is the peak value of the magnetic flux. In a similar manner, the RMS value of the secondary emf will be given by:

$$V_2 = 4.44 f N_2 \Phi_m \tag{7.12}$$

Example 7.2

Suppose a coil having 100 turns is wound on a core with a uniform cross-sectional area of 0.25 m². A 5 A, 60 Hz current is flowing into this coil. If the maximum magnetic flux density is 0.75 T, find the mmf and the voltage induced into the coil.

SOLUTION

The calculation starts with: $mmf = NI = 100 \times 5 = 500$ At
The induced voltage is computed by using Equation (7.11).

$$V = 4.44 \cdot 60 \cdot 100 \cdot 0.75 \cdot 0.25 \times 10^{-4} = 0.4995 \approx 0.5 \text{ V}$$

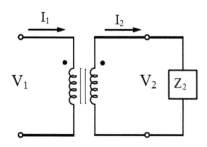

FIGURE 7.5   Load connected to an ideal transformer secondary.

By dividing Equation (7.11) into (7.12), the transformer voltage relationship Equation (7.8) is obtained. Consider now an arbitrary load (Z_2) connected to the secondary terminals of the ideal transformer as shown below.

The input impedance seen looking into the primary winding is given by:

$$Z_1 = \frac{V_1}{I_1} = \frac{aV_2}{\frac{I_2}{a}} = a^2 Z_2 \qquad (7.13)$$

The impedance seen by the primary voltage source of the ideal transformer is the secondary load impedance times the square of the transformer ratio (Figure 7.5). Using this property, the secondary impedance of the ideal transformer can be reflected to the primary. In a similar fashion, a load on the primary side of the ideal transformer can be reflected to the secondary.

$$Z_2 = \frac{V_2}{I_2} = \frac{\frac{V_1}{a}}{aI_1} = \frac{Z_1}{a^2} \qquad (7.14)$$

Another important property of an ideal transformer is derived from Equation (7.8):

$$V_1 I_1 = V_2 I_2 \qquad (7.15)$$

This is the power conservation, *the primary and secondary apparent powers (volt-amperes) are equal in an ideal transformer.*

Example 7.3

Determine the primary and secondary currents for the ideal transformer, supplied by 120 V, if the source impedance is $Zs = (18 - j4)$ Ω and the load impedance is $Z_2 = (2 + j1)$ Ω. The transformer ratio is a = 4.

SOLUTION

The secondary voltage is:

$$V_2 = \frac{V_1}{a} = \frac{120}{4} = 30 \text{ V (RMS)}$$

The load impedance seen in the primary side is:

$$Z_2' = a^2 Z_2 = (4)^2 (2 + j1) = 32 + j16 \ \Omega$$

The primary and secondary currents are computed as:

$$I_1 = \frac{V_1}{Z_S + Z_2'} = \frac{120\langle 0°}{18 - j4 + 32 + j16} = 2.33\langle -13.5° \text{ A (RMS)}$$

$$I_2 = aI_1 = 9.32\langle -13.5° \text{ A (RMS)}$$

While the primary and the secondary voltages are:

$$V_1 = Z_2'I_1 = (32 + j16)2.33\langle -13.5° = 83.50\langle 13.07° \text{ V (RMS)}$$

$$V_2 = \frac{V_1}{a} = \frac{83.50\langle 13.07°}{4} = 20.88\langle 13.07° \text{ V (RMS)}$$

7.2.1 POLARITY OF TRANSFORMER WINDINGS

The operation of a transformer depends on the relative orientation of the primary and secondary coils. We mark one of the terminals on the primary and secondary coils with a dot to denote that currents entering these two terminals produce magnetic flux in the same direction within the transformer core (as shown in Figure 7.6). If either coil orientation is reversed, the dot positions are reversed and the current and voltage equations must include a minus sign. With power or distribution transformers, polarity is important only if the need arises to parallel transformers to gain additional capacity or to hook up three single-phase transformers to make a three-phase bank. The way the connections are made affects angular displacement, phase rotation, and direction of rotation of connected motors. Polarity is also important when hooking up current transformers for relay protection and metering. Transformer polarity depends on which direction coils are wound around the core (clockwise or counterclockwise) and the leads. Transformers are sometimes marked at their terminals with polarity marks. Often, polarity marks are shown as white paint dots (for plus) or plus-minus marks on the transformer and on the nameplate.

More often, transformer polarity is shown simply by the American National Standards Institute (ANSI) designations of the winding leads as H1, H2 and X1, X2. By ANSI standards, if you face the low-voltage side of a single-phase transformer (the side marked X1, X2), the H1 connection will always be on your far left. If the terminal marked X1 is also on your left, it is subtractive polarity. If the X1 terminal is on your right, it is additive polarity. Additive polarity is common for small distribution transformers. A transformer is said to have additive polarity if, when adjacent high- and

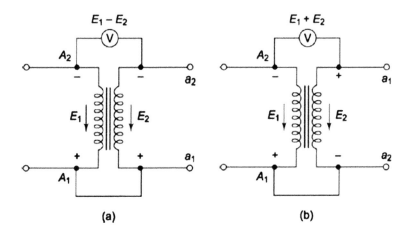

(a)   (b)

FIGURE 7.6   (a) Subtractive polarity, and (b) additive polarity.

low-voltage terminals are connected and a voltmeter placed across the other high- and low-voltage terminals, the voltmeter reads the sum (additive) of the high- and low-voltage windings. Figure 7.6 shows the high- and low-side voltage relationships for subtractive and additive polarity. It is subtractive polarity if the voltmeter reads the difference (subtractive) between the voltages of the two windings. If this test is conducted, use the lowest AC voltage available to reduce potential hazards. An adjustable AC voltage source is recommended to keep the test voltage low.

7.2.2 PRACTICAL (NON-IDEAL) TRANSFORMERS

We discussed ideal transformers in the previous chapter section. Certain assumptions made there are not valid in a practical transformer. In a practical transformer, the windings have resistances, not all windings link the same magnetic flux, permeability of the core material is not infinite, and the core losses occur when core material is subject to time-varying magnetic flux. In the analysis of practical or actual transformers, all these must be considered. In each of the transformer's configuration types, most of the flux is confined to the core and therefore links both windings. The windings also produce additional flux, known as leakage flux, which links one winding without linking the other. Although leakage flux is a small fraction of total flux, it plays an important role in determining the behavior of the transformer. In practical transformers, leakage is reduced by subdividing the windings into sections placed as close together as possible. In core-type construction, each winding consists of two sections, one section on each of the two legs of the core, the primary and secondary windings being concentric coils. In shell-type construction, variations of the concentric-winding arrangement may be used or the windings may consist of a number of thin "pancake" coils assembled in a stack with primary and secondary coils interleaved. A non-ideal transformer accounts for all of the loss terms that are neglected in the ideal transformer model. The individual loss terms are: (a) primary and secondary winding resistances (losses in the windings due to the resistance of the wires); (b) primary and secondary leakage reactances (losses due to flux leakage out of the transformer core); (c) core resistance (core losses due to hysteresis loss and eddy-current loss); and (d) magnetizing reactance (magnetizing current necessary to establish magnetic flux in the transformer core). The equivalent circuit diagram of a non-ideal transformer is shown in Figure 7.7. Given that the voltage drops across the primary winding resistance and the primary leakage reactance are typically quite small, the shunt branch of the core loss resistance and the magnetizing reactance (excitation branch) can be shifted to the primary input terminal. The primary voltage is then applied directly across the shunt impedance and allows the winding resistances and leakage reactances to be combined.

The parameters of the transformer equivalent circuit (Figure 7.7) can be determined experimentally from three tests: (1) a DC test in which the resistances of the primary and secondary windings can be measured; (b) an open-circuit test, where the secondary is left unconnected and the normal-rated voltage is applied to the primary; and (c) a short-circuit test, where the secondary terminals are short-circuited and a low voltage is applied to the primary, sufficient to circulate the

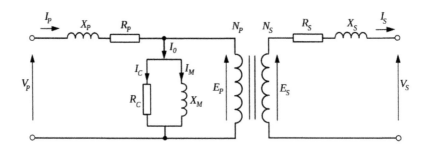

FIGURE 7.7 The equivalent circuit of a non-ideal transformer.

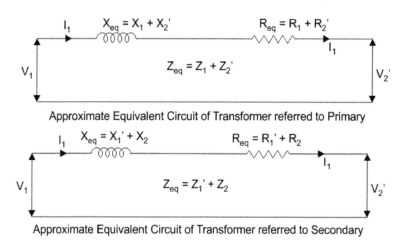

FIGURE 7.8 Approximate equivalent circuit of a transformer.

normal full-load current. Because of the nonlinear magnetic properties of iron, the waveform of the exciting current differs from the waveform of the flux; the exciting current for a sinusoidal flux waveform will not be sinusoidal. This effect is especially pronounced in closed magnetic circuits such as are found in transformers. In magnetic circuits where the reluctance is dominated by an air gap with its linear magnetic characteristic, such as is the case in many electric machines, the relationship between the net flux and the applied magneto-motive force (mmf) is relatively linear and the exciting current will be much more sinusoidal. A further approximation to the equivalent circuit can be made by eliminating the excitation branch. This approximation removes the core losses and the magnetizing current from the transformer model. The resulting equivalent circuit is shown in Figure 7.8. This is the so-called cantilever approximate circuit of a transformer. The *equivalent resistances* of the transformer, referred to as the primary side and the secondary side, respectively, are:

$$R_{eq-1} = R_1 + R_2' = R_1 + a^2 R_2 \tag{7.16}$$

$$R_{eq-2} = R_1' + R_2 = \frac{R_1}{a^2} + R_2 \tag{7.17}$$

While the equivalent reactances, are expressed in a similar manner as:

$$X_{eq-1} = X_1 + X_2' = X_1 + a^2 X_2 \tag{7.18}$$

$$X_{eq-2} = X_1' + X_2 = \frac{X_1}{a^2} + X_2 \tag{7.19}$$

When no confusion is likely to arise, the adjective "equivalent" and subscript "eq" are often omitted. In power engineering applications, it is convenient to specify the equivalent resistance and reactance in a way that is indicative of the voltage across them relative to the rated transformer voltage. *Percent resistance* of the transformer is defined as the voltage across the equivalent resistance referred to the primary (or secondary) when the primary or secondary windings are carrying the rated current, expressed as a percent of the rated primary (or secondary) voltage. The *percent reactance* is defined in a similar manner by inserting the reactance in place of resistance in the above definition. The percent values are the per-unit values, which we learned in the previous chapter, normalized values of specific quantities.

Example 7.4

A single-phase step-up transformer is rated at 1000 kVA, 60 Hz, and 11/110 kV. Transformer equivalent resistance and reactance are 1.452 Ω and 12.1 Ω, respectively. Compute the percent resistance and reactance, as seen from primary and secondary transformer sides. (The transformer rated voltage ratio is always its turn ratio and no-load voltage ratio, unless specified otherwise.)

SOLUTION

Transformer ratio is

$$a = \frac{V_P}{V_S} = \frac{11}{110} = 0.1$$

Rated primary current is:

$$I_P = \frac{1000 \text{ kVA}}{11 \text{ kV}} = 90.9 \text{ A}$$

Now, the primary percent equivalent resistance and reactance are:

$$R_{eq}(\%) = \frac{R_{eq}(\text{in } \Omega) \times \text{Rated Current}}{\text{Rated Voltage}} = \frac{1.452 \times 90.9}{11000} \times 100 = 1.2\%$$

$$X_{eq}(\%) = \frac{X_{eq}(\text{in } \Omega) \times \text{Rated Current}}{\text{Rated Voltage}} = \frac{12.1 \times 90.9}{11000} \times 100 = 10.0\%$$

The secondary resistance and reactance are computed as:

$$R_{eq(secondary)} = \frac{R_{eq(primary)}}{a^2} = \frac{1.452}{(0.1)^2} = 142.5 \ \Omega$$

$$X_{eq(secondary)} = \frac{X_{eq(primary)}}{a^2} = \frac{12.1}{(0.1)^2} = 1210 \ \Omega$$

In order to determine the secondary percent quantities, the rated secondary current is needed:

$$I_S = \frac{1000 \text{ kVA}}{110 \text{ kV}} = 9.09 \text{ A}$$

And the secondary percent (per-unit) resistance and reactance are:

$$R_{eq}(\%) = \frac{R_{eq}(\text{in } \Omega) \times \text{Rated Current}}{\text{Rated Voltage}} = \frac{145.2 \times 9.09}{110000} \times 100 = 1.2\%$$

$$X_{eq}(\%) = \frac{X_{eq}(\text{in } \Omega) \times \text{Rated Current}}{\text{Rated Voltage}} = \frac{1210 \times 9.09}{110000} \times 100 = 10.0\%$$

7.2.3 Transformer Voltage Regulation

For a given input (primary) voltage, the output (secondary) voltage of an ideal transformer is independent of the load attached to the secondary. As seen in the transformer equivalent circuit, the output voltage of a realistic transformer depends on the load current. Assuming that the current through the excitation branch of the transformer equivalent circuit is small compared to the current that flows through the winding loss and leakage reactance components, the transformer approximate equivalent circuit referred to the primary is shown below. The percentage voltage regulation (VR) is defined as the percentage change in the magnitude of the secondary voltage as the load current changes from the no-load to the loaded condition.

$$VR(\%) = \frac{|V_S|_{NL} - |V_S|_{FL}}{|V_S|_{NL}} \times 100 \tag{7.20}$$

The transformer equivalent circuit above gives only the reflected secondary voltage. The actual loaded and no-load secondary voltages are equal to the loaded and no-loaded reflected secondary values divided by the transformer turn ratio. Thus, the percentage voltage regulation may be written in terms of the reflected secondary voltages.

$$VR(\%) = \frac{|V_S'|_{NL} - |V_S'|_{FL}}{|V_S'|_{NL}} \times 100 \tag{7.21}$$

Here the prime quantities are the secondary voltages reflected in the primary of the transformer. According to the approximate transformer equivalent circuit, the reflected secondary voltage under no-load conditions is equal to the primary voltage, while the secondary voltage for the loaded condition is taken as the rated voltage, so:

$$|V_S'|_{NL} = V_P$$
$$|V_S'|_{FL} = |V_S'|_{rated}$$

Inserting the previous two equations into the percentage voltage regulation equation gives:

$$VR(\%) = \frac{|V_P| - |V_S'|_{rated}}{|V_S'|_{rated}} \times 100 \tag{7.22}$$

Example 7.5

Compute the voltage regulation of the transformer in Example 7.4 for: (a) unit power factor, (b) 0.8 lagging power factor, and (c) 0.8 leading power factor.

SOLUTION

The impedance of the transformer (Figure 7.8) is:

$$Z_S = R_S + jX_S = 145.2 + j1210 \ \Omega$$

a. For unit power factor

$$V_1' = V_2 + I_2 Z_2 = 110000 + 9.09\langle 0° \times (145.2 + j1210) = 111320 + j11100 \ V$$

and

$$|V_1'| = \sqrt{113206^2 + 11100^2} \cong 118,870 \ V$$

The voltage regulation, by using Equation (7.19) is:

$$VR = \frac{111870 - 110000}{110000} \times 100 = 1.7\%$$

b. For 0.8 lagging power factor

$$V_1^1 = V_2 + I_2 Z_2 = 110000 + 9.09 \langle -36.9° \times (145.2 + j1210) = 117805 + j8078 \text{ V}$$

and

$$|V_1'| = \sqrt{117805^2 + 8078^2} \cong 118,070 \text{ V}$$

The voltage regulation, by using Equation (7.19) is:

$$VR = \frac{118070 - 110000}{110000} \times 100 = 7.3\%$$

c. For 0.8 leading power factor

$$V_1^1 = V_2 + I_2 Z_2 = 110000 + 9.09 \langle 36.9° \times (145.2 + j1210) = 104355 + j9662 \text{ V}$$

and

$$|V_1'| = \sqrt{104355^2 + 9662^2} \approx 104800 \text{ V}$$

The voltage regulation, by using Equation (7.19) is:

$$VR = \frac{104800 - 110000}{110000} \times 100 = -4.7\%$$

7.2.4 Multiwinding Transformer

Transformers that have more than one secondary winding on the same core are known commonly as multiple winding or multi-winding transformers. A three-winding transformer has a primary winding and two secondary windings. The principle of operation of a multi-winding transformer is no different from that of an ordinary transformer. Primary and secondary voltages, currents, and turn ratios are all calculated the same; the difference this time is that we need to pay special attention to the voltage polarities of each coil winding, the dot convention marking the positive (or negative) polarity of the winding, when we connect them together. Multi-winding transformers can also be used to provide a step-up, a step-down, or a combination of both between the various windings. In fact, multiple winding transformers can have several secondary windings on the same core with each one providing a different voltage or current-level output. Consider a four-winding transformer in Figure 7.9.

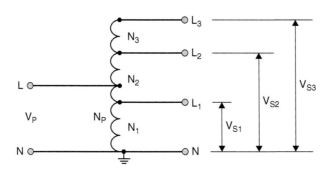

FIGURE 7.9   Multi-winding transformer.

Because the magnetic flux linking the fur windings with assumption of an ideal transformer is the same, we have the following voltage relationship:

$$\frac{V_P}{N_P} = \frac{V_{S1}}{N_1} = \frac{V_{S2}}{N_2} = \frac{V_{S3}}{N_3} \tag{7.23}$$

With the same ideal transformer assumption, the conservation of power in a multi-winding transformer is valid, so:

$$N_P I_P - N_1 I_1 - N_2 I_2 - N_3 I_3 = 0$$

Here I_P, I_1, I_2, and I_3 are the primary current and the currents in the secondary windings, respectively.

7.2.5 Transformer Ratings, Categories, Types, and Tap Changers

Transformer ratings are usually given on the so-called nameplate, which indicates the normal operating conditions. The nameplate includes the following parameters: primary-to-secondary voltage ratio, design frequency of operation, and apparent rated output power. Transformer ratings are related to the primary and secondary windings. The ratings refer to the power in kVA and primary/secondary voltages. A rating of 10 kVA, 1100/110 V means that the primary is rated for 1100 V while the secondary is rated for 110 V ($a = 10$). The kVA rating gives the power information. Power generated at a generating station (usually at a voltage in the range of 11 to 25 kV) is stepped up by a generator transformer to a higher voltage (220, 345, 400, or 765 kV) for transmission. It is one of the most important and critical components of a power system; usually it has a fairly uniform load. Generator transformers are designed with higher losses because the cost of supplying losses is cheapest at the generating station. Generator transformers are usually provided with off-circuit tap changers with a small variation in voltage (e.g., ±5%) because the voltage can always be controlled by generator field current. Unit auxiliary transformers are step-down transformers with primary connected to generator output directly. Its secondary voltage is of the order of 6.9 kV for fitting the power requirements of the auxiliary generating station equipment. Station transformers are required to supply auxiliary equipment during setting up of the generating station and subsequently during each start-up operation. The rating of these transformers is small, and their primary is connected to a high-voltage transmission line. This may result in smaller conductor size for HV winding, necessitating special measures for increasing the short-circuit strength. Interconnecting transformers are normally autotransformers used to interconnect two grids/systems operating at two different system voltages (say, 400 and 220 kV or 345 and 138 kV). They are normally located in the transmission system between the generator transformer and receiving end transformer, and in this case they reduce the transmission voltage (400 or 345 kV) to the sub-transmission level (220 or 138 kV). In autotransformers, there is no electrical isolation between primary and secondary windings, some volt-amperes are conductively transformed and the remaining is inductively transformed. Autotransformer design becomes more economical as the ratio of secondary voltage to primary voltage approaches unity. These are characterized by a wide tapping range and an additional tertiary winding, which may be loaded or unloaded. Unloaded tertiary acts as a stabilizing winding by providing a path for the third harmonic currents. Synchronous condensers or shunt reactors are connected to the tertiary winding, if required, for reactive power compensation. In the case of an unloaded tertiary, adequate conductor area and proper supporting arrangement are provided for withstanding short-circuit forces under asymmetrical fault conditions.

Example 7.6

Determine the turn ratio and the rated currents of a transformer from its nameplate data, 480 V/120 V, 48 kVA, and 60 Hz.

SOLUTION

Assuming ideal transformer, the transformer ratio is:

$$a = \frac{480}{120} = 4$$

The primary and the secondary currents are:

$$I_P = \frac{|S|}{V_P} = \frac{48000}{480} = 100 \text{ A}$$

$$I_S = \frac{|S|}{V_S} = \frac{48000}{120} = 400 \text{ A}$$

Receiving station transformers are basically step-down transformers reducing transmission/sub-transmission voltage to primary feeder level (e.g., 33 kV). Some of these may directly supply an industrial plant. Loads on these transformers vary over wider limits, and their losses are expensive. The farther is the location of transformers from the generating station, the higher the cost of supplying the losses. Automatic tap changing on load is usually necessary, and tapping range is higher to account for wide variation in the voltage. A lower noise level is desirable if they are close to residential areas. Distribution transformers are used to adjust the primary feeder voltage to the actual utilization voltage (~415 or 460 V) for domestic or industrial use. A great variety of transformers fall into this category due to many different arrangements and connections. Load on these transformers varies widely, and they are often overloaded. A lower value of no-load loss is desirable to improve all-day efficiency. Hence, the no-load loss is usually capitalized with a high rate at the tendering stage. Because very little supervision is possible, users expect the least maintenance on these transformers. The cost of supplying losses and reactive power is highest for these transformers. Classification of transformers as above is based on their location and broad function in the power system. Power distribution transformers for utility applications, single-phase, or three-phase service have fixed voltage ratings, with no way to adjust the transformer voltage ratio to allow for applications in which the system voltage is slightly off the nominal values. When a transformer is required to give a constant load voltage despite changes in load current or supply voltage, the turn ratio of the transformer must be altered. In such situations where the system voltage differs from the nominal, voltage taps are used to accomplish such tasks, via a tap-setting mechanism. The taps are usually on the high-voltage windings to adjust for the variations in the supply voltage. This is the function of a tap changer; the two basic types are: (a) *off-load*, and (b) *on-load*. To better understand tap operation, let's consider how a standard tap changer works by using a common example.

Example 7.7

A 13,800 V/4160 V transformer has five taps on the primary winding giving −5%, −2 1/2%, nominal, +2 1/2%, and +5% turns. If, on-load, the secondary voltage reduces to 4050 V, then which tap should be used to maintain 4160 V on-load (assuming the supply voltage remains constant)?

SOLUTION

To keep the secondary voltage at (or as close as possible to) 4160 V, either the primary supply voltage or the HV winding tap position must be altered. Examining the transformer relationship indicates that, in order to keep the equation in balance with primary voltage and secondary winding turns fixed, either V_2 or N_1 must be adjusted. Because the objective is to raise V_2 back to nominal, then N_1 must be reduced. To raise V_2 from 4050 V to 4160 V requires an increase in secondary volts of:

$$Tap = \frac{4160}{4050} = 1.027 \text{ or } 102.7\%$$

N_1 must be reduced to $1/1.027 = 0.974$. Therefore N_1 must be reduced by $(1 - 0.974) = 0.026$ or 2.6%. Reducing N_1 by 2.6% accomplishes the increase in secondary voltage output. The nearest tap to select is −2 1/2%, as specified in the example statement.

Most transformers associated with medium-level voltage distribution for stations have off-circuit tap changers. With this type of tap changer, the transformer has to be switched out of circuit before any tap changing can be done. The contacts on the tap changer are not designed to break any current, even the no-load current. If an attempt is made to change the tap positions while online, severe arcing results, which may destroy the tap changer and the transformer. On-load tap changers, as the name suggests, permit tap changing and hence voltage regulation with the transformer on-load. Tap changing is usually done on the HV winding for two reasons. First, because the currents are lower, the tap changer contacts, leads, etc., can be smaller. Second, as the HV winding is wound outside the LV winding, it is easier to get the tapping connections out to the tap changer. Figure 7.10 shows the connections for an on-load tap changer that operates on the HV winding of the transformer.

The tap changer has four essential features. Selector switches select the physical tap position on the transformer winding and, because of their construction, cannot and must not make or break the load current. The load current must never be interrupted during a tap change. Therefore, during each tap change, there is an interval where two voltage taps are spanned. Reactors (inductors) are used in the circuit to increase the impedance of the selector circuit and limit the amount of current circulating due to this voltage difference. Under normal load conditions, equal load current flows in both halves of the reactor windings and the fluxes balance out giving no resultant flux in the core. With no flux, there is no inductance and, therefore, no voltage drop due to inductance. However, a very small voltage drop will occur due to resistance. During the tap change, the selector switches are selected to different taps and circulating current flows in the reactor circuit. This circulating current creates magnetic flux and the resulting inductive reactance limits the flow of circulating current. The vacuum switch device performs the duty of a circuit breaker that makes and breaks current during the tap changing sequence. The bypass switch operates during the tap changing sequence, but at no time does it make or break load current, though it does make "before break" each connection.

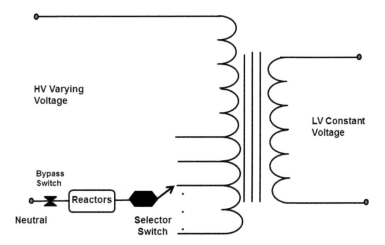

HV Varying
Voltage

LV Constant
Voltage

Bypass
Switch

Reactors

Neutral

Selector
Switch

FIGURE 7.10 High-voltage tap changer.

7.2.6 TRANSFORMER CONNECTIONS

Single-phase transformers can be connected in many configurations. Two single-phase can be connected in four different combinations provided that their polarities are observed. When transformer windings are connected in parallel, the transformers with the same voltage and polarity are paralleled. When connected in series, windings of opposite polarity are joined in one junction. Coils of unequal voltage may be series-connected with polarities either adding or opposing. In many sections of the power system, three winding transformers are used, with the three windings housed on the same core to achieve economic savings. Three-phase (three-phase or 3φ supplies) are used for electrical power generation, transmission, and distribution, as well as for all industrial uses. Three-phase supplies have many electrical advantages over single-phase power, and when considering three-phase transformers we have to deal with three alternating voltages and currents differing in phase-time by 120 degrees. A transformer cannot act as a phase-changing device and change single-phase into three-phase or three-phase into single-phase. To make the transformer connections compatible with three-phase supplies, we need to connect them together in a particular way to form a three-phase transformer configuration.

A three-phase transformer can be constructed either by connecting together three single-phase transformers, thereby forming a so-called three-phase transformer bank, or by using one pre-assembled and balanced three-phase transformer that consists of three pairs of single-phase windings mounted onto one single laminated core. The advantages of building a single three-phase transformer are that for the same kVA rating it will be smaller, cheaper, and lighter than three individual single-phase transformers connected together because the copper and iron core are used more effectively. The methods of connecting the primary and secondary windings are the same, whether using just one three-phase transformer or three separate single-phase transformers. The primary and secondary windings of a transformer can be connected in different configurations, as shown, to meet practically any requirement. In the case of three phase transformer windings, three forms of connection are possible: star (wye), delta (mesh), and interconnected star (zig-zag). The combinations of the three windings may be with the primary delta-connected and the secondary star-connected, or star-delta, star-star, or delta-delta, depending on the transformer's use. When transformers are used to provide three or more phases, they are generally referred to as poly-phase transformers. A three-phase transformer has three sets of primary and secondary windings. The manner in which these sets of windings are interconnected determines whether the connection is a star (Y) or delta (Δ) configuration. The three available voltages, which themselves are each displaced from the other by 120 electrical degrees, not only decide on the type of electrical connections used on both the primary and secondary sides, but also determine the flow of the transformer currents. With three single-phase transformers connected together, the magnetic flux in the three transformers differs in phase by 120 time-degrees. With a single three-phase transformer, there are three magnetic fluxes in the core differing in time-phase by 120 degrees.

The standard method for marking three-phase transformer windings is to label the three primary windings with capital (uppercase) letters A, B, and C, used to represent the three individual phases of RED, YELLOW, and BLUE (see Figure 7.11 for details). Secondary windings are usually labeled with small (lowercase) letters a, b, and c. Each winding has two ends normally labeled 1 and 2 so that, for example, the second winding of the primary has ends that will be labeled B_1 and B_2, while the third winding of the secondary will be labeled c1 and c2 as shown. We now know that there are four ways in which three single-phase transformers may be connected together between primary and secondary three-phase circuits. The configurations are delta-delta (Δ-Δ), star-star (Y-Y), star-delta (Y-Δ), and delta-star (Δ-Y). Transformers for high-voltage operation with star connections have the advantage of reducing the voltage on an individual transformer, reducing the number of turns required, and increased size of conductor, making the coil windings easier and cheaper to insulate than delta transformers. The delta-delta connection nevertheless has one big advantage over the star-delta configuration, in that if one transformer of a group of three should

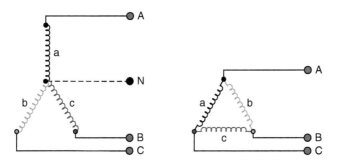

FIGURE 7.11   Star (Y) and delta (Δ) transformer connections.

become faulty or disabled, the two remaining ones will continue to deliver three-phase power with a capacity equal to approximately two-thirds of the original output from the transformer unit. One disadvantage of delta-connected three-phase transformers is that each transformer must be wound for the full-line voltage (in our example above 100 V) and for 57.7% line current. Greater numbers of turns in the windings and required insulation between turns necessitate larger and more expensive coils than the star connection. Another disadvantage with delta-connected three-phase transformers is that there is no "neutral" or common connection. In the star-star arrangement (Y-Y), each transformer has one terminal connected to a common junction, or neutral point with the three remaining ends of the primary windings connected to the three-phase main supply. The number of turns in a transformer winding for star connection is 57.7% of that required for delta connection. The star connection requires the use of three transformers, and if any one transformer becomes faulty or disabled, the whole group might become disabled. Nevertheless, the star-connected three-phase transformer is especially convenient and economical in electrical power distributing systems, in that a fourth wire may be connected as a neutral point (n) of the three star-connected secondaries as shown.

7.2.7   TRANSFORMER EFFICIENCY

The efficiency (η) of a transformer is defined as the ratio of the output power (P_{Out}) to the input power (P_{In}). The output power is equal to the input power minus the transformer losses. The transformer losses have two components: core loss (P_{Core}) and so-called copper loss (P_{Cu}) associated with winding resistances. The transformer efficiency in percent is given by:

$$\eta = \frac{P_{Out}}{P_{In}} \times 100 = \frac{P_{Out}}{P_{Out} + Losses} \times 100 = \left(1 - \frac{Losses}{P_{In}}\right) \times 100 \qquad (7.24)$$

or

$$\eta = \frac{P_{Out}}{P_{Out} + P_{Core} + P_{Cu}} \times 100 \qquad (7.25)$$

Assuming a relatively constant voltage source on the primary of the transformer, the core loss can be assumed to be constant and equal to power dissipated in the core loss resistance of the equivalent circuit for the no-load test. The copper loss in a transformer may be written in terms of both primary and secondary currents, or in terms of only one of these currents based on the relationship of Equation (7.8). The copper losses are a function of the load current, while the core losses depend on the peak core magnetic flux density, which in turns depends on the transformer applied voltage. The core losses are in fact constant because the supply voltage is constant. From the equivalent circuit, one can show that the transformer efficiency depends on the load current (I_2) and the load power factor (θ_2), expressed as:

$$\eta = \frac{V_2 I_2 \cos(\theta_2)}{V_2 I_2 \cos(\theta_2) + P_c + R_{2eq} I_2^2} \times 100 \qquad (7.26)$$

Example 7.8

Compute the efficiency of a transformer that has core losses of 1 kW, load current 17.7 A, and load phase angle 30°. The output voltage is 900 V, and the equivalent resistance from the secondary side is 1.2 Ω.

SOLUTION

The copper losses are:

$$P_{cu} = R_{2eq} I_2^2 = 1.2 \cdot (17.7)^2 = 375.95 \approx 376 \text{ W}$$

The output power is:

$$P_{out} = V_2 I_2 \cos(30°) = 900 \cdot 17.7 \cdot 0.866 = 13795.4 \text{ W}$$

Using Equation (7.25) the efficiency is:

$$\eta = \frac{V_2 I_2 \cos(\theta_2)}{V_2 I_2 \cos(\theta_2) + P_c + R_{2eq} I_2^2} = \frac{13795.4}{13795.4 + 1000 + 376} = 0.909 \text{ or } 90.9\%$$

By taking the derivative of the efficiency vs. load current, the transformer maximum efficiency can be determined. The condition for maximum efficiency is that the copper losses must be equal to the core losses. Power distribution transformers usually operate near maximum capacity over 24 hours and are taken out of the power network when they are not required. In order to account for the efficiency performance of a distribution transformer, a merit figure is used. This is "all-day" or "energy" distribution transformer efficiency, expressed as:

$$\eta_{All-day} = \frac{24\text{-hour energy output}}{24\text{-hour energy intput}} \qquad (7.27)$$

7.3   AC ELECTRIC MOTORS

Electric motors can be found in almost every production process today. Getting the most out of your application is becoming more and more important in order to ensure cost-effective operations. An electrical motor is an electromechanical device that converts electrical energy into mechanical energy. Electric motors are essentially inverse generators: a current through coils of wire causes some mechanical device to rotate. The core principle underlying motors is electromagnetic induction. By Ampere's law, the current induces a magnetic field, which can interact with another magnetic field to produce a force, and that force can cause mechanical motion. A motor is basically a generator run backwards (using current to produce motion rather than motion to produce current), and in fact the modern era of practical motors was initiated by accident when one DC generator was accidentally connected to another in 1873, producing motion and leading Zenobe Gramme to realize that his generators could also be used as motors. The first AC motors (synchronous and then induction) were invented by Tesla in the 1880s. All electrical motors exploit the force that is exerted on a current-carrying conductor placed in a magnetic field. The magnitude of the force depends directly on the current in the wire, the strength of the magnetic field, and the angle between the conductor and the magnetic field. The force on a wire of length l, carrying a current I, and exposed to a uniform magnetic flux density B throughout its length is given by:

$$F = BIl \qquad (7.28)$$

where F is in newton (N), B is in tesla (T), I in ampere (A), and l in meter (m). In electrical motors, we intend to use the high magnetic flux density to develop force on current-carrying conductors.

7.3.1 ELECTRIC MOTOR FUNDAMENTALS

Electric motors are estimated to now consume over 25% of US electricity use (though some estimates are even higher, to up to 50%, and over 20% of US total primary energy). While large electric motors can be extremely efficient at converting electrical energy to kinetic energy (efficiency higher than 90%), those efficiencies are only achieved when motors are well-matched to their loads. Actual efficiencies in normal usage practice in the US are substantially suboptimal (motors are often oversized for the loads they drive). Small electric motors are also inherently less efficient (more like 50% or so). Motor design and, even more importantly, motor choice and use practices are an important area of potential energy conservation. Before we can examine the function of a drive, we must understand the basic operation of the motor. It is used to convert the electrical energy, supplied by the controller, to mechanical energy to move the load. There are really two types of motors, AC and DC. The basic principles are alike for both. Magnetism is the basis for all electric motor operation. It produces the force required to run the motor. The two types of magnets are the permanent magnet and the electromagnet. Electromagnets have the advantage over permanent magnets in that the magnetic field can be made stronger. Also the polarity of the electromagnet can easily be reversed. When a current passes through a conductor, lines of magnetic force (flux) are generated around the conductor. The direction of the flux is dependent on the direction of the current flow. If you are thinking in terms of conventional current flow (positive to negative), then, using your right hand, point your thumb in the direction of the current flow and your fingers will wrap around the conductor in the same direction of the flux lines. There are basically two types of AC motors: synchronous and induction.

7.3.2 SYNCHRONOUS MOTORS

The synchronous motor has the special property of maintaining a constant running speed under all conditions of load up to full load. This constant running speed can be maintained even under variable line-voltage conditions. It is, therefore, a useful motor in applications where the running speed must be accurately known and unvarying. It should be noted that, if a synchronous motor is severely overloaded, its operation (speed) will suddenly lose its synchronous properties and the motor will come to a halt. The synchronous motor gets its name from the term synchronous speed, which is the natural speed of the rotating magnetic field of the stator. This natural speed of rotation is controlled strictly by the number of pole pairs and the frequency of the applied power. Like the induction motor, the synchronous motor makes use of the rotating magnetic field. In a synchronous machine, the rotor is magnetized and it runs at the same speed as the rotating magnetic field. Permanent-magnet rotors are common in small machines, so the machine structure is similar to that of the brushless DC motors. Unlike the induction motor, however, the torque developed does not depend on the induction currents in the rotor. Briefly, the principle of operation of the synchronous motor is as follows: a multiphase source of AC is applied to the stator windings and a rotating magnetic field is produced. A DC current is applied to the rotor windings and a fixed magnetic field is produced. The motor is constructed such that these two magnetic fields react upon each other causing

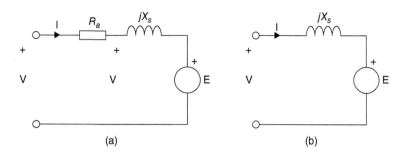

FIGURE 7.12 (a) Synchronous machine equivalent circuits, (b) approximate equivalent circuit.

the rotor to rotate at the same speed as the rotating magnetic field. If a load is applied to the rotor shaft, the rotor will momentarily fall behind the rotating field but will continue to rotate at the same synchronous speed. Once the rotor's north and south poles line up with the stator's south and north poles the stator current is reversed, thus changing the south- and north-pole orientation in the stator and the rotor is pushed again. This process repeats until the current in the stator stops alternating or stops flowing. In a three-phase motor, the stator magnetic flux rotates around the motor and the rotator actually follows this rotating magnetic field. This type of motor is called a synchronous motor because it always runs at synchronous speed (rotor and magnetic field of stator are rotating at exactly the same speed). Maximum torque is achieved when the stator flux vector and the rotor flux vector are 90° apart. Synchronous motors operate at synchronism with the line frequency and maintain a constant speed regardless of load without sophisticated electronic control. The synchronous motor typically provides up to a maximum of 140% of rated torque. These designs start like an induction motor but quickly accelerate from approximately 90% synchronous speed to rated synchronous speed. When operated from an AC drive they require boost voltage to produce the required torque to synchronize quickly after power application. A non-salient synchronous machine can be represented by a simple equivalent circuit, shown in Figure 7.12.

There is no fundamental difference between a synchronous motor and a synchronous generator. In a motor, the magnetic axis of the rotating magnetic field is ahead of the magnetic axis of the rotor, resulting in a positive torque that depends on the displacement between the two axes. In a synchronous generator, the displacement is reversed: the magnetic axis of the rotor is ahead of the magnetic axis of the rotating field, so the torque is negative. Most of the AC generators in electric power systems are synchronous machines. High-speed turbine generators normally have two poles. The rotor is made from a cylindrical steel forging, with the field winding embedded in slots machined in the steel. Apart from the slots for conductors, the active surfaces of the stator and rotor are cylindrical, so these are uniform air-gap or non-salient machines. Low-speed hydro generators have many poles. These are salient-pole machines, where the poles radiate like spokes from a central hub. The circuits in Figure 7.12 represent one phase of a three-phase synchronous machine. The voltage V is the phase voltage at the machine terminals, and the current I is the corresponding phase current. Other elements in the circuit have the following significance. The voltage E is termed the excitation voltage. It represents the voltage induced in one phase by the rotation of the magnetized rotor, so it corresponds to the rotor magnetic field. The reactance X_S is termed the synchronous reactance. It represents the magnetic field of the stator current in the following way: the voltage jX_SI is the voltage induced in one phase by the stator current. This voltage corresponds to the stator magnetic field. The resistance R_a is the resistance of one phase of the stator, or armature, winding. The resistance R_a is usually small in comparison with the reactance X_S, usually being neglected in most calculations from the equivalent circuit (Figure 7.12b). The voltage V represents the voltage induced in one phase by the total magnetic field, and is expressed as:

$$V = E + (R_a + jX_S)I \qquad (7.29)$$

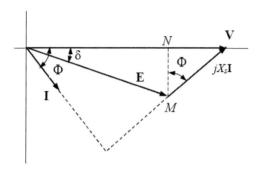

FIGURE 7.13   Synchronous motor phasor diagram.

The approximate equivalent circuit, shown in Figure 7.12b, is described by the simplified equation:

$$V = E + jX_S I$$

For operation as a motor, V leads E by an angle δ (the torque or power angle), as shown in Figure 7.13. The phase angle is now less than 90°, indicating a flow of electrical power into the machine. The developed torque is thus given by:

$$T_d = \frac{3VE}{\omega_S X_S} \sin(\delta) \tag{7.30}$$

This has a maximum value when the torque angle $\delta = 90°$, given by:

$$T_{dMax} = \frac{3VE}{\omega_S X_S}$$

Then the three-phase synchronous motor power is given by:

$$P_d = \frac{3VE}{X_S} \sin(\delta) \tag{7.31}$$

The magnitude of internal generated voltage induced in a given stator is:

$$E = K\Phi\omega \tag{7.32}$$

Because the magnetic flux, Φ, in the machine depends on the field current through it, the internal generated voltage is a function of the rotor field current. If the mechanical load on a synchronous motor exceeds T_{dMax}, the rotor is pulled out of synchronism with the rotating field, and it stalls. T_{dMax} is therefore known as the pullout torque. The synchronous motor torque characteristic has an important practical consequence. If the rotor loses synchronism with the rotating field, the load angle will change continuously. The torque will be alternately positive and negative, with a mean value of zero. Synchronous motors are therefore not inherently self-starting. Induction machines, discussed in the next chapter section, do not have this limitation, and the induction principle is generally used for starting synchronous motors. If the mechanical load is removed from an over-excited synchronous motor, then $\delta = 0$, and the phase angle is now 90°, so the machine behaves as a three-phase capacitor. In this condition, it is known as a synchronous compensator. The magnitude of the current, and hence the effective value of the capacitance, depends on the difference between E and V. The ability of a synchronous motor to operate at a leading power factor is extremely useful. It will be shown in the next section that induction motors always operate at a lagging power factor. Many industrial processes use large numbers of induction motors, with the result that the total load current is lagging. It is possible to compensate for this by installing an over-excited synchronous motor. This may be used to drive a large load such as an air compressor, or it may be used without a load as a synchronous compensator purely for power factor correction.

Example 7.9

A synchronous generator stator reactance is 190 Ω and the internal voltage (open circuit) generated is 35 kV. The machine is connected to a three-phase bus whose line-to-line voltage is 35 kV. Find the maximum possible output power of this synchronous generator.

SOLUTION

The line-to-neutral input voltage is:

$$V_{In} = \frac{V_{LL}}{\sqrt{3}} = \frac{35\ kV}{\sqrt{3}} = 20.2\ kV$$

The maximum power is when the torque angle is 90°, so from Equation (7.24):

$$P_{dMax} = \frac{3VE}{X_S} = \frac{3 \times 20.2 \times 20.2}{190} = 6.3\ MW$$

Efficiency of a synchronous motor is computed by using a well-known relationship. It is defined in the normal way as the ratio of useful mechanical output power P_{OUT} to the total electrical input power P_{IN}:

$$\eta = \frac{P_{OUT}}{P_{IN}} = 1 - \frac{Losses}{P_{IN}} \tag{7.33}$$

Example 7.10

A 1492 kW, unity power factor, three-phase, star-connected, 2300 V, 50 Hz, synchronous motor has a synchronous reactance of 1.95 ohm/phase. Compute the maximum torque in N-m that this motor can deliver if it is supplied from a constant frequency source and if the field excitation is constant at the value that would result in unity power factor at rated load. Assume that the motor is of cylindrical rotor type. Neglect all losses.

SOLUTION

Rated three-phase apparent power, at PF = 1 is $S_{3-\phi}$ = 1492 kVA, while the rated per-phase apparent power is S = 1492/3 = 497.333 kVA. The rate voltage per-phase is:

$$V_{ph} = \frac{V_{LL}}{\sqrt{3}} = \frac{2300}{\sqrt{3}} = 1327.906\ V$$

The rated per-phase current is:

$$I_{ph} = \frac{S}{V_{ph}} = \frac{497,333}{1327.906} = 374.52\ A$$

The induced per-phase voltage for unit power factor is:

$$E = \sqrt{V_{ph}^2 + \left(X_s I_{ph}\right)^2} = 1515.489 \text{ V}$$

The maximum power is computed using Equation (7.24) for a torque angle of 90°:

$$P_{Max} = \frac{EV_{ph}}{X_S} = \frac{1515.489 \times 1327.906}{1.95} = 1032.014 \text{ kW/phase}$$

The maximum per-phase torque computed is then:

$$\tau_{Max} = \frac{P_{Max}}{\omega_m} = \frac{1032,014}{2\pi 50} = 3285 \text{ N·m/phase}$$

The three-phase maximum torque of this synchronous motor is 9855 Nm.

In general, larger synchronous machines have higher efficiencies because some losses do not increase with machine size. Losses are due to rotor resistance, iron parts moving in a magnetic field causing currents to be generated in the rotor body, resistance of connections to the rotor (slip rings), stator resistance, magnetic losses (e.g., hysteresis and eddy-current losses), mechanical losses (windage, friction at bearings, friction at slip rings), and stray load losses, due to nonuniform current distribution.

Synchronous motors are usually used in large sizes because in small sizes they are more expensive as compared with induction machines. The principal advantages of using synchronous machine are as follows:

1. Power factor of synchronous machine can be controlled very easily by controlling the field current.
2. It has very high operating efficiency and constant speed.
3. For operating speed less than about 500 rpm and for high-power requirements (above 600 kW) a synchronous motor is cheaper than an induction motor.

In view of these advantages, synchronous motors are preferred for driving loads requiring high power at low speed; e.g., reciprocating pumps and compressor, crushers, rolling mills, pulp grinders. Synchronous motors are used for constant-speed, steady loads. For high power factor operations these motors are sometimes exclusively used for power factor improvement. These motors find application in driving low-speed compressors, slow-speed fans, pumps, ball mills, metal rolling mills, and process industries.

Example 7.11

A factory takes 600 kVA at a lagging power factor of 0.6. A synchronous motor is to be installed to raise the power factor to 0.9 lagging when the motor is taking 200 kW. Calculate the corresponding apparent power (in kVA) taken by the motor and the power factor at which it operates.

SOLUTION

Load (factory) power factor angle is:

$$\theta = \cos^{-1}(0.6) = 53.2°$$

Load active and reactive powers are:

$$P = S \cdot \cos(53.2) = 600 * 0.6 = 360 \text{ kW}$$
$$Q = S \cdot \sin(53.2) = 600 * 0.8 = 480 \text{ kVAR}$$

With the addition of the synchronous motor (200 kW) the overall factory power factor is 0.9 then:

$$\alpha = \tan^{-1}(0.9) = 42.0°$$
$$\tan(\alpha) = \frac{Q_{old} - Q_{SM}}{P_{old} + P_{SM}}$$

Solving for the new reactive power, we get:

$$Q_{SM} = Q_{old} - (P_{old} + P_{SM}) \cdot \tan(\alpha) = 480 - (360 + 200) \cdot \tan(42°) = 208.78 \text{ kVAR}$$

The synchronous motor apparent power is:

$$S_{SM} = \sqrt{P_{SM} + Q_{SM}} = 289.118 \text{ kVAR}$$

7.3.3 POLYPHASE INDUCTION MOTORS

An **induction** or **asynchronous motor** is an AC electric motor in which the electric current in the rotor needed to produce torque is obtained by electromagnetic induction from the magnetic field of the stator windings. An induction motor can therefore be made without electrical connections to the rotor as are found in universal, DC, and synchronous motors. An induction motor's rotor can be either wound type or squirrel-cage type. The electrical section of the three-phase induction motor as shown in Figure 7.14 consists of the fixed stator or frame, a three-phase winding supplied from the three-phase mains, and a turning rotor. There is no electrical connection between the stator and the rotor. The currents in the rotor are induced via the air gap from the stator side. Stator and rotor are made of highly magnetizable core sheet providing low-eddy current and hysteresis losses. The *stator winding* consists of three individual windings, which overlap one another and are offset by an electrical angle of 120°. When it is connected to the power supply, the incoming current will

FIGURE 7.14   State-of-the-art closed squirrel-cage three-phase motor.

first magnetize the stator. This magnetizing current generates a rotary field, which turns with synchronous speed N_S. The *rotor* in induction machines with squirrel-cage rotors consists of a slotted cylindrical rotor core sheet package with aluminum bars that are joined at the front by rings to form a closed cage. The rotor of three-phase induction motors sometimes is also referred to as an anchor. The reason for this name is the anchor shape of the rotors used in very early electrical devices. In electrical equipment, the anchor's winding would be induced by the magnetic field, whereas the rotor takes this role in three-phase induction motors. The stopped induction motor acts like a transformer shorted on the secondary side. The stator winding thus corresponds to the primary winding, the rotor winding (cage winding) to the secondary winding. Because it is shorted, its internal rotor current is dependent on the induced voltage and its resistance. The interaction between the magnetic flux and the current conductors in the rotor generates a torque that corresponds to the rotation of the rotary field. The cage bars are arranged in an offset pattern to the axis of rotation in order to prevent torque fluctuations.

Three-phase squirrel-cage induction motors are widely used in industrial drives because they are rugged, reliable, and economical. Single-phase induction motors are used extensively for smaller loads, such as household appliances like fans. Although traditionally used in fixed-speed service, induction motors are increasingly being used with variable-frequency drives (VFDs) in variable-speed service. VFDs offer especially important energy savings opportunities for existing and prospective induction motors in variable-torque centrifugal fan, pump, and compressor load applications. Squirrel-cage induction motors are very widely used in both fixed-speed and variable-frequency drive (VFD) applications. Variable voltage and variable-frequency drives are also used in variable-speed service. The induction motor operates much the same way that the synchronous motor does. It uses the same magnetic principles to couple the stator and the rotor. However, one major difference is the synchronous motor uses a permanent magnet rotor and the induction motor uses iron bars arranged to resemble a squirrel cage. As the stator magnetic field rotates in the motor, the lines of flux produced will cut the iron bars and induce a voltage in the rotor. This induced voltage will cause a current to flow in the rotor and will generate a magnetic field. This magnetic field will interact with the stator magnetic field and will produce torque to rotate the motor shaft, which is connected to the rotor. The torque available at the motor shaft is determined by the magnetic force (flux) acting on the rotor and the distance from the center of rotation that force is. The flux is determined by the current flowing through the stator windings. Another factor determining torque and another difference between the induction motor and the synchronous motor is slip. Slip is the difference between the stator magnetic field speed and the rotor speed. As implied earlier, in order for a voltage to be induced into a conductor, there must be a relative motion between the conductor and the magnetic lines of flux. Slip is the relative motion needed in the induction motor to induce a voltage into the rotor. If the induction motor ran at synchronous speed, there would be no relative motion and no torque would be produced. This implies that the greater the slip, the greater the torque. This is true to a limit, as we can see in the figure below. The above curve shows the speed/torque characteristics that the typical induction motor would follow, excited by a given voltage and frequency. We can see by this curve that the motor produces zero torque at synchronous speed because there is no slip. As we apply a load, the rotor begins to slow down, which creates slip. At about 10% slip (at the knee of the curve) we get maximum torque and power transfer from the motor. This is really the best place on the curve to operate the motor. Vector control (slip control) from a closed-loop drive system can be used to keep the motor operating at this optimum point on the curve. Vector control is implemented using a microprocessor-based system that has a mathematical model of the motor in memory and a position transducer on the motor to indicate rotor. The mathematical model allows the microprocessor to determine what the speed vs. torque curve the motor will follow with any applied voltage and frequency. This allows the system to control the slip in the motor to keep it operating at the knee of the speed/torque curve. This technology achieves extremely high performance. Now that we have a basic understanding of the operation of the motor, we can better understand the function and operation of the high performance drive. The RPM (rotation per minute) speed and the torque of an induction motor are given by the following equations:

$$N_S = \frac{120f}{P} \qquad (7.34)$$

and

$$T = K_{IM} I_{RMS} \qquad (7.35)$$

Here, T is the motor torque, K_{IM} is the torque constant, I_{RMS} is the RMS motor current, N_S is the motor synchronous (rpm), f is the frequency of stator current (power supply frequency), and P is the motor number of poles. The magnetic field produced in the rotor because of the induced voltage is alternating in nature. To reduce the relative speed, with respect to the stator, the rotor starts running in the same direction as that of the stator flux and tries to catch up with the rotating flux. However, in practice, the rotor never succeeds in "catching up" to the stator field. An essential feature of induction motors is the speed difference between the rotor and the rotating magnetic field, which is known as slip. There must be some slip for currents to be induced in the rotor conductors, and the current magnitude increases with the slip. It follows that the developed torque varies with the slip, and therefore with the rotor speed. The rotor runs slower than the speed of the stator field. This speed is called the base or mechanical speed (N_m). The difference between N_S (or angular synchronous velocity, ω_{syn}) and N_m (or actual angular velocity of the motor, ω_m) is called the slip, s and is expressed as:

$$s = \frac{N_S - N_m}{N_S} = \frac{\omega_{sun} - \omega_m}{\omega_{syn}} \qquad (7.36)$$

The slip varies with the load. An increase in load will cause the rotor to slow down or increase slip. A decrease in load will cause the rotor to speed up or decrease slip. The frequency of the rotor currents is proportional to the slip speed, and therefore proportional to s. When the rotor is stationary, $s = 1$, and the rotor frequency must be equal to the stator supply frequency, so we have the important result:

$$f_r = sf_s \qquad (7.37)$$

where f_r is the frequency of rotor currents and f_s is the stator supply frequency.

Example 7.12

An induction motor has four poles, the supply frequency is 60 Hz, and the actual speed of rotation is 1775 rpm. Determine the synchronous speed, the slip, and the rotor frequency.

SOLUTION

The synchronous speed is

$$N_S = \frac{120f}{P} = \frac{120 \times 60}{4} = 1800 \text{ RPM}$$

Then the slip is:

$$s = \frac{1800 - 1775}{1800} = 0.014 \text{ or } 1.4\%$$

By using Equation (7.32) the rotor frequency is:

$$f_r = 0.014 \times 60 = 0.833 \text{ Hz}$$

The rotating magnetic field exerts a torque T_d on the rotor, and does work at the rate $\omega_s T_d$. This represents an input of power to the rotor. The rotor revolves at an angular speed ω_r, and therefore does work at the rate $\omega_r T_d$, which represents the mechanical output power of the rotor. The difference between these two powers represents power lost in the resistance of the rotor. As in any electrical motor, there are mechanical losses in the motor, so the shaft torque T is less than T_d. We have the following set of rotor power relationships. If the rotor electromagnetic input is:

$$P_{em} = \omega_{syn} T_d \tag{7.38}$$

then the rotor power output is:

$$P_{rot} = \omega_r T_d = (1 - s)\omega_{syn} T_d \tag{7.39}$$

So the rotor copper loss is then given by the difference between Equations (7.38) and (7.39):

$$P_{Loss} = (\omega_{syn} - \omega_r)T_d = s\omega_{syn} T_d \tag{7.40}$$

Thus, a fraction $(1 - s)$ of the rotor electromagnetic input power is converted into mechanical power, and a fraction s is lost as heat in the rotor conductors. The quantity $(1 - s)$ is termed the rotor efficiency. Because there are other losses in the motor, the overall efficiency must be less than the rotor efficiency. For high efficiency, the fractional slip s should be as small as possible. In large motors, with power ratings of 100 kW or more, the value of s at full load is about 2%. For small motors, with power ratings below about 10 kW, the corresponding value is about 5%. When the rotor is stationary, the induction motor behaves as a three-phase transformer with a short-circuited secondary. When the rotor moves, the voltage induced in the rotor will depend on the relative motion, which can be represented by a simple change to the equivalent circuit: the secondary resistance is not constant, but depends on the fractional slip s. The circuit for one phase takes the form shown in Figure 7.15a.

The parameters in Figure 7.16a have the same significance as in a transformer: R_S is the stator winding resistance, x_S is the stator leakage reactance, representing stator flux that fails to link with the rotor, R_r is the rotor resistance referred to the stator, x_r is the rotor leakage reactance referred to the stator, R_C represents core loss, mainly in the stator, and X_m is the magnetizing reactance. With a transformer, it is possible to simplify the equivalent circuit by moving the shunt elements to the input terminals. This is a poor approximation with an induction motor, however, because the magnetizing reactance X_m is much smaller in comparison with the leakage reactances x_S and x_r. The reason for this is the presence of an air gap between the stator and the rotor, which increases the reluctance

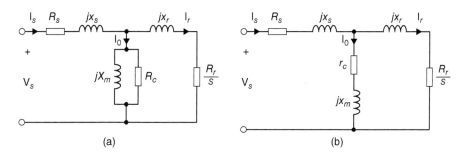

FIGURE 7.15   (a) Induction motor-equivalent circuit; and (b) modified equivalent circuit.

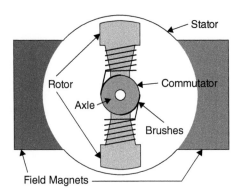

FIGURE 7.16 Schematic diagram of a DC motor.

of the magnetic circuit. The developed torque is obtained by equating the power absorbed in the resistance R_r/s to the rotor input power from Equation (7.38), giving the result:

$$T_d = \frac{3}{\omega_{syn}} \frac{R_r}{s} I_r^2 \text{ N} \cdot \text{m} \tag{7.41}$$

It is necessary to solve the equations of the equivalent circuit (Figure 7.15a) for the currents, I_S and I_r, and hence determine the torque from Equation (7.41). This process is simplified by transforming the equivalent circuit to the form shown in Figure 7.15b. Here, the parallel combination of X_m and R_C has been replaced by the series combination of x_m and r_C. The series elements in Figure 7.15b are related to the parallel elements in Figure 7.16a by the following equations:

$$r_C = \frac{X_m^2}{R_C^2 + X_m^2} R_C$$

$$x_m = \frac{R_C^2}{R_C^2 + X_m^2} X_m \tag{7.42}$$

The value of r_c depends on X_m, and therefore on the frequency. When the speed of an induction motor is controlled by varying the frequency, the resistance R_C is approximately constant. Under these conditions, r_C is proportional to the square of the frequency, so the modified equivalent circuit is less useful. In this section, however, the frequency is assumed constant, so the modified circuit can be used. Values of the stator and rotor currents are easily determined by first defining impedances as follows:

$$Z_S = R_S + jx_S, \quad Z_r = \frac{R_r}{s} + jx_r, \quad Z_m = r_C + jx_m \tag{7.43}$$

The parallel combination of the magnetizing branch and the rotor branch is:

$$Z_P = \frac{Z_m Z_r}{Z_m + Z_r} \tag{7.44}$$

So the stator and the rotor currents are now estimated as:

$$I_S = \frac{V_S}{Z_S + Z_P} \tag{7.45}$$

and

$$I_r = \frac{Z_P I_S}{Z_r} \tag{7.46}$$

Example 7.13

A four-pole 3.6 kW, wye-connected induction motor operates from a 50 Hz supply with a line voltage of 400 V. The equivalent-circuit parameters per phase are as follows:

$$R_S' = 2.27 \ \Omega, \quad R_r = 2.28 \ \Omega, \quad x_S = x_r = 2.83 \ \Omega$$
$$X_m = 74.8 \ \Omega, \quad r_C = 3.95 \ \Omega$$

If the full-load slip is 5%, determine: (a) the no-load current, (b) the full-load stator current, (c) the full-load rotor current, (d) the full-load speed in rev/min, and (e) the full-load developed torque.

SOLUTION

The per-phase voltage is:

$$V_p = \frac{400}{\sqrt{3}} = 231 \text{ V}$$

The motor impedances are:

$$Z_S = R_S + jx_S = 2.27 + j2.83 \ \Omega$$
$$Z_r = \frac{R_r}{s} + jx_r = \frac{2.28}{0.05} + j2.83 \ \Omega = 45.6 + j2.83 \ \Omega$$
$$Z_m = r_C + jx_m = 3.95 + j74.8 \ \Omega$$
$$Z_P = \frac{Z_m Z_r}{Z_m + Z_r} = 31.1 + j20.3 \ \Omega$$

a. The no-load current is:

$$I_0 = \frac{V_p}{|Z_S + Z_m|} = \frac{231}{|6.22 + j77.6|} = \frac{231}{77.9} = 2.97 \text{ A}$$

b. The full-load current and its magnitude are:

$$I_S = \frac{V_P}{Z_S + Z_P} = \frac{231}{33.4 + j23.1} = 4.68 - j3.23 \text{ A}$$
$$|I_S| = |4.68 - j3.23| = 5.69 \text{ A}$$

c. The full-load rotor current and its magnitude are:

$$I_r = \frac{Z_P I_S}{Z_r} = \frac{211 - j5.89}{45.6 + j2.83}$$
$$|I_r| = \frac{|211 - j5.89|}{\left|\frac{211 - j5.89}{45.6 + j2.83}\right|} = \frac{211.1}{45.7} = 4.62 \text{ A}$$

d. The synchronous speed and the supply angular speed are:

$$N_s = \frac{120f}{P} = \frac{120 \times 50}{4} = 1500 \text{ RPM}$$

$$N_m = N_s(1-s) = 1500(1 - 0.05) = 1425 \text{ RPM}$$

$$\omega = 2\pi f = 2 \times 50 \times \pi \approx 314 \text{ rad/s}$$

e. The developed torque is then:

$$T_d = \frac{3}{\omega_s} \frac{R_r}{s} I_r^2 = \frac{3 \times 2.28 \times (4.62)^2}{314 \times 0.05} = 37.2 \text{ N} \cdot \text{m}$$

Notice: The maximum torque is known as the *breakdown torque*. If a mechanical load torque greater than this is applied to the motor, it will stall. The torque is zero at the synchronous speed of 1500 rpm. When the rotor is stationary, the fractional slip is $s = 1$. When the rotor is running at the synchronous speed, the fractional slip is $s = 0$. The rotor frequency is:

$$f_r = sf = 0.05 \times 50 = 2.5 \text{ Hz}$$

Most induction motors are started by connecting them straight to the AC mains supply. This is known as direct online starting; however, it may result in a large starting current. Direct online starting may be unacceptable, either because the supply system cannot support such a large current, or because the transient torque could damage the mechanical system. Details about induction motor starting methods are discussed later in this book. The efficiency of an induction motor, as well as of any power system equipment, is of great importance to the user. It is defined in the normal way as the ratio of useful mechanical output power P_{OUT} to the total electrical input power P_{IN}:

$$\eta = \frac{P_{OUT}}{P_{IN}} = 1 - \frac{Losses}{P_{IN}} \tag{7.47}$$

Induction motor losses are considered to have five components as follows:

1. Stator I^2R loss (stator copper loss), P_{SCL}
2. Rotor I^2R loss (rotor copper loss), P_{RCL}
3. Core losses
4. Friction and windage loss (rotational loss): P_{w+f}
5. Stray load loss: P_{Stray}

The total loss is the sum of items 1 through 5. Core loss is the eddy-current and hysteresis loss in the magnetic core of the machine, mostly in the stator, which is represented by the resistance r_C. Friction and windage loss is the total mechanical power loss within the motor, from bearing friction and aerodynamic drag on the rotor. Stray load loss is an additional loss under load, which is not included in the other four categories. It may be attributed to departures from a purely sinusoidal winding distribution and to effects of the stator and rotor slot openings on the magnetic field distribution in the machine. The full set of power relationships for poly-phase induction motors is:

$$P_{IN} = \sqrt{3} V_{LL} I_L \cos\theta = 3 V_{ph} I_{ph} \cos\theta$$

$$P_{SCL} = 3 R_S I_S^2$$

$$P_{RCL} = 3R_r I_r^2$$

$$P_{AG} = P_{IN} - \left(P_{SCL} + \text{Core Losses}\right)$$

$$P_{Conv} = P_{AG} - P_{RCL}$$

$$P_{OUT} = P_{Conv} - \left(P_{w+f} + P_{Stray}\right)$$

The parameters of the induction motor equivalent circuit in Figure 7.15b are determined from three tests: (a) a DC measurement of the stator phase resistance, (b) no-load test, for efficiency determination, and (c) a locked-rotor (or blocked-rotor) test, where the rotor is prevented from revolving. These tests resemble the open-circuit and short-circuit tests for determining the equivalent-circuit parameters of the transformer.

7.4 DC ELECTRIC MACHINES

The armature of the motor is a loop of wire (current-carrying conductor), which is free to rotate. The field magnets are permanent or electromagnets with their N and S poles facing each other to set up the lines of flux in the air gap (see Figures 7.16 and 7.17 for details about DC motors). The armature is connected to the commutator, which rides along the brushes that are connected to a DC power source. The current from the DC power source flows from the positive lead, through the brush through one commutator section, through the armature coil, through the other commutator section, back to the negative lead. This current generates lines of flux around the armature and affects the lines of flux in the air gap. On the side of the coil where the lines of flux oppose each other, the magnetic field will be made weaker. On the side of the coil where the lines of flux are not opposing each other, the magnetic field is made stronger. Because of the strong field on one side of the coil and the weak field on the other side, the coil will be pushed into the weaker field, and, because the armature coil is free to rotate, it will rotate. The torque available at the motor shaft (turning effort) is determined by the magnetic force (flux) acting on the armature coil and the distance from the center of rotation that force is. The flux is determined by the current flowing through the armature coil and strength of the field magnets, so the DC motor torque can be expressed as:

$$T = K_{DC} \Phi_F I_a \tag{7.48}$$

Here, T represents the DC motor torque, K_{DC} is the torque constant, Φ_F is the magnetic field flux, and I_a is the motor armature current. The total emf induced in the motor by several such coils wound on the rotor can be expressed as:

$$E_b = K \Phi_F \cdot \omega_m \tag{7.49}$$

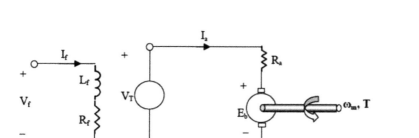

FIGURE 7.17 Separately excited DC motor.

where K is an armature constant, and is related to the geometry and magnetic properties of the motor, and ω_m is the rotor angular velocity (rad/s). The rotational (rpm) speed (N) of the motor is determined by the voltage applied to the armature coil and is expressed as:

$$N = \frac{V_t - R_a I_a}{K_V \Phi_F} \tag{7.50}$$

where V_t is the terminal (supply) voltage, K_V is the motor voltage constant, I_a is the armature current, and R_a is the armature resistance. The DC motor converts electrical power from the adjustable DC voltage source to rotating mechanical force and power. Motor shaft rotation and direction are proportional to the magnitude and polarity of the DC voltage applied to the motor. The electrical power (motor-developed mechanical power) generated by the machine is given by:

$$P_{developed} = E_b I_a = \omega_m T_{developed} = K \Phi_F \omega_m I_a \tag{7.51}$$

This is the power delivered to the induced armature voltage (counter-voltage) and is given by:

$$E_b I_a \,(\text{Electrical Power}) = \omega_m T_{developed} \,(\text{Mechanical Power}) \tag{7.52}$$

The following are the basic types of DC motors and their operating characteristics:

a. Shunt-wound motors have the field controlled separately from the armature winding. With constant armature voltage and constant field excitation, a shunt-wound motor offers relatively flat speed-torque characteristics.
b. The series-wound motor has the field connected in series with the armature. Although a series-wound motor offers high starting torque, it has poor speed regulation. Series-wound motors are generally used on low speed, very heavy loads.
c. The compound-wound DC motor utilizes a field winding in series with the armature in addition to the shunt field, to obtain a compromise in performance between a series- and a shunt-wound type motor. The compound-wound motor offers a combination of good starting torque and speed stability.
d. In a separately excited DC motor, the field winding is independent of the armature winding. Such DC motors offer flexible speed-torque control. They are often used as actuators. These type of motors are used in trains and for automatic traction purposes.
e. A permanent-magnet motor has a conventional wound armature with commutator and brushes. Permanent magnets replace the field windings. This type of motor has excellent starting torque, with speed regulation slightly less than that of the compound motor. Peak starting torque is commonly limited to 150% of rated torque to avoid demagnetizing the field poles. Typically these are low-horsepower.

DC machines can be classified according to the electrical connections of the armature winding and the field windings. The different ways in which these windings are connected lead to machines operating with different characteristics. The field winding can be either self-excited or separately excited; that is, the terminals of the winding can be connected across the input voltage terminals or fed from a separate voltage source (as in the previous paragraph). Further, in self-excited motors, the field winding can be connected either in series or in parallel with the armature winding. These different types of connections give rise to very different types of machines, as we will study in this section. In separately excited machines, the armature and field winding are electrically separate from each other, and the field winding is excited by a separate DC source. Applying KVL in the armature and field circuits of Figure 7.18:

$$V_T = E_b + R_a I_a \tag{7.53}$$

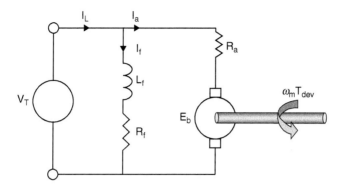

FIGURE 7.18   Shunt DC motor.

and

$$V_f = R_f I_f \tag{7.54}$$

where V_T is voltage applied to the armature terminals of the motor and R_a is the resistance of the armature winding, V_f is voltage applied to the field winding (to produce the magnetic field), R_f is the resistance of the field winding, and I_f is the current through the field winding. Note that the total input power of a separately excited DC motor is:

$$P_{IN} = V_f I_f + V_T I_a \tag{7.55}$$

In self-excited DC electrical machines, instead of a separate voltage source, the field winding is connected across the main voltage terminals. A schematic diagram of a shunt DC machine is shown in Figure 7.18. The armature and field windings are connected in parallel, and the same terminal voltage is across each of them. The total current (I_L) drawn from the power supply is:

$$I_L = I_f + I_a \tag{7.56}$$

while the total input power of a DC shunt motor is given by:

$$P_{IN} = V_T I_L \tag{7.57}$$

Voltage, current, and power equations are given in Equations (7.53), (7.55), (7.56), and (7.57).

Example 7.14

A 230 V, 10 HP DC shunt motor delivers power to a load at 1200 r/min. The armature current drawn by the motor is 200 A. The armature circuit resistance of the motor is 0.2 Ω and the field resistance is 115 Ω. If the rotational losses are 500 W, what is the value of the load torque?

SOLUTION

The back EMF induced in the armature is:

$$E_b = V_T - R_a I_a = 230 - 0.2 \cdot 200 = 190 \text{ V}$$

Power developed (in the rotor), and the power delivered to the load

$$P_{developed} = E_b I_a = 190 \times 200 = 38000 \text{ W}$$
$$P_{OUT} = P_{developed} - \text{Rotational Losses} = 38000 - 500 = 37500 \text{ W}$$

The angular velocity is:

$$\omega_m = \frac{2\pi N}{60} = \frac{2\pi 1200}{60} = 40\pi \ \text{rad/s}$$

The load torque is given by:

$$T_{Load} = \frac{P_{OUT}}{\omega_m} = \frac{37500}{40\pi} = 298.4 \ \text{Nm}$$

In a series DC machine, the field winding and armature winding are connected in series, carrying the same current. A series DC wound motor is also called a universal motor. It is universal in the sense that it will run equally well using either an AC or a DC voltage source. Reversing the polarity of both the stator and the rotor cancel out. Thus the motor will always rotate the same direction regardless of the voltage polarity. By applying KVL in Figure 7.19, the terminal voltage, armature current, back EMF, field, and armature resistance are related to the following equation:

$$V_T = E_b + \left(R_f + R_a\right)I_a \tag{7.58}$$

In a compound DC machine, when both series and shunt field windings are used, the motor is said to be compounded. In a compound machine, the series field winding is connected in series with the armature, and the shunt field winding is connected in parallel. Two types of arrangements are possible in compound motors: cumulative compounding, in which the magnetic fluxes produced by both series and shunt field windings are in the same direction (i.e., additive), and the machine is called cumulative compound. In differential compounding, where the two fluxes are in opposition, the machine is a differential compound. In both these types, the connection can be either short-shunt or long-shunt. In most applications, DC motors are used for driving mechanical loads. Some applications require that the speed remain constant as the load on the motor changes. In some applications, the speed is required to be controlled over a wide range. It is therefore important to study the relationship between torque and speed of the motor. The performance measure of interest is the speed regulation, defined as the change in speed as full load is applied to the motor, expressed as:

$$SR(\%) = \frac{N_{NL} - N_{FL}}{N_{FL}} \times 100 \tag{7.59}$$

where N_{NL} is the speed at no load, and N_{FL} is the speed when full load is applied. In order to effectively use a DC motor for an application, it is necessary to understand its characteristic curves. For every motor, there is a specific torque/speed curve and power curve. The relation between torque and speed is important in choosing a DC motor for a particular application. A separately excited

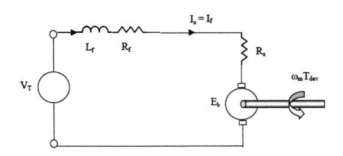

FIGURE 7.19 Series DC motor.

DC motor equivalent circuit is shown in Figure 7.14. From voltage and torque equations, we can get a relationship for developed torque, voltage, and magnetic flux for separately excited and shunt DC motors, expressed as:

$$T_{developed} = \frac{K\Phi}{R_a}(V_T - K\Phi\omega_m)$$ (7.60)

This equation shows the relationship between the torque and speed of a separately excited DC motor. If the terminal voltage V_T and flux Φ are kept constant, the torque-speed relationship is a straight drooping line. The torque is inversely proportional to the speed of the output shaft. In other words, there is a tradeoff between how much torque a motor delivers, and how fast the shaft spins. However, the motor load determines the final operating point on the torque curve. The DC series motor characteristics can be analyzed in the same way as the shunt motor. In series motors, the series field winding is connected in series with the armature (Figure 7.17). Therefore, the torque developed in the rotor can be expressed as:

$$T_{developed} = \frac{K_m V_T^2}{\left(R_a + R_f + K_m\omega_m\right)^2}$$ (7.61)

Here K_m is the series DC motor constant. From this equation, if the terminal voltage V_T is kept constant, the speed is almost inversely proportional to the square root of the torque. A high torque is obtained at low speed and a low torque is obtained at high speed. As power flows from DC motor input terminals to the output (shaft), some losses take place. Figure 7.17 shows the flow of power in a separately excited DC motor. Efficiency of the motor can be calculated as the ratio of output power to the total input power. For example, for a separately excited DC motor, the input power is given by the relation in Equation (7.55). The total copper loss is given by the power dissipated in the field and armature windings:

$$\text{Copper Losses} = \frac{V_f^2}{R_f} + R_a I_a^2 = R_f I_f^2 + R_a I_a^2$$ (7.62)

The developed power is the difference between input power, copper losses, and core losses (hysteresis and eddy-current losses). The output power and torque are less than the developed values because of rotational losses, which include friction and windage losses. Rotational power loss is approximately proportional to motor speed. The efficiency of the DC motor can be calculated as:

$$\eta = \frac{P_{OUT}}{P_{IN}} = \frac{P_{OUT}}{P_{OUT} + \text{Copper Losses} + \text{Core Losses} + \text{Rotational Losses}}$$ (7.63)

It is important to mention that the total input and output power can be calculated in many different ways using the power flow diagram of a specific type of DC, depending on the information given. Also note that the torque developed inside the rotor is different from the final (output) torque supplied to the load due to rotational losses. DC motors are typically rated in terms of: (a) rated voltage (the operating voltage on the input side of the motor), (b) rated power (power in horsepower [HP] or watts) that the motor is designed to deliver to the load (i.e., output power) for continuous operation, (c) rated speed (speed in revolutions per minute, denoted by r/min or rpm) for which the motor is designed to operate for continuous operation, and (d) rated load (the load which the motor is designed to carry for [theoretically] an infinite period of time). "Full load" or "rated load" operating condition refers to the operation of a motor when it is delivering rated power to the load. *Note*: A motor may not always operate at its rated power and/or speed. Operation above these values is not advisable due to overloading. (Note that 1 HP = 746 W.)

Example 7.15

A series-connected DC motor has an armature resistance of 0.5 Ω and field winding resistance of 1.5 Ω. In driving a certain load at 1200 rpm, the current drawn by the motor is 20 A from a voltage source of $V_T = 220$ V. The rotational loss is 150 W. Find the output power and efficiency.

SOLUTION

Total input power is given by:

$$P_{IN} = V_T I_a = 220 \times 20 = 4400 \text{ W}$$

The induced voltage is:

$$E_b = V_T - (R_f + R_a)I_a = 220 - (0.5 + 1.5)20 = 180 \text{ V}$$

Power developed in the armature can be calculated as:

$$P_{dev} = E_b I_a = 180 \times 20 = 3600 \text{ W}$$

Output power delivered to the load is:

$$P_{OUT} = P_{dev} - \text{Rotational Losses} = 3600 - 150 = 3450 \text{ W}$$

Therefore, efficiency can be calculated as:

$$\eta = \frac{P_{OUT}}{P_{IN}} = \frac{3450}{4400} = 0.784 \text{ or } 78.4\%$$

Example 7.16

A 440 V, 20 HP, 500 RPM, DC shunt motor has rotational losses of 780 W at rated speed. The armature resistance and the field resistance are 0.5 Ω and 220 Ω, respectively. The armature current is 38 A. Compute: (a) the electromagnetic torque, (b) the shaft (output) torque, and (c) the motor efficiency.

SOLUTION

The output power and the developed (electromagnetic) power are:

$$P_{OUT} = 20 \times 746 = 14920 \text{ W}$$
$$P_{em} = P_{OUT} + \text{Rotational Losses} = 14920 + 780 = 15700 \text{ W}$$

a. The electromagnetic torque is then: $\tau_{em} = \frac{P_{em}}{\omega} = \frac{15700}{2\pi\frac{500}{60}} = 300 \text{ Nm}$

b. The shaft (output) torque is: $\tau_{OUT} = \frac{P_{OUT}}{\omega} = \frac{14920}{2\pi\frac{500}{60}} = 285 \text{ Nm}$

c. Input current is:

$$I_L = I_a + I_f = I_a + \frac{V_T}{R_f} = 38 + \frac{440}{220} = 40 \text{ A}$$

The efficiency is then:

$$\eta = \frac{P_{OUT}}{P_{IN}} = \frac{P_{OUT}}{V_T \times I_L} = \frac{14920}{440 \times 40} = 0.847 \text{ or } 84.7\%$$

7.5 CHAPTER SUMMARY

Motors and transformers are the key driving force for industrial, commercial, and residential equipment and appliances. In industry, all types of linear or rotational force, movement, torque, and so on are applied largely by electrical motors. Industries are getting automated day by day, hence the use of electrical motors is increasing at the same pace. The power supply to any medium- or large-scale industry comes through a transformer as the utilities prefer to supply at higher grid voltage. Transformers are a common and indispensable component of power and electronics systems where they are used to transform voltages, currents, and impedances to appropriate levels for optimal use. Transformers offer us opportunities to investigate the properties of magnetic circuits, including the concepts of mmf; magnetizing current; and magnetizing, mutual, and leakage fluxes and their associated inductances. In both transformers and rotating machines, a magnetic field is created by the combined action of the currents in the windings. In an iron-core transformer, most of this flux is confined to the core and links all the windings. This resultant mutual flux induces voltages in the windings proportional to their number of turns and is responsible for the voltage-changing property of a transformer. In rotating machines, the situation is similar, although there is an air gap that separates the rotating and stationary components of the machine. Directly analogous to the manner in which the transformer core flux links the various windings on a transformer core, the mutual flux in rotating machines crosses the air gap, linking the windings on the rotor and stator. As in a transformer, the mutual flux induces voltages in these windings proportional to the number of turns and the time rate of change of the flux. A significant difference between transformers and rotating machines is that in rotating machines there is relative motion between the windings on the rotor and stator. An electromechanical energy-conversion device is essentially a medium of transfer between an input side and an output side. Three electrical machines (DC, induction, and synchronous) are used extensively for electromechanical energy conversion. Electromechanical energy conversion occurs when there is a change in magnetic flux linking a coil, associated with mechanical motion. The maximum portion of power that is consumed in any industry is by electrical motors. So efficiency is a great issue for an industry owner to think about. As the major consumer of energy, motors must be as highly efficient as possible. The efficiency of the transformer, through which all the power is consumed, must also be near 100 percent. So all personnel involved with industry decision-making must have knowledge regarding the energy efficiency issue for electrical motors and transformers. The basic principles of transformers and motors, and their operation, applications, and uses are also discussed in this chapter.

7.6 QUESTIONS AND PROBLEMS

1. Core losses of a transformer operated from a constant voltage supply are assumed to be constant and independent of the load. Why?
2. How much force will be created on a wire that is parallel to the magnetic field?
3. What is meant by leakage flux? How is it kept to a minimum?
4. What are two types of core construction used in transformers?
5. What is the function of a transformer?
6. Why is the capacity of a transformer rated in kilo-volt-amperes?
7. What are the three-phase transformer connections?
8. State the purpose of a tap changer in an electrical system.
9. What is the starting torque of a motor?
10. What determines the speed of a synchronous motor?
11. Explain how a synchronous motor can improve the power factor of a load with low lagging power factor.
12. What are some applications of synchronous motors?
13. What effect does the slip have on the rotor reactance of an induction motor?

14. Upon what factors does the torque of an induction motor depend?
15. What happens if the direction of current at the terminals of a DC series motor is reversed?
16. What happens when the load from a series motor is suddenly taken off?
17. What happens when a DC motor is connected across an AC supply?
18. How would the field winding inductance affect the operation of a DC motor under steady-state?
19. What happens if the direction of current at the terminals of a DC series motor is reversed?
20. A transformer is rated at 500 kVA, 60 Hz, and 2400/240 V. There are 200 turns on the 2400 V winding. When the transformer supplies a rated load, find (a) the ampere-turns of each winding, and (b) the current in each winding.
21. A transformer is made up of a 1200-turn primary coil and an open-circuited 80-turn secondary coil wound around a closed core of cross-sectional area 45 cm². The core material can be considered to saturate when the RMS applied flux density reaches 1.50 T. What maximum 60 Hz RMS primary voltage is possible without reaching this saturation level? What is the corresponding secondary voltage? How are these values modified if the source frequency is lowered to 50 Hz?
22. A single-phase transformer has 1200 turns on primary and 400 turns on secondary. The primary winding is connected to 240 V supply and the secondary winding is connected to a 6.40 kVA load. If the transformer is considered ideal, determine: (a) the load voltage, (b) the load impedance, and (c) the load impedance referred to the primary side.
23. A single-phase core-type transformer is designed to have primary voltage 33 kV and secondary voltage 6.6 kV. If the maximum flux density permissible is 1.24 Wb/m² and the number of primary turns is 1350, calculate the number of secondary turns and the core cross-sectional area when operating at 50 Hz frequency.
24. A transformer is to be used to transform the impedance of a 75 Ω resistor to an impedance of 225 Ω. Calculate the required turns ratio, assuming the transformer to be an ideal one.
25. Prove that the maximum efficiency of a transformer is when the core losses are equal to RI^2 copper losses (due to the winding resistance) if the secondary voltage and the power factor are assumed constant.
26. A single-phase step-up transformer is rated at 1800 kVA, 60 Hz, and 13.5/135 kV. Transformer-equivalent resistance and reactance are 1.560 Ω and 15.6 Ω, respectively. Compute the percent resistance and reactance, as seen from primary and secondary transformer sides.
27. The high-voltage side of a step-down transformer has 800 turns, and the low-voltage side has 100 turns. A voltage of 240 V is applied to the high side, and the load impedance is 3 Ω (low side). Find:
    a. The secondary voltage and current.
    b. The primary current.
    c. The primary input impedance from the ratio of primary voltage and current.
    d. The primary input impedance.
28. Compute the voltage regulation of the transformer in Problem 26 for: (a) unit power factor, (b) 0.85 lagging power factor, and (c) 0.85 leading power factor.
29. A single-phase transformer with a nominal voltage ratio of 4160 V to 600 V has an off-load tap changer in the high-voltage winding. The tap changer provides taps of 0, ±2½, and ±5%. If the low voltage is found to be 618 V, what tap would be selected to bring the voltage as close to 600 volts as possible?
30. Determine the rotor speed (rpm) of the following three-phase synchronous machines: (a) P = 4 and f = 60 Hz, (b) P = 12 and f = 50 Hz, and (c) P = 4 and f = 400 Hz.
31. If the torque angle of Example 7.9 is limited to 45 degrees, find the generator power output.
32. What is the speed (rpm) of a 30-pole, 60 Hz, 440 V synchronous motor? Is this motor classed as a high- or low-speed motor?

33. A synchronous motor with an input of 480 kW is added to a system that has an existing load of 720 kW at 0.82 lagging power factor. What are the new system's active power, apparent power, and power factor if the new motor is operated at: (a) 0.85 lagging power factor, (b) unit power factor, and (c) 0.85 leading power factor?

34. A three-phase induction motor, 50 Hz, has a synchronous speed of 1500 rpm and runs at 1450 rpm at full load.
    a. How many poles have the motor?
    b. What is the slip when it runs at full load?
    c. What is the rotor frequency?

35. A 208 V, 10 HP, four-pole, 60 Hz, Y-connected induction motor has a full-load slip of 5%. Find: (a) the motor synchronous speed, (b) the actual motor speed at rated load, (c) the rotor frequency, and (d) the shaft torque at rated load.

36. What is the rotor frequency of an eight-pole, 60 Hz, squirrel-cage motor operating at 850 rpm?

37. A 50 HP, 230 V, three-phase induction motor requires a full-load current of 130 A per terminal at a power factor of 0.88. What is its full-load efficiency?

38. A 480 V, 60 Hz, three-phase induction motor is drawing 60 A at 0.85 PF lagging. The stator copper losses are 2 kW, the rotor copper losses are 0.7 kW, the windage and friction losses are 0.6 kW, the core losses are 1.8 kW, while the stray losses are 0.2 kW. Find: the air gap power, the converted power, the output power, and the motor efficiency.

39. A two-pole, 50 Hz induction motor supplies 15 kW to a load at a speed of 2950 rpm.
    a. What is the motor's slip?
    b. What is the induced torque in the motor in Nm under these conditions?
    c. What will be the operating speed of the motor if its torque is doubled?
    d. How much power will be supplied by the motor when the torque is doubled?

40. Find the counter EMF of a DC shunt motor when the terminal voltage is 240 V and the armature current is 60 A. The armature resistance is 0.1 Ω. What is the developed power by this motor?

41. A 240 V DC shunt motor on no load runs at 900 rpm and takes 5 A. Compute rotational losses if the field resistance and the armature resistance are 240 Ω and 0.2 Ω, respectively.

42. A separately excited DC motor has a variable armature voltage supply that can vary from 50 V to 250 V. The field circuit is supplied from a 250 V constant voltage source, the field and the armature resistances are 250 Ω and 0.2 Ω, respectively. The motor runs at 1000 rpm at no load and takes 4 A from the armature supply when it is 250 V. Compute the rotational losses at 1000 rpm, and shaft power if the armature takes 50 A when the armature supply is 250 V.

REFERENCES AND FURTHER READINGS

1. P. Schavemaker and L. van der Sluis, *Electrical Power System Essentials*, John Wiley and Sons, 2008.
2. A. R. Daniels, *Introduction to Electrical Machines*, Macmillan Press, London, UK, 1976.
3. H. Majmudar, *Introduction to Electrical Machines*, Allyn and Bacon, Boston, MA, 1969.
4. O. Elegerd, *Basic Electric Power Engineering*, Addison-Wesley, 1977.
5. A. C. Williamson, *Introduction to Electrical Energy Systems*, Longman Scientific & Technical, 1988.
6. V. del Toro, *Electric Machines and Power Systems*, Prentice Hall, 1985.
7. R. D. Schultz and R. A. Smith, *Introduction to Electric Power Engineering*, John Wiley & Sons, New York, 1988.
8. B. S. Guru and H. R. Hiziroglu, *Electric Machinery and Transformers*, Oxford University Press, 2001.
9. J. J. Cathey, *Electric Machines: Analysis and Design Applying MATLAB*, McGraw-Hill, 2001.
10. C. I. Hubert, *Electric Machines – Theory, Operation, Adjustment and Control*, Prentice Hall, 2002.
11. J. R. Cogdell, *Foundations of Electric Power*, Prentice Hall, Upper-Saddle River, NJ, 1999.
12. S. J. Chapman, *Electrical Machinery and Power System Fundamentals* (2nd ed.), McGraw-Hill, 2002.

13. M. A. Pai, *Power Circuits and Electro-Mechanics*, Stipes Publishing L.L.C., Champaign, IL, 2003.
14. A. E. Fitzgerald and C. Kingsley, *Electric Machinery* (6th ed.), McGraw-Hill, 2003.
15. M. A. El-Sharkawi, *Electric Energy: An Introduction* (2nd ed.), CRC Press, 2009.
16. P. Schavemaker and L. van der Sluis, *Electrical Power System Essentials*, John Wiley and Sons, 2008.
17. M. E. El-Hawary, *Introduction to Power Systems*, Wiley – IEEE Press, 2008.
18. T. Gönen, *Electrical Machines with MATLAB* (2nd ed.), CRC Press, 2012.
19. J. L. Kirtley, *Electric Power Principles: Sources, Conversion, Distribution and Use*, Wiley, 2010.
20. F. M. Vanek, L. D. Albright, and L. T. Angenent, *Energy Systems Engineering*, McGraw-Hill, 2012.
21. J. J. Winders, *Power Transformers – Principles and Applications*, Marcel Dekker, Inc., 2002.
22. P. Sen, *Principles of Electric Machines and Power Electronics* (3rd ed.), Wiley, USA, 2014.
23. N. Mohan, *Electric Power Systems – A First Course*, Wiley, 2012.
24. M. R. Patel, *Introduction to Electrical Power Systems and Power Electronics*, CRC Press, 2013.
25. J. E. Fleckenstein, *Three-Phase Electrical Power*, CRC Press, 2016.

8 Building Energy Systems, Heat, and Air-Conditioning

8.1 INTRODUCTION AND THEORETICAL BACKGROUND REVIEW

A building is a complex system consisting of several clearly defined subsystems and functions, and relationships to its context, environment, and users. Buildings have the primary function to provide shelter for humans from the surrounding climate, creating appropriate comfort and conditions to support different activities. This is an illustration of the relationship that is essential to a building's whole scope and its role in human culture: A building can be viewed as an interface between humans and nature. Buildings typically last for several decades, usually even longer, so it is critical and very important to consider energy technologies and equipment that can be used to retrofit existing buildings as well as new buildings. The technological and energy-intensive solutions for providing comfort are currently being updated with more efficient and sustainable approaches. Heating, ventilation, and air-conditioning (HVAC) systems, critical to the building function and to provide comfort, are among the largest users of the energy, depending on the purpose and the activities performed in the building, where they are installed. Modern air-conditioning was created at the very beginning of the twentieth century, based on the works and ideas of Hermann Rietschel, Alfred Wolff, Stuart Cramer, and Willis Carrier. Cramer, a textile engineer in Charlotte, North Carolina, is credited with coining the term *air-conditioning* in 1906. Carrier is credited with integrating the scientific method, engineering, and business of this developing technology and creating the industry that is today known as heating, ventilating, and air-conditioning. Notice that the often-used acronym HVAC&R stands for heating, ventilating, air-conditioning, and refrigerating. The combination of these processes is equivalent to the functions performed by an air-conditioning system. Much modern energy equipment and many HVAC technologies can be used in both new and existing structures. Retrofits present unique challenges, and technologies focused on retrofits merit attention because of the large, existing stock and its generally lower efficiency. These include low-cost solutions such as thin, easily installed insulation; leak detectors; devices to detect equipment and systems problems (e.g., air conditioners low on refrigerants); and better ways to collect and disseminate best practices. Building energy uses depend on a combination of architecture and the design of energy systems, as well as on the operations and maintenance once the building is occupied. Buildings are quite sophisticated, complex, integrated, and interrelated systems, and different climates require different building designs, technologies, equipment, and operation methods. The performance and the value of any technology or equipment depend on the system in which they are embedded. Attractive lighting depends on the performance of the devices, converting the electricity to light, as well as on the window design, window and window-covering controls, occupancy detectors and monitoring, and the building's overall controls and auxiliary equipment. However, for example, as the light-fixture efficiency is increased, lighting controls have less net impact on the building energy use. In addition, the thermal energy released by lighting decreases, which then affects the building heating and cooling loads.

Heating, ventilation, and air-conditioning systems are responsible for meeting the requirements of an indoor environment, with regard to air temperature, humidity, and air quality by space heating or cooling. HVAC systems are an important part of the energy needs and uses in various industries and building sectors, comprising almost 50% of the total energy demands. A HVAC system operation is shown in Figure 8.1 diagram. HVAC systems are a component of the energy chain for conditioning a space using different energy sources and converting them into thermal energy to meet the required level of comfort. Such systems use fuels and electricity for operation. Fossil fuels are the most common source for heating, whereas electricity is almost the only source for cooling. HVAC systems are classified as either self-contained unit packages or as central systems. Unit package describes a single unit that converts a primary energy source (electricity or gas), providing final heating and cooling to the conditioned space. Examples of self-contained unit packages are rooftop HVAC systems, air-conditioning units for rooms, and air-to-air heat pumps. In central systems, the primary fossil fuel (e.g., natural gas) or electricity conversion takes place in a central location, with some form of thermal energy distributed throughout the building or facility. Central systems are a combination of central supply subsystem and multiple end-use subsystems, configuring several topologies and variations of combined central supply and end-use zone systems. The most commonly used combination is central hot and chilled water distributed to multiple fan systems. The fan systems use water-to-air heat exchangers called coils to provide hot and/or cold air for the controlled spaces. End-use subsystems can be fan systems or terminal units. If the end-use subsystems are fan systems, they can be single- or multiple-zone type. A zone is an area, e.g., a room or several rooms, conditioned by an HVAC terminal subsystem and monitored by a thermal sensor. Apart from cooling, electricity is also used to transfer fluids to enable thermal energy to reach the required spaces. Fans and pumps used in thermal generation and transfer processes are the major electric energy consumers in an HVAC system, being driven by electric motors. Electric motor energy efficiency is strongly influenced by the use of power electronics-based drives and advanced control methods. Apart from energy-use optimization by electric motors, there is a great potential for energy savings of HVAC systems as a whole, or in heating and cooling HVAC subsystems. This can be achieved by reducing the thermal energy demand by heating or cooling passive methods, advanced control, and the use of renewable energy to support heat and cold production. Effective design and analysis of HVAC systems to accomplish air-conditioning processes depends on knowledge of the properties of moist air and familiarity with the tools of thermal sciences and environmental physics. Heat transfer is the basis of HVAC system performances to heat or cool a space (Figure 8.1). Such transfers are performed in a set of heat exchangers (evaporators and condensers) that can be on their own or a part of a cool- or heat-generation device. Heat exchange processes are studied using methods of thermodynamics.

Refrigeration is defined as the process of achieving and maintaining a temperature below that of the surroundings, the aim being to cool products or spaces to the required temperature. One of the most important refrigeration applications is the preservation of perishable food products by storing them at lower temperatures. Refrigeration systems are also used extensively for providing thermal comfort through the air-conditioning, such as providing cooling and dehumidification in summer for personal comfort. Air-conditioning refers to the treatment of air so as to simultaneously control its temperature, moisture content, cleanliness, odor, and circulation, as required by occupants, a process, or products in the space. The refrigeration and air-conditioning subject has evolved out of the human needs for food and comfort. The development of the refrigeration and air-conditioning industry depended to a large extent on the development of refrigerants to suit various applications and the development of various system components. In older days, the main refrigeration purpose was to produce ice, which was used for cooling beverages, food preservation, refrigerated transport, etc. Modern refrigeration and air-conditioning have so many applications that they have become essential for humankind—without refrigeration and air conditioning the basic fabric of society is adversely affected. Of course, refrigeration is required for many applications other than air-conditioning, and

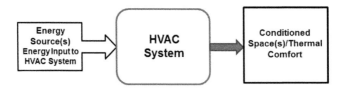

FIGURE 8.1 The energy flow to deliver thermal comfort.

air-conditioning also involves processes other than cooling and dehumidification. Refrigeration has become an essential part of the food chain, from post-harvest heat removal to the processing, distribution, and storage. Refrigeration is also essential for many chemical and processing industries, improving the standard, quality, precision, and efficiency of manufacturing processes. Industries such as oil refineries, petrochemical plants, and paper pulp facilities require very large cooling capacities. The requirements for each industry—with regards to process and equipment—are different; hence, refrigeration systems must be customized and optimized for each individual application.

Heat flows naturally from a hot area (higher temperature) to a cold area (lower temperature), and such heat transfer does not require energy use. Heat exchangers, evaporators, and condensers, the components of any HVAC system, are simple devices that normally consist of two fluids at different temperatures that flow through two different circuits. The hotter fluid transfers its heat to the colder fluid. After such a transfer, the hot fluid loses temperature and the cold one increases its temperature. However, the reverse process is not possible by itself: it needs the use of a thermal engine, a refrigerator. Thermal engines are devices that have a cyclic operation, removing heat from one region and injecting it into another. A fluid (called a refrigerant) is usually used as the medium to transfer such heat during its cycle. Refrigerators and heat pumps, as shown in Figure 8.2, work on such principles. The fluid (refrigerant) enters through a system compressor, as vapor, is compressed up, and achieves higher temperature. When it flows through the condenser, it cools and condenses, delivering its heat to the hot environment. At this low temperature of the system evaporator, it evaporates, absorbing heat from the refrigerated space. The cycle is completed when the fluid leaves the evaporator and flows into the compressor again. The objective of a refrigerator is to eliminate heat, Q_L, from a refrigerated space. In order to achieve this, it requires external work, W_{net}. The efficiency of a refrigerator is expressed in terms of the coefficient of performance (COP), which is the ratio of useful transferred (removed) heat (thermal energy) to the used work, as it will be presented in the next chapter section. A heat pump is another device that uses the same cycle as a refrigerator and can transfer heat from an area with low temperature to another area with high temperature. However, the purpose of a heat pump is to keep a space heated, which it achieves by absorbing heat from a cold environment and delivering that heat to the required space.

A building's energy consumption depends significantly on the demands of the indoor environment, which also affects the health, performance, and comfort of the occupants. It is commonly

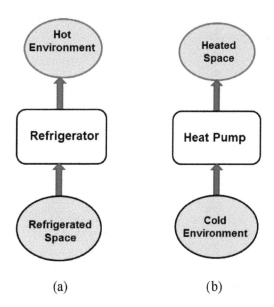

(a)   (b)

FIGURE 8.2 Energy flows in (a) refrigeration systems, and (b) heat pumps.

estimated that people in economically developed countries spend about 80% or even more of their time indoors (at home, at work, at school, or when commuting), suggesting that the indoor environmental quality (IEQ) has significant impacts on people's well-being, and on the energy consumption for heating, cooling, and air quality. In an effort to maintain the quality of the indoor environment, buildings are electromechanically conditioned to provide *constant*, *uniform*, and *comfortable* environments. IEQ requirements address four related aspects: indoor air quality, thermal environment, lighting, and acoustics, all of which affect the occupants' health, comfort, and performance. However, health, comfort, and productivity can be influenced by physiological, behavioral factors or social and organizational variables, not only by IEQ conditions. The most important variables that affect thermal comfort are the air temperature, the mean radiant temperature, the relative air velocity, the relative humidity (or the water content in the air), the type of activity undergone by people (which determines the heat production in the body), and their clothing (which offers thermal resistance to the heat transfer between the human body and the environment). Thermal conditions can affect not only occupant health, but also occupants' performance or work through several mechanisms and impacts. An unpleasant sensation of being too hot or too cold (thermal discomfort) can distract people from their work and disturb their feelings of well-being. This may lead to reduced concentration and decreased motivation to work. Indoor air quality (IAQ) is an important parameter that characterizes the indoor environment and is strongly related to the health of a building's occupants. The indoor air quality is defined as the level of the contaminants and pollutants in indoor air. The operation and the performance of the air ventilation system directly determine the IAQ of an indoor space. In addition, the air flow and paths inside the building have an important influence on the thermal comfort of its occupants, especially during the summer. So the impact of thermal comfort expectations and the impact of air ventilation on the building energy performance are critical. Recent standards, such as ANSI/ASHRAE Standard 55-2010 and EN-15251-2007, specify the combinations of the indoor thermal environmental factors and personal factors that produce the thermal environmental conditions that are acceptable to a majority of the occupants. Heat transfer is the basis of the calculations involved in the IEQ thermal aspects and the performances and design of an HVAC system in order to heat or cool a specific space. Heat exchange processes are studied using thermodynamics.

8.1.1 THERMAL ENGINEERING BASICS

Gas thermal energy results from the kinetic energy of microscopic molecule movements. Temperature is a body thermal energy characteristic, due to the internal motion of molecules. Two systems in thermal contact are in thermal equilibrium if they have the same temperature. In general, if a phase change (e.g., from solid to liquid) is not involved, the temperature of any material increases as it absorbs heat. The heat required to raise the system temperature, with specific heat, c, by an amount ΔT is expressed as:

$$\Delta Q = mc\Delta T \tag{8.1}$$

Example 8.1

The specific heat of water is 4180 J/(kg·°C). Calculate the energy required to increase the temperature of 1 kg of water 25°C.

SOLUTION

Using Equation (8.1) to solve for the heat gives:

$$Q = (1 \text{ kg}) \times \left[4180 \ J/\left(kg°C\right) \right] \times 25 \ °C = 1.045 \times 10^5 \ J$$

However, heat and temperature are different. Heat is the energy, while the temperature is the potential for heat transfer from a hot to a cold place. Materials with larger specific heat require larger heat amounts per unit of mass to raise their temperature by a given amount. Such materials can store larger thermal energy amounts per unit of mass for a small increase in temperature, and they find applications in energy storage. Understanding of the energy concept arises from the laws of thermodynamics, discussed in detail in Chapter 1. They are expressed as:

1. Energy is conserved, and *cannot be created or destroyed,* only transformed from one form to another.
2. Thermal energy (heat) cannot be transformed totally into mechanical work. Systems tend toward disorder, and in energy transformations, disorder increases. Entropy is a measure of disorder, meaning that some energy forms are more useful than others, with heat naturally flowing from a hot to a cold place.

Mathematically, the process of energy transfer is described by the first law of thermodynamics:

$$\Delta E = W + Q \qquad (8.2)$$

Here E is the internal energy, W is the work done by the system, and Q is the heat flow. By convention, Q and W are positive if heat flows into the system and work is done by the system. When the thermal efficiency of the system is high, the heat term in Equation (8.2) is ignored:

$$\Delta E = W \qquad (8.3)$$

This simplified equation is the one used to define the joule (J). Errors often occur when working with energy, power, heat, or work. Units and quantities are mixed up quite often. Wrong usage of these quantities and/or units can dramatically change the statements and cause misunderstandings. The thermal energy contained in hot material can be converted into other energy forms, which is the principle of heat-engine operation. In addition to the two laws of thermodynamic discussed above, there is a third one. It states that temperature of absolute zero cannot be attained. The third law's details and origin are not relevant to this textbook's topics, so they are not discussed further. If heat is moved from a hot reservoir to a cold reservoir, some of the thermal energy can be converted into mechanical work; the device doing this is called a *heat engine.* Examples of heat engines include steam turbines, car internal-combustion engines, and jet engines. Engine operation consists of removing heat from a hot reservoir at temperature, T_H, while some of the heat is deposited into a cold reservoir, at temperature T_C, and some is used to develop mechanical work. If the removed heat from the hot reservoir is Q_H, and the heat deposited into the cold reservoir is Q_C, then the relationship including the developed work is:

$$Q_H = Q_C + W \qquad (8.4)$$

The terms *energy efficiency* and *energy conservation* have often been used interchangeably in policy discussions, but they do have very different meanings. Energy conservation is reduced energy consumption through lower quality of energy services, e.g., lower heating levels, through turning down thermostat levels, reducing car speed limits, and capacity/consumption limits on appliances, often set by standards. Often it means "doing without" to save money or energy. Energy conservation is strongly influenced by regulations, consumer behavior, and lifestyle changes. Energy efficiency is simply the ratio of energy out to the energy input. It means getting the most out of every unit of energy. It is a technical process caused by stock turnover where old equipment is replaced by more efficient equipment. Measuring energy efficiency, particularly on a macro scale, is very difficult. *Efficiency* means often different things to the professions engaged in achieving it. To engineers, *efficiency* means a physical output/input ratio. To economists, *efficiency* means a monetary output/input ratio, and confusingly, the *efficiency* may refer to the economic optimality of a market transaction or process. Physically, energy efficiency is defined as:

$$\eta = \frac{P_{OUT}}{P_{IN}} = \frac{P_{IN} - Losses}{P_{IN}} \tag{8.5}$$

Example 8.2

An electric motor consumes 100 W of electricity to obtain 87 W of mechanical power. Determine its efficiency.

SOLUTION

Because power is the rate of energy utilization, efficiency can also be expressed as a power ratio. The time units cancel out, and we have:

$$\eta = \frac{Power\ Output}{Power\ Input} = \frac{87\ W}{100\ W} = 0.87\ or\ 87\%$$

Technical systems perform the energy conversions with various efficiencies. The ratio of Q_C and Q_H can show the ratio of the reservoir temperatures. For a heat engine, it can be written as:

$$\eta = 1 - \frac{Q_C}{Q_H} = 1 - \frac{T_C}{T_H} \tag{8.6}$$

The efficiency from involving cold and hot reservoir temperatures is more convenient because temperature is a much more easily measured quantity than heat. The efficiency as stated by Equation (8.6) is known as *ideal Carnot efficiency*, after the name of French engineer Sadi Carnot, and is the maximum efficiency attainable by a heat engine. Real heat engines typically operate at efficiencies that can be much less than the Carnot efficiency. The ideal Carnot efficiency is valid for ideal processes, taking place in either direction equally, existing in the real world only for limited cases. Time-reversible ideal processes require that net entropy of the S remain constant. An alternative definition of the second law of thermodynamics is that in any real process dS > 0. Working on the same thermal principles, a heat engine is the heat pump, which uses mechanical work to transfer heat from a cold reservoir to a hot one. Similarly, the conservation of energy principle requires that:

$$W + Q_C = Q_H \tag{8.7}$$

Example 8.3

What is the thermal efficiency of the most efficient heat engine that can run between a cold reservoir at 20°C and a hot reservoir at 200°C?

SOLUTION

Carnot engine has the ideal or maximum efficiency, so:

$$\eta = 1 - \frac{T_C}{T_H} = 1 - \frac{273.15 + 20}{273.15 + 200} = 0.38\ or\ 38\%$$

Heat pumps have practical applications for the heat transfer from cold areas/reservoirs to hot ones. They have applications in combined heat and power generation, in recovering wasted heat, or moving heat from the outside to the inside on a cold day. Although heat pumps may be economically attractive for heating purposes, careful considerations of cost, local climate, and other factors are necessary to assess their viability. Heat-pump performances are expressed through the *coefficient of performance* (COP), the ratio of heat deposited in the hot reservoir to the work done:

$$COP = \frac{Q_H}{W} \tag{8.8}$$

COP is a quantity greater than 1 (or, in percent, greater than 100%). Using the relationship between heat and temperature, COP can be expressed as:

$$COP = \frac{Q_H}{Q_H - Q_C} = \frac{T_H}{T_H - T_C} \tag{8.9a}$$

In the case of a refrigerator, the COP expression is:

$$COP = \frac{Q_C}{Q_H - Q_C} = \frac{T_C}{T_H - T_C} \tag{8.9b}$$

However, because they are heat engines, these devices are also limited by Carnot's theorem. The limiting value of the Carnot efficiency for these processes, with the equality theoretically achievable only with an ideal reversible cycle, means that the COPs of the actual systems are less:

$$COP_{Heating} < \frac{T_H}{T_H - T_C} \tag{8.10a}$$

and

$$COP_{Cooling} < \frac{T_C}{T_H - T_C} \tag{8.10b}$$

The same device used between the same temperatures is more efficient when considered as a heat pump than when considered as a refrigerator: $COP_{Heating} - COP_{Cooling} = 1$. This is because when heating, the work used to run the device is converted to heat, being added to the desired effect, whereas when cooling, the heat resulting from the input work is just an unwanted byproduct.

8.1.2   HEAT TRANSFER

Basically the three types of heat transfer are: *conduction, convection,* and *radiation*. Conduction is the thermal transfer process due to molecules' random motions. The average molecule energies are proportional to the temperature. The heat flow rate, in the steady-state along the length of bar, d, of cross-sectional area, A, with one end at higher temperature, T_1, and the other at lower temperature, T_2, is given by the Fourier law of heat conduction:

$$Q = kA \frac{T_1 - T_2}{d} \tag{8.11a}$$

where k is the thermal conductivity of the bar material. Its SI units are W/m·K, the imperial units of the thermal conductivity are BTU/(h·ft·°F), while accepted and used units are W/cm·K. Thermal conductivity is a well-tabulated material property for a large number of materials. Some values for familiar

TABLE 8.1

Thermal Conductivity at Room Temperature for Some Common Metals and Nonmetals

Material	k (W/m·K)
Silver (Ag)	420
Copper (Cu)	390
Aluminum (Al)	200
Iron (Fe)	70
Steel	50
Water (H_2O)	0.61
Air	0.06
Engine oil	0.15
Hydrogen (H_2)	0.18
Pyrex glass	1.09
Fiberglass	0.038
Brick	0.40-0.50
Wood	0.20
Cork	0.04

materials are given in Table 8.1; others can be found in the references. The thermal conductivity is a function of temperature, and the values shown in Table 8.1 are for standard (room) temperature.

Example 8.4

A steel bar of 2 m and a cross-sectional area of 10 cm² has at one end a temperature of 1200°C and 200°C at the other end. Calculate the heat thermal flow along the bar in the steady-state, ignoring the heat losses from the bar surface, if the steel thermal conductivity is 50 W·m^{-1}·K^{-8}.

SOLUTION

From Equation (8.11a) the heat flow along the bar is calculated as:

$$Q = 50 \times 10 \times 10^{-4} \frac{1200 - 200}{2} = 25 \text{ W}$$

The heat conduction rate through any medium depends on the medium geometry, thickness, material(s), and the temperature difference across the medium. It is well-known that wrapping a hot water tank with an insulating material reduces the rate of heat loss from the tank. The thicker the insulation, the smaller the heat loss. It is also known that a hot water tank loses heat at a higher rate when the surrounding temperature is lowered. Further, the larger the tank, the larger is the surface area and thus the rate of heat loss. Consider steady heat conduction through a large plane wall of thickness, L, and area A. The temperature difference across the wall is $\Delta T = T_2 \, T_1$. Experiments have shown that the rate of heat transfer $\dot{Q}$ through the wall is proportional to the temperature difference, ΔT, across the wall, the area, A, normal to the direction of heat transfer, and inversely proportional to the wall thickness, Δx. Thus, the rate of heat conduction through a plane layer is proportional to the temperature difference across the layer and the heat transfer area, but is inversely proportional to the thickness of the layer, expressed as:

$$\dot{Q}_{Cond} = -kA \frac{\Delta T}{\Delta x}$$

(8.11b)

Or in differential form as:

$$\dot{Q}_{Cond} = -k \cdot A \frac{dT}{dx} \tag{8.11c}$$

which is called *Fourier's law of heat conduction* after the French scientist, J. Fourier, who expressed this relationship in his heat transfer text in 1822. The *dT/dx* is the *temperature gradient*, the rate of change of *T* with *x*, at location *x*. The relation above indicates that the rate of heat conduction in a direction is proportional to the temperature gradient in that direction. Heat is conducted in the decreasing temperature direction, and the temperature gradient becomes negative when temperature decreases with increasing *x*. The heat transfer area *A* is always *normal* to the direction of heat transfer. The rate heat of conduction through an object, as specified in the relationships of Equation (8.11) refers to a steady-state process, in which there is no temporal change in the temperatures.

Example 8.5

An electric-based heated home roof is 5.85 m long, 10 m wide, and 0.28 m thick, and is made of a flat material layer, having thermal conductivity 0.70 W/m°C. The temperatures of the roof's inner and outer surfaces at night are 18°C and 4°C, respectively, for about 10-hour period. Determine the heat loss rate through the roof, during the night and the heat loss cost if the electricity cost is $0.075/kWh.

SOLUTION

The heat transfer through the roof is by conduction and the area of the roof is A = 5.85 × 10 = 48.5 m², then the steady rate of heat transfer through the roof is determined by Equation (8.16a):

$$\dot{Q} = 0.7 \times 58.5 \frac{18-4}{0.28} = 2047.5 \text{ W}$$

The amount of heat lost through the roof during a 10-hour period and its cost are determined from:

$$Q = \dot{Q} \times \Delta t = 2047.5 \times 10 = 20.475 \text{ kWh}$$

and

$$\text{Cost} = 20.475 \times 0.072 = \$1.4742$$

Convection represents the heat transfer through the fluid bulk motion, the actual substance movement or circulation taking place in fluids, such as water or air where the materials are able to flow. In a typical convective heat transfer, a hot surface heats the surrounding fluid, which is then carried away by fluid movement such as wind. The warm fluid is replaced by cooler fluid, which can draw more heat away from the surface. Because the heated fluid is constantly replaced by cooler fluid, the rate of heat transfer is enhanced. *Natural convection* (or *free convection*) refers to a case where the fluid movement is created by the warm fluid itself. The density of fluid decrease as it is heated; thus, hot fluids are lighter than cool fluids. Warm fluid surrounding a hot object rises, being replaced by the cooler fluid, and the result is a circulation of air above the warm surface. Much of the atmosphere and ocean heat transport is carried by convection. However, the atmospheric circulation consists of vertical as well as horizontal components, so both vertical and horizontal heat transfer occurs. Forced convection uses external means of producing fluid movement. Forced convection is what makes a windy, winter day feel much colder than a calm day with the same temperature. The heat loss from your body is increased due to the constant replenishment of cold air by the wind. Natural wind and fans are the two most common sources of forced convection. Considering a fluid of density ρ, temperature T, and moving with velocity v, the heat flow rate per unit of area is the product of the mass flow per unit area per second, ρv and the thermal energy per unit mass, cT:

$$\frac{Q}{A} = \rho vcT \tag{8.12}$$

When a cold fluid is forced to flow over a hot surface, the heat transfer rate in this forced convection from the surface to the fluid is higher than in the case of a stationary fluid. The temperature gradient at the surface is very large, the fluid layer above the surface is rapidly heated by thermal convection, and the heat transfer rate per unit the area is often expressed as:

$$\frac{Q}{A} = Nu \frac{k(T_S - T_\infty)}{L} \tag{8.13}$$

Here, T_S and T_∞ are the surface and the fluid temperatures, L is the characteristic length, k is the thermal conduction, and Nu is the dimensionless Nusselt parameter. The choice of L depends on the fluid-surface geometry, while the Nusselt parameter is a function two other nondimensional parameters, the Prandtl and Reynolds numbers, which are determined by the fluid mechanical and thermal properties, which are obtained from the empirical data and correlations. In technical and engineering applications, a slightly different relationship is used to estimate heat transfer through convection. Convection coefficient, h, in W/m²K is the measure of how effectively a fluid transfers heat by convection, being experimentally measured and influenced by factors such as the fluid density, viscosity, and velocity. The rate of heat transfer from a surface by convection is given by:

$$\dot{Q}_{Convection} = -h \cdot A(T_{Surface} - T_\infty) \tag{8.14}$$

where A is the surface area of the object, $T_{surface}$ is the surface temperature, and T_∞ is the ambient or fluid temperature.

Example 8.6

Determine coefficient for heat transfer between the outer surface of a 2.5 m long, 0.4 cm diameter electrical wire extending across a room at 15°C and the air in the room. Heat is generated in the wire as a result of resistance heating, and the surface temperature of the wire is measured to be 165°C in steady operation, and the wire voltage and current are 50 V and 2.4 A.

SOLUTION

The wire electric power is 120 W, and the wire (surface) area is 0.0314 m². From Equation (8.19), the convection coefficient is then:

$$h = \frac{\dot{Q}_{convection}}{-A(T_{Surface} - T_\infty)} = \frac{120}{-0.0314(15 - 165)} = 25.477 \text{ W/m}^2 \cdot °C$$

Radiative heat transfer represents the energy transfer through the electromagnetic radiation, which includes the heat transfer into the vacuum. The power per unit area transferred from a surface at temperature T is given by the Stefan-Boltzmann law:

$$P_{rad} = \varepsilon \sigma T^4 \tag{8.15a}$$

Here, ε is the surface emissivity (a dimensionless number ranging from 0 to 1, depending on the surface nature), $\sigma \approx 5.67 \times 10^{-8}$ W·m⁻²·K⁻⁴ is the Stefan-Boltzmann constant. Opaque surfaces absorb electromagnetic radiation from the environment. The surface absorptivity and emissivity are the same. The absorption rate per unit of area, for the environment temperature T_0 is:

$$P_a = \varepsilon \sigma T_0^4 \qquad (8.15b)$$

The net emission rate per unit of area per unit of time is then given by:

$$P_{rad} = P_e - P_a = \varepsilon \sigma \left(T^4 - T_a^4 \right) \qquad (8.16)$$

A blackbody is a surface that absorbs all incident electromagnetic radiation. The outer sun surface determines the electromagnetic radiation flux incident to the upper earth atmosphere. Another important relationship is the one describing mathematically the relationship between the temperature (T) of the radiating body and its wavelength of maximum emission (λ_{max}), the Wien's displacement law:

$$\lambda_{max} = \frac{C}{T} \qquad (8.17)$$

Here, C, the Wien's constant is equal to 2898 μm·K.

Example 8.7

Using Wien's law, estimate the maximum wavelengths of the sun and the earth, assuming that the sun's temperature is 6000 K, and the earth's temperature is 300 K.

SOLUTION

Applying Wien's law for the sun and earth temperatures, we found:

$$\lambda_{max}(Sun) = \frac{2898}{6000} = 0.483 \ \mu m$$

and

$$\lambda_{max}(Earth) = \frac{2898}{300} = 9.660 \ \mu m$$

Radiation is often identified by the effects it produces when it interacts with an object. We divide radiant energy into categories based on our ability to perceive it; however, all wavelengths of radiation behave in a similar manner. An important difference among the various wavelengths is that shorter wavelengths are more energetic. For example, the sun emits all forms of radiation, in varying quantities with over 95% of solar radiation emitted in wavelengths between 0.1 and 2.5 μm, with much of the energy concentrated in visible and near-visible parts of the electromagnetic spectrum. The visible light spectrum band, wavelengths between 0.4 and 0.7 μm, represents over 43% of the total energy emitted, while the rest lies in the infrared (IR), about 49%, and ultraviolent (UV), about 7%. In order to have a better understanding of how the sun's radiant energy interacts with the earth's atmosphere and land-sea surface, it is helpful to have a general understanding of basic radiation laws. However, this is beyond the scope of this book, and interested readers are directed to elsewhere in the literature. Radiation is the dominant energy transfer mode in power plant furnaces. The atmospheric effects on electromagnetic radiation transmission are very important in determining the earth's surface temperature.

One-dimensional conduction refers to special cases where there is only one spatial variable, e.g., the temperature varies in one direction only. Fourier's heat law is, in many ways, quite analogous to Ohm's law. A model used often to calculate the heat transfer through a one-dimensional system is called the *thermal circuit model*. This model simplifies the heat conduction analysis through composite materials, which is of great interest in building energy systems. In this model, each layer is replaced by an equivalent thermal resistor (similar as in the electric circuits) called the *thermal resistance*. For thermal conduction, the thermal resistance is expressed as:

$$R_{Thermal} = \frac{L}{k \cdot A}, \; (\text{K/W})$$ (8.18a)

So Fourier's heat law may be written, by using thermal resistance, as:

$$\dot{Q}_{Cond} = \frac{\Delta T}{R_{Thermal}}, \; (\text{W})$$ (8.18b)

where L is the thickness of the layer, k is the thermal conductivity of the layer, and A is the cross-sectional area. The thermal resistance of a slab having thickness L, thermal conductivity k, and area A is given by Equation (8.18a), and once the slab thermal resistance is calculated, then the rate of heat that flows through that slab for any given temperature condition can be computed easily. It is convenient to think in terms of thermal resistances once we begin combining materials. A wall is typically made by combining layers of different materials. When there is more than one layer in the composite, the total thermal resistance of the circuit must be calculated. The total thermal resistance for layers in series is simply the sum of the individual thermal resistances:

$$R_{Thermal-total} = \sum_{k=1} R_{Thermal,\,k}$$ (8.19a)

For layers on top of each other, thermal resistances are in parallel, and the total thermal resistance of the structure is computed as:

$$\frac{1}{R_{Thermal-total}} = \sum_{k=1} \frac{1}{R_{Thermal,\,k}}$$ (8.19b)

Example 8.8

Calculate the thermal resistance of a wall of a house that has a thickness of 0.35 m and an area of 12 m². The wall is constructed from a material (brick) that has a thermal conductivity of 0.55 W/m K.

SOLUTION

The conduction thermal resistance, Equation (8.18a), is:

$$R_{Th} = \frac{0.33}{12.8 \times 0.55} = 0.04972 \; \frac{K}{W}$$

The convection at the surface must also be expressed as a thermal resistor:

$$R_{Thermal-convection} = \frac{1}{h \cdot A}$$ (8.20)

Once the total resistance of a structure is found, the heat flow through the layers can be found by:

$$\dot{Q} = \frac{T_{Final} - T_{Inital}}{R_{Thermal-total}} \tag{8.21}$$

where $T_{initial}$ and T_{final} refer to the temperatures at the two ends of the thermal circuit (analogous to voltage difference in an electrical circuit) and q is the heat flow through the circuit (current).

Example 8.9

Consider a composite structure (panel) shown in the diagram below. Conductivities of the layer are: $k_1 = k_3 = 10$ W/mK, $k_2 = 20$ W/mK, and $k_4 = 10$ W/mK. The convection coefficient on the right side of the composite is 30 W/m²K. Calculate the total resistance and the heat flow through the composite. The vertical layer thicknesses are: $L_1 = L_3 = L_4 = 10$ cm, and $L_2 = 20$ cm; layer areas are A1 = A2 = A3 = 10 m²; and the horizontal layer top area is $A_4 = 0.4$ m², while its lateral area is 0.1 m².

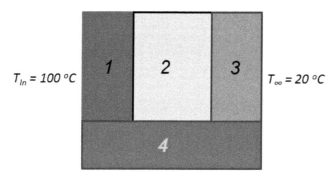

SOLUTION

First, the thermal circuit for the composite wall consists of three thermal resistances in series, all in parallel with the horizontal layer thermal resistance in the final series with the convection thermal resistance. The circuit spans between the two known temperatures, T_{In} and T_∞, and is shown here:

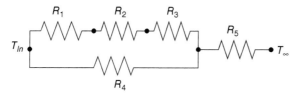

The layer thermal resistances, computed with Equations (8.19a,b), and the convective thermal resistance, computed with Equation (8.20) are:

$$R_{Th,1} = R_{Th,3} = 0.001 \text{ (K/W)}$$
$$R_{Th,2} = 0.001 \text{ (K/W)}$$
$$R_{Th,4} = 0.025 \text{ (K/W)}$$

and, the conduction thermal resistance, considering the composite panel total lateral area 10 + 0.1 = 10.1 m², is:

$$R_{Cond,5} = \frac{1}{30 \times 10.1} = 0.00033 \ (K/W)$$

Then the composite panel total thermal resistance is computed from the above circuit diagram as:

$$R_{Th-total} = \frac{R_4 \cdot (R_1 + R_2 + R_3)}{R_1 + R_2 + R_3 + R_4} + R_5 = \frac{0.025 \cdot (0.001 + 0.001 + 0.001)}{0.001 + 0.001 + 0.001 + 0.025} + 0.00033$$

$$= 0.003009 \ (K/W)$$

The heat transfer through the composite panel, Equation (8.20), is then:

$$\dot{Q} = -\frac{20 - 100}{0.003009} = 26586.906 \ W \ or \sim 26.6 \ kW$$

8.1.3 Heat Pumps and Cooling Systems

Heat pumps are one of the most underused means of conserving energy for heating and cooling buildings. However, heat-pump technology is considered to be one of the most sophisticated and beneficial engineering accomplishments of the twentieth century. Heat pumps are devices that use compressor-pump system(s) to extract heat from a low-temperature reservoir and to reject it at higher temperatures. They are, in essence, Carnot cycle applications, operating at the highest efficiency levels by employing transport heat that already exists, without generation, for both heating and cooling, by using only a small fraction of transferred energy. Such devices are often used in low-temperature geothermal schemes to maximize heat extraction from fluids, while their specific function in any scheme depends on the fluid temperature used. A heat pump is an environmentally friendly thermal engine that uses energy from the environment, e.g., solar energy accumulated by soil, ground formations, water reservoirs and storage systems, or borehole fluids. The basic heat pump components are: a *working fluid* or *refrigerant,* a *compressor,* two *heat exchangers, piping system, controls,* and *accessories,* which provide either heating or cooling to a building space or an industrial process. These devices form a closed circuit, and the refrigerant circulates through the entire cycle. Regardless of its source—outdoor air, ventilation air, surface or deep layers, groundwater or sewage—thermal energy is everywhere, ready to be transferred and converted into higher-value energy. However, only a small portion of excess energy released into the environment can be used. In heating mode, heat is extracted from a natural or waste heat source, transferred to the space while in cooling mode, removed from the building or industrial process, and discharged to a heat sink. The four basic types of heat pumps are: *air-to-air, water-to-air, water-to-water,* and *earth-to-air.* In an air-to-air heat pump system, heat is removed from building or space indoor air and rejected to the outdoor air during the cooling cycle, and the reverse happens during the heating cycle. Water can be used instead of air as the heat source or sink, depending on the unit mode: heating or cooling mode. Air-to-air heat pumps or air-source heat pumps (ASHPs) are typically rooftop units either completely packaged or split-package systems. Split-package heat pumps are designed with an air-handling unit located inside the conditioned space while the condenser and compressor are packaged for outdoor installation on the roof. ASHPs are best suited for mild climates and areas where natural gas is either unavailable or expensive. There are several heat pumps in industrial applications, divided into the following categories: vapor compression cycle or mechanical compression cycle, mechanical vapor recompression cycle, thermal vapor recompression cycle, absorption cycle, and chemical heat pumps. The steady-state performance of a heat pump cycle is evaluated by a coefficient called the coefficient of performance, defined (Equation [8.9]), as:

$$COP = \frac{Q_{Useful}}{P_{In}} \qquad (8.22)$$

where Q_{Useful} is the useful heat delivered by the heat pump and P_{In} is the high-grade (primary) energy input, used to transfer the useful heat. The basic principles of the industrial heat pumps are as follows: In water-source heat pumps, instead of air, water is used to transfer the heat between the building and the outside. Geothermal heat pumps (GHPs) use energy from the ground soil or groundwater as the thermal source or sink. In winter, a geothermal heat pump transfers thermal energy from the ground to provide space heating. In summer, the energy-transfer process is reversed—the ground absorbs heat from the conditioned space and cools the space air. The benefits of GHPs result from a nearly constant ground temperature year round, which is higher on average than winter air temperatures and lower on average than summer air temperatures. GHP energy efficiency is thus higher than that of conventional ASHPs; some are also more efficient than fossil fuel furnaces in the heating mode. The primary difference between an ASHP and a GHP is the investment in a ground loop for heat collection and rejection required for the GHP system. Whether or not the GHP is cost-effective relative to a conventional ASHP depends upon generating annual energy cost savings that are high enough for the extra cost of the ground loop.

A GHP is a heat pump that uses the earth's thermal capacity as an energy source to provide heat to a system or as an energy sink to remove heat from a system, cooling the system. The three main GHP types are: ground-coupled, ground-water, and hybrid GHPs. The ground-coupled GHPs are of two types: vertical closed-loop and horizontal closed-loop, based on the piping system shape. In the case of a moderate fluid temperature range (50°C to 70°C), the extracted heat depends primarily on the heat exchanger, and the heat pump connections to extract additional heat from the geothermal fluid. For low temperature range (less than 50°C, often 40°C or lower), heat extraction is not possible and the heat pump is connected to ensure all heat transfer. GHPs take advantage of the earth's relatively constant temperature at depths of a few meters to about 100 m (10 ft. to about 300 ft.). GHPs can be used almost everywhere in the world, as they do not have the requirements of fractured rock and water as are needed for a conventional geothermal reservoir. GHPs circulate water or other liquids through pipes buried in a continuous loop, either horizontally or vertically, under a landscaped area, parking lot, or any number of areas around the building, being one of the most efficient heating and cooling systems available. To supply heat, the system pulls heat from the earth through the loop and distributes it through a conventional duct system. For cooling, the process is reversed; the system extracts heat from the building and moves it back into the earth loop. It can also direct the heat to a hot water tank, providing another advantage, free hot water. GHPs reduce electricity use by 30%–60% compared with traditional heating and cooling systems, because the electricity that powers them is used only to collect, concentrate, and deliver heat, not to produce it.

Direct geothermal systems for heating and cooling typically comprise a primary circuit that exchanges heat with the ground, a heat pump that exchanges and enhances heat transfer between the primary circuit and the secondary circuit, and a secondary circuit that circulates heat within the building. Space heating systems usually require higher temperatures than that of the ground. At first glance, the use of the cooler ground to heat a building appears as a contravention of the second law of thermodynamics. Heat pumps overcome this apparent restriction by enhancing the ground-sourced energy with external work. Refrigerators are an everyday-life example of heat pumps. Figure 8.3 shows a GHP schematic diagram and its basic operating principle. Heat transfer occurs in fluids when they change temperature and/or phase, while the heat transfer associated with phase change is much greater than that corresponding to only temperature change, and heat pumps make use of the properties of refrigerants (which can change phase at suitable operating temperatures and pressures) to achieve efficient heat transfer.

The heat pump principle and operation consist of a cooled, liquid refrigerant that is pumped into a heat exchanger (evaporator), where it absorbs thermal energy from the ambience as a result of the temperature differences, then through the compressor to a condenser where the heat is released, as shown in the diagram of Figure 8.4. During this process, the refrigerant then changes state, becoming gas, and then the gaseous refrigerant is recompressed in the compressor, resulting in a temperature increase. A second heat exchanger (the condenser) transfers the thermal energy to the heating system where the refrigerant reverts to liquid, and the refrigerant pressure is reduced again in the expansion valve. Basically, the process is: (1) the liquid refrigerant absorbs heat from a heat source

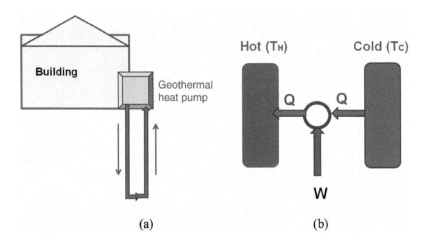

(a)   (b)

FIGURE 8.3   (a) Geothermal heat pump diagram, (b) its operating principle.

and evaporates, (2) the refrigerant, being cooler than the heat source, having a boiling point below the heat source temperature, the refrigerant gas then passes through the compressor, increasing its pressure and temperature, and (3) the compressor's hot and high-pressure refrigerant is hotter than the heat sink, so the heat flows from the refrigerant to the heat sink. At this higher pressure, the refrigerant gas condenses at a higher temperature than its boiling point. Thus when the refrigerant gas reaches the condenser, it condenses and releases heat. The hot, high-pressure liquid refrigerant then passes through an expansion valve that returns the pressure and temperature of the liquid to its original conditions prior to the cycle. For GHPs in heating mode, refrigerant evaporation occurs where the heat pump joins the primary (ground) circuit and condensation occurs where the heat pump joins the secondary (facility) circuit. Refrigerant evaporation cools the circulating fluid in the primary circuit and is then reheated by the ground. This process is reversed in cooling mode as refrigerant condensation heats the circulating fluid in the primary circuit, which is recooled by the ground. It is important to note that GHPs require energy (for the compressor and the pumps that circulate fluid) to move heat around the system. However, its energy input required is smaller compared to the heat output. Typically, a GHP produces around 3.5 kW to 5.5 kW of thermal energy for every 1 kW of electricity. Heat-pump efficiency increases as the temperature difference between the heat sink and source decreases. GHPs are thus more efficient than air-source heat pumps because the seasonally averaged ground temperature is closer to the desired ambient building temperature than the air is. In order to transfer heat from a heat source to a heat sink, heat pumps need external energy. Theoretically, the total heat delivered by a heat pump is equal to the heat extracted from the heat source, plus the amount of drive energy supplied to it. It is simply a heat engine running in

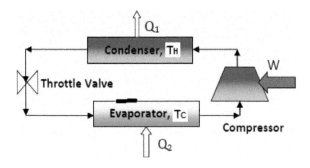

FIGURE 8.4   Building geothermal heat pump.

reverse mode (Figure 8.4), accepting heat Q from the sink at T_C (lower temperature), and rejecting the heat to the source at a higher temperature, T_H, by consuming work W, from an external source. From a thermodynamic point of view, a heat-pump thermal cycle is a reverse thermal engine cycle, so the input work provides the heat transfer to raise the temperature. The basic relationships governing heat pumps are the thermodynamics laws, being independent of the working fluid, cycle type, and the form of heat transfer. The heating efficiency of a heat pump is given by is its COP, as defined before:

$$COP = \frac{\text{Heat Output}}{\text{Input Work}} = \frac{Q_H}{W} < \frac{T_H}{T_H - T_C} \tag{8.23}$$

However, it is usual to define overall efficiencies for actual (real) heat engines and heat pumps, according to the Carnot cycle, as:

$$W = \eta \cdot Q_H$$

with $\eta < \eta_C$, for heat engines, and $\eta > \eta_C$ for heat pumps, where η_C is the Carnot efficiency, operating between the same temperatures. The previous equation is usually rewritten for heat pumps as:

$$W = \frac{Q_H}{COP_{HP}} = \frac{Q_C}{COP_{RF}} \tag{8.24}$$

The COP_{HP} is based on heat output Q_H from a heat pump or vapor recompression system, and COP_{RF} is the coefficient for a refrigeration system based on heat absorbed Q_C from the process. COPs are defined for real (actual) systems, with temperature expressed in Kelvin units (the absolute temperature) as:

$$COP_{HP} = \frac{\text{Desired Output}}{\text{Required Input}} = \frac{\text{Heating Effect}}{\text{Input Work}} = \frac{Q_H}{W} = \eta_{mech} \frac{T_H}{T_H - T_C} \tag{8.25a}$$

and

$$COP_{RF} = \frac{\text{Desired Output}}{\text{Required Input}} = \frac{\text{Cooling Effect}}{\text{Input Work}} = \frac{Q_C}{W} = \frac{Q_H - W}{W} = COP_{HP} - 1 \tag{8.25b}$$

Hence, the heat pump as a low-temperature lifting device ($T_C \rightarrow T_H$) gives higher COP_{HP}, which is inversely proportional to the temperature stretch and a large amount of upgraded heat per unit power. However, lower values of T_C reduce COP_{RF}, so refrigeration systems need more power per unit for upgraded heat as the absolute temperature falls. With the above understanding, one can start analyzing the GHP problems. Theoretically, heat pumping can be achieved by many more thermodynamic cycles and processes. These include Stirling cycle, single-phase cycles (e.g., with air, CO_2, or noble gases), solid-vapor absorption systems, hybrid systems (combining the vapor compression and absorption cycle), and electromagnetic and acoustic processes. Some of these systems are close to entering the market or have reached technical maturity, and they can become significant in the future. All heat pumps fall into two categories, i.e., either based on a vapor compression, or on an absorption cycle. Heat pumps are used for refrigeration/cooling below-ambient temperature, or as a heat-recovery system.

Example 8.10

Compare the heating efficiencies (maximum COP) of the same heat pump installed in New Orleans, Louisiana, and in Cleveland, Ohio.

SOLUTION

In New Orleans, the climate is milder with higher average temperatures; we can assume that T_H (summer) is about 20°C (or 70°F) and that T_C (winter season) is about 5°C (or about 40°F). In Cleveland, assume that T_H is the same, but that T_C (the outside temperature) is much lower, on average about –10°C (or about 14°F). Because the heat pump is used as a heater, after the conversion to absolute temperatures, the maximum COP at each of the two locations is calculated as:

$$\text{Cleveland: } COP_{max} = \frac{T_H}{T_H - T_C} = \frac{293}{293 - 263} = 9.77$$

$$\text{New Orleans: } COP_{max} = \frac{T_H}{T_H - T_C} = \frac{293}{293 - 278} = 19.53$$

The most important GHP benefit is that GHPs use about one-third or even over one-half less electricity than conventional heating and cooling systems with good potential for energy consumption reductions in any area. The cooling efficiency is defined as the ratio of the heat removed to the input energy, or the energy efficiency ratio (EER). Good GHP units must have a COP of 3 or greater and an EER of 13 or greater.

8.2 BUILDING ENERGY SYSTEMS

Modern buildings consume a large fraction of generated electricity, having a significant impact on utility and grid operations and management. The ability of buildings to shift the energy demands from peak periods to low-demand periods can greatly reduce the cost and pollutant emissions by allowing utilities to reduce the needs for their least efficient, higher-cost operations, and most polluting peak power plants. On-site generation, energy storage technologies, and coordinating building energy systems with other buildings and utilities can lower overall costs and emissions, while increasing the utility grid reliability. Providing a comfortable and healthy indoor environment is the core function of a building energy system, accounting for about 30% of total building energy use. New and smart technologies for heating, cooling, and ventilation not only can achieve higher efficiencies and energy conservation targets, but also improve the way building systems meet occupant needs and preferences by providing extended control, reducing unwanted temperature variations, and improving indoor air quality. Opportunities for improvements are separated into the following categories: good building design; improved building envelope, including roofs, walls, and windows; improved equipment for heating and cooling air and removing humidity; thermal energy storage that can be a part of the building structure or separate equipment; improved sensors and control systems for optimizing system performance. Both building designs and the selection of equipment depend on the climate where the building operates. Building walls, foundation, roof, and windows are coupling the exterior environment with the building interior environment in complex and dynamic ways. The insulating properties of the building envelope and construction quality together control the way heat and moisture flows into or out of the building. The color of the building envelope and other optical properties govern how solar energy is reflected and how thermal energy (heat) is radiated from the building. Windows bring sunlight and the sun's energy into the building. About 50% of the heating load in residential buildings and 60% in commercial buildings results from flows through walls, foundations, and the roof. Virtually the entire commercial cooling load comes from energy entering through the windows (i.e., solar heat gain). The bulk of residential cooling results from window heat gains although infiltration also has a significant role.

Water heaters, refrigerators, washers, and dryers are major energy consumers and are responsible for about 20% of all building energy use. Many of the technologies designed to improve whole-building energy performance can also be used to increase the efficiency of these appliances. For example, water-heating efficiency can be improved using advanced heat pumps, low-cost variable-speed motors, thin insulation, and other improved designs. Improved insulation and other strategies can reduce the losses from lengthy hot-water distribution systems in commercial buildings and large homes. Water heaters with storage tanks are good candidates for load shifting and providing other services important for optimizing electric utility performance with improved controls, energy management, and communications technologies, and to ensure that these approaches are designed for the optimum size ranges needed for appliances. Significant gains have been made in refrigerator performance over the past decades, but these gains have been partially offset by the increasing number of refrigerators and freezers used per household. Improvement in heat pumps, advanced thermal cycles, heat exchangers, and thin, highly-insulating materials (e.g., vacuum insulation) can lead to major performance gains. Further gains are possible by using separate compressors optimized for freezers and refrigerator compartments and using variable-speed drives and new sensors and controls to reflect ambient temperatures and react to signals from utilities. Until recently, clothes dryers were untouched by the technical advances transforming markets for other building equipment, but this is changing rapidly. Clothes dryers now on the market use heat pumps to circulate heated air over clothing in a drum, pass the air over a heat exchanger cooled by the heat pump, condense the water out of the air, and then reheat and recycle the air. Because air is recycled, there is no need for an air vent. These appliances operate at lower temperatures (thus are gentler to clothes) and reduce utility peaks because their peak electric demands are one-fifth that of conventional dryers. The technology is attractive for designs that provide washing and drying in the same front-loading unit.

In order to improve the overall building operations, improved thermal comfort, building energy conservation, and efficient usage, it is necessary to view as a whole-building system the building, its subsystems, and the ways that they interact. However, energy conservation and efficiency are not the primary objectives for a building. Rather, the main objectives for a building include its functions and aspects, such as basic provision of shelter; space; and comfort for different activities, such as living, working, and interacting. Other factors are secondary but important, such as aesthetics, integration to the landscape and environment, and cultural settings. In an integral view, the aesthetics, technologies, and functionalities work together for higher comfort levels, better life conditions, and sustainability for the best building design approach. The system view is also appropriate for studying buildings from an energy perspective, conservation, and savings. For distinguishing the building system, basic characteristics and essential building functions must be well-defined and analyzed. The primary building function is to provide shelter for the humans from the surrounding climate, and to create appropriate space with comfort and safety for various activities. In short, a building system's primary function is to provide appropriate space for human activities in a physical shelter from the natural environment. To accomplish this function the building system needs to fulfill several demands, depending on the specific activities the building houses, such as: protection from cold, wind, rain, heat excess, humidity, moisture, or snow; appropriate and appealing space; structural support for shelters and spaces; daylight, heating, cooling, night lighting, fresh air, cold and hot water supplies; access and connections to the electricity and communication grids; and safety and monitoring.

The *building envelope* is the boundary that divides and separates the inner and the outer spaces, satisfying the first building demand as discussed in the previous paragraph. The foundation, the walls, and the roof are all part of the building envelope, which are key factors for *passively* achieving energy efficiency within the building system. The last building system demands are fulfilled by the building subsystems. These include: the heating, cooling, and ventilation systems; the cold and hot water systems; and the electric (building electric service), communication, internet, phone, security, and monitoring systems. By optimizing these building subsystems, *active* energy efficiency and conservation can be obtained. This division between passive and active approaches and building subsystems is not always very clear or strictly necessary for building system energy optimization. From an integrated point of view, a building system is composed of its constructional and spatial structures, surrounded by the environment, and housing the humans. For the characterization of building interactions to those surrounding systems, a number of subsystems are used, as shown in Figure 8.5, and they are described in the next paragraphs. In order to make the building

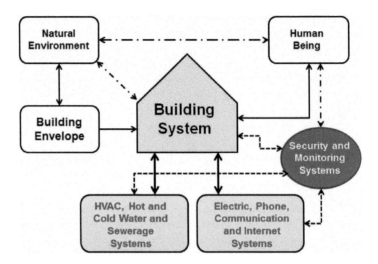

FIGURE 8.5   An overview of the building system.

respond to the shifting and changing characteristics of the surroundings and climate, a broad and comprehensive integration between the humans, environment, and building system and subsystems must be achieved. The characteristics of the surroundings, climate, and environment that are considered into the building design and operation include: daily, seasonal, and yearly temperature; solar insulation,; humidity and wind variations and characteristics; topography; vegetation; landscape; altitude; climate changes; and other buildings and structures.

Building response to external conditions is quantified by a set of building codes and provisions. One of the most important factors is the highest allowed *U-value* for the building as a whole, describing how much thermal power is transmitted through unit area of the structure at one degree temperature difference, measured in W/m^2K. Usually and traditionally, the energy codes set figures for every building element, such as walls, the floor, the roof, and the windows separately. However, this approach is a poor integral view, where the building parts and elements are considered separate systems rather than as components of the whole building. By introducing an overall U-value demand, there is more flexibility to the building design, element selection, and overall energy system operation. For operation and energy management, a building consists of several subsystems that are coordinated and managed to fulfill the building characteristics and functionalities. The most general description of the building system consists of the supporting, enclosing, and maintaining building systems. The enclosing system is represented by the building envelope for its thermal properties. The maintaining systems are further divided into the heating, ventilation, cold and hot water, electric, monitoring, and communication systems.

The most active subsystem for maintaining the building and providing comfort is the thermal system. The heating system is usually divided into three main components: the heating source(s), the distribution subsystem, and the space heaters. The space heaters can be radiators, convectors, or a grid of pipes or electric conductors integrated into the floors. Floor heating systems have gained preference during the last decades for their comfort and flexibility. After the optimum and appropriate temperature, the second factor for achieving comfort is the air quality. Without natural and/ or artificial ventilation, the air in a populated area gets increasing concentrations of odors and CO_2 from respiration. Ventilation systems are likely the most debated in the low-energy building design, such as the use of only natural ventilation without electricity use, but also without recovering any heat from ongoing air. For hygiene, drinking, and cooking, the water system (grid) is an essential part in any standard building. Water temperature must be controlled in prescribed ranges for both cold and hot water. Energy losses in water systems must be reduced to increase overall building system efficiency. The electric network in a building is essential for lighting, household appliances, computers, TVs, cable, and internet. For electrical heated buildings, the electric network is an essential part of the heating and water subsystems. The energy efficiency for electrical equipment is critical for optimum building energy usage. The building envelope is the most critical building subsystem, constituting the boundary between exterior and interior, consisting of the outer walls (with windows and doors), floor, and roof. The building envelope composition determines the amount of the gained and lost thermal energy and expresses the building in relation to its context (façade). However, with new and more efficient HVAC and electrical technologies and equipment, the building envelope has lost some of its importance as the main component of the building thermal balance mix.

8.2.1 BUILDING THERMAL BALANCE, COOLING, AND THERMAL LOAD CALCULATIONS

Heating and cooling load calculations are performed to estimate the required capacity of heating and cooling systems to maintain the required conditions and comfort in indoor spaces. Accurate load calculations have a direct impact on energy efficiency, occupant comfort, indoor air quality, and building durability. The load calculation is the first step of the HVAC design; a full HVAC design involves much more than just the load calculation. The loads modeled by the heating and cooling load calculation process dictate the equipment selection and duct design to deliver conditioned air to the rooms of the house. To estimate the required cooling or heating capacities, the most complete and full information available is necessary, regarding the design indoor and outdoor conditions; building specifications; conditioned space specifications, such as the occupancy, activity level, various appliances and equipment used; and any special requirements of the particular application. For comfort applications, the required indoor conditions are fixed by the criteria and characteristics of thermal comfort, while for industrial or commercial applications, the required indoor conditions are fixed by the specific processes being performed or the products being stored. Heating load calculations are carried out to estimate the heat loss from the building in winter to determine the required heating capacities. Usually during winter months the peak heating load occurs before sunrise and the outdoor conditions do not vary significantly throughout the winter season. In addition, internal heat sources such as occupants or appliances are beneficial as they compensate for part of the heat losses. As a result, the heat load calculations are carried out assuming steady-state conditions (no solar radiation and steady outdoor conditions) and neglecting internal heat sources. This is a simple but conservative approach, leading to a small heating capacity overestimation. For accurate estimation of heating loads, the thermal capacity of the walls and internal heat sources needs to be taken into account, making the problem more complicated. For estimating cooling loads, unsteady state processes need to be considered, as the peak cooling load occurs during the daytime and the outside conditions vary significantly throughout the day due to solar radiation. In addition, all internal sources add on to the cooling loads, so neglecting them leads to underestimation of the required cooling capacity and to the failure to maintain the required indoor conditions. Thus cooling load calculations are inherently more complicated as they involve solving unsteady equations with unsteady boundary conditions and internal heat sources. Building load calculations are usually performed by using software applications, most of them using data and algorithms developed by ASHRAE. Manual calculations are used for simpler buildings or primary designs. Such methods are based on the estimate methods of the heat-transfer processes. The conduction heat transfer is determined by the building interior and exterior temperature differences. Inside temperature is set by the personal comfort or manufacturing processes, and by the weather conditions (outside air temperature, humidity, and wind velocity). Conduction depends on the insulating quality of the walls, roof, and floor, expressed by the R-value (the thermal resistance to heat transfer, hr·ft$^{2.0}$F/Btu or m^{2}K/W). Envelope elements are made of several layers, and the heat flows through each layer in sequence, so the insulation value of an assembly is the sum of the R-values of each component. The heat conduction of assemblies is characterized by the U-factor, the reciprocal of the R-value. The units of U-factor are W/m²·K in SI units and BTU/h·ft²·°F in Imperial System units, with a 0.176 conversion. The heat transfer, Q, the temperature difference, ΔT, the area, A, and the U-value are related by:

$$Q = U \times A \times \Delta T = UA \cdot (T_{In} - T_{Out}) \tag{8.26}$$

Equation (8.31) implies at first look, that the relationship between Q and AT is linear. Whereas this is approximately so over limited ranges of temperature difference for which U is nearly constant, in practice U may well be influenced by the temperature difference and the absolute value of the temperatures. If it is required to know the area needed for the transfer of heat at a specified rate, the temperature difference ΔT, and the value of the overall heat-transfer coefficient must be known. Thus the U value(s) calculation is a key requirement in any design problem in which heating and cooling are involved. Large parts of heat-transfer studies are therefore devoted to the evaluation of this coefficient. The value of this coefficient depends on the mechanism by which heat is transferred, on the fluid dynamics of both the heated and the cooled fluids, on the properties of the materials through which the heat must pass, and on the geometry of the fluid paths. In solids, heat is normally transferred by conduction; some materials such as metals have a high thermal conductivity, whilst others such as ceramics have a low conductivity. Transparent solids like glass also transmit radiant energy particularly in the visible part of the spectrum. Liquids also transmit heat readily by conduction, though circulating currents are frequently set up and the resulting convective transfer may be considerably greater than the transfer by conduction. Many liquids also transmit radiant energy. Gases are poor heat conductors and circulating currents are difficult to suppress, so convection is therefore more important than conduction in a gas. The above equation is used for heat-load calculations, to estimate the heat losses through walls, roofs, or windows. However, in HVAC calculations, various equations considering the effects of the sun, humidity and the outside temperature are used. In building heat transfer, different types of energy transport are effective, and often, heat is transported by different modes to or from the same place. Energy that reaches a point via different paths and modes may be added up for the heat balance. Primary heat transport modes are conduction, convection, and radiation, as discussed before. In buildings, heat is also transported by the following mechanisms, which basically belong to the convective mode: transfer of latent heat by transport of water or water vapor, thermal energy associated with the air replaced in a building by ventilation or by air leakage (infiltration), and thermal energy associated with domestic water and combustion air (including flue gases), and fluids feeding heat pumps. For any building there exists a balance point at which the solar radiation (Q_{Solar}) and internal heat generation rate (Q_{Int}) exactly balance the heat losses from the building. Thus from sensible heat balance equation, at the thermal (internal and external heat flows) building balanced conditions:

$$(Q_{Solar} + Q_{Int})_{sensible} = UA \cdot (T_{In} - T_{Out}) \tag{8.27}$$

where UA is the product of overall heat-transfer coefficient and heat-transfer area of the building, T_{In} is the required indoor temperature, and T_{Out} is the outdoor temperature. From the above equation, the outside temperature at balanced condition ($T_{Out-Bal}$) is given by:

$$T_{Out-Balance} = T_{In} - \frac{(Q_{Solar} + Q_{Int})_{sensible}}{UA} \tag{8.28}$$

Example 8.11

A model building is represented by a rectangular shape of 5 m × 5 m × 10 m. If no heat is lost into the solid and with $U = 0.35$ W/(m²K), and the inside and outside temperature difference is 20°C, what is the total heat loss?

SOLUTION

Applying Equation (8.26), the building heat is:

$$Q = UA \cdot (T_{In} - T_{Out}) = 0.35 \times (5 \times 5 \times 10) \times 20 = 1750 \text{ W}$$

If the outdoor temperature is greater than the balanced outdoor temperature given by the above equation, i.e., when $T_{Out} > T_{Out\text{-}Bal}$, then there is a need for cooling the building, while when the outdoor temperature is less than the balanced outdoor temperature, i.e., when $T_{Out} < T_{Out\text{-}Bal}$, then there is a need for heating the building. When the outdoor temperature equals the balanced outdoor temperature, i.e., when $T_{Out} = T_{Out\text{-}Bal}$, there is no need for either building cooling or heating. For residential buildings (with fewer internal heat sources), the balanced outdoor temperature may vary from 10°C to 22°C, meaning that if the balanced outdoor temperature is 18°C, then a cooling system is required when the outdoor temperature exceeds 18°C (or about 68°F). This implies that buildings need cooling not only during summer but during spring and fall as well. If the building is well-insulated (small UA factor) and/or internal loads are large, then from the energy balance equations, the balanced outdoor temperature reduces leading to an extended cooling season and a shortened heating season. Thus, a smaller balanced outdoor temperature implies higher cooling requirements and smaller heating requirements, and vice versa. For commercial buildings with large internal loads and relatively smaller heat transfer areas, the balanced outdoor temperature can be as low as 2°C, implying a longer cooling season and a smaller heating season. If there are no internal heat sources and if the solar radiation is negligible, then the heat balance equation, $T_{Out\text{-}Bal} = T_{In}$, implies that if the outside temperature exceeds the required inside temperature (21°C for comfort) then there is a need for cooling; otherwise there is a need for heating. Thus depending upon the specific conditions of the building, the need for either a cooling system or a heating system varies. This also implies a need for optimizing the building insulation depending upon outdoor conditions and building heat generation to use free cooling provided by the environment during certain periods.

Example 8.12

A building has a U-value of 0.75 W/m²·K and a total exposed surface area of 400 m². The building is subjected to an external load of 2.0 kW and an internal load of 1.0 kW, sensible only. The required internal temperature is 21°C, whether a cooling system is required or a heating system is required when the external temperature is 5°C. How will the results change, if the U-value of the building is reduced to 0.35 W/m²·K?

SOLUTION

From energy balance, Equation (8.28), the balance temperature is:

$$T_{Out\text{-}Balance} = 21 - \frac{(2.0 + 1.0) \times 1000}{0.75 \times 360} = 11.10 \text{ °C}$$

Since the outdoor temperature at balance point is greater than the external temperature, the building requires heating. When the U-value of the building is reduced to 0.35 W/m²·K, the new balanced outdoor temperature is given by:

$$T_{Out\text{-}Balance} = 21 - \frac{(2.0 + 1.0) \times 1000}{0.35 \times 360} = 23.81 \text{ °C}$$

Since now the outdoor temperature at balance point is smaller than the external temperature, the building now requires cooling. The above example shows that adding more insulation to a building extends the cooling season and reduces the heating season.

Usually, the heating and cooling load calculations involve using systematic, stepwise (algorithmic) procedures to arrive at the required system capacity by taking into account all the building energy flows. In practice, a variety of methods ranging from simple rules-of-thumb and empirical methods to complex *transfer function methods (TFMs)* are used to arrive at building loads. For example, typical rules-of-thumb methods for cooling loads specify the required cooling capacity based on the floor area or occupancy. Such rules-of-thumb are useful in preliminary estimation of the equipment size and cost, but their main conceptual drawback is the presumption that the building design does not make any difference. Thus such rules for a badly designed building are the same as for a good design. Accurate load estimation methods involve a combination of analytical methods and empirical results obtained from actual observation and measurement data, for example the use of *cooling load temperature difference* (CLTD) for estimating material heat gain and the use of *solar heat gain factor* (SHGF) for estimating the heat transfer through fenestration. These methods are widely used by HVAC engineers because they yield reasonably accurate results and estimations can be carried out manually. Over the years, more accurate methods that require the use of computers have been developed for estimating cooling loads, e.g., the *transfer function methods*. Because these methods are expensive and time-consuming, they are generally used for estimating cooling loads of large commercial or institutional buildings. ASHRAE suggests different methods for estimating cooling and heating loads based on the specific applications and building designation, such as for residences, for commercial or industrial buildings, schools, etc.

Heating and cooling load calculations rely on the size and type of construction for each component of the building envelope, as well as the heat given off by the lights, people, and equipment inside the house. If a zoned heating and cooling system is used, the loads in each zone should be calculated separately. The total building cooling load consists of heat transferred through the building envelope (walls, roof, floor, windows, doors, etc.) and heat generated by occupants, equipment, and lights. The instantaneous rate of heat gain is the rate at which heat enters into and/or is generated within a space at a given instant. The heat gain is classified by the manner in which it enters the space, as: solar radiation through transparent surfaces such as windows; heat conduction through exterior walls and roofs; heat conduction through interior partitions, ceilings, and floors; heat generated within the space by occupants, lights, appliances, equipment, and processes; loads as a result of ventilation and infiltration of outdoor air; and other miscellaneous heat gains. Sensible heat is the heat which a substance absorbs, and while its temperature goes up, the substance does not change state. Sensible heat gain is directly added to the conditioned space by conduction, convection, and/ or radiation. Note that the sensible heat gain entering a conditioned space does not equal the sensible cooling load during the same time interval because of the stored heat in the building envelope. Only the convective heat becomes cooling load instantaneously. Sensible heat load is the total of the heat transmitted through floors, ceilings, and walls; occupants' body heat; heat generated by appliances and lighting devices; solar heat gains through glass and transparent materials; infiltration of outside air; and air introduced by ventilation. Latent heat gain occurs when moisture is added to the space either from internal sources (e.g., vapor emitted by occupants and equipment) or from outdoor air as a result of infiltration or ventilation to maintain proper indoor air quality. Latent heat load is the total of the moisture-laden outside air from infiltration and ventilation; occupant respiration and physical activities; and moisture from equipment and appliances. The heat received from the heat sources (e.g., conduction, convection, solar radiation, lighting, people, equipment) does not go immediately to heating the room air. Only some portion of it is absorbed by the air in the conditioned space instantaneously leading to a minute change in its temperature. Most of the radiation heat, especially from the sun, lighting, and people, is first absorbed by the internal surfaces, which include the ceiling, floor, internal walls, and furniture. Because of the large but finite thermal capacity of the roof, floor, walls, etc., their temperature increases slowly due to absorption of radiant heat. The radiant portion introduces a time lag and also a decrement factor depending upon the dynamic characteristics of the surfaces. To maintain a constant humidity ratio, water vapor must condense on a cooling apparatus at a rate equal to its rate of addition into the space. An adjustment of Equation

(8.26) consisting of the summation of the conductive heating load of each envelope elements can be used, being expressed as:

$$Q_{env} = \sum_{j=1}^{N_{env}} U_j A_j \left(T_{In} - T_{Out} \right)$$ (8.29)

Here: Q_{env} is the heating load (W) (kcal/h), A_j is the area of each envelope element (m²), U_j is the overall element heat-transfer coefficient (W/m²K) (kcal/m²h°C), T_{In} is the indoor design air temperature (K, or °C), and T_{Out} is the outdoor design air temperature (K, or °C). The difference between T_{In} and T_{Out} is called the cooling load temperature difference (CLTD). For sunlit surfaces, the CLTD is obtained from CLTD tables, e.g., ASHRAE manuals. Adjustment to the values obtained from the table is needed if actual conditions are different from those based on which the CLTD tables are prepared. For surfaces that are not sunlit or that have negligible thermal mass (such as doors), the CLTD value is simply equal to the temperature difference across the wall or roof. For example, for external doors, the CLTD value is simply equal to the difference between the outdoor and indoor dry bulb (thermometer) temperatures, $T_{Out} - T_{In}$. Dry bulb (DB) temperature is the ambient air temperature as measured by a standard thermometer, thermocouple, or resistance temperature device. DB temperature is used in combination with globe thermometer temperature and air velocity to calculate mean radiant temperature. For interior air-conditioned rooms surrounded by non-air-conditioned spaces, the CLTD of the interior walls is equal to the temperature difference between the two spaces. If an air-conditioned room is surrounded by other air-conditioned rooms, with all of them at the same temperature, the CLTD values of the walls of the interior room will be zero. Estimation of CLTD values of floors and roofs with false ceilings could be tricky. For floors standing on ground, one has to use the temperature of the ground for estimating CLTD. However, the ground temperature depends on the location and varies with time. ASHRAE suggests suitable temperature difference values for estimating heat transfer through ground. If the floor stands on a basement or on the roof of another room, then the floor CLTD values are the temperature difference across the floor (i.e., temperature difference between the basement or room below and the conditioned space). This discussion also holds good for roofs that have non-air-conditioned rooms above them. For sunlit roofs with false ceilings, the U-value may be obtained by assuming the false ceiling to be an air space. However, the CLTD values obtained from the tables may not exactly fit the specific roof, so designers must use their own judgment to select the suitable CLTD values. In summary, cooling load calculations involve the estimate of each of the components described above from the available data. In this chapter, the cooling load calculations are carried out based on the ASHRAE CLTD/CLF methods. For more advanced methods, such as TFM, the reader should refer to ASHRAE or other references.

Simple load calculation methods involve the steps described here. Heat transfer through opaque surfaces, which is basically a sensible heat-transfer process, such as walls, roof, floor, and doors, is given by Equation (8.30) with the appropriate temperature-difference adjustments. However, solar effects must be considered in the estimations of heat gains through windows and skylights. These heat loads are considered in two parts: conduction and solar transmission to that transparent area. Solar load through transparent building areas can be estimated by the equation used for radiant sensible loads from the transparent or translucent elements such as window glass, skylights, and plastic sheets, and the conduction to the material, as recommended by ASHRAE, is:

$$\dot{Q} = U \times A \times TD + A \times SHGF \times SC \times CLF$$ (8.30)

Here, TD is the temperature difference, A is area of transparent glass (window), roof, and/or wall transparent section, calculated from building plans, measured or given in the building data, SHGF is the solar heat gain coefficient (ASHRAE tables), SC is the shading coefficient, and CLF is the cooling load factor (ASHRAE tables, see also Appendix B tables). The SHGF represents the solar heat amount that enters a clear single-pane window at a given time (day and year) and specific orientation. The SC coefficient is a property of a glazing material and accessories (e.g., blinds, draperies), being a measure of the shading effectiveness of a glazing product. The SC is the ratio of solar heat admitted in comparison what is admitted by a clear glass, having shading coefficient 1.0. CLF factors are used to account for the fact that building thermal mass creates a time lag between heat generation from internal sources and the corresponding cooling load. CLF factors are presented in a set of tables that account for number of hours the heat has been on, thermal mass, type of furnishing or window shading, type of floor covering, number of walls, the room air circulation, the solar time, and the facing direction. CLF accounts for the fact that all the radiant energy that enters the conditioned space at a particular time does not instantly become a part of the cooling load. Notice that the unshaded area has to be obtained from the dimensions of the external shade and solar geometry. SHGF and SC are obtained from ASHRAE tables based on the orientation of the window, location, month of the year, and the type of glass and internal shading device. The CLF values for various surfaces have been calculated as functions of solar time and orientation and are available in the form of tables in ASHRAE handbooks. As solar radiation enters the conditioned space, only a small portion of it is absorbed by the air of the conditioned space instantaneously leading to an insignificant change in its temperature. Most of the radiation is first absorbed by the space's internal surfaces, e.g., ceiling, floor, internal walls, furniture. Due to the large but finite thermal capacity of internal surfaces, their temperature increases slowly due to solar radiation absorption. As the surface temperature increases, heat transfer takes place between these surfaces and the conditioned space air. Depending upon the thermal capacity of the wall and the outside temperature, some of the absorbed energy due to solar radiation may be conducted to the outer surface and may be lost to the outdoors. Only that fraction of the solar radiation that is transferred to the air in the conditioned space becomes a load on the building; the heat transferred to the outside is not a part of the cooling load. Thus it can be seen that radiation heat transfer introduces a time lag and also a decrement factor depending upon the dynamic characteristics of the surfaces. Due to the time lag, the effect of radiation will be felt even when the source of radiation, in this case the sun, is removed. The CLF values for various surfaces have been calculated as functions of solar time and orientation and are available in the form of tables in ASHRAE handbooks.

Example 8.13

A house has two windows facing east, four windows facing south, and two windows facing west. The windows are identical 1.5 m by 1.4 m. Assuming that the window U-factor is 3.18 W/m²·K, the SC is 0.65, and SHGF is 94.6 W/m² for the east- and west-facing windows, and 631 W/m² for the south-facing windows on a specific day at the house location, estimate the heat gain through the house windows, and the total equivalent temperature for east- and west-facing windows is 10°C and 42°C for the south-facing ones, estimate the total window heat gain.

SOLUTION

The area of a window is 2.10 m², and applying Equation (8.29) for a total of seven windows, the heat gain through windows is:

$$Q_{south} = 3.18 \times 3 \times 2.10 \times 32 + 3 \times 2.10 \times 0.65 \times 631 = 3225 \text{ W}$$

And for the east- and west-facing windows the heat gain is:

$$Q_{east} = Q_{west} = 3.18 \times 2 \times 2.10 \times 32 + 2 \times 2.10 \times 0.65 \times 94.6 = 814.9 \text{ W}$$

And the total gain through all house windows is:

$$Q_{Total} = 814.9 + 3225 + 814.9 = 4854.9 \text{ W}$$

Ventilation air is the amount of outdoor air required to maintain the indoor air quality for the occupants (e.g., ASHRAE Standard 62, specifying the minimum ventilation requirements) and needed to make up for the air that is leaving the space due to equipment exhaust, exfiltration, and pressurization.

$$\dot{Q}_{Sensible} = 1.08 \times CFM \cdot (T_{outdoor} - T_{CC}) \tag{8.31a}$$

$$\dot{Q}_{Latent} = 4840 \times CFM \cdot (W_{outdoor} - W_{CC}) \tag{8.31b}$$

and

$$\dot{Q}_{Total} = 4.8 \times CFM \cdot (h_{outdoor} - h_{CC}) \tag{8.31c}$$

Here, *CFM* is the ventilation airflow rate (found in ASHRAE tables), $T_{outdoor}$ is the outside dry bulb temperature, °F, T_{CC} is the dry bulb temperature of air leaving the cooling coil (AC system), °F, $W_{outdoor}$ is the outside humidity ratio, lb (water)/ lb (dry air), W_{CC} is the humidity ratio of air leaving the cooling coil, lb (water)/lb (dry air), $h_{outdoor}$ is the outside/inside air enthalpy, Btu per lb (dry air), and h_{CC} is the enthalpy of air leaving the cooling coil Btu per lb (dry air). In addition to this, sensible and latent heat transfer to the building also occurs due to heat transfer and air leakage in the supply ducts. A safety factor is usually provided to account for this depending upon the specific details of the supply air ducts. Heat transfer due to infiltration consists of both sensible as well as latent components. The sensible heat transfer rate due to infiltration is given by:

$$\dot{Q}_{SH,\,Infiltration} = \dot{m}_o C_{p,m} (T_o - T_i) = \rho_o \dot{V}_o (T_o - T_i) \tag{8.32a}$$

and

$$\dot{Q}_{LH,\,Infiltration} = \dot{m}_o h_{lg} (T_o - T_i) = \rho_o \dot{V}_o h_{lg} (T_o - T_i) \tag{8.32b}$$

In addition to this, sensible and latent heat transfer to the building also occurs due to heat transfer and air leakage in the supply ducts. A safety factor is usually provided to account for this depending upon the specific details of the supply air ducts. The load due to heat transfer through the envelope is called an external load, and all other loads are called internal loads. The percentage of external versus internal loads varies with building type, local climate, and building design. The total building cooling load consists of both sensible and latent load components. The sensible load affects dry bulb temperature, while the latent load affects the moisture content of the conditioned space. Buildings may be classified as *externally loaded* or *internally loaded*. In externally loaded buildings, the cooling loads are mainly due to heat transfer between the surroundings and the internal conditioned space. Because the surrounding conditions are highly variable in any given day, the *cooling load of an externally loaded building varies widely*. In internally loaded buildings, the cooling load is mainly due to internal heat-generating sources such as occupants or appliances or processes. In general, the heat generation due to internal heat sources may remain fairly constant, and because the heat transfer from the variable surroundings is much less compared to the internal heat sources (Figure 8.6), the *cooling load of an internally loaded building remains fairly constant. Infiltration* is the uncontrolled passage of outdoor air into a building through unintended leaks in the building envelope (e.g., cracks between wall sections, wall-floor connections, corners, the roof-wall interface, around windows and doors). The *exfiltration* process is the opposite of infiltration. Infiltration and exfiltration are driven by air-pressure differences that exist between the inside of a building relative to the building outside across its envelope. These air-pressure differences are the result of natural forces (e.g., wind and temperature) and a building's geometry and structure, HVAC system design, and envelope tightness. *Ventilation* is the process of "changing" or replacing the air in any space to provide high indoor air quality (i.e., to control temperature, replenish oxygen, or remove

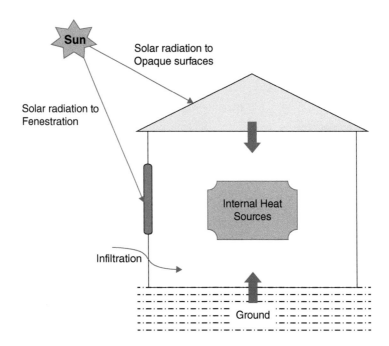

FIGURE 8.6 Building cooling load components.

moisture, odors, smoke, heat, dust, airborne bacteria, and carbon dioxide). Ventilation is used to remove unpleasant smells and excessive moisture, introduce outside air, to keep interior building air circulating, and to prevent stagnation of interior air. Ventilation includes both the exchange of air to the outside as well as circulation of air within the building. It is one of the most important factors for maintaining acceptable indoor air quality in buildings. Methods for ventilating a building may be divided into mechanical/forced and natural types.

The external loads consist of heat transfer by conduction through the building walls, roof, floor, doors, etc., and heat transfer by radiation through fenestration such as windows and skylights. All these are sensible heat transfers. In addition to these, the external load also consists of heat transfer due to infiltration, which consists of both sensible as well as latent components. The heat transfer due to ventilation is not a load on the building but a load on the system. The internal loads consist of sensible and latent heat transfer due to occupants, products, processes and appliances, as well as sensible heat transfer due to lighting and other equipment. Internal cooling loads consist of the following: (1) sensible and latent loads due to people; (2) sensible loads due to lighting; (3) sensible loads due to power loads and motors (equipment, elevators, pumps, fans, and other machinery); and (4) sensible and latent loads due to appliances. An internal load calculation is the *area of engineering judgment*. Internal loads are sometimes about 60% of the load. However, these data are generally the least amount of information available at the design stage, and therefore generic and technical rules are most often employed to fix the variables. The equations used in estimating internal loads are separated into several categories, such as: people, building equipment, lighting, appliances, motors and power loads, then separated in two categories, heat gains from the driving equipment inside the conditioned space, installed outside of the space, and heat gain when driven equipment is located outside the space to be conditioned with the motor inside the space or air stream, and the case of the heat gain from the HVAC system itself. The heat gains from people, lighting systems, appliances, power loads, and motors are expressed by the following equations:

$$Q_{Sensible-people} = N \times QS \times CLF \tag{8.33a}$$

$$Q_{Sensible-people} = N \times QL \tag{8.33b}$$

Here, N is the number of people in the conditioned space, QS and QL are the sensible and latent heat gain from occupancy (as specified in ASHRAE tables), CLF is the cooling load factor (as given in ASHRAE tables; notice that CLF is equal to 1.0, if the operation is 24 hours or if cooling is off at night or during weekends). The lighting system operation results in sensible heat gain, being expressed by:

$$Q_{light} = 3.41 \times W \times UF \times BFA \times CLF \tag{8.34}$$

where, W is the installed lamp system wattages, the input from the electrical lighting plan or lighting load data, UF is the lighting utilization (use) factor, as appropriate (the full description was given in Chapter 6 of this book), and again CLF is given in ASHRAE fundamentals. In calculations, as mentioned above CLF is equal to 1.0, for 24-hour operation or if the cooling is off at night or during weekends. The heat gain from the appliances is given by:

$$Q = 3.41 \times W \times F_U \times F_{rad} \times CLF \tag{8.35}$$

Here, W is again the installed power rating of appliances in watts (ASHRAE tables or the manufacturer's data; for computers, monitors, printers, and miscellaneous office equipment, see the ASHRAE tables), F_U is the usage factor, F_{rad} is the radiation factor, (both given in ASHRAE fundamentals, Chapter 28), CLF is cooling load factor, by hour of occupancy (ASHRAE fundamentals). Note that CLF is 1.0, if operation is 24 hours or if cooling is off at night or during weekends. Power loads and motor heat gain contributions are separated into three classes, as discussed before. Heat gain of power-driven equipment and motors when both are located inside the space to be conditioned, is expressed as:

$$Q = 2545 \times \frac{P_{motor}}{\eta_{eff}} \times FUM \times FLM \tag{8.36a}$$

Here, P_{motor} is the horsepower rating from electrical power plans or manufacturer's data, the power is expressed in watts in the case of the SI units, η_{eff} is the equipment motor efficiency, FUM, is the motor utilization factor (normally = 1.0), and FLM is the motor load factor (normally = 1.0). Note, $FUM = 1.0$, if the operation is 24 hours. While in SI units, the motor power in watts, the motor heat gain is expressed by:

$$Q = \frac{P_{motor}}{\eta_{eff}} \times FUM \times FLM \tag{8.36b}$$

Heat gain, in the second case of the power loads and motors, when driven equipment is located inside the space to be conditioned space and the motor is outside the space or the air stream, with notations, given above is expressed, by the imperial system and in SI units, respectively as:

$$Q = 2545 \times P \times FUN \times FLM \tag{8.37a}$$

and

$$Q = P \times FUM \times FLM \tag{8.37b}$$

In the third case of the power loads and motors, the heat gains of when driven equipment is located outside the space to be conditioned space and the motor is inside the space or air stream, for both using the Imperial System and SI units, with previous notations, are expressed by:

$$Q = 2545 \times P_{motor} \times \frac{1 - \eta_{eff}}{\eta_{eff}} \times FUM \times FLM \tag{8.38a}$$

and, respectively:

$$Q = P_{motor} \times \frac{1 - \eta_{eff}}{\eta_{eff}} \times FUM \times FLM \qquad (8.38b)$$

The last internal load is related to the HVAC itself, and the heat gain from the HVAC system is separated into two parts. First, the supply fan heat load, because the supply and/or return fans that circulate or supply air to the space add heat to the space or system depending on the location relative to the conditioned space. The heat added may take one or all of the following forms. Instantaneous temperature rises in the air stream due to fan drive inefficiency. Temperatures rise in the air stream when the air is brought to static equilibrium and the static and kinetic energy is transformed into heat energy. The location of the fan and motor relative to the cooling coil and space being conditioned determines how the heat is added to the system. If the fan is downstream of the cooling coil (draw-through), then the fan heat load is added to the space-cooling load. If the fan is upstream of the cooling coil, then the fan heat load is added to the system cooling coil load. The thermal energy is calculated, by using Imperial System units, by:

$$Q = 2545 \times \frac{P}{\eta_{eff1} \cdot \eta_{eff2}} \qquad (8.39a)$$

and

$$Q = \frac{P}{\eta_{eff1} \cdot \eta_{eff2}} \qquad (8.39b)$$

Here, P is the horsepower rating from electrical power plans or manufacturer's data, or in watts in SI units, 2545 is the conversion factor for converting horsepower to BTU per hour, η_{eff1} is the full-load motor and drive efficiency, and η_{eff2} is the fan static efficiency.

Then the total building load is simply the summation of the external and the internal loads, both sensible and latent loads. Usually a 10% safety margin is added, but it depends on how accurate are the inputs. The final load is then used to size the HVAC equipment. HVAC equipment is rated in BTU/h, but is commonly expressed in tonnage in the United States and in joules in most other countries. Remember that a BTU (British thermal unit) is the amount of heat needed to raise 1 pound of water 1 degree Fahrenheit. A *ton* of cooling load is actually 12,000 BTU per hour heat extraction equipment. The term *ton* comes from the amount of cooling provided by 2000 pounds, or 1 ton, of ice. Traditionally, cooling loads are calculated based on worst-case scenarios. Cooling loads are calculated with all equipment and lights operating at or near nameplate values, occupant loads assumed to be at a maximum, and extreme outdoor conditions assumed to prevail 24 hours per day. Real occupant loads are seldom as high as design loads. In detailed designing, the internal and external loads are individually analyzed, because the relative magnitude of these two loads has a bearing on equipment selection and controls. Analysis of this breakup provides an idea of how much each component of the building envelope contributes to the overall cooling load and what can be done to reduce this load. Reducing solar heat gain through windows is clearly one of the key areas.

8.3   HEATING, VENTILATION, AND AIR-CONDITIONING SYSTEMS

HVAC systems range widely, from the simplest hand-stoked stove used for comfort heating, to large-facility centralized systems, to the extremely reliable total (full) air-conditioning systems found in submarines and space shuttles. Cooling equipment varies from small domestic units to refrigeration machines that are 10,000 times the size, which are used in industrial processes and the food industry. Depending on HVAC system complexity and requirements, the HVAC designer must consider several issues and requirements beyond simply keeping temperatures comfortable. Basically, the HVAC system service covers air circulation, temperature and humidity control, and IEQ level control. The needs for ventilation and air-conditioning vary widely depending on the requirements of the indoor environment. Exact predictions of the cooling and heating load, proper HVAC system sizing, and optimal control are important to minimize energy consumption. Factors that affect cooling loads are the external climate such as outdoor temperature, solar radiation, and humidity. Local climatic conditions are important parameters for the energy efficiency of buildings, and the energy building consumption depends strongly on the climatic conditions. HVAC system performances change with design in building HVAC applications, taking into account the climatic conditions resulting in better comfort and energy-efficient buildings. Calculation of building thermal load is essential to find the best air-conditioning equipment and air-handling unit, and optimum HVAC system size, in order to achieve comfort and good air distribution in the conditioned zone. The concepts and fundamentals of air-conditioner sizing are based on the building heat gains and/or losses. The heat gains and losses must be balanced by the heat removal or addition, to get the desired indoor comfort. The heat gain and heat loss through a building depend on:

a. The temperature difference between outside temperature and the indoor desired temperature.
b. The construction type and the amount and the quality of insulation of the building ceiling, floor, and walls. Buildings built from materials with higher thermal conductivity (U-value) use more energy than those built of materials with lower thermal conductivity.
c. Shade and shade control, which have significant effects on the heating and cooling loads. Two identical buildings with different orientation with respect to the direction of sun rise and fall may require different air-conditioner sizing.
d. The room size and the surface area of the walls. The larger the surface area, the more heat can be lost or gained through it.
e. Infiltration. Door gaps, cracked windows, and chimneys are the doorways for air to enter from outside into your living space.
f. The occupants and their activity types.
g. Activities and other equipment within a building (e.g., cooking, hot bath, gymnasium, or laboratory).
h. Amount of lighting in the room. High-efficiency lighting fixtures generate less heat.
i. Number of items of power equipment inside the space, e.g., oven, washing machine, computers, TV sets; all contribute to heat.

HVAC system efficiency, performance, durability, and cost depend on the optimum matching of the size to the above factors and requirements. Often designers use simple methods for sizing HVAC systems. Such methods are useful in preliminary equipment sizing. The main drawback of such rules-of-thumb methods is the presumption that the building design does not make any difference. Thus the rules for a poorly designed building are basically the same as for a good design. Improper design and improper installation of the HVAC system have negative impacts on personal comfort and energy bills; this can create dangerous conditions that may reduce comfort, degrade indoor air quality, or even threaten the occupants' health. HVAC systems are responsible for meeting the indoor environment requirements and good living and working conditions, such as: air temperature, humidity and air quality by heating or cooling the indoor spaces, being an important part of the energy demands in the industry and building sectors, and requiring up to 50% of the total building energy demands. Notice that the building sectors, demands, and shares of electricity uses have grown dramatically over the past decades from about 25% of US annual electricity consumption in the 1950s to about 40% in the early 1970s to more than 75% by 2010. Without significant increases in building system efficiency, the total US electricity demand would have grown much more rapidly over this period. HVAC systems are related to the energy code and standard requirements, so knowledge of the most common HVAC systems and equipment, along with energy-related components and controls is critical for energy conservation, efficiency, and reduction in the overall building energy uses, regardless of the designation. An HVAC system consists of components and equipment to condition the air, transport it to the conditioned space, and to control the indoor environmental parameters of a specific space within required limits. HVAC system controls are the information link between varying energy building demands and the usual demands for indoor environmental conditions. Without a good control system, the most expensive, most thoroughly designed HVAC system can be a failure, by simply not controlling indoor conditions to provide comfort. Factors to the design of a HVAC system include:

1. Sizing the system for the specific building heating and cooling load;
2. Proper selection and proper installation of controls;
3. Correctly charging the unit with the proper amount of refrigerant;
4. Sizing and designing the layout of the ductwork or piping for maximizing energy efficiency; and
5. Insulating and sealing all ductwork.

Air-conditioning is a combined process, performing simultaneous functions, such as the air-conditioning and transporting to introduce the air into the conditioned space, to providing heating and cooling from its central plant or rooftop units. It also controls and maintains the temperature, humidity, air movement, air cleanliness, noise level, and pressure differential in a space within predetermined limits for the comfort and health of the occupants of the conditioned space or for the purpose of product processing. As previously discussed, the term *air-conditioning*, properly used, means the total control of temperature, moisture in the air (humidity), supply of outside air for ventilation, filtration of airborne particles, and air movement in the occupied (conditioned) space. Seven main processes are required to achieve full air-conditioning, listed and explained here. The processes are:

1. *Heating* is the process of adding thermal energy (heat) to the conditioned space for the purposes of raising or maintaining the temperature of the space.
2. *Cooling* is the process of removing thermal energy (heat) from the conditioned space for the purposes of lowering or maintaining the temperature of the space.
3. *Humidifying* is the process of adding water vapor (moisture) to the air in the conditioned space for the purposes of raising or maintaining the moisture content of the air.
4. *Dehumidifying* is the process of removing water vapor (moisture) from the air in the con-ditioned space for the purposes of lowering or maintaining the moisture content of the air.
5. *Cleaning* is the process of removing particulates, dust, biological contaminants, etc., from the air delivered to the conditioned space for the purposes of improving or maintaining the air quality.
6. *Ventilating* is the air-exchanging process between the outdoors and the conditioned space for the purposes of diluting the gaseous contaminants in the air and improving or main-taining air quality, composition, and freshness. It can be achieved either through natural ventilation or mechanical ventilation. Natural ventilation is driven by natural drafts, such as when you open a window. Mechanical ventilation can be achieved by using fans to draw air in from outside or by fans that exhaust air from the space to outside.
7. *Air movement* is the process of circulating and mixing air through conditioned spaces in the building for the purposes of achieving the proper ventilation and facilitating the ther-mal energy transfer.

The requirements and importance of the seven processes vary. In a climate that stays warm all year, heating may not be required at all or very little, while in a cold climate the warm summer periods may be so infrequent as to make cooling unnecessary. In dry desert climates, dehumidification may be redundant, while in hot, humid climates the dehumidification may be the most important HVAC design aspect. In summary, the main HVAC system design objectives are as follows: (1) control of temperature, humidity, air purity, and correct pressurization to avoid contamination; (2) provide comfort and healthy indoor environment of office buildings, educational buildings, cinemas, libraries, auditoriums, multiplexes, shopping centers, hotels, and other public places; and (3) provide special air filtration to remove bacteria, promote high indoor quality, and avoid cross contamination. The air-conditioning process involves both cooling the air and removing moisture. The usual approach does both using vapor-compression heat pumps. Smaller systems, including most residential systems, move conditioned air while most large commercial buildings use central chillers to cool water and transfer heat from water to air closer to the occupied spaces. Dehumidification is the process of removing the water from air, typically achieved by inefficiently cooling moist air until the water vapor condenses and then reheating the air to the needed temperature. HVAC system efficiency improvements involve efforts to improve the heating or cooling air process efficiency, and technology that efficiently removes the moisture from air. Air-conditioning systems perform the following functions: provide the required cooling and heating, supply air-conditioning (heat or cool, humidify or dehumidify, clean and purify), attenuate objectionable HVAC equipment-produced noise, distribute the conditioned air, contain sufficient outdoor air to the conditioned space, and control and maintain the indoor environmental parameters, such as temperature, humidity, cleanliness, air movement, sound level, and pressure differential between the conditioned space and surroundings, within predetermined limits. Parameters such as the size and the occupancy of the conditioned space, the indoor environmental parameters to be controlled, the quality and the effectiveness of control, and the cost involved determine the various types and arrangements of components used to provide appropriate characteristics. Air-conditioning systems are usually classified according to their application, use, and designation as: comfort air-conditioning systems and process air-conditioning systems. Comfort air-conditioning systems provide occupants with a comfortable and healthy indoor environment in which to carry out their activities. Process air-conditioning systems provide needed indoor environmental control for manufacturing, product storage, or other research and development processes. Air-conditioning systems can also be classified according to their construction and operating characteristics and functionalities.

Detailed discussions of the HVAC equipment and designs are beyond the chapter scope; a brief presentation is included and necessary. HVAC systems are used in different building types, such as industrial, commercial, residential, and institutional buildings, and their size, type, characteristics, and designs vary accordingly. HVAC systems are classified according to the required processes, which include heating, cooling, and ventilation, as well as humidification and dehumidification. These processes are achieved by using suitable HVAC equipment such as heating systems, air-conditioning systems, ventilation fans, and dehumidifiers. The HVAC systems need a distribution system to deliver the required amount of air with the desired environmental condition. The distribution system mainly varies according to the refrigerant type and the delivering method such as air-handling equipment, fan coils, air ducts, and water pipes. System selection depends on three main factors including the building configuration, the climate conditions, and owner (user) desires. The HVAC's basic components that deliver conditioned air to satisfy thermal comfort for the space and to achieve indoor air quality are: (a) mixed-air plenum and outdoor air control, (b) air filter, (c) supply fan, (d) exhaust or relief fans and an air outlet, (e) outdoor air intake, (f) ducts, (g) terminal devices, (h) return air system, (i) heating and cooling coils, (j) self-contained heating or cooling unit, (k) cooling tower, (l) boiler, (m) control subsystem, (n) water chiller, and (o) humidification and dehumidification subsystem.

The main types of HVAC systems are central systems and decentralized or local systems. System types depend on the primary equipment location, such as: centralized systems, which condition the entire building as a whole unit, or decentralized systems, which separately condition specific zones of a building. Therefore, the air and water distribution systems are designed based on system classification and the location of the primary equipment. The criteria mentioned above should also be applied in selecting between two HVAC types. The four main subsystems for any HVAC system are the primary equipment, the space requirements, the air distribution, and the piping network. Primary equipment includes heating equipment such as steam boilers and hot water boilers to heat buildings or spaces, air delivery equipment as packaged equipment to deliver conditioned ventilated air by centrifugal fans, axial fans, and plug or plenum fans, and the refrigeration equipment that delivers cooled or conditioned air into a space. It includes cooling coils based on water from water chillers or refrigerants from a refrigeration process. Space requirements are essential in shaping and setting an HVAC system to be central or local, its components and configuration. Equipment rooms are needed, because the total mechanical and electrical space requirements range between 4% and 9% of the gross building area. HVAC equipment and refrigeration equipment require several facilities to perform their primary tasks of heating and cooling the building. The heating equipment requires boiler units, pumps, heat exchangers, pressure-reducing equipment, control air compressors, and miscellaneous equipment, while the refrigeration equipment requires water chillers or cooling water towers for large buildings, condenser water pumps, heat exchangers, air-conditioning equipment, control air compressors, and miscellaneous equipment. The design of equipment rooms to host both pieces of equipment should consider the size and the weight of equipment, the installation and maintenance of equipment, and the applicable regulations to combustion air and ventilation air criteria. Fan rooms contain the HVAC fan equipment and other miscellaneous equipment. The rooms should consider the size of the installation and removal of fan shafts and coils, the replacement, and maintenance. Access to the equipment room for maintenance and keeping the equipment in excellent condition are critical for savings on maintenance costs. Air distribution considers ductwork that delivers the conditioned air to the desired area in as direct, quiet, and economical way as possible. All ductwork and piping should be insulated to prevent heat loss and save building energy. It is also recommended that buildings should have enough ceiling spaces to host ductwork in the suspended ceiling and floor slab and to be used as a return-air plenum to reduce the return ductwork.

8.3.1 TYPES OF AIR-CONDITIONING SYSTEMS

In institutional, commercial, and residential buildings, air-conditioning systems are usually for the health and comfort of the occupants, often called *comfort air-conditioning systems.* In industrial and manufacturing buildings (facilities), air-conditioning systems are provided for product processing, or for the health and comfort of workers as well as processing, and are called *processing air-conditioning systems.* Depending on their size, construction, and operating characteristics, air-conditioning systems can be classified as the following. *Individual room systems* usually employ either a single, self-contained, packaged room air-conditioner (installed in a window or through a wall) or separate indoor and outdoor units to serve an individual room. Self-contained, packaged means factory assembled in one package and ready for use. *Space-conditioning* or *space systems* have their air-conditioning, cooling, heating, and filtration performed predominantly in or above the conditioned space. Outdoor air is supplied by a separate outdoor ventilation system. *Unitary packaged systems* are the ones installed with either a single self-contained, factory-assembled packaged unit (PU) or two split units: an indoor air handler, normally with ductwork, and an outdoor condensing unit with refrigeration compressor(s) and condenser. In a packaged system, air is cooled mainly by direct expansion of refrigerant in coils and heated usually by a gas (or oil) furnace or electric heating. The unit may also function as a reverse cycle and provide heat (heat pump). *Central (hydronic) systems* use chilled water or heating hot water from a central plant to cool and heat the air at the coils in an air-handling unit (AHU). For energy transport, the heat capacity of water is about 3400 times greater than that of air. Central systems are built-up systems assembled and installed on the site. Both central and space-conditioning systems consist of the following components or subsystems. An air system is also called an air-handling system or the air side of an air-conditioning or HVAC&R system. Its function is to condition the air, distribute it, and control the indoor environment according to requirements. The primary equipment in an air system is an air-handling unit; both of these include fan, coils, filters, dampers, humidifiers (optional), supply and return ductwork, supply outlets and return inlets, and controls. Water systems include chilled water, hot water, and condenser water systems. A water system consists of pumps, piping work, and accessories. The water system is sometimes called the water side of a central or space-conditioning system. Central plant refrigeration and heating systems are designed for the following tasks. The refrigeration system in the central plant of a central system is usually in the form of a chiller package with an outdoor condensing unit. The refrigeration system is also called the refrigeration side of a central system. A boiler and accessories make up the heating system in a central plant for a central system, and a direct-fired gas furnace is often the heating system in the air handler of a rooftop packaged system. Control systems, discussed in chapter 11, usually consist of sensors, a microprocessor-based direct digital controller (DDC), a control device, control elements, personal computer (PC), and communication network.

HVAC systems consist of equipment, distribution network(s), and terminals for delivering thermal energy to condition a space either collectively or individually. The interaction of these during their performance makes energy balance assessment a complex task, significantly affecting the system performances and characteristics. HVAC systems contain a large variety of components that include equipment for generating heat and cool, air-handling units, distribution networks, and terminals. The more services are demanded, the more complex the system. A central HVAC system can serve one or more building thermal zones, and its major equipment is located outside of the served zone(s) in a suitable central location whether inside, on top of, or adjacent to the building. Central systems must condition zones with their equivalent thermal load. Central HVAC systems have several control points such as thermostats for each zone. The thermal energy transfer medium can be air or water or both, which represent as all-air systems, air-water systems, or all-water systems. The central systems include water-source heat pumps and heating and cooling panels, central humidifier, reheat coil, cooling coil, preheat coil, mixing box, filter, and outdoor air. The thermal energy-transfer medium through the building delivery systems is air. All-air systems can be subclassified based on the zone as single-zone or multizone, airflow rate for each zone as constant air volume and variable air volume, terminal reheat, and dual duct. A single-zone system consists of an air-handling unit, a heat source and cooling source, distribution ductwork, and appropriate delivery devices. The air-handling units can be fully integrated where heat and cooling sources are available or separate where heat and cooling source are detached. The integrated package is most commonly a rooftop unit connected to ductwork to deliver the conditioned air into several spaces with the same thermal zone. The main advantage of single-zone systems is simplicity in design and maintenance and low first cost compared to other systems. However, its main disadvantage is serving a single thermal zone when improperly applied. In a single-zone all-air HVAC system, one control device such as a thermostat located in the zone, controls the operation of the system. In a multi-zone all-air system, individual supply air ducts are provided for each zone in a building. Cold air and hot (or return) air are mixed at the air-handling unit to achieve the thermal requirement of each zone. A particular zone has its conditioned air that cannot be mixed with that of other zones, and all multiple zones with different thermal requirements demand separate supply ducts. Multi-zone all-air HVAC systems consist of an air-handling unit with parallel flow paths through cooling coils and heating coils and internal mixing dampers. It is recommended that one multi-zone unit can serve up to 12 zones because of physical restrictions on duct connections and damper size. If more zones are required, additional air handlers may be used. The multi-zone system advantage is that it adequately conditions several zones without energy waste associated with a terminal reheat system. Some spaces require different airflow of supply air due to changes in thermal loads. Therefore, a variable-air-volume (VAV) all-air system is the suitable solution for achieving thermal comfort. The previous four types of all-air systems are constant-volume systems. The VAV system consists of a central air-handling unit, providing air to the VAV terminal control box located in each zone to adjust the supply air volume. Temperature of supply air of each zone is controlled by the supply-air flow rate. The main disadvantage is that the controlled airflow rate can negatively impact adjacent zones with different or similar airflow rate and temperature. In an all-water system, heated and cooled water is distributed from a central system to conditioned spaces. This type of system is relatively small compared, because it uses pipes as distribution containers and the water has higher heat capacity and density than air, which requires a lower volume to transfer heat, and convectors. However, all-water cooling-only systems are unusual, such as valance units mounted in the ceiling. The primary type that is used in buildings to condition the entire space is a fan-coil unit. A fan-coil unit, installed vertically or horizontally, is a considerably small unit used for heating and cooling coils, circulation fan, and proper control unit. The fan-coil unit can be placed in the room or exposed to occupants, so it is essential to have appropriate finishes and styling. For central systems, fan-coil units are connected to boilers to produce heating and to water chillers to produce cooling to the conditioned space. Air-water systems are introduced as a hybrid system to combine both advantages of all-air and all-water systems. The volume of the combination is reduced, and outdoor ventilation is produced to properly condition the desired zone. The water medium is responsible for carrying the thermal load in a building by 80%–90% through heating and cooling water, while the air medium conditions the remainder.

Heating technologies comprise boilers, furnaces, and unit heaters, which typically use fossil fuels. Cooling technologies include chillers, cooling towers, and air-conditioning equipment, which usually consume electricity. Compression systems using gas or other cooling systems such as absorption chillers, adsorption chillers, or desiccant systems are alternatives to electric chillers. Distribution networks include air-handling units, fans, and pumps, which mainly consume electricity. The most common HVAC systems for buildings are all-air and all-water systems. All-air systems are usually used when cool generation is mainly required, including a boiler as a generator of heat and a water-cooled chiller connected to a cooling tower as a generator of cool. In such systems, outdoor air enters the air-handling unit, which includes a fan, filters to clean the air, coils for heating or cooling the air that passes through them, and humidifiers, if required. After going through the air-handling unit, the conditioned air is delivered to the spaces by a network of supply-air ducts and a parallel network of return-air ducts transports exhausted air. Most all-air HVAC systems have a single duct system that provides either heat or cool to the conditioned spaces. Two types of heating systems are most common in a new home: forced-air or radiant, with forced-air being used in the majority of homes. The heat source is either a furnace, which burns a gas, or an electric heat pump. Furnaces are generally installed with central air conditioners. Heat pumps provide both heating and cooling. Some heating systems have an integrated water heating system. One of the most common systems is the forced-air heating and cooling system. These systems use a central furnace plus an air conditioner, or a heat pump. In a typical system, several of these components are combined into one unit. Forced-air systems utilize a series of ducts to distribute the conditioned heated or cooled air throughout the home. A blower, located in a unit called an air handler, forces the conditioned air through the ducts. In many residential systems, the blower is integral with the furnace enclosure. Radiant heating systems typically combine a central boiler, water heater, or heat pump water heater with piping, to transport steam or hot water into the living area. Heating is delivered to the rooms in the home via radiators or radiant floor systems, such as radiant slabs or underfloor piping.

8.4   INDUSTRIAL HEATING PROCESS METHODS

Industrial sectors use about one-third of the total annual final energy consumed in the United States, about 28% in the European Union, with similar figures in many other developed and developing countries. Process heating is essential in most industrial sectors, including those dealing with products made from metal, plastic, rubber, concrete, glass, and ceramics; the energy sources involved in this process are quite varied. There are several options and approaches for industrial processes involving heating, melting, annealing, drying, distilling, separating, coating, drying, etc. The traditional existing heating solutions use direct fuel burning (oil, gas, by-products or waste products), hot air, steam, water, etc. On the other hand, the last several decades have witnessed the rapid expansion of different electric-based process heating systems, the so- called electro-heating technologies that assure the necessary heat by transforming the electrical energy into thermal energy. The practical implementation of these systems is based on direct, indirect, or hybrid heating methods. Direct heating methods generate heat directly within the processed material, by using one of the following techniques: an electrical current through the material, inducing an eddy current into the material, or exciting atoms and/or molecules within the material with electromagnetic radiation (e.g., microwaves). Indirect heating methods use one such method to heat a special element, while the heat is transferred to the material by either conduction, convection, radiation, or a combination of them, depending on the work temperature and the type of heating equipment. Hybrid systems use a combination of process heating systems based on different energy sources or different heating methods based on the same energy source. Electric-based heating process applications are in large part based on the huge progress in power electronics. The widespread industrial implementation of power electronics was based on their high performances and reduced prices, largely contributing to the successful application of electricity in heating processes. The main contribution was the possibility of changing the frequency of industrial distribution networks; indeed, the enhancement of heating process efficiency requires that power can be supplied to the application at an appropriate (depending on the heating method) controlled frequency.

Various techniques and methods are used in modern industrial processes for transferring the supplied electrical energy into thermal energy to heat up the material: resistance heating (direct, indirect), infrared radiation, induction, dielectric heating, etc.; all of these methods have both advantages and drawbacks, so the choice often has to be made on a case-by-case basis. In some cases, electric-based technologies are chosen for their unique technical capabilities, as the application cannot be used economically without an electric-based system; in other cases, the price of natural gas (or other fossil fuels) and electricity is the deciding factor. However, the electro-heating technologies compete with concurrent technologies using fossil fuels, reducing an industry's investment and operating costs, energy costs, and primary energy requirements. For some industrial applications, electric-based technologies are the most commonly used; in others, these are only used in certain applications. It is important to mention that in all cases, the efficiency and performance are considered together, always based on a lifecycle cost analysis. Electro-heating technologies comprise high-power heating processes that are powered through electrical energy and cover a large percentage of industrial electricity consumption. In regard to the whole industrial process, the implementation of electro-heating fulfills the following expectations:

1. It reduces the energy required to produce goods (improved energy efficiency).
2. It reduces greenhouse gases.
3. It improves product quality, while increasing productivity.
4. Electric-based process heating systems are controllable.
5. It improves the quality of the work environment.

Different electric-based process heating systems assure the necessary heat by transforming the electrical energy into thermal energy transmitted to the material to be treated. In the case of direct heating methods (e.g., direct resistance, dielectric heating, induction), the heat is generated within the material; for indirect methods (e.g., indirect resistance, indirect induction, infrared, electric arc, laser) energy is transferred from a heat source to the material by conduction, convection, radiation, or a combination of these techniques. In most processes, an enclosure is needed to isolate the heating process and the environment from each other. The enclosure reduces thermal losses and assures the containment of radiation (e.g., microwave or infrared), the confinement of combustion gases and volatiles, the material containment, the control of the atmosphere surrounding the material, and combinations of these. The most important electro-heating technologies further presented are resistance heating (direct or indirect), infrared radiation, induction, dielectric heating, and arc furnaces. Depending on the process heating application, system sizes, configurations, and operating practices differ widely throughout industry. It is not uncommon that electro-heating technologies compete with concurrent technologies using fossil fuels, reducing an industry's investment and operating costs, energy costs, and primary energy requirements. For some industrial applications, electric-based technologies are the most commonly used; in others, these are only used in certain niche applications. It is important to mention that in all cases, efficiency and performance must be considered together, always based on a lifecycle cost analysis. Electro-heating technologies use electric currents or electromagnetic fields to heat a large variety of materials (e.g., metals, ceramics, natural fibers, polymers, foodstuffs). Currently they are associated with many industrial processes involving heating, melting, annealing, drying, distilling, separating, coating, drying, etc. Most electro-heating installations can be very accurately controlled and/or process materials in controlled atmosphere enclosures. These attributes guarantee a better quality product, less material and energy wasted, and reduced operation time, that is energy savings, reduced costs, and reduced pollutant and CO_2 emissions. In high-temperature applications, the electro-heating installations are generally more energy efficient than their alternatives (furnaces based on direct fuel burning). The optimal efficiency of an electric furnace can reach up to 95% process efficiency, while the equivalent for a gas furnace is only in the 40% to 80% range. They are the only solutions for technological processes involving very high temperatures.

Electric-based process heating systems to assure the necessary heat to industrial processes, e.g. material treatment. In direct heating methods, the heat is generated within the material, while for the indirect methods, it is transferred from a heat source to the material by conduction, convection, radiation, or a combination of them. In most processes, an enclosure is needed to isolate the heating process and the environment from each other. Depending on the process, operation type, system sizes and configurations, heating processes differ widely throughout industry. Resistance heating is the simplest (but also the oldest) electric-based method of heating and melting metals and nonmetals. The efficiency of this technique can rise up to close to 100%, while working temperatures can exceed 2000∘C, so it can be used for both high-temperature and low-temperature applications. Direct resistance heating involves passing an electric current directly through the product to be heated, causing an increase in temperature, being an example of the joule effect. Infrared radiation heating is practically a variant of indirect resistive heating; it uses radiation emitted by electrical resistors, usually made of nickel-chromium or tungsten, heated to relatively high temperatures. Electric infrared processing systems are used by many manufacturing sectors for heating, drying, curing, thermal-bonding, sintering, and sterilization applications. Induction heating consists of applying an alternating magnetic field created by an inductance coil (inductor) to an electrically conducting object. The variable (oscillating) magnetic field produces an electric current (called an induced current) that flows through this body and heats it by the joule effect. In addition to the heat induced by eddy currents, magnetic materials also produce heat through the hysteresis effect: magnetics naturally offer resistance to the rapidly alternating electrical fields, and this causes enough friction to provide a secondary source of heat. Induction heating uses the same principle as a power transformer.

8.5 CHAPTER SUMMARY

This chapter provides a comprehensive overview of building heating, ventilation, air-conditioning, and energy systems, their characteristics, and functionalities. The general aspects of building heat transfer and heating systems are included and presented. Thermal and cooling load estimate methods, building envelope, and building subsystems are also discussed and introduced. Heating has one of its main functions, besides providing shelter to humans of any building, to provide the occupants' required and appropriate comfort. Air-conditioning is one of the major applications of refrigeration, making living conditions more comfortable, hygienic, and healthy in offices, workplaces, and homes. Air-conditioning involves cooling and dehumidification in summer months;

this is essentially done by refrigeration. It also involves heating and humidification in cold climates, which is conventionally done by a boiler unless a heat pump is used. The major applications of refrigeration can be grouped into the following four major equally important areas: food processing, preservation, and distribution; chemical and process industries; special refrigeration applications; and comfort air-conditioning. Space heating with heat pumps has a big advantage over fossil-fuel-based heating, due to the lack of combustion processes. Heating takes place automatically by the program entered in the machine, it depends on outdoor temperature changes. From a comfort point of view, a heat pump heating system is comparable to a centralized heating system. A heat pump provides the ability to install environmentally clean space heating. While operating, it does not emit pollutants, which harm the environment, human health and cause the greenhouse effect. Heat pumps also allow help to preserve the rapidly shrinking land resources and at the same time considerably reduce heating costs. Cooling load calculations may be used to accomplish one or more of the following objectives: (a) provide information for equipment selection, system sizing and system design, (b) provide data for evaluating the optimum possibilities for load reduction, and (c) permit analysis of partial loads as required for system design, operation, and control.

This chapter presents in some detail the types of HVAC systems. The ultimate purpose of an HVAC control system is to control zone temperature (and secondarily air motion and humidity) to conditions that assure maximum comfort and productivity of the occupants. From a controls viewpoint, the HVAC system is assumed to be able to provide comfort conditions if controlled properly. HVAC systems have several requirements including primary equipment such as heating equipment, cooling equipment, and delivery equipment; space requirements such as HVAC facilities, equipment room, and vertical shaft; air distribution; and piping. Types of HVAC systems can be divided into central HVAC systems and local HVAC systems. This classification depends on zone types and the location of HVAC equipment.

8.6 QUESTIONS AND PROBLEMS

1. What are the major components of a building energy system?
2. Briefly describe the major types of air-conditioning systems.
3. List the most important factors affecting indoor environment quality.
4. Briefly describe the main functions of HVAC systems.
5. What is the driving force for heat transfer?
6. What are the mechanisms of heat transfer? How are they distinguished from each other?
7. State in your own words the differences between a refrigerator and a heat pump.
8. Consider a compound slab, consisting of two materials having thicknesses, L_1 and L_2, and the thermal conductivities, k_1 and k_2, respectively. If the outer temperatures are T_2 and T_1, find the heat transfer rate through the slab in a steady state. Numerical application: $L_1 = 20$ cm, $L_2 = 30$ cm, $k_1 = 50$ W·m^{-1}·K^{-1}, and $k_2 = 60$ W·m^{-1}·K^{-1}.
9. A heat engine operates between a temperature of 20°C and 150°C. What is the efficiency of the engine if it works at 75% of the maximum possible Carnot efficiency?
10. A refrigerator operates between a temperature of 21°C and −8°C. What is the COP of the refrigerator if it works at 83% of the maximum possible Carnot COP?
11. Which of the following statements are true?
    a. Infiltration load is a part of the building load.
    b. Infiltration load is not a part of the building load.
    c. Infiltration rate increases as the pressure difference across the building decreases.
    d. Infiltration rate is uncontrollable.
12. What is the coefficient of performance of a refrigerator that uses 150 W to remove 1350 W from a cold chamber? What is the coefficient of performance of the same device if it is used as a heat pump?

13. Compare the heating efficiencies (maximum COP) of the same heat pump installed in Phoenix, Arizona, and in Portland, Maine.

14. Describe the relationship between the temperature of a radiating body and the wavelengths it emits.

15. True or false: The outdoor temperature at balance point decreases as the amount of insulation increases.

16. Briefly explain the physical mechanism of each heat-transfer mode.

17. Define thermal conductivity and briefly explain its significance in heat transfer.

18. How does heat conduction differ from convection?

19. Consider heat transfer through a windowless wall of a house on a winter day. Discuss the parameters that affect the rate of heat conduction through the wall.

20. A rectangular slab, 2 cm thick, is measured to be 100°C on one side and 96.2°C on the other side. The slab is 20 cm by 20 cm. Calculate the rate of heat transfer through the slab if the conductivity of the slab is 170 W/m·K.

21. Consider two house walls that are identical except that one is made of 15 cm, thick wood, while the other is made of 25 cm thick brick. Through which wall is the house losing more heat in winter?

22. If the earth has no atmosphere, its radiation emission would be lost quickly to outer space, making its temperature about 33 K cooler. Calculate the rate of radiation emitted and the wavelength of the maximum radiation emission for the earth at 255 K.

23. How much heat (thermal energy) is needed to warm 250 L of water from 20°C to 80°C?

24. Estimate the heat loss through a 250 ft.² wall with an inside temperature of 68°F and an outside temperature of 35°F. Assume the exterior wall is composed of 2 "of material having a 'k' factor of 0.75, and 2" of insulation having a conductance of 0.15.

25. Calculate the U-value for a layered wall assembly composed of three materials: (1) Plywood, 3/4 inch thick (R_1 = 0.94); (2) Expanded polystyrene, 2 inches thick (R_2 = 8.00); and (3) Hardboard, 1/4 inch thick (R_3 = 0.18). Assume the resistance of inside still air is R_{indoor} = 0.68 and resistance of outside air at 15 mph wind velocity is $R_{outdoor}$ = 0.17.

26. An industrial-process furnace is constructed with 0.20 m of fire brick, 0.10 m of insulating brick, and 0.20 m of building brick. The inside temperature is 1350 K and the outside temperature is 335 K. If the thermal conductivities are as shown in the following diagram, estimate the heat loss per unit area and the temperature at the junction of the fire brick and the insulating brick.

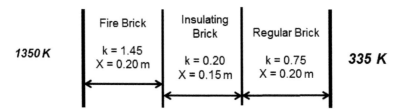

27. The inner and outer surfaces of a 4.5 m by 7.2 m brick wall of thickness 35 cm and thermal conductivity 0.695 W/m·°C are maintained at temperatures of 20°C and 5°C, respectively. Determine the rate of heat transfer through the wall, in W.

28. Determine the range for the rate of heat loss through a 1.2 m by 1.8 m window of a house that is maintained at 20°C when the outdoor air temperature is 8°C. Assume the value of the window U-factor ranges from about 1.25 W/m² °C (or 0.22 BTU/h·ft²·°F) for low-coated, argon-filled, quadruple-pane windows to 6.25 W/m²·°C (or 1.1 BTU/h·ft²·°F).

29. Consider two walls of a house that are in identical areas, except that one is made of 0.1 m thick wood, while the other is made of 0.25 m thick brick. Through which wall does the house lose more heat in winter?

30. Consider steady heat transfer between two large parallel plates at constant temperatures of 305 K and 210 K that are 10 cm apart. Assuming the surfaces to be black (emissivity $\varepsilon = 1$), steady operating conditions exist. There are no natural convection currents in the air between the plates, determine the heat transfer rate between the plates per unit area assuming the gap between the plates is: (a) filled with atmospheric air (thermal conductivity 0.0219 W/m·°C), (b) evacuated condition, and (c) filled with a superinsulation that has a thermal conductivity of 0.00003 W/m·°C.

31. The walls of a house are filled with glass wool insulation. The wall is 5 inches thick. Assuming that the average temperature on the inside wall is 75°F and outside it is about 10°F. The total walls of the house are 15 feet high and 540 feet wide, and the thermal conductivity is 0.04 W/m·°C. What is the rate of heat transfer through this wall?

32. The inner and outer surfaces of a 0.5 cm thick 1.5 m by 2 m window glass in winter are 10°C and 3°C, respectively. If the thermal conductivity of the glass is 0.78 W/m·°C, determine the amount of heat loss, in kJ, through the glass over a period of 12 hours. What would the answer be if the glass were 1 cm thick?

33. Estimate the coefficient of performance of a refrigerator that consumes 1000 W of power to remove heat at a rate of 5.2 BTU per second.

34. Briefly describe the main advantage of a geothermal heat pump over the other heat pump types.

35. The flow rate of water that enters and leaves a cooling chamber is 1 kg/s. What is the heat rate that must be extracted from the water to reduce its temperature by 20°C?

36. A model building is represented by a rectangular shape of 4 m × 6.5 m × 18 m. If no heat is lost into the solid and with $U = 0.4$ W/(m²K), and an inside and outside temperature difference of 22 °C, what is the total heat loss in 24 hours?

37. A building has a U-value of 0.75 W/m²·K and a total exposed surface area of 400 m². The building is subjected to an external load (only sensible) of 2.1 kW and an internal load of 1.35 kW (sensible). The required internal temperature is 22°C, whether a cooling system is required or a heating system is required when the external temperature is 5°C. How will the results change, if the U-value of the building is reduced to 0.36 W/m²·K?

38. Why is it necessary to ventilate buildings? What is the effect of ventilation on energy consumption for heating in winter and for cooling in summer?

39. Briefly describe the negative effects of a poorly designed HVAC system.

40. List and briefly describe the main objectives of HVAC system design.

41. List and briefly discuss the main HVAC types.

42. A typical classroom that contains 28 people is to be air-conditioned using window air-conditioning units of 5 kW cooling capacity. A person at rest may be assumed to dissipate heat at a rate of 350 kJ/h. There are 12 light bulbs in the room, each with a rating of 100 W. The rate of heat transfer to the classroom through the walls and the windows is estimated to be 13,500 kJ/h. If the room air is to be maintained at a constant temperature of 21°C, determine the number of window air-conditioning units required.

43. List and briefly describe the HVAC processes and functions.

44. Briefly describe a central HVAC system. Do the same for HVAC zonal and multizone systems.

45. List the methods of industrial heating processes.

REFERENCES AND FURTHER READINGS

1. *ASHRAE Handbook Fundamentals*, ASHRAE Publishing, 2017, ISBN-10: 193920058X; ISBN-13: 978-1939200587

2. F. Bueche, *Introduction to Physics for Scientists and Engineers*, McGraw-Hill, 1975.

3. E. L. Harder, *Fundamentals of Energy Production*, Wiley, NY, 1982.

4. T. D. Eastop and D. R. Croft, *Energy Efficiency for Engineers and Technologists*, Longman, Harlow, UK, 1990.
5. A.W. Culp, Jr., *Principles of Energy Conversion* (2nd ed.), McGraw-Hill, New York, 1991.
6. A. F. Mills, *Heat Transfer* (2nd ed.), Prentice Hall, New Jersey, 1999.
7. J. F. Kreider (ed.), *Handbook of Heating, Ventilation and Air Conditioning*, CRC Press, 2001.
8. R. J. Ribando, *Heat Transfer Tools*. New York, McGraw-Hill, 2002.
9. F. Kreith and W. Z. Black, *Basic Heat Transfer*, Harper & Row, New York, 1980.
10. D. T. Allen and D. R. Shonnard, *Green Engineering*, Prentice Hall, Upper Saddle River, NJ, 2003.
11. F. P. Incropera and D. P. DeWitt, *Introduction to Heat Transfer*, Wiley, New York, 2002.
12. J. P. Holman, *Heat Transfer* (10th ed.), New York: McGraw-Hill, 2010.
13. R. A. Ristinen and J. J. Kraushaar, *Energy and Environment*, Wiley, Hoboken, NJ, 2006.
14. E. L. McFarland, J. L. Hunt, and J. L. Campbell, *Energy, Physics and the Environment* (3rd ed.), Cengage Learning, 2007.
15. F. J. Trost and I. Choudhury, *Design of Mechanical and Electrical Systems in Buildings*, Pearson, 2004.
16. A. Von Mayer, *Electric Power Systems – A Conceptual Introduction*, Wiley-IEEE Press, 2006.
17. D. R. Patrick and S. W. Fardo, *Electrical Distribution Systems* (2nd ed.), CRC Press, 2008.
18. F. Kreith and D. Y. Goswami (eds.), *Energy Management and Conservation Handbook*, CRC Press, 2008.
19. M. A. El-Sharkawi, *Electric Energy: An Introduction* (2nd ed.), CRC Press, 2009.
20. A. Sumper and A. Baggini, *Electrical Energy Efficiency – Technology and Applications*, Wiley, 2012.
21. G. Petrecca, *Energy Conversion and Management – Principles and Applications*, Springer, 2014.
22. R. Fehr, *Industrial Power Distribution* (2nd ed.), Wiley, 2015.
23. A. C. Metaxas, *Foundations of Electro-heat*, John Wiley & Sons Ltd., 1996.
24. D. Comsa, *Industrial Electro-heating Installations*, Editura Tehnica, Bucharest, Romania, 1986.
25. M. Ungureanu, M. Chindris, and I. Lungu, *End Use of Electricity*, EDP Press, Bucharest, Romania, 2000.
26. B. Stein and J. S. Reynolds, *Mechanical and Electrical Equipment for Buildings*, Wiley, NY, 2000.
27. R. R. Janis and W. K. Y. Tao, *Mechanical and Electrical Systems in Buildings* (5th ed.), Pearson, 2013.
28. B. Everett and G. Boyle, *Energy Systems and Sustainability: Power for a Sustainable Future* (2nd ed.), Oxford University Press, 2012.
29. W. T. Grondzik, and A. G. Kwok, *Mechanical and Electrical Equipment for Buildings* (12th ed.), Wiley, 2015.
30. W. E. Glassley, *Geothermal Energy: Renewable Energy and the Environment* (2nd ed.), CRC Press, 2015.
31. J. F. Kreider (ed.), *Handbook of Heating, Ventilation, and Air Conditioning*, CRC Press, Boca Raton, FL, 2000.
32. R. McDowall, *Fundamentals of HVAC Systems*, McGraw-Hill, 2006.
33. R. W. Haines and M. E. Myers, *HVAC Systems Design Handbook* (5th ed.), McGraw-Hill, 2010.
34. M. W. Earley (ed.), *NFPA 70, National Electrical Code (NEC) Handbook*, 2014.
35. R. Belu, *Industrial Power Systems with Distributed and Embedded Generation*, The IET Press, 2018.

9 Distributed Generation and Energy Storage

9.1 INTRODUCTION TO DISTRIBUTED GENERATION

Energy sustainability is the cornerstone to the health and competitiveness of the industries in today's global economy. It is more than being environmentally responsible, meaning the ability to utilize and optimize multiple sources of secure and affordable energy for enterprises, and then continuously improve the utilization through systems analysis, energy diversification, conservation, and intelligent use of these resources. Distributed generation (DG) is not a new phenomenon and application. Prior to the advent of alternating current (AC) and large-scale steam turbines, during the beginnings of the power industry, most, if not all, energy requirements (e.g., heating, cooling, lighting, motive power) were supplied at or near their point of use. There are several definitions of distributed generation, and in fact, there is not clear consensus as to what constitutes distributed generation. There is not a single standard definition nor an agreement as to what can be included in distributed generation and distributed energy resource concepts. One common definition is that a DG system is a generating unit serving a customer on-site or providing support to a distribution network, connected to the grid at distribution level. DG usually refers to small-scale systems that generate electricity and often heat close to the point where the energy is actually used. Distributed energy resources (DER) and dispersed or distributed generation systems are becoming more important to the electricity generation mix. DG is typically viewed as small-scale generation that is used on-site and/or connected to power and thermal distribution networks, installed usually at strategic points of the electric power system or locations of load centers. DG can be used in an isolated way, to supply consumer power demands, or integrated into the grid. DG technologies can run on renewable energy resources, fossil fuels, or waste heat. Equipment ranges in size from less than a kilowatt to tens of megawatts, in order to meet all or part of customers' power needs. Historically, the type of technologies employed has varied, but these are generally limited to small power units. Recently, renewable resources such as solar photovoltaic (PV) systems, small power-hydro units, and wind energy conversion systems are included in the DG mix. However, small-scale, fossil-fired generation is still seen as primarily providing reliable backup power in the event of grid interruptions. A grid-connected device for electricity storage can also be classified as a DER system, often being called a distributed energy storage system (DESS). By means of an interface, DER systems can be managed and coordinated within a smart grid. DESS units can enable the collection of energy from several sources and may lower environmental impacts and improve security of supply. With distributed generation in the building sector, referred to the on-site generation, often electricity from renewable energy systems is widely available today.

DG technologies include, among others, diesel and Stirling engines, microturbines, fuel cells, PV, and wind energy systems. The International Energy Agency (IEA) considers that *dispersed generation* includes distributed generation plus wind power and other generation, either connected to a distribution network or completely independent of the grid. In addition to providing a definition for distributed generation, the IEA has also provided nomenclature for other dispersed, distributed, or decentralized energy resources. For example, *distributed power* includes *distributed generation* combined with *energy storage technologies* such as flywheels, fuel cells, or compressed air storage. *Distributed energy resources* include *DG units* plus demand-side functionalities, while *decentralized power* refers to a system of distributed energy resources connected to a power distribution network.

Main DG features include: not centrally planned and mostly operated by independent power producers or consumers, not centrally dispatched (although virtual power plants, where DG units are operated as one single unit, infringe on this definition), smaller than 50 MW (although some sources consider certain systems up to 300 MW to be classed as DG), connected to the electricity distribution network, which usually refers to the part of the network that has an operating voltage of 240/400 V up to 110 kV. Notice that most renewable energy systems are also distributed generation systems, although large-scale hydro, offshore wind parks, and co-combustion of biomass in fossil-fuel power plants are exceptions. Combined heat and power generation (CHP), also referred to as cogeneration, indicates the joint generation and use of electricity and heat. Usually, a part of the electricity is used locally and the remainder is fed into the grid. The heat, on the other hand, is always used locally, as heat transport is costly and involves relatively large losses. Generally, distributed generation based on fossil fuels is also cogeneration as the local use of the *waste heat* is an important benefit of DG. Typical uses of DG are:

1. Domestic (micro-generation: electricity and heat);
2. Commercial (building related: electricity and heat);
3. Greenhouses (process related: electricity, heat, and carbon dioxide for crop fertilization);
4. Industrial (process related: electricity and steam);
5. District heating (building related: electricity and heat through heat distribution grid); and
6. Grid power (only electricity to the grid).

The main reasons why central, rather than distributed, electricity generation still dominates current electricity production are the economy of scale, efficiency, fuel capability, lifetime, system resilience, and robustness. Increasing the generation unit size increases the efficiency and decreases the cost per MW. Even where a large power plant is based on several smaller units of the same size, the facility cost per MW will be lower. However, the advantage of economy of scale is decreasing; small units are benefiting from continuing technological developments, while large units are already fully developed. Fuel capability is another reason to keep building large power plants. Coal, especially, is not economically suitable for DG, but it is the most abundant fossil fuel with steady suppliers all over the world and a stable price, if compared to oil and natural gas prices. Additionally, with lifetimes up to 50 years, large power plants will remain the prime electricity source for many years to come. Another reason is the efficient use of the heat that is always generated when electricity is generated, increasing the overall plant efficiency. As heat must be used locally, the need for distributed generation to be close to the demand point is obvious. Other DG benefits include improved security of supply, avoidance of overcapacity, peak load reduction, reduction of grid losses, and network-related benefits, such as the distribution network infrastructure cost deferral, power quality support, and reliability improvement. Disadvantages of DG, beyond those mentioned earlier, are the costs of connection, operation, metering, and balancing. This chapter presents an overview of the key issues concerning the structure, configurations, integration of distributed and dispersed generation systems, role of thermal energy storage systems, and applications. A synopsis of the main challenges and issues that must be overcome in the process of DG and DER applications and integration is presented. DER systems typically use renewable energy sources, including small hydro, biomass, biogas, solar power, wind power, and geothermal power, playing an increasingly important role in electric power distribution. A detailed presentation and descriptions of the distributed generation, cogeneration, and distributed energy resource and types, characteristics, and performances are the objectives of this chapter.

9.1.1 DG and DER Reasons, Benefits, and Technologies

Due to the large variations in the definitions used in the literature, the following different issues have to be discussed to define distributed generation more precisely: the purpose, location, technology, ratings, power delivery area, environmental impact, mode of operation, ownership, and DG penetration level. DG can be applied in many ways, and some examples are listed below:

1. It may be more economic than running a power line to remote locations.
2. It provides primary power, with the utility providing backup and supplemental power.
3. It can provide backup power during utility system outages, for critical facilities.
4. For cogeneration, where waste heat can be used for heating, cooling, or steam. Traditional uses include large industrial facilities with high steam and power demands, such as universities and hospitals.
5. It can provide higher power quality for electronic equipment.
6. For reactive supply and voltage control, by injecting or absorbing reactive power, it controls the voltage.
7. For network stability in using fast-response equipment to maintain a secure transmission system.
8. For system black-start to start generation and to restore a portion of the utility system without outside support after a system collapse.

No single DG technology can accurately represent the full range of capabilities and applications or the scope of benefits and costs associated with DG and DER applications. Some of these technologies are used for many years, especially reciprocating engines and gas turbines. Others, such as fuel cells and microturbines, are relatively new technological developments. Several DG technologies are now commercially available, and some are expected to be introduced or substantially improved within the next few years. *Diesel* and *gas reciprocating engines* are well-established commercial DG technologies. Industrial-sized diesel engines can achieve fuel efficiencies in excess of 40% and are relatively low-cost per kilowatt. While nearly half of the capacity is ordered for standby use, the demand for units for continuous or peak use has also been increasing. The technology is also suitable for backup generation, as it can be started up quickly and without the need for grid-supplied power. *Gas turbines*, originally developed for jet engines, are now widely used in the power industry and for micro-CHP applications. The electric utility industry uses simple-cycle gas turbines as units to serve peak load and they generally tend to be larger in size. Such turbines have the same operating characteristics as reciprocating engines in terms of start-up and the ability to start independently of grid-supplied power, making them suitable as well for backup power needs. This technology is also often run in combined heat and power applications, which can increase overall thermal efficiency. Capital costs are on par with natural gas engines with a similar operating and levelized cost profile. Small industrial gas turbines in the range of 1 MW to 20 MW are commonly used in combined heat and power applications. They are particularly useful when higher-temperature steam is required than can be produced by a reciprocating engine. The maintenance cost is slightly lower than for reciprocating engines, but so is the electrical conversion efficiency. Gas turbines can be noisy. Emissions are somewhat lower than for engines, and cost-effective NOx emission-control technology is commercially available. *Microturbines* extend the gas turbine technology to units of smaller size. A microturbine (MT) is a Brayton cycle engine using atmospheric air and natural gas fuel to produce power. The technology was originally developed for transportation applications, but is now finding a place in power generation. One of the most striking technical characteristics of microturbines is their extremely high rotational speed, up to 120,000 rpm. Individual units usually range from 30 kW to 200 kW but can be combined into systems of multiple units. This technology takes simple-cycle gas technology and scales it down to capacities of 50 kW to 100 kW. The installed costs are greater than for gas turbines, and the efficiencies are lower; however, it is much quieter than a gas turbine with a much lower emissions profile than a gas turbine. The possibility also exists for microturbines to be used in CHP applications to improve overall thermal efficiencies. Low combustion temperatures can assure very low NOx emission levels, making less noise than a comparable size engine. Natural gas is expected to be the most common fuel, but flare gas, landfill gas, or biogas can also be used. The key barriers to MT usage include maintenance cost, questionable part load efficiency (manufacturer data varies), limited field experience, use of air bearings is desirable to reduce maintenance but air-filtration requirements are stringent, and high-frequency noise that is produced but is relatively easy to control. *Fuel cells* are compact, quiet DC power generators that use hydrogen and oxygen to make electricity. The transportation sector is the major potential market for fuel cells, and car manufacturers are making substantial investments in research and development. Power generation, however, is seen as a market in which fuel cells could be commercialized much more quickly. Fuel cells can convert fuels to electricity at very high efficiencies (up to 60%), compared with conventional technologies. As there is no combustion, other noxious emissions are low. Fuel cells can operate with very high reliability and so could supplement or replace grid-based electricity. Only one fuel cell technology for power plants, a phosphoric acid fuel cell (PAFC), is currently commercially available. Three other types of fuel cells, namely molten carbonate (MCFC), proton exchange membrane (PEMFC), and solid oxide (SOFC), are the focus of intensive research and development. Fuel cell technology is fairly new and can run at electrical efficiencies comparable to other mature technologies. Fuels cells have the highest capital cost among fossil-fired technologies and consequently have the highest levelized costs.

There are four major types of renewable energy technologies discussed here, with good building DG and DER potential: solar photovoltaic (PV), geothermal energy, bioenergy, and wind energy. Each of these technologies is intermittent in that it is dependent upon the sun, river flows, or wind. Consequently, these technologies are not suitable for backup power, but also have no fuel costs and have a zero emissions profile. However, the capital costs vary significantly among the technologies and operating conditions over the year affect their respective levelized costs. *Photovoltaic systems* are a capital-intensive, renewable technology with very low operating costs. Photovoltaic (PV) cells directly convert sufficiently energetic photons in sunlight to electricity. Because sunlight is a diffuse resource, large array areas are needed to produce significant power. However, offsetting this is the zero cost of the fuel itself. Today, there is a PV market worldwide on the order of 100 MW per year. They generate no heat and are inherently small-scale. These characteristics suggest that photovoltaic systems are best-suited to household or small commercial applications, where power prices on the grid are highest. The key barriers to PV usage include: the price of delivered power exceeds other DG resources; subsidies exist in some states that make PV-produced power competitive, and temporal match of power produced to load is imperfect; and batteries or other systems are often needed. Wind energy generation is rapidly gaining a share in electricity supply worldwide. Wind power is sometimes considered to be DG, because the size and location of some wind farms make it suitable for connection at distribution voltages. Different from PV, which changes solar energy to electricity directly, a solar thermal electric system first collects and concentrates the solar energy using a receiver. The receiver contains a fluid, such as molten salt, that circulates in the receiver and heats an engine, such as a Stirling or Brayton cycle machine, to produce power. Solar thermal electricity involves two pieces: a concentrator and an engine/receiver. Different concentrators have been developed to collect the solar energy in different applications. Operating costs are very low, for both PV and solar-thermal systems, as there are no fueling costs. Geothermal energy is defined as a natural heat flow from the earth. Geothermal resources are being exploited around the world in a variety of applications and resources. A geothermal system consists of a heat source, permeable rock, and an inflow of water. The heat source is mostly an intrusion of magma that is close to the earth's surface. When the water inflow is heated through the heat source, the hot water or steam can be trapped in the permeable and porous rocks, forming a geothermal reservoir. Geothermal reservoir temperature normally increases with an increase in depth into the earth's crust. Geothermal heat is used directly, without involving a power plant or a heat pump, for a variety of applications such as space heating and cooling, food preparation, hot spring bathing and spas (e.g., balneology), agriculture, aquaculture, greenhouses, and industrial processes. Geothermal heat pumps (GHP) take advantage of the earth's relatively constant temperature at depths of about 10 ft. to 300 ft. for providing thermal energy to buildings and industrial processes. GHPs can be used almost everywhere in the world, as they do not share the requirements of fractured rock and water, as are needed for a conventional geothermal reservoir. GHPs circulate water or other liquids through pipes buried in a continuous loop, either horizontally or vertically, under a landscaped area, parking lot, or any number of areas around the building.

Distributed generation can provide a multitude of services to both utilities and consumers, including standby generation, peak shaving capability, peak sharing, base-load generation, or combined heat and power that provide for the thermal and electrical loads of a given site. Less well-understood benefits include ancillary services, reactive power support, voltage support, network stability, black start, spinning reserve, and others—which may ultimately be of more economic benefit than simple energy for the intended load. DG technologies can have environmental benefits ranging from truly green power (i.e., photovoltaics) to significant mitigation of one or more pollutants often associated with coal-fired generation. Natural gas-fired DG turbine generators, for example, release less than one-quarter of the emissions of sulfur dioxide (SO_2), less than 1/100th of the nitrogen oxides (NO_x), and 40% less carbon dioxide (CO_2) than many new coal-boiler power plants. One of the potential benefits is to operate DG in conjunction with combined heat and power (CHP) applications, which improves overall thermal efficiency. On a stand-alone electricity basis, DG is most often used as backup power for reliability purposes, but can also defer investment in the transmission and distribution network, avoid network charges, reduce line losses, defer the construction of large generation facilities, displace more expensive grid-supplied power, provide additional sources of supply in markets, and provide environmental benefits.

9.1.2   Energy Conservation and Efficiency in Buildings

Energy efficiency, energy conservation, and renewable energy technologies are the pillars of a sustainable energy policy. There several reasons to pursue energy conservation and efficiency and renewable energy use, such as increasing and diversifying the energy supply, making the existing energy supply last longer, or having a cleaner environment. Greater energy efficiency and higher levels of energy conservation are critical sustainable development issues. For example, increased energy efficiency and energy conservation provide to any country significant economic, environmental, and energy security benefits. Higher energy efficiency translates directly into lower costs of energy, products, and services, and higher life standards. Energy efficiency and conservation methods can also prolong equipment life. The two concepts, closely related, have different connotations. Efficiency involves using less energy to achieve the same results, while conservation means using less energy even if there may be a compromise on the results. An energy balance is a set of relationships accounting for all the energy produced and consumed, matching the system inputs and outputs, over a given time period. The system can be anything from a whole country or a specific area, to a factory process. An energy balance is usually made with reference to a year, though it can also be made for consecutive years to show variations over time. Energy balances provide overviews; they are basic energy-planning tools for analyzing current and projected energy use. The overviews aid resource management, indicating options for energy saving, or for policies of energy pricing and redistribution, etc. An energy-conversion device or equipment, as represented in Figure 9.1, has the efficiency defined as the ratio of the energy output to the energy input:

$$\text{Equipment (Device) Efficiency} = \frac{\text{Useful Energy Output}}{\text{Energy Input}} \qquad (9.1)$$

The meaning of the word *useful* depends strongly on the device or equipment purpose. Energy efficiency of a device, equipment, or process is a quantitatively dimensional value between 0 and 1, the larger it is the higher the efficiency. The concept of efficiency embodies the laws of thermodynamics. On the other hand, even though almost all of the energy on our planet comes directly from the sun, we are not yet able to harness large-scale solar energy directly and efficiently. Instead we have,

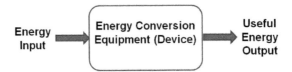

FIGURE 9.1 Schematic diagram of energy conversion equipment (device).

in a large measure, to rely on the chemical energy of fossil fuels for most of our energy needs. The problem with chemical energy is that it is a potential energy form and must be converted to other energy forms before it can be used. The only way to exploit the stored solar energy is to release it by burning fossil fuels, through combustion processes and eventually convert it into mechanical, electrical, or other energy forms. The conversion processes of the stored energy involve two or more conversion devices, each with their own efficiency, so it is useful to introduce *system efficiency*. A system, consisting of two or more devices or subsystems, means a well-defined space in which two or usually more than two energy conversions take place. The efficiency of a system is equal to the product of efficiencies of the individual devices or subsystems, and the concept is shown in the following example.

Example 9.1

Calculate the overall efficiency of a power plant if the efficiencies of the boiler, steam turbine, generator, and step-up transformer are 90%, 42%, 96%, and 98%, respectively.

SOLUTION

The overall (system) efficiency of the power plant is the product of the conversion subsystem efficiencies.

$$\eta_{Power\ Plant} = \eta_{Boiler} \times \eta_{Turbine} \times \eta_{Generator} \times \eta_{Transformer} = 0.90 \times 0.42 \times 0.96 \times 0.98$$
$$= 0.3556\ or\ 35.56\%$$

Note that the system efficiency is lower than any one of the efficiencies of the individual components of the system. In the case of this electric power plant, only 35% of the chemical energy input is converted to electricity, while the rest is lost to the environment, mostly as heat. However, if part of the wasted heat is recovered, via CHP methods the overall power plant (system) efficiency can be significantly improved.

The building sector accounts for about 40% of the total energy used in the United States, European Union, Canada, and most other developed countries. However, at the same time, the building sector has a documented cost-effective savings potential of up to 80%, which can be effected over the next 40 years. In order to ensure energy conservation and savings, while at the same time using renewable energy in an optimal way, building-integrated energy technologies are required. Energy demands in the commercial, industrial, residential, and utility sectors vary on a daily, weekly, and seasonal basis. These demands are matched by various energy-conversion systems, operating synergistically. Peak hours are the most difficult and expensive to supply, being met by conventional gas turbines or diesel generators, which are reliant on costly oil or natural gas. Energy storage provides alternative ways to supply peak energy demands or to improve the operation of DG, solar, and wind facilities. Energy storage plays significant roles in energy conservation. In processes where there is a large recoverable wasted energy, the storage can result in savings of premium fuels. Thermal energy storage (TES) is one of the key technologies for energy conservation, being of great importance in energy sustainability, and a well-suited technology for hot water, heating, and cooling thermal applications, which can be easy embedded into distributed generation, contributing significantly to overall system efficiency improvement. TES deals with the stored energy by cooling, heating, melting, solidifying, or vaporizing a material, while the thermal energy becomes available when the process is reversed. Large TES systems are employed for applications ranging from solar hot water storage to building air-conditioning systems. However, advanced TES technologies have only recently been developed to a point that they have significant impacts on modern industries. TES appears to be an important solution to correcting the mismatch between energy supply and demand. TES can contribute significantly to meet the needs for more efficient, environmentally benign energy use. TES is a key component of many thermal systems, and a good TES should allow minimal thermal losses, leading to energy savings, while permitting the highest reasonable extraction efficiency of the stored thermal energy.

Thermal energy storage is a key component of many successful building and industrial energy systems, minimizing thermal energy losses, enhancing energy savings, and permitting the highest appropriate extraction efficiency of the stored energy. The design and selection criteria for TES systems are examined. Further, energy-saving techniques and applications are discussed and highlighted with illustrative examples. TES is considered by many to be an *advanced energy technology*, and there has been increasing interest in using this essential technology for thermal applications such as hot water, space heating, cooling, air-conditioning, and so on. TES systems have enormous potential for permitting more effective use of thermal energy equipment and for facilitating large-scale energy substitutions. The resulting benefits of such actions are especially significant from an economic perspective. In general, a coordinated set of actions has to be taken in several energy-system sectors for the maximum TES potential benefits to be realized. TES appears to be the best means of correcting the mismatch that often occurs between the supply and demand of thermal energy. The first step of a TES project is to determine the energy load profile of the building. Parameters influencing the building demand and load profile are the building use, internal loads, and climatic conditions. The following steps are to determine the type and amount of storage appropriate for the particular application, the effect of storage on system performance, reliability, and costs, and the storage systems or designs available. It is also useful to characterize the TES types in relation to their storage duration. Short-term storage is used to address peak loads lasting from a few hours to a day in order to reduce the system size, taking advantage of the energy-tariff daily structures. Long-term storage is used when waste heat or seasonal energy loads can be transferred with a delay of a few weeks to several months. Related to the energy storage amount required, it is important to avoid both undersized and oversized systems. Undersizing the systems results in poor performance levels, while oversizing results in higher initial costs and energy waste if more energy is stored than is required. The effect of TES on the overall energy system performance should be evaluated in detail. The economic justification for storage systems requires that the annualized capital and operating costs for TES be less than those required for primary generating equipment supplying the same service loads and periods. TES systems are an important element of many energy-saving programs in a variety of sectors: residential, commercial, industrial, utility, and transportation. TES can be employed to reduce energy consumption or to transfer an energy load from one period to another. The consumption reduction can be achieved by storing excess thermal energy that would normally be released as waste, such as heat produced by equipment, appliances, lighting, and even by occupants. Energy-load transfer is achieved by storing energy at a given time for later use, and can be applied for either heating or cooling capacity. The consumption of purchased energy can be reduced by storing waste or surplus thermal energy available at certain times for use at other times. The demand of purchased electrical energy can be reduced by storing electrically produced thermal energy during off-peak periods to meet the thermal loads that occur during high-demand periods. There has been an increasing interest in the reduction of peak demand or transfer of energy loads from high- to low-consumption periods. The use of TES can defer the need to purchase additional equipment for heating, cooling, or air-conditioning applications and reduce equipment sizing in new facilities. The relevant equipment is operated when thermal loads are low to charge the TES, and energy is withdrawn from storage to help meet the thermal loads that exceed equipment capacity.

Energy storage density, in terms of the energy per unit of volume or mass, is an important factor for optimizing solar ratio (how much solar radiation is useful for the heating/cooling purposes), efficiency of appliances (solar thermal collectors and absorption chillers), and energy consumption for space heating or cooling room consumption. Therefore, the possibility of using phase-change materials (PCMs) in solar system applications should be investigated. PCMs might be able to increase the energy density of small-sized water storage tanks, reducing storage volume for a given solar fraction or increasing the solar fraction for a given available volume. It is possible to consider thermal storage on the hot and/or cold side of the building, allowing the hot water storage from the collectors and the auxiliary heater to be supplied to the generator of the absorption chiller (in cooling mode) or directly to the users (in heating mode). The latter allows the storage of cold water produced by the absorption chiller to be supplied to the cooling terminals inside the building. It is usual to identify three situations as "hot," "warm," and "cold" storage based on the different temperature ranges. Typically, a hot tank may work at 80°C to 90°C, a warm tank at 40°C to 50°C, and a cold tank at 7°C to15°C. While heat storage on the hot side of solar plants is always present because of heating and/or domestic hot water (DHW) production, cold storage is justified in larger plants. Cold storages are used not only to gain economic advantages from lower electricity costs (in the case of electric compression chillers) depending on the time of day but also to lower the cooling power installed and to allow more continuous operation of the chiller. The use of thermal storage, initially, could not provide effective backup but helped the system to thermally stabilize. Consequently, thermal storage found use in solar-assisted thermal systems. Since then, studying thermal energy storage technologies as well as the usability and effects of both sensible and latent heat storage in numerous applications increased, leading to a number of reviews. These reviews focused only on one side (cold or hot) or component of the system or one of its integral mechanisms.

9.2 COGENERATION AND MICRO-COMBINED HEAT AND POWER

The cogeneration principle or concept has long been known; a number of cogeneration units were already supplying heat and electricity to houses, commercial, and industrial facilities at the end of the nineteenth century. Power plants create large amounts of heat in the process of converting fuel into electricity. On average in a utility-sized power plant, about two-thirds of the energy content of the input fuel is converted to heat. Conventional power plants discard this waste heat, and by the time electricity reaches the average user, only about 30% of the energy remains. DG systems, due to their load-appropriate size and siting, enable the economic recovery of this heat. An end user can generate both thermal and electrical energy in a single CHP system located at or near its facility, with efficiencies exceeding 90%. Cogeneration, or combined heat and power production, is the process of producing both electricity and usable thermal energy, heat, or cooling at high efficiency and near the user. It thus has three defining elements: simultaneous electricity and heat production, a performance criterion of high overall efficiency, and a locational criterion concerning the proximity of the energy conversion unit to a customer site. While the discussion on micro-cogeneration, or micro-CHP, has only recently gained momentum, the technological roots of micro-cogeneration go back to the early development of steam and Stirling engines in the eighteenth and nineteenth century, respectively. Today, several technologies are capable of providing cogeneration services, such as reciprocating engines, gas turbines, Stirling engines, and fuel cells. But, in principle, the exhaust heat from any thermal power plant, such as gas combined-cycle power plants or coal power plants, can be used for cogeneration. Advances in the technology, as well as a general trend towards smaller unit sizes of power plants, have led to an increased interest in small units, with the hope of ultimately developing units that can provide electricity and heat for individual buildings. This is what we call micro-cogeneration, which we define as the simultaneous generation of heat, or cooling, energy and power in an individual building, based on small energy-conversion units below 50 kWe.

Combined heat and power systems have been used by energy-intensive industries (e.g., pulp and paper, petroleum, or food industries) to meet their steam and power needs for more than 100 years. They can be deployed in a wide variety of sizes and configurations for industrial, commercial, and residential users. CHP strategies can even be used with utility-sized generation, usually in conjunction with a district energy system. CHP systems can also involve nonelectric power production, or the electricity can be used only internally. CHP systems capture the heat energy from electric generation for a wide variety of thermal needs, including hot water, steam, and process heating or cooling. Whereas the heat produced is used for space and water heating inside the building, the electricity produced is used within the building or fed into the utility grid. CHP systems can be designed to use one or more types of fuel: fossil fuel, biomass, biofuels, and waste heat. Such systems are fuel cells, microturbines, combustion engines, steam-cycle turbines, Stirling engines, and gasification digesters. Successful development of a micro-CHP system for residential applications has the potential to provide significant benefits to users, customers, manufacturers, and suppliers of such systems and, in general, to the nation as a whole. The benefits to the ultimate user are a comfortable and healthy home environment at an affordable cost, potential utility savings, and a reliable supply of energy. Manufacturers, component suppliers, and system integrators will see growth of a new market segment for integrated energy products. The benefits to the nation include significantly increased energy efficiency, reduced consumption of fossil fuels, and pollutant and CO_2 emissions from power generation, enhanced security from power interruptions, as well as enhanced economic activity and job creation. An integrated micro-CHP energy system provides advantages over conventional power generation, because the energy is used more efficiently by means of efficient heat recovery. In summary, integrated micro-CHP system solutions represent an opportunity to address all of the following requirements at once: conservation of scarce energy resources, reducing the pollutant emissions into our environment, increasing the overall industrial processes, and assuring comfort for home owners. The micro-CHP technological core is an energy-conversion unit that allows the simultaneous production of electricity and heat in very small size units. In addition to this core, further technology components are involved in micro-cogeneration systems, such as advanced grid interfaces, including possible metering and control devices.

A conversion technology serves to convert chemical energy of a fuel into *useful energy forms*, i.e., electricity and heat. A number of different conversion technologies have been developed that have residential CHP applications. The conversion process can be based on combustion and subsequent conversion of heat into mechanical energy, which then drives a generator for electricity production (e.g., reciprocating engines, Stirling engines, gas turbines, or steam turbines). Alternatively, it can be based on direct electrochemical conversion from chemical energy to electrical energy (i.e., fuel cells). Other processes include photovoltaic conversion of radiation (e.g., thermo-photovoltaic devices) or thermoelectric systems. In principle, most conventional cogeneration systems can be downscaled for micro-cogeneration applications. However, some of them have yet to be successfully implemented for very small applications. Quantitative performance and costs of a broad range of existing technologies allow a detailed techno-economic assessment of potential CHP system solutions from a technology-neutral perspective, existing under-development or envisioned technologies. A prime mover in a micro-CHP system is the equipment used to generate electricity and the waste heat is recovered downstream. The prime mover characteristics, including performance and cost, are discussed in detail here. The prime movers to be evaluated include micro-steam and gas turbines, Stirling engines, reciprocating engines, and various fuel cell types. Fuel cells are discussed in the energy storage section of this chapter.

Reciprocating engines are based on conventional piston-driven internal combustion engines. The two types of reciprocating engines are gasoline engines running on the Otto cycle and diesel engines running on the diesel cycle. A gasoline engine is ignited using an electric spark. It has relatively low combustion temperature, thus lower efficiency. However, it is quieter than a diesel engine. A diesel engine is ignited when the diesel mist/air mixture reaches a high temperature when the mixture is compressed in the cylinder. Because of the higher compression ratio, diesel engines usually have higher combustion temperature and higher efficiency. They are usually heavier than gasoline engines for the same power, and traditionally are used in heavy-duty applications such as trucks or ships, although some small passenger cars do use a diesel engine. For micro-CHP applications, typically, spark-ignition engines are used. In an Otto engine, a fuel, for instance natural gas, is mixed with air and compressed in a cylinder, ignited by an externally supplied spark. The mechanical energy produced by this combustion is then used to drive a generator. The exhaust heat as well as the heat from the lubricating air cooler and the jacket water cooler of the engine are recovered using heat exchangers, and then supplied to the heating system. The electrical efficiency of reciprocating engines, defined as net electrical energy output divided by the energy of the natural gas input, depends strongly on the electrical capacity of the system. Thermal efficiency depends on the system and its level of heat integration. With combined electrical and thermal efficiency, the overall efficiency varies between 80% to above 90%. Similar to electrical efficiency, capital costs per kW_{el} depend on the electrical capacity of the system. IC engines for micro-CHP ideally take natural gas or heating oil as fuel. Gasoline-fired IC engines are close to natural gas engines, and heating oil fired engines are essentially diesel engines, because home heating oil is very close to diesel oil in composition. Reciprocating engines have several advantages for micro-CHP. First, they are a mature and well-understood technology. They can be designed for different fuels including gasoline, diesel, natural gas, or landfill gas. Large engines can run more than 20 years, although smaller engines have somewhat shorter lives. The efficiency of a reciprocating engine is around 25% to 45%, which is higher than steam engines and current Stirling engines. Also, reciprocating engines have a shorter startup time than external combustion engines. One of the limitations of reciprocating engines is the frequent maintenance.

Unlike spark-ignition engines, for which combustion takes place inside the engine, Stirling engines generate heat externally, in a separate combustion chamber. In the Stirling engine, developed in 1816 by Robert Stirling, a working gas (for instance helium or nitrogen) is, by means of a displacer piston, moved between a chamber with high temperature and a cooling chamber with very low temperature. On the way from the hot to the cold chamber, the gas moves through a regenerator, consisting of wire, ceramic mesh, or porous metal, which captures the heat of the hot gas and returns it to the gas as the cold gas moves back to the hot chamber. Stirling engines can be designed in different configurations, distinguished by the position and number of pistons and cylinders and by the drive methods (cinematic and free-piston). A Stirling cycle has two isothermal processes and two isochoric (constant-volume) processes. The two main types of Stirling engines are *displacer type* and *two-piston type*. The theoretical efficiency of the Stirling engine is equal to that of the Carnot engine, the highest possible of all heat engines. Stirling engines generally are small in size, ranging from 1 kW to 25 kW, although some can be up to 500 kW. Due to the fact that fuel combustion is carried out in a separate burner, Stirling engines offer high fuel flexibility, in particular with respect to biofuels, and, because of the continuous combustion, lower emissions. In principle, other heat sources, such as concentrated solar irradiation, can be used. Stirling engines have the potential to reach higher overall electrical, plus thermal, efficiencies. Their electrical efficiencies, however, are only moderate, up to 20% seasonal average. Small Stirling engines are designed for low cost; consequently, they achieve lower electrical efficiencies than larger units, typically around 10% to 12%. Durability and cost are two major shortcomings of current Stirling-engine technology. Their durability challenges include shaft seal, piston rings, and bearings leakage; minimization of material stress and corrosion in the high-temperature region; and problems with abrasive particles generated at the piston rings. Stirling engines are expected to have long life and low maintenance.

The Rankine cycle is a thermodynamic cycle that is used to generate electricity in major power stations. In this cycle, superheated, high-pressure steam is generated in a boiler, and then expanded in a steam turbine, driving a generator to convert the work into electricity. The remaining steam has low pressure and is then condensed. Condensed fluid is recycled to the boiler. The water-steam Rankine cycle is mainly used in large scale (up to hundreds of MW scale) because Rankine cycle efficiency is roughly proportional to the working fluid temperature and pressure, and high fluid temperatures can be achieved in a boiler by external combustion of fuels. This type of cycle is called a topping power generation cycle. If the heating source temperature is relatively low, e.g., 400°C, steam cannot reach very high pressure, and organic working fluids can be used to achieve higher pressures and conversion efficiencies. Therefore the organic Rankine cycle (ORC) is more often used in a bottoming cycle, based on the fact that the cycle is at the bottom stage of an energy system to recover the waste heat. Although typical Rankine cycle systems are in the order of MW or above, some small-scale systems have capacity as low as 50 kWe, with some Rankine cycle engines in the order of 1 kWe to 10 kWe, developed for micro-CHP applications. The Rankine cycle for micro-CHP is less expensive than most other prime mover technologies and is likely to be a competitive prime mover technology. Rankine-cycle engines have a design life of 30 years. Small-scale Rankine-cycle engines are also a type of external combustion engine, which can use a conventional burner, easing the electric-thermal load balancing in micro-CHP. Much of the system can be sourced from the standard equipment list, thus saving equipment cost. The major disadvantage of Rankine-cycle engines is that the efficiency is low. Small-scale Rankine engines typically do not have super-heaters or reheaters as in large power plants. Therefore, the electric conversion efficiency of small engines is much lower, around 10%, down from about 40% in large power stations. However, this does not constitute a problem for micro-CHP because matching the thermal and electric loads for a typical residence can be achieved even if the micro-CHP prime mover does not operate at high efficiency.

Microturbines are small combustion turbines with an electrical capacity of 25 to 500 kW. They are designed mainly to meet the load for small or mid-size commercial customers as standby power, peak shaving, or cogeneration. The technologies used in microturbines are similar to large combustion turbines, except that they are less sophisticated and thus have relatively lower efficiency compared to larger turbines. Certain microturbines are equipped with heat recuperators, which lead to higher electric efficiency of around 20% to 30%. Microturbines without a recuperator have efficiencies of only about 15%.

For power-generation purposes or commercial CHP, it is desirable to have as high efficiency as possible. Current development work focuses on improving the mechanical design for higher efficiency and lower equipment cost. The microturbine equipment cost is from $700 to $1100/kW and the installation is another $500/kW roughly. The total microturbine cost is higher than the internal combustion engine cost. Current microturbine research efforts include the use of different fuels (such as landfill gas), heat recovery and cogeneration design, application to vehicles, and integrated fuel cell/microturbines. The electric capacity of current microturbines, usually 25 kW or above, is too high to be in a residential micro-CHP unit. For residential purposes, a prime mover should be of the order of 1 kW. Microturbines are favored because they have lower NOx emission compared to IC engines, due to lower combustion temperature. Also microturbines have smaller weight than IC engines of the same power. They have fast response, in the order of seconds, and, therefore are good candidates for backup power. Microturbines are compact in size and are easy to maintain. The major disadvantages of this technology include the high cost ($700 to $1100/kW), relatively shorter life (~10 years), and high operation and maintenance costs, which are around $0.005–0016/kWh. The efficiency of microturbines is not very high, although this is enough or more than enough for residential micro-CHP because of the high thermal-electric load ratio. Finally, microturbine power output decreases with elevation and temperature. This is not desirable for applications in certain locations and in hot weather.

Thermally activated technologies (TAT) typically refers to thermally activated cooling (adsorption and absorption chillers), dehumidification (solid and liquid desiccant systems), heating (heat recovery exchangers and boilers), or power generation (organic Rankine cycle bottoming devices). TAT devices can use the waste heat that leaves a prime mover in a CHP system so that the overall energy utilization is increased. An absorption chiller operates on the absorption cooling cycle, which involves a pair of working fluids: an absorbent and a refrigerant. There are two dominant working fluid pairs currently: lithium bromide (Li-Br) solution-water and ammonia-water or ammonia aqueous refrigeration (AAR). Li-Br-water has been used mainly for space cooling applications and cannot produce cooling loads below the freezing point. Ammonia-water can produce cooling loads with temperatures typically at –30°C. Therefore, AAR can be used for refrigeration purposes and heat pumps in winter. In an absorption cycle, refrigerant is separated from the absorbent when heated in the generator. The refrigerant vapor then condenses in the condenser, releasing the latent heat. The liquid refrigerant enters a low-pressure space after an expansion valve. It evaporates there by absorbing energy from the fluids to be cooled, thus creating useful cooling. The vapor is then absorbed by the condensed absorbent solution, which becomes diluted and then pumped back to the generator to complete the cycle. If there are a high-stage generator and a low-stage generator in the cycle, the system can be designed to generate more refrigerant vapor by utilizing part of the refrigerant latent heat that would otherwise be removed by the cooling water. The COP can then be improved dramatically, from 0.7 to about 1.3. Such a chiller is called a double-effect chiller. If there are three stages of generators, it is called a triple-effect chiller. Theoretically, the higher the number of stages, the higher the chiller COPs. Because the heating source fluid is not in direct contact with the chiller fluids, the absorption chillers can be heated with different heating sources without significantly changing their major structure. Chillers heated by hot water, steam, direct gas, or liquid fuel firing as well as waste heat have been developed. Absorption chillers can also provide heating when the cooling water to the chiller is disconnected. In that case, the heat is carried from the heating source in the generator to the evaporator so that the chiller serves as a boiler. Although it seems that the chiller components are not fully used in heating mode, this configuration saves a boiler, additional heating coils, and piping. Absorption chillers have the advantage that they are driven by heat, not electricity, reducing the peak power load in summer if the cooling would otherwise be provided by an electricity-driven vapor compression chiller. An absorption chiller does not have moving parts, other than small liquid pumps, consuming only a small electricity amount. It also is quieter than typical vapor compression chillers. In winter, the chiller can serve as a boiler thus saving partial heating equipment costs. It usually has a long life, around 15–25 years. As a thermally driven device, it can utilize the heat from a prime mover. The disadvantages of absorption chillers include: (1) They are roughly twice as expensive as a vapor compression chiller with the same capacity, (2) Absorption chillers use a large amount of working fluid and, therefore, have slow transient response. Chiller start-up and shutdown processes take much longer times than those for vapor compression chillers. (3) Absorption chillers need a cooling tower to operate, which is not desired by some owners of some buildings, such as supermarkets. It is also difficult for regions where water supply is an issue. For micro-CHP in a residential market, the absorption chiller must be very small in capacity, which is less cost-effective than larger chillers. An adsorption chiller is different from an absorption chiller in that the absorbent is not a liquid, but a solid substance that can adsorb the refrigerant vapor. Typical working adsorbent-absorbent pairs are silica-gel-water and methanol/carbon. In an adsorption chiller, there is no continuous adsorbent flow so that the chiller is designed to have two chambers and the chiller is to run intermittently. During operation, water is brought into the evaporator and evaporates, generating the useful cooling load in the cooling circuit. The evaporated water is adsorbed on the receiver, the silica-gel, and then the adsorbed water is de-adsorbed with the supply of thermal energy. The de-adsorbed water vapor is condensed in the condenser cooling cycle. After the cycle-time expiration, the machine switches over by means of valves.

Dehumidification can be realized in two ways: (1) use a vapor compression chiller, and (2) use a desiccant system in which the solid or liquid desiccant absorbs the water vapor. When the humid air passes through the cold coil, water vapor condenses so that the supply air is dehumidified. The two types of solid desiccant systems are: a passive system and an active system. The passive desiccant system utilizes the waste heat from the building exhaust air, while the active system utilizes a heater or, preferentially, waste heat from either a condenser or CHP system to regenerate the desiccant material. The simplest and most common design is a rotating desiccant-coated honeycomb wheel in which approximately 75% of the wheel is exposed to the air stream to be conditioned while the remaining 25% portion is used to regenerate the desiccant. The most commonly used material is typically silica gel. Other materials such as zeolites, molecular sieves, activated aluminas, and synthetic polymers have also been used. Liquid desiccant systems provide dehumidification by spraying the liquid into an incoming air stream to remove moisture. Both solid desiccant and liquid desiccant systems have been used for decades to control the humidity level in buildings. Solid desiccant systems offer several benefits to the building operator/owner. These benefits include reduced electrical energy consumption, reduced condensation and growth of molds, and downsizing of cooling equipment. Additional benefits resulting from downsized equipment include chiller reduced housing space and lower initial costs. Limitations include an added higher cost of the desiccant equipment and limited application potential depending on the building geographical location and conditioning needs, and potential desiccant lifetime degradation due to clogging from foreign particles in the air stream. Prevention requires the use of particle filters, which will increase the overall life cycle cost of the system. Liquid desiccants have several advantages over solid desiccants. First, liquid desiccants have a higher water-holding capacity than solid desiccants, thus providing deeper drying capabilities. Second, further improvement in indoor air quality (other than moisture removal inhibiting mold growth) is accomplished through the disinfecting attributes of liquid desiccants. Another advantage is the fact that the liquid desiccants have the capability of being used as a potential energy-storage media. On the other hand, there are several limitations to this technology. Specifically, there is a higher complexity factor (i.e., more components than a solid desiccant system), resulting in higher initial costs and reduced reliability confidence. Also, because liquid desiccants are usually toxic and corrosive, carryover concerns arise. Carryover occurs when small droplets of the liquid desiccant are entrained with the process air. As a result, human health issues arise as well as the possibility of increased maintenance costs due to corroding HVAC ducting. Another limitation of liquid desiccants relates to the potential desiccant lifetime degradation due to chemical reactions with, for example, sulfur compounds in the air stream.

In a house, a boiler usually provides domestic hot water for washing or space heating by burning liquid fuel or natural gas. Although this is one of the simplest and most primitive energy-conversion methods, i.e., from fuel chemical energy to heat, it is the most important and widely used method of energy utilization for a house, both in developing or developed countries. Often single-family houses use a furnace for space heating and a gas or electric boiler for domestic hot water. Some houses use a combined furnace for space heating and hot water. If a micro-CHP is to be installed, it is desirable to use the micro-CHP waste heat for both space heating and hot water. A heat recovery exchanger/boiler will be needed instead of a furnace/hot water heater. Here we assume the heat recovery exchanger/boiler is a flue gas-liquid water heat exchanger. If the house uses forced-air heating, air will be heated by the water coil. If a hydronic heating system is used, the hot water from the heat recovery equipment may be used directly. From a heat-transfer standpoint, there is no difference between a direct-fired boiler and a heat recovery boiler. The difference lies in that a heat recovery boiler usually is subject to lower gas temperature than a direct-fired one. In a direct-fired boiler, the flue gas temperature is high enough so that bare tubes can be used to provide enough heating load, although the efficiency of heat utilization and the cost of the boiler are lower. For waste heat recovery, depending on the temperature of the waste heat, fin tubes may be needed to recover enough heat. In micro-CHP applications, the heat recovery design should focus on the following aspects: (1) Integration with the prime mover or burner as well as related thermal/electric load balancing and control. For example, in a Stirling engine based micro-CHP system, the engine and the heating load both get the heat from the burner. The burner fuel flow rate needs to consider both the electric and heating load, which is more complex than ordinary boilers; (2) Extent of heat recovery from the flue gas and the effects of potential water vapor condensation. In case of condensation, both material corrosion and mold development would likely necessitate raising the gas, leaving temperature above the dew point. The benefit of heat recovery through heat exchangers is that the technology is simple. Typical houses have large heating and/or hot water loads so that a heat recovery exchanger fits well into the micro-CHP blueprint. If there is enough waste heat, extra hot water generated can be used for absorption/adsorption cooling. More importantly, the heat recovery exchanger cost, which is essentially a gas-liquid heat exchanger, is smaller compared to other technologies, such as heat-driven cooling or dehumidification. The only disadvantage appears to be the fact that this is a very common technology and there is less competitive advantage except for load control strategies.

9.3 ENERGY STORAGE SYSTEMS

Energy storage, designed to provide support for both long-term applications and dynamic performance enhancement, can provide better balancing between the electricity demand and the supply, allowing increased asset utilization, facilitating renewable energy penetration, and improving the flexibility, reliability, and overall efficiency of the grid. Almost all renewable energy sources are characterized by generation variability, intermittency and discontinuity. Their energy generation is not controlled by the system operator, making difficult their integration into power systems. *After completing this chapter, readers should understand the role, importance, configurations, and topologies of energy storage systems, operation principles, characteristics, performances, and operation of major energy storage systems used in power systems, buildings, and industrial facilities.* Another benefit is that readers will be able to understand the critical role and necessity of energy storage systems in power and renewable energy systems, the differences between large-, medium-, and small-scale energy storage systems, and how a system is selected on specific applications based on system characteristics and performances. Major energy storage technologies discussed in this chapter are: compressed air energy storage, pumped hydropower storage systems, batteries, flywheels, hydrogen energy storage, fuel cells, super-capacitors, and superconducting energy storage systems. Thermal energy storage systems are covered in detail in another book chapter. This chapter provides comprehensive reviews of energy storage technologies and gives an up-to-date comparative summary of their performances, characteristics, and applications. Over the past few decades, many innovative ideas have been explored in the energy storage areas, ranging in size, capacity, and design complexity. Some of them are designed for large-scale power system applications, others for small- or medium-scale renewable energy systems, while still others are designed for short-term energy storage ride-through capabilities for critical infrastructure (military, hospitals, or communication facilities). Energy storage has become an enabling technology for renewable energy applications and for enhancing power quality and power system stability, having a great potential to improve power grids and to provide alternatives to conventional energy generation. Some major constraints for increasing renewable energy penetration is their availability and intermittency, which can be addressed through energy storage. The energy storage system choice depends on specific requirements; usually several energy storage systems are used in order to increase capacity and improve supply security. The aim of energy storage is to provide the basis for development of new energy storage possibilities in buildings, industrial, and commercial facilities and in power distribution. Among the parameters used in comparisons of various energy storage technologies are efficiency, energy capacity and energy density, run time, costs, response time, lifetime in years and cycles, self-discharge rate, and technology maturity. The most common energy storage technologies for medium- and low-scale applications are batteries, fuel cells, flywheels, capacitors, and superconductive energy storage systems.

Energy storage technology has been in existence for a long time and has been utilized in many forms and applications, from a flashlight to spacecraft systems. Today energy storage is used to make electric power systems more reliable as well as make broader renewable energy use a reality. Energy storage systems (ESS) become critical in enabling renewable energy applications, providing the means to make non-dispatchable resources into dispatchable ones. In order to match power demand, in the renewable energy intermittent and non-dispatchable context and with fairly predictable electrical demand, the energy storage systems are critical, allowing decoupling of generation from consumption, reducing the needs for constant monitoring and energy demand predictions. Energy storage also provides economic benefits by allowing plant energy generation reduction to meet average demands rather than peak power demands. The current status shows that several drivers are emerging and are spurring growth in energy storage demands, such as: renewable energy growth, increasingly strained grid infrastructure as new lines lag well behind demands, or microgrid emergence as part of smart grid architecture, higher energy supply reliability and security, and improved and optimized building and industrial process energy use and efficiency. However, issues regarding the optimal integration (operational, technical, and market) of energy storage into the electric grid are still not fully developed, tested, and standardized. ESS integration and further energy-conversion unit development, including renewable energy, must be based on the existing electric infrastructure, requiring optimal integration of energy storage systems. Energy storage becomes a critical factor that can solve the problems described above. A renewable energy system with its corresponding energy storage system can behave as a conventional power plant, at least for time intervals in the order of half an hour up to a day, depending on the storage capacity. Electricity generated from renewable sources can rarely provide immediate response to demand as these sources do not deliver a regular supply easily adjustable to consumption needs. Thus, the growth of this decentralized production means greater network load stability problems and requires energy storage, as a potential solution. However, lead batteries cannot withstand high cycling rates, nor can they store large amounts of energy in a small volume. Energy storage is also a crucial element in energy management from renewable sources, allowing energy to be released into the grid during peak hours when it is more valuable. In the conventional integration and operation planning process of bulk power plants, normally a top-down strategy, coming from an energy-consumption point of view down to a stepwise detailed description is used. In this strategy, the planning horizon is subdivided into long-, medium-, and short-term planning tasks. In the distributed generation (DG) case, the technical boundaries are initially more important for the planning process to get an economical optimal supply configuration under stable operation conditions. When this is not taken into account, potential conflicts may arise between the energy supplier and the operator of electricity distribution networks, especially the ones based on renewable energy sources. Therefore, the need arises for a planning and integration strategy, which already includes clear specifications and definitions of the system inherent in forefront. This means, the process of system modeling and structuring for optimal integration of energy storage systems and distributed energy resources (DERs) must already include ESS and DER control functions, limits, and boundaries. These requirements lead to bottom-up strategy starting from distributed generation up to the centralized conventional electrical supply system. With a bottom-up strategy, clear definitions of supply and control functionality between ESS, DER, and the electrical grid can be derived.

There are several well-established electricity storage technologies, as well as a large number in development, offering significant application potential. Economically viable energy storage requires conversion of electricity or energy storage into other energy forms, which can be converted back to electricity when needed. All storage methods need to be feasible, efficient, and environmentally safe. Energy storage systems can be separated into four major classes: *mechanical, electrical, thermal,* and *chemical* based systems. Each class contains several technologies with specific characteristics and applications. Mechanical storage includes pumped hydropower storage, compressed air energy storage, and flywheels. Electrochemical storage includes batteries, hydrogen-based energy storage, and fuel cells. Here also is included thermochemical energy storage, such as solar-hydrogen, solar-metal techniques, etc., not discussed here. Electromagnetic energy storage includes super-capacitors and superconducting magnetic energy storage. Thermal energy storage includes two broad categories: low-temperature and high-temperature energy storage. Figure 9.2 summarizes the most common energy storage technologies. Despite the opportunity offered by energy storage to increase the stability and reliability of intermittent energy sources, there were only very few installed energy storage systems (batteries, pumped hydropower storage, compressed air energy storage, and thermal storage) with a capacity exceeding 100 MW. The opportunities for significant improvement in development of energy storage are strong, with the most appropriate applications to power quality management, load shifting, and energy management.

ESS integration and further developments must be based on the existing electric supply system infrastructure, requiring optimal integration of energy storage systems. Renewable energy systems

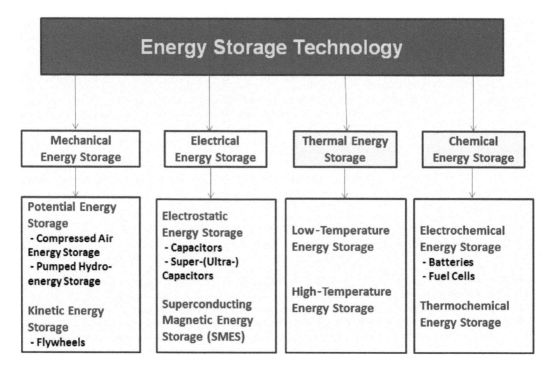

FIGURE 9.2 Classification of the major energy storage technologies.

with optimum energy storage can behave as conventional power plants, at least for short time intervals from a half-hour to a day, depending on the energy storage system capacity. Electricity generated from renewable energy sources often are not providing immediate response to energy demands, as these sources do not have capabilities to easily adjust to the consumption needs, being in this regard low-inertia systems. Significant growth of decentralized power production may result in network stability and other grid issues, making the energy storage as one of the potential solutions. Energy storage is also crucial in the energy management from renewable energy sources (e.g. wind or solar energy), allowing energy, stored during off-peak periods to be released into the grid during peak hours when it is more valuable. In the conventional integration and operation planning process of bulk power plants, usually a top-down strategy is used, which is, coming from an energy consumption point of view down to a stepwise detailed description. In this strategy, the planning horizon is usually subdivided into long-, medium- and short-term planning tasks. The time scale in each planning step is a compromise between optimum operation, the number of technical and economical boundaries. Planning strategy is usually driven by economic considerations in a unidirectional electrical power supply chain. In these planning strategies, the detailed control functionality can be figured out only when the system model in the planning approach is detailed enough. This means the detailed accuracy of the model inside of each planning stage (long-, medium-, short-term, quasi-stationary, or dynamic) defines the optimal layout of those conventional supply systems. In the distributed generation (DG) case, the technical boundaries are important for the planning process to get an economical optimal supply under stable operation conditions. When this is not taken into account, a conflict potential between the energy supply and the operation of electricity networks, especially the ones based on renewable energy sources may exist. Therefore, needs arise for a planning and integration strategy that includes clear system specifications and definitions, meaning a system structure for optimal integration of energy storage and distributed energy resources (DERs), which include ESS and DER control functions, limits, and boundaries. These requirements lead to a bottom-up strategy starting from distributed generation up to the centralized conventional electrical supply system, from which a clear definition of supply and control functionality between ESS, DER, and the electrical grid is derived.

9.3.1 Energy Storage Functions and Applications

Electrical energy can be stored in kinetic, potential, electrochemical, or electromagnetic energy forms, which can be transferred back into electrical energy when required or needed. The conversion of electrical energy to different forms and back to electrical energy is done by conversion systems. Energy generated during off-peak periods can be stored and used to meet the loads during peak periods when energy is more expensive, improving the power system economics. Compared to conventional generators, the storage systems have a faster ramping rate to respond to load fluctuations. Therefore, ESSs can be a perfect spinning reserve, providing a fast load following and reducing the need for spinning reserves from conventional generation plants. ESSs were initially used only for load-leveling applications. Today, they are seen as tools to improve the power quality and stability, to ensure a reliable and secure power supply, and to black start the power system. Breakthroughs that dramatically reduce ESS costs could drive revolutionary changes in power system design and operation. Peak-load problems could be reduced, stability could be improved, and power quality disturbances could be reduced or even eliminated. Storage can be applied at the power plant level, in support of the transmission system, at various distribution system points, and on particular appliances and equipment on the customer side. Figure 12.2 shows how the new electricity value chain is changing, supported by the integration of energy storage systems. ESSs in combination with advanced power electronics can have a great technical role and can lead to many benefits. ESSs by their nature are suitable for particular applications, primarily due to their potential storage capacities. Therefore, in order to provide a fair comparison of various energy storage technologies, they are grouped together based on the size of power and storage capacity. The four categories are: devices with large power (>50 MW) and storage (>100 MWh) capacities; devices with medium power (1–50 MW) and storage (5–100 MWh) capacities; devices with medium power or storage capacities but not both; and finally, small-scale energy storage systems, with power less than 1 MW and capacity less than 5 MWh. Storage systems such as pumped-hydroelectric energy storage (PHES) have been in use since 1929 primarily for daily load leveling. As the electricity sector is undergoing significant changes, energy storage is a realistic option for: (1) electricity market restructuring; (2) integrating renewable resources; (3) improving power quality; (4) aiding the shift towards distributed energy; and (5) helping networks operate under more stringent environmental requirements. ESSs are needed by the conventional electricity-generating industry, because unlike any other successful commodities, the conventional electricity-generating industries have little or no storage facility. The electricity transmission and distribution systems are operated for the simple one-way transportation from usually large power plants to consumers, meaning that electricity must always be used precisely when produced. However, the electricity demand varies considerably emergently, daily, and seasonally, while the maximum demand may only last for very short periods, leading to inefficient, overdesigned, and expensive plants. ESS allows energy production to be decoupled from its supply, self-generated, or purchased. Having large-scale electricity storage capacity available over any time, system planners would need to build only sufficient generating capacity to meet average electrical demand rather than peak demands. Therefore, ESS can provide substantial benefits including load following, peaking power and standby reserve, as well as increasing the net efficiency of thermal power sources while reducing harmful emissions. Furthermore, ESS is regarded as an imperative technology for distributed energy resource systems. The traditional electricity value chain consists of five links: fuel/energy source, generation, transmission, distribution, and customer-side energy service as shown in Figure 9.3. Supplying power when and where it is needed, ESS is on the brink of becoming the "sixth link" by integrating the existing segments and creating a more responsive market.

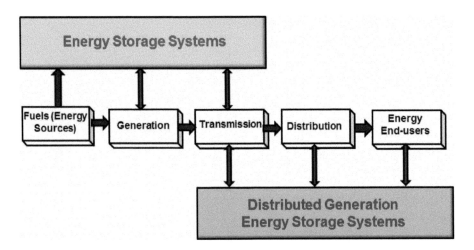

FIGURE 9.3 New chain of electricity, with energy storage as the sixth dimension.

ESS performances include cycle efficiency, cost per unit capacity, energy density, power capacity, lifecycle, and environmental effects including end-of-life disposal cost. An ideal ESS system is one exhibiting the best possible performances so that it will have the minimum amortized (dollar or environmental) cost during its whole lifetime. Unfortunately, no single ESS type can simultaneously fulfill all the desired characteristics of an ideal ESS system, and thus minimize the amortized lifetime cost of ESS. Capital cost, one of the most important criteria in the ESS design, can be represented in the forms of cost per unit of delivered energy ($/kWh) or per unit of output power ($/kW). Capital cost is an especially important concern when constructing hybrid energy storage (HES) systems. Such systems often consist of several ESS elements with relatively low unit cost (e.g., lead-acid batteries) and ESS elements with relatively high unit cost (e.g., super-capacitors). The overall HES system cost is minimized by allocating the appropriate mix of low-cost vs. high-cost ESS elements while meeting other constraints such as cycle efficiency, total storage capacity, or peak output power rate. The ESS cycle efficiency is defined by the "roundtrip" efficiency, i.e., the energy efficiency for charging and then discharging. The cycle efficiency is the product of charging efficiency and discharging efficiency, where charging efficiency is the ratio of electrical energy stored in an ESS element to the total energy supplied to that element during the entire charging process, and discharging efficiency is the ratio of energy derived from an ESS element during the discharging process to the total energy stored in it. Charging-discharging efficiency is significantly affected by the charging/discharging profiles and the ambient conditions. The ESS state of health (SOH) is a measure of its age, reflecting the ESS general condition and its ability to store and deliver energy compared to its initial state. During the ESS lifetime, its capacity or "health" tends to deteriorate due to irreversible physical and chemical changes taking place. The term *replacement* implies the discharge, meaning the use, until the ESS is no longer usable (its end-of-life). To indicate the rate at which SOH is deteriorating, the lifecycle may be defined as the number of ESS cycles performed before its capacity drops to a specific capacity threshold, being one of the key parameters and gives an indication of the expected ESS working lifetime. The lifecycle is closely related to the replacement period and full ESS cost. The self-discharge rate is a measure of how quickly a storage element loses its energy when it simply sits on the shelf, being determined by the inner structure and chemistry, ambient temperature and humidity, and significantly affects the sustainable energy storage period of the given storage element.

Deferred from the conventional power system, which has large, centralized units, DERs are installed at the distribution level, close to consumers, and generate lower power typically in the range of a few kW to a few MW. The electric grid is undergoing changes to become a mixture of centralized and distributed subsystems with higher and higher DER penetration. However, more drastic load fluctuations and emergent voltage drops are anticipated, due to smaller capacity and higher line fault probability than in conventional power systems. ESS is a key solution to compensate for power flexibility and provide a more secure power supply, being also critically important to the integration of intermittent renewable energy. The renewable energy penetration can displace significant amounts of energy produced by large conventional power plants. A suitable ESS could provide an important (even crucial) approach to dealing with the inherent RES intermittency and unpredictability as the energy surplus is stored during periods when generation exceeds the demand and then used to cover periods when the load demands are greater than the generation. Future development of RES technologies is believed to drive down energy storage costs. Nonetheless, the widespread deployments face the fundamental difficulty of intermittent supplies, requiring demand flexibility, backup power sources, and enough energy storage for a significant time. For example, the ESS applications to enhance wind energy generation are: (1) transmission curtailment, mitigating the power delivery constraint due to insufficient transmission capacity; (2) time-shifting, firming, and shaping of wind-generated energy by storing it during the off-peak interval (supplemented by power from the grid when wind generation is inadequate) and discharging during the peak interval; (3) forecast hedge mitigating the errors in wind energy bids prior to required delivery, reducing the price volatility and mitigating consumer risk exposure to this volatility; (4) grid frequency support through the energy storage during sudden, large decreases in wind generation over a short discharge interval; and (5) fluctuation suppression through stabilizing the wind farm generation frequency by suppressing fluctuations (absorbing and discharging energy during short-duration variations in output).

Advanced electric energy storage technologies, when utilized properly, would have environmental, economic, and energy diversity advantages to the system. These include:

1. *Matching electricity supply to load demand:* energy is stored during periods when production exceeds consumption (at lower cost possible) and the stored energy is utilized at periods when consumption exceeds production (at higher cost level), so the electricity production need not be scaled up and down to meet demand variations. Instead, production is maintained at a more constant and economic level. This has the advantage that fuel-based power plants (i.e., coal, oil, gas) are operated more efficiently and easily at constant production levels, while maintaining continuous power to the customer without fluctuations.
2. *Reducing the risks of power blackouts:* energy storage technologies have the ability to provide power to smooth out short-term fluctuations caused by interruptions and sudden load changes. If applied properly, real long-term energy storage can also provide power to the grid during longer blackouts.
3. *Enabling renewable energy generation:* solar and wind energy systems are intermittent sources, expected to produce about 20% of the future electricity, generating during off-peak, when the energy has a low financial value, and the energy storage can smooth out their variability, allowing the electricity to dispatch at a later time, during peak periods, making them cost-effective, and more reliable options.
4. *Power quality:* it may cause poor operations or failures of end-user equipment. Distribution networks, sensitive loads, and critical operations suffer from outages and service interruptions, leading to financial losses to utility and consumers. Energy storage, when properly engineered and implemented, can provide electricity to the customer without any fluctuations, overcoming the power quality problems such as swells/sags or spikes. A summary of energy storage applications and requirements is given in Table 9.1.

The electricity supply is deregulated with clear separations between generation, transmission system operators, distribution network operators and supply's companies. Energy storage applications

TABLE 9.1

Energy Storage Applications in Power Systems

Applications	Matching Supply & Demand	Providing Backup Power	Enabling Renewable Technology	Power Quality
Discharged Power	1 MW – 100 MW	1MW – 200 MW	20 KW – 100 MW	1 KW – 20 MW
Response Time	< 10 Min.	< 10 ms (Quick) < 10 min. (Conventional)	< 1 s	< 1 s
Energy Capacity	1 – 1000 MWh	1 – 1000 MWh	10 KWh – 200 MWh	50 – 500 kWh
Efficiency Needed	High	Medium	High	Low
Life Time Needed	High	High	High	Low

to power distribution can benefit the customer, supply's company, and generation operator in several ways. Major areas where energy storage systems are finding application applied are:

1. Voltage control, supporting a heavily loaded feeder, providing power factor correction, reducing the need to constrain DG, minimizing on-load tap changer operations, mitigating flicker, sags and swells.
2. Power flow management, by redirecting power flows, delay network reinforcement, reduce reverse power flows, and minimize losses.
3. Restoration, by assisting voltage control and power flow management in a post-fault reconfigured network.
4. Energy market, through the arbitrage, balancing market, reduced DG variability, increased DG yield from non-firm connections, replacing the spinning reserve.
5. Commercial and regulatory, by assisting the compliance with energy security standard, reducing customer minutes lost, while reducing generator curtailment.
6. Network management, by assisting islanded networks, supporting black starts, switching ESS between alternative feeders at a normally open point.

It is evident that developing a compelling EES installation at power distribution level in today's electricity market with present technology costs is difficult if value is accrued from only a single benefit. EES systems can contribute significantly to meeting the needs for more efficient and environmentally benign energy use in buildings, industry, transportation, and utilities. Overall the ESS often uses results in significant benefits such as: reduced energy costs and consumption, improved air and water quality, increased operation flexibility, reduced initial and maintenance costs, reduced size, more efficient and effective utilization of the equipment, fuel conservation, by facilitating more efficient energy use and/or fuel substitution, and reduced pollutant emissions.

9.3.2 Types of Energy Storage Systems

Energy storage technologies can be classified in four categories depending on the type of energy stored: Mechanical, electrical, thermal, and chemical energy storage technologies each offer different opportunities, but also have distinct disadvantages. In the following section, each method of electricity storage is assessed and the characteristics of each technology, including overall storage capacity, energy density (energy stored per kilogram), power density (energy transfer time rate per kilogram), and roundtrip efficiency of energy conversion are compared. Energy storage can be defined as the conversion of electrical energy from a power network into a form in which it can be stored until converted back to electrical energy. Energy storage systems are currently characterized by (a) disagreement on the role and design of energy storage systems, (b) common energy storage uses, (c) new available technologies still under demonstration and illustration, (d) no recognized planning tools/models to aid understanding of storage devices, (e) system integration including power electronics must be improved, and (f) it seems small-scale storage will have great importance in the future. However, several energy storage systems are available nowadays with different characteristics, capabilities, and applications. The next subsections include a comprehensive presentation of various energy storage technologies.

9.3.2.1 Compressed-Air Energy Storage

Compressed-air energy storage (CAES) has been an established energy storage technology in the grid operation since the late 1970s. Energy is stored mechanically by compressing the air, and the energy is released to the grid when the air is expanded. CAES is achieved at high pressures (40–70 bars), at near-ambient temperatures. This means less volume and a smaller storage reservoir. Large caverns made of high-quality rock deep in the ground, ancient salt mines, or underground natural gas storage caves are the best options for CAES, as they benefit from geostatic pressure, which facilitates the containment of the air mass (see Figure 9.4). Studies have shown that the air could be compressed and stored in underground, high-pressure piping (20–100 bars). A CAES facility may consist of a power-train motor that drives a compressor (to compress the air into the cavern), high-pressure and low-pressure turbines, and a generator. In a gas turbine (GT), 66% of the energy used is required to compress the air. Therefore, the CAES pre-compresses the air using off-peak electrical power, taken from the grid to drive a motor (rather than using GTs) and stores it in large storage reservoirs. CAES may use the peaks of energy generated by renewable energy plants to run a compressor that compresses the air into underground reservoirs or surface vessel/piping systems. The compressed air is used, combined with a variety of fuels in combustion turbines, to generate electricity during peak demand. The energy storage capacity depends on the compressed-air volume and maximum storage pressure. When the grid is producing electricity during peak hours, the compressed air is released from the storage facility, so instead of using expensive gas to compress the air, cheaper off-peak electricity is used. However, when the air is released from the reservoir, it must be mixed with small amounts of gas before entering the turbine, to avoid air temperature and pressure issues. If the pressure using air alone was high enough to achieve a significant power output, the temperature of the air would be far too low to be tolerated by the materials and connections. The amount of gas required is very small, so a GT working with CAES can produce three times more electricity than operating alone, using the same amount of natural gas. The reservoir can be human-made, which is an expensive choice, or CAES locations can usually use suitable natural geological formations, such as: salt caverns, hard-rock caverns, depleted gas fields, or an aquifer selected to suit specific requirements. In a salt cavern, fresh water is pumped into the cavern and left until the salt dissolves and saturates the fresh water. Then the salted water is transferred at the surface to remove salt, and the cycle is repeated until the required cavern volume (reservoir) is created. This process

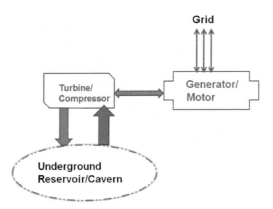

FIGURE 9.4   Diagram of typical compressed-air energy storage.

is expensive and can take up to 2 years to complete. Hard-rock caverns are even more expensive, about 60% higher than salt caverns. Finally, aquifers cannot store the air at higher pressures, having relatively lower energy capacities. CAES efficiency is difficult to estimate, especially when it is using both electrical energy and natural gas, with estimated efficiencies based on the compression and expansion cycles in the rage of 68% to 75%. CAES systems with typical capacities between 50 MW to 300 MW are used for large- and medium-scale applications. Their lifetimes are far longer than those of existing gas turbines and the charge/discharge ratio is dependent on the compressor and reservoir size and pressure.

With assumptions of ideal gas and isothermal processes, the energy stored by compressing m amount of gas at constant temperature from initial pressure, P_i, to final pressure, P_f, is given by:

$$E_{fi} = -\int_{P_i}^{P_f} VdP = mR_g \ln\left(\frac{P_i}{P_f}\right) = P_f V_f \ln\left(\frac{P_i}{P_f}\right) = P_f V_f \ln\left(\frac{V_i}{V_f}\right) \tag{9.2}$$

Here, air is assumed an ideal gas whose specific heat is constant, R_g, the ideal gas constant (8.31447 J·K^{-1}mol^{-1}), V_i, and V_f are the initial and final volumes of the compressed air. Compression power, P_C, depends on the air flow rate, Q (the volume per unit time), and the compression ratio (P_f/P_i), expressed as:

$$P_C = \frac{\gamma}{\gamma - 1} P_i \cdot Q\left[\left(\frac{P_f}{P_i}\right)^{\frac{\gamma}{\gamma-1}} - 1\right] \tag{9.3}$$

where γ is the ratio of air specific heat coefficients ($\gamma \approx 1.4$), P_i, P_f are the initial and final pressure, the atmospheric and compressed state, respectively. A 290 MW CAES (300,000 m^3 and 48,000 Pa) was built in 1978 in Handorf, Germany, and a 110 MW CAES (540,000 m^3 and 53,000 Pa) was built in 1991 at McIntosh, Alabama. Other CAES projects are implemented or in process to be built in Canada, the United States, and the European Union, using different technologies, geological structures, and approaches at various power capacities.

Example 9.2

A CAES has a volume of 450,000 m^3, and the compressed-air pressure range is from 75 bars to atmospheric pressure. Assuming isothermal processes and an efficiency of 30%, estimate the energy and power for a 3-hour discharge period.

SOLUTION

From Equation (9.2), assuming the atmospheric pressure is 1 bar, the energy is:

$$E = 45 \cdot 10^4 \times 75 \cdot 10^5 \ln\left(\frac{75}{1}\right) \approx 1457.2 \cdot 10^4 \text{ MJ}$$

The average power output is then

$$P = \frac{\eta \cdot E}{\Delta t} = \frac{0.3 \times 1457.2 \cdot 10^{10}}{2 \times 3600} \approx 404.76 \text{ MW}$$

Compression of a fluid or air in CAES generates heat, while after the decompression the air is colder. If the heat generated during compression can be stored and used again during decompression, the overall system efficiency improves considerably. CAES storage can be achieved in *adiabatic, diabatic,* and *isothermal* processes. An adiabatic storage has basically no heat exchange during the compression-expansion cycle, with heat being stored in fluids, such as oil or molten salt solutions, while in diabatic storage the heat is dissipated with intercoolers. In an isothermal compression and expansion, the operating temperature is basically maintained constant (or rather quasi-constant) through heat exchange with the environment, being practical only for low power levels. Some heat losses are unavoidable, with the compression process not being truly isothermal. For an isothermal process (a reasonable assumption), the maximum energy that can be stored and released is given by:

$$E_{fi} = nRT \ln\left(\frac{V_i}{V_f}\right) \tag{9.4}$$

Here T is the absolute temperature (K), and n is the number of moles of the air in the reservoir.

Example 9.3

Determine the maximum stored energy if a mass of 2900 kg of air is compressed isothermally at 300 K from 100 kPa to 1500 kPa, assuming a heat loss of 33 MJ and ideal gas.

SOLUTION

With the ideal gas assumption, molar mass for air 29 kg/Kmol and R 8.314 kJ/kmol·K, the number of moles of air, is:

$$n = \frac{2900}{29} = 100 \text{ kmol}$$

The mechanical energy stored, by using Equation (9.4), is then:

$$E_{fi} = -nRT \ln\left(\frac{V_i}{V_f}\right) = -100 \times 8.314 \times 300 \cdot \ln\left(\frac{100}{1500}\right) = 675,441.9 \text{ kJ}$$

Net stored energy is then:

$$E_{net} = E_{fi} - \text{Heat Losses} = 675441.9 - 33000 = 642,441.9 \text{ kJ}$$

CAES systems, the only very-large-scale energy storage other than PHES, have fast reaction time with plants usually able to go from 0% to 100% in less than 10 minutes, 10% to 100% in approximately 4 minutes, and from 50% to 100% in less than 15 s. As a result, it is ideal for acting as a large sink for bulk energy supply, being also able to undertake frequent start-ups and shutdowns. CAES do not suffer from excessive heat when operating on partial load, as traditional gas turbines. These flexibilities mean that CAES can be used for ancillary services such as load following, frequency regulation, and voltage control. As a result, CAES has become a serious contender in the wind-power energy storage market. A number of possibilities are being considered, such as integrating a CAES facility with several wind farms within the same region or area, so the excess off-peak power from these wind farms could be stored in a CAES facility. CAES advantages are high energy and power capacity and long lifetime, while the major disadvantages are low efficiencies, some adverse environmental impacts, and difficulty of siting.

9.3.2.2 Electrochemical Energy Storage

Chemical energy storage can be further classified into electrochemical and thermochemical energy storage. Electrochemical energy storage refers to conventional batteries, such as lead-acid (LA), nickel-metal hydride, and lithium-ion (Li-ion), and flow batteries, such as zinc/bromine (Zn/Br), vanadium redox, and metal air batteries. Electrochemical energy storage is also achieved in fuel cells (FCs), most commonly hydrogen fuel cells, but also direct-methanol, molten carbonate, and solid-oxide fuel cells. Batteries and fuel cells are the most common energy storage devices used in power systems and in several other applications. In batteries and fuel cells, electrical energy is generated by conversion of chemical energy via redox reactions at the anode and cathode. As reactions at the anode usually take place at lower electrode potentials than at the cathode, the terms *negative* and *positive* electrode are used. The more negative electrode is designated the anode, whereas the cathode is the more positive one. The difference between batteries and fuel cells is related to the locations of energy storage and conversion. Batteries are closed systems, with the anode and cathode being the charge-transfer medium and taking an active role in the redox reaction as "active masses." In other words, energy storage and conversion occur in the same compartment. Fuel cells are open systems where the anode and cathode are just charge-transfer media and the active masses undergoing the redox reaction are delivered from outside the cell, either from the environment, for example, oxygen from air, or from a tank, for example, fuels such as hydrogen and hydrocarbons. Energy storage (in the tank) and energy conversion (in the fuel cell) are thus locally separated. However, to date the only portable and rechargeable electrochemical energy storage systems (with the exception of capacitors) are batteries.

Basic configuration of a battery or a fuel cell consists of two electrodes and an electrolyte, placed together in a special container and connected to an external device (a source or a load). Usually a battery consists of two or more electric cells joined together, in series, parallel, or series-parallel configurations in order to achieve the desired voltage, current, and power. The cells, consisting of positive and negative electrodes joined by an electrolyte, convert chemical energy to electrical energy. The chemical reaction between the electrodes and the electrolyte generates DC electricity. In the case of secondary or rechargeable batteries, the chemical reaction is reversed by reversing the current and the battery is returned to a charged state. A battery, by definition, is one or more electrically connected electrochemical cells having terminals/contacts to supply electrical energy. Battery types are of two forms, *disposable* or *primary batteries* and *rechargeable* or *secondary batteries*. A primary battery is a cell, or group of cells, for the generation of electrical energy intended to be used until exhausted and then discarded; they are assembled in the charged state, while the discharge is the primary process during operation. A secondary battery is a cell or group of cells for the generation of electrical energy in which the cell, after being discharged, may be restored to its original charged condition by an electric current flowing in the direction opposite to the flow of current when the cell was discharged. Other terms for this type of battery are rechargeable battery or accumulator. As secondary batteries are usually assembled in the discharged state, they have to be charged first before they can undergo discharge in a secondary process. Mature rechargeable battery chemistries are: (a) lead-acid (LA), (b) nickel-cadmium (Ni-Cd), (c) nickel-metal hydride (Ni-MH), (d) lithium-ion (Li-ion), (e) lithium-polymer/lithium metal (Li-polymer), (f) sodium-sulfur (Na-S), (g) sodium-nickel chloride, and (h) lithium-iron phosphate. Flow batteries, a different type, will be discussed later. Facing growing demands from electric vehicles and portable consumer products, companies are spending significant funds to research new battery technologies, such as zinc-based chemistries and silicon as a material for improving battery properties and performances. Higher energy density and life cycle, environmental friendliness, and safer operation are among the general design research targets for secondary batteries. Primary batteries are a reasonably mature technology, in terms of chemistry, but still there is research to increase the energy density, reduce self-discharge rate, increase the battery life, or to improve the usable temperature range. To complement these developments, many semiconductor manufacturers continue to introduce new integrated circuits for battery power management. Other chemical energy storage options, such as thermochemical energy storage, although promising, are still at the infant development stages and are not included in this entry.

Chemical energy storage is usually achieved through accumulators or batteries, characterized by a double function of storage and release of electricity by alternating the charge-discharge phases. They can transform chemical energy generated by electrochemical reactions into electrical energy and vice versa, without very little harmful emissions or noise, and require little maintenance. A wide range of battery technologies is in use, and their main assets are their energy densities (up to 2000 Wh/kg for lithium-based types) and technological maturity. Chemical energy storage devices (batteries) and electrochemical capacitors (ECs) are among the leading ESS technologies today. Both technologies are based on electrochemistry, and the fundamental difference between them is that batteries store energy in chemical reactants capable of generating charges, whereas electrochemical capacitors store energy directly as electric charges. Although the electrochemical capacitor is a promising technology for electrical energy storage, especially considering its high power capability, the energy density is too low for large-scale energy storage. The most common rechargeable battery technologies are lead-acid, sodium-sulfur, vanadium redox, and lithium-ion types. A battery comprises several electrochemical cells, with each cell comprising an electrolyte, positive, and negative electrodes. During discharge, electrochemical reactions at the two electrodes generate an electron flow through an external circuit. Vanadium-redox batteries have good prospects because they can scale up to much larger storage capacities and show great potential for longer lifetimes and lower per-cycle costs than conventional batteries requiring refurbishment of electrodes. Lithium-ion batteries also display very high potential for large-scale energy storage. A battery consists of one or more electrochemical cells, connected in series, in parallel or series-parallel configuration in order to provide the desired voltage, current, and power. The anode is the electronegative electrode from which electrons are generated to do external work. The cathode is the electropositive electrode to which positive ions migrate inside the cell and electrons migrate through the battery external electrical circuit. The electrolyte allows the flow of ions and electrons, from one electrode to another, being commonly a liquid solution containing a dissolved salt, and must be stable in the presence of both electrodes. Current collectors allow the transport of electrons to and from the electrodes, being typically made of metals, and must not react with the electrode or electrolyte materials. Cell voltage is determined by the chemical-reaction energy occurring inside the cell. The anode and cathode are, in practice, complex composites, containing, besides the active material, polymeric binders to hold together the powder structure and conductive diluents such as carbon black to give the whole structure electric conductivity so that electrons can be transported to the active material. In addition, these components are combined to ensure sufficient porosity to allow the liquid electrolyte to penetrate the powder structure and permit the ions to reach the reacting sites. During the charging process, the electrochemical reactions are reversed via the application of an external voltage across the electrodes.

Today, battery technologies range from the mature and long-established lead-acid system to various more recent and emerging systems and technologies. The newest technologies are attracting an increased interest for possible use in power systems, having achieved market acceptance and uptake in consumer electronics in the so-called 3Cs sector—cameras, cellphones, and computers. Batteries have the potential to span a broad range of energy storage applications, in part due to their portability, ease of use, large power-storage capacity (100 W up to 20 MW), and ease of connecting in series-parallel combinations to increase their power capacity for specific applications. Major battery advantages include: standalone operation, no need to be connected to an electrical system, easy to expand; typical disadvantages are: expensive, limited life cycle, and maintenance. These systems can be located in any place and be rapidly installed, have less environment impacts than other ESS technologies, and can be located in a building (or similar facility) near the demand point. Battery energy storage system (BESS) connected to a power system uses an inverter to convert the battery DC voltage into AC grid-compatible voltage. These units present fast dynamics with response times near 20 ms and efficiencies ranging from 60% to 80%. The battery temperature change during charge and discharge cycles must be controlled because it affects its life expectancy. Depending on the battery and cycle, a BESS can require multiple charges and discharges per day. The battery cycle is normal while the discharge depth is small, but if the discharge depth is high, the battery cycle duration could be degraded. The expected useful life of a Ni-Cd battery is 20,000 cycles if the discharge depth is limited to 15%. Examples of large-scale BESSs installed today are: 10 MW (40 MWh) Chino system in California and 20 MW (5 MWh) in Puerto Rico. Their main inconvenience, however, is their relatively low durability for large-amplitude cycling. They are often used in emergency backup, renewable-energy system storage, etc. The minimum discharge period of electrochemical accumulators rarely reaches less than 15 min. However, for some applications, power up to 100 W/kg, even a few kW/kg, can be reached within a few seconds or minutes. As opposed to capacitors, their voltage remains stable as a function of charge level. Nevertheless, between a high-power recharging operation at near-maximum charge level and nearing full discharge, the voltage can easily vary by a ratio of two.

9.3.3 Battery Operation Principles and Battery Types

Batteries convert the chemical energy contained in their active materials into electrical energy through an electrochemical oxidation-reduction reversible reaction. Battery fundamental principles and operation, regardless of the battery type, can be explained by using the so-called galvanic element or electrochemical cell. A galvanic cell consists of three main components: the *anode*, the *cathode*, and the *electrolyte*. In the case of a galvanic cell, the electrons needed for conduction are produced by a chemical reaction. From thermodynamics, we know that the work done by the electrochemical cell comes at a cost, which in turn implies that the chemical reactions taking place within the cell must lead to a decrease in free energy. In fact, for a reversible process at constant temperature and pressure, the maximum work done by the system, W, equals the free energy (Gibbs free energy) change $-\Delta G$. The work performed when transporting an electric charge e (in C) through a potential difference E (in V) is simply the product of eE. Here we can express this work on a per-mole basis. The total charge carried by 1 mole of positively charged ions of valence +1 is 96,487 C and this number is denoted by F, Faraday's constant. Thus, the work produced by the electrochemical cell is:

$$W = -\Delta G = \xi F \cdot E \rightarrow E_{max} = \frac{-\Delta G^0}{\xi F} \tag{9.5}$$

where ξ is the valence of the ions produced in the chemical reaction. The electric potential difference across the cell electrodes, E, is the electromotive force, or EMF, of the galvanic cell. It is clear from Equation (9.5) that in order to have higher work and potential, we need to find reactions with the highest driving force, the free energy change $-\Delta G$.

Example 9.4

Estimate the maximum output voltage that a Zn-Cu electrochemical cell can generate.

SOLUTION

The Zn-Cu reaction taking place in this electrochemical cell is:

$$Zn + Cu^{2+} \rightarrow Cu + Zn^{2+}$$

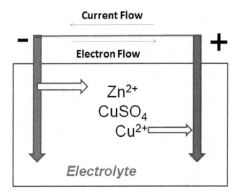

FIGURE 9.5 Galvanic element diagram (consisting of Zn and Cu electrodes, and CuSO₄ electrolyte). Current flows are also shown here.

By using Equation (9.6) and from the data tables of the book appendixes (for Zn-Cu cell ξ is equal to 2, and ΔG^0 is equal to 216,160 kJ/kmole), the maximum voltage is:

$$E_{max} = \frac{-\Delta G^0}{\xi F} = \frac{216,160 \text{ kJ/kmole}}{2 \times 96,500.0} = 1.12 \ V$$

To understand the role of each of the electrochemical cell components and their operation it is best to refer to specific examples. If, for example, a galvanic element made of a zinc (Zn) electrode, a copper (Cu) electrode, and an electrolyte (CuSO₄, for example), the two electrodes are electrically connected (see diagram of Figure 9.5), the Zn^{2+} ions flow from the Zn electrode, through electrolyte, while the electrons migrate from Zn, and eventually combine with Cu^{2+} ions, residing into the electrolyte and form copper atoms, increasing the Cu electrode volume. The direction of electrons is determined by the potential difference between metal and electrolyte, $\Delta\phi_{metal-electrolyte}$, the electron transport is in such way that the metal (Zn, here) separates the electrons and ions, because it has a lower metal-electrolyte potential difference than that of the Cu electrode against electrolyte. The metal (and electrode) with higher potential is serving as positive electrode (and cell terminal) of a galvanic element. The metals are usually arranged in voltaic sequences, in a way that all metals (e.g., Fe, Cd, Ni, Pb, Cu) in the right side of a certain metal (e.g., Zn) do for a positive pole in a combination of electrodes, chosen based on the metal-electrolyte potential difference values of Table 9.2. For example, Zn is a negative pole with respect to Fe, Cd, Ag, or Au electrodes, while Li electrodes form a negative pole with respect to K, Na, Mg, Zn, or Fe electrodes. The potential difference (external voltage) between the cell (galvanic element) terminals is the voltage difference, for this galvanic element, existing between Zn electrode and electrolyte (CuSO₄) and between Cu electrode and electrolyte. Notice that it is not possible to directly measure the potential difference between the metal electrode and electrolyte; only the metal-metal potential difference can be measured.

TABLE 9.2

Metal Voltaic Series

Metal electrode	Li	K	Na	Mg	Zn	Fe	Cd
$\Delta\phi_{metal-electrolyte}$ (V)	−3.02	−2.92	−2.71	−2.35	−0.762	−0.44	−0.402
Metal electrode	Ni	Pb	H₂	Cu	Ag	Hg	Au
$\Delta\phi_{metal-electrolyte}$ (V)	−0.25	−0.126	0.0	+0.345	+0.80	+0.86	+1.50

Example 9.5

Determine the potential differences between Li-Au, Li-Ag, and Pb-H$_2$, using the values in Table 9.2.

SOLUTION

The required potential differences are:

$$\Delta\phi_{Li-Au} = -3.02 - 1.50 = -4.52 \text{ V}$$
$$\Delta\phi_{Li-Au} = -3.02 - 0.86 = -3.88 \text{ V}$$

And

$$\Delta\phi_{Ni-Cu} = -0.25 - 0.345 = -0.595 \text{ V}$$

Lead-acid batteries (Pb-acid/LA) are the most common electrical energy storage device used at the present, especially in transportation, renewable energy, and standalone hybrid power systems. Their success is due to their maturity (research has been ongoing for about 150 years), low cost, long lifespan, fast response, and low self-discharge rate. They are used for both short- and long-term applications. A lead-acid battery works somewhat differently than series galvanic elements, employing the positive and negative charge polarizations. Lead and lead-dioxide are good electrical conductors, the electrolyte contains aqueous ions (H$^+$ and SO$_4^{-2}$), and the conduction electrolyte mechanism is via migration of ions through diffusion or drift. Two lead plates of a cell are immersed in a dilute sulfuric acid solution, each covered with a PbSO$_4$ layer, basically consisting of spongy lead anode and lead-acid cathode, with lead as the current collector. The lead-acid batteries for electric vehicles use a gel rather than a liquid electrolyte, in order to withstand the deep cycle that is required here, and for that reason are more expensive than regular ones. The sulfuric acid combines with the lead and lead oxide to produce lead sulfate and water, and the electrical energy is released during the process. The overall lead-acid battery reaction is:

$$Pb + PbO_2 + 2H_2SO_4 \rightleftarrows 2PbSO_4 + 2H_2O \tag{9.6}$$

This reaction proceeds from left to right during battery discharge, and in the opposite direction during the battery-charging process. During the discharging process, the electrons flow from the cathode to anode (current is flowing in the opposite direction), until the chemical process is reversed and PbSO$_4$ forms on the anode and cathode, while the anode is covered with PbO$_2$ that gets reduced as the discharging process progresses. The voltage across the two lead plates is $e_0 = 2.02$ V. If six sets for lead plates are connected in series, then the total terminal voltage is $E^0 = 6 \cdot e_0 = 12.12$ V. The efficiency of a lead-acid battery is from 80% to 90% of the charged energy. During discharging, the electrolyte gradually loses the sulfuric acid and becomes more dilute, while during the charging process, the electrolyte reverts to lead and lead dioxide, and also recovers its sulfuric acid and the concentration increases.

The two lead-acid battery types are: (a) flooded lead-acid (FLA), and (b) valve-regulated lead-acid (VRLA). From an application point of view, lead-acid batteries are split into stationary, traction, and car batteries. Stationary batteries ensure uninterruptible electric energy supply, in case of power system failure, undergoing usually only a few cycles, so their lifespan is about 20 years. Traction batteries, used for power supply of electric vehicles, electro-mobiles, industrial tracks, etc., work in deep-cycle charging-discharging regimes with a lifespan of about 5 years (about 1000 charge-discharge cycles). Automotive (car) batteries, used to crank car engines, can supply short and intense discharge current, and also support car electrical devices when the engine is not running. FLA batteries, described above, are made up of two electrodes (lead plates), immersed in a mixture of water (65%) and sulfuric acid (35%). VRLA batteries have the same operating principle, with the difference that they are sealed with pressure-regulating valves, to prevent the air entering the cells and hydrogen venting. VRLA batteries operate with the help of an internal oxygen cycle, emitted in the later charging stages and during overcharging on the positive electrode, traveling through a gas separator to the cathode where it is reduced to water. This oxygen cycle makes the cathode potential less negative, so the rate of hydrogen evolution decreases. Part of the generated electricity is used by the internal recombination oxygen cycle and is converted to heat. VRLA batteries have lower maintenance costs, and less weight and space, but these advantages are coupled with higher initial costs and shorter lifetime. During discharge, lead sulfate is the product on both electrodes. Sulfate crystals become larger and difficult to break up during recharging, if the battery is over-discharged or kept discharged for a prolonged time period. Hydrogen is produced during charging, leading to water losses if overcharged. Their popularity is a result of their wide availability, robustness, and reasonably low cost. The disadvantages of lead-acid batteries are their weight, low specific energy and specific power, short cycle life (100–1000 cycles), high maintenance requirements, hazards associated with lead and sulfuric acid during production and disposal, and capacity drop at low temperatures. FLA batteries have two primary applications: starting and ignition, short bursts of strong power (e.g., car engine batteries), and deep-cycle, low steady power over long durations. VRLA batteries are very popular for backup power, standby power supplies in telecommunication centers, and uninterruptible power supply. Among the major reasons for failure of lead-acid batteries include: positive plate expansion, positive mass fractioning, water loss, acid stratification, incomplete charging causing active mass sulfating, positive grid corrosion, and negative active mass sulfating.

Nickel-cadmium (Ni-Cd) and nickel-metal hydride (Ni-MH) batteries use nickel oxyhydroxide for the cathode and metallic cadmium as the anode with potassium hydroxide as an electrolyte. Other possible systems are Ni-Fe, Ni-Zn, and Ni-H$_2$. These types of batteries were very popular between 1970 and 1990 but have been largely superseded by Ni-MH due to their inferior cycle life, memory effect, energy density, and toxicity of the cadmium in Ni-Cd, compared to Ni-MH batteries. Ni-MH also have the advantage of improved high-rate capability (due to the endothermic nature of the discharge reaction), and high tolerance to over-discharge. Ni-MH use nickel oxyhydroxide for the cathode, with a potassium hydroxide electrolyte and a hydrogen-absorbing alloy, usually lanthanum and rare earths, serving as a solid source of reduced hydrogen that can be oxidized to form protons. Ni-Cd and Ni-MH batteries are mature technologies, relatively rugged with higher energy density, lower maintenance requirements, and higher cycle life than lead-acid batteries. They are, however, more expensive than lead-acid batteries, with limitations on long-term cost reduction potential due to material costs. In addition, Ni-Cd batteries contain toxic components (cadmium) making them environmentally challenging. With a higher energy density, longer cycle life, and low maintenance requirements, Ni-Cd batteries have a potential edge over lead-acid batteries. The drawbacks of this technology include the toxic-heaviness of cadmium and higher self-discharge rates than lead-acid batteries. Also, Ni-Cd batteries may cost up to ten times more than a lead-acid battery, making it quite a costly alternative. Ni-Cd batteries are produced in a wide commercial variety from sealed free-maintenance cells (capacities of 10 mAh to 20 Ah) to vented standby power units (capacities of 1000 Ah or higher), longer cycle life, overcharge capacity, high charge-discharge rates, almost-constant discharge voltage, and they can operate at low temperature. Depending on construction, Ni-Cd batteries have energy densities in the range 40 Wh/kg to 60 Wh/kg, and cycle life above several hundred for sealed cells to several thousand for vented cells. There are two main construction types of Ni-Cd batteries. The first type uses pocket plate electrodes, in vented cells, where the active material is found in pockets of finely perforated nickel-plated sheet steel, and positive and negative plates are separated by plastic pins or ladders and plate edge insulators. The second type use sintered, bonded, or fiber plate electrodes, for both vented and sealed cells. Active material is distributed within the pores. The electrolyte is an aqueous solution of KOH, with 20%–28% weight concentration and density of 1.18 to 1.27 g/cm^3 at 25°C, with 1%–2% of LiOH usually added to minimize the coagulation of the NiOOH electrode during the charge-discharge cycling. An aqueous NaOH electrolyte may be used in cells operating at higher temperatures. The overall Ni-Cd cell discharge reaction is:

$$2NiOOH + Cd + 2H_2O \rightarrow 2Ni(OH)_2 + Cd(OH)_2 \qquad (9.7)$$

The Ni-Cd cell voltage is $E^0 = +1.30$ V. Ni-Cd batteries are designed as positive limited oxygen cycle use, and the amount of water decreases during discharge. The oxygen evolved at anode during charging diffuses to the cathode area and reacts with cadmium, forming $Cd(OH)_2$, and carbon dioxide from air reacts with KOH in electrolyte to form K_2CO_3 and $CdCO_3$ at the cathode, increasing internal resistance and reducing the capacity of Ni-Cd batteries. Sealed Ni-MH cells have as active negative material hydrogen absorbed in a metal alloy that increases its energy density and makes it more environmentally friendly compared with Ni-Cd cells. However, these cells have higher self-discharge rates and are less tolerant to overcharge than Ni-Cd cells. The hydrogen absorption alloys are made of two metals (the first metal absorbs hydrogen exothermically and the second endothermically), and serve also as catalysts for the absorption of hydrogen into the alloy lattice. A hydrolytic polypropylene separator is used in Ni-MH cells. The electrolyte used in Ni-MH cells is potassium hydroxide, and the cell voltages range from 1.32 V to 1.35 V. Their energy density is 80 Wh/kg (25% higher than that of Ni-Cd cells), life cycle over 1000 cycles, and a high self-discharge rate of 4%–5% per day. Ni-MH batteries are used in electric vehicles, consumer electronics, phones, medical instruments, and other high-rate and longer life-cycle applications. The overall Ni-MH cell reaction during discharge cycle is:

$$NiOOH + MH \rightarrow Ni(OH)_2 + M \tag{9.8}$$

Ni-Dc and Ni-MH cells both suffer from memory effects, a temporary but reversible capacity reduction caused by shallow charge-discharge cycling. After a shallow cycling, there is a voltage step during discharge, and the cell remembers the depth of the shallow cycling. The voltage reduction depends on the number of preceding shallow cycles and the discharge current. However, the cell capacity is not affected. If the cell is fully discharged and then charged, a deep discharge shows a normal discharge curve.

Lithium ion (Li-ion) batteries are largely cobalt- or phosphate-based. In both embodiments, lithium ions flow between the anode and cathode to generated current flow. Li-ion batteries have a high energy-to-weight ratio (density), no memory effect, light weight, high reduction potential, low internal resistance, and low self-discharge rates. Prices may be high and increasing penetration may push prices higher as limited lithium resources are depleted. Lithium ion battery technology has progressed from developmental and special-purpose status to a global mass-market product in less than 20 years. The technology is especially attractive for low-power and portable applications because Li-ion batteries offer high-power densities, typically 150–200 Wh/kg and generally acceptable cycle life, about 500 cycles, very low self-discharge rate (less than 10% per month) and high voltage, about 3.6 V. The operation of lithium ion cells involves the reversible transfer of lithium ions, between electrodes, during charge and discharge. During charging, lithium ions move out (de-intercalate) from the lithium metal oxide cathode and intercalate into the graphite-based anode, with the reverse happening during the discharge reaction. The non-aqueous ionic conducting electrolyte takes no part in the reaction except for conducting the lithium ions during the charge and discharge cycles. Notwithstanding their significant advantages, lithium ion systems must be maintained within well-defined operating limits to avoid permanent cell damage or failure. The technology also possesses no natural ability to equalize the charge amount in its component cells. This, and the closely defined operational envelope of lithium ion batteries, essentially dictates the use of relatively sophisticated management systems. Most commercial Li-ion cells have anodes of cobalt oxide, or manganese oxide. The negative electrode is made of carbon, in graphite form (light weight and low price) or an amorphous material with a high surface-area. The electrolyte is made of an organic liquid (ether) and a dissolved salt, and the positive and negative active mass is applied to both sides of thin metal foils, aluminum on the positive side and copper on the negative. A microporous polymer is used as a separator between the positive and negative electrode. The positive electrode reaction in a Li-ion cell is:

$$LiCoO_2 \rightarrow Li_{1-x}CoO_2 + xLi^+ + xe^- \tag{9.9a}$$

And the negative electrode reaction is:

$$xLi^+ + xe^- + C_6 \rightarrow Li_xC_6 \tag{9.9b}$$

Here the index x moves on the negative electrode from 0 to 1, and on the positive electrode from 0 to 0.45. Li-ion batteries are manufactured in coin format, cylindrical, and prismatic shapes. On the other hand, the application of the technology to larger-scale systems is relatively limited to date, although various developments are emerging with potential for automotive, power utility, submersible, and marine sectors. Lithium-polymer (Li-pol) batteries employ the polymer property (containing a hetero-atom, i.e., oxygen or sulfur) to dissolve lithium salts in high concentrations, leading at higher temperatures to a good electric conductivity (0.1 S/m at 100°C), allowing such polymers to be used as electrolyte for lithium batteries. Polymer electrolyte, not being flammable, is safer than liquid electrolyte.

9.3.3.1 Battery Fundamentals, Parameters, and Electric Circuit Models

Battery or cell capacity ($CAP(t)$) means an integral of current, $i(t)$ over a defined time period, as:

$$CAP(t) = \int_0^t i(t)dt \tag{9.10}$$

The above relationship applies to either battery charge or discharge, meaning the capacity added or capacity removed from a battery or cell, respectively. The capacity of a battery or cell is measured in milliampere hours (mAh) or ampere hours (Ah). This basic definition is simple and straight; however, several different forms of capacity relationship are used in the battery industry. The distinctions between them reflect differences in the conditions under which the battery capacity is measured. Standard capacity measures the total capacity that a relatively new, but stabilized production cell or battery can store and discharge under a defined standard set of application conditions, assuming that the cell or battery is fully formed, that it is charged at standard temperature at the specification rate, and that it is discharged at the same standard temperature at a specified standard discharge rate to a standard end-of-discharge voltage (EODV). The standard EODV is subject to variation depending on discharge rate. When the application conditions differ from standard ones, the cell or battery capacity changes, so the term *actual capacity* includes all nonstandard conditions that alter the amount of capacity the fully charged new cell or battery is capable of delivering when fully discharged to a standard EODV. Examples of such situations might include subjecting the cell or battery to a cold discharge or a high-rate discharge. That portion of actual capacity that is delivered by the fully charged new cell or battery to some nonstandard EODV is called available capacity. Thus, if the standard EODV is 1.6 V/cell, the available capacity to an EODV of 1.8 V/cell would be less than the actual capacity. Rated capacity is defined as the minimum expected capacity when a new, fully formed, cell is measured under standard conditions. This is the basis for C rate (defined later) and depends on the standard conditions used, which may vary depending on the manufacturers and the battery types. If a battery is stored for a period of time following a full charge, some of its charge will dissipate. The remaining capacity that can be discharged is called retained capacity. In most of the practical engineering applications, the battery capacity, C, for a constant discharge rate of I (in A) is:

$$C = I \times t \tag{9.11}$$

It is clear from the above equation that the capacity of a battery is reduced if the current is drawn more quickly. Drawing 1 A for 10 hours does not take the same charge from a battery as running it at 10 A for 1 hour. Notice that the relationship between battery capacity and discharge current is not linear, and less energy is recovered at faster discharge rates. This phenomenon is particularly important for electric vehicles, as in this application the currents are generally higher, with the result that the capacity might be less than is expected. It is important to be able to predict the effect of current on capacity, both when designing electric vehicles, or when designing and making instruments to measure the charge left in a battery, the so-called battery fuel gauges. The best way to do this is by using the Peukert model of battery behavior. Although this model is not very accurate at low currents, for higher currents it models battery behavior well enough. The starting point of this model is that there is a capacity, called the Peukert capacity, which is constant, and is given by the equation:

$$C_{Pkt} = I^k \cdot t \qquad (9.12)$$

where C_{Pkt} is the amp-hour capacity at a 1 A discharge rate, I is the discharge current in amperes, t is the discharge time, in hours, and k is the Peukert coefficient, typically with values from 1.1 to 1.3.

Example 9.6

A lead-acid battery has a nominal capacity of 50 Ah at a rate of 5 h, and a Peukert coefficient of 1.2. Estimate the battery Peukert capacity.

SOLUTION

This battery of a capacity of 50 Ah is discharged at a current of:

$$I = \frac{50 \text{ Ah}}{5 \text{ h}} = 10 A$$

If the Peukert coefficient is 1.2, then the Peukert capacity is:

$$C_{Pkt} = 10^{1.2} \cdot 5 = 79.3 \text{ Ah}$$

The capacity of a battery, sometimes referred to as C_{load} or simply C, is an inaccurate measure of how much charge a battery can deliver to a load. It is an imprecise value because it depends on temperature, age of the cells, state of the charge, and the rate of discharge. It has been observed that two identical, fully charged batteries, under the same circumstances, will deliver different charges to a load depending on the current drawn by the load. In other words, C is not constant and the value of C for a fully charged battery is not an adequate description of the characteristic of the battery unless it is accompanied by additional information, *rated time of discharge,* with the assumption that the discharge occurs under a constant-current regime. Usually, lead-acid batteries are selected as energy storage for building DC microgrids, because of relatively low cost and mature technology. The energy storage systems are usually operated by current closed-loop control, while the storage power is controlled by a supervision unit, which calculates the corresponding power reference. The storage state of charge (SoC) must be respected to its upper (maximum) and lower (minimum) SoC limits, SoC_{max} and SoC_{min} respectively, to protect the battery from overcharging and overdischarging, as given in Equation (12.15), below. SoC is then calculated with Equation (12.16), where SoC_0 is the initial SoC at t_0 (initial time), CREF is the storage nominal capacity (Ah) and V_S is the storage voltage.

$$SOC_{min} \leq SOC(t) \leq SOC_{max} \qquad (9.13)$$

and

$$SOC(t) = SOC_0 + \frac{1}{3600 \times V_S \times CREF} \int_{t_0}^{t} (P_{SC} - P_{SD}(t)) dt \qquad (9.14)$$

Accurate battery models are required for the simulation, analysis, and design of energy consumption of electric vehicles, portable devices, or renewable energy and power system applications. The major challenge in modeling a battery is the nonlinear characteristics of the equivalent circuit parameters, which depend on the battery state of charge, and require complete and complex experimental and/or numerical procedures. The battery itself has internal parameters, which need to be taken care of for modeling purposes, such as internal voltage and resistance. All electric cells in a battery have nominal voltages that give the approximate voltage when the cell is delivering electrical power. The cells can be connected in series to give the overall voltage required by a specific application.

Example 9.7

A battery bank consists of several cells, connected in series. Assume that the cell internal resistance is 0.012 Ω and the cell electrochemical voltage is 1.25 V. If the battery bank needs to deliver 12.5 A at 120 V to a load, determine the number of cells.

SOLUTION

With cells in series, the number of cells can be estimated as:

$$N(cells) = \frac{V_L}{V_{OC}(cell)} = \frac{120}{1.25 - 12.5 \times 0.012} = 109.09$$

We choose 110 cells, rounding off to the upper integer, meaning a higher terminal voltage than 120 V. However, the terminal voltage decreases over the battery bank lifetime.

The three basic battery models used most in engineering applications are: the ideal, linear, and Thevenin models. The battery ideal model, a very simple one, is made up only by a voltage source (Figure 9.6a), and ignores the internal parameters. Widely used in applications, the battery linear model (Figure 9.6b) consists of an ideal battery with open-circuit voltage, V_0, and an equivalent series resistance, R_S, while V_{Out} represents the battery terminal (output) voltage. This terminal voltage is obtained from open-circuit tests as well as from load tests conducted on a fully charged battery. Although this model is quite widely used, it still does not consider the varying characteristics of the internal impedance of the battery with the varying state of charge (SoC) and electrolyte concentration. A more complete, accurate, and complex model is the Thevenin model (Figure 9.6c), which, in addition to the open-circuit voltage and internal resistance, consists of an internal capacitance, C_{in}, and overvoltage resistance, R_O. The capacitor C_{in} accounts for the capacitance of the parallel plates and resistor R_O accounts for the nonlinear resistance offered by the plate to the electrolyte. All the elements in this model are assumed to be constants. However, in reality, they depend on the battery conditions. Thus, this model is also not the most accurate, but is by far the most widely used. In this

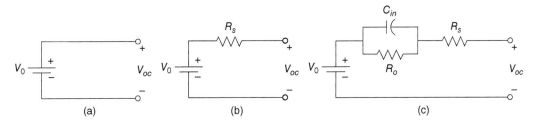

FIGURE 9.6   Battery electric diagrams, steady-state conditions: (a) battery ideal model, (b) battery linear model, and (c) Thevenin battery model.

view, a new approach to evaluate batteries is introduced. Any battery, regardless of the type, in the first approximation (for steady-state operation) works as a constant voltage source with an internal (source/battery) resistance, by considering the battery linear model as shown in Figure 9.6b. These parameters, the internal voltage (V_{OC}), and resistance (R_S) are dependent on the discharged energy (Ah) as:

$$V_{OC} = V_0 - k_1 \times DoD \tag{9.15a}$$

and

$$R_S = R_0 + k_2 \times DoD \tag{9.15b}$$

Here, V_{OC}, known also as the open-circuit or electrochemical voltage, decreases linear with depth of discharge, DoD, while the internal resistance, R_S increases linear with DoD. V_0 and R_0 are the values of the electrochemical (internal) voltage and resistance, respectively, when the battery is fully charged, DoD is 0, and when fully discharged DoD is 1.0. The constants, k_1 and k_2, are determined from the battery test data, through curve fitting or other numerical procedures. The depth of discharge (DoD) is defined from the battery state of charge (SoC), as:

$$DoD = \frac{\text{Ah drained form battery}}{\text{Battery rated Ah capacity}} = 1 - SoC \tag{9.16}$$

While, SoC (as defined in Equation [9.15]) is computed, for practical application by a much simpler relationship, that can be estimated from the battery monitoring (test data) as:

$$SoC = 1 - DoD = \frac{\text{Ah remaining in the battery}}{\text{Rated Ah battery capacity}} \tag{9.17}$$

Notice, the battery terminal voltage is lower and the internal resistance is higher in a battery partially discharge state (i.e., any time when $DoD > 0$). All electric battery cells have nominal voltages, which gives the approximate voltage when the battery cell is delivering electrical power. Cells can be connected in series to give the required voltage. The terminal voltage, V_L, of a partially discharged battery, with notation of Figure 9.7 is expressed as:

$$V_L = V_0 - I \cdot R_S = V_0 - k_1 \times DoD - I \cdot R_S \tag{9.18}$$

The load delivered power is I^2R_L, the battery internal loss is I^2R_S, dissipated as heat inside the battery. In consequence, as the battery discharges, the internal resistance, R_L, increases, and more heat is generated.

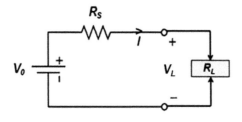

FIGURE 9.7 Electric diagram of battery load, by using the steady-state battery linear model.

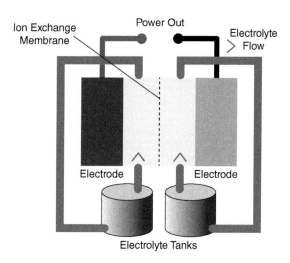

FIGURE 9.8 Flow battery schematic diagram.

Example 9.8

A 12 V lead-acid car battery has a measured voltage of 11.2 V when delivering 40 A to a load. What are the load and the internal battery resistances? Determine its instantaneous power and the rate of sulfuric acid consumption.

SOLUTION

For the battery equivalent circuit of Figure 9.8, the voltage across the load is:

$$V_L = V_0 - I \cdot R_S$$

and the internal battery and load resistances are:

$$R_S = \frac{12.0 - 11.2}{45} = 0.02 \ \Omega$$

$$R_L = \frac{V_0}{I} - R_S = \frac{12}{40} - 0.02 = 0.3 - 0.02 = 0.28 \ \Omega$$

The power delivered by the battery is:

$$P = V_L \times I = 11.2 \times 40 = 448 \ \text{W}$$

In order to perform the analysis of a battery cell, let us consider a reversible battery cell with terminal potential difference V, discharging a current I through a variable load resistance. The electron flow through the load produces the instantaneous electrical power, $I \cdot V$. At low current levels the cell voltage is close to the cell EMF, which is the open-circuit voltage, and the external work done and power output are small. To estimate the sulfuric acid (reactant) consumption rate, we suppose that the battery consumes N_{Rct} moles of reactant per second, and the released electrons in reaction flow through electrodes, and external circuit is proportional to the reaction rate, jN_{Rct}, the rate of flow of electrons from the cell (in mole per second), while j ($j = 2$ for lead-acid batteries) is the number of moles of electrons released per mole of reactant. Suppose that the chemical reaction in the battery consumes reactant at an electrode at a rate of N_C moles per second. Electrons released in the reaction flow through the electrode to the external load at a rate proportional to the rate of reaction, $j_{rct} \cdot N_C$, where j_{rct} is the number of moles of electrons released per mole of reactants. For example, for a lead-acid battery, two moles of electrons are freed to flow through the external load for each mole of lead reacted, j_{rct} is equal to 2. Thus $j_{rct} \cdot N_C$, is the cell electron flow rate, in moles of electrons per second. There are 6.023×10^{23} electrons per gram-mole of electrons, and each has a charge of 1.602×10^{-19} C, and their product, F, is the Faraday constant and is equal to 96,488 C/g-mol. The electric current from a cell may then be related to the rate of reaction in the cell as:

$$I = j N_{Rct} F \ \text{(A)} \tag{9.19}$$

F is the Faraday constant equal to the Avogadro number times the electron charge:

$$F = 6.023 \times 10^{23} \cdot 1.609 \times 10^{-19} = 96,488 \ \text{C/g-mole}$$

The instantaneous power delivered by the cell is then:

$$P = j N_{Rct} \cdot F \cdot V \tag{9.20}$$

The battery cell electrode and electrolyte materials and cell design determine the maximum cell voltage. Equations (9.19) and (9.20) show the nature and rate of chemical reactions that are controlling the cell current and maximum power output of a cell. Moreover, it is clear that the store of consumable battery reactants sets a limit on battery capacity.

Example 9.9

For the battery of previous example, 40 A current, estimate the acid consumption rate.

SOLUTION

For the battery in this example, I equal to 40 A, the sulfuric acid consumption rate, calculated from Equation (9.20) is:

$$N_{Rct} = \frac{I}{jF} = \frac{40}{2 \times 96488} = 2.0728 \times 10^{-4} \ \text{g-mole/s}$$

Because the sulfuric acid is consumed at the same rate at both anode and cathode, the above value is multiplied by 2, and it can be expressed in kg/hour, by using the sulfuric acid molecular weight:

$$N_{Rct} = 2 \times 2.0728 \times 10^{-4} \cdot 10^{-3} \cdot 98 \times 3600 = 0.1463 \ \text{kg/hr}$$

9.3.3.2 Summary of Battery Parameters
1. *Cell and battery voltage:* All battery cells have a nominal voltage, which gives the approximate voltage when delivering power. Cells are connected in series to give the overall battery voltage required. Figure 9.8 shows one of the equivalent battery electric circuits.
2. *Battery charge capacity (Ahr):* The most critical parameter is the electric charge that a battery can supply, and it is expressed in ampere hour (Ahr). For example, if the capacity of a battery is 100 Ahr, the battery can supply 1 A for 100 hours. The storage capacity, a measure of the total electric charge of the battery, is an indication of the capability of a battery to deliver a particular current value for a given duration. Capacity is also sometimes indicated as energy storage capacity, in watt-hours (Wh).
3. *Cold-cranking-amperage (CCA):* For a battery composed of nominal 2 V cells, cold-cranking-amperage is the highest current (A) that the battery can deliver for 30 seconds at

a temperature of 0°F and still maintain a voltage of 1.2 V per cell. The CCA is commonly used in automotive applications, where the higher engine starting resistance in cold winter conditions is compounded by reduced battery performance. At 0°F the cranking resistance of a car engine may be increased more than a factor of two over its starting power requirement at 80°F, while the battery output at the lower temperature is reduced to 40% of its normal output. The battery output reduction is due to the decrease of chemical reaction rates with temperature decreases.

4. *Storage capacity:* A measure of the total electric charge of the battery, storage capacity is usually quoted in ampere hours rather than in coulombs (1 A·hr = 3600 C), being an indication of the capability of a battery to deliver a particular current value for a given duration. Thus a battery that can discharge at a rate of 5 A for 20 hours has a capacity of 100 A·hr. Capacity is also sometimes indicated as energy storage capacity, in watt-hours. The energy stored in a battery depends on the *battery voltage*, and the *charge* stored. The SI unit for energy is joule (J); however, this is an inconveniently small unit, and in practical application watt-hour (Wh) is used instead. The energy is expressed in Wh as:

$$\text{Energy (Wh)} = V \times A \cdot hr \qquad (9.21)$$

5. *Specific energy* is the amount of electrical energy stored in a battery per unit battery mass (kg), and in practical applications is expressed in Wh/kg.

6. *Energy density* is the amount of electrical energy per unit of battery volume, expressed in practical applications and engineering in Wh/m^3.

7. *Charge (Ahr) efficiency* of actual batteries is less than 100%, and depends on the battery type, temperature, rate of charge, and varies with the battery state of charge. An ideal battery will return the entire stored charge to a load, so its charge efficiency is 100%.

8. *Energy efficiency* is another important parameter, defined as the ratio of the electric energy supplied by the battery to the electric energy required to return the battery at its state before discharge. In other words, it is the ratio of the energy delivered by a fully charged battery to the recharging energy required to restore it to its original state of charge.

9. *Self-discharge rate* refers to the fact that most battery types are discharging, when left unused, the self-discharge, an important battery characteristic, meaning that some batteries must be recharged after longer periods. The self-discharge rate varies with battery type, temperature, and storage conditions.

10. *Battery temperature, heating, and cooling.* Although most battery types run at ambient temperatures, there are battery types that need heating at start and then cooling when in use. For some battery types, performances vary with temperature. The temperature effects, cooling, and heating are important parameters that designers need to take into consideration.

11. Most rechargeable batteries have a limited *number of deep cycles* of 20% of the battery charge, in the range of hundreds or thousands cycles, which also determines the **battery life**. The number of deep cycles depends on the battery type, design details, and the ways and conditions that a battery is used. This is a very important parameter in battery specifications, reflecting the battery lifetime, and its cost.

9.3.3.3 Flow Batteries and Special Battery Types

A flow battery is a type of rechargeable secondary battery in which energy is stored chemically in liquid electrolytes. Secondary batteries use the electrodes as an interface for collecting or depositing electrons and as a storage site for the products or reactants associated with the battery's reactions. Consequently, their energy and power densities are set by the electrodes' size and shape. Flow batteries, on the other hand, store and release electrical energy by means of reversible electrochemical reactions in two liquid electrolytes. Their electrolytes contain dissolved electro-active species that flow through a power cell, converting chemical energy to electricity. So, flow batteries are fuel cells that can be recharged. From a practical point of view, the storage of reactants is very important. Gas storage in fuel cells requires large, high-pressure tanks or cryogenic storage, prone to thermal self-discharge. Flow batteries do offer some advantages over secondary batteries, such as: the system capacity is scalable by simply increasing the solution amount, leading to cheaper installation costs as the systems get larger, the battery can be fully discharged with no ill effects, and it has little loss of electrolyte over time. Flow battery flow cells (or redox flow cells) convert electrical energy into chemical potential energy through a reversible electrochemical reaction between two electrolyte solutions. In contrast to conventional batteries, they store the energy in the electrolyte solutions, and the power and energy ratings of redox flow cells are independent variables. Their power rating is determined by the active area of the cell stack assembly and their storage capacity by the electrolyte quantity. Figure 9.8 depicts a flow cell energy storage system, having two compartments, one for each electrolyte, physically separated by an ion-exchange membrane, basically two flow loops. Electrolytes flow into and out of the cell through separate manifolds and undergo chemical reaction inside the cell, with ion (proton) exchange through the cell membrane and electron exchange through the external electric circuit. The electrolytes' chemical energy is turned into electrical energy and vice versa. All flow batteries work in the same way but vary in the chemistry of electrolytes. Because the electrolytes are stored separately and in large containers (with a low surface-area-to-volume ratio), flow batteries show promise to have some of the lowest self-discharge rates of any energy storage technology available. The development activities have been centered on four principal electrochemistry combinations for flow batteries: vanadium-vanadium, zinc-bromine, polysulfide-bromide, and zinc-cerium, although others are under development, too. Their major advantages are high power capacity and long life; while the main disadvantages are low energy density and efficiency. Flow batteries can be used for any energy storage applications, including load leveling, peak shaving, and renewable energy integration. Under normal conditions, flow batteries can charge or discharge very fast, making them useful for frequency response and voltage control. However, poor energy densities and specific energies remand these battery types to utility-scale power shaping and smoothing, although they might be adaptable for distributed-generation use. Nevertheless, the Tennessee Valley Authority released a finding of no significant impact for a proposed 430 GJ facility and deemed it safe. Three types of flow batteries are closing in on commercialization: *vanadium redox, polysulfide bromide*, and *zinc bromide*. Production of flow cell-based energy storage systems proceeds at a slow pace, via the activities of a relatively small number of developers and suppliers. The two costs associated with flow batteries are: the power and the energy costs, as they are independent of each other. However, installations to date have principally used the vanadium redox and zinc-bromine. The polysulfide-bromide system was developed for grid-connected, utility-scale storage applications at power ratings from 5 MW upwards.

Zinc-bromine (Zn-Br) batteries are a type of redox flow battery, operating by using a pump system that circulates reactants through the battery. One manufacturer, ZBB Energy, builds Zn-Br batteries in 50 kWh modules made of three parallel connected 60-cell battery stacks. The modules are rated to discharge at 150 A at an average voltage of 96 V for 4 hours. The battery stack design allows for individual stacks to be replaced instead of the entire module. Zinc-bromine battery life time is rated at 2500 cycles. The electrochemical charging and discharging reaction is reversible and nondestructive, meaning it is capable of 100% depth of discharge. The Zn-Br battery uses less toxic electrolytes compared to lead-acid batteries, making them a more environmentally friendly choice. Zinc-bromine flow batteries lack technological maturity, with only a few real installations. The overall chemical reaction during discharge:

$$Zn + Br_2 \rightarrow ZnBr_2, \quad \text{and with the theoertical voltage: } E_0 = 1.85 \text{ V} \tag{9.22}$$

During the discharge product of reaction, the soluble zinc bromide is stored, along with the rest of the electrolyte, in the two loops and external tanks. During charge, bromine is liberated on the positive electrode and zinc is deposited on the negative electrode. Bromine is then mixed with an organic agent to form a dense, oily liquid poly-bromide complex system. It is produced as droplets and these are separated from the aqueous electrolyte on the bottom of the tank in a positive electrode loop. During discharge, bromine in the positive electrode loop is again returned to the cell electrolyte in the form of a dispersion of the poly-bromide oil. Vanadium redox batteries have the advantage of being able to store electrolyte solutions in non-pressurized vessels at room temperature. Vanadium-based redox flow batteries hold great promise for storing electric energy on a large scale (theoretically, they have infinite capacity).

Vanadium redox battery flow battery (VRB) technology was pioneered at the University of New South Wales, Australia, and has shown potential for long cycle life and energy efficiencies of over 80% in large installations. A vanadium redox battery is another type of a flow battery in which electrolytes in two loops are separated by a proton exchange membrane (PEM). The VRB uses compounds of the element vanadium in both electrolyte tanks. The electrolyte is prepared by dissolving vanadium pentoxide (V_2O_5) in sulfuric acid (H_2SO_4). The electrolyte in the positive electrolyte loop contains $(VO_2)^+ - (V^{5+})$ and $(VO)^{2+} - (V^{4+})$ ions, the electrolyte in the negative electrolyte loop, V^{3+} and V^{2+} ions. Chemical reactions proceed on the carbon electrodes, while the reaction chemistry at the positive electrode is:

$$VO_2^+ + 2H^+ + e^- \leftrightarrow VO^{2+} + H_2O, \text{ with } E_0 = +1.00 \text{ V} \tag{9.23a}$$

And at the negative electrode, the reaction is:

$$V^{2+} \leftrightarrow V^{3+} + e^-, \text{ with } E_0 = -0.26 \text{ V} \tag{9.23b}$$

Under the actual VRB cell conditions, an open-circuit voltage of 1.4 V is observed at 50% state of charge, while a fully charged cell produces over 1.6 V at open-circuit, and a fully discharged cell of 1.0 V. The extremely large capacities possible for VRB batteries make them well-suited in large grid applications, where they could average out the production of highly variable wind and/or solar power sources. Their extremely rapid response times make them suitable for uninterruptible power system applications, where they can be used to replace lead-acid batteries. Disadvantages of vanadium redox batteries are low energy density of about 25 Wh/kg of electrolyte, low charge efficiency (need to use pumps), and a high price.

Polysulfide-bromide (PSB) battery technology was developed in Canada for utility-scale storage applications at power ratings from 5 MW upwards. A PSB battery utilizes two salt solution electrolytes, sodium bromide (NaBr) and sodium polysulfide (Na_2S_x). PSB electrolytes are separated in the battery cell by a polymer membrane that only passes positive sodium ions. The chemical reaction at the positive electrode is:

$$NaBr_3 + 2Na^+ + 2e^- \leftrightarrow 3NaBr \tag{9.24a}$$

And at the negative electrode the reaction is:

$$2Na_2S_2 \leftrightarrow Na_2S_4 + 2Na^+ + 2e^- \tag{9.24b}$$

This technology is expected to attain energy efficiencies of approximately 75%. Although the salt solutions themselves are only mildly toxic, a catastrophic failure by one of the tanks could release highly toxic bromine gas. Nevertheless, the Tennessee Valley Authority released a finding of no significant impact for a proposed 430 GJ facility and deemed it safe. PSB batteries have a very fast response time; they can react within 20 ms if electrolyte has retained charged in the stacks (of cells).

Sodium-Sulfur (Na-S) battery technology was originally developed by the Ford Motor Company in the 1960s. It contains sulfur at the positive electrode and sodium at the negative electrode.

These electrodes are separated by a solid beta alumina ceramic. Through an electrochemical reaction, electrical energy is stored and released on demand. Na-S batteries have an operating temperature between 300°C and 360°C. The primary manufacturer of Na-S batteries, NGK Insulators, LTD, builds the batteries in 50 kW modules that are combined to make MW power battery systems. Na-S batteries exhibit higher power and higher energy density, higher columbic efficiency, good temperature stability, long cycle life, and low material costs. Their energy density is approximately three times that of traditional lead-acid batteries, with a high DC conversion efficiency of approximately 85%, making them an ideal candidate for use in future DC distribution networks. Na-S batteries can be used for a wide variety of applications including peak shaving, renewable energy grid integration, power quality management, and emergency power units. They have the ability to discharge above their rated power, which makes them ideal to operate in both a peak shaving and power quality management environment.

9.3.4 FUEL CELLS AND HYDROGEN ENERGY

Fuel cells are an interesting alternative for power generation technologies because they have higher efficiencies and very low environmental effects. A fuel cell is an electrochemical-conversion device that has a continuous supply of fuel such as hydrogen, natural gas, or methanol, and an oxidant such as oxygen, air, or hydrogen peroxide. It can have auxiliary parts to feed the device with reactants as well as a battery to supply energy for start-up. In conventional power generation systems, fuel is combusted to generate heat and then heat is converted to mechanical energy before it can be used to produce electrical energy. The maximum efficiency that a thermal engine can achieve, when it operates at the Carnot cycle, is related to the ratio of the heat source and sink absolute temperatures. Fuel cell operation is based on electrochemical reactions and not fuel combustion, by avoiding the chemical energy conversion into mechanical energy, the thermal phase enables fuel cells to achieve higher efficiency than that of conventional power generation technologies. A fuel cell can be considered as a "cross-over" of a battery and a thermal engine, resembling an engine because theoretically it can operate as long as it is fed with fuel. However, its operation is based on electrochemical reactions, resembling batteries providing significant advantages for fuel cells. On the other hand, batteries are devices that when their chemical energy is depleted, they must be replaced or recharged, whereas fuel cells can generate electricity as long as they are fueled. However, fuel cells resemble rechargeable batteries, while their theoretical open-circuit voltage is given by Equation (9.5). The open-circuit voltage of a single fuel cell is about 1.2 V. Hydrogen has been preferred for quite a long time as an environmentally friendly and powerful energy storage medium and fuel. Among the most important advantages of using hydrogen as an energy storage medium are: it is the lightest element, very stable compound, on a volumetric basis can store several times more energy than compressed air, reacts easily with oxygen to generate energy, forms water, harmless to the environment, can be easily used in fuel cells, and has a long industrial application history. However, hydrogen has a few disadvantages as an energy storage medium: flammable and explosive, requiring special containers and transportation; highly diffusive, due to its specific energy content; and being the lightest element high-pressure and large containers must be used for significant mass storage. Hydrogen (H_2) can be produced with electrolysis, consisting of an electric current applied to water separating it into components O_2 and H_2. The oxygen has no inherent energy value, but the higher heat value (HHV) of the resulting hydrogen can contain up to 90% of the applied electric energy. This hydrogen can then be stored and later combusted to provide heat or work, or to power fuel cells. Compression to a storage pressure of 350 bar, the value usually assumed for automotive technologies, consumes up to 12% of the hydrogen's HHV if performed adiabatically, although the loss approaches a lower limit of 5% in a quasi-isothermal compression. Alternatively, the hydrogen can be stored in liquid form, a process that costs about 40% of HHV, using current technology, and that at best would consume about 25%. In the context of the hydrogen economy, hydrogen will be an energy carrier, rather than a primary energy source. It may be generated through electrolysis or other industrial processes, using the energy harnessed by wind or solar power conversion systems, or chemical methods, and used as fuel with almost no harmful environmental impacts. However, the establishment of the hydrogen economy requires that the current hydrogen storage and transportation issues are solved, and suitable materials for storage are readily available. A hydrogen economy may lead to widespread use of renewable-energy-based power generation, avoiding the use of expensive and environmentally harmful fossil fuels. Whether a hydrogen economy evolves in the near or far future is strongly dependent on technological advances in storage and transportation.

9.3.4.1 Fuel Cell Principles and Operation

Fuel cells are electrochemical devices that produce electricity from paired oxidation or reduction reactions, being in some ways batteries with flows/supplies of reactants in and products out. Invented in about 1840, fuel cells are hardly a new idea, but they are really making their mark as a power source for electric vehicles, space applications, and consumer electronics, especially in the last part of the twentieth century, and their time is about to come. A battery has all of its chemicals stored inside, and it converts those chemicals into electricity. This means that a battery eventually "goes dead" and you either throw it away or recharge it. With a fuel cell, chemicals constantly flow into the cell so it never goes dead—as long as there is a flow of chemicals into the cell, the electricity flows out of the cell. Fuel cells, similar to batteries, exhibit higher efficiency at partial load than at full load and with less variation over the entire operating range, having good load-following characteristics. Fuel cells are modular in construction with consistent efficiency regardless of size. Reformers, however, perform less efficiently at part load so that overall system efficiency suffers when used in conjunction with fuel cells. Fuel cells, like batteries, are devices that react chemically and instantly to changes in load. However, fuel cell systems are comprised of predominantly mechanical devices, each of which have their own response time to changes in load demand. Nonetheless, fuel cell systems that operate on pure hydrogen tend to have excellent overall response. Fuel cell systems that operate on reformate using an on-board reformer, can be sluggish, particularly if steam reforming techniques are used. Fuel cells are distinguished from secondary rechargeable batteries by their external fuel storage and extended lifetime. Fuel cells have the advantages of high efficiency, low emissions, quiet operations, good reliability, and fewer moving parts (only pumps and fans to circulate coolant and reactant gases) over other energy-generation systems. Their generation efficiency is very high in fuel cells with higher power density, lower-vibration characteristics. A fuel cell is a DC voltage source, operating at about 1 V level. However, this might be set to change over the next 20 or 30 years. The basic principle of the fuel cell is that it uses hydrogen fuel to produce electricity in a battery-like device, as discussed later. The fuel cell basic chemical reaction is:

$$2H_2 + O_2 \rightarrow 2H_2O + \text{Energy} \tag{9.25}$$

The reaction products are thus water and energy. The sole reaction product of a hydrogen-oxygen fuel cell is water, an ideal product from a pollution standpoint. In addition to electrons, heat is also a reaction product. This heat must be continuously removed, as it is generated, in order to keep the cell reaction isothermal. A fuel cell-based vehicle can be described as zero-emission, running off a fairly normal chemical fuel (hydrogen), a very reasonable energy can be stored, and its range is quite satisfactory, thus offering the only real prospect of a silent zero-emission vehicle with a range and performance broadly comparable with internal combustion engine vehicles. It is not surprising then that for many years, there have been those who have seen fuel cells as a technology that shows great promise, and could even make serious inroads into the internal combustion engine domination. Fuel cells must overcome quite a few problems and challenges before they become a commercial reality as a vehicle power source or other power applications. The main problems and issues are listed here. Fuel cells are currently far more expensive than IC engines, and even hybrid IC/electric systems, so cost reduction is a priority. Water management is another important and difficult issue with automotive fuel cells. The thermal management of fuel cells is actually rather more difficult than for

IC engines. Hydrogen is the preferred fuel for fuel cells, but hydrogen is very difficult to store and transport. There is also the vital question of, "Where does the hydrogen come from?" These issues are difficult and important, with many rival solutions. However, there is great hope that these problems can be overcome, and fuel cells can be the basis of less environmentally damaging transport. We have seen that the basic principle of the fuel cell is the release of energy following a chemical reaction between hydrogen and oxygen. The key difference between this and simply burning the gas is that the energy is released as an electric current, rather than heat. The four main parts of a fuel cell are: *the anode, the cathode, the catalyst,* and *the proton exchange membrane (PEM).* The catalyst is a special material that facilitates the reaction of oxygen and hydrogen, made usually of platinum powder very thinly coated onto carbon paper or cloth. The catalyst is rough and porous so that the maximum surface area of the platinum can be exposed to the hydrogen or oxygen. The platinum-coated side of the catalyst faces the PEM. The electrolyte is the proton exchange membrane, being a specially treated material that only conducts positively charged ions, and the membrane blocks the electrons. In order to understand the fuel cell operation, the separate reactions taking place at each electrode must be considered. These important details vary for different types of fuel cells, but if we start with a cell based on an acid electrolyte, we shall consider the simplest and the most common type. At the anode of an acid electrolyte fuel cell, the hydrogen gas ionizes, releasing electrons and creating H^+ ions:

$$2H_2 \rightarrow 4H^+ + 4e^- \tag{9.26}$$

During this reaction, energy is released. At the cathode, oxygen reacts with electrons taken from the electrode, and H^+ ions from the electrolyte, to form water.

$$O_2 + 4H^+ + 4e^- \rightarrow 2H_2O \tag{9.27}$$

In order for these reactions to proceed continuously, electrons produced at the anode must pass through an electrical circuit to the cathode. Also H^+ ions must pass through the electrolyte, so an acid, having free H^+ ions, serves this purpose very well. Certain polymers can also contain mobile H^+ ions. These reactions may seem simpler, but they are not in normal circumstances. Also, the fact that hydrogen has to be used as a fuel is a disadvantage. To solve these and other problems, many fuel cell types have been researched. The different types are usually distinguished by the electrolyte used, though there are other important differences as well. In fuel cells, similar to batteries, the electrode reactions are surface phenomena, occurring at a liquid-solid or gas-solid interface, and therefore proceed at a rate proportional to the exposed solid areas. For this reason, porous electrode materials are used, often porous carbon impregnated or coated with a catalyst to speed the reactions. Thus, because of microscopic pores, the effective area of the electrodes is very large. Any phenomenon that prevents the gas from entering the pores or deactivates the catalyst must be avoided in order for a fuel cell to function effectively over longer periods. Because of the reaction rate-area relation, fuel cell current and power output increase with increased cell area. The surface power density (W/m^2) is an important parameter in comparing fuel cell designs, and the fuel cell power output can be scaled up by increasing its surface area. The electrolyte acts as an ion transport medium between electrodes. The passage rate of the positive charge through the electrolyte must match the rate of electron arrival at the opposite electrode to satisfy the physical requirement of electrical neutrality of the discharge fluids. Impediments to the ion transport rate through the electrolyte can limit current flow and the power output. Thus care must be taken in design to minimize the length of ion travel path and other factors that retard the ion transport. The theoretical maximum energy of an isothermal fuel cell (or other isothermal reversible control volume) is the difference in free energy (Gibbs) functions of the cell reactants and the products. The drop in free energy is mainly associated with Equation (9.28), expressing free energy of the chemical potential, e.g., of the H^+ ions dissolved in the electrolyte, which relate to the maximum (open-circuit) voltage of a fuel cell through Equation (9.5). The maximum voltage of a fuel cell is about 1.23 V, as estimated in the example below. The maximum work output at which a fuel cell is able to perform is given by the decrease in its free energy, ΔG. The fuel cell conversion efficiency, η_{fc}, is defined as the electrical energy output per unit mass (or mole) of fuel to the corresponding heating value of the fuel consumed. The total energy drawn from the fuel, given by the maximum energy available from the fuel in an adiabatic steady-flow process, is the difference in the inlet and exit enthalpies, ΔH. The thermal fuel cell efficiency is expressed as:

$$\eta_{fc}(thermal) = \frac{\Delta G}{\Delta H} \tag{9.28}$$

Example 9.10

Determine the open-circuit voltage, maximum work, and the thermal efficiency for a direct hydrogen-oxygen fuel cell at standard reference conditions (see Table B.11, Appendix B). Consider two cases when the product is liquid water and vapor.

SOLUTION

The maximum work of the fuel cell is the difference of the free energy of the products and the free energy of the reactants. The maximum work for the liquid-water product and the water-vapor product are, respectively:

$$W_{max}(liquid) = G_r - G_p(l) = 0 - (-237,141) = 237,141 \text{ kJ/kg} \cdot mole$$
$$W_{max}(vapor) = G_r - G_p(v) = 0 - (-228,582) = 228,582 \text{ kJ/kg} \cdot mole$$

The theoretical ideal one-circuit voltages, in the two cases, are computed with Equation (9.6), as:

$$V_{max}(liquid) = \frac{W_{max}(l)}{nF} = \frac{237141 \times 10^3}{2 \times 96400 \times 10^3} = 1.2299 \approx 1.23 \text{ V}$$
$$V_{max}(vapor) = \frac{W_{max}(v)}{nF} = \frac{228582 \times 10^3}{2 \times 96400 \times 10^3} = 1.1855 \approx 1.19 \text{ V}$$

The change in the enthalpies for water-liquid product and water-vapor product (Table B.11, Appendix B) are: 285,830 kJ/kg·mole and 241,826 kJ/kg·mole, respectively. The thermal maximum efficiencies in the two cases, computed with Equation (9.28), are:

$$\eta_{fc}(liquid) = \frac{\Delta G}{\Delta H} = \frac{237141}{285830} = 0.83 \text{ or } 83\%$$
$$\eta_{fc}(vapor) = \frac{\Delta G}{\Delta H} = \frac{228582}{241826} = 0.95 \text{ or } 95\%$$

However, the actual efficiencies are lower than in this example, while the actual open-circuit voltage (potentials) of 0.7 V to 0.9 V are typical for most of fuel cells, regardless of type. Losses or inefficiencies in fuel cells, often called polarizations, are reflected in a cell voltage drop when the fuel cell is under load. Three major types of polarization are: (a) the *ohmic polarization* due to the internal resistance to the motion of electrons through electrodes and of ions through the electrolyte; (b) the *concentration polarization* due to the mass transport effects relating to diffusion of gases through porous electrodes and to the solution and dissolution of reactants and products; and (c) the *activation polarization*, which is related to the activation energy barriers for the various steps in the oxidation-reduction reactions at the electrodes. The net effect of these and other polarizations is a decline in terminal voltage with increasing current drawn by the load. This voltage drop is reflected in a peak in the fuel cell power-current characteristic. The high thermal efficiencies of the fuel cells are based on the operation of the fuel cell as an isothermal, steady-flow process. Although heat must be rejected, the transformation of chemical energy directly into electron flow does not rely on heat rejection in a cyclic process to a sink at low temperature as in a heat engine, which is why the Carnot efficiency limit does not apply. However, the inefficiency associated with the various polarizations and the incomplete fuel utilization tends to reduce overall cell conversion efficiencies to well below the thermal efficiencies predicted in the example above. Fuel cells, like other energy-conversion systems, do not function in exact conformance with simplistic models or without inefficiencies. Defining the voltage efficiency, η_V, as the ratio of the terminal voltage to the theoretical EMF, V_T/EMF, and the current efficiency, η_I, as the ratio of the cell electrical current to the theoretical charge flow associated with the fuel consumption, the overall fuel cell conversion efficiency is related to the thermal efficiency by this relationship:

$$\eta_{fc} = \frac{V_T}{EMF} \times \frac{I}{jN_{Rct}F} \times \eta_{th} = \eta_V \cdot \eta_I \cdot \eta_{th} \tag{9.29}$$

Example 9.11

The actual DC output voltage of a hydrogen-oxygen fuel cell is 0.80 V. Assuming its current efficiency 1.0, what is the fuel cell actual efficiency?

SOLUTION

From the previous example, the fuel cell thermal efficiency is 0.83, and the voltage efficiency is equal to 0.80/1.23 = 0.65. Then the actual fuel cell efficiency, Equation (9.29), is:

$$\eta_{fc} = \eta_V \cdot \eta_I \cdot \eta_{th} = 0.65 \times 1.0 \times 0.83 = 0.54 \text{ or } 54\%$$

9.3.4.2 Fuel Cell Types and Applications

An important element of fuel cell design is that, similar to the large battery systems, they are built from a large number of identical units or cells. Each has an open-circuit voltage on the order of 1 V, depending on the oxidation-reduction reactions taking place in the cell. Fuel cells are usually built in sandwich-style assemblies called *stacks*, while the fuel and oxidant cross-flow through a portion of the stack. A fuel cell stack can be configured with many groups of cells in series and parallel connections to further tailor the voltage, current, and power produced. The number of individual cells contained within one stack is typically greater than 50 and varies significantly with stack design. The basic components that comprise a fuel cell stack include the electrodes and electrolyte with additional components required for electrical connections and/or insulation and the flow of fuel and oxidant through the stack. These key components include current collectors and separator plates. The current collectors conduct electrons from the anode to the separator plate. The separator plates provide the electrical series connections between cells and physically separate the oxidant flow of one cell from the fuel flow of an adjacent cell. The channels in the current collectors serve as distribution pathways for the fuel and oxidant. Often, the two current collectors and the separator plate are combined into a single unit called a bipolar plate. Electrically conducting bipolar separator plates serve as DC transmission paths between successive stack cells. This modular type of construction allows research and development of individual cells and engineering of fuel cell systems to proceed in parallel. The preferred fuel for most fuel cell types is hydrogen. Hydrogen is not readily available, however, but the infrastructure for reliable extraction, transport, distribution, refining, and/or purification of hydrocarbon fuels is well-established. Thus, fuel cell systems that have been developed for practical applications to date have been designed to operate on hydrocarbon fuels. In addition to the fuel cell system requirement of a fuel processor for operation on hydrocarbon fuels, a power conditioning, and for grid-connection to supply an AC load, an inverter is also needed. Stationary and mobile applications use five major types of fuel cells. All these fuel cell types have the same basic design as mentioned above, but with different chemicals used as the electrolyte. These fuel cells are:

1. Alkaline fuel cell (AFC);
2. Phosphoric acid fuel cell (PAFC);
3. Molten carbonate fuel cell (MCFC);
4. Solid oxide fuel cell (SOFC); and
5. Proton exchange membrane fuel cell (PEMFC).

Regardless of type, all these fuel cells require fairly pure hydrogen fuel to run. However, large amounts of hydrogen gas are difficult to transport and store. Therefore, a reformer is normally equipped inside these fuel cells to generate hydrogen gas from liquid fuels such as gasoline or methanol. Among these five types of fuel cell, PEMFC has the highest potential for widespread use. PEMFC is getting cheaper to manufacture and easier to handle. It operates at relatively low temperatures when compared with other types of fuel cells. AFC systems have the highest efficiency and therefore have been used for more than 30 years to generate electricity in spacecraft systems. However, AFC requires very pure hydrogen and oxygen to operate; thus, the running cost is very expensive. As a result, AFCs are unlikely to be used extensively for general purposes, such as in vehicles and in our homes. In contrast, the MCFC and the SOFC are specially designed to be used in power stations to generate electricity in large-scale power systems. Nevertheless, there are still a lot of technical and safety problems associated with the use of MCFC and SOFC fuel cells in long-term applications. Apart from these fuel cell types, a new type of fuel cell, the direct methanol fuel cell (DMFC), is under vigorous ongoing research and may soon be coming to the market. This type of fuel cell has the same operating mechanism as PEMFC, but instead of using pure hydrogen, it is able to use methanol directly as the basic fuel. A reformer is therefore not essential in this fuel cell system to reform complex hydrocarbons into pure hydrogen. Several companies around the world are presently working on DMFC to power electronic equipment. The DMFC appears to be the most promising alternative electric source to replace the batteries used in portable electronics such as mobile phones and laptop computers. Although the development of fuel cell technologies is fast, fuel cells have not yet reached their potential level of commercial success due to high material costs (Platinum electrode) and market barriers.

9.3.4.3 Flywheel Energy Storage

A flywheel energy storage (FES) system converts electrical energy supplied from a DC or three-phase AC power source into kinetic energy of a spinning mass or converts kinetic energy of a spinning mass into electrical energy. Basically, a flywheel is a disk with a certain amount of mass that can spin, holding kinetic energy. Modern high-tech flywheels are built with the disk attached to a rotor in upright position to prevent gravity influence. They are charged by a simple electric motor that simultaneously acts as a generator in the process of discharging. When dealing with efficiency, however, it gets more complicated: As stated by the rules of physics, flywheels will eventually have to deal with friction during operation. Therefore, the challenge to increase that efficiency is to minimize friction. This is mainly accomplished by two measures: the first one is to let the disk spin in a vacuum, so there will be no air friction; and the second one is to bear the spinning rotor on permanent and electromagnetic bearings so it basically floats. The spinning speed for a modern single flywheel reaches up to 16000 rpm and offers a capacity up to 25 kWh, which can be absorbed and injected almost instantly. These devices are comprised of a massive or composite flywheel coupled with a motor-generator and special brackets (often magnetic), set inside a housing at very low pressure to reduce self-discharge losses. They have a great cycling capacity (a few 10,000 to a few 100,000 cycles) determined by fatigue design. For electrical power system applications, high-capacity energy flywheels are needed. Friction losses of a 200 ton flywheel are estimated at about 200 kW. Using this hypothesis and an efficiency of about 85%, the overall efficiency would drop to 78% after 5 hours, and 45% after 1 day. Long-term energy storage with flywheels is therefore not foreseeable. A FES device is made up of a shaft, holding a rotor, rotating on two magnetic bearings to reduce friction. These are all contained within a vacuum to reduce aerodynamic drag losses. Flywheels store energy by accelerating the rotor/flywheel to a very high speed and maintaining the energy in the system as kinetic energy, and release the energy by reversing the charging process so that the motor is used as a generator. As the flywheel discharges, the rotor slows down until eventually coming to a complete stop. The rotor dictates the amount of energy that the flywheel is capable of storing. Due to their simplicity, FES systems have been widely used in commercial small-power units (about 3 kWh) in the range of from 1 kW and 3 hours to 100 kW and 3 seconds. Energy is stored as kinetic energy using a rotor:

$$E = \frac{1}{2}J\omega^2 \tag{9.30}$$

where J is the momentum of inertia and ω is the angular velocity. The moment of inertia is given by the volume integral taken over the product of mass density, ρ, and squared distance r^2 of mass elements with respect to the axis of rotation:

$$J = \int \rho r^2 \, dV$$

However, in the case of regular geometries, the momentum of inertia is given by:

$$J = kmr^2 \tag{9.31}$$

Here, m is the flywheel total mass, and r the outer radius of the disk. By including the momentum of inertia expression, Equation (9.30) then becomes:

$$E = \frac{1}{2}kmr^2\omega^2 = \frac{1}{2}k(\rho\Delta V)r^2\omega^2 \tag{9.32}$$

Here, ΔV is the increment of the volume. The inertial constant depends on the shape of the rotating object. For thin rings, $k = 1$, while for a solid uniform disk, $k = 0.5$. A flywheel rotor is usually a hollow cylinder with magnetic bearings to minimize the friction. The rotor is located in a vacuum pipe to decrease the friction even more. The rotor is integrated into a motor/generator machine that allows energy flow in both directions. The energy storage capacity depends on the mass and shape of the rotor and on the maximum available angular velocity.

Example 9.12

A flywheel is a uniform circular disk of diameter of 2.90 m, 1500 kg, and is rotating at 5000 rpm. Calculate the flywheel kinetic energy.

SOLUTION

In Equation (9.32), $k = 0.5$, so:

$$E = \frac{1}{2}kmr^2\omega^2 = \frac{1}{2} \times 0.5 \times 1500 \times \left(\frac{2.90}{2}\right)^2 \times \left(\frac{2\pi 5000}{60}\right)^2 = 2.16142 \times 10^8 \, J = 216.142 \, MJ$$

There are two topologies, *slow* flywheels (with angular velocity below 6000 rpm) based on steel rotors, and *fast* flywheels (below 60,000 rpm) that use advanced material rotors (carbon fiber or glass fiber) that present a higher energy and power density than steel rotors. The flywheel designs are modular and systems of 10 MW are possible, with efficiency of 80%–85%, with a useful life of 20 years. Advances in rotor technology have permitted high dynamic and durability of tenths of thousands of cycles. These characteristics make them suitable for power quality applications: frequency deviations, temporary interruptions, voltage sags, and voltage swells. FES applied to the renewable energies trend is to combine them with other technologies, like micro-CAES or thermal energy storage. Flywheels store power in direct relation to the mass of the rotor, but to the square of its surface speed. The amount of energy that can be stored by a flywheel is determined by the material design stress, material density, and total mass, as well as flywheel shape factor K. It is not directly dependent on size or angular speed because one of these can be chosen independently to achieve the required design stress. Material properties also govern flywheel design and therefore allowable K values. In order to take maximum advantage of the best properties of highly anisotropic materials, the flywheel shape is such that lower K values have to be accepted compared with those normally associated with flywheels made from isotropic material. However, at first sight from Equations (9.30) through (9.32), we are tempted to maximize spatial energy density by increasing the product of the three factors ω^2, ρ, and the squared distance r^2 of the relevant mass elements with respect to the axis of rotation. This approach, however, cannot be extended to arbitrarily high spatial energy densities, because the centrifugal stresses caused by the rotation also roughly increase with the product of these three factors. To avoid fragmentation of the flywheel, a certain material-dependent tensile stress level must not be exceeded. Consequently, the most efficient way to store energy in a flywheel is to make it spin faster, not by making it heavier. The energy density within a flywheel is defined as the energy per unit mass:

$$\frac{W_{KIN}}{m_{FW}} = 0.5 \cdot v_l^2 = \frac{\sigma}{\rho} \tag{9.33}$$

where W_{KIN} is the total kinetic energy in joules (J), m_{FW} is the mass of the flywheel in kg, v_l is the linear velocity of the flywheel in m/s, σ is the specific strength of the material in Nm/kg; and ρ is the density of the material in kg/m³. For a rotating thin ring, therefore, the maximum energy density is dependent on the specific strength of the material and not on the mass. The energy density of a flywheel is normally the first criterion for the selection of a material. Regarding specific strength, composite materials have significant advantages compared to metallic materials. Table 9.3 lists some flywheel materials and their properties. The burst behavior is a deciding factor for choosing a flywheel material. However, not all the stored energy during the charging phase can be used during

TABLE 9.3
Typical Materials Used for Flywheels and Their Properties

Material	Density (kg/m³)	Strength (MN/m²)	Specific Strength (MNm/kg)
Steel	7800	1800	0.22
Alloy (AlMnMg)	2700	600	0.22
Titanium	4500	1200	0.27
GFRP*	2000	1600	0.80
CFRP*	1500	2400	1.60

* GFRP, glass fiber-reinforced polymer; CFRP, carbon fiber-reinforced polymer

the discharging phase. The useful energy per mass unit (the energy released during the discharge) is expressed as:

$$\frac{E}{m} = \left(1 - s^2\right) K \frac{\sigma}{\rho} \qquad (9.34)$$

where s is the ratio of minimum to maximum operating speed, usually taken to be 0.2.

Example 9.13

Compute the energy density for a steel flywheel.

SOLUTION

Using Equation (9.33) and the values of the density and specific strength for steel, from Table 9.3, the flywheel energy density is:

$$\frac{W_{KIN}}{m_{FW}} = \frac{0.22 \times 10^6}{7800} = 28.21 \, \text{J/kg}$$

The power and energy capacities are decoupled in flywheels. In order to obtain the required power capacity, you must optimize the motor/generator and the power electronics. These systems, so-called low-speed flywheels, usually have relatively low rotational speeds, approximately 10,000 rpm and a heavy rotor made from steel. They can provide up to 1650 kW, for very short times (up to 120 s). To optimize the flywheel storage capacities, the rotor speed must be increased. These high-speed flywheels spin on a lighter rotor at much higher speeds, with prototype composite flywheels claiming to reach speeds in excess of 100,000 rpm. However, the fastest flywheels commercially available spin at about 80,000 rpm. They can provide energy up to 1 hour, with a maximum power of 750 kW. Over the past years, flywheel efficiency has improved up to 80% or higher. Flywheels have an extremely fast dynamic response, a long life (~ 20 years), require little maintenance, and are environmentally friendly. As the storage medium used in flywheels is mechanical, the unit can be discharged repeatedly and fully without any damage to the device. Flywheels are used for power quality enhancements, uninterruptable power supply, and to dampen frequency variation of the electrical output from wind turbines. The stored energy in flywheels has significant destructive potential when released uncontrolled, with safety being a major disadvantage. Efforts are made to design rotors that are less harmful when they break. However, even with careful design, a composite rotor still can fail dangerously. The safety of a flywheel system is not related only to the rotor. The housing enclosure, and all components and materials within it, can influence the result of a burst significantly, so low-loss and long-life bearings and suitable flywheel materials need to be developed. Some new materials are steel wire, vinyl-impregnated fiberglass, and carbon fiber. However, the major advantages of flywheels include high power density, nonpolluting, high efficiency, long life (over 20 years), and operation independent from extreme weather conditions. Their major disadvantages are safety, noise, and high-speed operations leading to wear, vibration, and fatigue.

9.4 THERMAL ENERGY STORAGE

Thermal energy storage (TES) technologies store thermal energy for later use as required, rather than at the time of production. They are therefore important counterparts to intermittent renewable-energy generation and also provide a way to use the waste process heat and reduce the energy demand of buildings, facilities, and industrial processes. The use of TES in buildings, industrial facilities, and processes in combination with distributed generation, space heating, hot water, and space cooling is an important way to conserve and to efficiently use energy. A variety of TES techniques have been developed, including building thermal mass utilization, phase change materials (PCM), underground thermal energy storage, heat pumps, and energy storage tanks. TES systems can be either centralized or distributed systems. Centralized TES systems are used in district heating or cooling systems, large industrial plants, large CHP units, or renewable energy plants. Solar thermal systems are often applied in residential and commercial buildings to capture solar energy for water and space heating or cooling. In such cases, TES systems can reduce the energy demand during peak times. The economic performance depends significantly on the specific application and operational needs, and the number and frequency of storage cycles. TES systems for cooling or heating are used where there is a time mismatch between the demand and the economically most favorable energy supply. TES can provide short-term storage for peak shaving as well as long-term storage for the introduction of renewable and natural energy sources. Sustainable buildings need to take advantage of renewable and waste energy to approach ultra-low-energy buildings. TES systems enable the collection and preservation of excess heat for later utilization. Practical situations where TES systems are often installed are solar energy systems, geothermal systems, DG units, and other energy-conversion systems where heat availability and peak use periods do not coincide. The three basic types of TES systems are *sensible heat storage, latent heat storage,* and *thermochemical heat storage.* Energy storage by causing a material temperature to change is sensible heat storage, for which the system performance depends on the storage material specific heat, and if the volume is important, on the density. Sensible heat storage systems usually use rocks, ground, oil, or water as the storage medium. Latent heat storage systems store energy in PCMs, with the thermal energy stored when the material changes phase, e.g., from a solid to a liquid. Latent heat storage systems use the fusion heat, needed or released when a storage medium changes phase by melting, solidifying, liquefaction, or vaporization. Thermochemical energy storage is based on chemical reactions in inorganic substances. The TES choice depends on the required storage time period, e.g., day-to-day or seasonal, and outer operating conditions. The specific heat of solidification or vaporization and the temperature at which the phase change occurs are design criteria. Both sensible and latent heat types may occur in the same storage material. Notice that the oldest TES form involves harvesting ice from lakes and rivers, and storing it in well-insulated warehouses for use throughout the year for almost all tasks that mechanical refrigeration satisfies today (e.g., preserving food, cooling drinks, and air-conditioning).

   The need for thermal energy storage is often linked to cases where there is a mismatch between thermal energy supply and energy demand, when intermittent energy sources are utilized, and for compensation of the solar fluctuations in solar heating systems. Possible technical solutions to overcome the thermal storage needs may be the following: building production overcapacity, using a mix of different supply options, adding backup/auxiliary energy systems, only summer-time utilization of solar energy, and short-/long-term thermal energy storage. In traditional energy systems, the need for thermal storage is often short-term and therefore the technical solutions for thermal energy storage may be quite simple, and for most cases water is used as the storage medium. Large-volume sensible heat systems are promising technologies with low heat losses and attractive prices. Sensible heat-based thermal energy storage is based on the temperature change in the material and the unit storage capacity (J/kg) is equal to the medium heat capacitance times the temperature change. Phase-change-based energy storage means the material changes its phase at a certain temperature while heating the substance, then heat is stored in the phase change. Reversing, heat is dissipated when at the phase-change temperature it is cooled back. The storage capacity of the phase-change materials is equal to the phase-change enthalpy at the phase-change temperature plus the sensible heat stored over the whole energy storage temperature range. Sorption or thermochemical reactions provide thermal storage capacity, based on the endothermic chemical reaction (when a chemical reaction absorbs more energy than it releases) principle:

$$AB + \text{Heat} \Leftrightarrow A + B \qquad (9.35)$$

TABLE 9.4

Examples of Materials Suitable for Thermal Energy Storage

Thermal Storage Method	Material	Thermal Storage Method	Material	Thermal Storage Method	Material
Sensible heat	Water Ground Rocks Ceramics	Phase change	Inorganic salts Organic and inorganic compounds Paraffin	Thermochemical reactions	*Working Fluid:* Water, ammonia, hydrogen, carbon dioxide alcohols *Sorption Materials:* hydroxides, hydrates, ammoniates, metal hydrides, carbonates, alcoholates

The heat used for a compound AB to be broken into components can be stored separately, and later by bringing the A and B compounds together to form again the AB compound, the heat is released.

The energy storage capacity is the reaction heat or the reaction free energy. TESs based on chemical reactions have negligible losses whereas sensible heat storage systems dissipate the stored heat to the environment and need to be isolated to perform efficiently. Materials are the key issues for thermal energy storage. A large range of different materials can be used for thermal storage, as shown by Table 9.4 or Table B.12 of Appendixes. One of the most common storage media is water. TES in various solid and liquid media or materials is used for solar water heating, space heating, and cooling as well as high-temperature applications such as solar-thermal power generation. Important parameters in a storage system include the storage duration, energy density (or specific energy), and charging and discharging cycle (storage and retrieval) characteristics. However, the energy density is a critical factor for the size and application of any energy storage system. The rate of charging and discharging depends on thermo-physical properties such as thermal conductivity and design of TES systems. TES systems involve the storage of energy by material cooling, heating, melting, solidifying, or vaporizing, and the thermal energy becomes available when the initial process is reversed. The materials that store heat are typically well-insulated. A primary disadvantage of a TES system, similar to other energy storage technologies, is the large initial investments required to build the energy storage infrastructure. However, it has two primary advantages: (1) the energy-system efficiency is improved with the implementation of a thermal energy storage system (CHP is approximately 85% to 90% efficient while conventional power plants are only 40% efficient or lower), and (2) these techniques have already been implemented with good results. On the negative side, thermal energy storage does not improve flexibility within the transportation sector like hydrogen energy storage systems, but this is not a critical issue. TES does have disadvantages, but these are small compared to its advantages. Due to the efficiency improvements and maturity of these systems, it is likely that they will become more prominent, enabling the utilization of intermittent renewable energy, but also to maximize fuel use within power plants. These systems have already been put into practice with promising results. Therefore, it is evident this technology can play a crucial role in future energy and power systems. TES operation and system characteristics are based on thermodynamics and heat transfer principles and laws. There are two major TES types for storing thermal energy, sensible heat storage and latent heat storage. The first consists of changing the temperature of a liquid or solid, without changing its phase. Thermal energy quantities differ in

TABLE 9.5

Thermal Capacities at 20°C for Some Common TES Materials

Material	Density (kg/m3)	Specific Heat (J/kg·K)
Aluminum	2710	896
Brick	2200	837
Clay	1460	879
Concrete	2000	880
Glass	2710	837
Iron	7900	452
Magnetite	5177	752
Sandstone	2200	712
Water	1000	4182
Wood	700	2390

temperature, and the energy required E to heat a volume V of a substance from a temperature T_1 to a temperature T_2 is expressed by the well-known relationship:

$$E = m \int_{T_1}^{T_2} C dT = mC(T_2 - T_1) = \rho VC(T_2 - T_1) \tag{9.36}$$

where C is the specific heat of the substance, m is the mass, and ρ is its density. The energy released by a material as its temperature is reduced, or absorbed by a material as its temperature is increased, is called the sensible heat. The second type of energy storage implies the phase change. The ability to store sensible heat for a given material is strongly dependent on the value of ρC. For example, water has a high value, is unexpansive, however being a liquid must be contained in a better and more expensive container than the one for a solid material. For high-temperature sensible heat, TES (i.e., in the range of 100°C), iron and iron oxide have very good characteristics comparable to water, low oxidization in high-temperature liquid or air flow, with moderate costs. Rocks are unexpansive sensible heat TES materials; however, the volumetric thermal capacity is half that of water. Some common TES materials and their characteristics are listed in Table 9.5. Latent heat is associated with the changes of material state or phase changes, for example from solid to liquid. The amount of energy stored (E) in this case depends upon the mass (m) and latent heat of fusion (λ) of the material:

$$E = m \cdot \lambda \tag{9.37}$$

The storage operates isothermally at the melting point of the material. If isothermal operation at the phase-change temperature is difficult to achieve, the system operates over a range of temperatures T_1 to T_2 that includes the melting point. The sensible heat contributions have to be considered in the top of latent heat, and the amount of energy stored is given by:

$$E = m \left[\int_{T_1}^{T_{melt}} C_{Sd} dT + \lambda + \int_{T_{melt}}^{T_2} C_{Lq} dT \right] \tag{9.38}$$

Here, C_{Sd} and C_{Lq} represent the specific heats of the solid and liquid phases and T_{melt} is the melting point. It is relatively straightforward to determine the value of the sensible heat for solids and liquids, being more complicated for gases. If a gas restricted to a certain volume is heated, both the temperature and the pressure increases. The specific heat here is the specific heat at constant volume, C_v. If, instead the volume is allowed to vary and the pressure is fixed, the specific heat at constant pressure, C_p, is obtained. The ratio $\gamma = C_p/C_v$ and the fraction of the heat produced during compression can be saved, affecting the energy storage system efficiency. TES specific applications determine the method used. Among others, considerations include storage temperature range, storage capacity, having a significant effect on the system operation, storage heat losses, especially for long-term storage, charging and discharging rate, initial and operation costs. Other considerations include the suitability of container materials, the means adopted for transferring the heat to and from the storage, and the power requirements for these purposes. A figure of merit that is used occasionally for describing the performance of a TES unit is its efficiency, which is defined by Equation (9.39). The time period over which this ratio is calculated would depend upon the nature of the storage unit. For a short-term storage unit, the time period would be a few days, while for a long-term storage unit it could be a few months or even 1 year. For a well-designed short-term storage unit, the value of the efficiency should generally exceed 80%.

$$\eta = \frac{T_{max} - T_{min}}{T_{charging} - T_{min}} \tag{9.39}$$

where T_{min}, T_{max} are the maximum and minimum temperatures of the storage during discharging respectively, and $T_{charging}$ is the maximum temperature at the end of the charging period. Heat losses to the environment between the discharging end and the charging beginning periods, as well as during these processes, are usually neglected. Two particular problems of thermal energy storage systems are the heat exchanger design and in the case of phase-change materials, the method of encapsulation. The heat exchanger should be designed to operate with as low a temperature difference as possible to avoid inefficiencies. In the case of sensible heat storage systems, energy is stored or extracted by heating or cooling a liquid or a solid, which does not change its phase during this process. A variety of substances have been used in such systems, such as: (a) liquids (water, molten salt, liquid metals, or organic liquids), and (b) solids (metals, minerals, or ceramics). In the case of solids, the material is invariable in porous form and the heat is stored or extracted as gas or a liquid flowing through the pores or voids. For incompressible type of thermal storage, for example the ones using heavy oils or rocks, the maximum work that can be produced is given in terms of specific heat capacity, C and mass:

$$W_{max} = m\left[C(T_{str} - T_{amb}) + CT_{amb} \ln\left(\frac{T_{str}}{T_{amb}}\right)\right] = \rho V\left[C(T_{str} - T_{amb}) + CT_{amb} \ln\left(\frac{T_{str}}{T_{amb}}\right)\right] \tag{9.40}$$

Here, T_{str} and T_{amb} are the storage material temperature and ambient temperature (in Kelvin units), respectively, m is the storage material mass, ρ the storage material density, and V the volume. The storage materials (water, steam, molten salt, heavy oil, or solid rocks) are at temperatures significantly higher than the ambient one, so the heat is continuously lost from the thermal storage, regardless of the insulation quality. Given enough time, the stored energy, if not used, is dissipated. For this reason TES are suitable for short-term or intermediate-period applications rather than long-term ones. The total rate of heat transfer, q, from a TES reservoir depends on the overall heat transfer coefficient, C_{trsf} the reservoir instant temperature, T, the ambient temperature and the reservoir total surface, A_{tot}, expressed as:

$$q = C_{trsf} \cdot A_{tot} \cdot (T - T_{amb}) \tag{9.41}$$

A number of TES applications are used to provide building or facility heating and cooling, including *aquifer thermal storage* (ATS) and *duct thermal storage* (DTS). However, these are heat-generation techniques rather than energy storage techniques. An aquifer is a groundwater reservoir, consisting of highly water-permeable materials such as clay or rocks, having large volumes and high thermal storage capacities. When heat extraction and charging performances are good, high heating and cooling powers are achieved by such systems. The energy that can be stored in an aquifer depends on the local conditions (allowable temperature changes, thermal conductivity, and groundwater flows). An aquifer storage system is used for short to medium storage periods, daily, weekly, seasonal, or mixed cycles. In terms of storing energy, there are two primary thermal energy storage options. One option is a technology used to supplement building air-conditioning. Thermal energy storage can also be used very effectively to increase the energy system capabilities to facilitate the penetrations of renewable energy sources, which can be increased. Unlike other energy storage systems that enable interactions between the electricity, heat, and transport sectors, thermal energy storage combines only the electricity and heat sectors with one another. By introducing district heating into an energy system, then electricity and heat can be provided from the same facility to the energy system using CHP plants. This brings additional flexibility to the system, enabling larger renewable energy penetrations.

Example 9.14

A residence requires 72 kWh of heat on a winter day to maintain a constant indoor temperature of 21°C. (a) How much solar collector surface area is needed for an all-solar heating system that has 20% efficiency? (b) How large does the storage tank have to be to provide this much energy? Assume the average solar energy per square meter and per day for the area is 6.0 kWh/m²/day.

SOLUTION

a. Daily thermal energy per unit of area converted into thermal energy is:

$$\text{Thermal Energy} = \frac{6.0 \times 0.20}{1.0} = 1.20 \text{ kWh/m}^2/\text{day}$$

The minimum converter area is then:

$$A_{Collector} = \frac{72}{1.2} = 60 \text{ m}^2$$

b. If we are assuming the storage medium, water, the most common storage medium in residential applications, heat capacity of water is 1 kcal/kg/°C, and the temperature difference is that between the hot fluid and the cold water going into the storage tank, about 40°C. Therefore, the required mass of water for a day's worth of heat is:

$$\text{Tank Mass} = \frac{72}{1.116 \times 10^{-3} \times 40} = 1612.9 \text{ or} \approx 1613 \text{ kg}$$

9.5 EMBEDDING MICRO-COGENERATION INTO THE BUILDING ENERGY SYSTEM

CHP systems can provide cost savings as well as substantial pollution and emissions reductions for industrial, institutional, and commercial users. However, an important prerequisite for the large-scale introduction of micro-cogeneration into the building and facility energy systems is its compatibility with the existing and future energy system. This involves different aspects, which are investigated briefly here, with the potential impact of micro-cogeneration on supply security being a particular important one. Selecting the right CHP or micro-CHP technology for a specific application depends on many factors, including amount of power needed, the duty cycle, space constraints, thermal needs, emission regulations, fuel availability, utility prices, and interconnection issues. Designing a technically and economically feasible CHP system for a specific application requires detailed engineering and site data. Engineering information should include electric and thermal load profiles, capacity factor, fuel type, and performance characteristics of the prime mover. Site-specific criteria such as maximum noise levels and footprint constraints must be taken into account. On the other hand, during the past two decades, security of the energy supply has become an increasingly important issue, having significant impacts on the adaptation of CHP solutions. In the near future, the security of both electricity and the fuel supply will confront new challenges, such as high costs of power blackouts in digital economies, growing demands for highly reliable electricity and energy supplies, higher integration of renewable energy sources, increased rate of the decommission of conventional power plants, energy supply security, and volatile prices of fossil fuels. With distributed generation increasing, the number of central regulating power plants is decreasing. In the long run, the risk to fuel supply security is best mitigated by enhancing energy efficiency and increasing the proportion of renewable energy sources and distributed generation. DG units, particularly combined heat and power systems, are regarded as a way forward in the mitigation of the risks to energy supply security and controlling the energy costs. However, embedding distributed power plants into electricity and heat networks is also a challenge, because they affect loads and currents in the grid. The potential impact of micro-cogeneration plants on security and the reliability of the energy supply must be carefully studied and addressed. Last but not least, micro-CHP units reduce transmission losses, because the generated electricity is consumed directly at the site of generation.

Selecting the most appropriate CHP technology for a specific application depends on several factors, such as: the amount of power needed, the duty cycle, space constraints, thermal needs, emission regulations, fuel availability, utility prices, and interconnection issues. The prime movers in a micro-CHP system generate electricity and the prime mover waste heat is used to drive the thermally activated equipment. Thermally activated technology (TAT) usually denotes heat-driven cooling (absorption and adsorption cooling), dehumidification (solid and liquid desiccant systems), bottoming power generation (organic Rankine cycle), or waste heat recovery equipment. The various prime mover and TAT technologies described earlier are usually evaluated using the metrics and weighting factors required by the application. The weighting factors are assigned subjectively and are intended to provide a relative measure of the importance of each metric with respect to application in a CHP system. Designing a technically and economically feasible CHP system for a specific application requires detailed engineering and site data. Engineering information should include electric and thermal load profiles, capacity factor, fuel type, and performance characteristics of the prime mover. Site-specific criteria such as maximum noise levels and footprint constraints must be taken into account. By definition, CHP systems imply the simultaneous generation of two or more energy products that function as a system. One of the most important factors in the analysis of CHP feasibility is obtaining accurate information and data on the electric and thermal loads. This is particularly true in situations where CHP systems are not allowed to export electricity to the grid. Such applications are usually load-following applications where the prime mover must adjust its electric output to match the demand of the end user while maintaining zero output to the grid. Thermal load profiles can consist of hot water use, low- and high-pressure steam consumption, and cooling loads. The shape of the electric load profile and the distribution between minimum and maximum values largely dictate the number, size, and type of prime movers. It is recommended that electric and thermal loads be monitored and analyzed if such information is not readily available. After identifying the optimum candidate micro-CHP equipment and devices, each technology is evaluated based on various performance and cost indicators and metrics. This evaluation is not intended to provide a definitive figure of merit for each technology, but rather to provide a simple means of differentiating the various technologies. The results of this process provide a first-order filter that is used to identify the technologies that have the greatest potential for achieving high-value impact as energy storage devices in micro-CHP systems. Capacity factor (CF) is a key indicator of how the capacity of the prime mover is utilized during operation, being a useful means of indicating the overall economic viability of the CHP system, defined as:

$$CF = \frac{\text{Actual Used (or Generated) Energy}}{\text{Prime-mover Peak Capacity} \times 8760} \qquad (9.42)$$

Lower capacity factors are indicative of peaking applications that derive economic benefits generally through the avoidance of high demand charges, while a higher capacity factor is desirable for most CHP applications to obtain the greatest economic benefit. Higher capacity factors effectively reduce the fixed system unit costs ($/kWh) and help to maintain its competitiveness with grid-supplied power.

For example, gas turbines are usually selected for applications with relatively constant electric load profiles in order to minimize cycling the turbine or operating the turbine for a large percentage of hours at part-load conditions where efficiency declines rapidly, being ideal for industrial or institutional end users with 24-hour operations or where export to the grid is intended. Most commercial and industrial end users have varying electric load profiles, i.e., high peak loads during the day and low loads at night after business hours. On the other hand, reciprocating engines are a popular choice for commercial CHP due to good part-load operation, ability to obtain an air-quality permit, and availability of size ranges that match the load of many commercial and institutional end users. Reciprocating engines exhibit high electric efficiencies, meaning that there is less available rejected heat, often compatible with the thermal requirements of end users. Thermal demand of commercial end users often consists of winter hot water and/or low-pressure steam demands and summer cooling demands. Heat from the prime mover is often used in a single-stage absorption chiller, allowing the CHP system to operate continuously throughout the year while maintaining a good thermal load without the need to waste the heat. The thermal requirements of the end user may dictate the feasibility of a CHP system or the selection of the prime mover(s). For example, gas turbines offer the highest-quality heat that is often used to generate power in a steam turbine. Gas turbines reject heat almost exclusively in their exhaust gas streams. The high temperature of this exhaust can be used to generate high-pressure steam or lower-temperature applications such as low-pressure steam or hot water. Some of the fuel cell technologies, e.g., MCFC and SOFC, also provide high-quality rejected heat comparable to a gas turbine. Reciprocating engines and the commercially available PAFC produce a lower grade of rejected heat. Heating applications that require low-pressure steam or hot water are most suitable, although the exhaust from a reciprocating engine can generate steam up to 100 psig. Reciprocating engines typically have higher efficiency than gas turbines in the same output range and are a good fit where the thermal load is low relative to electric demand. Reciprocating engines can produce low- and high-pressure steam from their exhaust gas, with the water temperatures up to 210°F, so the heat is recovered in the form of hot water. After the system and components are analyzed and selected based on the factors and criteria previously described, the overall system structure is analyzed and configured. The overall objective of the system synthesis and analysis effort is to characterize a range of micro-CHP systems that satisfy customer requirements. These characteristics consist of performance, economics, emissions, reliability, nuisances, and operation. The first three are correlated and strongly depend on the system configuration (architecture), design, and operation. In order to quantify the economics and performance and to quantify the value of each micro-CHP system, analysis tools are needed. Such tools must have the following characteristics:

1. Building load and climate models for representative building types and climate zones.
2. Thermal performance models of the micro-CHP systems, their constituent subsystems, and alternate conventional (non-CHP) systems to produce detailed energy utilization profiles.
3. Electricity and natural gas rate models for the desired markets to derive projections of operating costs.

4. Capital cost data and models for complete life cycle cost analysis for the energy-conversion equipment.

5. Emission models to evaluate the micro-CHP emission, as well as that of conventional systems.

All the candidates discussed and analyzed thus far for micro-CHP technologies, can potentially be used in a micro-CHP system. The synthesis of system solutions, i.e., their configuration, structure, or architecture, and their operation strategy to meet customer needs require careful study and analysis. In its simplest embodiment and structure, a micro-CHP system includes a prime mover to produce power and a heat exchanger to recover heat. However, a large number of possible configurations exist, due to the specificities of several prime-mover technologies in terms of capacity, power, power conversion (DC-AC), efficiency, turn-down operation, or waste heat grade. Moreover, the system operation, aiming to optimize the system functional requirements, such as maximizing system economic value, is another complexity level for a particular system. Finally, the performance, economics, and emission characteristics of micro-CHP systems are compared to the performance of conventional house HVAC equipment and systems. The operating and life cycle costs of the conventional system are used as a baseline to calculate the payback period, cash flow, and emissions. The metrics used to compare the performance of the different micro-CHP systems are the payback period with respect to investment and baseline system definition, cash flow, and emissions compared to a conventional, non-CHP system, as well as favorable features such as automatic switching at power outage or provision of additional backup power for lighting and appliances. The micro-CHP system configuration usually consists of a combination of the following main subsystems and components:

- A prime mover to generate electricity that partially meets household electrical load.
- A heat recovery unit to recover the prime-mover waste heat to meet hot water and heating loads.
- An electrical power interface with the grid to meet electrical loads and ensure power quality.
- A process air subsystem to handle air circulation to the house.
- A supplementary burner to provide additional energy for space heating and hot water, if the waste heat is not sufficient.
- A thermally activated chiller such as an absorption or adsorption chiller to meet the cooling load.
- A vapor compression lag chiller to meet cooling loads, if a thermally activated chiller is not available.

9.6   CHAPTER SUMMARY

Distributed generation continues to be an effective energy solution under certain conditions and for certain types of customers, particularly those with needs for emergency power, uninterruptible power, and combined heat and power. However, for the many benefits of DG to be realized by electric system planners and operators, electric utilities will have to use more of it. The cogeneration technology, including the micro-CHP, delivers two useful products: heat and electricity. Consequently, the successful implementation of cogeneration or micro cogeneration depends not only on the development of the electricity market, but also eminently on the developments of the heat and overall energy markets and technologies. In the context of growing concern over scarcity of energy resources, ever increasing pressure to mitigate environmental impacts, and in order to comply with the need for more secure comfort for home or building owners, integrated micro-CHP system solutions represent an opportunity to answer, at least in part, all these requirements. Successful development of a micro-CHP system for residential applications has the potential to provide significant benefits to users, customers, manufacturers, and suppliers of such systems and in general, to the nation as a whole. The benefits to the ultimate user are a comfortable and healthy home environment at an affordable cost, potential utility savings, and a reliable supply of energy. Manufacturers, component suppliers, and system integrators will see growth of a new market segment for energy products. The benefits to the nation include significantly increased energy efficiency, reduced consumption of fossil fuels, reduced pollutant and CO_2 emissions from power and heat generation, enhanced security from power interruptions, as well as enhanced economic activity and job creation. An integrated CHP energy system provides advantages over conventional power generation, because the energy is used more efficiently through means such as heat recovery. This chapter identifies the characteristics, possible applications, strengths, and weaknesses of the different distributed, dispersed generation, as well as the energy storage concepts and technologies. ESSs are the key enabling technologies for transportation, building energy systems, conventional and alternative energy systems for industrial processes and utility applications. In particular, the extended applications of energy storage enable the integration and dispatch of renewable energy generation and are facilitating the emergence of smarter grids with less reliance on inefficient peak power plants. Energy storage will play a critical role in an efficient and renewable energy future; much more so than it does in today's fossil-based energy economy. Major power and energy storage systems are reviewed in this chapter. It is concluded that ESSs can contribute significantly to the design and optimal operation of power generation and smart grid systems, as well as improved energy security and power quality, while reducing the overall energy costs. A variety of options are available to store intermittent energy until it is needed for electricity production. Mature energy storage technologies can be used in several applications, but in other situations, these technologies cannot fulfill the application requirements. Thus, new storage systems have appeared, passing new challenges that have to be solved by the research community. A micro-CHP system produces electricity and heat, which may not always be equal to the required electric and heating loads in a house. Economics are the key driver in the introduction of any new technology, and micro-CHP offers the potential for significant energy cost savings and substantially higher profits for energy suppliers. CHP offers a customer enhanced reliability, operational and load management flexibility (when also connected to the grid), ability to arbitrage electric and gas prices, and energy management (including peak shaving and possibilities for enhanced thermal energy storage). The value of these benefits will depend on the characteristics of the facility, the form and amount of energy it uses, load profile, rate tariffs, prices of electricity and gas, and other factors. If the generation amount by the micro-CHP is smaller than the house load requirements, the house needs to purchase electricity from the grid and/or burn more fuel using a burner. If the generation from the micro-CHP unit is more than the required house loads, means must be available to handle the surplus such as sending the electricity back to the grid or dissipating the extra heat to the environment. To conclude, if implemented at suitable locations, micro-cogeneration could form a noteworthy contribution to the overall energy supply.

9.7   QUESTIONS AND PROBLEMS

1. Briefly describe the distributed generation (DG), distributed energy resources (DER), and micro combined heat and power generation (micro-CHP) concepts.
2. List the main distributed generation types.
3. List the main distributed generation and distributed energy resources benefits.
4. What are the benefits of energy storage systems?
5. What are the major problems with using: (a) compressed-air energy storage; (b) fuel cells; and (c) flywheels?
6. Briefly describe micro combined heat and power generation benefits and advantages.
7. Which of the following is *not* a method to store energy? (a) battery, (b) motor, (c) compressed air, (d) flywheel, and (e) all of the above store energy.
8. List the benefits of electricity energy storage for power grid operation.

9. Classify and explain compressed-air storage systems.

10. List the essential criteria for comparing energy storage systems.

11. List a few of the potential applications for energy storage options.

12. What gases do we use to feed a fuel cell to produce electricity? Can you name any by-products that may have been produced by fuel cells?

13. Explain the differences between shallow- and deep-cycle batteries. List the main characteristics and applications for each type.

14. List the major advantages and disadvantages of fuel cells.

15. List the main combined heat and power technologies.

16. Briefly describe the main micro-CHP prime movers.

17. What are the major features of a Stirling engine?

18. The lead-acid battery (a secondary type cell) can be recharged and reused for a few years after which it cannot be recharged and reused. Can you think of reasons why it cannot be used any further?

19. What energy storage options are available for solar energy applications?

20. How much kinetic energy does a flywheel (steel disk of diameter 6 in., thickness 1 in.), speed of rotation 30,000 rpm have?

21. Estimate the energy of the compressed-air energy storage (CAES) facilities at Handorf, Germany and McIntosh, Tennessee, United States.

22. A compressed-air energy storage (CAES) has a volume of 500,000 m^3, and the compressed-air pressure range is from 80 bars to 1 bar. Assuming isothermal process and an efficiency of 33% estimate the energy and power for a 3-hour discharge period.

23. If the power generated by compressed-air energy storage (CAES) of 300,000 m^3, in a 2-hour discharge period from 66 bars to the atmospheric pressure is 360 MW, what is the system efficiency?

24. An old salt mine, having a 15,000 m^3 storage capacity, has been selected for pressurized air storage at 33 bar. If the temperature during the filling/charging phase is 175°C, assuming an isothermal process and an efficiency of 35%, estimate the energy and average power for a 3-hour discharge period.

25. Assuming a flow rate of 36 m^3/s, what is the power capacity of the compressed-air energy storage (CAES) in Example 8.2?

26. An underground cavern of volume 35,000 m^3 is used to store compressed-air energy isothermally at 300 K. Determine the maximum stored energy by air compression from 100 kPa to 1750 kPa, assuming heat loss of 63,000 kJ. Hint: One mol of air occupies 22.4 L.

27. Determine the maximum available stored energy if 1450 kg of air is compressed from 100 kPa to 1600 kPa at 27°C, assuming isothermal condition and a heat loss of 27.5 MJ.

28. A flywheel has a weight of 20 kg, 8 m diameter, an angular velocity of 1200 rad/s, a density of 3200 kg/m^3, and a volume change of 0.75 m^3. Evaluate for k equal to 0.5 and 1.05 how much energy is stored in this flywheel.

29. Very high-speed flywheels are made of composite materials. Two flywheels have the following characteristics: (a) a ring with radius of 2.5 m, mass of 100 kg, and speed of 25,000 rpm; and (b) a solid uniform disk, with same mass, radius, and running at the same speed of rotation. How much energy is stored in each system?

30. If the speed of rotation in the previous problem decreases to 2500 rpm during discharge phase, what is the useful energy density for this device?

31. Estimate the energy stored in a CAES with capacity of 250,000 m^3, if the reservoir air pressure is 60 bar, and the overall system efficiency is 33%. Estimate also the total generated electricity if the air is discharged over a period of 90 minutes.

32. A flywheel constructed in a toroidal shape, resembling a bicycle wheel, has a mass of 300 kg. Assuming all mass is concentrated at 1.8 m, what is its rpm to provide 800 kW for 1 min.? In your opinion, is this system physically possible?

33. Compute the kinetic energy density of flywheels made of alloy, glass fiber-reinforced polymer (GFRP), and carbon fiber-reinforced polymer (CFRP).
34. What is the significance of depth-of-discharge (DoD), and how does this parameter affect the battery lifecycle?
35. A battery with an internal resistance of $0.02\ \Omega$ per cell needs to deliver a current of 25 A at 150 V to a load. If the cell electrochemical voltage is 1.8 V, determine the number of cells in series.
36. A battery stack, used as an electrical energy storage system, is designed for 20 MW peak power supply for a duration of 4 hours. This energy storage system uses 600 Ah batteries operating at 420 V DC. Estimate the stack minimum number of batteries and the current in each during peak operation.
37. For a lead-acid battery having a nominal Peukert capacity of 60 Ah, assuming 1.2 Peukert coefficient, plot the capacity for different discharge rates and for different currents.
38. The voltage of a 12.5 V lead-acid battery is measured as 10.5 V when delivering a current of 50 A to an external (load) resistance. What are the load and the battery internal resistances? What is the battery voltage and current through a load of $7.5\ \Omega$?
39. An automobile storage battery with an open-circuit voltage of 12.8 V is rated at 280 Ah. The internal resistance of the battery is $0.25\ \Omega$. Estimate the maximum duration of current flow and its value through an external resistance of $1.85\ \Omega$.
40. Under standard (normal) operation conditions a fuel cell, hydrogen-oxygen type, generates a 3.5 A current and 0.85 V. What are the fuel cell internal resistance and its energy efficiency?
41. A 2.45 MW hydrogen-oxygen fuel cell stack, designed to produce electrical energy for a small electric network (microgrid) has an efficiency of 74%. Calculate the flow rates (kmol/s) of hydrogen and oxygen, and the fuel cell stack losses.
42. A hydrogen-oxygen fuel cell has a liquid water as product when it operates at 0.825 V. What is the electrical energy output in kJ/kg·mole of hydrogen, and the cell efficiency?
43. A hydrogen-oxygen fuel cell stack produces 60 kW of DC power at an efficiency of 63%, with water vapor as product. What is the hydrogen mass flow rate, in g/s, and the cell voltage?
44. Briefly describe the criteria for selecting the configuration, size, and structure of a micro-CHP.

REFERENCES AND FURTHER READINGS

1. F. Kreith and D. Y. Goswami (eds.), *Energy Management and Conservation Handbook*, CRC Press, 2008.
2. J. P. Holman, *Heat Transfer* (10th ed.), McGraw-Hill, New York, 2010.
3. R. A. Ristinen and J. J. Kraushaar, *Energy and Environment*, Wiley, Hoboken, NJ, 2006.
4. E. L. McFarland, J. L. Hunt, and J. L. Campbell, *Energy, Physics and the Environment* (3rd ed.), Cengage Learning, 2007.
5. F. J. Trost and I. Choudhury, *Design of Mechanical and Electrical Systems in Buildings*, Pearson, 2004.
6. M. A. El-Sharkawi, *Electric Energy: An Introduction* (2nd ed.), CRC Press, 2009.
7. A. Sumper and A. Baggini, *Electrical Energy Efficiency – Technology and Applications*, Wiley, 2012.
8. IEEE Standard 1547, IEEE Standard for Interconnecting Distributed Resources with Electric Power Systems, Standards Coordinating Committee 21, The Institute of Electrical and Electronics Engineers, (IEEE), 1547-2003, 28 July 2003.
9. EPRI 1023518, *Green Circuits: Distribution Efficiency Case Studies*, Electric Power Research Institute, Palo Alto, CA, 2011.
10. EPRI 1024101, *Understanding the Grid Impacts of Plug-In Electric Vehicles (PEV): Phase 1 Study— Distribution Impact Case Studies*, Electric Power Research Institute, Palo Alto, CA, 2012.
11. D. Linden and T. B. Reddy, *Handbook of Batteries* (3rd ed.), McGraw-Hill, USA, 2002.
12. C. A. Vincent and B. Scrosati, *Modern Batteries*, Anthony Rowe Ltd., 2003.

13. F. A. Farret and M. G. Simões, *Integration of Alternative Sources of Energy*, New York: Wiley-Interscience, 2006.
14. R. O'Hayre, et al., *Fuel Cell Fundamentals*, Wiley, 2006.
15. V. Quaschning, *Understanding Renewable Energy Systems*, Earthscan, 2006.
16. D. Y. Goswami and F. Kreith, *Energy Conversion* (1st ed.), CRC Press, 2007.
17. F. Kreith and D. Y. Goswami (eds.), *Energy Management and Conservation Handbook*, CRC Press, 2008.
18. H. Chen, T. N. Cong, W. Yang, C. Tan, Y. Li, and Y. Ding, Progress in electrical energy storage system: A critical review, *Progress in Natural Science*, vol. 19, p. 291–312, 2009.
19. B. K. Hodge, *Alternative Energy Systems and Applications*, Wiley, 2010.
20. A. Sumper and A. Baggini, *Electrical Energy Efficiency – Technology and Applications*, Wiley, 2012.
21. B. Everett and G. Boyle, *Energy Systems and Sustainability: Power for a Sustainable Future* (2nd ed.), Oxford University Press, 2012.
22. R. Belu, Renewable Energy: Energy Storage Systems in *Encyclopedia of Energy Engineering & Technology*, vol. 2 (eds: Dr. Sohail Anwar, R. Belu, et al.), CRC Press/Taylor and Francis, 2014.
23. R. A. Dunlap, *Sustainable Energy*, Cengage Learning, 2015.
24. V. Nelson and K. Starcher, *Introduction to Renewable Energy (Energy and the Environment)*, CRC Press, 2015.
25. G. Petrecca, *Energy Conversion and Management - Principles and Applications*, Springer, 2014.
26. R. Fehr, *Industrial Power Distribution* (2nd ed.), Wiley, 2015.
27. B. Stein and J. S. Reynolds, *Mechanical and Electrical Equipment for Buildings*, Wiley, New York, 2000.
28. R. R. Janis and W. K. Y. Tao, *Mechanical and Electrical Systems in Buildings* (5th ed.), Pearson, 2013.
29. W. T. Grondzik and A. G. Kwok, *Mechanical and Electrical Equipment for Buildings* (12th ed.), Wiley 2015.
30. R. Belu, *Industrial Power Systems with Distributed and Embedded Generation*, The IET Press, 2018.

10 Building-Integrated Renewable Energy and Microgrids

10.1 INTRODUCTION

Electrical energy demands keep growing at an increasing pace, together with a significant increase in distributed generation penetration as one of the ways to fulfill these increased energy demands. Distributed energy resources (DER) and renewable energy sources (RES) have and are expected to have even more significant roles in the energy and power industries. DER systems include, among others: photovoltaics, wind turbines, microturbines, small hydropower units, geothermal energy systems, fuel cells, and internal combustion (IC) engines. In addition, energy storage systems (ESS) are used to harness excess electricity produced by the most efficient electricity generators during off-peak and low-demand periods. The harvested energy can be fed into the grid, when it is needed, eliminating or reducing the use of high-cost peak-load generators. Inclusion of the energy storage in the distributed generation system can provide dispatchability of renewable energy sources, such as PV, wind energy, and solar-thermal energy systems, which have no dispatchability on their own. Modern power systems generate and supply electricity taking into account the following considerations. At level 1, electricity generation is produced in large power plants, usually located close to the primary energy source (e.g., coal mines or water reservoirs), usually far away from large consumer centers. At level 2, electricity is delivered to the customers using a large passive but complex power distribution infrastructure, involving high-voltage (HV), medium-voltage (MV), and low-voltage (LV) electric networks. The power distribution networks (level 3) are designed to operate mostly radially. However, other configurations are used, especially in urban or suburban areas. The power flows only in one direction, from HV levels down to customers. Nowadays, the technological advancements, environmental policies, and expansion of electrical markets, are promoting significant changes in the electricity sector. New technologies allow electricity to be generated in smaller-sized plants and in DG units, located in the MV and LV grid sections, closer to the customers (Figure 10.1). Moreover, the increasing use of RES in order to reduce the environmental impact of power generation leads to the development and application of new electrical energy supply schemes (Figure 10.1). In this architecture, the generation is not exclusive at the generation end, but shifted also in transmission or distribution sections. Hence, some of the energy demand is supplied by the centralized generation and another part is produced by distributed generation.

To meet the increased energy demands along with environmental and natural resource preservation, the power grids must incorporate RES and DG units to a greater degree than today. Furthermore, due to deregulations and increased dependability, today's energy systems show an increased brittleness. To cope with the issues of flexibility and increased brittleness, new supporting information infrastructures and intelligent components of energy systems are needed. On the other hand, autonomous off-grid electrification using on-site RES generation has proven to be capable of delivering reliable and acceptable quality electricity for rural locations, islands, or remote areas. Microgrids are described as a self-contained set of local generation, power distribution assets, protection and control units, and smart loads that can operate in either utility grid-connected or islanded modes. In addition to providing reliable power supply, microgrids can provide a wide range of ancillary services, e.g., voltage support, frequency regulation, power factor correction, harmonic control, spinning and non-spinning reserves. Microgrids are distributive in their nature, by including DG and RES units, conventional generation, energy storage units, protection systems, critical and noncritical loads, and other elements. In order to achieve a MG coordinated performance within

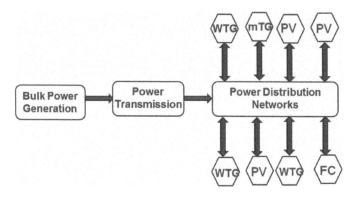

FIGURE 10.1   Distributed energy resources on the power distribution network.

the scope of power distribution, it is required to perform distributed or cooperative control. MG control and management components perform sensory, communication, management, operation, and control tasks, to achieve MG required reliability levels, being that the microgrids possess self-configurable networks for communication among the MG components, leading to increased needs for information, control, real-time, automated, interactive technologies, smart metering, and provision of timely information and control options to the consumers.

DER systems are not new in many respects; however, they are receiving increased attention because of their ability to be used in CHP applications, to provide peak power, demand reduction, backup power, improved power quality, and ancillary services to the power grid. A wide range of power generation technologies is currently in use or in the development phase, including small gas turbines and microturbines, small-size steam turbines, fuel cells, small-scale hydropower units, new types of photovoltaics, solar-thermal energy, wind turbines, energy storage systems, etc. The DER benefits to the transmission and distribution (T&D) include system loss reductions, enhanced service reliability and quality, improved voltage regulation, and relieved T&D congestion. However, DER integration into an existing power distribution system has several impacts on the system, with protection of the power system being one of the major issues. The integration of DG into grids has several associated technical, economical, and regulatory issues, which must be addressed. Adding CHP capabilities to DER systems or facilities can increase the overall system efficiencies to as much as 80%, as reported in the literature, a dramatic improvement over producing electricity and heat separately, which generally has efficiencies of about 45% or lower. The improved efficiency by adding CHP results by meeting the thermal loads with the waste heat produced by the electricity generation. Customers with substantial thermal loads and high electricity prices realize higher gains from investing in CHP systems. In microgrids, for example, the inclusion of CHP can provide additional economic benefits to make DER economically attractive to customers who otherwise may not have enough incentive to join the microgrid. The process of converting the input energy form to the final output form generally comprises a number of intermediate conversions. Any energy-conversion process usually includes several conversion stages; each process phase has its conversion efficiency, and the overall system efficiency is a product of the individual efficiencies. Another way to combat the efficiency problem, is to harness the waste heat from the process and to use it else-where. For instance, the exhaust gases from a boiler in an agro-processing facility, containing a sub-stantial amount of heat, can be used via a heat exchanger to dry the product, to generate electricity, or for other purposes. Reducing both the wasted energy and the input energy increases the overall system energy efficiency. This idea is the basis of cogeneration or CHP, where the waste heat from electricity generation is used as process heat in a factory. Rather than allowing the waste heat to run off into the atmosphere, the heat is used via a heat exchanger to provide hot water for an industrial process, to dry products, or to generate electricity. The overall efficiency of a cogeneration system can be 80% or higher.

The chapter's sections and subsections focus on a review of thermal engineering basics, thermal energy storage systems, microgrid concepts, architecture and structure, micro-combined heat and power generation, and cogeneration concepts, structure, and applications. Other chapter sections are dedicated to energy conservation and energy efficiency. A chapter section is also dedicated to microgrid challenges and issues, such as control, protection and islanding. We conclude with a summary and relevant concept discussions, end-of-chapter questions and problems, and critical references.

10.2 BUILDING-INTEGRATED RENEWABLE ENERGY SYSTEMS

Fast-evolving cities are one of the major drivers into the transformation of the global energy system in the twenty-first century. Cities already account for nearly two-thirds of global energy use and an even larger share of energy-related emissions in developed countries and even in many developing countries (Dincer and Rosen, 2007). Therefore, developing effective energy alternatives for buildings, electricity, heating, cooling, and the provision of hot water is imperative. One way to reduce fossil-fuel dependence is the use of renewable energy, which is generally environmentally benign (Gevorkian, 2010). In many countries, RES and, in particular, solar water heating and solar thermal systems, are used extensively. Meeting building electrical and thermal demands can be achieved through such systems, following standard building energy-saving measures. Designers, architects, and construction engineers already have a number of sustainable technologies to choose from: premium high efficient thermal insulation materials and types, advanced HVAC equipment and installations, passive solar architecture featuring climate-conscious building orientation and advanced glazing and daylighting options, active solar thermal technologies for space heating and domestic hot water, and energy-efficient lighting and appliances. All these measures and technologies have already significantly reduced the thermal energy requirements of buildings. However, such approaches and technologies in turn have increased the share of electricity in the energy balance of the building sectors. It is expected that the increasing demand for renewable energy systems, such as solar thermal systems (STSs), photovoltaics (PVs), building-integrated wind turbines, and geothermal energy systems, will play a pivotal role, as they contribute directly to the heating and cooling of buildings and the provision of electricity and domestic hot water. Meeting building electrical and thermal loads will be primarily achieved through an extensive use of renewable energy systems, following standard building energy-saving measures, such as good insulation or advanced glazing systems. These systems are typically mounted on building roofs with no attempt to incorporate them into the building envelope, creating aesthetic challenges, space availability issues, and envelope integrity problems. However, switching to renewable energy sources means rethinking the entire urban energy landscape, from buildings, to transportation, to industry and power systems. It means integrating the energy supply and demand across the board, through smart and intelligent technologies, rigorous planning, and holistic decision-making. Technologies such as rooftop solar thermal and photovoltaic power generation, distributed energy storage, and electric mobility are becoming an integral part of city power systems, with key roles in balancing the demand and supply of electricity. Buildings are now increasingly both consumers and producers, switching between these functions at different times of day.

On a global level, installations of wind power plants, solar PV power plants, and rooftop systems have reached record highs. In building heating and cooling, however, the progress has been slower. Options range from decentralized renewable energy production, supplying energy within the direct vicinity of buildings, to centralized renewable energy production, in which energy is generated elsewhere and then distributed to buildings via energy networks. Decentralized options can include solar thermal collectors, solar PV panels, biomass boilers, and modern cookstoves using bioenergy (mainly used in developing countries). Centralized options include using renewable energy applications to generate heat or for cooling supplied to buildings through district energy networks, and renewable energy systems, which can be used for lighting, appliances, and heating or cooling.

Renewable energy options for heating (space and water) consist of decentralized building equipment, and centralized power generation. Decentralized solid biofuel-fired boilers (e.g., wood pellets and chips) are a mainstream technology. Solar thermal systems have been used for decades for water heating, and to some extent also for space heating. District heating networks using bioenergy have been in operation for a long time in many cities, mainly in Europe. The heat is typically produced in large boilers known as heat plants, and then it is transported across cities to households, commercial buildings, and industrial facilities. Other renewable energy resources are also used, such as geothermal in Iceland and industrial excess waste heat in various cities in Europe. The use of geothermal energy is growing in certain cities, as well as building-integrated solar-thermal systems. While more of this electricity can be generated by renewable energy, cooling can also be supplied by renewable options integrated in the building infrastructure. This includes solar cooling systems, such as absorption chillers, adsorption chillers, and desiccant cooling systems. Absorption chillers use a refrigerant to cool the environment, and account for over 70% of installed systems today.

Advantages of building integration of RES are that space on the building for the RES installation replaces a conventional building component or used available space, which increases the building's economic viability. At times, resolutions are still being sought for aesthetic and architectural challenges of building integration, such as rainwater sealing and protection from overheating (avoiding increased cooling loads during summer). However, the extra thermal energy that is available can also be used for the heating of the building in winter. Because RES systems depend on latitude, with respect to façade application and solar incidence-angle effects, these factors must be considered as countries near the equator have higher incidence angles (the sun is higher in the sky) but more energy is available compared with the higher latitudes. The building integration of RES can fundamentally change the technologies that affect residential and commercial buildings throughout the world. There are many vivid discussions about the advantages and disadvantages of adopting RES towards achieving zero-energy or nearly zero-energy buildings. Among the most important advantages are:

1. Offer local generation of heat and electricity so transmission losses are minimized.
2. Usually RES generation is friendly to the environment.
3. The building owner can install high-tech systems by taking advantage of the various subsidies.
4. The building energy consumption expenditure is minimized or it is not existent (for 100% coverage).
5. The overall value of a property increases tremendously.
6. The building can be of a higher class concerning energy performance certificates.
7. As the energy performance certificate and the class of the building are related to the amount of rent requested, a higher income can result for the owner.
8. There may be substantial income to the building owner by selling the electricity produced to the grid.

The disadvantages are also important, and include:

1. The initial expenditure for the building is higher.
2. The owner usually has to pay the RES expenditure and later get the money back from subsidies.
3. Problems with RES installation on the building structure and with respect to the required space.
4. Most RES systems require periodic maintenance, which creates extra costs for the building owner.
5. On existing buildings, there may be disruption of existing services.
6. Building integration may require specialist training for the installers.

10.2.1 Building-Integrated Photovoltaic and Solar Thermal Systems

Among the renewable energy resources, solar energy is the most sustainable energy because of its ubiquity, abundance, and sustainability. The systems employed in buildings include: PVs, STS systems, and solar collectors. PV systems can supply the electricity required to the building or the generated electricity can be fed to the grid. The latter is usually preferred as the system does not require energy storage and takes advantage of higher electricity rates that can be obtained by selling the electricity to the grid. STS can supply thermal energy for space heating and cooling, and hot water for the needs of the building. Both PV and STS are expected to take a leading role in providing the electrical and thermal energy needs, respectively, as they can contribute directly to the building electricity, heating, cooling, and domestic hot water requirements. PV modules are solid-state devices that convert solar radiation directly into electricity with no moving parts, require no fuel, and create virtually no pollutants over their life cycle. The conversion process takes place in a solar or PV cell, usually made of silicon, although new materials are also used. PV cells are encapsulated and connected as a group into larger DC electrical units, PV modules, panels, and arrays. To capture the maximum amount of sunlight, PV arrays are installed outdoors, so they need to be durable. The state-of-the art modern PV technology can be characterized by:

1. PV modules are technically well-proven with an expected service time of at least 30 years.
2. PV systems have successfully been used in thousands of small and large applications.
3. PV is a modular technology and can be employed for power generation from milliwatt to megawatt facilitating dispersed power generation in contrast to large central stations.
4. PV electricity is a viable and cost-effective option in many remote site applications where the cost of grid extension or maintenance of conventional power supply systems would be prohibitive.
5. PV technology is universal: the PV modules feature a *linear* response to solar radiation.

Standard PV modules are designed to achieve the maximum energy yields at the lowest cost. They are mostly glass-film laminates, with or without aluminum frames. The frame improves the module strength and rigidity and helps mountings. Frameless PV modules can be mounted into special profile systems. Typical crystalline cells are square, so they fit within the module with a minimum of gaps and thus of wasted space, with sizes of 10 cm by 15 cm. The cells are arranged in a variety of patterns to make a PV module. A typical standard module consists of 36 to 216 cells and has a peak power specification of 100 W_p to 300 W_p. The PV cells are arranged in four to eight rows, resulting in rectangular forms. Strings of 36 or 72 cells are connected in series. Larger modules use parallel connections of two or three strings. There is a limited number of standard module sizes because their dimensions are determined by the size, number, and layout of PV cells within the module. When designers integrate modules into a façade, they choose to either design within the constraints of existing sizes or to specify a custom size. Semitransparent and translucent PV modules present a wide range of possibilities to combine electricity generation with natural lighting, creating interesting light effects. In a glass-glass laminate, some light passes through, referred to as semitransparency or light-filtering, being very useful in overall building lighting. PV modules have a dark appearance because they are designed to reflect a minimum of light, in order to produce maximum electricity output. Monocrystalline silicon PV cells are typically black, grey, or blue, while polycrystalline silicon cells are usually medium or dark blue. The appearance of thin-film amorphous silicon cells is uniform, with a dark matte surface, while colors also include grey, brown, and black. By varying the thickness of the antireflection coating, other colors can be obtained, such as the multicolored polycrystalline cells, at the expense of PV cell efficiencies. Flexible PV cells are a relatively new product that allows attractive building-integrated options. Curved modules with a minimum radius of 0.9 m can be fabricated from crystalline PV cells by embedding the cells between curved sheets or curving finished modules. Thin-film modules are permanently flexible and rollable when deposited onto malleable substrates. Flexible and curved modules are not laminated in hard glass but on versatile materials (e.g., metal and synthetic foils, synthetic resin, glass-textile membranes).

One PV cell produces only about 3 W at 0.6 or 0.7 V DC. For higher power and voltage, 30 or more identical PV cells are connected in series to form a PV module. The PV module is the basic unit for building integration. Under full illumination, a 36-cell module has an output voltage of about 17 V DC across its two output contacts. The PV module also performs the essential encapsulation role, protecting cells against mechanical stress, weathering, and humidity. Only by lamination are the PV modules able to operate for the required, often guaranteed, 20 to 25 years, even in harsh environments. The so-called balance-of-system components include all components needed for a PV system to operate or to be connected to the grid, e.g., inverters, batteries, enclosures, disconnect switches, charge controllers, monitors and meters, wiring, and connectors. In both grid-connected and off-grid solar PV modes, the PV panels are at the top of the electricity-production process. The most common way to compare PV modules is by their peak-power specifications, expressed in watts-peak (W_p). This rating is made at a well-defined set of conditions known as standard test conditions (STC): the actual temperature of the PV cells (25°C or 77°F), the intensity of radiation (1.0 kW/m²), the spectral distribution of the light (air mass 1.5 or AM 1.5, the spectrum of sunlight is filtered by passing through 1.5 thicknesses of the atmosphere). STC correspond to noon on a clear sunny day with the sun about 60° above the horizon, the PV module directly facing the sun, and an air temperature of 0°C (32°F). In reality, these conditions occur very rarely. When the sun shines with the specified intensity, the cell temperature is then 25°C. For such reasons, the so-called nominal operating cell temperature (NOCT) is often specified. NOCT is determined for an irradiance level of 800 W/m², an ambient temperature of 20°C, and a wind velocity of 1.0 m/s. The PV system electrical output is set by the solar energy received at the location. A useful quantity is the average daily insolation, often in energy units of kWh (not to be confused with electrical energy, which uses the same unit). Daily insolation varies through the year, increasing with day length and sun altitude. The insolation over the span of a whole year is at maximum when the PV module has a tilt about 20° less than the local latitude angle, and where the northern hemisphere is oriented south. The main factors that determine the total electrical output from a PV installation over a year are: the annual average daily insolation, the tilt and orientation of the PV array, whether there will be any overshadowing or partial shading, and extent of solar heating (or ineffectiveness of ventilation cooling), the efficiency of balance-of-system components, and the efficiency of the PV module, which is material-dependent.

A building-integrated photovoltaics (BIPV) system consists of integrating photovoltaics modules into the building envelope, such as the roof or the façade, simultaneously serving as building envelope material and power generator. BIPV systems can provide savings in materials and electricity costs, reduce use of fossil fuels and emissions, and add architectural interest to the building. Installations of solar PV technologies on building rooftops are common in some parts of the world. The vast majority of these systems are composed of modules that are mounted off the surfaces of roofs using different types of racking hardware. System designs are most influenced by PV performance considerations, and aesthetics are often secondary. Product designs and intended functionality create inherent cost differences between PVs and BIPVs. BIPV devices often include additional materials such as flashing to ensure buildings are protected from a wide range of weather conditions. On the other hand, most BIPV products reduce installation costs by eliminating common PV mounting hardware such as struts and clips, and associated labor costs. BIPV modules may also install more quickly than incumbent PV modules. While the majority of BIPV systems are interfaced with the available utility grid, BIPV may also be used in stand-alone, off-grid systems. One of the benefits of grid-tied BIPV systems is that, with a cooperative utility policy, the energy storage system is essentially free. It is also 100% efficient and unlimited in capacity. Both the building owner and the utility are benefiting with a grid-tied BIPV. The on-site production of solar electricity is typically greatest at or near the time of a building's and the utility's peak loads. The solar contribution reduces energy costs for the building owner while the exported solar electricity helps support the utility grid during the time of its greatest demand. A photovoltaic system is constructed by assembling a number of individual collectors called modules electrically and mechanically into an array. BIPV is the integration of a PV system into the building envelope. The PV modules serve the dual function of a building outer layer, by replacing conventional building envelope materials, and power generator. By avoiding the cost of conventional building materials, the incremental PV costs are reduced and the building life-cycle cost is improved. That is, BIPV systems often have lower overall costs than PV systems requiring separate, dedicated mounting systems. A complete BIPV system includes:

a. the PV modules (which might be thin-film or crystalline, transparent, semi-transparent, or opaque);
b. a charge controller, to regulate the power into and out of the battery storage bank, if included;
c. a storage system, the utility grid in grid-connected mode or usually batteries in stand-alone systems;
d. power conversion equipment including inverters to convert the PV modules' DC output to AC compatible with the utility grid;
e. backup power supplies such as diesel generators (usually employed in stand-alone systems); and
f. appropriate support and mounting hardware, wiring, and safety disconnects.

BIPV systems can either be interfaced with the available utility grid or they may be designed as stand-alone systems, in either case reducing the demands on the conventional utility generators, and quite often significantly reducing the overall pollutant emissions. A second solar energy is the concentrating solar power (CSP) technology. Such systems concentrate the solar energy to produce high-temperature heat that is then converted into electricity or used for space heating, to provide steam and/or hot water to buildings and industrial processes. The solar resources for generating power from parabolic CSP systems are very plentiful and can provide sufficient electric power for the entire country if sufficient number systems are installed. CSP systems can be sized from 2 to 10 kW or could be large enough to supply grid-connected power of up to 200 MW. Some existing systems use thermal storage during cloudy periods and are combined with natural gas generation, resulting in hybrid power plants that provide grid-connected dispatchable power. The three most advanced CSP technologies currently in use are parabolic troughs (PTs), central receivers (CRs), and dish engines (DEs). CSP technologies are considered one of the most efficient, reliable, and environmentally benign power plants. Solar energy can be used for example directly in cooling systems. The basic principle of absorption chillers is gasification of lithium-bromide or ammonia, when exposed to heat. Heat could be from fossil-fuel burners, geothermal energy, passive solar water heaters, or microturbines. The combined use of passive solar and natural gas-fired media evaporation has given rise to a generation of hybrid absorption chillers that can produce a large tonnage of cooling energy by the use of solar- or geothermal-heated water. A class of absorption that commonly uses LiBr is available commercially, with solar power as its main source of energy. Another solar power cooling technology makes use of solar desiccant-evaporator air-conditioning, which reduces outside air humidity and passes it through an ultra-efficient evaporative-cooling system. This system uses an indirect evaporative processes and minimizes air humidity, which is quite effective technology in coastal and humid areas.

10.2.2 Residential and Rooftop Wind Turbines

Wind is a converted form of solar energy, because the sun heats different regions at different rates, air has mass, and when in motion it contains the kinetic energy of that motion. Parts of that energy can be converted into other forms such as mechanical or electrical. In a mechanical windmill, the energy is used directly by machinery, while in electrical wind turbines the mechanical energy is converted into electrical energy. The wind energy is free, clean, and renewable, does not emit pollutants, requires only a small plot of land, can provide power to remote areas not connected to the power grid, comes in a large power ranges, and is low-cost and low-maintenance after initial setup. The main disadvantages for electrical wind turbines are that they can be damaged by thunderstorms, the blades can hit birds or bats, ice can be thrown by the turbine, noise, or at many locations there is not sufficient wind. The wind turbine power output depends on turbine size and wind speed through the rotor. For example, a 10 kW turbine can generate about 10 MWh annually with wind speeds averaging 12 mph, enough to power a typical household. A 5 MW turbine generates enough power to power about 1500 households if sited at a good wind energy location. Wind turbines are often clustered together as a wind farm. Such a cluster of turbines can provide power to the electrical grid. Wind energy is used as a source of electric power in more than 50 countries. It is estimated that 72 TW of world wind power potential is commercially viable, while the average global power consumption is about 15 TW.

The two main types of wind turbines are vertical-axis wind turbine (VAWT) and horizontal-axis wind turbine (HAWT), or propeller-style. Nearly all of the utility-scale (100 kW capacity and larger) wind turbines in global markets are of the HAWT type (Figure 10.2). The size and use of wind turbines can vary considerably. Land-based wind turbines have rotor diameters that range from 50 m to 100 m. The towers typically are about the same height as the rotor diameter. Offshore turbines tend to have larger rotors, up to 130 m in diameter. For residential or small businesses, rotor diameters of 10 m or less mounted on towers of 30 m in height or less are typical. In the case of HAWT type, the wind flows over the blades causing a lift force on each blade and a torque on the rotor axis. A gear box connects low- and high-speed shafts, driving the generator, increasing the rotational speeds to 1000 to1800 rpm. Geography, terrain, and local conditions can greatly affect the wind power. Knowing this information prior to setting up a wind turbine is critical. Calculating the average power from a wind turbine is performed by a simple equation:

$$P_{WT} = 0.5 \cdot C_{WT} \cdot \rho \cdot A \cdot v^3 \tag{10.1}$$

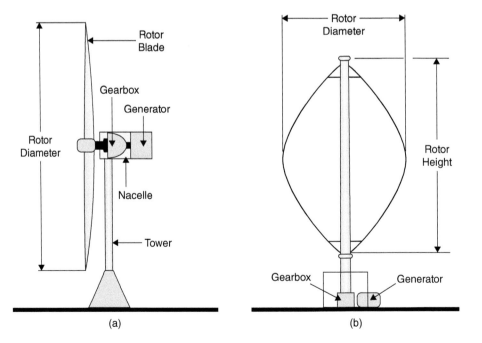

(a)   (b)

FIGURE 10.2   Wind turbine schematics: (a) horizontal-axis and (b) vertical-axis wind turbines.

Here, ρ is the air density (kg/m³), A is the wind turbine rotor swept area (m²), v is the wind speed (m/s), and C_{WT} is the turbine capacity factor. Equation (10.1) indicates the importance of wind speed in power generation because power generation increases proportionally as wind increases to the third power. Knowing the power density allows the wind turbines to be placed in good wind energy locations. The power coefficient, C_{WT}, the percentage of power received by the wind turbine through the rotor swept area, is the ratio of the wind turbine power output to the available wind power, with the maximum value 16/27, the so-called Betz limit. Wind velocity changes rapidly with time and space. In tune with these changes, the power and energy available from the wind also vary. The variations may be short time fluctuations, day-night variations, or weekly, monthly, seasonal, yearly, or multi-decadal variations. Statistical descriptions, tables, or histograms are useful tools to characterize and assess the wind regime. However, for a number of theoretical reasons, it is more convenient to model wind speeds and directions by continuous mathematical functions rather than a table of discrete values. Several probability distribution functions are fitted with wind velocity field data to identify suitable statistical distributions for representing wind regimes. Once a probability distribution function is found, the wind regime and wind power characteristics can be computed. Examples of commonly used probability distributions in wind engineering are the Rayleigh and Weibull functions, given here:

$$f_{RL}(v) = \frac{\pi}{2}\frac{v}{c^2}\exp\left[-\frac{\pi}{4}\left(\frac{v}{c}\right)^2\right] \tag{10.2a}$$

and

$$f_{WB} = k\frac{v^{k-1}}{c^k}\exp\left(-\left(\frac{v}{c}\right)^k\right) \tag{10.2b}$$

The Weibull distribution is a function of two parameters: the shape parameter, k, and the scale factor, c, defining the shape or steepness of the curve and the mean value of the distribution. These coefficients are adjusted to match the wind data at a particular site. Notice the Rayleigh distribution is a special case of the Weibull distribution where k is equal to 2. These probability distribution functions are the most commonly used for wind energy analysis and assessment. Another important factor is the height of the turbine rotor. One of the major reasons wind turbine costs are so high is because the higher altitude the turbine is located, the higher the wind velocity, which in turn increases the turbine power. Wind speed increases with height, so higher elevation sites can offer greater wind resources than comparable lower ones, making it advantageous to site wind turbines at higher elevations. However, the decrease of air density with height can reduce the output power, due to decreases in air density, so less power is produced by a particular turbine at higher elevations. Output power and the power curve depend on air density. Air density is a function of atmospheric pressure, temperature, humidity, elevation, and acceleration due to gravity. Wind turbulence has a strong impact on the power output fluctuation of a wind turbine, being even more critical in build environments. Heavy turbulence may generate large dynamic fatigue loads acting on the turbine and thus reduce the expected turbine lifetime or result in turbine failure. In selecting wind farm sites, knowledge of wind turbulence intensity is crucial for the stability of wind power production, turbine control, and proper wind turbine design. Variations of the wind speed $v(z)$ at a height z can be calculated directly from the wind speed $v(z_{ref})$ at height z_{ref} (usually the standard measurement level, 10 m) by using the logarithmic law (the Hellmann exponential law) expressed by:

$$\frac{v(z)}{v_0} = \left(\frac{z}{z_{ref}}\right)^\alpha \tag{10.3}$$

where, $v(z)$ is the wind speed at height z, v_0 is the speed at z_{ref} (usually 10 m height, the standard meteorological wind measurement level), and α is the friction coefficient, power low index, or Hellman index. When the exact value of the friction coefficient is not available, 1/7 is commonly used.

Example 10.1

Using Rayleigh distribution and $\mu = 5.57$ m/s, calculate the probability the wind speed is between 4 and 10 m/s. The probability the wind is in the range is the difference between the probability of wind at the maximum value and the probability at the minimum value.

SOLUTION

From Equation (10.2a) the required probabilities are estimated as:

$$P\left(\text{Wind Speed} \le 4 \text{ m/s}\right) = 1 - \exp\left[\left(\frac{\pi}{4}\right)\left(\frac{4}{5.57}\right)^2\right] = 0.3333 \text{ or } 33.33\%$$

$$P\left(\text{Wind Speed} \le 10 \text{ m/s}\right) = 1 - \exp\left[\left(\frac{\pi}{4}\right)\left(\frac{10}{5.57}\right)^2\right] = 0.9205 \text{ or } 92.05\%$$

Therefore, the probability that the wind speed is in the range 4 to 10 m/s is 92.05 – 33.33 = 58.72%.

The integration of wind energy systems in the urban environment has often been advocated because it represents a yet unexploited potential, energy is produced close to where it is needed, and urban wind energy is complementary to solar energy so both could thus be combined. The most common disadvantage is that small wind turbines are less efficient and economically viable, mean wind speed in urban environments being lower and turbulence higher. Siting is an important aspect of wind turbine installation because it plays a major role in performance. In order to determine the feasibility of rooftop or residential wind turbines, the placement is critical. The available wind resource plays a major role in finding good sites with proper wind conditions, where the wind turbines function effectively. The common cut-in speed, or the minimum wind speed for a wind turbine to produce power, ranges from 1.8 m/s (4 mph) to 4 m/s (9 mph) for small wind turbines. If the average wind speed in a particular location falls below the cut-in speed, the location is not suitable for wind energy. Compared to rural locations, suburban areas have wind speeds, 13% to 20% lower. Rooftop structures need to ensure that a wind turbine installation does not compromise the rooftop or building integrity. A number of rooftop attributes should be considered when installing wind turbines, such as roof material, support, and durability. It is important to acknowledge the different loads caused by a rooftop wind turbine: static and dynamic loading. *Static loading* is the dead weight of the turbine, tower, and foundation. *Dynamic loading* is created by the wind on the turbine, adding stress to the foundation. The varying torque caused by the moving blades also causes vibration, which needs to be considered in design. The capacity factor of a wind turbine tends to be lower in urban areas due to higher turbulence and other factors. Turbulence and erratic changes in direction and wind speed occur when there are obstructions in the wind path, such as buildings, trees, etc. To avoid this problem, some studies suggest that the roof where the turbines are located should be approximately 50% higher than any surrounding objects, a difficult requirement to meet. Wind turbines can also generate noise, depending on the model and wind regime, which can be an issue for wind turbines sited in residential areas. Wind turbines are often visible from a long distance off because of their towers. Some people dislike the appearance of wind turbines because they believe they spoil the scenery. Rooftop wind turbines present challenges concerning aesthetics in urban environments. Even though the wind turbines considered for rooftop or residential applications are smaller in size, it is quite likely that they are seen either from the street or from neighboring buildings. This can be a problem if enough people oppose wind turbine placement due to aesthetic reasons.

A number of choices or options are considered in several exploratory studies to provide a first assessment of the potential for urban wind energy in an area or at any location. The first decision is to consider only vertical axis wind turbines (VAWTs) of the Darrieus type or horizontal axis wind turbines (HAWT). The second decision concerns the type of building and residential integration of the wind energy system. The three categories of possibilities are: (1) siting stand-alone wind turbines in urban locations; (2) retrofitting wind turbines onto existing buildings; and (3) full integration of wind turbines together with architectural form. The performance of systems in the first category has been reported to be very site-specific. A number of interesting third-category systems have been contemplated but not included in the assessment in this report due to the additional construction costs involved and the limited size of the wind turbines that can be integrated in these systems. The most straightforward solution is to place wind turbines on mast on top of high-rise buildings, high enough so that the wind turbine is situated outside the area of separated flow above the roof for all possible wind directions, in amplified wind speed zone, above the separated flow. A variety of roof-integrated systems has been devised, all belonging to the third category. Such systems consist of a special roof configuration in which one or more VAWTs can be embedded. Three important disadvantages of this type of roof-integrated wind turbines are: (1) extra costs for the special roof structure; (2) the increased flow resistance, reducing the wind speed and the power output; and (3) the limited space and height to add multiple wind turbines. Façade-integrated wind turbines are inspired by the amplified wind speed that can occur around building corners. Nevertheless, when integrated in the corner itself, they can be sited in the areas of separated flow or flow stagnation when oriented to the approaching wind. For the other wind directions, they will almost always be situated in the area of separated flow or in the building wake. In addition, often Savonius turbines are used, being characterized by low power coefficients and efficiencies, but less expensive and easy to install.

10.2.3 BUILDING GEOTHERMAL ENERGY SYSTEMS

Geothermal energy comes from Greek words: *thermal* meaning heat and *geo* meaning earth, being the energy contained as heat in the earth's interior. The earth gives us the impression that it is dependably constant, because over the human time scale, little seems to change. However, the earth is a dynamic entity, with time scales spanning from seconds for earthquakes, years that volcanoes appear and grow, over millennia that landscapes slowly evolve, to over millions of years the continents rearranging on the planet's surface. The energy source to drive these processes is heat, with a constant flux from every square meter of the earth's surface. The average heat flux is 87 mW/m^2 or for the total global surface area of 5.1×10^8 km^2, the heat flux is equivalent to about 4.5×10^{13} W. For comparison, it is estimated that the total power used by all human activity in 1 year is approximately 1.6×10^{13} W, the earth's heat having the potential to significantly contribute to the world's energy needs. This heat, the source of geothermal energy, is contained in the rock and fluid inside the earth's layers. Its origin is linked with the earth's internal structure and composition, and the associated physical processes. Despite the fact that it is present in huge, inexhaustible quantities in the earth's crust or deeper layers, it is unevenly distributed, seldom concentrated, and most often at depths too great to be exploited industrially. Geothermal resources can reasonably be extracted at costs competitive with other energy forms, confined to the earth's crust areas where heat flow is higher than in surrounding areas, and heat the water contained in permeable rocks in the reservoirs. Usually resources with the highest energy potential are concentrated on the plate boundaries, where significant geothermal activity exists, such as hot springs, stem vents, fumaroles, or geysers. Active volcanoes are also a geothermal activity, on a particularly and more spectacular large scale.

Geothermal heat is used directly for a variety of applications such as: space heating and cooling, food preparation, hot spring bathing, balneology, agriculture, aquaculture, greenhouses, and industrial processes. Uses for heating and bathing are traced back to ancient Roman times. From about 5.4×10^{27} J of the earth continents' available thermal energy, about a quarter is estimated to be available at depths shallower than 10 km. In order to be useful directly, the thermal energy temperature must be above ambient surface temperatures and be easily and efficiently transferred to the designed premises. Such conditions are satisfied in places where there are warm or hot springs and in locations where high thermal gradients exist. Such sites are quite restricted in their distribution, being concentrated in regions where there is recent volcanic activity or where the rifting of continents occurs. For these reasons, a relatively small fraction of the heat contained within the continents can be economically employed for geothermal direct-use applications. The geothermal energy resources that are readily available are not well-assessed and their complete distribution map of such resources is not available. The main geothermal energy direct uses are: (1) swimming, bathing and balneology, (2) space heating and cooling including district heating, (3) agricultural, aquaculture, and industrial applications, and (4) geothermal (ground-source) heat pumps. The annual growth rate for direct-use is about 8.3%, with the largest annual energy increase in the GHP technology. The overall global geothermal energy direct use is estimated close to 75 GWt, while the contribution of shallow reservoirs for small commercial, industrial, or domestic applications is quite difficult to estimate. The capacity factors for such applications are in the range of 15% to 75%. Direct use of geothermal energy in the United States is about 12.5 GWt, China 32.0 GWt, and European Union about 20 GWt. The oldest direct use of geothermal energy is for bathing and therapy by using the hot spring or the surface heating hot water. Another traditional direct use of geothermal energy is space heating. The hot water, typically at 60°C or higher from geothermal reservoirs, is pumped into the building heating system, through the heat exchanger to provide heat to the building. The water is then re-injected into the geothermal reservoir for reheating. For direct use of geothermal energy, a well is drilled into the geothermal reservoir, and there are pumps to bring the hot water to the surface.

Direct use of geothermal energy includes the hydrothermal resources of low to moderate temperatures, providing direct heating in residential, commercial, and industrial sectors, including, among others: space, water, greenhouse, and aquaculture heating; food dehydration; laundries; and textile processes. These applications are commonly used in many countries. Unlike geothermal power generation, direct-use applications use heat directly to accomplish a broad range of purposes. The temperature range of these applications is from 10°C to about 150°C. Given the ubiquity of this temperature range in the shallow subsurface, these types of geothermal applications have the potential to be installed almost everywhere. Geothermal resources are also used for agricultural production, to warm greenhouses, to help in cultivation or for industrial purposes, including drying fish, fruits, vegetables, and timber products; washing wool; dying cloth; manufacturing paper; and in milk industry. Geothermal heated water can be piped under sidewalks and roads to keep them from icing during cold weather, can extract gold and silver from ore, and can even be used for refrigeration and ice-making. Geothermal, ground-source heat pumps have the largest energy use and installed capacity worldwide, accounting for 70.95% of the installed capacity and 55.30% of the annual energy use. The installed capacity is 50,000 MWt and the annual energy use is 325,028 TJ/yr., with a capacity factor of 0.21 (in the heating mode). The energy use reported for heat pumps is deduced from the installed capacity, based on an average coefficient of performance (COP) of 3.5, which allows for 1 unit of energy input (electricity) to 2.5 units of energy output, for a geothermal component of 71% of the rated capacity. The cooling load was not considered a geothermal effect; however, it has a significant role in the use of fossil fuels and pollutant emission reductions.

Thermal equilibrium is achieved when coexisting systems reach the same temperature, with heat being spontaneously transferred from a hot to a colder body. This fundamental principle upon which all direct-use applications rely is also the mechanism of the unwanted heat losses. The ability to manage heat transfer by minimizing unwanted losses and maximizing useful heat is required to set and operate an efficient direct-use application. The heat-transfer processes are conduction, convection, radiation, and evaporation. Heat transfer by conduction occurs at a microscopic level through the exchange of vibrational energy by atoms and molecules, while at the macroscopic level this process is manifest as changes in temperature when two bodies at different temperatures are placed in contact with each other at a certain time t_1. If T_1 and T_2 represent the initial temperatures of bodies 1 and 2, respectively, and T_3 is the equilibrium temperature they eventually achieve at time t_2.

Note that T_3 is not halfway between T_1 and T_2, reflecting the effect of heat capacity of each body material. In this example, the heat capacity of body 1 must be higher than that of body 2. Conductive heat transfer is described by the relationship:

$$\frac{dQ_{cnd}}{dt} = k \cdot A \cdot \frac{dT}{dx}$$ (10.4)

where dQ_{cnd}/dt (J/s or W) is the rate at which heat transfer occurs by conduction over the area A, k is the thermal conductivity (W/m·°K), and dT/dx is the temperature gradient over the distance x (m). Equation (10.4) is the Fourier's law of heat conduction. Equation (10.4) indicates that the heat transfer rate increases by increasing the area over which heat transfer occurs or by decreasing the distance. The thinner the plate, the greater is the temperature gradient across the plate and hence the greater the heat loss rate. Table 10.1 lists the thermal conductivities of some common materials that may be used in direct-use applications. Notice, there is a difference of more than two orders of magnitude in the rate of heat loss, but that, for many common materials, the difference is more than four orders of magnitude. Although large temperature changes are not associated with most direct-use applications, it is worth mentioning that thermal conductivity is a temperature-dependent material characteristic. Thus, accurate computation of heat transfer rates must take into account the temperature dependence of k. It is so important to know the thermal conductivities and spatial geometry of the materials that are used for a direct-use application, or which are encountered when constructing a facility. Incorrect data on these parameters result in seriously undersizing thermal insulation, inadequately sizing piping, and underestimating the rate of heat loss to the environment, all of which can seriously compromise the efficient operation of a direct-use system.

Heat transfer by convection is a complex process that involves the movement of mass that contains a quantity of heat. Convection represents the heat transfer due to the bulk motion of a fluid. Convective heat transfer also occurs at interfaces between materials, as when air is in contact with a warm pool of water or is forced to flow at high velocity through a heat exchange unit. In such cases, buoyancy effects, the flow characteristics, the boundary layers, the effects of momentum and viscosity, and the surface properties and shape of the geometry of the flow pathway influence the heat transfer. For example, we assume that cool air at temperature T_2 moves over a body

TABLE 10.1
Thermal Conductivities of Selected Materials Used in Direct-Use of Geothermal Energy

Material	Thermal Conductivity (W/m·°K)
Aluminum	202
Copper	385
Iron	73
Carbon steel	43
Marble	2.90
Magnesite	4.15
Quartz	6.50
Glass	0.78
Concrete	1.40
Sandstone	1.83
Air	0.0240
Water	0.5560
Water vapor	0.0206
Ammonia	0.0540

of warm water at temperature T_1. Viscous and frictional forces act to slow the air movement near the water surface, forming a boundary layer, which is a region where a velocity gradient develops between the interface and the main air mass that has velocity v. At the interface, the velocity approaches zero. The characteristics of the boundary layer are dependent upon the fluid properties, velocity, temperature, and pressure. Heat is transferred by diffusive processes from the water surface to the fluid at the near-zero velocity boundary layer base, causing its temperature to approach that of the water, T_1, resulting in a temperature gradient, in addition to the velocity gradient, between the main mass of moving fluid and the water–air interface. This thermal gradient becomes the driving force behind thermal diffusion that contributes to heat transfer through this boundary layer. Advective transport of heated molecules provides an additional mechanism for heat to move through the boundary layer and into the main air mass, resulting in an increase in the temperature of the air. The rate at which convective heat transfer occurs follows Newton's law of cooling, which is expressed here as:

$$\frac{dQ_{cnv}}{dt} = h \cdot A \cdot dT \tag{10.5}$$

where dQ_{cnv}/dt is the rate at which heat transfer occurs by convection, h is the convection heat transfer coefficient (J/s-m^2-K), A is the exposed surface area (m^2), and dT is the temperature difference between the warm boundary and the overlying cooler air mass, $(T_1 - T_2)$. Values for h are strongly dependent on the properties of the materials involved, the pressure and temperature conditions, the flow velocity and whether flow is laminar or turbulent, surface properties of the interface, the geometry of the flow path, and the orientation of the surface with respect to the gravitational field, h being variable and specific to a given situation. Determining values for h requires geometry-specific experiments, or access to functional relationships that have been developed in analog cases. For example, convective heat loss from a small pond over which air is flowing at a low velocity can be reasonably accurately represented by:

$$\frac{dQ_{cnv}}{dt} = 9.045 \cdot v \cdot A \cdot dT \tag{10.6}$$

Here v is the velocity of the air and the effective units of the coefficient, 9.045, are kJs/m^3·h°C. In the ideal case, radiation heat transfer is represented by considering a so-called ideal blackbody. A blackbody radiator emits radiation that is strictly and absolutely dependent only on temperature, so the wavelength of the emitted radiation is strictly inversely proportional to the temperature. At room temperature, for example, an ideal blackbody would emit primarily infrared radiation while at very high temperatures the radiation would be primarily ultraviolet. However, real materials emit radiation in more complex ways that depend upon both the surface properties of an object and the physical characteristics of the material of which the object is composed. In addition, when considering radiative heat transfer from one object to another, the geometry of the heat source, as seen by the object receiving the radiation, must also be taken into account. For most considerations involving radiative heat transfer in direct geothermal energy applications, interfaces are commonly flat plates or enclosed bodies in a fluid, the geometrical factor of minimal importance, and the following relationship is used:

$$\frac{dQ_{rad}}{dt} = \varepsilon \cdot \sigma \cdot A \cdot \left(T_2^4 - T_1^4\right) \tag{10.7}$$

where ε is the emissivity of the radiating body (ε equals 1.0 for a perfect blackbody), at temperature T_1, σ is the Stefan-Boltzmann constant, equals 5.669 10^{-8} W/m^{2}K^4, T_1 and T_2 are the respective temperatures of the two involved objects, and A is its effective surface area.

Heat transfer through evaporation can be an efficient energy transport mechanism. The factors that influence evaporation rate are temperature and pressure of the vapor, overlying the evaporating fluid, the exposed area, the fluid temperature, the equilibrium vapor pressure, and the wind velocity. Each of these properties is relatively simple to formulate individually; however, the evaporation process is affected by factors similar to those that influence convective heat transfer, which makes them quite complex. A complicating factor is that the boundary layer behavior with respect to the partial pressure varies both vertically away from the interface and along the interface due to turbulent flow mixing, so the ambient vapor pressure is not represented rigorously. Moreover, the evaporation rate is affected by the temperature gradient above the interface, which is affected by the boundary layer properties, which in turn, influence the equilibrium vapor pressure. Because the gradient is the driving force for diffusional processes, the rate at which diffusion transports water vapor from the surface is affected by local temperature conditions as well as the fluid velocity. These complications have led to an empirical approach for establishing evaporation rates in which various functional forms are fit to data sets that span a specific range of conditions. One such common heat rate relationship due to evaporation used in direct geothermal energy use is:

$$\frac{dQ_{evp}}{dt} = \frac{a \times (P_{water} - P_{air})^b \times H_w}{2.778 \times e^{-7}} \tag{10.8}$$

where, P_{air} and P_{water} are the water vapor saturation pressures (kPa) at the water and the air temperatures, respectively, H_w is the water enthalpy (kJ/kg), a and b, two empirical constants, determined by the velocity of the fluid moving over the interface (m/s), that often are calculated as:

$$a = 74.0 + 97.97 \times v + 24.91 \times v^2$$
$$b = 1.22 - 0.19 \times v + 0.038 \times v^2$$

Example 10.2

Estimate the convection, conduction, radiation, and evaporation heat transfer rates for a rectangular water pool, constructed of concrete walls 15 cm in the ground, having the dimensions, 15 m by 40 m and 2 m depth, assuming the external air temperature 20°C, the wind speed is 5 m/s and the pool has a temperature of 37°C.

SOLUTION

Applying Equations (10.5), (10.6), (10.7) and (10.8) the heat transfer rates due to conduction, convection, radiation, and evaporation are computed as follows. The total area for conduction heat transfer is calculated considering the lateral walls and the bottom:

$$A_1 = 2 \times (15 + 40) \times 2 + 15 \times 40 = 880 \text{ m}^2$$

$$\frac{dQ_{cnd}}{dt} = 1.4 \frac{W}{m \cdot K} \times 880 \text{ m}^2 \times \left(\frac{17}{0.2 \text{ m}}\right) = 104720 \text{ W(J/s)}$$

Convection and radiation heat losses are taking place only on the pool upper (open) surface.

$$\frac{dQ_{cnv}}{dt} = 9.045 \cdot 5(\text{m/s}) \cdot (15 \times 40) \text{ m}^2 \cdot (37 - 20)(°C) = 461295 \text{ J/s (W)}$$

$$\frac{dQ_{rad}}{dt} = 0.99 \times 5.669 \times 10^{-8} \frac{W}{m^2 K^4} \times 600 \text{ m}^2 (310^4 \text{ K}^4 - 293^4 \text{ K}^4) = 62807.1 \text{ W(J/s)}$$

And the rate of evaporation heat loss, assuming the partial pressure of 3.7 kPa and 1.23 kPa is then:

$$a = 74.0 + 97.97 \times 5 + 24.91 \times 5^2 = 1186.6$$
$$b = 1.22 - 0.19 \times 5 + 0.038 \times 5^2 = 1.22$$

$$\frac{dQ_{evp}}{dt} = \frac{1186.6 \times (3.7 - 1.23)^{1.22} \times 1}{2.778 \times e^{-7}} = 14411639 \text{ W(J/s)}$$

The amount of heat, Q_{Load}, required to operate the function of the designed installation depends upon the specific process, the operation size, and the external conditions. Assuming that load Q_{Load} is constant over time, then the geothermal resource must be of sufficient temperature and flow rate to satisfy:

$$\frac{dQ_{Geothermal}}{dt} > \frac{dQ_{Load}}{dt} + \frac{dQ_{Losses}}{dt} \tag{10.9}$$

From the discussion of heat transfer mechanisms, the total heat losses, Q_{Losses}, that need to be accounted for in any application are the sum of all relevant heat loss mechanisms, conduction (Q_{cnd}), convection (Q_{cnv}), radiation (Q_{rdt}), and evaporation (Q_{evp}), expressed as:

$$\frac{dQ_{Losses}}{dt} = \frac{dQ_{cnd}}{dt} + \frac{dQ_{cnv}}{dt} + \frac{dQ_{rad}}{dt} + \frac{dQ_{evp}}{dt} \tag{10.10}$$

This equation represents the heat loss that is assumed for the application operating conditions. The heat, Q_{Load}, required to perform the function of the designed installation (load) depends upon the specific processes and the operation size. For most applications it is likely that seasonal variability influences the total heat losses through changes in the air temperature and other seasonal weather variables, such as wind or radiation. For this reason, the design load concept was developed, the most severe condition set a facility is likely to experience, maximizing the heat losses. Hence, when evaluating the feasibility of a potential direct-use project, it is important to establish whether the resource is sufficient to meet the maximum demanding conditions that are likely to be encountered. In instances where an abundant resource is available, it may be suitable to size the facility in such a way to meet all probable energy demands. In other instances, possibly for reasons of economics, it may turn out to be sufficient to design the facility such that the geothermal resource meets the demand of some maximum percentage of probable events, while the remainder is addressed with other supplemental energy sources. However, direct heating is more efficient than electricity generation and places less demanding temperature requirements on the heat resources. Heat may come from cogeneration with a geothermal electrical plant or from smaller wells or heat exchangers buried in shallow ground. As a result, geothermal heating is economical over a much greater geographical range than geothermal electricity. Where natural hot springs are available, the heated water can be piped directly into radiators. If the ground is hot but dry, earth tubes or downhole heat exchangers can collect the heat. But even in areas where the ground is colder than room temperature, heat can still be extracted with a geothermal heat pump more cost-effectively and cleanly than it can be produced by conventional furnaces. These devices draw on much shallower and colder resources than traditional geothermal techniques, and they frequently combine a variety of other functions, including air-conditioning, seasonal energy storage, solar energy collection, and electric heating. Geothermal heat pumps can be used for space heating essentially anywhere in the world.

Example 10.3

For the pool of Example (10.2) estimate the geothermal inflow rate from a 50°C reservoir to keep the pool temperature at 37°C.

SOLUTION

The total heat losses, due to conduction, convection, radiation, and evaporation are given by Equation (10.10):

$$\frac{dQ_{Losses}}{dt} = 104720 + 461295 + 62807 + 14411639 = 1638929 \text{ W(J/s)}$$

The geothermal inflow rate is computed as:

$$q_{inflow} = \frac{dQ_{Losses}/dt}{C_P \left(T_{Geothermal} - T_{Water}\right)} = \frac{1638929}{4183.3 \times (45 - 37)} = 489.7 \text{ kg/s}$$

Notice the evaporation heat transfer losses are the most important for heat losses for the open ponds.

10.2.3.1 District Heating

Approximately 5.4×10^{27} J of geothermal energy is available, of which nearly a quarter is available at depths shallower than 10 km. In order to be useful directly, this heat must be significantly above ambient surface temperatures and easy to be transferred efficiently. The basic district heating require- ments are a warm geothermal fluid source, a pipe network for fluid distribution, a control system, and a disposal or a reinjection system. These conditions are usually satisfied in places where hot springs emerge at the surface, or in locations where higher thermal gradients allow shallow drilling to access heated waters. Such sites are relatively restricted in distribution, being concentrated in regions where volcanic activity is present or where rifting of continents has happened. For these reasons, a relatively small fraction of the large amount of heat contained within the crust can be economically employed for direct-use applications. However, for example, space heating accounted for about 60,000 TJh/yr of the total 273,372 TJh/yr of energy consumed through direct-use applications, which is the third worldwide largest geothermal direct use. The majority of such heating systems involve district sys- tems in which multiple users are linked into a network that distributes heat to users (Figure 10.3). District heating systems distribute hydrothermal water through piping systems to buildings.

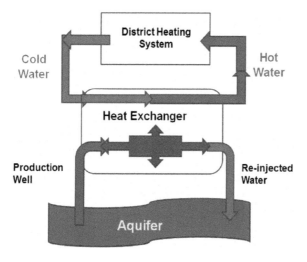

FIGURE 10.3 Block diagram of a district heating system.

Such system design requires matching the distribution network size to the available resource. The resource attributes are the sustainable flow rate (usually this is required to be between 30 and 200 kg/s, depending on the size of the district heating system and the temperature of the resource) and the temperature of the resource. The three typical components of a district heating system are a production facility, a mechanical system, and a disposal system. In a production system, the well is needed to bring the thermal water/heat energy from the geothermal reservoir, the mechanical system delivers the hydrothermal water/heat energy to the process, and the disposal system is a medium that receives the cooled geothermal fluid, e.g., a pond, river, or an injection-fluid system. The geothermal power (Q_{Geo}) that must be provided from a resource, must be higher than the thermal demand, as given by:

$$Q_{Geo} > m \times C_P \times (T_{Geo} - T_{Rtn}) \tag{10.11}$$

Here m is the mass flow rate (kg/s), C_P is the constant pressure heat capacity of the fluid (J/kg-K), T_{Geo} and T_{Rtn} are the temperature (K) of the water from the geothermal source and temperature of the return water after it has been through the network, respectively. In summary, the district heat basic requirements and components are a source of geothermal fluid, a network of pipe to distribute the heated fluid, a control system, and a disposal or reinjection system. System design requires matching the size of the distribution network to the available geothermal resource and operation constraints. The geothermal water passes through heat exchangers, with heat being transferred to a cleaner secondary fluid, often fresh water, distributed via insulated pipes to the district, where it transfers the energy to a group of houses and/or commercial facilities. Fluid valves control the amount of geothermal water at heat exchangers and balance the heating demands with the supply. Geothermal water does not exit the heat exchangers, dissolved gases or solids are re-injected into reservoir, with almost no environmental impact. The resource characteristics are a sustainable flow rate (usually this is required to be between 30 and 200 kg/s, depending on the size of the district heating system and the temperature of the resource) and the temperature of the resource. The geothermal power, P_G, provided from a resource is estimated from:

$$P_G = q_m C_P (T_{GW} - T_{RW}) \tag{10.12}$$

where q_m is the mass flow rate (kg/s), C_P is the constant pressure heat capacity of the fluid (J/kg-°K), and T_{GW} and T_{RW} are the temperature (K) of the water from the geothermal source and temperature of the return water after it has been through the network, respectively. The heat demand, Q_{Load}, imposed on the system, is a function of time. During the day the load can vary by up to a factor of 3, being affected by the climatic variability. This equation makes clear that the only controllable variable affecting P_{GW} is the return water temperature, T_{RW}, because the other variables are set by the natural system. Maximizing the temperature drop across the network thus becomes a means to increase the power output of the system. However, how this is addressed depends upon the operating mode that can be employed for the system.

Example 10.4

Assuming that a geothermal well provides a flow rate of 300 kg/s with the difference of temperature of water between the geothermal source and return of 75°C, what is the geothermal power of this application? If the total losses are 30% of the geothermal power, what is the maximum load power?

SOLUTION

The geothermal power of the well is given by Equation (10.12), as:

$$P_G = 300 \times 4180 \times 80 = 100.32 \text{ MW}$$

The maximum usable power is:

$$Q_{Load(Max)} = (1 - 0.30) \times 100320000 = 70.224 \text{ MW}$$

10.2.3.2 Geothermal Heat Pumps

Heat pump technology is considered to be one of the most beneficial engineering application, being in essence, Carnot cycles, operating at highest efficiency levels, by employing transport heat that already exists, without generation, for both heating and cooling, while using only a small fraction of transferred energy. Such devices are often used in low-temperature geothermal schemes to maximize heat extraction from fluids, while their specific function in any scheme depends on the used geothermal fluid temperature. A geothermal heat pump (GHP) is a heat pump that uses geothermal energy for its operation. The three most common types of GHP are: ground-coupled GHPs, ground-water GHPs, and hybrid GHPs. Ground-coupled GHPs are of two types: vertical closed-loop and horizontal closed-loop. For a moderate fluid temperature range (50°C to 70°C), the extracted heat depends on the heat exchanger and the heat pump connection to extract maximum geothermal fluid heat. For low-temperature ranges (up to 40°C), heat extraction is not possible and a heat pump is used for heat transfer. GHPs take advantage of the deeper ground constant temperature up to about 100 m (about 300 ft.). GHPs are versatile systems and can be used almost everywhere in the world. GHPs circulate a liquid, usually water through pipes buried in a continuous loop, either horizontally or vertically, under any area around the building, being one of the most efficient heating and cooling systems available. To supply heat, the GHPs extract heat from the earth, distributing it through the building duct system. For cooling, the process is reversed; the system extracts heat from the building and moves it back into the earth loop. GHPs can reduce the electricity use up to 60% compared with conventional heating and cooling systems.

Direct geothermal systems for heating usually consists of a primary circuit that exchanges heat with the ground, and secondary circuit that circulates heat within the building. Space heating systems require higher temperatures than the ground ones, and a heat pump can overcome this restriction by enhancing the ground energy with external electrical or mechanical energy.

The geothermal heat pump operation as described in Chapter 8 absorbs thermal energy from the ground, then through the compressor and condenser (using electricity) where it releases the heat into the building (Figure 10.4). For GHPs in heating mode, refrigerant evaporation occurs where the heat pump joins the primary circuit and condensation occurs where the heat pump joins the secondary circuit. Refrigerant evaporation cools the circulating fluid in the primary circuit, which is then reheated by the ground. GHPs require energy input (for the compressor and to the pumps that circulate the fluid) to move heat around the system. However, this energy input required is

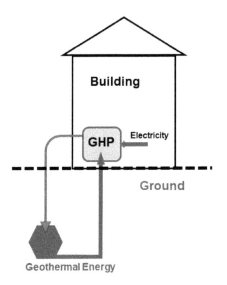

FIGURE 10.4 Heat pump schematic and operating diagram.

small compared to the heat output. Heat pump efficiency also increases as the temperature difference between the heat sink and source decreases. GHPs are thus more efficient than air-source heat pumps when the seasonally averaged ground temperature is closer to the desired ambient building temperature than the air. In order to transport heat from a heat source to a heat sink, heat pumps need external energy. Theoretically, the total heat delivered by the heat pump is equal to the heat extracted from the heat source, plus the amount of drive energy supplied to it. A GHP accepts heat Q from the sink at T_C, rejects heat Q into the building, at a higher temperature, T_H, consuming work W, from an external source (electricity). The heating efficiency of a heat pump is given by its coefficient of performance (COP). However, it is typical to define overall efficiencies for actual (real) heat engines and heat pumps, according to the Carnot cycle, as:

$$W = \eta \cdot Q_H \qquad\qquad (10.13)$$

with $\eta < \eta_C$, for heat engines, and $\eta > \eta_C$ for heat pumps, where η_C is the Carnot efficiency, operating between the same temperatures. The previous equation is usually rewritten for heat pumps as:

$$W = \frac{Q_H}{COP_{HP}} = \frac{Q_C}{COP_{RF}} \qquad\qquad (10.14)$$

Here COP_{HP} is the heat pump coefficient of performance based on heat output Q_H from a heat pump or vapor recompression system, and COP_{RF} is the coefficient for a refrigeration system based on heat absorbed Q_C from the process, as discussed in Chapter 8. The heat pump as a low-temperature lifting device ($T_C \rightarrow T_H$) gives higher COP_{HP} inverse proportional to the temperature difference. The most important benefits of GHPs are that they use about one-third or even over half less electricity than conventional heating and cooling systems, having a good potential for energy savings in any area. The cooling efficiency is defined as the ratio of the heat removed to the input energy, or the energy efficiency ratio (EER). Good GHP units must have a COP of 3 or greater and an EER of 13 or greater.

10.3   BUILDING MICROGRIDS

Microgrid concepts are not new; in fact, the first power grids were more or less organized as microgrids (Belu, 2014a,b; Belu, 2018). However, as new technologies come into place to employ renewable energy, energy storage and more efficient electricity production methods coupled with the flexibility of power electronics and control, microgrids seem a suitable approach for extracting the maximum benefits for owners, consumers, and the power grid. Microgrids may be a prospective power system that addresses RES technologies accompanying necessary growing deployment of distributed energy resources, micro-CHP and small-scale renewable energy sources. Microgrids, consisting of various DG units and loads, are viewed as small controllable power distribution subsystems, having characteristics such as the ability to operate in parallel or in isolation from the electrical grid, capabilities to improve service and power quality, reliability, and operational optimality. Microgrid operation, either interconnected to or islanded from the grid, depends on factors like planned disconnection, grid outages, or economical convenience. A typical microgrid configuration consists of a group of radial feeders, a point of common coupling, critical and noncritical loads, and microsources. These may include: microturbines, fuel cells, PV systems, solar thermal arrays, fuel cells, and wind turbine installations. Also, the storage, load control, power and voltage regulation, and heat recovery units need to be grouped together into microgrids. A microgrid is a self-contained subset of indigenous generation, distribution system assets, protection and control capabilities, and end-user loads that may be operated in either a utility-connected mode or in an isolated (islanded from the utility) mode. In addition to providing a reliable electric power supply, microgrids are also capable of providing a wide array of ancillary services such as voltage support, frequency regulation, harmonic cancellation, power factor correction, and spinning and nonspinning reserves. A microgrid may be intrinsically distributive in nature, including several DGs, both renewable and conventional sourced, energy storage elements, protection systems, end-user loads, and other elements. In order to achieve a coordinated performance of a microgrid (or several microgrids) within the scope of a utility, it is required to perform distributed or cooperative control. Agent technology is one of the techniques for achieving the objectives of distributed control of microgrids. An agent is a software or hardware entity that exhibits characteristics of autonomy, self-organization, decentralization, and limited purview so as to progress the entire system toward a common goal such as in cooperative distributed problem solving. Agents in microgrids are expected to perform sensory, communication, and actuation (control) tasks. In order to achieve reliable levels of distributed control of microgrids, it is imperative that the microgrid possess a self-configurable sensor network that aids in communication among the constituent agents. Microgrids seem to be quite suitable to address these challenges.

Microgrids may be a prospective framework that addresses technical concerns accompanying necessary growing deployment of RES by DG technologies to meet the increasing requirements for power quality, grid resilience, and reliability. Microgrids are usually embedded in power distribution systems and often within customer commercial or industrial facilities by incorporating modern controls that enable them to operate with a degree of autonomy from the traditional macrogrid. The growing requirements for energy services can be met using a collage of approaches, technologies, and solutions. DER systems located near local loads, which may include DG and distributed energy storage (DES), can provide several benefits if properly operated. Integration of various DG technologies with the utility power grid is an important pathway to a clean, reliable, secure, and efficient energy system for developed economies with established levels of quality and reliability of electrical service. A microgrid is usually created by connecting a local group of small power generators using advanced censoring, controls, communications, and protection techniques. The development of the microgrid based on the DG together with the opportunities offered by renewables, could have a major effect on the way rural electrification is approached, both in developed and developing countries. Microgrids are expected to enable new opportunities for customers and business stakeholders, by using local energy resources. DER units consist of distributed generation, energy storage, or demand response, all having impacts on the power balance. One MG purpose is to improve power distribution reliability by reducing the amount and duration of interruptions with island operation functionality. A microgrid operates safely and efficiently within its local power distribution network, and is capable of islanding when needed. MG design and operation demand new skills and technologies, while power distribution containing several DER units may require considerable control capabilities. While not strictly compliant with the above definition, small isolated power systems are usually considered microgrids. Usually, microgrids are spanned on the power distribution, supplying AC loads through AC electric infrastructure, being the AC microgrids. However, recently, DC microgrids exploiting a DC electric infrastructure have gained attention due to some specific advantages. In either case, the energy sources are integrated into the microgrid by using power electronic interfaces. In AC microgrids, RES and DG units are integrated to the grid through AC-AC power converters (inverters) while in DC microgrids these sources are connected through DC-DC power converters. In either case, the converters are equipped with controls to meet the network objectives, e.g., active and reactive power flow, voltage, frequency, or power control.

In a typical MG architecture (Figure 10.5), the electric network is often of radial type, connected to the power distribution through a separation device, and the point of common coupling (PCC) is usually a static switch. Each feeder has a circuit breaker and a power flow controller. The MG structure comprises LV network(s), controllable and noncontrollable loads, microsources, energy storage units, and hierarchical management and control schemes supported by communication, used to monitor and control microsources, storage units, and loads. The MG multilevel

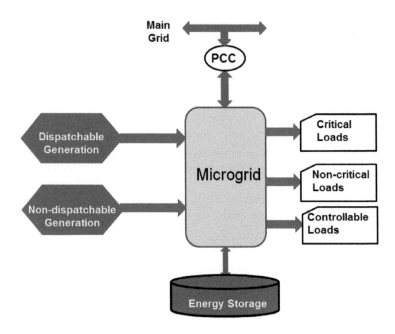

FIGURE 10.5 Microgrid architecture and structure.

control system consists of an MG central controller and a second control level, controlling the loads and energy storage units and microsources, exchanging information with central control, managing the MG operation. The exchanged data amount is usually small, e.g., messages containing controller set points, information requests to microsource and storage unit controllers on the active and reactive powers, voltage levels and messages to control MG switches. MG design, operation, and power distribution with high-DER penetration require considerable control capabilities. Power distribution is evolving from currently passive to an active network, in the sense that decision-making and control are distributed, while the power flow is bidirectional. This power network eases DG, RES, demand side, and energy storage integration, creating opportunities for novel types of equipment and services, all of which are required to conform to new protocols and standards. The main function of active power distribution is to efficiently link generation with consumer demands, allowing both to decide how best to operate in real-time. Active power distribution requires the implementation of radically new system concepts. Microgrids, characterized as *smart grid building blocks*, are perhaps the most promising approach. The MG organization is based on the control capabilities over the network operation, allowing the MG to operate when isolated from the main grid, in case of faults or other external disturbances or natural disasters, thus increasing the supply quality and grid resilience.

Microgrid bases are the DER, DG, and energy storage units and demand response, all of which have impact on the power balance. One MG purpose is to improve power distribution reliability by reducing the periods of interruptions, via its island operation capability. It is useful to understand the MG definition variations for the detailed MG studies. Microgrid can be: (1) a power distribution part, having islanding capability, reducing the power outages; (2) an integration platform for the supply-side (energy sources) and demand-side resources (energy storage and controllable loads) located in the local distribution; and (3) an autonomous power system able to provide power and all critical services, the islanded operation being a special emergency case. This list shows that the MG definition is still incomplete; however, in summary microgrids are entities that can provide improved and more reliable energy distribution, enabling easier RES and DG integration. Microgrids may also be considered as self-managed and self-serviced power systems where islanded operation is only a special case. Differences in the MG definitions are explained by the operation drivers, e.g., the operating environment and network structure variations. For example, the fact that electricity transmission and distribution are regulated as monopoly businesses has impact. In addition to variations into the MG functionalities, the ways by which the wanted functionalities are achieved vary significantly. Basically the MG term describes a network with features: (1) system has islanding capability, being normally grid-connected; (2) DER units produce a large part of the consumed electricity; and (3) power flow is bidirectional. There are still uncertainties related to microgrids, such as: what are the stakeholder roles, how fast is the new technology implemented, and whether microgrids are a cost-effective way to limit outages. The MG future is highly dependent on the smart grid features, e.g., fast, reliable, and inexpensive communication, cost-effective energy storage, or DER implementation. Microgrids have been extensively investigated in a number of research and demonstration projects with considerable work undertaken on their design and operation. However, to date, microgrids have not been implemented widely in electricity systems, and a robust commercial justification or business case for their use has yet to be developed.

The last decades have seen an increased interest in DC or hybrid (AC and DC) power networks, for reasons such as: cost, efficiency, energy conservation, reduced losses, increased DG and RES uses, supply reliability, security, or higher power quality demands. DC networks are used in commercial applications, for practical reasons, such as: easier DC machine speed control, compatibility with digital electronics or more reliable and simpler power systems, using DC sources and batteries. If DC electric systems will gain interest, there will be a transition period when both AC and DC are used in parallel, while the existing loads, without significant modifications, operate independent of the voltage type. It is also of interest to study within which voltage ranges existing loads can operate with DC power systems. For example, resistive loads operate with DC in the full tested range without any changes. Problems may arise with load switches, which are not designed to interrupt DC currents. On the other hand, an electronic interface is needed to connect DC supplies with an AC system. The interface design has great significance on the DC or AC power system operation. A well-designed interface needs to provide a controllable DC-link voltage, higher power quality, and transient performance during faults and disturbances, and have also lower losses and cost. Moreover, bidirectional power flow capability is desired if generation is present in the DC system, to transfer power to the AC system during low-load or high-generation conditions in the DC system. Finally, galvanic isolation is required to prevent a current path between the AC and the DC systems in the case of faults. Commercial LV power systems often have sensitive nonlinear loads, which must be protected from disturbances and outages to operate correctly. A DC microgrid or nanogrid can be preferred over an AC microgrid in cases where there is sensitive electronic equipment and the power sources are connected through power electronic interfaces, having the advantages of simpler and more efficient power electronic interfaces. In a DC microgrid, energy storage, power sources, and loads are normally interconnected through DC busses.

For decades, chemical plants, refineries, military installations, college campuses, and other large facilities structured as microgrids, have had the ability to generate and manage their own electricity needs, while remaining connected to the grid for supplemental or emergency needs. Today, affordable and locally produced electricity is encouraging MG expansion, having the potential to revolutionize power grids by increasing efficiency and environmental sustainability. Microgrids offer an efficient energy system, based on collocating sources and loads that can operate independently in case of outages or energy crises. The MG concept, a major part of the smart grid, assumes an aggregation of loads and DG units operating as a single entity to provide power and heat. The MG critical elements are: (1) *embedded distributed energy resources*; (2) *advanced energy storage*; and (3) *flexible demand*. Most MGs combine loads, sources, and storage units, allowing the intentional islanding operation, while trying to reuse the available waste heat and energy; improving efficiency, power quality and reliability; minimizing the overall energy consumption; reducing pollutant emissions; or allowing cost-efficient electricity infrastructure replacement. MGs link several DERs into a small power network serving some or all of the customer energy needs, having the potential to accommodate new electricity demands, shifting the current electricity paradigm to one in which local, clean energy resources supplement the power grid. However, MGs raise several challenges, such as protection and control. MG potential advantages are: more reliable energy supply to customers in the event of a major disturbance, reduced transmission losses, preventing network congestions, enhancing system reliability and stability. A main MG advantage is *economic savings*, through the MG energy portfolio. The major microgrid benefits are:

- **Sustainability:** The microgrid portfolio enables a hedge against fuel cost increases.
- **Stewardship:** The MG enables deep penetration of renewables, missions reduction, green marketing.
- **Reliability:** The microgrid actively controls the network for better reliability.

As a result of the major operation differences between large power grids and microgrids, new approaches of the unit commitment, control, protection, management, and economic dispatch must be developed. The major differences between these two models are outlined in this chapter. The RES and DG penetration in microgrids is much higher than in conventional power grids. For this reason, it is more difficult to forecast and predict the MG power outputs. Moreover, if the microgrid is working in connected mode, it could sell to or buy electricity from the grid depending on the prices. Although in large power grids it is possible to store energy, the large part of generated power is instantly consumed. However, due to the reduced size of a MG, a large part of the produced energy is often stored, which is a reason for the MG's extended use of distributed storage systems. Unlike the microgrids, committing a generation unit in large power systems means bringing online a large power generator, a task more difficult and time-consuming than in the case of small generators. Furthermore, small MG generators have more flexibility constraints in relation to the minimum time that a unit must be down or up. These limits are set to avoid equipment fatigue and to prevent excessive maintenance and repair costs due to frequent unit cycling. Another major difference between large power systems and microgrids is that in the latter case, the number of generators involved may easily vary throughout time, so a unit-commitment algorithm, working well with a limited number of generators may not be able to handle a larger number of units, so the MG characteristics make it necessary to modify the existing unit-commitment algorithms. The MG control system is composed of several levels, and specific characteristics, assigning the active and reactive powers to each microgrid element according to the load demands, by fixing the voltages and frequencies references whereas the controllers respond to the voltage and frequency deviations.

10.3.1 Microgrid Concepts and Architecture

A microgrid, defined as an aggregation of electrical generation, storages, and loads, can take the form of shopping center, industrial park, college campus, or an apartment building. To the utility, a microgrid is an electric load that can be controlled; this load can be constant, can increase at night when electricity is cheaper, or can be held at zero during grid stress. The MG concept supersedes all the advantages of single-source DG or hybrid power systems. Moreover, it also includes all the grid advantages, at mini-scale, having also capabilities for combined heat and power (CHP) generation and integrated cogeneration units. The use of waste heat through CHP, a versatile approach to save energy and improve energy-system efficiency, implies integrated energy systems, delivering both electricity and useful heat. CHP processes can convert as much as 90% of their fuel into useable energy. To maximize system efficiency, the energy sources need to be placed closer to the heat load rather than the electrical load since it is easier and more efficient to transport electricity over longer distances. Small local energy sources can be sited optimally for heat uses, so a distributed system, such as MG integrated with DGs is very pro-CHP. A microgrid combined with a power electronic interface is a self-sufficient network, usually with autonomous control, communication, and protection. It is capable of providing capacity support to the grid while in grid-connected mode, and with capacity in excess of coincident peak demand. So, the microgrids comprise low-voltage

(LV) distribution systems with integration of DER units together with distributed energy storage units, and controllable and smart loads that behave as a coordinated entity networked by employing advanced power electronic conversion and control capabilities. It has been proposed that one solution to the reliability and stability issues is to take advantage of MG technologies. A microgrid offers three major advantages over traditional electricity supply systems:

1. Application of combined heat and power technology;
2. Opportunities to tailor the quality of power delivered to suit the requirements of end users; and
3. Creates a favorable environment for energy efficiency and small-scale RES investments.

Distributed energy sources have the potential to increase system reliability and power quality due to the decentralization of supply. The integration of RES units into the power system provides unique challenges. Due to the RES intermittent nature, central generation is required to provide the base power supply and backup power when RES are not generating power. Systems with intermittent sources can experience similar problems as those with large, intermittent loads. Distributed generation can ease the burden of high RES penetration by filling in when intermittent generation is low and by smoothing transmission system loading. A conventional architecture (Figure 10.6) has a microgrid connected to a larger system and a disconnect switch that *islands* the DG units to protect sensitive loads. A major MG factor is the disconnect switch enabling the microgrid to maintain compliance with current standards, needed to realize the high reliability and power quality that microgrids offer. Small generators have a lower inertia and are better at automatic load following and helping to avoid large standby charges that occur when there is only a single large generator. Having multiple distributed generators available makes the chance of an all-out failure less likely, especially if there are storage and backup generation capable of being quickly and easily connected to the system. This configuration of multiple independent generators creates a peer-to-peer network, with no master controller that is critical to the MG operation. A peer-to-peer system implies that the microgrid can continue to operate with the loss of any component or generator. With the loss of one source, the grid regains all its original functionality with the addition of a new source, if available. This ability to interchange generators and create components with plug-and-play functionality is

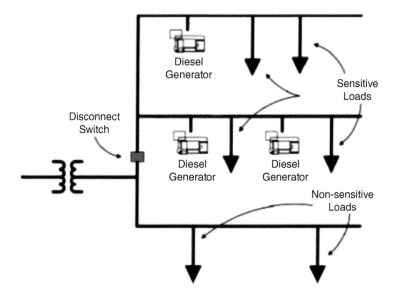

FIGURE 10.6 A typical microgrid configuration with diesel gensets and loads.

one requirement of microgrids. The concept can be extended to allowing generators to sit idly on the system when there is more electrical capacity than necessary. As the load on the system increases, additional generators would come online at a predetermined set-point necessary to maintain the correct power balance. Intelligent devices could sense when there is extra generation capacity on the system and would disconnect and turn off generators to save fuel and increase machine efficiencies.

The microgrid of Figure 10.6 normally operates in a grid-connected mode through a substation's transformer, being expected to provide sufficient generation capacity, controls, and operational strategies to supply at least an overall load portion, after being disconnected from the distribution system, remaining operational as an autonomous (islanded) entity. The existing power utility practice often does not permit accidental islanding and automatic resynchronization of a microgrid, primarily due to the human and equipment safety concerns. However, the high penetration of DER units potentially requires provisions for both islanded and grid-connected operation modes and smooth transition between the two (i.e., islanding and synchronization transients) to enable the best use of MG resources. DER units, in terms of their interface with a microgrid, are divided into two groups. The first group includes rotary units, interfaced as in conventional grids, while the second group consists of electronically coupled units, using power electronic converters to provide coupling with the host system. The control concepts, strategies, and characteristics of power electronic converters, as the interface for most DG and DES units, are significantly different from those of conventional rotating machines. Therefore, the control strategies and dynamic behavior of a microgrid, particularly in an autonomous mode of operation, can be noticeably different from that of a conventional power system. Figure 10.7 shows a schematic diagram of the microgrid building blocks that include load, generation/storage, electricity, and thermal grids, implying two control levels, i.e., component-level and system-level controls. Both DG and DES units are usually connected at either medium- or low-voltage MG levels. A DG unit usually comprises a primary energy source, an interface, and switchgear at the unit point of connection (PC). In a conventional DG unit (e.g., a synchronous generator driven by a reciprocating engine or an induction generator driven by a fixed-speed wind turbine) the rotating machine: (1) converts the power from the primary energy source to the electrical power, and (2) acts as the interface medium between the source and the microgrid. For an electronically coupled DG unit, the coupling converter: (1) can provide another layer of conversion and/or control (e.g., voltage and/or frequency control), and (2) acts as the interface medium with the microgrid. The input power to the interface converter from the source side can be AC at fixed or variable frequency or DC type. The microgrid side of the converter is at the

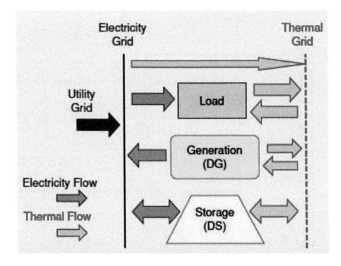

FIGURE 10.7   A general representation of microgrid building blocks.

TABLE 10.2
DER and DG Units' Interface and Control

DER and DG Type	Primary Energy Source	DER Interface	Power Flow Control
Conventional DG	Reciprocating engines Small hydro Fixed-speed wind turbine	Synchronous or Induction generators	Power flow and/or wind turbine control
Nonconventional DG	Microturbines, variable-speed wind turbines, solar PVs, fuel cells	Power electronics Converters	DG control, speed, output, frequency-voltage control
Energy storage	Batteries, ultra-capacitors, flywheels, TESs	Power electronics Converters	DG control, state-of-charge, speed, output, frequency-voltage control

frequency of either 50 or 60 Hz. Table 10.2 outlines typical interface configurations and methods for power flow control of DG and DS units for the widely used primary energy sources and storage media, respectively. It should be noted that in addition to the two basic types of DG and DES units, a DER unit can be of a hybrid type; i.e., a unit that includes both a *primary energy source* and *energy storage*. A hybrid DER unit is often interfaced to the host microgrid through a converter system that includes bidirectional AC-DC and DC-DC power converters. In terms of power flow control, a DG unit is either a dispatchable or a nondispatchable unit. The output power of a dispatchable DG unit can be controlled externally, through set points provided by a supervisory control system. A dispatchable DG unit is either a fast-acting or a slow-response unit. An example of a conventional dispatchable DG unit is one using a reciprocating engine as its primary energy source, controlled by a governor setting the output power, voltage, and frequency. In contrast, the output power of a nondispatchable DG unit is normally controlled based on the optimal operating condition of its primary energy source. For example, a nondispatchable wind unit is normally operated based on the maximum power tracking concept to extract the maximum possible power from the wind regime. Thus, the output power of the unit varies according to the wind conditions. To maximize output power of a renewable energy-based DG unit, normally a control strategy based upon maximum point of power tracking (MPPT) is used to deliver the maximum power under all viable conditions. IEEE Standards Coordinating Committee 21 supports the development of IEEE P1547.4 Draft Guide for Design, Operation, and Integration of Distributed Resource Island Systems with Electric Power Systems, providing approaches and practices for MG design, operation, and integration, covering the ability to disconnect and reconnect to the grid. Intended for designers, operators, and equipment manufacturers, the guide covers DER interconnection and microgrids. Its implementation expands the DER and MG benefits by enabling improved reliability, based on the IEEE 1547-2003 standard specifications.

IEEE Standards Coordinating Committee 21 supports the development of *IEEE P1547.4 Draft Guide for Design, Operation, and Integration of Distributed Resource Island Systems with Electric Power Systems*. It will provide alternative approaches and good practices for the design, operation, and integration of the microgrid and cover the ability to separate from and reconnect to part of the Area EPS while providing power to the islanded local power systems. The guide covers the DER, interconnection systems, and participating microgrids. Its implementation expands the benefits of DER by enabling improved EPS reliability and building on the requirements of *IEEE 1547-2003*. Technical challenges include the design, acceptance, and availability of low-cost technologies for microgrids. Significant changes in the regulatory and operational climate of electric utilities and the emergence of DG and RES systems have opened new opportunities for on-site power generation, located at user sites where the electricity and heat meet the customer demands with an emphasis on supply security, reliability, power quality, and less harm to the environment. The DER includes generators, energy storage, load control, and, for certain systems, advanced power electronic inter-faces, while the MG concept assumes their aggregation, as a single entity to provide power and heat, an integrated platform of supply-side generation, energy storage, demand resources, and control-lable loads, located typically in a local power distribution. A microgrid is typically located at the LV grid sections with installed generation capacity in the MW ranges, although there are exceptions, with a microgrid located in MV section for interconnection or other purposes. Most MG energy sources are power electronic interfaced to provide the required flexibility and to ensure operation as a single aggregated entity. The control flexibility allows the MG to present itself to the main power system as a single controllable entity, meeting local needs for supply reliability and security, being proposed as one of the solutions to grid reliability and stability issues. MGs offer three major advan-tages over the traditional electricity supply involving large generation stations and long transmission through complex electric networks:

- Application of combined heat and power, as well as any heat and energy waste recovery technology;
- Opportunities to tailor the power quality delivered to suit the end-user requirements; and
- Creating a favorable environment for energy efficiency, DG and RES investments, and operation.

The traditional power grid is dominated by unidirectional power flows and production based on large synchronous generators. The developments to the RES and DG generation led to new aspects such as energy storage units located in the microgrids, bidirectional power flows, and "prosumers," i.e., customers who are using and producing energy. The challenges to RES and DG integration can be faced through the MG approach. MG definition includes clearly defined electrical boundaries, local control or flexible loads, DG and RES integration issues, secured power supply, challenges of bidirectional power flows, demand-side participation, and intermittent generation. The MG can use the waste heat through micro-CHP, implying an integrated energy system, delivering both electric-ity and thermal energy, having the capacity of converting over 80% of its primary fuel into useable energy. Small local energy sources can be sited optimally for heat uses, so the distributed energy sys-tems, integrated with DGs are very pro-CHP. Power quality and reliability are used in quantifying the levels of electrical service. Both scheduled and unscheduled outages affect the end-user service availability, increasing the dependence on on-site backup power supplies, which is quite often very expensive. The power quality degradation has more subtle, but important, effects as well. Voltage sags, harmonics, and imbalances are triggered by switching events and by faults. While power quality events do not lead to electrical losses, they may degrade the end-use processes, affecting equipment operation, performances, and durability. DERs have the potential to increase system reli-ability and power quality due to supply decentralization. Increases in reliability levels are obtained if DG units operate autonomously in transient conditions, when there are outages or disturbances upstream in the electrical supply. DG units located close to the load deliver electricity with minimal losses, bringing higher values than power coming from large, central generators through grid trans-mission and distribution infrastructures. With the RES and DGs, the fossil-fuel dependency and the fluctuating prices, pollutant emissions, and operation costs are minimized. If, in addition, DG and consumption are integrated into a single system, the power supply reliability increases signifi-cantly. The importance and quantification of the MG benefits must be recognized and fully incorpo-rated within the technical, commercial, and regulatory framework. However, under the present grid codes, all DG types must be shut down during power outages. This is precisely when on-site energy sources offer the greatest value to both the owners and the society. A microgrid can optimize the following aspects: power quality and reliability, sustainability, economic benefits, and flexible dual operation. MG structure varies a lot and variables such as the generator type and number, and grid topology impact the final solution. A MG definition is related to its functionalities, capabilities, and operation, and not all microgrid functionalities need to be implemented in all microgrid operations.

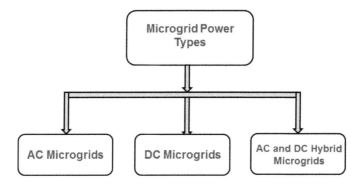

FIGURE 10.8 Microgrid classification based on the power type (AC, DC, or hybrid).

All of the MG variables include uncertainties, making it hard to predict the final, integrated MG model. The MG capabilities can also vary according to the needs. However, the MG main idea is islanding capability, which can reduce outages and offer independence for customers. In terms of power type, the microgrid can be classified as an AC power system, a DC power system, or a hybrid system, as shown in Figure 10.8. When used, each microgrid type presents advantages and disadvantages, and the application can set the MG type.

The DC microgrid presents several operational advantages, such as: most DG systems employed in microgrids are DC supply, such as PV units and fuel cells. Storage devices have a DC output voltage, so connecting them to the DC microgrid requires only a voltage regulator, as compared to an AC microgrid, which additionally needs to synchronize the system by matching the voltage magnitude, the phase, and the frequency to the grid. Most DC microgrid connected loads are conventional ones, e.g., electronic devices, TVs, computers, lights, variable-speed drives, and appliances, reducing or even eliminating the needs for multiple power conversions, such as AC-to-DC or DC-to-AC, as is required for AC microgrids. The DC microgrids do not use transformers, making them more efficient, smaller in size, and reliable. There is no reactive power flow in a DC microgrid, and the voltage control is concerned with only active power flow. In an AC microgrid, the voltage control is related to the reactive power flow at the same time injecting the active power. On the other hand, an AC microgrid has a facility to use existing infrastructure from the utility grid, the nature of its power system and its compatibility with the utility grid. When using an AC microgrid, there is no requirement to reconfigure loads or the building power system of the supply. This implies that AC loads are connected directly to the AC microgrid without any power conversion through an AC-DC converter interface. Also, it contributes to the utility grid stability by offering reactive power support for balancing and ancillary services, being fully compatible with the utility grid.

10.3.2 Microgrid Features and Benefits

Theoretically, a microgrid can be formed in any power grid section, based on the control methods and/or on different voltage levels as long as there is enough local power generation and control capability. The MG operation principle remains the same despite the variations. The MG main configurations are separated into four classes: (1) island microgrid, (2) customer microgrid, (3) LV microgrid, and (4) MV microgrid. The first type is the stand-alone microgrid providing electricity to a customer or a small community outside the grid, e.g., islands, isolated farms, or villages far away from the grid. The second type refers to the LV customer microgrid, such as a farm or a detached building with its own DER units to provide the needed electricity, operating usually in parallel with the utility grid. However, in case of a grid fault, the microgrid operates in island mode. In such cases, the microgrid needs energy storage to maintain a constant electricity supply. The third MG type, the LV microgrid, consists of a group of customers, anything from a few consumption points to the whole LV network fed by a transformer. Power production of this MG type is based on several

DER units of different type. In the MV microgrid, larger generation units are connected to the MV network section. Such microgrids can consist of part of or the entire HV or MV substation output. MV microgrids offer an opportunity for wind parks or other large RES facilities to produce electricity during utility grid interruption. The basic principles, concepts, and protocols are the same in all MG types, but when it comes to more details, there are separating factors. In addition to technical issues, separating factors are related to the market structure and the grid ownership. The solution must be feasible for all business parties, and the biggest challenge, related to all microgrids, is how to do things economically. However, there also are several variables related to MG operation and management. In addition to the differences in voltage levels, there are differences in the number and type of generators, or the type and nominal power of the DER units. Higher numbers of generators increase the system complexity, requiring communication to enable the needed control capabilities. On the other hand, higher numbers of generators increase the reliability because the system does not collapse if one unit is not working.

A microgrid is usually connected to the grid at a PCC, appearing to the grid as a single controllable entity. The connection switch connects the microgrid to the distribution system (see Figure 10.9), enabling high DG and RES penetration, without requiring power distribution redesign or reconfiguration. A major MG feature is to ensure stable operation during faults and network disturbances, realized by opening the static switch, disconnecting the microgrid from the grid. DG generations and loads are separated from the power distribution to isolate them during faults and/or malfunctions, or by intentionally disconnecting when the grid power quality falls below certain standards. In such an event, the DG units maintain the prescribed voltage and frequency levels while sharing the power. Microgrids major desired features are:

- Accommodates a wide variety of generation options: distributed, intermittent, or dispatchable.
- Empowers consumers to actively manage their energy use and to reduce the energy costs.
- Allows plug-and-play functionality, for switching to a suitable operation mode either into the grid connected or islanded operation, providing voltage and frequency protection during islanded operation and capability to resynchronize safely the MG connection to the grid.

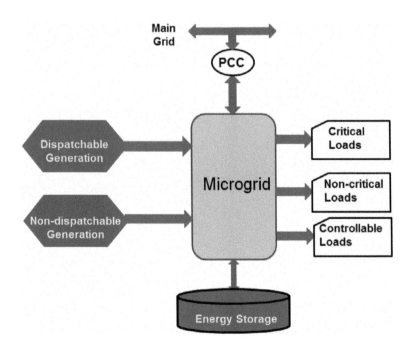

FIGURE 10.9 Microgrid structure and architecture, with point of common connection to the grid.

- MGs can operate without grid connection, during islanding mode, while all loads are supplied and shared by DG units, providing solutions to supply power in emergencies and power shortages.
- Often microgrids are equipped with thermal energy units capable of waste heat recovering, an inherent by-product of fossil-based electricity generation, or energy conversion, through CHP systems.
- MGs can service a variety of loads (residential, office, industrial parks, and commercial facilities).
- Provides power quality at the twenty-first century standards, as required by the end users.
- MGs can provide self-healing, anticipating and instantly responding to system problems in order to avoid or mitigate power outages and power quality problems.
- Tolerant of attacks, by mitigating and standing resilient to physical and/or cyber-attacks.
- Enables a competitive, open energy market, through real-time information and lower transaction costs.
- Asset optimization, through IT capabilities, minimizing operation and maintenance costs.

In summary, the three main MG benefits are: (1) increased power system flexibility; (2) opportunities to tailor the power quality to the end-user requirements; and (3) favorable to the environment to establish higher energy efficiency and small-scale energy generation investments. Microgrids can power buildings, neighborhoods, or entire cities, even if the grid suffers severe outages, providing uninterrupted power supply to customers during unexpected power outages, e.g., natural disasters and severe human-induced faults in the utility grid, an important feature in the case of critical loads, such as hospitals, first responder units, or military facilities. Less essential loads can be switched off to increase the withstand periods. Reduction of grid interaction results in improving the self-consumption. Microgrids can reduce and control the electricity demand and mitigate grid congestion, lowering the electricity prices and reducing the peak power requirements. In remote areas where a grid is not available, a microgrid can reduce the investment costs for substations, transmission lines, or other infrastructure components. Microgrids, with advanced control technology, can generate electricity from renewable energy sources at costs only a little higher to those of the conventional grid. The price is also lowered, due to lower transmission losses and less sophisticated transmission equipment, because the grid interaction is shorter, reducing the electricity bills for customers. Self-consumption and self-sufficiency coefficients are parameters characterizing the generation unit performances. The self-consumption coefficient is defined as the ratio of the internal consumed energy to the total generated RES or DG energy:

$$SC = \frac{\text{DG or RES Internally Consumed Power}}{\text{Total Generated Power}} \quad (10.15)$$

The self-sufficiency coefficient is determined as the ratio of the internal consumed energy to the total load (electricity demand):

$$SS = \frac{\text{DG or RES Internally Consumed Power}}{\text{Total Electricity Demand}} \quad (10.16)$$

Flexibility is a feature of a power system that is easiest to be identified in the portfolio of any power station. The inclusion of MGs makes a power network more efficient and flexible. Optimal dispersion allows generators to be smaller and deeply embedded with the energy demands. Unscheduled outages and malfunctions are usually more disruptive and threatening to people and property, while power quality deteriorations have mixed and less dramatic effects. While the ideal is rarely achieved in practice, the MG paradigm can in theory provide a universal level of power quality to every load. The MGs and the technology selection on the demand side hold a unique advantage, absent from the macrogrid. The RES units on both demand and supply sides have a chance at being even-handed and considered, without the macrogrid issues. Because MG penetration into the grids is expected to be moderate, the initial unit commitment schedule of the centralized power generation is not expected to change dramatically in the near future. However, there are changes in the economic dispatch of the most expensive (critical network units), the network losses and pollutant emissions. In order to successfully integrate RES and DG units and to fully operate microgrids, several technical issues and challenges must be addressed to ensure that the present reliability levels are not significantly affected, and the DG or RES benefits are fully harnessed. In this regard, the main issues are: the schedule and dispatch of RES units under supply and demand uncertainty and the determination of appropriate levels of reserves, reliable and economical MG operation with high penetration RES levels in stand-alone operation modes; the design of appropriate demand-side management schemes to allow customers to react to the power grid changes; the new market models, allowing competitive participation of intermittent energy units, and providing appropriate incentives for the investments, development of appropriate protection schemes at the distribution level to account for bidirectional power flows, voltage, and frequency control methods to account for the power-electronics-interfaced DG and RES units; and the development of market and control mechanisms that exhibit a plug-and-play feature to allow for seamless integration of DERs and DG units.

Example 10.5

A microgrid consists of four 15 kW wind turbines, 36 PV panels of 345 W each, and four battery banks with enough capacity to store all the excess energy, and the power management, control, and protection units. The microgrid supplies power to a remote atmospheric research facility having a daily power demand of 13.5 kW. Each wind turbine and PV panel has a capacity factor of 0.23 and 0.26, respectively. Each wind turbine uses internally 0.25 kW of generated power, while each PV panel 7.5 W. The battery banks, power management, control, and protection use 1.5% of the output generated power to operate. Estimate the MG self-consumption and self-sufficiency coefficients.

SOLUTION

The average daily wind turbine output is 13.8 kW, and PV panels 1.8 kW, an overall average daily generated power of 15.6 kW. The total internally used power by wind turbines and PV panels is then 0.52 kW. The overall power output is then 15.08 kW. The power used by the battery banks, power management, control, and protection units is 0.237 kW. The MG self-consumption coefficient is then:

$$SC = \frac{0.52 + 0.237}{15.6} = 0.0485 \text{ or } 4.85\%$$

and the MG self-sufficiency coefficient is:

$$SS = \frac{0.52 + 0.237}{13.5} = 0.0561 \text{ or } 5.61\%$$

A microgrid is simply defined as an aggregation of generators, energy storage units, and loads that can take the form of a shopping center, industrial park, campus, or large building. To the utility, a microgrid is a controllable electric load, which can be constant, can increase at night when the electricity is cheaper, or can be held at zero during grid stresses. A MG supersedes all the combined advantages of individual DG sources, while including at a smaller scale all the grid advantages. Microgrids are self-sufficient energy networks with autonomous control, communication, and protection, being capable of providing support to the transmission network in grid-connected mode and with capacity in excess of coincident peak demand. Microgrids behave as a coordinated entity networked by advanced power electronic interfaces and control capabilities, being proposed as one solution to the grid reliability and stability issues. One of the major MG aims is to combine the benefits of renewable energy, low-carbon generation technologies, and highly efficient CHP systems. The choice of a distributed generator mainly depends on the climate and the area topology. Sustainability of a microgrid depends on the energy scenario, strategy, and policy, varying from region to region. Microgrids can potentially improve the technical performance of the local power distribution grid through: the reduction of energy losses due to the decreased line power flows, voltage variation mitigation through reactive power control and constrained active power dispatch, or peak loading relief of constrained network devices through selective scheduling of MS outputs; and enhancement of supply reliability via partial or complete islanding during loss of the main grid. The waste-heat recovery through CHP implies an integrated energy system, such as a microgrid delivering both electricity and heat, with the small energy sources sited optimally for heat utilization, distributed systems, integrated with distributed generation are very pro-CHP systems. Systems with intermittent energy sources can experience similar problems as systems with large and intermittent loads, but the DG can ease the burden by filling in when intermittent generation is low and by smoothing the transmission loading. A common MG architecture and structure were shown in the diagrams of Figures 10.8 and 10.9. This microgrid is connected to a larger system and a disconnect switch that *islands* the DG units to protect sensitive loads. A major MG component is the disconnect switch enabling the microgrid to maintain compliance with the standards, requiring high reliability and power quality. Small electrical generators have a lower inertia and are better at automatic load following and help avoid large standby charges, occurring when there is only a large generator.

The presence of multiple DG units makes the chance of an all-out failure very unlikely, especially when energy storage and backup generation units are available. Such configurations create peer-to-peer networks with no need for a master controller to the MG operation. A peer-to-peer system implies that the microgrid continues to operate, regardless of the loss of a component or a generator. With one source lost, the microgrid regains its original functionality by new energy-source additions, if available. The ability to interchange generators and components with plug-and-play capability is an MG main requirement. The concept can be extended to allow generators to sit idly when there is more capacity than needed. As the load increases, additional generators come online at a predetermined set point to maintain the power balance. Intelligent control can sense the extra generation capacity and turn off the generators, increasing the overall efficiencies. The microgrid, as shown in Figure 10.8, operates in a grid-connected mode. However, it is also expected to provide sufficient generation capacity, controls, and operational strategies to supply at least a load portion when disconnected from the grid, remaining operational as an autonomous entity. The existing utility practice does not permit accidental islanding and automatic resynchronization, due to safety reasons. However, MG operation needs new provisions for both islanded and grid-connected operation modes, and smooth transition between them, to enable the best use of MG resources. Table 10.2 lists the control concepts, strategies, and characteristics of power electronic converters, as the DG and RES interface media are significantly different from those of the rotating machines. Therefore, the MG control strategies and dynamic behavior, particularly in autonomous operation mode, can be very different from those of a conventional power system.

10.3.3 Microgrids vs. Virtual Power Plants

DG systems are used to displace energy from conventional generating plants but not to change their capacity. Small DG units are not visible to power system operators and are controlled to maximize energy from renewable energy sources or in response to the heat demands of the host site, not to provide additional capacity for the power system. A *virtual power plant* (VPP) is defined as a cluster of DERs, collectively operated by a central control entity. A VPP can replace a power plant, while providing higher efficiency and flexibility, leading to very large-generation margins, avoiding underutilization of assets and low operating efficiencies. The VPP concept has been developed to increase DG visibility and control. In this way, a very large number of small generation units can be aggregated, in order to take part in the markets for energy and ancillary services. In a VPP, the DG units together with controllable, responsive loads are aggregated into a controllable entity, visible to the power system operator, and can be controlled to support the grid operation and can effectively trade in energy markets. Such an entity behaves in a power system in a similar manner as a large transmission-connected generation station. Although the microgrid and the VPP appear to be similar concepts, they differ in some distinct ways, such as: the *locality*, in a microgrid, DERs are located within the same distribution network, aiming to satisfy primarily local energy demand, while in VPPs, DERs are not necessarily located on the same network, being coordinated over a wider area. The VPP-aggregated production participates in trading in the energy markets. The *size* differences—installed capacities of microgrids are typically smaller (from few kW to several MW), while the VPP's power ratings can be much larger. Customer interests in microgrids focus on local consumption, while VPP systems deal with consumption only as a flexible energy resource, participating in aggregate power trading. Following on from the VPP definition as a commercial entity that aggregates different generation, storage or flexible loads, regardless of their locations, the technical VPP has been proposed, which also takes into account local network constraints. In any case, VPPs tend to ignore local consumption, except for DSI, while MGs acknowledge local power consumption and give end consumers the choice of purchasing local generation or generation from the upstream energy market. This leads to better controllability of microgrids, where both the supply and the demand resources of a microgrid can be simultaneously optimized, leading to better DG profitability.

Through VPP aggregation, individual DG units become visible, gain access to energy markets, and maximize revenue opportunities, and system operation benefits from the effective DG use and increased operation efficiency. When operating alone, DG units, even in a large number, have no sufficient capacity, flexibility, or controllability to allow them to effectively take part in system management and energy market activities. A VPP represents a portfolio of DG units, through which smaller electric units can take part in power system operation. A VPP aggregates the capacity of diverse DG units, creating a single operating entity from the parameters characterizing the generators. A VPP is characterized by a set of parameters, such as: scheduled output, ramp rates, and voltage regulation capability or reserve. Furthermore, as the VPP also includes controllable loads, parameters such as demand-price flexibility and load recovery patterns characterize a VPP. A VPP performs similarly to a large generating unit (Figure 10.10). As a VPP is composed of several DG units of various technology operating pattern and availability, its characteristics vary significantly during operation. Furthermore, as the VPP is connected to various power distribution points, the network characteristics (topology, impedances, and constraints) affect the VPP characteristics. A VPP facilitates trading in the wholesale energy markets, providing services to support transmission

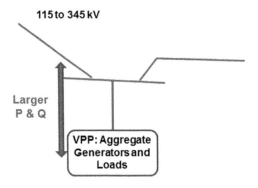

115 to 345 kV

Larger P & Q

VPP: Aggregate Generators and Loads

FIGURE 10.10   Aggregate VPP, larger active and reactive power transfer and control.

system management through various reserve types, frequency, and voltage regulation. In the VPP concept, the activities of market participation, system management, and support are described as *VPP commercial* and *technical* activities. For the DG units in a VPP portfolio, the approach reduces the imbalance risks associated with individual unit operation and provides the benefits of resource diversity and increased capacity through aggregation. DG units can benefit from scale economies in the market participation to maximize revenue. A VPP can still represent distributed generation from various places, but aggregation of resources occurs by location, resulting in a set of generation and load portfolios defined by geographic location. The technical VPP provides DG visibility to system operators, allowing the DG units to contribute to the system management and to facilitate the use of controllable loads to provide system balancing at the lowest cost. The technical VPP aggregates controllable loads, generators, and networks within a single electric geographical area and models its responses. A hierarchy of technical VPP aggregation can be created to characterize the DG operation connected to the grid low-, medium-, and high-voltage sections, but at the grid interfaces the VPP presents a single profile representing the whole local network.

10.3.4 MICROGRID STRUCTURE AND OPERATION

Microgrids consist of several basic operation technologies, including: DG and DES units, critical and noncritical loads, interconnection switches, control subsystems, and protection devices. Basically an MG structure consists of five major components: energy microsources, loads, energy storage unit(s), control and protection subsystems, and a point of common coupling. These five aggregate components are usually connected to a LV power distribution. In order to provide synchronization and control, the mode of operation can be determined by the PCC. Distributed energy storage units are used in MGs where the generation and loads are not matching, providing a bridge to meet power requirements, and enhancing its performance in three ways: (1) It stabilizes and permits DG units to run at a constant and stable power output, despite load or generation fluctuations. (2) It provides ride-through capability when there are dynamic variations of primary energy supply. (3) It permits DG units to seamlessly operate as dispatchable units. The MG control is designed to safely operate the system in grid-connected and stand-alone modes. The control may be based on a central controller or imbedded as autonomous parts of each DG unit. In islanded operation mode, MG must control the local voltage and frequency, providing instantaneous real power differences between generation and loads, the difference between generated and the actual reactive power consumed by the load, and protecting the microgrid. In the grid-connected mode, the microgrid can be considered a controllable load, or it can supply power and act like a generator from the grid point of view. In the islanded mode, the energy generation, storage, load control, and power quality control are implemented in a stand-alone system approach. The overall energy storage denotes the total DES units within a smart microgrid that are under control of the microgrid energy management system. The microgrid concept was initially favored for its ability to provide fault isolation and ease of DG handling. Within the context of the smart grid, the MG definition is evolving into the smart microgrid, and information and communication technologies are becoming integrated to load control tasks and also being used for energy trading among communities.

A traditional MG architecture consists of several energy sources and loads, with a microgrid connected to a larger system and a disconnect switch that *islands* the DG units to protect sensitive loads. In terms of energy security, multiple small generators are more efficient than relying on a single large generator. Small generators have a lower inertia and are better at automatic load following and help to avoid large standby charges that occur when there is only a single large generator. If multiple DG units are available, the chance of an all-out failure is less likely, especially if the backup generation can be quickly and easily connected to the grid. The configuration with multiple independent generators are setting a peer-to-peer network that it is critical to the MG operation. This functionality gives the benefit of placing generation close to the load, further increasing efficiency by reducing transmission losses. The concept can be extended to allow the generators to sit idly on the system

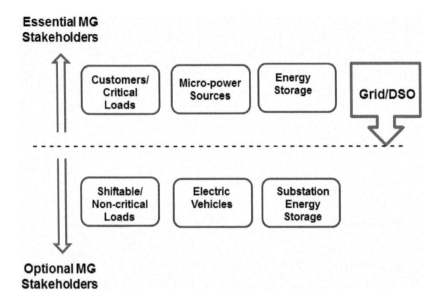

FIGURE 10.11 Microgrid stakeholders, configuration, and components.

when there is more electrical capacity than necessary, then when the load increases, additional generators would come at a predetermined set point necessary to maintain the correct power balance. A typical MG structure and components, with large distributed energy units, as in Figure 10.11, include a power-supplying network, several power sources, loads, and energy storage units. The distributed energy sources, energy storage, and loads can be of any type, all being connected with the distributed network. The MG-generated power may be higher than, equal to, or lower than the loads, with large variations. In cases of larger generated power than total loads, the excess is stored or supplied to the grid, and when the generation is less than the loads, the energy is supplied by the energy storage and/or grid. A microgrid is most likely to be set on a small, dense group of contiguous geographic sites that exchange electrical energy and maybe heat. The MG generators and loads are located and coordinated to minimize the costs of electricity and heat, while operating safely and maintaining power balance and quality. Microgrids manage to move the power quality control closer to the end uses, matching the end-user needs effectively, therefore, improving the overall efficiency of electricity supply. As microgrids become more prevalent, the grid power quality standards can ultimately be matched to the purpose of bulk power delivery. Within microgrids, loads and energy sources can be disconnected from and reconnected to the utility grid with minimal disruption to the loads. The required flexible load can be met under controlled operation at higher energy efficiency, by providing both power and heat. Key technical issues raised by microgrids include: *protection, energy management, load control,* and *power electronic interfaces.* Besides the LV network, loads, generators, and energy storage, a microgrid that has a hierarchical-type management and control schemes is supported by a communication infrastructure to monitor and control microsources and loads. The level of technical benefits from a microgrid, however, depends strongly on two factors: the optimality of DER allocation and the degree of coordination among components. Just as effective planning of DER dimensioning and interconnection decisions can maximize unit contribution to system performance, unguided penetration of oversized units at weak grid points can create more technical problems than benefits.

In summary, all microgrids consist of DER units, power conversion equipment, communication system, controllers, and energy management to obtain flexible energy management. DER units

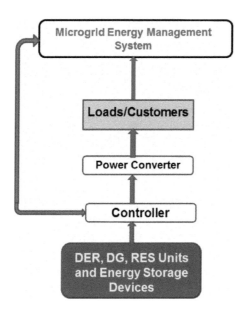

FIGURE 10.12 Microgrid operation flow and key components.

involving DG and RES, distribute energy storage to meet energy demand. If the DER units are producing DC or AC voltage with other magnitude and frequency than the grid, the power electric interfaces are needed. A communication system monitors and controls the MG information, interconnecting the MG elements, ensuring management and control. The energy management system is used for data gathering and device control, state estimates and reliability evaluation of the system. It also predicts the output power from the renewable energy systems, load forecasting, and planning. Presented in Figure 10.12, the basic MG architecture shows microgrid structure, control, and power flows. The power distribution networks can be classified in three types: DC line, 60 Hz, or 50 Hz AC line frequency, or in the case of special microgrids, such as ships, aircrafts, data, and telecommunication centers, high-frequency AC (HFAC) lines. However, most DERs generate DC power, and the DC distribution system has no power quality issues. The DC current from DERs is transformed to the AC signal by inverters and then transmitted to the load side. There are several ways to connect DER units in a MG system. The use of HFAC signals and lines in a microgrid is a new concept, still at the developmental stage. In an HFAC microgrid, the DGs are connected to a common bus, and the generated electricity is transformed and HF by power converters and is transmitted to the load side, and converted back to 50 Hz or 60 AC by an AC-AC power converter. The load is connected to the distribution network, which can guarantee an effective interaction between the microgrid and the grid. At higher frequency, the harmonics are filtered, thus limiting power quality problems. A main disadvantage is that HFAC increases the power loss.

For power control and protection, communication systems are very important. The basic communication methods so far used in the existing communication test-beds are: power-line carrier, broadband over power line, leased telephone line, global system for mobile communication, LAN, WAN or Internet (TCP/IP), wireless radio communication, optic fiber, WiFi 802.11b, WiMAX 802.16, and ZigBee/IEEE 802.15.4 (for advanced metering infrastructure). A microgrid must be able to import or export energy from the grid, control the power flows, and balance the voltage bus level. To achieve these objectives, small generators, storage devices, and loads have to be controlled. Usually, the distributed energy sources or energy storage devices use power electronic interfaces for MG connections. The principal control functions for non-interactive control methods are active power, reactive power, voltage, and frequency control. The droop control method is often used for interactive or distributed control. In interactive control, a power electronic converter has two separate operation modes. If it is connected to the grid it follows the grid inverter, acting as a current source; if the microgrid is in the islanded mode, it acts as a voltage source. A compromise between fully centralized and decentralized control is the hierarchical control scheme. In the context of power systems, hierarchical control includes three control levels: primary, secondary, and tertiary. These control levels differ in their response speed, operating time frame, and infrastructure requirements, e.g., need for communication. Microgrids are not considered a really new concept since historically the first power systems were small-scale grids, and such networks already existed in remote areas where grid interconnection is not possible due to technical or economic reasons. Nevertheless, combustion-based electric generators, which are fully deterministic and dispatchable, have been so far the most common generation choice. The main challenge is to make microgrids ensure the system operation without relying on fossil fuels but through an efficient coordination of different zero-carbon-emission technologies. Although microgrids may have arbitrary configurations, some elements are usually present, such as renewable energy sources, energy storage, and controllable generation units. However, it should be understood that high RES integration, in spite of many environmental advantages, raises some technical concerns that must be solved in order to ensure system reliability. The most relevant challenges in microgrid management and control include:

1. **Intermittent power:** The power output of the renewable energy sources is variable and is determined by external factors, such as weather and different hours of the day. Therefore, there are situations where the power balance is not feasible. To overcome this issue, the microgrid is equipped with energy storage units that are charged when there is power availability and discharged when a load peak occurs. However, the energy storage units are not fully controllable sources because they depend on their state parameters.

2. **Bidirectional power flows:** Power distribution feeders were designed for unidirectional power flows. However, the introduction of DERs to low-voltage levels can cause reverse power flows, given, for instance, the energy storage presence that can either absorb or deliver power, leading to complications in protection coordination, undesirable power flow patterns, fault current distribution, or voltage control.

3. **Low-inertia:** Unlike power systems where a high number of synchronous generators ensures large system inertia, microgrids are characterized by a low-inertia characteristic as most DG sources are controlled through power electronics converters. This interface is necessary since many micro-generation units generate DC power or not synchronous AC power, such as wind turbines. Therefore, power converters, such as inverters are needed. Although such interfaces enhance the dynamic performance, the lack of synchronous and high-inertia rotating generators makes system control more critical as relevant voltage or frequency deviations can occur, especially when the microgrid is not supported by the host grid.

4. **Uncertainty:** This is another issue for proper system coordination because neither MG generation sources nor loads are deterministic systems. Indeed, even though load profiles and weather forecasts are often available, their reliability is controversial. This factor is more critical in microgrids than in large power systems due to the reduced number of loads and the high correlation variations of the available energy resources, limiting the averaging effect that a large electrical system has.

All these issues may be overcome through the presence of a supervising, advanced, and properly designed control system that is in charge of the coordination of all microgrid systems. It has to ensure that reliability is never compromised, especially in islanded operation, and it could also take into account the economic factors for efficient management of energy resources.

10.4   DC MICROGRIDS AND NANOGRIDS

AC microgrids are mostly used because the existing electrical grid is an AC system. However, a small section of an existing AC grid, e.g., a residential community, can be converted to an AC microgrid through the installation of sufficient DG units along with an isolating switch at the grid interface for islanding purposes. Microgrids can have different configurations and can be composed of networks of different natures forming hybrid AC-DC systems interconnected through power electronic converters. A DC microgrid is more suitable for new developments in rural areas, commercial facilities, or residential buildings. However, the DC microgrid concept can also be applied for existing installations since existing AC systems are usually three-phase systems, consisting of at least three wires, the number of wires needed for a DC network: positive, negative, and ground. Nevertheless, the conversion to a DC microgrid requires considerable equipment retrofitting as well as appropriate power electronic converters. DC power systems are already used in industrial and commercial applications, for practical reasons, such as: the versatile speed control of DC electrical machines and the possibility to build reliable, simple, power networks by using directly connected batteries, fuel cells, and DC energy sources. Today's communication and data centers are supplied with 48 V DC through power converters connected to the AC grid. A low-voltage DC microgrid or nanogrid can be used to supply sensitive loads, combining the advantages of using a DC supply for electronic loads with using local generation units. The smaller losses due to fewer power conversion steps result in reduced operation costs. To ensure reliable operation of a LV DC microgrid, well-designed control and protection systems are needed. In the case of power outages, the loads are supplied from batteries or fuel cells connected to MG DC buses. In some cases, a standby diesel generator can also be used to support the network. A similar solution is also used for power supply of the control and protection equipment in power plants and substations. However, higher voltage levels (110 or 220 V DC) are used for longer distances between sources and loads, and higher power ratings. Moreover, an LV DC system is well-suited for PV systems and fuel cells, both DC systems. Microturbines, small hydropower stations, and wind turbines usually generate AC with variable frequency or at a different frequency than the grid, and hence such settings need an AC-DC-AC converter to meet the grid frequency requirement. These energy sources can benefit a DC system connection, since DC-AC power converters are removed or replaced by simpler, cheaper, more robust and efficient DC-DC power converters. Battery units can be directly connected without any converters, resulting in reduced costs and losses. Compared to an AC system, in which frequency, phase, and reactive power control are required, a DC microgrid requires only the balancing of the active power in the system for proper voltage regulation. Due to the presence of the AC-DC converter, a DC microgrid does not increase the short-circuit capacity of the main grid at the interconnection point, and the interconnection converter can be used to mitigate power quality problems, e.g., harmonics, and voltage fluctuations, resulting in a higher power quality. In addition, a DC network prevents the propagation of power system disturbances between the MG and grid.

   Standards describing component modeling and calculation methods are necessary in order to analyze any power network. Available standards today for LV DC systems are IEEE Standard 399-1997 and IEC 61660, both covering load flow and short-circuit calculations of DC auxiliary power systems. These are used, for example, in power plants and substations. Loads in these standards are modeled as constant-resistance, constant-current, or constant-power loads, depending on the load characteristic. These models are adequate for load-flow calculations and simplified short-circuit calculations. Speed control of DC electrical machines is obtained by changing the supplying voltage or the magnetic flux. In the early beginnings of the electric networks for DC machines, a variable resistor in series with the machine, or a variable resistor in the excitation circuit was used, a simple solution, but with higher losses and a poor speed-torque characteristic. The use of power electronics enables a better, faster, and more precise control of both AC and DC machines. Today's DC machines are found in traction applications or in industrial drive systems.

Although AC systems made a big breakthrough in the beginning of the twentieth century, DC solutions have been adopted by a number of new applications such as distributed generation, energy storage, electric vehicles, hybrid electric vehicles, electric ships, and HV DC transmission. Besides a proper control system, a well-functioning protection system is needed to ensure reliable DC microgrid operation. It can be designed by using the techniques already used in existing protection systems for higher power level DC power networks, for protection of generating stations and traction power systems. However, these DC networks utilize grid-connected rectifiers with current-limiting capability during DC faults. An LV DC microgrid must be connected to an AC grid through power converters with bidirectional power flow capability; therefore, a different protection system is required.

DC microgrids can have different configurations with different RES and DG units, affecting the system control and protection in certain ways. It is desirable to design a control strategy that is applicable to any microgrid configuration, with only minor changes. All components of a DC microgrid configuration are interconnected through power electronic converters. On the microgrid DC side, the power converters are responsible for maintaining the grid voltage within reasonable limits. For this purpose, a power-based droop control solution is often used to control the DC voltage fast, and to establish power sharing between the converters connected to the DC network. Apart from integrating more RES units, the DC microgrid concept has other advantages and applications. For example, DC microgrid architecture facilitates the design of ultra-available power sources for critical loads, such as hospitals, security, and first-response stations, or data centers. It has been proven that, for critical loads, DC systems offer a higher availability over AC systems (at least two orders magnitude higher). On top of that, DC facilitates integrating the majority of modern electronics since all of them are DC. Coming back to renewable energy sources, a few advantages are worth mentioning when using DC microgrids:

- Reduced system power losses, by reducing the number of the AC-to-DC conversions;
- Loads are supplied through the distribution line when there is a main grid blackout, and having higher local availability, through local power sources and energy storage in redundant architectures;
- There is no need to synchronize distributed generators;
- Fluctuations of generated power and the loads can be compensated through energy storage modules; and
- The system does not require long transmission lines, or high-capacity lines.

Along with so many benefits, the MG DC concept raises a couple of difficulties. As mentioned before, DC systems do not experience harmonic issues because the fundamental frequency of a DC system is 0 Hz and an integral of multiple of 0 Hz does not exist. The functions of a LV DC microgrid protection system are to detect, isolate, and clear the faults quickly and accurately, in order to minimize the effects of disturbances. Its design depends on a number of factors, such as the fault types that can occur, their consequences, and type of protection devices required; the need for backup protection; detection methods; the measures designed to prevent faults; and, finally, the measures to prevent incorrect protection system operation. Possible fault types in the DC microgrid are pole-to-pole and pole-to-ground faults. Pole-to-pole faults have low fault impedance, while pole-to-ground faults are characterized as either low-impedance or high-impedance faults. The location of the faults can be on the bus or one of the feeders, inside the sources or the loads. The main difference between an LV DC microgrid and other existing LV DC networks is the type of power converters used to interconnect the DC system with the AC grid. For example, power converters used in DC auxiliary power systems for generating stations and substations and traction applications are designed to have a power flow only from the AC side to the DC side. Therefore, it is also possible to design power converters to be able to handle faults on the DC side by limiting the current through them. However, the power flow between an LV DC microgrid and an AC grid must be bidirectional, requiring a different type of power converter, and it may not be possible to limit the current through the converter during a fault in the DC microgrid or nanogrid. During a fault, all energy sources and storage units connected to the DC microgrid or nanogrid are contributing to the total fault current. The fault current from each DER unit is determined by design and the total fault impedance. The converters used in the LV DC microgrid have a limited steady-state fault-current capability due to their semiconductor switches. However, they can provide a fault current with a high amplitude and a short duration from their DC-link capacitors. Energy storage (e.g., lead-acid batteries) can provide large steady-state fault currents, but in contrast to the power converters, they have a longer rise time. The components within the DC microgrid must be protected from both overloads and short-circuits. Depending on the component sensitivity, various solutions exist. Power converters are very sensitive to overcurrents, and if they are without internal current-limiting capability, they require very fast protection. Examples of such devices are fuses, hybrid circuit breakers, and power electronic switches. Batteries and loads do not require fast protection, and therefore simpler and cheaper devices are used. To achieve selectivity in the DC microgrid, it is necessary to coordinate the protection devices. Feeders and loads are preferably protected by fuses because they are simple and cheap, and it is easy to obtain selectivity.

DC nanogrids are considered elements of a microgrid, often referred to as small microgrids, meaning that they can be interconnected to form a larger microgrid. Nanogrids can also be separated from a microgrid and function independently with their own voltage, phase, and frequency from DC to kHz.AC, but more often are referred only to DC networks. Nanogrids play a different role to microgrids and/or main grids in the power system hierarchy. For example, by connecting multiple nanogrids, a microgrid can be formed. This introduces an alternative approach to the conventional microgrid. Interconnecting nanogrids facilitate the ability to increase the overall power supply. As the nanogrid structure is often confined to a single building, the technical objectives, hardware, and software often vary from that of a microgrid or of the main grid. A nanogrid allows a power structure to be obtained at a relatively low cost compared to microgrids. A nanogrid is different from a microgrid, through some microgrids can be developed for single buildings, being usually interfaced with the utility grid. However, a nanogrid, regardless of whether or not a utility grid is present, is mostly an autonomous DC-based system that digitally connects individual devices to one another, to power generation, and energy storage units within the building, and is viewed as a single power domain, for voltage, capacity, reliability, administration, and price. Nanogrids include internal energy storage. The local generation operates as a special type of nanogrid, while a building-scale microgrid can be as simple as a network of nanogrids without any central entity. The nanogrid is conceptually similar to a ship, car, or aircraft, which all house their own insulated networks powered by RES and energy that can support electronics, lighting, and internet communications. Uninterruptible power supplies also perform a similar function in buildings during grid disturbances or malfunctions. Essentially, a nanogrid allows that most devices plug into power sockets and connect to the nanogrid, which balances the energy supply with the demand from those individual loads. A nanogrid assumes and requires digital communication among its entities, the DC power network, being only intended for use within or between adjacent buildings. Building-scale microgrids are built on a configuration of nanogrids and pervasive communication. The system capacity ranges from a few kilowatts to hundreds of kilowatts, very often in the lower power ranges. There are lots of potential benefits to structuring the local DC networks or DC power distribution in this way, as suggested in the literature. Conversion losses would be cut, investments in inverters and breakers would be reduced, and device-level controls would enable a much more nimble way to match generation or storage capabilities with demand. The building can theoretically be immune to problems more likely to be encountered with a local microgrid or the broader centralized grid. Theoretically, such localized, autonomous power systems could eventually be scaled up or down without almost any interference with the utility grid.

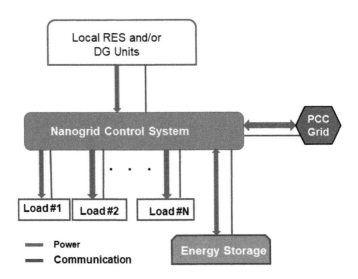

FIGURE 10.13   A typical nanogrid architecture.

The components of a nanogrid consist of a controller, gateway, load, and optional storage, as shown in Figure 10.13. The typical load sizes are in hundreds of watts and, but at times can follow in the tens of watts range. The controller is considered the core or the authority. It controls the loads, power flows, and communication and manages the energy storage. Energy storage can be installed internally or through a second nanogrid used solely to relieve the primary nanogrid and act specifically as storage. Gateways can be considered one-way or two-way, with a capacity limits. These gateways consist of two components, including communication and power exchange. The communication portion should be considered generic giving it the ability to run across physical layers. The power exchange is a component that focuses on defining the various amounts of voltages and capacities. Figure 10.13 shows a schematic of a typical nanogrid and its main components. The nanogrid control strategy should be extended to consider the combined consumption and production of a cluster of houses to further reduce the effects of RES generation intermittency. Nanogrid control is usually divided into two categories; supply-side management and demand-side management. Supply-side management focuses on controlling the nanogrid power supplies and energy storage (such as PV, small-scale wind turbines, geothermal heat pumps, battery banks, fuel cells) to ensure the demand (load) is met and/or the state of charge (battery banks) is optimized. This is an important aspect of nanogrid control as often multiple energy sources exist and their integration needs to be balanced in such a way that a specific source can be selected to supply the nanogrid (e.g., with a grid-tied PV system, it is favorable to supply the loads with the PV first, before supplying the unmet load with the grid). On the other hand, demand-side management manipulates the load to meet the characteristics of the overall energy supply, balancing the nanogrid power.

DC nanogrids employ many technologies, but the subject that dominates the DC nanogrid literature is the power converter topologies. Power converters are responsible, within the nanogrid, for manipulating voltages to meet the requirements of a specific task. This is typically used, but not limited to interfacing the nanogrid energy sources with the system (DC nanogrid) power bus and eventually the main grid (at PCC), along with interfacing the nanogrid loads. The common categories of power converters used in nanogrids are DC-DC, DC-AC, and AC-DC. The converters can perform other tasks within the nanogrid, such as maximum point of power tracking (MPPT) of a renewable source or charge controlling a battery bank. Charge controllers make use of converters to regulate charging and ensure battery banks are not overcharged, which lengthens the life of a battery bank. The goal of MPPT is to address the nonlinearities presented to a system by a RES unit (e.g., PVs or wind turbines).

10.4.1 Control of Low-Voltage DC Microgrids

The DC nature of emerging renewable energy sources (e.g., PV panels) or energy storage units (e.g., batteries, fuel cells, ultra-capacitors) efficiently lends itself toward a DC microgrid paradigm that avoids redundant conversion phases. Many of the loads are electronic DC loads and even some conventional AC loads, e.g., induction machines appear as DC loads when controlled by inverter-fed drive systems. Given the intermittent nature of electric loads, energy sources must be dynamically controlled to supply the load power demand at any moment, while preserving a desired voltage at consumer terminals. Sources may reflect a variety of rated powers, being desirable to share the total load demand among these sources in proportion to their rated power, so-called proportional load sharing. This approach prevents overstressing of sources and helps to expand the lifetime of MG power-generating units. While the source voltages are the sole variables controlling power flow, they must be tightly managed to also ensure a desirable voltage regulation. DC microgrids are also shown to have about two orders of magnitude more availability compared to their AC counterparts, making them ideal candidates for critical applications. Moreover, DC microgrids can overcome disadvantages of AC systems, e.g., transformer inrush current, frequency synchronization, reactive power flow, and power quality issues. Given the desire for developing DC microgrids, control algorithms must be tailored to account for the individual behavior of the entities, forming a DC system, and their interactive behavior. The development of distributed control techniques that are inspired by the operation of DC systems provides global voltage regulation and accurate load sharing through minimal communication. Proliferation of power electronics loads in DC distribution networks shifts the load consumption profiles from the traditional constant impedance loads to electronically driven loads with variable power profiles. Such fast-acting consumption patterns can destabilize the entire distribution network, given their weak nature due to the lack of the damping and generation inertia. Hardware-centric approaches focus on placement of energy storage devices or power buffers for source decoupling, load, and power distribution network dynamics. Control-centric approaches are the alternative solutions. Commercial LV power systems often have a large amount of sensitive nonlinear loads, which must be protected from disturbances and outages in order to operate correctly. Examples of such loads are data and communication systems, access and safety systems, and HVAC equipment. A way to ensure reliable power supply is to install uninterruptible power supplies (UPSs). These UPSs are used to protect the sensitive loads from transients or short-time interruptions or other power disturbances. Microgrids or DC nanogrids are well-suited to protect sensitive loads from power outages and power disturbances. An isolated power system with high reliability can be obtained by utilizing the local energy sources together with fast protection systems.

Besides the advantages, the DC microgrids avoid some of the disadvantages of AC systems, e.g., transformer inrush currents, frequency synchronization, reactive power flow, phase unbalance, and power quality issues. A DC microgrid is an interconnection of DC sources and DC load through a transmission and distribution network. Given the intermittent nature of electric loads, the energy sources must be dynamically controlled to provide load power demand at any moment, while preserving a desired voltage at consumer terminals. While source voltages are the sole variables controlling power flow, they must be tightly managed to also ensure a desirable voltage regulation. A hierarchical structure is widely used to control DC sources. Such structure includes primary, secondary, and tertiary levels, where the primary is the highest and the tertiary is the lowest level. Droop control is an adaptive voltage positioning method in circuit design. The droop control principle is to linearly reduce the DC voltage reference with increasing output current. By involving adjustable voltage deviation, which is limited within the acceptable range, current sharing among multiple converters can be achieved. In most cases, the current sharing accuracy is enhanced by using larger droop coefficient. However, the voltage deviation increases accordingly, and hence, the common design criterion is to select the largest droop coefficient while limiting the DC voltage deviation at the maximum load condition, as expressed below. The controller uses the droop

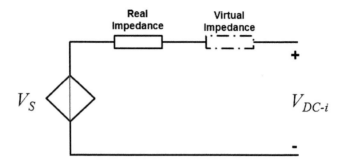

FIGURE 10.14 Thévenin equivalent circuit model of droop-controlled interface converter.

mechanism to handle proportional load sharing; therefore, a virtual resistance, R_D, is introduced to the output of the energy source. While the load sharing benefits from this virtual resistance, it is not physical impedance, thus, does not cause any power losses. In this stage, the voltage controllers inside each source follow the voltage reference generated by the droop mechanism, as:

$$V^*_{0-i} = V_{\text{Ref}} - R_{D-i} \cdot i_{0-i} \qquad (10.17)$$

The droop voltage deviation is then expressed as:

$$\Delta V_{DC} = |V_{\text{Ref}} - V_{DC}| \le \Delta V_{DC-\text{max}} \qquad (10.18)$$

Where V^*_{0-i} is the reference voltage for the inner-loop voltage controller (converter) #i, R_{D-i} is the droop coefficient of converter #i, V_{Ref} is the MG-rated voltage, and i_{0-i} is the source output current, ΔV_{DC} and $\Delta V_{DC-\text{max}}$ are the voltage deviation and its maximum value. In steady-state, given low distribution line resistances, all terminal voltages converge to the same value. Given identical rated voltages used at all energy sources, the droop terms, R_{D-i} and i_{0-i} share identical values, implying that the total load is shared among the energy sources inverse to their droop coefficients. By choosing the droop coefficients in inverse proportion to the source power ratings, the droop control successfully manages proportional load sharing. The interface converter with droop control is modeled by using a Thévenin equivalent circuit (see Figure 10.14). This virtual resistance allows additional control flexibility of the DC microgrids. Notice that the linear droop technique cannot ensure low voltage regulation and proportional current sharing. To achieve acceptable voltage regulation at full load and to ensure proportional current sharing, nonlinear and adaptive droop techniques are being proposed and studied.

Example 10.6

The reference voltage of a DC microgrid is 360 V (DC), the droop coefficient is 0.50 Ω, and source output current is 15 A. Estimate the controller voltage and the voltage deviation.

SOLUTION

Applying Equations (10.18) and (10.19), the required parameter values are:

$$V^*_{0-i} = 360 - 0.5 \cdot 15 = 352.5 \text{ V}$$

and

$$\Delta V_{DC} = |360 - 352.5| = 7.5 \text{ V}$$

Despite its benefits, the droop method suffers from poor voltage regulation and load sharing, particularly when the line impedances are not negligible, which frequently happens in DC microgrid configurations. Voltage drop caused by the virtual impedance in droop mechanism, and voltage mismatch among power converters are the main reasons. To eliminate voltage deviation induced by droop mechanism, a voltage secondary control loop is often applied. The concept of secondary control, under the name of automatic generation control, is used in large power systems to address the steady-state frequency drift caused by the generation droop characteristics. It is usually implemented via a slow, centralized PI controller with low bandwidth communication. In AC microgrids, however, the *secondary control* term is used not only for frequency regulation, but also for voltage regulation, load power sharing, grid synchronization, and power quality issues. To eliminate voltage deviation induced by droop control, a voltage secondary control loop is often applied to a DC system, assigning a proper voltage set point for primary control of each converter to achieve global voltage regulation. The secondary control coefficient (δV_t^κ) changes the voltage reference of local unit(s) by shifting the droop lines, regulating the voltage to the nominal value:

$$V_{DC-i} = V_{Ref} - R_{D-i} \cdot i_{0-i} + \delta V_t^\kappa \tag{10.19}$$

where V_{Ref} is the microgrid reference voltage, V_{DC-i} is the local voltage set point for *ith* converter, i_{0-i} is the output current injection, and R_{D-i} is the droop coefficient. In the islanded operation mode, the global reference voltage, V_{Ref}, is usually the rated MG voltage. However, in the grid-connected mode, a new reference voltage is set by the tertiary control in order to exchange power between grid and microgrid. On the other hand, proper current sharing is a highly desirable feature in microgrid operation, in order to prevent overloading the converters. In droop-controlled DC microgrids, load power is shared among converters in proportion to their rated power. Because voltage is a local variable across the microgrid, in practical applications where line impedances are not negligible, droop control itself is unable to provide accurate current sharing among the sources. In other words, the line impedances incapacitate the droop mechanism in proportional sharing of the load. To improve current sharing accuracy, another secondary control loop is often employed. This current regulator generates another voltage correction term, δV_i^{cr} to be added to the droop mechanism. The droop correction term forces the system to accurately share the currents among the MG components according to the converter power rates. In an alternative approach, the current sharing module updates the virtual impedance, to properly manage the current sharing, and the droop correction term generated by the secondary controller, δR_{D-i}, adjusts the droop control mechanism. The relationships of these approaches are expressed by the relationships given below. Notice that although the secondary control ensures a proportional current sharing, it can inversely affect the voltage regulation. Therefore, there is an inherent trade-off between these two control objectives, i.e., voltage regulation and current sharing.

$$V_{DC-i} = V_{Ref} - R_{D-i} \cdot i_{0-i} + \delta V_t^{cr} \tag{10.20a}$$

and

$$V_{DC-i} = V_{Ref} - \left(R_{D-i} - \delta R_{D-i} \right) \cdot i_{0-i} \tag{10.20b}$$

10.4.2   DC Microgrid Protection

The protection of DC grids, in general, and the protection methods for DC microgrids and DC nanogrids, in particular, are considered one of the biggest challenges in power engineering fields. Unlike in AC systems, the current has no zero crossing that extinguishes arcs, resulting in high fault currents in DC electricity systems. Moreover, series arcing can be an issue when high power loads are unplugged. Special plugs with leading pins are one possible solution, but selective load side arc detection could also be implemented. Furthermore, coordination with frequency-based backup protection would need to be standardized. Traditional protection schemes in the LV grid rely on high short-circuit currents, a radial system, and unidirectional power flow for selectivity. Because these three elements are not usually present in DC distribution systems, new short-circuit protection strategies are needed. Advances in power electronics have led to a reduction of capacitor size and consequently their contribution to short-circuit currents. Oversizing the power converters is an expensive approach, sometimes employed for AC systems. A small DC nanogrid in islanded operation might not be able to produce high enough short-circuit currents, even with the oversized power converters. Because high fault currents are not inherently desirable, a new low short-circuit current protection philosophy is required. Low short-circuit currents allow for solid-state breakers to be used, enabling fast fault clearing and avoiding arcing. Fast selectivity in meshed microgrids with bidirectional power flow is therefore essential and an important research challenge. Current-limiting inductors, limiting the rate of change of the current, are needed to be used. The significant impact of these limiting inductors on the control system needs to be taken into account. Inrush currents and ramp rates must be specified in order to allow fast fault discrimination. Grounding is another important topic to be considered—DC can cause corrosion if it flows through metallic structures in the environment for an extended period. High impedance grounding schemes have often been used in DC microgrids (e.g., data centers), because they sustain operations during a single ground fault. However, selectivity is not possible; therefore, this is not feasible for DC distribution grids. Solid grounding in one point allows for selective protection for residual ground currents; however, if there are multiple grounding points, ground currents would flow. Multiple grounding points would be needed in order to be able to island individual nanogrids. The development of advanced grounding schemes is therefore fundamental. Preliminary research indicates that capacitive grounding could be an interesting alternative, as it provides low impedance for fault transients and blocks DC currents.

According to the fault characteristics, the DC microgrid fault types are pole-to-pole fault and pole-to-ground fault. The pole-to-ground faults are the most common in industrial systems. Usually, the fault impedances of pole-to-pole faults are low. However, the fault impedance of pole-to-ground faults can be either low or high. On the other hand, the fault types can be bus-fault and feeder-fault based on the fault location. Electronic equipment is vulnerable and the tolerance for overcurrent is finite. The fault is much more severe when the fault location is closer to the energy sources; therefore, the bus fault is critical for the whole system. Faults inside VSCs and batteries may cause a pole-to-pole short-circuit fault, and these are terminal faults that generally cannot be quickly cleared. In such cases, the devices must be replaced and using fuses could be a proper choice. The fault characteristics obtained from detailed analyses help to define protection operation time and assess the effect of any proposed protection schemes. The pole-to-pole fault is the most typical type in DC microgrids. The pole-to-pole fault response can be depicted in three stages: capacitor discharge stage, diode freewheel stage, and grid-side current feeding stage. Protection devices commercially available for LV DC systems are fuses, molded-case circuit breakers (MCCB), LV power circuit breakers, and isolated-case circuit breakers. Such models are specially designed for DC, but most of them can be used in both AC and DC applications. A fuse consists of a fuse link and heat-absorbing material inside a ceramic cartridge. When the current exceeds the fuse limit, the fuse link melts and an arc is formed. In order to quench the arc, the arc voltage must exceed the system voltage. This is done by stretching and cooling the arc. There is no current zero in a DC system that helps to interrupt the fault current. The fuse voltage and current ratings are given in RMS values, so they are valid for both AC and DC networks. A MCCB consists of a contractor, a quenching chamber, and a tripping device. When an MCCB is tripped the contacts begin to separate, and an arc is formed. The arc is forced into the quenching chamber by air pressure and magnetic forces. The quenching chamber consists of multiple metal plates, designed to divide the arc into several smaller arcs, which increases the total arc voltage, decreases the arc temperature, and the arc in most cases extinguishes. To improve the voltage withstanding capability, multiple poles can be connected in series. Molded-case circuit breakers are usually equipped with a thermal-magnetic tripping device.

Their voltage and current ratings are given in RMS values. The magnetic tripping senses the current instantaneous value, meaning that the rated current for DC is about 1.42 higher than for the AC case. The problems associated with fuses and circuit breakers in LV DC systems include larger time constants and longer breaker operation time. By utilizing power electronic switches such as gate-turn-on thyristors, the operation speed decreases and the inductive current interruption capability is increased. However, such a solution has higher losses than a mechanical switch. Therefore, a combination of one mechanical switch and one power electronic switch is often proposed to solve these issues. Grounding is a critical issue for DC microgrids protection. Different grounding options come with different fault characteristics and influence the configuration and setting of the protection. The purpose of grounding designs is to facilitate the ground-fault detection, minimize the stray current, and ensure personnel and equipment safety. DC microgrids can be grounded with either high-resistance or low-resistance, even ungrounded. The ungrounded mode has been highly recommended especially in LV applications. In that case, the common-mode voltage could not be high enough to pose a threat to personnel and equipment safety. Meanwhile the system could operate continuously when a single phase-to-ground fault occurs. However, a possible second ground fault at another pole may result in a line-to-line fault and do severe damage to the whole system. Solid grounding has rarely been adopted because of the corrosion caused by stray current. Compared with low-resistance grounding, high-resistance grounding limits the ground-fault current so that the system can keep operating during a fault, but the detection and location of the fault becomes much more difficult. Moreover, the point of the system selected to be grounded could be the midpoint, the positive, or the negative pole of the common DC link.

Similar to an AC power system, the IEC 60364 standard determines the grounding strategies.

Similar to the AC power systems, the protection schemes designed for DC systems are divided into non-unit and unit protection techniques. All the desirable characteristics such as reliability, selectivity, speed, performance, economics, and simplicity have to be taken into consideration when designing and setting a protection system. Non-unit protection is not able to protect a distinct zone of the power system and operates directly when the threshold is exceeded. Meanwhile, non-unit protection schemes have inherent advantages for coordinating the whole protection system. Non-unit protection realizes fault discrimination in a DC microgrid system by analyzing the current, voltage, current and voltage changing time rates (di/dt, dv/dt), and the impedance response in a range of the fault. Some protection systems are designed based on the fault current natural characteristics and its first- and second-order derivatives. Various faults are easily discriminated with the derivatives of the currents. The selectivity of non-unit protection methods was on the basis of complex setting values and proper time delays. Given that, current-limiting methods were presented to release the time coordination tension. Because of the additional cost of the crucial communication and relay devices, the implementation of the unit protection method is closely restricted. But the development of the smart grid and microgrid may call for increased investments in sensors and communication infrastructures within the distribution systems to achieve advanced automatic network monitoring and management. Apparently the deployment of these devices will provide the opportunity to promote the application of unit protection schemes. Unit protection supports a clear zone and never responds to an external circuit fault. In comparison with non-unit protection, unit protection does not have backup protection to the adjacent elements in the system, thus it is common that non-unit protection is deployed alongside the unit protection to act as a backup protection.

10.5 CHAPTER SUMMARY

The sun, our only primary source of energy, emits continuous energy as electromagnetic radiation at an extremely large and relatively constant rate. All of earth's renewable energy sources are generated from solar radiation that can be converted to useful energy by using various technologies. Direct solar energy has long been used in many ways, not only to keep us warm but also to manufacture construction materials such as mud bricks. Wind and solar regime knowledge and characteristics are important for assessment and analysis of wind or solar energy potential for an area or location, exploitation of wind or solar energy, design, management, or operation of conversion systems, as well as in other engineering and technology branches, including building industries. Wind energy is a converted form of solar energy, like most other energy forms. Photovoltaics and solar-thermal energy systems have found increased applications in building energy sectors. Renewable energy systems, such as wind and solar energy or geothermal systems, can be integrated into buildings or operated inside the building structures or near buildings and significantly reduce the building energy demands and uses. Photovoltaic devices are rugged and simple in design, requiring very little maintenance. Their biggest advantage is their construction as stand-alone systems to give outputs from microwatts to megawatts. Many modern products incorporate PV cells or modules in order to operate independently of other electrical supplies. Electricity produced from photovoltaic (PV) systems has a far smaller impact on the environment than traditional methods of electrical generation. There are also several applications and technologies for direct solar energy uses for cooling and/or to provide heat, hot water, and steam for buildings and industrial processes. Solar energy and wind energy can be effectively used and even combined to produce electrical power by photovoltaics (PV) and wind turbines (WT) respectively and to be employed for space heating, cooling, and to provide hot water or steam to buildings and industrial processes. Geothermal resources can be considered renewable on the time-scales of societal systems. Commercial geothermal energy exploitation is primarily a heat-mining operation rather than tapping an instantly renewable energy source, such as solar, wind, or biomass energy. Geothermal systems can be classified by a variety of criteria, such as the nature of heat transfer (conductive vs. convective), the presence or absence of recent magmatism or volcanic activity, the particular geologic setting (e.g., type of volcanic environment) or tectonic setting (e.g., type of plate boundary or intra-plate geologic hot spot), and, of course, temperature (low-, moderate-, and high-enthalpy systems). Although power generation captures much of the attention of the geothermal energy field, direct-use is much more widely applied as sub-power-generating temperatures of fluids are widespread and can be developed with less cost than power plants. Indeed, one of the important attributes of geothermal systems is their wide use over a cascading range of temperatures. Even where no hot fluids or rocks are present, the earth acts like a thermal bank, where heat can be stored during the summer and withdrawn during the winter. In regions characterized by hot summers and cold winters, geo-exchange systems can significantly reduce energy consumption using conventional fossil-fuel sources.

The term *microgrid* is typically used from the LV networks of the (smart) grid with an island operation capability. Microgrids allow the integration of renewable energy sources as the most optimal and reliable sources of energy in the power networks. Potential applications for microgrids are shopping or office centers, university campuses or high schools, hospitals, power for essential and critical services (police, fire, water treatment facilities), farms located far from the grid, remote power networks (islands, rural areas, isolated villages), suburbs and blocks not connected to centralized district heating systems, and military installations and facilities. The use of a microgrid with island operation capability as part of future smart grid architecture allows the reliability benefit of DER to be realized as well as the energy efficiency aspects to be fulfilled. In developing countries and rapidly industrializing economies, the microgrid concept can be an interesting option to meet the local challenges with electricity distribution and generation, which are quite different from those for post-industrial economies. In developing economies, the reach of the transmission grid is geographically limited and it is not profitable to build more transmission grid due to lack of purchasing power and low levels of average consumption among the unserved populations. Instead, small remote hybrid systems based on operating diesel engine generators together with solar and/or wind power systems have been used as separate island grids. The microgrid concept highlights three essential microgrid features: local loads, local micropower sources, and intelligent control units. A microgrid can be defined as an electrical network of small modular distributed generators, including also energy storage devices and controllable loads. Most microgrid small-size generators are connected to the network through power electronic interfaces. A microgrid can operate in grid-connected or islanded mode, increasing the reliability of energy supplies by disconnecting from the grid in the case of network faults or reduced power quality. It can also reduce transmission and distribution losses by supplying loads from local generation and form a benign element of the distribution system. An LV DC microgrid can be preferable to an AC microgrid, where most of the sources are interconnected through a power electronic interface and most loads are sensitive electronic equipment. The advantage of an LV DC microgrid is that loads, sources, and energy storage then can be connected through simpler and more efficient power-electronic interfaces. To achieve a higher quality of service (e.g., voltage regulation, power flow control), communication-based higher control layers must be applied to these small power systems.

10.6 QUESTIONS AND PROBLEMS

1. List the major benefits of microgrids.
2. What is a virtual power plant? Briefly describe its attributes.
3. What are the major divers of smart grid development?
4. List and briefly describe the advantages and disadvantages of building-integrated renewable energy source (RES) systems.
5. List the main components (subsystems) of a building-integrated photovoltaic (BIPV) system.
6. Briefly describe the criteria for installing a wind turbine in a building environment.
7. Briefly describe the photoelectric effect.
8. Why can a microgrid be one of the main smart grid contributors?
9. Indicate whether the following statements are true or false, and justify your answers: (a) A solar cell converts solar radiation directly into electricity. (b) A window in your city that is facing south receives as much as three times more solar radiation than a window facing north.
10. Why do we need to collect photovoltaic cells in series, in parallel, or in series-parallel configurations?
11. What are a solar module and a photovoltaic array?
12. What is the maximum theoretical power that can be produced by a horizontal axis wind turbine of 4.5 m diameter in a wind of 4 m/s, 7.5 m/s, 10 m/s, and 13.5 m/s?
13. Calculate the wind speed distribution for a Weibull distribution for $c = 7.5$ m/s and $k = 2.7$.
14. A wind turbine is about to be installed at a site in which the average wind speed is 6.75 m/s. Assuming that the wind is modeled by a Rayleigh probability distribution, plot the probable wind-speed distribution throughout the year. What is the probability of having a wind speed of 12 m/s?
15. Calculate the factor for the increase in wind speed if the original wind speed was taken at the standard level of 10 m. New heights are at 40 m, 60 m, and 80 m. Use the power law with an exponent of 0.20.
16. What are the major reasons for using DC electric networks and power?
17. How do direct use of geothermal energy applications work?
18. How much geothermal energy is used in the United States? How much geothermal resources could potentially be used in the United States? How does geothermal energy benefit local economies?
19. How does geothermal energy benefit developing countries?
20. What geological areas are most likely to have higher heat flows?
21. If a geothermal heat pump has a coefficient of performance of 3.50, estimate the electrical energy needed for heating (mid-November to mid-April) for a 200 m^2 house in Detroit, Michigan.
22. Repeat the problem above for cooling season (June–August).
23. What are the three features defining the microgrid concept?

24. List the attributes differentiating a virtual power plant from a microgrid.
25. What are the major types of microgrids?
26. Define in your own words the distributed generation concept.
27. List the microgrid internal and local control functionalities.
28. Briefly discuss the microgrid protection issues.
29. List the main features of a microgrid.
30. Briefly describe microgrid main stakeholders.
31. List and briefly discuss the microgrid protection methods.
32. Briefly describe the microgrid control methods.
33. Briefly describe the microgrid functionalities.
34. A microgrid consists of four 30 kW wind turbines, 36 PV panels of 265 W each, and four battery banks with enough capacity to store all the excess energy, and the power management, control, and protection units. The microgrid supplies power to a small remote hotel with daily energy demands of 18.5 kW. Each wind turbine and PV panel has a capacity factor of 0.26 and 0.28, respectively. Each wind turbine is using internally 0.375 kW of the generated power, each of the PV panel 9.5 W, while the battery banks, power management, control and protection are using 1.8% of the output generated power to operate. Estimate the microgrid self-consumption and self-sufficiency coefficients.
35. List and briefly discuss the major microgrid applications.
36. The reference voltage of a DC microgrid is 360 V (DC), the droop coefficient is changing from 0.75 to 0.25 Ω, and source output current is 15 A. Estimate the minimum and maximum controller voltage and the voltage deviation.

REFERENCES AND FURTHER READINGS

1. B. Sorensen, *Renewable Energy – Its Physics, Engineering, Environmental Impacts, Economics & Planning* (3rd ed.), Elsevier/Academic Press, 2004.
2. F. A. Farret and M. G. Simões, *Integration of Alternative Sources of Energy*, Wiley-Interscience, New York, 2006.
3. I. Dincer and M. Rosen, *Energy: Energy, Environment and Sustainable Development*, Elsevier, New York, 2007.
4. J. Andrews and N. Jelley, *Energy Science, Principles, Technologies, and Impacts*, Oxford University Press, 2007.
5. M. Kaltschmitt, W. Streicher, and A. Wiese, *Renewable Energy: Technology, Economics and Environment*, Springer, 2007.
6. P. Sen, *Principles of Electric Machines and Power Electronics* (2nd ed.), Wiley, New York, 1996.
7. R. H. Lasseter, *Microgrids*, in Proceedings IEEE Power Engineering Society Winter Meeting, vol. 1, New York, NY, pp. 305–308, January 27–31, 2002.
8. R. H. Lasseter and P. Paigi, MicroGrid: A conceptual solution, in *IEEE Annual Power Electron Specialists Conference*, Vol. 1, pp. 4285–4290, 2004.
9. R. H. Lasseter, Microgrids and distributed generation, *J. Energy Eng.*, Vol. 1, pp. 1–7, 2007.
10. J. A. P. Lopes, C. L. Moreira, and A. G. Madureira, Defining control strategies for microgrids islanded operation. *IEEE Trans. on Power Systems*, Vol. 21, pp. 916–924, 2006.
11. S. Chowdhury, S. P. Chowdhury, and P. Crossley, *Microgrids and Active Distribution Networks, IET Renewable Energy Series 6*, London IET, 2009.
12. D. Salomonsson, L. Soder, and L. Sannino, Protection of low-voltage DC microgrids, *IEEE Transactions on Power Delivery*, Vol. 24(3), pp. 1045–1053, 2009.
13. S. Roberts, *Building Integrated Photovoltaics*, Birkhäuser Architecture, 2009.
14. Z. Xiao, J. Wu, and N. Jenkins, An overview of microgrid control, *Intelligent Automation and Soft Computing*, Vol. 16(2), pp. 199–212, 2010.
15. H. J. Laaksonen, Protection principles for future microgrids, *IEEE Transactions on Power Electronics*, Vol. 25(12), pp. 2910–2918, 2010.
16. N. Jenkins, J. B. Ekanayake, and G. Strbac, *Distributed Generation*, The IET Press, 2010.
17. P. Gevorkian, *Alternative Energy Systems in Building Design*, McGraw-Hill, 2010.

18. J. Casazza and F. Delea, *Understanding Electric Power Systems: An Overview of Technology, the Marketplace, and Government Regulation* (2nd ed.), Hoboken, NJ: John Wiley & Sons, 2010.
19. R. Teodorescu, M. Liserre, and P. Rodriguez, *Grid Converters for Photovoltaic and Wind Power Systems*, John Wiley & Sons, 2011.
20. R. Belu, Wind energy conversion and analysis, in *Encyclopedia of Energy Engineering and Technology* (ed. Sohail Anwar et al.), Taylor and Francis, 2013 (DOI: 10.1081/E-EEE-120048430/27 pages).
21. A. Ruiz-Alvarez, A. Colet-Subirachs, F. Alvarez-Cuevas Figuerola, O. Gomis-Bellmunt, and A. Sudria-Andreu, Operation of a utility connected microgrid using an IEC 61850-based multi-level management system, *IEEE Trans. on Smart Grids*, Vol. 3(2), pp. 858–865, 2012.
22. S. Sinisa, H. Campbell, and J. Haris, *Urban wind energy*, Earthscan, 2009.
23. M. Yazdanian, G. S. Member, and A. Mehrizi-sani, Distributed control techniques in microgrids, *IEEE Trans. Smart Grid*, Vol. 5(6), pp. 2901–2909, 2014.
24. N. Hatziargyriou (ed.), *Microgrids Architectures and Control*, Wiley, 2014.
25. R. Belu, Microgrid concepts and architectures, in *Encyclopedia of Energy Engineering & Technology* (Online) (ed. S. Anwar, R. Belu, et al.), CRC Press/Taylor and Francis, 2014a (DOI: 10.1081/E-EEE-120048432/22 pages).
26. R. Belu, Microgrid protection and control, in *Encyclopedia of Energy Engineering & Technology* (Online) (ed. S. Anwar, R. Belu), CRC Press/Taylor and Francis, 2014b (DOI: 10.1081/E-EEE-120048432/22 pages).
27. R. A. Higgins, *Energy Storage* (2nd ed.), Springer, USA, 2016.
28. S. Roberts and N. Guariento, *Building Integrated Photovoltaics - A Handbook*, Birkhäuser, Basel – Boston – Berlin, 2010.
29. J. Siegenthaler, *Heating with Renewable Energy*, Cengage Learning, 2017.
30. R. Belu, *Industrial Power Systems with Distributed and Embedded Generation*, The IET Press, 2018.

11 Building Automation, Control, Communication, and Monitoring

11.1 INTRODUCTION

A building is a dynamic and complex system with a variety of physical processes interacting with each other and with the environment. From the control viewpoint, it is a system with several multivariant, dynamic subsystems with linear or nonlinear behaviors and complex interactions. Environmental and occupancy changes increase the complexity of the management and control operations. Occupants impose control goals related to thermal and visual comfort and indoor air quality, influencing the building processes impacting indirectly on the control functions of the different processes (HVAC, lighting, safety, etc.). Environmental concerns also require that lighting, building, and HVAC control systems play increased roles in energy-use reductions, without impeding occupant comfort. Lighting, heating, and air-conditioning are often the largest electrical loads in office and residential buildings. However, in an office building, the cost of lighting energy consumption remains low when compared to personnel costs. Thus, its energy-saving potential is often neglected. For example, on average, the office buildings dedicate about 50% of their electricity for lighting, whereas the share of electricity for lighting is about 30% in hospitals, 15% in factories and schools, and about 10% in residential buildings. Developments in building interior environment control have been in a step-by-step process with different environmental aspects usually being dealt with separately. HVAC control is a major component of building interior environment control, providing the required and expected occupant comfort. Building control, management, and monitoring systems pose significant challenges as complex interconnected subsystems, with interactions between occupants and automation subsystems, and time-varying operating conditions depending on usage and weather conditions. With typically large numbers of occupants and diverse usage of space in commercial, industrial, and large residential buildings, automated building management, monitoring, and control that coordinates multiple subsystems and equipment are becoming essential and critical. These subsystems and equipment include HVAC, lighting, power distribution, transportation, communication, plumbing, access and evacuation control, security and monitoring, and fire protection. As sensors, smart and embedded controllers, microprocessors, and communication become more powerful, complex, sophisticated, and less expensive, it is now feasible, and even essential, to model the building dynamics and coordinate various building control subsystems to achieve energy efficiency, occupant comfort, and safety.

Building control systems (BCSs), building automation systems (BASs) and building management systems (BMSs) are the control and management structures, consisting of integrated hardware and software entities, designed to monitor and control the indoor climatic conditions, occupant safety, and security.

Originally, almost all BCSs consisted only of heating, ventilation, and air-conditioning control, and disregarded lighting and safety control. The evolution of electromechanical and electronic equipment used in residential, commercial, and industrial facilities has been a continuous progress over time, with potential to reduce building energy and operation costs, while improving occupant comfort and worker productivity, providing the expected environmental conditions, indoor air quality, HVAC system performances, and opportunities for improved building management. Such structures monitor and control HVAC, lighting, life safety, security, and other building subsystems. These systems are divided into four sections: applications, hardware, communications, and oversights. BASs or BMSs consist of subsystems installed in buildings to manage building services and functionalities, coordinating several electrical and mechanical devices and equipment interconnected in a distributed manner by means of underlying control networks. They can be deployed in industrial infrastructures such as factories, in enterprise buildings and malls, hospitals, or even in houses. Building automation has been receiving greater attention due to its potential for reducing energy consumption, while facilitating building operation, monitoring, management, and maintenance, and improving occupant satisfaction. These systems achieve such potential by employing a wide range of sensors (e.g., temperature, CO_2, toxic gases or bad odor concentrations, zone airflow, lighting levels, occupancy levels), providing information that enables decision-making regarding how the building equipment is controlled, and reducing operation costs while maintaining occupant comfort. The management and control of a building and a process should be efficient, effective, and robust, for example, in creating optimized and efficient services, or in effective management of heat and electricity.

For large buildings such as schools, malls, hospitals, large apartment buildings, and office facilities, the overall operational costs are significant, and BCSs can provide great savings potential. However, the role of building automation and control systems has been broadened beyond reducing building energy and maintenance costs. Additionally, the number of devices and their capabilities used by the BCSs are increasing due to cheap hardware costs and advances in wireless technologies. BCSs is on its way to becoming a prime example of a distributed control system equipped with sensors, monitoring devices, and actuators for data acquisition, service automation, and management; BCS structures will soon be a major part of building infrastructure. Machine-to-machine networking allows every connected component to be internet-addressable. On the other hand, the BCS's proper access control becomes a matter of concern when a security issue can have unwanted consequences, ranging from a privacy breach to a life-threatening situation (e.g., compromised ventilation system). In BCS structures, the intended information flow depends not only on the rules governed by centralized control but also on the building environment. BCS services are provided by several subsystems (e.g., lighting at one floor or others, heating services in various building spaces). BCSs are highly complex networks, because services provided by each subsystem may have their own communication architecture and protocols. Effective access control cannot be designed without considering the BCS functional architectures, communication topologies, and protocols. Modern BCS systems are distributed systems oriented to building service computerized control and management, tending to separate the automation logic from the user interface through service-oriented entities, providing flexible access from different platforms and locations. The management can be carried out by humans or by computers. However, automation can help to correct human errors, avoid unnecessary energy consumption, assist and guide users to the optimal facility management and control, helping to save energy, reduce operation costs, and improve environmental sustainability.

11.2 BUILDING AUTOMATION, CONTROL, AND MANAGEMENT

Modern buildings are becoming increasingly complex and technologically advanced, as demands on services keep increasing. A modern building is expected to provide conditions and services with high security, efficient energy use, adaptability, and expected comfort and convenience. For buildings with complex requirements due to the building's hosted activities, such as a hospital, a school, or a mall, the services provided are even more advanced and the requirements and expectations are higher. Many of these services benefit from communicating with each other, sharing functions and data, and being monitored together in an integrated system. Building control has traditionally been organized into autonomous subsystems such as lighting, HVAC, and security. As buildings are automated, these independent systems are interconnected to deliver integrated services. Interconnection among these systems is enabled by equipping each subsystem with data communication functions, and computerized control and management of building services. Such integrated structures can control and monitor building mechanical and electrical equipment, HVAC, lighting, power and energy distribution, fire protection, and building security. Over the years, BCS units have evolved tremendously. They initially relied on pneumatic (air-compressed) controls, but back in the 1980s, these systems were replaced by computers, microprocessors, and direct digital controls (DDC), subsequently leading to the introduction of standardized building protocols. In recent decades, some energy-monitoring systems have been integrated with building domotic systems, with the goal of giving information about energy consumption and controlling the user-comfort parameters. To control and monitor building services in an efficient way, more-advanced building automation systems (BASs) are required, including smart and adaptive monitoring, control, safety, efficient control strategies, remote services, etc. However, experience has shown that such approaches also have drawbacks, such as a higher competence level required, larger risk of becoming dependent on services from one company, privacy concerns, higher investment costs, etc.

Control and automation (an equipment reacting without operator intervention, to satisfy pre-set conditions) provide the intelligence (smartness) of building mechanical and electrical systems, enabling specific equipment or entire systems to operate precisely and reliably for comfort, safety, and efficient energy use. Tasks are accomplished by controlling several parameters or properties (e.g., temperature, pressure, humidity, flow rate, speed, light, time), lighting, and transfer media (e.g., air or water, related equipment, and systems). Control systems can be electric, electronic, pneumatic, direct digital, or a combination of these. Electric controls employ usually line voltage (120 V or 220 V), often even low-voltage (12 V to 24 V), which is more sensitive for performing basic control functions. Less frequently used today, pneumatic controls use 5 psi to 30 psi compressed air and receiver controllers with force-balance mechanisms, adjusting output pressure to actuators in response to the input pressure to produce the desired result (effect). Electronic control systems use similar controller concepts. A computer-based control system is installed in buildings to control and to monitor the building equipment such as ventilation, lighting, power systems, fire systems, and security systems, consists of software and hardware, with the software configured in a hierarchical manner by using specific protocols. The BMS and BCS market, a growing segment of the larger building-efficiency industry, is gaining momentum as an alternative, cheaper means for end users to implement energy-efficiency applications in commercial buildings. Conceived out of the market demand for smaller, less-expensive, and strictly energy-related automation and management systems for commercial buildings, the BMS market includes both specialized and broad-based solutions. Such solutions range from reactive energy-efficiency optimization software to predictive supply- and demand-side energy management architectures. A BMS is most common in a large building, and its core function is to manage the building environment and to control temperature, carbon dioxide levels, and humidity within the building. A core function in most BMS systems is controlling heating and cooling by managing the systems that distribute this air throughout the building (e.g., by operating fans or opening/closing). For large buildings, the automation process would cover many functions of building operation, including the following: HVAC, lighting, security and access control, life safety (e.g., fire services), vertical transportation (e.g., lifts and escalators), energy management and power distribution, tenant telecommunications, internet, communications within public spaces, and building structure monitoring. Nowadays, efforts to make buildings smarter focus on cutting costs by streamlining building operations. Building automation is crucial to these efforts, because it could reduce operating costs through effective monitoring and optimization.

BCSs comprise actuators, sensors, and hardware modules. Actuators react to signals closing circuits or varying the electric loads, which are physical devices such as a window blind or a ceiling lamp. Sensors convert a physical reality into a signal that can be measured (Cook and Das, 2005). Some devices do fit into both groups due to their sensing and actuating capabilities, but they are usually perceived as two different virtual devices: one device for sensing and one for actuating. Actuators are often confused with electrical loads or with the physical device they are driving. For example, a "blind actuator" is more precisely a motor actuator (attached to a blind). Actuators and sensors are attached to input or output ports of hardware modules that generate electric signals according to digital output commands and create readings from input signals. The interaction between devices is coordinated often by controllers. In building control systems, a controller usually consists of application-specific hardware with embedded software that continually controls physical actuators (e.g., lights, blinds) depending on the feedback given by monitored inputs, such as light or occupation sensors or by receiving commands from the system. A variety of commercial monitoring and control systems that utilize electronic controls and remote sensors are available. The remote sensors may be located anywhere from the controller, depending on the application. Highly reliable and accurate sensors are available for temperature, humidity, light level, carbon monoxide level, and static pressure measurements, the most commonly monitored variables in building and commercial facilities. Depending on the configuration, such systems are used on individual houses to provide electromechanical thermostatic control, to multiple sensor systems that can control temperature, humidity, static pressure, evaporative coolers, supplemental heaters and/or brooding systems, lights, and ventilation rate with readouts in a facility. Depending on the configuration and structure, distributed monitoring and control systems can provide the functions as dedicated electronic controls, or they can be linked together to a central computer that monitors and records events and conditions at several locations. Conditions can usually be monitored on the dedicated controller at each house or at the central monitoring point, which can be located anywhere from several hundred feet to miles from the houses being controlled, or monitoring can be done remotely via telecommunication. Environmental condition monitoring units are usually logged at a central location. Alarms for out-of-range variables can usually be indicated (visually or audible) at both the individual and central locations. Failure of one or more of the dedicated control systems or the communication link between the central location and the remote buildings does not affect any of the other systems in a multiple-house system.

HVAC system controls are the information link between varying energy demands on a building system and the (usually) approximately uniform demands for indoor environmental conditions. Without a properly functioning control system, the most expensive, best designed HVAC system will be a failure, by not controlling the indoor conditions to provide comfort. Heating, cooling, and indoor air quality vary significantly with time, so the amount of heating, cooling, or ventilation should be controlled as variables change. The designated space parameters must be monitored to keep them in prescribed limits through the building control, management, and monitoring systems, which are critical components in achieving these goals and objectives, and proper occupant comfort. HVAC system designers face considerable challenges when designing a control system that is both energy efficient and reliable. In order to do that, the BCSs deliver multiple and critical functions, such as:

1. Control of the building's environment is primarily delivered through the automated control of the HVAC system and its individual components (e.g., air handlers, coils, fans, pumps, chillers, boilers). The other important aspects include the control of lighting and indoor air quality.
2. Energy management aims to minimize the energy costs through extended monitoring and smart HVAC operation with respect to occupancy, weather, and prices of electricity, gas, or other energy sources.
3. Monitoring of facility assets has the objective of detecting performance problems of HVAC equipment and addressing them early, before causing bigger issues.
4. Security and access control help to minimize risks related to security breaches and improve situational awareness, through access control, surveillance, perimeter protection, and risk mitigation.
5. Fire and life safety ensure that people and assets are protected from fires and other risks, which is accomplished by fire- and smoke-detection systems, sprinkler supervision, and emergency communication.

All these functions address a variety of operational goals and objectives, such as reducing energy consumption and maintenance and life-cycle costs, ensuring tenants' comfort and compliance with regulations (e.g., on the minimum required volume of fresh air in a given building), minimizing safety and security risks, and facilitating active participation in demand response or related energy-trading schemes enabled by smart grid technologies. The architectural complexity of modern BASs largely depends on the subsystems deployed. A BAS controls and monitors building services, and such systems are built up in several configurations. Real systems usually have several, but not all, of the features and components described before, in a specific structure or configuration. When defining BASs, it is advantageous to divide the system into levels. However, there is no standard procedure to defining and naming these levels in a system. In this chapter, the functions of all parts in a BAS are defined in a structured approach, starting with the so-called *field level*. The field level consists of all devices that physically control or detect the building functions, e.g., actuators, motion sensors, smoke detectors, valves, dampers, fans, card readers, motors, sprinklers, switches, or specific equipment. Most of these devices do not have *intelligence* on their own. They either send their status or react to control signals. In the most basic control system, the field-level devices are not connected to anything and are controlled manually. For simple automatic control, a control device can be connected to a sensor (e.g., a light switch connected to a motion sensor). Such solutions work adequately in many applications, but for advanced building services, advanced control systems are usually needed. This can be achieved by connecting the field-level devices to an advanced controller in the *automation level* through *field networks*. The field-level devices act as slaves, whereas the automation-level controllers are the masters. Some field-level devices connect to BASs in other ways. The main purpose with an automation network is to connect the actuators, sensors, and other field-level devices to a programmable logic controller, microcontroller, BAS central computer, or remote terminal unit. The physical connection between the two layers is of four types: hard-wired, bus system type, power line, or wireless.

11.2.1 CONTROL TYPES FOR BUILDING EQUIPMENT AND INSTALLATIONS

Basically the two types of controls are open-loop and closed-loop (Figure 11.1). *Open-loop control* (feed-forward control) is a system with no feedback (i.e., there is no way to monitor if the control system is working effectively). In open-loop control, the controller often operates an actuator or

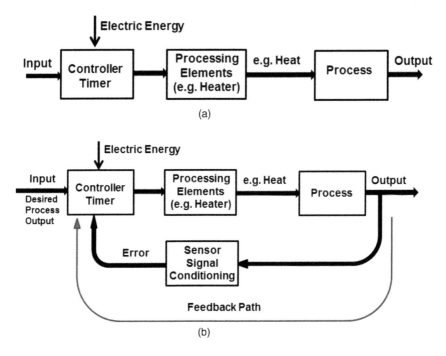

FIGURE 11.1   (a) Open-loop control system; (b) closed-loop control system.

switch through a timer, as shown in Figure 11.1. The operator sets a timer that operates the electrical circuit to the electric heating element. Once the system reaches the desired temperature, the timer *closes* the switch so that the temperature inside modulates around a set point. The open-loop configurations do not monitor or measure the condition of their output signal, as there is no feedback. However, the function of any control or electronic system is to automatically regulate the output and keep it within the system's desired input value or *set point*. If the system input changes for whatever reason, the output of the system must respond accordingly and change itself to reflect the new input value. Therefore, an open-loop system is expected to faithfully follow its input command or set point regardless of the final result, having no knowledge of the output condition, there is no error self-correction when the preset value drifts, even if this results in large deviations from the preset value. Another disadvantage of open-loop systems is that they are poorly equipped to handle disturbances or changes in conditions, which may reduce their ability to complete the desired task. *Closed-loop control systems* use feedback where a portion of the output signal is fed back to the input to reduce errors and improve stability. In a *closed-loop system*, the controller responds to an error in the controlled variable, a comparison of the sensed parameters is made with respect to the set parameters, and the corresponding signals are generated accordingly. Closed-loop systems are designed to automatically achieve and maintain the desired output condition by comparing it with the actual condition. It does this by generating an error signal, which is the difference between the output and the reference input. In other words, a closed-loop system is a fully automatic control system in which its control action is dependent on the output in some way. The main advantage of closed-loop feedback control is its ability to reduce system sensitivity to external disturbances, giving the system a more robust control as any changes in the feedback signal result in compensation by the controller. The main characteristics of closed-loop control are: reducing errors by automatically adjusting the system input, improving the system stability, increasing or reducing the system sensitivity, enhancing its robustness against external process disturbances, and producing a reliable and repeatable performance.

HVAC and many lighting controls are typically closed-loop systems. Closed-loop control systems are usually of two categories: two-position controls and continuous controls. The control elements or components include sensors, controller units, and controlled devices (e.g., valve actuator, valve, switches, and drives). The simplest form of controller is direct-acting, comprising a sensing element that transmits power to a valve through a capillary, bellows, and diaphragm. The measuring system derives its energy from the process under control without amplification by any auxiliary source of power, which makes it simple and easy-to-use. The most common example is the thermostatic radiator valve, which adjusts the valve by liquid expansion or vapor pressure. Direct-acting thermostats have little power and have some disadvantages, but their main advantage is individual and inexpensive emitter control. Direct-acting thermostatic equipment gives gradual movement of the controlling device and may be said to modulate. Sensors measure the controlled medium and provide a controller with information concerning changing conditions in an accurate and repeatable manner. The HVAC control variables are temperature, pressure, flow rate, and relative humidity. Positioning of sensors is critical for an accurate control. In sensing space conditions, the sensing device must not be exposed to direct solar radiation or located on surfaces that can give false readings such as an improperly insulated external wall. In pipework or ductwork, sensors must be arranged in a way such that the device active part is fully immersed in the sensed fluid. Often, averaging device combinations are used to give average values of the measured variable (i.e., in large spaces with several sensors, an averaging signal is important for the controller).

Various types of sensors produce different types of signals. Analog sensors are used to monitor continuously changing conditions, while a digital sensor provides a two-position open or closed signal (ON or OFF), providing the controller with a discrete signal (i.e., open or closed contacts). Some electronic sensors use an inherent attribute of their material (e.g., wire resistance) to provide a signal and can be directly connected to the electronic controller. For example, a sensor that detects pressure requires a transducer or transmitter to convert the pressure signal into a voltage or current usable by the electronic controller. Sensors can measure other temperatures, time of day, electrical demand condition, or other variables that affect the controller logic. Other sensors input data that influence the control logic or safety, including airflow, water flow, and current, fire, smoke, or high/low temperature limit. Sensors are an extremely important part of the control system and can be a weak link in the chain of control. Electronic sensors for temperature sensing in modern HVAC systems are usually resistance temperature detectors (RTDs), in which resistance varies with temperature. The least expensive and most common RTD is the thermistor, which uses a solid-state device. A thermistor has a high-reference resistance, typically 1 kΩ at 0 DC and a high change in resistance per degree change in temperature. Low-cost thermistors tend to drift (get out of calibration), especially when subjected to thermal cycling, so frequent calibration is required. Special caution is needed to ensure that the replacement thermistor has the same characteristics as the initial thermistor. Despite these disadvantages, thermistors are popular because they are low-cost devices that can be used with lower-quality electronics. Platinum RTDs, using pure-platinum wire resistors, retain their calibration indefinitely. Typically, a platinum RTD has a reference resistance of 100 Ω, making the resistance of the leads significant, and three- or four-wire leads are used to compensate. Higher-quality electronic circuits are needed because the resistance of platinum RTDs changes only slightly with temperature. A recent development is the thin-film platinum RTD, a temperature-sensitive resistor made by a substrate deposition of a thin platinum film. The device uses very little platinum but has a higher reference resistance of 1 kΩ or more. Thermocouples are seldom used in HVAC electronic control. However, there is one thermocouple application: small residential gas-fired heating systems. A special thermocouple called a thermopile (several thermocouples wired in series) is inserted in the gas burner pilot flame. The flame heat generates enough electric current to power a special gas valve through a bimetal thermostat.

Electronic pressure sensors can use the same sensing elements as pneumatic and electric devices while providing an electronic-level signal. Peculiar to electronic circuitry is the strain gauge where a small solid-state device is connected to the diaphragm of a pressure sensor. When it is distorted by pressure changes, the device resistance varies, and small distortions produce significant changes so that very small pressure changes can be measured, as low as a few hundredths of a water column inch. Bulb and capillary elements are used where temperatures are to be measured in ducts, pipes, tanks, or similar locations remote from the controller. The bulb is filled with liquid, gas, or refrigerant depending on the required temperature range. Fluid expansion in the heated bulb exerts a pressure, transmitted by the capillary to the diaphragm, where it is translated into movement, calibrated in temperature. RTDs operate on the principle that the electrical resistance of a metal changes predictably and in an essentially linear and repeatable manner with changes in temperature. RTDs have a positive temperature coefficient (resistance increases with temperature). Common materials used in RTD sensors are Balco wire (an annealed resistance alloy with a nominal composition of 70% nickel and 30% iron), copper, platinum, 10 K thermistors, and 30 K thermistors. Thermistors are temperature-sensitive semiconductors that exhibit a large change in resistance over a relatively small temperature range. The two main thermistor types are positive temperature coefficient (PTC) and negative temperature coefficient (NTC). NTC thermistors exhibit the characteristic of resistance falling with increasing temperature. These are most commonly used for temperature measurement. Unlike RTDs, for temperature-resistance, the thermistor characteristic is nonlinear, and cannot be characterized by a single coefficient. Manufacturers commonly provide resistance-temperature data in tables or diagrams. Linearizing the resistance-temperature correlation may be accomplished with analog circuitry, or by the application of mathematics using digital computation.

Several types of electronic sensors for humidity are in use. Synthetic fabrics that change dimensions with humidity are still in use; they have poor accuracy and require frequent calibration. A good device for measuring the dew point uses a tape impregnated with lithium chloride and wound with two wires, connected to a power supply. The lithium chloride absorbs moisture from the atmosphere, creating an electrical circuit, heating the system until in balance with the ambient moisture. The resulting temperature is measured, which is the dew point. The device is quite accurate but requires frequent maintenance. Solid-state humidity sensors use polymer film elements that give resistance or capacitance changes with the relative humidity. One of the most accurate dew-point sensors is the chilled-mirror type. A polished mirror is provided with a small thermoelectric cooling system and a light beam is reflected from the mirror to a photo cell. When the mirror is cooled to the ambient dew-point temperature, moisture condenses, changing the mirror from a specular to a diffuse reflector. The resulting change in light reflectivity serves as feedback to a circuit, controlling the temperature of the mirror at the dew point. The mirror temperature (the dew-point temperature) is measured by a platinum RTD sensor. The only maintenance required is periodic cleaning of the mirror. By measuring the dry bulb temperature, the relative humidity can be estimated. A method that uses resistance changes to determine relative humidity employs a layer of hygroscopic salt, such as lithium chloride, deposited between two electrodes. The salt absorbs and releases moisture, a function of relative humidity, causing the sensor-resistance changes, detected by an electronic circuit connected to the sensor. In a procedure using capacitance changes to determine the relative humidity, the capacitance changes between two conductive plates separated by moisture-sensitive material such as polymer plastic are employed. The capacitance decreases as the material between plates absorbs water, and an electronic circuit detects the changes. To overcome any hindrance of the material's ability to absorb and release moisture, the two plates and electric lead wires can be on one side of the polymer plastic and a third sheet of extremely thin conductive material on the other side of the polymer plastic forms the capacitor. This third plate allows moisture to penetrate and to be absorbed by the polymer, thus increasing sensitivity and response level. Usually the capacitance humidity sensors are combined with a transmitter to produce a higher-level voltage or current signal. Key considerations in the transmitter sensor combinations include range, temperature limits, end-to-end accuracy, resolution, long-term stability, and interchangeability. Capacitance-type relative humidity sensor/transmitters are capable of measurement from 0% to 100% relative humidity with application temperatures from −40°F to 200°F, with typical tolerances of ±1%, ±2%, and ±3%. Capacitance sensors are affected by temperature such that accuracy decreases as temperature deviates from the calibration temperature.

Liquid flow rate is usually estimated by measuring fluid velocity in a duct or pipe and multiplying by the cross-sectional duct or pipe area, at the measurement location. Common methods used to measure liquid flow include differential pressure measurements across a restriction to flow (orifice plate, flow nozzle, venture), vortex shedding sensors, positive displacement flow sensors, turbine-based flow sensors, magnetic flow sensors, ultrasonic flow sensors, and target flow sensors. A concentric orifice plate is the simplest differential pressure type meter that constricts the flow of a fluid to produce a differential pressure across the plate. The result is a high pressure upstream and a low pressure downstream that is proportional to the square of the flow velocity. An orifice plate usually produces a greater overall pressure loss than other flow elements. An advantage of this device is that cost does not increase significantly with pipe size. A Venturi tube meter consists of a rapidly converging section, increasing the flow velocity and hence reducing the pressure, then returning to the pipe's original dimensions, by a diverging *diffuser* section. The discharge can be calculated by measuring the pressure differences. This is an accurate flow-measurement method, with very small energy losses. Venturi tubes exhibit a very low pressure loss compared to other differential pressure meters, but they are also the largest and most expensive. Venturi tubes are usually restricted to those of low pressure drops and high-accuracy readings, being widely used in large-diameter pipes. Flow nozzles may be viewed as a variation on the Venturi tube. The nozzle opening is an elliptical restriction in the flow but with no outlet area for pressure recovery. Pressure taps are located approximately a half pipe diameter downstream and one pipe diameter upstream. The flow nozzle is a high-velocity flowmeter used where turbulence is high, such as steam flows at higher temperatures. The pressure drop of a flow nozzle falls between that of the Venturi tube and the orifice plate (30% to 95%). The turndown (ratio of the full range of the instrument to the minimum measurable flow) of differential pressure devices is generally limited to 4:1. With the use of a low-range transmitter in addition to a high-range transmitter or a high-turndown transmitter and appropriate signal processing, this can sometimes be extended to as great as 16:1 or more. Permanent pressure loss and associated energy cost is often a major concern in the selection of orifices, flow nozzles, and Venturi tubes. In general, for a given installation, the permanent pressure loss will be highest with an orifice-type device, and the lowest with a Venturi. Benefits of differential pressure instruments are their relatively low cost, simplicity, and proven performance. Turbine- and propeller-type meters operate on the principle that fluid flowing through the turbine or propeller induces a rotational motion that can be related to the fluid velocity. Turbine- and propeller-type flowmeters are available in full-bore, line-mounted versions and insertion types where only a portion of the flow being measured passes over the rotating element. Turbine flowmeters are commonly used where good accuracy is required for critical flow control or measurement for energy computations. Insertion types are used for less critical applications. Insertion types are often easier to maintain and inspect because they can be removed for inspection and repaired without disturbing the main piping.

11.2.2 LIGHTING CONTROL

Inappropriate operation and improper control of lighting systems are some of the main reasons that a large portion of electric light is delivered to spaces where no one is present, or for which there is adequate daylight, leading to energy waste, higher operation costs, and even often to improper lighting conditions. However, human requirements and the quality of the working and living environment are expressed in terms of thermal and visual comfort. The optimal conditions of thermal comfort are best described as a neutral perception of the interior environment, where occupants do not feel the need for changes to warmer or colder conditions. Visual comfort is not described easily, being perceived as receiving a message, rather than referring to a state of interior environment neutral perceptions. Aspects such as daylighting, glare, luminance ratios, light intensity, and contact to the outside have influences on our visual comfort perception. On the other hand, energy consumption can be significantly reduced by the optimum control of operating hours and/or the illuminance level. Appropriate lighting controls offer the ability for systems to adjust their characteristics to the existing specific interior conditions. To fulfill the comfort and energy-efficiency requirements, measures to reduce lighting energy and control lighting must be implemented. However, this is not sufficient—lighting energy management has to provide the optimal lighting levels by using the most efficient light source suitable for each application, and providing light only when and where it is needed. This can be achieved by using lighting adaptive-control strategies and smart lighting control. The main objective of these systems is to reduce energy use, while providing a productive visual environment, which includes: providing the *right amount* of light, providing that light *where* it is needed, and providing the lighting *when* it is needed. In fact, lighting control depends on the considered space. Thus, it is necessary to define the following factors: the lighting needs (e.g., level of illumination, ambience), the task zone/area/space (e.g., position, size, disposition), occupation time, and the user-control needs and requirements. Lighting control continuously evolves due to the constant flux in requirements for visual comfort and the increasing demands for lighting energy savings. Quite often, these needs are not even clearly identified. A system objective evaluation requires defining and understanding the performance parameters, as well as the baseline conditions with which the performance is compared. Common performance parameters include: visual performance and comfort, building energy use, cost effectiveness, maintenance, easy operation, flexibility (versatility), building constraints, system stability, and integration. An optimal system performance needs not only to perform well with respect to electricity savings, but also to be accepted by the end user. Oftentimes, within the visual comfort limits, it is difficult to define precisely the needs and priorities of the occupant(s).

The lighting installation task is to create optimum visual conditions for any specific situation, enabling the space users to perform the necessary visual tasks and move safely from room to room. Furthermore, it should also take into consideration the aesthetic and psychological effects of lighting, i.e., provide an aid for orientation, accentuate architectural structures, and support the architectural statement. For example, even the simplest of lighting tasks and requirements cannot be fulfilled by a single lighting concept. The lighting needs to be adjusted to meet the changing conditions in the environment, e.g., conditions for night lighting are different from the supplementary lighting required during the day. Notice that a minimum level of illumination may be desirable for the security of personnel or monitoring of the building and its contents for the detection of intruders. Changes from daylight, to full artificial lighting, then to security illumination can be achieved with either manual or automatically timed operating switches. The design of a control scheme for an occupied space may include a minimum number of luminaires, switched on from a time switch or by the occupant, to provide safe access and acceptable light levels. The requirements of a lighting installation are substantially different when it comes to space usage, e.g., the change of events in a multipurpose hall, the change of exhibitions in a museum, or even the way a standard office space can be transformed into a conference room. To meet such changing requirements, a lighting installation must be able to be controlled, dimmed, and switched in a variety of ways to create different light scenes. The prerequisite is that luminaires or groups of luminaires can be switched separately and their brightness controlled, so that the illuminance level and lighting quality in specific areas of the space can be adjusted to particular situations. The result is an optimum pattern of switched luminaires and brightness levels, a light scene that complies with the use of the room or creates specific ambient conditions. If large numbers of luminaires are to be controlled accurately, it is a good idea to store the light scenes electronically so that any scene may be called up as required.

Lighting controls can be manual or automatic, or combinations of both, and they are also expected to be *user-friendly*. There are several lighting system control technologies, ranging from simple (e.g., installing manual switches in proper locations) to sophisticated and smart control systems (e.g., installing occupancy sensors). The lighting control choice has a significant impact on the total lighting energy use. For example, by using advanced automatic lighting controls, a large fraction of energy (up to 35%) can be saved; such options can be very cost-effective. Energy used for electric lighting depends on the degree to which the lighting controls reduce or turn off the lights in response to: the availability of daylight, changes in visual tasks, occupancy schedules, regular maintenance, and cleaning practices. Advances in the technology and cost reductions of lighting controls, combined with economic and environmental considerations, have led to the increased use of advanced lighting controls, particularly in the commercial, industrial, and public sectors, where lighting represents an important share of total energy costs. Nowadays, automatic and smart lighting controls are used in only a fraction of cases where it would make good economic sense to install them. Consequently, lighting control offers important energy-saving opportunities. Currently, various control strategies are implemented, with their suitability depending on the specific applications and patterns of energy usage. All can be framed into one of the following two major types: ON/OFF control approach and lighting-level control strategies. For the first type, controls offer only the ability to turn on and off the existing lighting system; by using the level control, the luminous flux emitted by lamps may be adjusted continuously or in steps (usually, two or three). All existing lighting-control strategies can be implemented as either the manual or automatic type. Notice that the energy-efficiency levels of lighting control depend strongly on the implemented strategies.

The standard manual, single-pole switch was the first energy-conservation device; being the simplest device, it also provides the fewest options. The first lighting-control level, also the most widely used, is the manual switch to put on or off an individual luminaire or a group of luminaires. A *single-pole, double-throw switch* directs the current in either of two directions. It is used to alternately turn on two different luminaires or lighting devices with a single switch action, such as a safelight and the general light in a darkroom, to two positions, designed ON and OFF. A *three-way switch* controls an electrical light from two locations, allowing the circuit to use one of two alternate paths to complete itself. A *four-way switch* controls a lighting circuit from three locations, while a *five-way switch* controls a lighting circuit from four locations. Chapter 4 gives full descriptions of the electrical switches. For control from many different locations, low-voltage switching systems are used. A manual-control approach is not robust enough with respect to energy efficiency, relying solely on behaviors of occupants who may or may not be concerned about energy savings. If switches are far from room exits or are difficult to locate, the occupants likely will not bother to operate them. Fortunately, some ballast producers offer the opportunity to use dimmers in combination with energy-efficient lamps to overcome the issue. Another meaningful option for upgrading lighting controls is when existing lighting systems are designed such that all lights are controlled from one switch, although not all lights need to be activated simultaneously. Time clocks can be used to control lights in a space when their operation is based on a fixed operating schedule. Time clocks are available in several types. However, regular check-ups are needed to ensure that the time clock is controlling the system properly. However, after a power loss, electronic timers without battery backups can get off schedule and cycle at the wrong times, requiring maintenance to reset isolated time clocks.

Automatic lighting control systems are also known as *responsive illumination control methods.* Despite representing mature technologies, they were adopted only slowly in buildings, offering still a huge potential for building and facility cost-effective energy savings. Lighting-control strategies can provide additional cost savings through real-time pricing and load shedding. Reducing lighting power during electricity peak-use periods when energy rates are at the highest can also be achieved through a lighting management system. The management system enables the integration of the lighting systems with other building services such as heating, cooling, and ventilation, in order to achieve a global energy approach for the whole building, in particular for a green building or an energy-producing building. One of the simplest lighting automatic control methods is the automatic ON/OFF control. Energy consumption can be reduced by controlling the number of operating hours, namely by turning off the light when it is not needed; all spaces that are not used continuously need to have automatic switching, allowing lights to be turned off when the space is not in use. The automatic ON/OFF control can also use occupancy detection or timers. Automatic lighting level controls are usually a combination of photo sensors and electronic ballasts with dimming functions: sensors continuously measure the illuminance level, while ballasts automatically adjust the lumen output of artificial light sources. Such an automatic dimming system has the advantages of compensating for wasted power due to lamp lumen depreciation and luminaire dirt depreciation, adjusting the lighting levels according to activity levels, and responding to variations in daylight availability (see the next subsection). Usually, electronic ballasts with dimming functions operate linear fluorescent lamps, and many of these products start the lamps at any dimmer setting. Most dimmable ballasts have separate low-voltage control leads that can be grouped together to create control zones. The dimmer modules used in high-power lighting systems are suitable for interfacing with time clocks, photocells, or computers and may assure dimming ranges from 100% to 1%. The control often uses DC voltage (0 to 10 V), but dimming ballasts designed for AC control signals are also available on the market. Different control standards provide the interface platform, in particular for equipment and devices of the same manufacturer; in order to guarantee the exchangeability of dimmable electronic ballasts from different manufacturers, an international standard called Digital Addressable Lighting Interface (DALI) is now available. It was developed to overcome the problems associated with the analog 1 V to 10 V control interface and provides a simple digital communication between intelligent components in a local system (Zumtobel Lighting Handbook, 2004 and 2018). Energy efficiency of lighting control systems depends on the strategies implemented, availability of technology, financial resources, and facility designation. These factors are discussed briefly in the next paragraphs.

Predicted occupancy control strategy (POCS) is often used to reduce the operating hours of a lighting installation. It generates energy savings by turning lighting on and off on a preset daily time schedule. Schedules usually vary on a daily basis according to building occupancy. By automatically turning off lights at a preset time, the POCS assists building operators and facility managers to avoid having the lighting on during unoccupied hours, e.g., at night and during weekends. Different schedules can be programmed for different areas of the building based on occupant needs. A *time scheduling control strategy* enables switching on or off automatically based on time schedules and occupancy patterns for different zones. Twenty-four-hour timers allow occupants to set certain times for lighting. The timer is set to switch lighting on during occupancy. This type of control strategy can be used in applications where building occupancy patterns are predictable and follow daily and weekly schedules, such as classrooms, meeting rooms, and offices. The *dusk or dawn control strategy* is a type of predicted occupancy strategy based on sunrise and sunset, calculated for every building location. Light is switched on automatically when it gets dark, and off when there is enough daylight. This control type is not often applied for indoor lighting, but it is very efficient for atriums with good daylight availability or for glazed corridors linking buildings. This strategy is not necessarily achieved with an outdoor-daylight sensor. The on and off hours can be provided by a scheduler. *Real occupancy control strategy* (ROCS) limits the operation time of the lighting system based on the occupancy time of a space. In contrast to the POCS method, ROCS does not operate on a pre-established time schedule. The system detects when the room is occupied and then turns the lights on. When there is no activity in the room, it considers the room as unoccupied and turns the lights off. To prevent the system from turning the lights off while the space is still occupied, a delay time (typically 10 to 15 minutes) can be programmed. ROCS methods are best used in applications where occupancy does not follow a set schedule and is not predictable. Common ROCS applications include private offices, corridors, stairwells, conference rooms, library stack areas, storage rooms, and warehouses, with an energy-use reduction up to 50%. It depends on the level of detection, the place of the sensor, the coupling with daylight-harvesting, and of course the movements of the occupants.

11.2.2.1 Lighting Control Components and Effectiveness

A *lighting controller* is an electronic device used in a building to control the operation of one or several light sources or luminaires simultaneously. The majority of lighting controllers can control dimmers, which, in turn, control the light intensities. Some controllers can also control lighting, according to specific schedules or programs. Lighting controllers communicate with the dimmers and other devices in the lighting system via an electronic control protocol (DALI, ZigBee, etc.). The most common protocol used for lighting today is the DALI model. Controllers vary in size and complexity depending on the building types and sizes. The purpose of lighting controllers is to combine the control of the lights into an organized, easy-to-use system, reducing energy uses. A *sensor* is a device that measures or detects a real-world condition, e.g., motion or light level, and converts the condition into an analog or digital signal. Sensor specifications include performance factors (range, accuracy, repeatability, sensitivity, drift, linearity, and response time), and practical and economic considerations (costs, maintenance, compatibility with other components and standards, environment, and sensitivity to noise). Illuminance sensors indicate the illuminance level in their detection area. They are used to measure indoor illuminance (e.g., on a working plane) and outdoor illuminance (e.g., on the building roof). Illuminance sensors are mostly used to switch or to dim luminaires. Some illuminance sensors enable day or night detection. They can be used in integrated control strategies, particularly if solar protections are involved. Illuminance sensors command the control system to dim or to switch on/off according to the daylight level. Illuminance sensors have to be placed so that they measure the light levels that are representative of the space.

A *timer* is a device that automatically turns on electric lighting when it is needed and turns it off when it is not needed. Timers range in complexity from simple integral timers to microprocessor-based timers that can be programed for a sequence of events for a specific time period. With a simple integral timer, the lighting unit is switched on and held energized for a preset time period, usually within a range of a few minutes to several hours. An electromechanical time clock is driven by an electric motor, with contacts actuated by mechanical stops or arms affixed to the clock face. Electronic time clocks provide programmable selection of many switching operations and typically provide control over a several-day period. Electromechanical and electronic time clocks have periods from 24 hours to 7 days and often include astronomical correction to compensate for seasonal changes. *Occupancy sensors* (motion sensors) automatically switch luminaires on and off to reduce energy use, in response to the presence or absence of occupants in a space. Electrical consumption is reduced by limiting the number of hours that luminaires remain in use. Occupancy is sensed by one of four methods: audio, ultrasonic, passive infrared, or optical. Occupancy sensors can be mounted in several ways: they can be recessed or surface-mounted on ceilings, corners, or walls, and they can replace wall switches. The floor area covered by individual sensors can range from 150 sq. ft. in individual rooms, offices, or workstations, to 2000 sq. ft. in large spaces, by adding more sensors. Occupancy sensors can be used in combination with manual switches, timers, daylight sensors, dimmers, and central lighting control systems. Careful product selection and proper sensor location are critical to avoid the annoying inconvenience of false responses to movement by inanimate objects inside the room or people outside the entrance to the room.

Photo sensors (also called daylight sensors) use electronic components that transform visible radiation from daylight into an electrical signal, which is then used to control electric lighting. A photo sensor comprises different elements that form a complete system. The word *photocell* refers only to the light-sensitive component inside the photo sensor. The term *photo sensor* is used to describe the entire product, including the housing, optics, electronics, and photocell. The photo sensor output is a control signal that is sent to a device that controls the quantity of electric light. The control signal can activate two modes of operation: a simple on-off switch or relay, or a variable-output signal sent to a controller that continuously adjusts the output of the electric lighting. Different photo sensors are manufactured for indoor or outdoor use. In the northern hemisphere, photo sensors used in outdoor applications are usually oriented to the north. This orientation ensures more constant illumination on the sensor because it avoids the direct-sunlight contribution. Outdoor illuminance sensors measure the outdoor illuminance level. They can be combined with the lighting control so that indoor luminaires can be controlled by dimming or switching. Day-night sensors enable the comparison of outdoor illuminance with a predefined threshold in order to trigger actions on outdoor lighting (street lighting) or closing of shutters. They were developed primarily for street lighting and are usually very robust devices. Presence sensors detect the presence of occupants by detecting their movements. The most common sensors used in the building sector are passive infrared (PIR) sensors that react to variations of infrared radiations due to movements of persons. These sensors are usually equipped with Fresnel lenses that define the zone of detection. PIR sensors are usually distinguished into two types: movement sensors and occupancy sensors. Multifunction PIR sensors can integrate up to four functions listed below: occupancy detection, indoor illuminance sensor (level of illuminance for the switch-on of the lamps), infrared sensor, and timer (turn the lamps off after a specific delay). Active infrared (AIR) devices use infrared technology consisting of an infrared diode, which constantly or episodically sends infrared radiation into the controlled area. A receiver monitors the reflected radiation levels. The non-appearance of a reflected ray or a modification of its properties (wavelength or amplitude) indicates that a change occurred in the detection zone. Ultrasonic presence (UP) sensors send out inaudible ultrasound waves, while at the same time, another device is scanning for sound waves that are reflected at a specific rate. If a change in the reflected wave is detected, it indicates that something or someone has moved or is moving into the detection zone (area). Some sensor products in use combine the two technologies (e.g., PIR and ultrasonic presence detections). Such devices are called *passive dual technology sensors*.

Actuators are used for automation in all kinds of technical and industrial processes and systems. However, actuators are more commonly used in HVAC or water systems than n lighting control systems. Depending on their type of supply, actuators may be classified as pneumatic, hydraulic, or electric actuators. They measure or detect real-world conditions, such as displacement or light level, and convert the condition into an analog or digital representation. The *switch* is the most common interface between the lighting system and the occupant. The switch can integrate several modes: ON-OFF switch, and the timer for switch-off. Switching hardware is relatively simple and very cost-effective. Switching is appropriated in single-occupant spaces where light level changes are generated by the behavior of that occupant (when the occupant switches the lights on or when the lights are switched on by an occupancy sensor). For multiple-occupant spaces, automatic ON-OFF switching must be used with care. An automatic control causing unexpected changes in the light level, while a space is occupied, may confuse or annoy the occupants. Switching systems that automatically change the lights according to daylight should be used only in spaces where daylight levels are very high during most of the day. In such cases, the lights are off most of the day so the occupants are not bothered by cycling. Switching may also be used when occupants are transient or performing noncritical tasks. Switching systems are appropriate for atria, corridors, entryways, warehouses, and transit centers, especially when there is abundant daylight. A *dimmer* provides variation in the intensity of an electric light source. Full-range dimming is a continuous lighting intensity variation from maximum to zero. All dimming systems operate on one of two principles for restricting the flow of electricity to the light source: varying the voltage or varying the length of time that the current flows during each alternating current cycle. Dimming systems adapt the light levels gradually, reducing power and light output gradually over a specified range. Dimming can make large energy savings. However, dimmers are more expensive compared to switching devices. Dimming can be achieved through two modes: continuous dimming and step-by-step dimming. Continuous dimming consists of a continuous adaptation of the source light, which is function of external information. Often, the continuous dimming type is achieved through a DC control command on the ballast of the luminaire (discharge lamp) or through a transformer (e.g., halogen lamp). Some manufacturers have adopted a standard analog 0 V to 10 V dimming protocol, allowing ballasts from different manufacturers to be used. Step-by-step dimming is a way to control the light output of luminaires based on a limited number of configurations. The rated dimming levels are based on information generated by the controller, received by the actuator, and transmitted to the light source. The number of dimming steps is defined by the protocol used. A DALI-based dimming system is an example of this kind of step-by-step dimming (256 dimmed levels). Switching systems perform very well in climates with stable sky conditions, such as Arizona, while dimming systems are more appropriate to save energy in climates with variable sky conditions, such as Northern Europe. Historically, resistance dimmers were the first dimming method, used mainly in theaters and auditoriums for the first half of the twentieth century. A resistance dimmer (rheostat) controls the voltage by introducing into the circuit a variable resistance, controlling voltage and the lamp intensity. A major drawback to this kind of dimmer is that the portion of the current that would otherwise produce light is instead converted to heat. *Autotransformer dimmers* avoid these problems by using an improved method of dimming. Instead of converting the unused portion of the current into heat, the autotransformer *changes* the standard-voltage current into low-voltage current, with only a 5% power loss. Autotransformer dimmers are widely available in sizes up to many thousands of watts. Solid-state dimmers are predominant today; they use the second of the two methods of limiting current flow. A power control device, such as a silicon-controlled switch (SCS) under 6 kW, or a silicon-controlled rectifier (SCR) over 6 kW, allows electric current to flow at full voltage, but only for a portion of time. This causes the lamp to dim just as if less voltage were being delivered.

Wireless sensors (e.g., presence sensor, illuminance sensor) and actuators (e.g., dimmers, switches) are based on old existing concepts as far as their intrinsic functionalities. It is the way the control is achieved that makes them more attractive as the wireless operation. Their main advantage is their flexibility (no cabling). Thus, installation and operational costs can be reduced significantly, while flexibility is increased. That is why they are used more and more in refurbishing and open-space areas. Wireless sensor networks in commercial and industrial buildings are exposed to an increasing number of interference sources, like Wi-Fi nodes, microwaves, Bluetooth devices, and other types of wireless-sensor networks. Most wireless-sensor networks perform well in a *clean* RF environment after initial installation. The challenge is to maintain good performance when there is another deployment of supplementary RF sources. The building shell can be integrated in several types of active or passive components that impact directly on heat/cool energy or illuminance level. These components can be placed on the roof or facade. The evolution of building automation and direct digital control in the last three decades has been aided by the development of communication technologies in buildings. Today, BASs offer two main possibilities to integrate the control of different equipment/applications in a building, namely: *proprietary systems*, and *open systems*. A proprietary system is a closed system that is developed by a single manufacturer/contractor. In such systems, the initial costs may be lower and easy to set up, but the building owner is locked into products of a single manufacturer and cannot easily take advantage of innovative technologies. An open system is one where standards are developed, published, and maintained by an independent recognized organization body (e.g., CEN, ISO, ASHRAE/ANSI) or an industrial alliance (de facto standard). Any change to the system requires the comments of users and industry professionals before being approved by the standards organization or the alliance. Open communication protocols are now being used for equipment integration from different manufacturers.

Lighting effectiveness is usually expressed in illumination level, quality, and energy usage. Illumination levels and quality were discussed in Chapter 6. Energy consumption is computed by using appropriate power and energy relationships. As previously mentioned, lighting and lighting control represent significant contributions to the energy consumption of buildings. In order to estimate the lighting energy consumption and related impact of controls, the simplified equations, as presented in standards and codes, e.g., EN 15193, can be used. Such relationships relate the total energy for lighting, E_{tot}, the amount of energy consumed in period, t, by the luminaires and lighting units when operating, and parasitic loads when the luminaires are not operating, in a room or zone, $E_{L,t}$, the energy used for illumination, the amount of energy consumed in interval, t, by the luminaires and lighting units to fulfill the required illumination function and purpose in the building, $E_{P,t}$, the luminaire parasitic energy consumption, the parasitic energy consumed in the same period, by the charging circuits of emergency lighting and by the standby control system controlling the luminaires and lighting units. All these energy quantities are expressed in kWh. The lighting energy balance equation is then:

$$E_{tot} = E_{L,t} + E_{P,t} \tag{11.1}$$

The illumination energy is computed, as specified in the standard, by using this simplified relationship:

$$E_{L,t} = \sum \frac{(P_N \times F_C) \cdot (t_D \times F_O \times F_D + t_N \times F_O)}{1000}, \quad \text{(kWh)} \tag{11.2a}$$

The estimation of the parasitic energy $(E_{P,t})$ required to provide the energy for emergency lighting units and for standby energy for lighting controls in the building is estimated by using the following equation:

$$E_{P,t} = \sum \frac{(P_{P,c} \times |t_{Yr} - (t_D + t_N)|) + (P_{Emr} \times t_{Emr})}{1000}, \quad \text{(kWh)} \tag{11.2b}$$

Here, t_D is the daylight operating hours, t_N is the non-daylight operating hours, P_N is the total installed (nominal) lighting power, measured in W, F_D is the daylight dependency factor, a factor relating the usage of the total installed (nominal) lighting power to daylight availability in the room or zone, F_O is the occupancy dependency factor, a factor relating the usage of the total installed lighting (nominal) power to the occupancy period in the room or zone, F_C is the constant illuminance factor, a factor relating to the usage of the total installed (nominal) power when constant illuminance control is in operation in the room or zone, t_{Yr} is the standard year time, time taken for one standard year to pass, 8760 h, t_{Emr} is the emergency lighting charge time—the operating hours during which the emergency lighting and batteries are being charged in hours, $P_{P,c}$ is the total installed parasitic power of the controls in the room or zone, the input power of all control systems in luminaires in the room or zone, P_{Emr} is the total power of the emergency units, both measured in W. Reduced energy use is possible by adjusting the different elements of these equations, for example: the installed (nominal) power can be reduced by using low consumption and more efficient light sources and efficient control systems (electronic ballasts, DC power converters, etc.). Daylight dimming leads to an important energy reduction by smartly adjusting the light flux according to the daylight levels, through the F_D parameter, operating hours can be reduced by adjusting lighting according to predicted or real occupation strategies through the F_O parameter and the amount of working hours (t_N and t_D). In fact, only a fraction of a building lighting system is required at any given time. Lights frequently are left on in unoccupied places or unused rooms, where there is no need for lighting. Through the changes of the t_N, t_D, and F_O values, energy savings can be calculated.

Example 11.1

If in a medium-size office space the time for daylight usage is 2250 hours per year, time for non-daylight usage is 280 hours per year, occupancy factor is 0.9, the constant illuminance factor is 0.93, and daylight dependency factor is 0.8, and the time when parasitic power is used is the same as non-daylight usage time, estimate the total annual energy use. The lighting system power is 7200 W, the parasitic power is 15% of the installed power, and the emergency power is 75 W, being used 96 hours per year.

SOLUTION

The parasitic power is 1080 W. Applying Equations (11.2a) and (11.2b), the total energy used for illumination and parasitic energy are:

$$E_{L,t} = \frac{(P_N \times F_C) \cdot (t_D \times F_O \times F_D + t_N \times F_O)}{1000} = \frac{(7.2 \times 0.93) \cdot (2250 \times 0.9 \times 0.8 + 280 \times 0.9)}{1000} = 12534.9 \text{ kWh}$$

and

$$E_{P,t} = \frac{(1080 \times |8760 - (2250 + 280)|) + (75 \times 96)}{1000} = 6735.6 \text{ kWh}$$

The total energy consumption in this office space is: 19,270.5 kWh.

The strategies previously described can be applied in almost any building type. They can be stand-alone systems or part of a fully interoperable lighting management system (LMS). An LMS can schedule the light operations in any area, room, or zone within the building, or monitor occupancy patterns and adjust lighting schedules as required or needed. The LMS gives facility managers the ability to remotely control building lighting energy consumption. It also enables the facility manager to perform load-shedding strategies in case of high electricity demand in the building. The operation costs are thus reduced as the control strategy has turned off or dimmed some lights or lighting components during peak-use periods. The LMS also can give an efficient way to control light units, being able to manage lights in one zone in an independent way. An additional advantage of LMS is their ability to monitor the operation of the lighting systems such as the number of operating hours in a given area, or the number of times the lights are switched on. Using this information, maintenance operations like relamping (actions to replace a burned-out lamp) can be scheduled. In case of an implemented building management system (BMS), the LMS can be combined with heating, ventilation, air-conditioning, security, etc. This type of integrated management system allows sharing resources, e.g., actuators, sensors, routers.

In terms of lighting control, there are three levels of integration that are often distinguished for indoor lighting control. The first level (level 1) takes into account the artificial lighting alone. The second level (level 2) takes into account artificial lighting and its control by external information like daylighting, occupancy, etc. The third level (level 3) takes into account artificial lighting dealing with artificial lighting plus external interaction with external systems (elements), such as HVAC systems and/or blinds. In the integration level 1 (artificial lighting only), for example, the user controls the artificial lighting through a manual switch or dimmer. This solution is one of the most commonly used methods in buildings, consisting of only a switch for a lamp or a group of lamps. In level 2 (artificial lighting control based on external information) integration, for example, an illuminance sensor and an occupancy sensor can be combined with a manual switch-dimmer in order to increase the visual comfort of the occupant. For each sensor, a priority level is set. This system allows artificial lighting control to a manual switch (on/off) or dimming with a high priority level, by an occupancy sensor with an intermediate priority level, and by an illuminance sensor (in order to assume a constant light level) with a low priority level. The energy-saving potential of this solution is similar as daylight harvesting plus occupancy sensor. In level 3 (artificial lighting, daylight, and HVAC) integration type, for example, a full integration of the lighting system with the HVAC system and the blinds system to increase the occupant visual and thermal comfort is used. Such an integrated system allows the control of the artificial lighting, daylighting (with blinds), and HVAC systems. Additional sensors are included with their own priority level, such as: a manual temperature set-point button with a high priority level, a manual switch blind button with a high priority level, an indoor temperature sensor and wind speed sensor with an intermediate priority level, and a glare sensor with a low priority level. The communication scheme of this integration level system is rather complicated because of the multiple interactions between the sensors and the controllers. It is important to note that this kind of integration is not designed for buildings that consume large amounts of energy (cheaper solutions can be more relevant and less expensive). The equipment sharing is an important issue to achieve a proper integration of the control strategies (levels 1 through 3). The lighting control architecture supports the implementation of the defined strategies, usually organized in four levels: lighting service, lighting unit (plant), lighting zone/area, and lighting device. The lighting service level deals with the overall lighting management system, being the lighting backbone. A lighting unit (plant) as an analogy to an HVAC central plant deals with the control of central technical areas. It often appears at each building floor. A lighting zone (area) deals with the different interactions in a zone (a zone can be a room or a set of rooms). Finally, a lighting device is the terminal device, controlling the visual comfort of a specific area. The key points of lighting architecture definition are the position of the actuators, switches, and sensors.

Local, single-room systems typically consist of one control station with switches or manual controllers that control the power. The dimmable wattage is limited only by the capacity of the system. These local systems are easily expanded to multiple rooms and customized to offer many combinations of manual, preset, assigned, and time-clock control. They can incorporate energy-reduction controls such as occupancy sensors and photo sensors, and can handle some emergency power functions.

Whole-building systems use local or small modular dimmers, a central computer, and master control stations to control all of the luminaires in a home or commercial building. Such systems can also operate other electrical systems, such as motorized shades, fans, air-conditioning, heating, and audio systems, and they interface easily with burglar alarms, smart building systems, and other electrical control systems.

In centralized systems, a microprocessor processes the data, determines the required change, and initiates action to complete the change. More sophisticated processors can respond to a number of complex lighting conditions in the space, collect power and energy-use data, and supply summary reports for building management and tenant billing. Processors range in complexity from a microchip in a controller to a large computer. The central processor unit receives all inputs, analyzes the data, and then sends instructions to controllers located throughout a facility, allowing coordinated control of all system elements. In distributed processing, the ongoing decision-making is left to local processors, but a central processor is supervising the entire system, with the advantage that the entire system does not fail if any one processor does, and only the local processor has to be reprogrammed to accommodate changes.

11.2.3 HVAC CONTROL METHODS

Temperature control in an air-conditioning system that uses air as a delivery medium may use any of the following approaches and procedures. The basic constant-volume and variable-temperature method consists of varying the temperature of air supplied to the space while keeping the airflow rate constant. The constant-temperature methods consist of varying the airflow rate while keeping the temperature constant for air supplied to the space. Varying the airflow rate and changing the temperature for air supplied to the space is the variable-volume and temperature approach. The variable volume-reheat approach happens by varying both the supply-air temperature and flow rate where the airflow rate is varied down to a minimum value, then energy input for coil reheating is controlled to vary the supply-air temperature. Humidity control in a conditioned space is performed by controlling the water vapor amount present in the space air. When relative humidity at the desired temperature set point is too high, dehumidification is often required to reduce the amount of water vapor in the air for humidity control. Similarly, when relative humidity at the desired temperature set point is too low, humidification is required to increase the amount of water vapor in the air for humidity control. Because relative humidity varies significantly with dry bulb temperature, it is important to state both the dry bulb temperature and the relative humidity. Commonly used dehumidification methods include: surface dehumidification on cooling coils simultaneous with sensible cooling, sprayed coil dehumidifier with indirect cooling coils, and direct dehumidification with desiccant-based dehumidifiers. Humidification is not usually required in an HVAC system but, when required, it is provided by a humidifier. Commonly used humidification methods include: water spray humidifier, steam grid humidifier, and steam pan humidifier. Dehumidification is usually done at the same time as the sensible cooling by a surface dehumidification process on the system cooling coils, either indirect cooling using chilled water or other heat-transfer medium or direct-expansion refrigerant evaporator coils. Dehumidification in low dew-point process systems may be done in a separate dehumidification unit. Humidity relationships in HVAC systems are expressed in percent relative humidity (%RH).

Temperature control in a space is done by a temperature controller, with a thermostat being the most common device, set to the desired temperature value or the set point. A temperature deviation, or offset, from the set point causes a control signal to be sent to the controlled device at the HVAC system component that is being controlled, providing needed energy to adjust the space temperature. If the conditioned space temperature is controlled by a heat-exchange process to supply air from a heating or cooling coil, the temperature-control signal will cause a change in the flow of the coil cooling or heating medium. With a chilled water or heated water coil, the temperature controller may position a water valve to vary the flow rate of heated or chilled medium through the coil or may position face and bypass dampers at the coil to vary the air passing through the air side of the coil to that which bypasses the coil and is not conditioned. Automatic control valves used to control water flow through a water coil may be either two-way or three-way type and may be positioned in either two-position or modulating sequence. Valves used to control steam flow through a coil are two-way type and may be positioned in either two-position or modulating sequence. Figure 11.2 illustrates a basic control loop for a room heating system. In this example, the thermostat assembly contains both the sensor and the controller. The purpose of this control loop is to maintain the controlled variable (room air temperature) to some desired value, the set point. Heat energy necessary to accomplish heating is provided by the radiator and the controlled device is the two-way motorized or solenoid valve, which controls the flow of hot water to the radiator.

The system humidity controller, commonly called a humidistat, is located in the conditioned area, preferably adjacent to the thermostat, to ensure that the ambient temperature is that which the humidity is to be based upon. The space humidity controller is set at the desired relative humidity set point. A change in relative humidity from that set point causes a control signal to be sent to the controlled component. The control of an electrically heated steam humidifier is similar to valve-controlled, with electric contactors being the controlled devices. When variations of supply-air

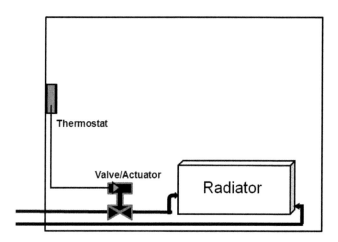

FIGURE 11.2   Conditioned-space temperature control.

volume are used to control the space temperature, the temperature controller may cycle the fan motor in ON-OFF control sequence, may modulate the fan motor speed, or damper the airflow, such as through volume-control dampers in air terminal units. In *variable air volume* (VAV) systems, the supply-air volume delivered to the space varies as the temperature controllers on individual terminal units position each of the modulating dampers on individual terminal units. The central station air-handling unit fan operates continuously and the fan performance varies to maintain duct static pressure within specified limits. Pressure control in variable-volume air-distribution systems utilizes a pressure controller set for the desired pressure set point. A deviation from the set point causes a control signal to the controlled device at the controlled component. Airflow control is done by several different methods or combinations of methods, such as ON-OFF fan control, variable volume control, terminal reheat, terminal bypass, and terminal induction. Control of air quality is done by several methods or combinations of them depending on the contamination degree, such as odor dilution with outside ventilating air, filtration of particulate matter with air filters, filtration of gaseous contaminants with odor-adsorbent or odor-oxidant filters, and local control of gaseous and particulate contaminant emission by using a local exhaust with exhaust hoods over processes.

11.2.4   HVAC CONTROL INFRASTRUCTURE

An air-conditioning or HVAC&R (heating, ventilation, air-conditioning, and refrigeration) system consists of a set of components and equipment arranged in sequential order to heat or cool, humidify or dehumidify, clean and purify, attenuate the objectionable system components and equipment noise, transport the conditioned air and recirculate air to the conditioned space, and control and maintain a prescribed indoor or enclosed environment at optimum energy use. The principal functions of HVAC systems and control systems are presented and discussed here. An HVAC system functions to provide a controlled environment in which these parameters are maintained within desired and specified ranges: temperature, humidity, air distribution, and indoor air quality. In order to accomplish this task, the automatic temperature control (ATC) system must be designed so as to directly control the first three parameters. The fourth parameter, indoor air quality, is influenced by the first three but may require separate control methods that are beyond the scope of this book. In summary, HVAC and building control systems are designed to maintain comfortable conditions in the space by providing the desired cooling and heating outputs, while factors which affect the cooling and heating outputs vary. Other important goals are to maintain comfortable conditions with the minimum possible energy use, and to operate the HVAC system so as to provide a healthy environment for occupants and safe conditions for equipment. Building types in which air-conditioning systems are used can be classified as: institutional (e.g., hospitals, schools, governmental buildings), commercial (e.g., offices, stores, shopping centers, large apartment buildings), residential (e.g., single-family and multifamily low-rise buildings of three or fewer stories above-grade), and manufacturing or industrial (e.g., industrial facilities, power plants). Controls of HVAC systems are similar in many ways, as are those for any other industrial processes. Residential HVAC units using electricity are usually heat pumps, refrigeration systems, and electrical heaters. Heaters range from electric furnace types, small radiant heaters, duct heaters, and strip or baseboard heaters, embedded into the floor or ceiling heating systems. Efficiency for heating is usually high because there are no stack or flue losses, and the heater transfers heat directly into the living space. Cooling systems range from window air-conditioning to central refrigeration or heat pump systems. Evaporative coolers are also used in some climates. Major operational and maintenance strategies for existing equipment include: system maintenance and cleanup, thermostat calibration and setback, microprocessor or computer controls, time clocks, night cooldown, improved and advanced controls and operating procedures, heated or cooled volume reduction, and the reduction of infiltration and exfiltration losses. System maintenance is an obvious but often neglected energy-saving tool. Dirty heat-transfer surfaces decrease efficiency. Clogged filters increase pressure drops and pumping power. Inoperable or malfunctioning dampers can waste energy and prevent proper operation of the system.

The infrastructure for monitoring and control of HVAC systems is the most commonly implemented part of any BAS, and it is also perhaps the largest and most complicated building electromechanical system because of the variety of control and monitoring devices involved and the multiple ways they can affect a building operation. Heating and cooling are performed by specific equipment and devices, such as refrigeration devices, boilers, furnaces, and fuel and/or electricity delivering subsystems. Heat or cooling fluid transfers are performed by air diffusers, pumps, grilles, water piping systems, and radiators. Delivery methods vary greatly in their ability to maintain the required space conditions, through the complexity of their operation, maintenance, control and monitoring subsystems, and in their structure or energy usage. Spaces with similar heat or cooling load characteristics require either heating or cooling, but not both at the same time, meaning that the HVAC services can be delivered by single systems that do not need to have simultaneously heating and cooling capabilities or functionalities. While most of commercial, institutional, and industrial facilities control their HVAC systems via a direct digital control (DDC) system, a few small and old HVAC systems still use nondigital control. For any HVAC system, the performance requirements and the design of the controls are interdependent. Inadequate control system design, inadequate commissioning, and inadequate documentation and training for the building staff often create problems and poor operational control of HVAC systems. Therefore, the design of the HVAC&R controls is an iterative process, performed as part of the overall HVAC design process from the beginning phase of the HVAC&R system design. Detailed operational sequences and control point definitions are the most important components to the control system design. The HVAC control system is typically distributed across three building areas. The HVAC equipment and their controls are located in the main mechanical room. Equipment includes chillers, boiler, hot water generator, heat exchangers, pumps, etc. The weather maker or the air-handling units (AHUs) may heat, cool, humidify, dehumidify, ventilate, or filter the air and then distribute that air to a section of the building. AHUs are available in various configurations and can be placed in a dedicated room called a secondary equipment room or may be located in an open area such as a rooftop. Individual room controls depend on the HVAC system design. The equipment includes fan-coil units, variable air-volume systems, terminal reheat, unit ventilators, exhausters, zone temperature/humidistat devices, etc.

HVAC system controls are the information link between varying energy and conditioned space demands and requirements on building energy systems and the usually quasi-uniform demands for indoor environmental conditions. Without a properly functioning control system, the most expensive, modern, and best designed HVAC system is a failure, by simply not properly controlling the indoor conditions to provide the expected comfort. The HVAC control system is expected to have certain capabilities, such as: sustain a comfortable building interior environment, maintain acceptable indoor air quality, be as simple and inexpensive as possible while meeting the HVAC system operation criteria reliably for the system lifetime, provide efficient HVAC system operation under all conditions, complying with the commissioning requirements, including the building, equipment, and fully document the control systems so that the building staff successfully operates and maintains the HVAC system. An HVAC process is controlled by three elements, which are critical components of any HVAC system: sensor(s) and/or other input device(s), a controller unit, and controlled output device(s), setting the conditioned space parameters. Sensors measure the controlled space parameters and/or other control inputs in the most accurate and repeatable manner. Common HVAC sensors measure the space temperature, pressure, relative humidity (RH), airflow state, etc. Other variables or space parameters that can impact the controller logic may also be measured. Examples include other temperatures, time-of-day, or the current electrical demand condition. Additional input information (sensed data) that influences the control logic may include the status of other parameters (e.g., airflow, water flow, or electric current) or safety-related information (e.g., fire, smoke, high/low temperature limits, or any other physical parameters). Sensors are the first, and the weakest, link in any HVAC control chain. The controller processes the data provided by the sensor(s) and/or other devices, applies the control logic to the data, and generates the output action signal(s). The signal(s) may be sent directly to the controlled device or to other control subsystems, for further processing and finally to the controlled device. The controller function is to compare its input(s) with a set of instructions, such as set point, throttling range, and action, and then generate an appropriate output signal, *the control response(s)*. Typical control responses are of the following types: *two-position control, floating control, proportional control, and proportional-integral control*.

Controller equations establish the relationship of output signals to input signals for various types of controllers on electronic and pneumatic systems when set up for direct action or reverse action, using set-up parameters including throttling range, proportional band, and ratio. The controller equations provide a method for predicting the output values of ATC system controllers for various input values as required in the control calibration. The controller equations apply to both electronic and pneumatic control systems. Electronic and pneumatic-type ATC systems are analog-type systems, where control voltage and control pressure are analogous to the sensed values of the media controlled by the system, temperature, humidity, or pressure. In electronic controllers, the actual controlled variable value, provided by the sensor, is related to the controlled variable set-point value, the midpoint of the controller voltage range. The value difference, between measured and set point, is multiplied by the controller sensitivity to give a voltage difference that is added or subtracted from the midpoint voltage to give the predicted output voltage value for the measured value, as expressed by:

$$V_{Out} = V_{STP} + VR \frac{T_{Meas} - T_{STP}}{TR} \; (V_{DC}) \tag{11.3}$$

V_{Out} is the output voltage from the controller, V_{DC}, V_{STP} is the voltage at set point or the voltage corresponding to set-point temperature, e.g., 7.5 V on 6 to 9 V system, T_{STP} is the set-point temperature, °F or °C, T_{Meas} is the measured temperature of the controlled medium, °F or °C, TR is the throttling range, °F or °C of the controller, and VR is the voltage range of the controller as temperature changes through throttling range, e.g., 3 V on 6 to 9 V system. The basic controller equation is also used when controlling other variables besides the temperature, such as relative humidity and static pressure. Throttling ranges are generally assigned rather than calculated in electronic control systems, as in pneumatic systems. The amount of change in the controlled variable that will take place during the control process must be considered in the assignment of a numerical value of throttling range to be set up on a controller. The throttling range of a controller is the value of the specific change in the controlled variable that will cause the output signal of the controller to go from maximum to minimum or vice versa. Throttling range is a single value that is assigned in programming the controller and is stated as units of the controlled variable, such as °F, °C, % RH, or inches water column ("wc).

Example 11.2

For a direct-acting controller, the set-point temperature is 72°F, the set-point voltage is 7.5 V (DC), the throttling range 12°F, the voltage range is 5 volts, and the space temperature, T, is 76°F. Calculate the output voltage of the controller.

SOLUTION

From Equation (11.3), the output voltage is:

$$V_{Out} = 7.5 + 3\frac{76-72}{12} = 8.5 \text{ V}$$

In the basic pneumatic controller relationship, the controlled variable actual value, measured by a sensor, is related to the controlled variable set-point value, the controller pressure range midpoint. The difference in value between measured value and the set-point or midpoint value is multiplied by the controller sensitivity rate to give a pressure difference. This difference is added to or subtracted from the midpoint pressure to give the predicted output pressure value for the measured value. The same equations are used for the calculation of set-up parameters for other variables, such as relative humidity and static pressure.

The basic pneumatic controller equation is expressed as:

$$P_{Out} = P_{STP} + PR\frac{T_{Meas} - T_{STP}}{TR} \tag{11.4}$$

where, P_{Out} is the output or branch pressure from the controller, psig, P_{STP} is the set-point pressure or the pressure corresponding to set-point temperature, e.g., 8 psig for 3 psig to 13 psig system or 9 psig for 3 psig to 15 psig system, T_{STP} is the set-point temperature, °F, T_{Meas} is the measured temperature of the controlled medium, °F, TR is the throttling range, °F, of the controller, and PR is pressure range of the controller as temperature changes through throttling range, e.g., 10 psig on 3 to 13 psig system or 12 psig on 3 to 15 psig system. A pneumatic controller with integral sensor usually has a throttling range adjustment, while a pneumatic controller with a remote sensor does not have a throttling range adjustment, being set up by the use of a proportional band adjustment. Throttling range and proportional band are closely related: The *throttling range* is used with sensors having variable sensing ranges of sensed medium, while the term *proportional band* (PB) is used with sensors having fixed sensing ranges, which must be related to sensor pressure span. The basic pneumatic controller equation used on controllers with integral sensors is modified for use on controllers with remote sensors to use *PB* as follows:

$$P_{Out} = P_{STP} + PR\frac{T_{Meas} - T_{STP}}{PB \times P_{SP}} \tag{11.5}$$

where *PB* is the proportional band, % of sensor span, P_{SP} is the sensor span, expressed in °F, % RH, or "wc for output variation from 3 to 15 psig for 12 psig span system or from 3 to 13 psig for 10 psig span system. The sensitivity of the controller is defined as the ratio of the controller pressure range over the throttling range, expressed in psig/°F. *Controller sensitivity* represents the change in output of a controller per unit change in the controlled variable, while *sensor sensitivity* is the change in the measured variable in a remote sensor or transmitter per unit change in the sensed medium. The value of the PB is calculated for a specific sensor that is to be used on a controller with an assigned throttling range as follows:

$$PB = \frac{TR}{P_{SP}} \times 100 \qquad (11.6)$$

Example 11.3

A pneumatic controller has 78°F set point, 6°F throttling range, and a 3 psig to 13 psig pressure range system. Determine the output pressure of this controller at 77°F, and its sensitivity.

SOLUTION

The pressure range is 13 − 3 = 10 psig, and midpoint pressure of this controller is 8 psig. The output pressure at 77°F is calculated with Equation (11.4), and the sensitivity with Equation (11.6), respectively.

$$P_{Out} = 8 + 10\frac{77-78}{6} = 6.33 \text{ psig}$$

and

$$Sesnitivity = \frac{10}{6} \approx 1.67 \text{ psig/°F}$$

In a two-position control, the value of an analog or variable input is compared with the programmed instructions (values) and a digital (two-position) control output is generated. The instructions involve the definition of the upper and lower limits for the measured variable. The output control changes its value as the input is out of these limit values. Two-position control can be used for simple control loops (e.g., temperature control) or limit control, such as *freeze-status*, designed to shut down an AHU when the unit-mixed air temperature is too low. The input can be any measured variable including temperature, RH, pressure, current, liquid level, etc. Time can also be the input to a two-position control response. This control response provides a schedule for operations and functions, like an old-fashioned time clock. A common application of two-position control is the residential heating system. For example, if a set point of 70°F (or 21°C) and a 4°F (or 2°C) differential range are set, the thermostat (the *controller*) energizes the heating system when the space temperature is less than 68°F (20°C) and turns it off when the space temperature is above 72°F (22°C). In the floating control, a control response produces two possible digital outputs (DOs) based on a change in the variable input. One output increases the signal to the controlled device, while the other output decreases the signal to the controlled device. This control response also involves an upper and lower limit with the output changing as the variable input crosses these limits. There are no standards for defining these limits, but the terms *set point* and *deadband* are common. The set point sets a midpoint and the deadband sets the difference between the upper and lower limits. For this control response to be stable, the sensor must sense the effect of the controlled device movement very rapidly. Floating control does not function well where there is significant thermodynamic lag in the control loop. Fast airside control loops respond well to floating control. Proportional control response produces an analog or variable output change in proportion to the change of a varying input. In this control response, there is a linear relationship between the input and the output. Set point, throttling range, and action typically define this relationship. There is a measured variable unique value that corresponds to the full range of the controlled device and a unique value that corresponds to zero travel on the controlled device. The change in the measured variable, causing the controlled device to move from fully closed to fully open is the *throttling range*, being the total change in the controlled variable, requiring the actuator or controlled device to move between its limits. It is within this range that the control loop controls, assuming that the system has the capacity to meet the requirements. The action sets the slope of the control response. With proportional control, the final control element moves to a position proportional to the deviation of the value of the controlled variable from the set point, in a linear manner. The final control element is seldom in the middle of its range because of the linear relationship between the position of the final control element and the value of the controlled variable. The set point is typically in the middle of the throttling range, so there is usually an offset between the control point and the set point. Proportional control adjusts the controlled variable in proportion to the difference between the controlled variable and the set point. The following equation defines the behavior of a proportional control loop:

$$u(t) = U_0 + K_P \cdot e(t) \tag{11.7}$$

Here, $u(t)$ is the controller output, U_0 is the constant value of controller output when no error exists (usually the mid-range point), K_P is the proportional control gain (it determines the rate or proportion at which the control signal changes in response to the error), and $e(t)$ is the error, for example, in the case of the steam coil, $e(t)$ is the difference between the air temperature set point, T_{Set} and the sensed supply-air temperature, T_{Sensed}, expressed by:

$$e(t) = T_{Set} - T_{Sensed} \tag{11.8}$$

Example 11.4

If a system steam heating coil has a heat output that varies from 0 to 30 kW as the outlet air temperature varies from 30°C to 50°C in an industrial process, what is the coil gain and what is the throttling range? Find the equation relating the heat rate at any sensed air temperature to the maximum rate in terms of the gain and set point, assuming steady-state operation.

SOLUTION

The control variable throttling range is $\Delta T_{max} = 50°C - 30°C = 20°C$, assuming that 0 kW corresponds to 30°C and 30 kW to 50°C, the proportional gain ratio is

$$K_P = \frac{\dot{Q}_{max} - \dot{Q}_{min}}{\Delta T_{max}} = \frac{30 - 0}{20} = 1.5 \text{ kW/°C}$$

In the assumption of linear dependence, the average air temperature (40°C) occurs at the average heat rate (15 kW). By inserting the numerical values in relationship, as one of Equation (11.7), we are getting:

$$\dot{Q} = 1.5 \frac{kW}{°C}(40 - T_{Sensed}) + 15 \text{ kW}$$

In the assumption of a linear characteristic the average air temperature of the throttling range occurs at the average heat rate, as in the previous example, the straight-line equation is:

$$\dot{Q} = K_P (T_{Set} - T_{Sensed}) + \frac{\dot{Q}_{max} - \dot{Q}_{min}}{2} \tag{11.9}$$

Notice that the quantity $(T_{Set} - T_{Sensed})$ is the error $e(t)$ and a nonzero value indicates that the set temperature is not met. A proportional control system, as one presented here, requires the presence of an error signal to fully open or fully close the system valve. *Integral control* is often added to proportional control to eliminate the offset inherent in proportional-only control, resulting in a proportional plus an integral control, the so-called PI control. Initially, the corrective action produced by a PI controller is the same as for a proportional-only controller. After the initial period, a further adjustment due to the integral term reduces the offset to zero. The rate at which this occurs depends on the timescale of the integration. In equation form, the PI controller is modeled by this relationship:

$$V_{PI}(t) = V_0 + K_P \cdot e(t) + K_I \int e(t) \cdot dt$$

(11.10a)

Here, K_I is the integral gain constant, having the units of reciprocal time and is the number of times that the integral term is calculated per unit time, the reset rate; reset control is an older term used by some to identify integral control. Notice that the PI control loops do not perform well when set points are dynamic, where sudden load changes occur, or if the throttling range is small. The integral term means that the controller output can be affected by the error signal integrated over time and multiplied by the integral gain constant. Note that the errors may be positive or negative; therefore, the integral term may be plus or minus, its effect is that the controller output will continue to change as long as any error persists, and the control offset will be eliminated. Although PI mode has long been used in the process control industry, it is somewhat new to HVAC. Some pneumatic controllers and most electronic controllers use PI mode. For derivative control modes, still another term is added to the control equation:

$$V_{PID}(t) = V_0 + K_P \cdot e(t) + K_I \int e(t) \cdot dt + K_d \frac{de(t)}{dt}$$

(11.10b)

Here, K_d is the derivative gain constant The derivative term provides additional controller output related to the rate of change of the controlled variable. A rapid rate of change in the error increases the absolute value of the derivative term, and a small rate of change decreases the value. The sensitivity of the control logic is now a function of both the proportional gain K_P and the integral gain K_I. Just as with proportional control, it is possible to have unstable control if the gains are too high; they must be tuned for the application. Proportional integrative derivative control (PID) is used to reduce overshoot when a rapid response is desired (requiring a high proportional gain). In most HVAC systems, derivative control adds unneeded complexity. However, because most HVAC system responses are relatively slow, the value of derivative control in most HVAC applications is minimal. Including the differential term may complicate the tuning process, and cause unstable responses. For these reasons, derivative control logic is normally not used in most field HVAC applications. For most control loops, a smooth transition by the controlled variable to the set point is desired and required. A properly adjusted PID controller can achieve this response without the need for derivative control action. For example, a PID controller that works well in practice in most situations without knowing in-depth control theory is shown in Figure 11.3. This controller is expected

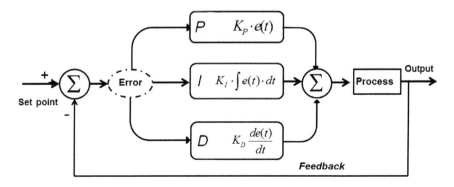

FIGURE 11.3 Typical implementation of a PID control scheme.

to have good stability and its performance should not be greatly compromised by small differences in plant, process, or operating conditions. Its error variable should converge to a small number, preferably 0, i.e., little oscillations even with some large disturbances. Implementing this type of PID controller can be done with analog components, but a microcontroller is much more flexible and preferable, interactive commands can be used to set K_P, K_I, K_D parameters.

11.2.5   HVAC CONTROL STRATEGIES AND TYPES

The simplest control in HVAC system is cycling or on/off control to meet part-load conditions. If a building needs only half the energy that the system is designed to deliver, the system runs for about 10 minutes or so, turns off for 10 minutes, and then cycles on again. As the building load increases, the system runs longer and its off-period is shorter. One problem faced by this type of control is short-cycling, which keeps the system operating at the inefficient condition and quickly wears out the component. A furnace or air conditioner takes several minutes before reaching steady-state performance. A compromise that allows adequate comfort without excessive wear on the equipment is modulation or proportional control, as described before. Under this concept, if a building calls for half the rated capacity of the chiller, the chilled water is supplied at half the rate or in the case of a heating furnace, fuel is fed to the furnace at half the design rate: The energy delivery is proportional to the energy demand. While this system is better than cycling, it also has its problems. Equipment has a limited turn-down ratio. A furnace with a 5:1 turn-down ratio can only be operated above 20% of rated capacity. If the building demand is lower than that, cycling would still have to be used. An alternate method of control under part-load conditions is staging. Several small units (e.g., four units at 25% each) are installed instead of one large unit. When conditions call for half the design capacity, only two units operate. At 60% load, two units are base-loaded (run continuously), and a third unit swings (is either cycled or modulated) as needed. To prevent excessive wear, sequencing is often used to periodically change the unit being cycled. Usually, the HVAC control system is distributed across the system components, e.g., chillers, boiler(s), heat exchangers, pumps, etc., with various purposes and designation, while individual room controls depend on the HVAC system design.

The most popular control system for large buildings historically has been pneumatics, which can provide both on-off and modulating control. Pneumatic actuators are described in terms of their spring range. Common spring ranges are 3 to 8 psig (21 to 56 kPa), 5 to 10 psig (35 to 70 kPa), and 8 to 13 psig (56 to 91 kPa). Compressed air with an input pressure can be regulated by thermostats and humidistats. By varying the discharge air pressure from these devices, the signal can be used directly to open valves, close dampers, and energize other equipment. Relatively inexpensive copper or plastic tubing carries the control signals around the building. The pneumatic system is very durable, is safe in hazardous areas where electrical sparks must be avoided, and, most importantly, is capable of modulation, or operation at part-load condition. Pneumatic controls use clean, dry, and oil-free compressed air, both as the control signal medium and to drive the valve stem with the use of diaphragms. Instrument-quality compressed air is more suitable for controls rather than industrial-quality and requires drying to a dew point low enough to satisfy the application. When compared to electronic systems, pneumatic controls are less reliable and noisier. Direct digital control (DDC) is the most common deployed control system today. The sensors and output devices (e.g., actuators, relays) used for electronic control systems are usually the same ones used on microprocessor-based systems. The distinction between electronic control systems and microprocessor-based systems is in the handling of the input signals. *In an electronic control system, the analog sensor signal is amplified, and then compared to a set point or override signal through voltage or current comparison and control circuits. In a microprocessor-based system, the sensor input is converted to a digital form, where discrete instructions (algorithms) perform the process of comparison and control.* Most subsystems, from VAV boxes to boilers and chillers, now have an onboard DDC system to optimize the performance of that unit. A communication protocol known as BACnet is a standard protocol that allows control units from different manufacturers to pass data to each other.

Combinations of controlled devices are possible. For example, electronic controllers can modulate a pneumatic actuator. Also, proportional electronic signals can be sent to a device called a transducer, which converts these signals into proportional air pressure signals used by the pneumatic actuators. These are known as electronic-to-pneumatic (E-P) transducers.

Direct digital control of an HVAC system is a method of monitoring and controlling HVAC system performances by collecting, processing, and sending information using sensors, signal conditioning, actuators, and microprocessors, or in the case of large facilities, by using computers. DDC is the concept or theory of HVAC system control that uses digital controls, while physically, DDC encompasses all the devices used to implement this control method: a whole group of DDC controllers/microprocessors, actuators, sensors, and other needed devices. Temperature measurements for DDC applications are made by three principal methods: thermocouples, resistance temperature detectors (RTDs), and thermistors.

A point represents any input or output device used to control the overall or specific performance of equipment or output devices related to the equipment. BASs are continuously evolving to efficiently address new challenges and enable flawless and cost-effective operation of high-performance buildings. The whole set of technologies deployed into buildings has a direct influence on BAS, their architectures, complexities, structures, and functions. The most important technology trends over the last decade relate to the increasing use of cloud technologies and data analytics, the prevalence of the Internet of Things (IoT) paradigm, and the growing emphasis on user experience and comfort. Cloud computing and data analytics have made significant progress over the last years in many domains and they continue to create impact in building applications. Their capabilities to collect and analyze data from multiple and potentially heterogeneous data sources and move them to a cloud repository allows the implementation of powerful applications that may provide insights into building operations. Cloud connectivity and real-time processing will enable the data to become fluid versus static with vast new opportunities. The sophistication of new and more powerful building analytics will likely improve from visualization and reporting dashboards to fault-detection and diagnostics too, and ultimately, applications in predictive maintenance and holistic dynamic optimization of buildings. Currently deployed building analytics can better inform facility managers about deviations from the expected energy consumption, likely HVAC equipment faults (Kreider, Rabl, and Curtiss, 2001) and underperforming controllers. IoT paradigms enable the connections of building automation components to IT networks to improve the interoperability and connectivity of control and monitoring devices. In summary, the controller's function is to compare its input(s) with a set of instructions, such as set point, throttling range, and action and then produce an appropriate output signal.

In HVAC terminology, a *zone* is an area or a space for which the temperature is controlled by a single thermostat. Zones must not be confused with rooms, while several rooms can be controlled by a single thermostat and are part of a zone. There are several types of zones, differentiated based on what is to be controlled, and the variability of what is to be controlled. The most common control parameters are: thermal (temperature), humidity, ventilation, operating periods, freeze protection, pressure, and importance. The most common reason for zoning is the variation in thermal loads. Basically, a zone is a portion of a building that has thermal loads that differ in magnitude and timing sufficiently from other areas so that separate portions of the secondary HVAC system and control system are needed to maintain expected comfort. The term *zone* is also applied to areas of humidity control. Having specified the zones, the designer must select the thermostat (and other sensors, if used) location. A wide variety of systems are available for building applications, differing with respect to cost, comfort, complexity, energy efficiency, maintenance, and flexibility. A simple residential system provides hot or cold air, and the supply temperature in either cooling or heating mode is basically constant. Controls are usually set to allow fan operation only when heating or cooling, and the supply-air volume depends on the fan operation time. The single-zone system is the simplest and most common type of all-air system, consists of an air handling unit with heating and cooling coils controlled in sequence in response to the room conditions as indicated by the room thermostat.

In cooler, dryer climate zones, the heating coil is typically installed upstream of the cooling coil and the single-zone system operates as a changeover system, providing heating or cooling, but not both simultaneously. However, in hot, humid climate zones, the heating coil must be installed downstream of the cooling coil and controlled to allow the use of reheat (simultaneous heating and cooling) when humidity control is needed. In commercial buildings, fans usually operate continuously to provide ventilation when the building is occupied. This operation mode is classified as *constant volume*. Single-zone constant volume is the simplest system, with air being supplied at a constant rate. Single-zone reheat, a constant-volume system, is an air-conditioning type used when humidity control is important. Such systems are characterized by high energy usage. Complex buildings require several zones of control to accommodate load variations among the building spaces. Often systems use air-terminal devices to serve individual spaces or groups of spaces, with each terminal corresponding to a control zone. Systems of constant-volume terminal-reheat multizones consist of AHUs containing cooling and reheating coils that chill or warm the air supplied to various zones, controlling the temperature in each zone. Reheat terminals (boxes) are modular devices available in different sizes and types. Constant-volume terminal-reheat systems are flexible, reconfigurable in agreement with space changes and/or destinations, characterized also by excellent humidity control. Constant-volume dual ducts send the warm and chilled air to a pair of ducts to the served area. Such systems are able to control multiple zones, each served by individual boxes. Mixing boxes are controlled by dampers, varying the quantity of warm and chilled air in response to the zone thermostat. Such systems are flexible and reconfigurable by adding or rearranging the mixing boxes on the main trunk-ducts. Multizone systems are very similar in operation to dual-duct systems, but the mixing of cold and warm air is done by dampers located in the AHU rather than at terminals located in the served areas. Both dual-duct and multizone systems waste energy by mixing warm and cold air. Multizone systems have the added disadvantage of inflexibility with regard to changes; rearranging and adding zones involve considerable duct-work and unit modifications. Single-zone variable air volume (VAV) systems, are an energy-efficient option used for cooling variable supplied air quantities, by the AHU in response to the space thermostats. Such systems use cooling coils to produce chilled air at a constant temperature, with air flow being controlled in response to the space thermostat. They are suitable for small and simple buildings. Multiple-zone VAV systems use a VAV air-handling unit to supply chilled air to a main trunk-duct to multiple VAV terminals, serving multiple zones. Each terminal controls the air in response to the served space thermostat. VAV terminals are inexpensive and energy-efficient. VAV systems provide only cooling, which is an important disadvantage, so buildings with heating loads must combine VAV systems with other systems. VAV reheat terminals are equipped with electric or hot water heating coils, similar to constant-volume reheat terminals, except for the air flow control and heating and cooling set by the space request.

11.3 MONITORING, SAFETY, AND SECURITY SYSTEMS

Modern home automation systems (HAS) are used in small buildings, particularly in residential areas, where a building control system (BCS) is used in large buildings such as offices, hotels, hospitals, malls, supermarkets, swimming pools, and schools). HASs control and monitor the lighting system (with presence detection and crepuscular outside-light control), HVAC, energy management system (monitoring and operating household appliances, other electric devices, or energy smart-metering devices), security and access-control system (e.g., antiburglary, video-surveillance), safety alarm system (e.g., fire detection, alarming), remote surveillance system and other functions like window blinds, communication equipment (e.g., intercom system, telephone, public address audio system), elevators, inside and outside irrigation (e.g., sensors, weather-station feedback), water supply system, audio/video (e.g., on/off, volume, other controls), de-icing system (e.g., gutters, stairs, alleys, terraces), garage and exterior gates control. Both HAS and BCS concern local, central, and remote access to building devices and their functions. Central monitoring and control equipment that can be accessed with remote wireless or wired devices can significantly reduce costs. However,

BCS and BMS were traditionally concerned only with the control of HVAC, as well as lighting and shading systems. They originated in a time where security was considered an afterthought at best. In recent decades, the trend in BCS technologies, system structure, and functionalities is to integrate security-critical services that were formerly provided by isolated subsystems. Security must no longer be neglected. The core application area of BCSs is the environmental control handled by the conventional services, HVAC, lighting, and shading. Other application-specific services are often implemented by separated systems. This is particularly true for safety-critical (e.g., fire, smoke, or alarm systems) and security-critical (e.g., intrusion alarm or access control systems) services, which are typically provided by proprietary stand-alone automatic control systems. Often only a loose coupling is implemented with the core BCS for visualization and alarm management. However, current trends are toward more integration. To activate all inherent synergies among these diverse systems, an integration of security and safety-critical systems is a major topic in research and technology development. The promised benefits range from lower (life cycle) costs to increased and improved functionalities and capabilities. Consider, for example, the possibility of sharing the data originating from one sensor in multiple application domains in parallel. This reduces the investment, operation, and maintenance costs and facilitates the management and, in particular, configuration of an integrated BCS, where a multitude of different management solutions can be replaced with a unified view and a single central configuration access point. Of note, all BCS manufacturers continue to strive to integrate all technology within a facility. Today, this extends to the functionality of business practice and services that include the security function.

Buildings may have operational problems due to degraded equipment, failed sensors, incorrect installation, poor maintenance, and improperly implemented controls. Currently, most problems related to building systems are detected through complaints from occupants or alarms provided by BCS units. Detection and diagnosis can be performed automatically by integrating the experience required to detect and diagnose operational problems into software tools that take advantage of existing sensors and control systems. These tools are not designed to replace the people who operate the building systems, but to help them improve the functioning of those systems. The automatic start-up and diagnosis technologies for systems and building equipment are expected to reduce the repair time, save costs, and act on problems and improve the functioning of the building, through the automatic and continuous detection of performance problems and maintenance requirements that are communicated to the building operators or managers, who can then perform the necessary corrective actions.

11.3.1 Fire Protection, Safety, and Smoke Management

Fire services installations and equipment, or fire protection systems, are all installations or equipment manufactured, used, or designed for the purpose of extinguishing, attacking, preventing, limiting, or giving warning of any fire. They may be fixed or portable, and either automatic or manual in operation. Fire is a rapid combustion process of some materials in the presence of oxygen, producing flames and heat, usually smoke and toxic gases. Fire products have two components, thermal elements, producing flames and heat, and nonthermal elements, producing smoke and toxic gases. Smoke is always associated with fire. A fire protection system is used to extinguish the fire and also to remove the smoke. According to NFPA classification, fires have four classes: Fires of ordinary combustible materials (Class A), fires from flammable liquids (Class B), fires involving energized electric equipment (Class C), and fires of combustible metals (Class D). Fire hazards are categorized by NFPA as: light (low) hazards, ordinary (moderate) hazards, and high (extra) hazards. Fire protection systems must be designed to be effective for each category of hazards. Fire protection starts with the design and construction of a fire-safe building, equipped with the most effective and appropriate fire protection system.

Fire alarm systems consist of different types of sensors such as fire detectors (automatic) or push-button alarms (non-automatic). Automatic and non-automatic sensors are grouped into detector groups.

Information coming from sensors of a detector group via a dedicated communication path is processed at the control and indicating equipment. According to the information received, the actuators are activated or not. Common actuators are fire doors or smoke-extraction devices. Moreover, a remote alarm to the fire department (control unit) can be transmitted. A fire protection system includes equipment, devices, sensors, control units, wiring, and piping subsystems to detect fire and smoke and to prevent and suppress the fire or smoke if it occurs, with the primary objective being to protect and save lives and property. Other objectives are to minimize interruptions of services due to the fire and/or smoke and to reduce and to minimize the losses caused by such events. A fire protection system is used not only to prevent and extinguish fire, but also to remove smoke and toxic gases. Fire safety decisions today are dominated by building codes, associated standards of practice, and insurance considerations, because they have been a part of the building process for nearly a century and have demonstrated success in reducing fire losses. The thought process for fire-related decisions is heavily influenced by experiences and interpretations of codes and standards. Most building codes specify the types of construction methods and materials that can be used for building occupancies.

In the event of fire and smoke presence, detected by either occupants or by automatic systems, appropriate measures must be taken in agreement with the event severity, such as activating the building alarm system, evacuating the building if needed, and notifying the fire department. Fire and smoke detection and signaling devices can be addressable or non-addressable. The former ones have the capability to be addressed and identified individually so the system can immediately identify the device location and type and initiate the associated subsystem(s) with that specific address. A manual alarm device, usually protected by glass, is basically an electric switch designed for the activation of fire alarms, such as: red and/or flashing lights, bells, etc. Thermal detectors are temperature-activated sensors, designed to meet various conditions that initiate the fire alarm when the temperature in the proximity area reaches a predetermined set point. Such detectors are used to provide property, but not life, protection. The commonly used temperature detectors are: *fixed-temperature type, rate-of-rise type,* or a *combination of these types.* Fixed-temperature detectors are either self-restoring, consisting of normally open contacts held by bimetallic elements that close when the ambient temperature reaches a fixed setting, or nonrestoring, which usually contain a fusible element that holds open the contact that melts if the temperature reaches the threshold and closes the alarm electric circuit. Rate-of-rise detectors react to the rising temperature rates. A sealed and slightly vented air chamber quickly expands if the temperature in the device vicinity is quickly rising, closing the alarm electric contacts. Such detectors react much more quickly than fixed-temperature detectors. The combination type reacts to both a fixed temperature set point and the temperature rising rates, often operating on the differential expansion principle between two metals or on the expansion rate of the air chamber. The combination type is the most desirable fire detector. Smoke detectors are designed to respond faster than fire detectors, if the fire-generated smoke is within their detectability limits. The most common smoke detector types are: *photoelectric smoke detectors* and *ionization smoke detectors.* Photoelectric smoke detectors operate on a light-scattering principle; if smoke is present in the detector chamber, a light beam generated by a light-emitting diode is scattered, which is detected by a photosensitive diode that activates the alarm electric circuit. Ionization smoke detectors operate on the principle of changing the air electric conductivity in the detector chamber if smoke is present, a current is flowing, and the alarm is activated. Ionization-type smoke detectors are the most sensitive smoke detectors. Signal devices are audio or video systems used to alert and inform the occupants of emergency situations and include bells, buzzers, flashing lights, sirens, horns, etc. Flow detectors are devices indicating or initiating an alarm if water is flowing in the fire-suppression system. Flow detectors monitor even unoccupied buildings for safety reasons.

Fire detection and alarm systems are key features and components of any building's fire prevention and protection strategy. Some major fire service installations are used in high-rise commercial buildings and/or residential buildings. Examples include audio/visual advisory systems, automatic actuating devices, automatic fixed installations other than water, emergency generators, emergency lighting, exit signs, fire alarm systems, fire detection systems, fire hydrant/hose reel systems, firefighter lifts, pressurization of staircases, sprinkler systems, and static or dynamic smoke extraction systems. A comprehensive discussion of all fire services, detection systems, fire sensors, equipment, installations, or fire protection and prevention methods is beyond the scope of this chapter. A fire alarm system serves primarily to protect life and secondarily to prevent property loss. Because buildings vary in occupancy, flammability, type of construction, and value, a fire alarm system must be tailored to the needs of each specific facility. The three basic parts of a fire alarm system are: signal initiation, signal processing, and alarm indication. For automatic fire alarm systems, the signal initiation can be actuated by fire and smoke detectors and/or water flow switches. The alarm signal is processed by some sort of control equipment, which in turn activates audible and visible alarms and in some cases, alerts a central fire station. To design the detection portion of a fire alarm system, it is necessary to determine where fire detectors should be placed in order to respond within the goals established for the system. Several detector types can respond to a fire, so it may be necessary to develop various candidate system designs, using combinations of detector types in order to optimize the system's performance and cost. In the incipient stage, the combustion products comprise a significant quantity of microscopic particles, from 0.01 to 1.0 micron, which are best detected by ionization-type detectors. Upon the flame stage, flame detectors can be of two types, namely ultraviolet radiation detectors or infrared radiation detectors. Fires burn openly and produce great heat, incandescent air, and smoke. Detectors should respond to heat; they are referred to as heat-actuated, thermal, thermostatic, or simply temperature detectors. A fixed-temperature detector responds when the air temperature surrounding the detector rises to a certain level. The rate-of-rise detector responds to a certain rate of rise of temperature of the surrounding air. A detector that responds when either of the two conditions occurs is available as well. A fire signature is defined as some measurable or sensible phenomenon present during combustion. The present practice in designing fire detection systems using heat detectors is to space the detectors at intervals equal to spacing intervals established by tests at Underwriters Laboratories (UL) Inc. Most codes require that detectors be spaced at intervals equal to the UL or FMRC spacing. NFPA 72®, National Fire Alarm Code, 2007 edition, requires that the installed spacing be less than the listed spacing to compensate for high ceilings, beams, and air movement. The best possible location for a heat detector is directly over the fire. If specific hazards are to be protected, the design should include detectors directly overhead or inside of the hazard. In areas without specific hazards, detectors should be spaced evenly across the ceiling. Spacing between detectors, S_{dtc}, is calculated as a function of its distance, d, from the center of the protected area, as:

$$S_{dtc} = \sqrt{2} \times d \qquad (11.11)$$

In building fires, smoke often flows to locations remote from the fire, threatening life and damaging property. NFPA 92 standard defines a smoke-control system as an engineered system that includes methods and physical mechanisms of smoke control that can be used individually or in combination to modify and control smoke flows. Conventional approaches to smoke control use physical mechanisms to prevent people from coming into contact with smoke for extended periods. Recent approaches consist of smoke contact-effect evaluations with the intent of providing a tenable environment for occupants. The physical smoke control mechanisms are: compartmentation, dilution, pressurization, airflow, and buoyancy. For a very long time, compartmentation has been the most effective way to control the spread of fire and smoke. Dilution can occur naturally, when smoke flows away from the fire and mixes with air, flowing naturally or forced by fan(s). However, dilution is not recommended in current technical practice. Many smoke-control systems use mechanical fans to control smoke. Pressure differences across a barrier can control smoke movement. Airflow has been used extensively to control smoke flow during fires in subways, railroads, and highway tunnels.

Smoke management or smoke-control design is a relatively new concept in the building industry. The concept of smoke management is to provide the building occupants time to safely evacuate the building during a smoke or fire event. Based on an analysis of egress time (typically 20 to 30 minutes), the HVAC equipment is used *to control the migration of smoke* to allow *tenable* conditions to exist for this time period. Although opinions differ on how to design these smoke-control systems, design guidelines have been established that should be adhered to by HVAC engineers and designers in order to provide the current standard of professional care in the case of smoke/fire events. There are essentially two different types of smoke-management systems. The first type is the so-called *zoned smoke control*, which is typically required for high-rise and institutional buildings (e.g., hospitals, jails, and prisons). The second type is *smoke evacuation*, used for malls, atria, and large spaces. The approaches are different but both share the mutual goal of providing tenable conditions for a specific time period, allowing for safe evacuation of the building. Some main requirements of smoke management are:

1. Maintain tenability (egress) from the building for a specific time interval for all occupants during a smoke or fire event, thus reducing or even avoiding any deaths and injuries;
2. Reduce damage and property loss caused by smoke damage;
3. Aid firefighters and assist with purging the building of smoke once the fire has been extinguished;
4. Maintain tenability for a minimum period of time, not smoke-free egress paths; and
5. Limit the migration of smoke from the origin of the smoke source/fire for a specific time; however these systems are not intended to allow for continual occupancy of the building during a fire event.

11.3.2 BUILDING ACCESS CONTROL

Access control by definition includes authentication and authorization of the requests to perform some actions in the system. Subjects are defined as the entities that perform actions in the system and objects are defined as the resources on which the action is being performed. It is worth noting, mainly in the field of BCS, that subjects can be both human users and other objects. The role of authentication in access control is mainly to further facilitate the authorization of the subjects. Sometimes, especially in the field of BCS, one-way authentication is not sufficient (e.g., a device in BCS can be programmed to execute some action on a certain kind of devices which could only be verified with the help of authentication). Following are the few classical access-control models that are frequently used to incorporate access control to various degrees depending on the requirements. *Discretionary access control* is when the access policies for the resource are decided by the owner of the resource. The discretionary access control model boils down to two requirements: every object in the system has a subject as its owner, and the owner decides the control to its resources through access rights and permissions to other subjects.

Mandatory access control is access control that allows access to a resource, *if and only if* there exists a policy that allows a given user to access the resource. The access policies are designed centrally and enforced to the whole system thereafter. Mandatory access control rules are often based on security labels on subjects (called clearance) and objects (called classification). Any central security system includes at least six major components, namely the sensors, signal processor, power source, system status display, alarm transmission and reception, and a zone-status indicator.

11.4 BUILDING ACOUSTIC AND NOISE CONTROL

In architecture, building operation, and management one of the many goals of the designers, managers, and operators is to create rooms, building spaces, and buildings that maximize sound performance, meaning that unwanted sounds should be prevented, and desired sounds should be enhanced. Few places on the planet are free from unwanted sounds or noise. Noise from many outdoor sources assails our hearing, invading our homes and workplaces: traffic, aircraft, loud music, construction equipment, etc. Noise within the workplace (e.g., from office equipment, telephones, ventilating systems, unwanted conversation in the next cubicle) distracts us from our work and makes us less productive.

Noise from within the home, from appliances, upstairs footsteps, TV sound traveling from room to room, keeps our homes from being the restful refuges they ought to be. Noise in the classroom impedes the learning process and threatens our children's educational experience. Noise can frustrate and impede speech communication. It can imperil us as we walk or drive city streets. It can be a physical health hazard as well: Exposure to higher noise levels, above a certain threshold, can cause permanent hearing loss. In summary, unwanted sounds or noises include distractions, sounds loud enough to damage hearing, and sound leakage that could affect privacy. Noise is usually described as an unwanted sound, very often with unpleasant effects. Control of noise is characterized and considered by the *source* (noise origin), the *path* (the means by which the generated noise reached the receiver), and the *receiver* (where the noise ended up). Noise control consists of the treatment of one or a combination of these three factors, (i.e., source, path, and receiver). However, noise control usually focuses on the first two factors. Special acoustical and noise-control considerations are required in many building spaces, such as theaters, auditoriums, classrooms, churches, lecture halls, restaurants, museums, display spaces, music practice rooms, recital halls, conference centers, and boardrooms. However, noise control is a general building design, operation, management, and occupant comfort issue.

To understand the material contained in this section, it is useful to review some basic concepts and definitions. A *sound* is a form of mechanical energy transmitted by vibration of the molecules of whatever medium the sound is passing through. The speed of sound in air is approximately 1130 ft/s or 340 m/s. The number of cycles per second made by a sound wave is termed its *frequency*. Frequency is expressed in Hertz (Hz). The heard sound is usually radiated in all directions from a vibrating medium. Sound waves can travel through any media, e.g., air, water, wood, masonry, or metal. Depending on the media through which it travels, sound is either airborne or structure-borne. Most of the sounds we hear, however, are a combination of several different frequencies. Perceptible acoustical sensations are usually classified into two broad categories, which are defined here. (a) A *sound* represents a disturbance in an elastic medium that results in an audible sensation. Noise is by definition an *unwanted sound*. (b) A *vibration* represents a disturbance in a solid elastic medium, which may produce a detectable elastic motion. Although this differentiation is useful and common in presenting acoustical concepts, in fact sound and vibration are often interrelated, with a sound being often the result of acoustical energy radiation from vibrating structures, and sometimes a sound can force structures to vibrate. Acoustical energy can be fully characterized by the determination of three parameters: *magnitude or energy level*, which is a measure of the intensity of the acoustical energy, *frequency* or *spectral content*, a description of an acoustical energy with respect to frequency composition, *time (temporal) variations*, which is a description of how the acoustical energy varies with respect to time. *Sound pressure*, *sound power*, and *sound intensity level* are basic acoustic quantities (parameters). The measurement parameters and units for each of these quantities are used to evaluate and characterize sounds and vibrations. Any sound or vibration is characterized by frequency, measured in cycles per second (Hz). The normal human hearing frequency range extends from a low frequency of about 20 Hz up to a high frequency of about 15,000 Hz or even higher, up to 20,000 Hz for some people. Frequency characteristics are important for the following reasons: people have different hearing sensitivity to different sound frequencies (generally, people hear better in the upper frequency region of 500 to 5000 Hz and are therefore more annoyed by loud sounds in this frequency region); high-frequency sounds of high intensity and long duration contribute to hearing loss; different pieces of electrical and mechanical equipment produce different amounts of low-, middle-, and high-frequency noise; and the noise control materials and treatments vary in their effectiveness as a function of frequency (e.g., low-frequency noise is more difficult to control, noise control

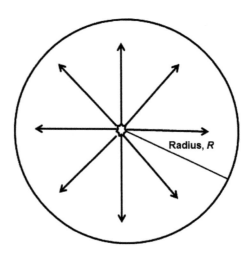

FIGURE 11.4   Radiation pattern of a point sound source.

methods perform better at higher frequencies). Frequency of sound is often segmented in *octave bands*, or sometimes in *one-third octave bands*. The upper frequency limit of an octave is twice of the lower limit, with the engineering preferred octave bands centered at 62.5, 125, 250, 500, 1000, 2000, 4000, and 8000 Hz. The wavelength (λ) of the sound wave is an important parameter in determining the behavior of sound waves. Wavelength is computed by multiplying the wave speed with period or divided by frequency.

Sound intensity is a scientific concept distinct from the more subjective *loudness*. It is defined as power of sound wave per unit area (Wm^{-2}). For example, sound intensity from a watch ticking is around 10^{-11} Wm^{-2}, but a jet taking off generates about 10^2 Wm^{-2}. Sound intensity (I) is a function of the mean square of the sound pressure (p^2), the fluid density, here the air density, (ρ), and speed of the sound in the fluid, here the sound speed into the atmosphere (c), as expressed in Equation (11.12). If a point sound source is emitting sound in all direction (as illustrated in Figure 11.4), then the total sound power is the sound intensity multiplied by the area ($4\pi R^2$, a sphere for the point source).

$$I = \frac{p^2}{\rho \cdot c}$$ (11.12)

Example 11.5

A plane sound wave is transmitted through air, speed of sound is 346.1 m/s, air density is 1.225 kg/m³ at 25°C (298.2 K or 778°F) and the atmospheric pressure is 101.3 kPa (14.7 psia). The sound wave has an acoustic pressure of 0.20 Pa. Determine the acoustic intensity for the sound wave.

SOLUTION

Using Equation (11.12), the sound wave intensity is:

$$I = \frac{(0.2)^2}{1.225 \cdot 346.1} = 94.346 \times 10^{-6} \ \text{W/m}^2$$

The product of the fluid density times the sound speed is the so-called acoustic impedance. Sound intensity level (*IL*), sound power level (L_{PW}), and sound pressure level (*SPL*) are parameters expressed in dB, all being related to acoustic power. Difference of powers of sound wave between the watch ticking and a jet is 10^{13} times. This scale is not comprehensive enough to tell the sound loudness. A logarithmic scale is used to measure sound intensity, *sound intensity level*, defined as:

$$IL = 10 \cdot log\left(\frac{I}{I_0}\right), \quad \text{in dB} \tag{11.13}$$

Here, IL is the sound intensity level (dB), I is the measured sound intensity (W·m^{-2}), and I_0 is the standard (reference) sound intensity (Wm^{-2}), the softest sound intensity that a human ear can hear. The standard reference of I_0 is 10^{-12} Wm^{-2}. Sound intensity is sound power per unit area, and like sound power, is not audible. Sound intensity is also a vector quantity, that is, it has both a magnitude and direction. Like sound power, sound intensity is not directly measurable, but sound intensity can be obtained from sound pressure measurements. The level basic unit in acoustics is the *decibel* (dB), expressing the ratio of two quantities that are proportional to power. In acoustics, the term *level* is used to designate that quantity, the reference value, which is either stated or implied. Notice that noise is a subjective and relative perception, but sound intensity level is a scientific measure of the power of sound. Even if a sound source has a relatively low sound-intensity level, some individuals might find certain sound frequencies annoying because the human ear reacts differently to different sound frequencies (pitches) and the hearing ability of individuals is subjective. Sound-intensity level can only be used as a standard to identify the sound volume that could cause hearing damage. It does not represent the exact perception of the sound to individuals. The decibel level is equal to 10 times the common logarithm of the power ratio, expressed as:

$$dB = 10 \cdot log\left(\frac{W_{ref}}{W}\right) \tag{11.14}$$

In this equation W is the absolute value of the power under evaluation and W_{ref} is the absolute value of a power reference quantity expressed in the same units. If the power W_{ref} is the accepted standard reference value, the decibels are standardized to that specific reference value. The reference power, in both SI and English unit systems, is 10^{-12} W. In acoustics, the decibel is used to quantify sound pressure levels that people hear, sound power levels radiated by sound sources, the sound transmission loss through a wall, and in other uses, such as simply a noise reduction of 15 dB (a reduction relative to the original sound level). Decibels are always related to logarithms to the base 10, so the notation 10 is usually omitted. It is important to realize that the decibel is in reality a dimensionless quantity (somewhat analogous to percent); therefore, when using decibels, reference needs to be made to the quantity under evaluation and the reference level. It is also important to note that the dB level is determined by the magnitude of the power level absolute value. That is, if the magnitude of two different power levels differs by a factor of 100 then the decibel levels differ by 20 dB. The sound power level (L_{PW}) in decibels is given by:

$$L_{PW} = 10 \cdot log\left(\frac{W}{W_{ref}}\right) = 10 \cdot log\left(\frac{I \times A}{W_{ref}}\right) \tag{11.15}$$

Here, W and W_{ref} are again the sound power and reference sound power in W, and A is the surface area. The conversion between sound intensity level (in dB) and sound power level, L_W (in dB), is expressed by:

$$L_W = 10 \cdot log\left[A \cdot \left(\frac{I}{I_0}\right)\right] \tag{11.16}$$

Here, A is the area over which the average intensity is determined in square meter (m^2). The above equation can also be expressed, if the area is in SI units (m^2) or English units (sq. ft.), as:

$$L_W = L + 10 \cdot log(A) \tag{11.17a}$$

and, respectively

$$L_W = L + 10 \cdot log(A) - 10 \tag{11.17b}$$

Example 11.6

If a point sound source in Figure 11.4 emits an acoustical power of 10^{-3} W, what is the sound power level and the sound intensity 5 m from the source?

SOLUTION

For a point source, the sound intensity is the sound power divided by the area of the sphere. The sound power level is calculated with Equation (11.16), as:

$$L_W = 10 \cdot log \left(\frac{10^{-3}}{10^{-12}} \right) = 90 \ dB$$

The area of the sphere with 5 m radius is:

$$I = \frac{10^{-3}}{4 \times \pi \times 5^4} = 3.185 \times 10^{-6} \ \text{W/m}^2$$

Notice, that if the area A completely closes the sound source, the above equations provide the total source sound power level. However, care must be taken to ensure that the intensity is representative of the total area. This is done by using the area weighted intensity or by logarithmically combining individual L_W values. The ear responds to sound pressure. Sound waves represent small pressure oscillations, above and below atmospheric pressure. These pressure oscillations impinge on the ear, and sound is heard. A sound level meter (instrument) is also sensitive to sound pressure. The sound pressure level (the abbreviation SPL is often used in practice), in dBs, is defined by:

$$SPL = 10 \cdot log \left[\left(\frac{p}{p_{ref}} \right)^2 \right] \tag{11.18a}$$

or

$$SPL = 20 \cdot log \left(\frac{p}{p_{ref}} \right) \tag{11.18b}$$

where p is the absolute sound pressure and p_{ref} is the reference pressure. Unless otherwise stated, the pressure, p, is the effective root mean square (RMS) sound pressure. Equations (11.18a) and (11.18b) are both correct; however, the sound pressure level is recommended to be computed by using Equation (11.18b). This is because when combining sound pressure levels, in almost all cases, it is the square of the pressure ratios (i.e. $(p/p_{ref})^2$) that should be summed, not the pressure ratios (i.e., not p/p_{ref}). This is also true for sound pressure level averaging. Sound pressure level, expressed in dBs, is the logarithmic ratio of pressures where the reference pressure is 20 µPa (the sound pressure at the threshold of hearing), the SI unit of pressure, $1\ \text{Pa} = 1\ \text{N/m}^2$. This reference pressure represents the faintest sound that can be heard by a young, sensitive, undamaged human ear when the sound occurs in the frequency region of maximum hearing sensitivity, about 1000 Hertz (Hz). A 20 µPa pressure is 0 dB on the sound pressure level scale. In the strictest sense, a sound pressure level includes the reference pressure base, such as 85 dB relative to 20 µPa. However, in normal practice, the reference pressure is omitted, but it is nevertheless implied.

Example 11.7

Compute the sound pressure level for the threshold of pain, 20 Pa, and a metalworking plant or workshop, 2 Pa.

SOLUTION

By using Equation (11.18b), the SPL for the cases are:

$$SPL = 20 \cdot log\left(\frac{20}{0.00002}\right) = 120\ \text{dB}$$

and

$$SPL = 20 \cdot log\left(\frac{2}{0.00002}\right) = 100\ \text{dB}$$

Sound intensity cannot be measured directly, but a reasonable approximation can be made if the direction of the energy flow can be determined. Under free field conditions where the energy flow direction is predictable (outdoors for example), the magnitude of the sound pressure level (L_p) is equivalent to the magnitude of the intensity level (L). This results because, under these conditions, the intensity (I) is directly proportional to the square of the sound pressure (p^2). This is the key to the relationship between sound pressure level and sound power level. This is also the reason that when two sounds combine, the resulting sound level is proportional to the log of the sum of the squared pressures (i.e., the sum of the individual p^2), not the sum of the pressures (the p values). Put another way, when two sounds combine, it is the intensities that add, not the pressures. Sound pressure levels can be used for evaluating the effects of sound with respect to sound level criteria. Sound pressure level data taken under certain installation conditions cannot be used to predict sound pressure levels under other conditions unless significant modifications are in place. Implicit in these modifications is a sound power level calculation. However, the sound power level is an absolute measure of the acoustical energy produced by an acoustic source. Sound power is not audible like sound pressure, but the two items are directly related: The way in which the sound power is radiated and distributed determines the sound pressure level at a specified location. The sound power level, when correctly estimated, is an indication of the sound radiated by the source and is independent of the room containing the source. Sound power level data can be used to compare sound data submittals more accurately and to estimate sound pressure levels for a variety of room or space conditions. Thus, there are technical needs for the higher quality sound power level data. Two notable limitations regarding sound power level data are: sound power cannot be measured directly but is calculated from sound pressure level data, and source directivity characteristics are not necessarily determined when the sound power level data are obtained. Notice also that recent advances in measurement techniques have resulted in equipment that determines sound intensity directly, both magnitude and direction, so the sound intensity measurements can be conducted in complex environments, where free field conditions do not exist and the relationship between intensity and pressure is not as direct. Vibration level, L_V, is analogous to sound pressure level, and is defined by:

$$L_V = 20 \cdot log\left(\frac{a}{a_{ref}}\right) \qquad (11.19)$$

where the parameter a, is the absolute level of the vibration, and a_{ref} is the reference vibration. In the past, different measures of the vibration amplitude have been utilized, including, peak-to-peak (p-p), peak (p), average and root mean square (RMS) amplitude. Unless otherwise stated, the vibration amplitude, a, is the root mean square (RMS). When an electrical or mechanical device or equipment is operating at a constant speed and often has some repetitive mechanisms that produce strong sounds, those sounds are usually concentrated at the principal frequency of device or equipment operation. Examples are: a fan or propeller blade passage frequency, the gear-tooth contact frequency or timing belt, the motor whining frequencies, the internal combustion engine firing rate, the impeller blade frequency of a pump or compressor, or the transformer hum. The frequencies, designated as *discrete frequencies* (*pure tones*), are when the sounds are clearly tonal in character, and their frequencies are calculable. The principal frequency is known as the fundamental, and most sounds also contain harmonics of the fundamental, the integer multiple of the fundamental frequency. For example, in a gear train, where gear tooth contacts occur at the rate of 200 per second, the fundamental frequency is 200 Hz, and it is likely that the gear would also generate sounds at 400, 600, 800, 1000, 1200 Hz, and so on for possibly 10 to 15 harmonics. Considerable sound energy is concentrated at these discrete frequencies. Such sounds are more noticeable and sometimes more annoying because of their presence. Discrete frequencies can be located and identified within a general background of broadband noise (noise that has all frequencies present, e.g., the roar of a jet aircraft or the water noise in a cooling tower) with the use of narrowband filters that can be swept through the full frequency range of interest, being very useful for noise and sound control.

Sound response is an individual perception, being quite different from person to person; however, there are some generally accepted sound perceptions useful to quantify the noise level. For example, a 10 dB increase in sound level is perceived as twice as loud, a 5 dB change in sound pressure is perceptible, while a 3 dB change in sound pressure is hardly perceptible. A low-frequency sound, below 500 Hz, is perceived less loud than a high-frequency sound when both are experienced at the same level. In general, human hearing is the most acute at frequencies around 1000 Hz. Overall sound levels, which measure a composite of all sounds usually over human hearing, are usually reported as A-weighted (in dBA), linear (in dBL or just dB), and C-weighted (in dBC). A-weighting, adjusting sound measurements to discriminate against the low frequency, much as the ear does in hearing, is the most prevalent measure, being commonly used in sound level measurements. The long-term average voice level for communication in normal voice, in the absence of background sound, is 60 dBA, and at 85 dBA background sound, conversation in normal voice became strained. A 120 dBA sound approaches the threshold of physical pain. Correlations between the experience and sound levels are tabulated or given in diagrams.

There are three basic elements in any noise-control system, as illustrated in Figure 11.5: the sound source, the path through which the sound travels, and the sound receiver. In many situations or cases, there are several sound sources, various sound paths, and more than one receiver. However, the basic principles of noise control remain the same as for the simple case of Figure 11.5. The objective of noise-control programs is to reduce the noise at the receiver level. This can be accomplished by modifying the source, path, receiver, or any combination of these three elements. The purpose of noise control can be to prevent hearing loss for personnel, allowing effective face-to-face communication or phone conversation, or to reduce noise so that neighbors of the facility will not become intensely annoyed with the sound emitted by the plant. Modifications at the sound

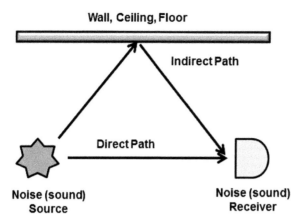

FIGURE 11.5 The three components of a common noise system.

source are usually considered to be the best noise-control solution. Components of a noise source may be modified to effect a significant change into its noise emission. Modifying the noise path is often used when modification of the noise source is not possible, not practical, or not economically feasible. For noise sources located outdoors or indoors, the transmission path may be modified by placing a wall or barrier between the source and receiver. An effective, although expensive, noise-control procedure is to enclose the sound source in an acoustic enclosure or enclose the receiver in a personnel booth. The human ear is the usual noise receiver, and there are quite limited modifications that can be done for the person's ear. Hearing protectors (earplugs or acoustic muffs) can be effective in preventing noise-induced hearing loss in an industrial environment.

Sound transmitted in a room is a function of the sound source (e.g., HVAC fans, ducts, diffusers) and the acoustical absorption properties of the room. In a room, as shown in Figure 11.5, there is a direct path and one or more reflected paths. The sound field composed from all reflected sounds is the so-called reverberating field. Free-field theory states that the sound pressure level decreases by 6 dB per doubling the distance from a point source. When sound is transmitted from one room to another, the sound level is reduced by a quantity called the *noise reduction* (NR), the difference in dBs between the sound pressure level in the source room and the sound pressure level in the receiving room. When a sound falls on a wall, some energy is reflected and some is absorbed by the wall. The absorptive properties of the material are defined by the absorption coefficient, α, defined as the ratio of the energy absorbed by the surface to the incident energy to the surface. The values of α range from 0.01, almost all energy reflected, to 1.0 for full absorption, are dependent on the sound wave frequency. Sound pressure level is a special case of the sound intensity level, the sound propagation from a point source radiating in a spherical pattern, and can be generalized for propagation in a free field, for the English system, as:

$$SPL = L_W + 10 \cdot \log\left(\frac{1}{4\pi d}\right) + 10.3 \tag{11.20a}$$

and, for the SI system as:

$$SPL = L_W + 10 \cdot \log\left(\frac{1}{4\pi d}\right) + 0.3 \tag{11.20b}$$

Here, d is the distance from the point source expressed in the system unit. The two major sources of noise in architecture and buildings are: the *airborne noise*, or sound waves transmitted from a source to a receiver through the air (atmosphere), and the *structure-borne noise* caused by vibration from within the building due to footsteps or machines. Air-borne sound radiates from a source directly into and travels through the air. Structure-borne sound travels through solid materials usually in direct mechanical contact with the sound source, or from an impact on that material. The vibration is directly transferred through the building's structures and materials. However, all structure-borne sound must eventually become airborne sound in order to be heard. The sound of an airplane's engine is a kind of exterior airborne sound that may affect communities near airports. Subways and metropolitan operating systems are quiet, but the vibration of the train rail generates exterior structure-borne noise. In a gymnasium, cheering spectators may cause interior airborne noise to the surrounding rooms, while the athletes playing a game can generate interior structure-borne noise. Sound waves travel from the source to the receiver via multiple paths. Altering the paths of sound transmission is a common way to control noise in architecture. When sound strikes on a surface, part of the sound reflects off the wall surface back into the space, and part of it is transmitted through the wall to the space beyond. *Flanking transmission* of sound is the propagation of sound wave via building elements. Notice that an air gap can stop the transmission of sound effectively. When sound wave travels through air or materials, the transmitted sound intensity level is reduced due to sound power loss. All materials exhibit sound-insulating properties. Airborne sound transmission loss is a measure of the degree to which a material or construction can block or reduce transmission of sound from one area to another. The degree to which a material or construction is effective at blocking airborne sound is expressed as its sound transmission loss (STL) value. They only vary in their level of acoustic performance, which is measured in terms of *the sound transmission loss* or *sound reduction index (SRI)*, in dBs, defined by:

$$SRI = 10 \cdot log\left(\frac{P_{in}}{P_{tr}}\right) \tag{11.21}$$

Here, P_{in} is the incident power to one side of the sound barrier, wall, door, window, etc. (W), and P_{tr} is the transmitted power (W). Notice that, when two partitions or sound barriers are completely separated and isolated from one another, their ability to reduce sound is increased. Sound reduction index is used for the purpose of choosing the appropriate sound insulating performance for a building, indicating the sound energy reduction, when a sound wave passes through a material. An operable double-glazed window, for example, provides horizontally offset openings that allow natural ventilation while preventing the direct propagation of traffic noise. The narrow path between the double window panes dissipates the sound energy and lowers the noise levels. Noise barriers can redirect the noise paths away from receivers. They can be artificial (e.g., walls) or natural (e.g., a forest). One common application for noise barriers is to prevent traffic sounds from penetrating nearby neighborhoods or habitats. Quite often, the noise barriers are integrated with the building design. Two common ways to control noise are by reducing transmitted sound pressure level and through the redirection of the sound transmission paths to the receivers. The noise generated by a source can be prevented from reaching a worker by means of an obstacle to its propagation, conveniently located between the source and receiver, the concept of sound isolation. Ideally the obstacle could isolate the noise completely, but in practice, some of the noise always passes through a noise barrier. The amount by which the noise is reduced by the obstacle, in dB, depends on the material noise reducing properties (its transmission loss) and the acoustic properties of the room into which the noise is being transmitted. The transmission loss is defined as: $TL = 10 \cdot log10^{\tau}$, where the transmission coefficient, τ, is defined as the ratio of transmitted-to-incident energy (on the obstacle). If the receiving space is outdoors in a *free field*, the noise reduction is equal to the transmission loss (ignoring, for now, the transmission of sound around the edges of the partition). If the receiving space is indoors, the noise reduction is given by:

$$NR = TL - 10 \cdot log\left(\frac{A_{wall}}{S\overline{\alpha}}\right) \tag{11.22}$$

Here, A_{wall} is the surface area of the partition and $S\bar{\alpha}$ is the absorption of the receiving space. It can be seen from the above equation that the performance of the partition in reducing noise levels is improved as the absorption is increased in the receiving room. Transmission loss through a partition depends on the type of material of which it is made and it varies as a function of frequency. For usual industrial noise, the transmission loss through a partition increases by about 6 dB for each doubling of its weight per unit of surface area. Therefore, the best sound-isolating materials are compact, dense, and heavy.

Example 11.8

The sound pressure level on one side of a 3 m × 5 m wall is measured 95 dB in the 500 Hz octave band. If the wall transmission loss is 35 dB in this band and in the receiving room absorption is 100 m², what will be the sound pressure level in the receiving room?

SOLUTION

The noise reduction, computed with Equation (11.22) is:

$$NR = 35 - 10 \cdot \log\left(\frac{3 \times 5}{100}\right) = 35 - (-8.2) = 43.2 \text{ dB}$$

So the sound pressure level, in dBs in the receiving room is: 95 – 43.2 = 51.8 dB.

Room criteria (RC) and noise criteria (NC) in buildings are two widely recognized criteria used in the evaluation of the suitability of intrusive mechanical equipment noise into indoor occupied spaces. Speech interference level (SIL) is used to evaluate the adverse effects of noise on speech communication. NC curves have been used to set or evaluate suitable indoor sound levels resulting from the operation of building mechanical equipment. These curves give sound pressure levels (SPLs) as a function of the octave frequency bands. The NC curves within this total range may be used to set desired noise-level goals for almost all normal indoor functional areas. In a strict interpretation, the sound levels of the mechanical equipment or ventilation system under design should be equal to or lower than the selected NC target curve in all octave bands in order to meet the design goal. In practice, however, an NC condition may be considered met if the sound levels in no more than one or two octave bands do not exceed the NC curve by more than one or two decibels. RC curves, like NC curves, are currently being used to set or evaluate indoor sound levels resulting from the operation of mechanical equipment. RC curves differ from NC curves in three important aspects. First, the low-frequency range has been extended to include the 16 and 31.5 Hz octave bands. Secondly, the high-frequency range at 2000 and 4000 Hz is significantly less permissive, and the 8000 Hz octave band has been omitted because most mechanical equipment produces very little noise in this frequency region. Thirdly, the range over which the curves are defined is limited from RC 25 to RC 50 because: (1) applications below RC 25 are special-purpose, requiring an expert consultation, and (2) spaces above RC 50 require less concern for the background sound quality and the NC curves are more applicable. The noise SIL is the arithmetic average of the SPLs of the noise in the 500, 1000, and 2000 Hz octave bands. Once the sound source characteristics have been determined, then the sound level at any location within an enclosed space can be estimated. Outdoors, in the *free acoustic field* (no reflecting surfaces, except the ground), the sound pressure level (SPL) decreases at a rate of 6 dB for each doubling of distance from the source. In an indoor situation, however, all the enclosing surfaces of a room confine the sound energy so that they cannot spread out indefinitely and become dissipated with distance. Sound and noise transmission can be reduced by containment or absorption. Containment implies an enclosed space with sound barriers all around, preventing basically the sound transmission. This is not as simple as it sounds, requiring complex and expensive measures. Massive barriers contain the most frequencies, but low frequencies may be transmitted. Lightweight barriers transmit more frequencies and may have a fairly high natural frequency. The best barriers combine mass with sound-absorbing materials. For all enclosure types, the difficulties are posed by openings or penetrations through the barrier, such as openings, duct, and pipe. These are sound leaks, often conveying sound as though there were an actual opening. Notice that openings, even small holes or cracks, greatly limit a partition's noise-reduction characteristics.

11.5 SELECTION OF LIGHTING AND HVAC SYSTEMS

The selection of energy, lighting, heating, ventilating, and air-conditioning equipment and system structure and configuration is a complex process because of the wide variety of available systems and factors, such as climate patterns, energy policies, or regulations, etc. Choosing between a simple solution and an advanced alternative depends on the building size, facility destination and configuration, grouping of the building units, available technologies, and budget, among other factors. Lighting, energy, and HVAC systems control indoor climate, facilitate proper indoor lighting, and provide energy, hot water, and heat when and where needed. This means achieving a comfortable temperature, optimum humidity levels, illumination levels, needed electricity, and clean air via heating, cooling, humidification, air purification, lighting, ventilation, and electricity, heat, and hot water distribution. The proper choice of HVAC, lighting, or energy system depends on several factors: the size of the space that needs air, heat, energy, and lighting control, the size of the indoor space available to house equipment, the type of energy needed to fuel it, and its efficiency and effectiveness. However, the design, equipment selection, and system configuration is determined by building and owner requirements for controllability and function. For example, in HVAC these include: zoning, part- and low-load operation, and areas of varying operating schedules. Most commonly, poor control results not from poor controls, but from poor HVAC, lighting, or monitoring system design. It is strongly recommended that the designer review the HVAC, lighting, control, monitoring, or access equipment and device specifications to determine what controls/safeties are specified with each piece of HVAC or lighting equipment (AHUs, packaged units, chillers, VFDs, packaged pumping sets, luminaires, lamp types, and so on). It may be obvious that some of these controls or safeties should be eliminated and instead provided by the control system. All control systems should be documented with written sequences, parts lists, data sheets, damper and valve schedules, control drawings, lighting drawings, control sequence lists, marked site plan locations for all controls and remote devices, and points lists where applicable. All control systems require periodic maintenance, adjustments, calibration checks, and testing in order to stay in operation properly. Most often, these are performed on an annual or semi-annual basis at minimum.

HVAC, lighting, and even monitoring systems are used at almost all places (e.g., shopping mall, commercial complex, hotel, hospital, office, or low- or high-rise residential buildings). HVAC, monitoring, and indoor lighting and access control systems have become an essential part of almost all building services. A typical person in modern society spends about 90% of day time indoors, so it is not surprising that providing a healthy, comfortable indoor environment has become a vital factor in the current economy (McQuiston et al., 2006, or Pita, 2012). HVAC, lighting, monitoring, or access control systems are available in many shapes, characteristics, and variations. Even though the objective is common, i.e., to control the indoor environment, HVAC, lighting, monitoring, or access control systems differ significantly according to different applications, building, and ground conditions. So decision-making is a very critical function to select the right type of HVAC, monitoring, or lighting system. HVAC, lighting, or design engineers must consider various factors like application, special requirements of the process, space available, load variations, operating reliability, building structure and shape, energy conservation, equipment types, system classifications, occupant expectations and requirements, existing building installations, total cost, etc., while selecting the proper equipment and installations. For example, HVAC systems are broadly classified as all-air systems, all-water systems (hydronic), air-water systems, central systems, unitary systems, and single-zone or multizone systems, each with specific characteristics, suitable for that specific building, with advantages and disadvantages. The benefits and advantages of HVAC central system designs include: larger size equipment has often higher quality, efficiency, and durability; the system maintenance is concentrated; noise can be removed from the zone; and diversity allows lower installed capacity. The benefits for the distributed system designs include: thermal storage can be used; easy to provide zoning, room control, and direct control by occupants; easier to set independent scheduling for energy savings and conservation; usually such systems have lower initial investment, capital costs, and shorter lead time for equipment; no need for dedicated maintenance staff, allowing the use of a service contract; and quite often equipment can be installed on roof or other space that is less useable for other items. In the case of lighting, the choice of light sources has a decisive influence on the qualities of a lighting installation. This applies first and foremost to the technical aspects of the lighting, the costs for control subsystem that may be required, and the possibility of incorporating a lighting control system. The operating costs for a lighting installation depend almost entirely on the choice of lamps. It also applies similarly to the light quality that the lighting designer is aiming to achieve, e.g., the choice of luminous color to create the atmosphere in specific spaces, the quality of color rendering, or the brilliance and modeling necessary for display lighting. Lighting depends not only on the use of a specific lamp type; it is the result of the correlation of lamp, luminaire, and illuminated environment. Nevertheless, the majority of lighting qualities can be achieved only with the correct choice of light source. The decision to select a particular light source is dependent on the criteria defined by the required lighting effect and basic conditions pertaining to the project, cost, structure, existing installations, and occupant requirements.

11.6   CHAPTER SUMMARY

The development of BCS is affected by many technical, managerial, human, social, and economic factors. In fact, the trends of BCSs and intelligent building technologies in the past two decades were driven mainly by the advancement and blooming of information technology and communication networks. By integrating and applying Web-based control, field bus capabilities, and wireless and mobile technologies, the modern BCS is able to enhance its functionality and cost-effectiveness. Modern BCS technologies are much better and sophisticated, at costs that have been reduced substantially. It is expected that further enhancements on interoperable systems, building information diagnosis, building simulation and performance monitoring, together with innovative approaches to apply pervasive and ubiquitous computing to BCS are also possible. To achieve intelligent facility management and enhance the overall building performance, it is necessary to develop and implement the BCS in an effective manner, by considering appropriate technologies, actual project requirements, and imaginations. HVAC technology is considered one of the greatest engineering achievements of the twentieth century. It is ubiquitous and almost a necessity for occupant comfort, health, safety, and productivity. Air-conditioning is a process that simultaneously conditions the air, distributes it combined with air ventilation to the conditioned space, while at the same time controls and maintains the space's temperature, humidity, air movement, air cleanliness, sound level, and pressure differential within predetermined limits for the health and comfort of the occupants, for product processing, or both. The chapter began with some general discussion on building automation, control, building management systems, control types, electric controls, pneumatic controls, electronic (analog electronic), and direct digital controls. Next, lighting and HVAC control types and characteristics are discussed in some detail. Each type has a niche where it is a very good choice, but there is a general trend toward DDC controls. This chapter also introduced the HVAC control requirements, functionalities, and the important features of properly designed control systems for HVAC applications. Sensors, actuators, and signal processing and control methods have been described and presented. The first step in designing a building security system is to conduct a thorough survey of the entire facility for an existing building or a review of the plans of a new building. Then, a security philosophy can be developed by considering the degree of security required. One chapter section focuses on noise control, with a good presentation of acoustic and noise theory, concepts, and parameters. This is an introduction to the fundamentals of acoustics and noise control in buildings, not an in-depth treatment, but it introduces the readers, students, and designers to some important principles and terminology.

11.7 QUESTIONS AND PROBLEMS

1. What are the main functions of an energy management and control system?
2. List the HVAC main components and briefly describe the HVAC system major operational and maintenance strategies.
3. Briefly describe open-loop and closed-loop control systems. List the main advantages and disadvantages of each system.
4. What are the typical types of closed-loop control?
5. Repeat the calculation of Example 11.1, if the lighting system is replaced with a more efficient one, having nominal power of 5600 W and overall parasitic power of 350 W. (The rest of the parameters remain the same.)
6. Briefly describe the advantages of integrated building automation systems.
7. In a school building, the time for daylight usage is 2200 hours per year, time for non-daylight usage is 960 hours per year, occupancy factor is 0.85, constant illuminance factor is 0.90, daylight dependency factor is 0.8, and time when parasitic power is used is 280 hours. The lighting system power is 7200 W for classrooms, 1800 W for laboratory rooms, and 960 W for gym hall; the parasitic power is 720 W of the installed power; and on average the emergency power of 180 W is used 108 hours per year. Estimate the total annual energy use.
8. By using Equations (11.1), (11.2a), and (11.2b), and appropriate values for the parameters, estimate the annual lighting energy consumption in a work space or school building.
9. What are the primary objectives of a fire protection system?
10. What are the three elements of an HVAC process?
11. List and briefly describe the major types of HVAC control.
12. Which three components are commonly involved in a control system? How is the control system categorized?
13. What is the main difference between an open-loop control system and a closed-control system? Use a block diagram to elaborate.
14. A single-input direct-acting electronic controller has set-point temperature of 72°F, throttling range of 4.5°F, set-point voltage of 7.5 V, and voltage range of 3.6 V.
    a.   What is the output voltage of the controller at a temperature of 74°F?
    b.   If this controller is to control a normally closed damper, what is the position of the damper when output voltage of the controller is 6 V DC (full-voltage range)?
    c.   What temperature corresponds to that voltage?
15. List and briefly describe the HVAC control capabilities and expectations.
16. List the most common performance parameters that characterize a lighting control system.
17. In your own words, describe the manual lighting control strategy and options.
18. What are the advantages of HVAC digital control?
19. Estimate the output pressure of the pneumatic controller of Example 11.2, when the measured temperature is 83°F.
20. If the controller of Example 11.2 has a 72°F set point with an 8°F throttling range assigned, calculate the output pressures for measure temperatures of 71°F, 78°F, and 81°F, and the controller sensitivity.
21. If the set-point temperature is 72°F, the set-point pressure is 8.5 psig, the throttling range 12°F, the pressure range 12 psig, and the space temperature, T, is 76°F, calculate the controller output pressure.

22. If a system steam heating coil has a heat output that varies from 0 to 60 kW as the outlet air temperature varies from 35°C to 75°C in an industrial process, what is the coil gain and what is the throttling range? Find the equation relating the heat rate at any sensed air temperature to the maximum rate in terms of the gain and set point, assuming a steady-state operation. Plot the heat vs. sensed temperature.

23. What are the main advantages of variable air volume HVAC control methods?

24. What are the major advantages of automatic lighting control methods?

25. List and briefly describe the main components of lighting control systems.

26. What are the main components of lighting control architecture?

27. Briefly describe the lighting control levels.

28. Briefly describe the purpose of smoke-control methods.

29. Briefly describe the advantages of lighting control.

30. What is a manual fire alarm system? How does it operate?

31. Which is a feature of an energy-efficient lighting control system?
    a. All luminaires remain switched on during the work day as frequent starting uses more energy.
    b. Lamp deterioration increases with frequent switching, so lighting should remain on continuously.
    c. Occupancy sensors are programmed to keep lights on for half an hour after occupants leave.
    d. Timed switch-off controller minimizes lighting use after occupation ceases.

32. List and briefly describe the most common fire and smoke detectors.

33. Compute the sound pressure level for loud rock music 2 Pa, average street noise 0.06 Pa, average office noise 0.02 Pa, quiet residential street 0.006 Pa, and very quiet home radio 0.002 Pa.

34. Repeat the estimate of the sound intensity of Example 11.5, but of air density of 0.95, 1.05, 1.10, and 1.15 kg/m^3, as in the case of a drier climate.

35. Repeat the estimates of Example 11.6, but for an acoustic power emitted by the point sound source of 10^{-2} W and at a distance of 12 m. Compute also the intensity level.

36. Briefly describe the types of noise propagation in a building.

37. Repeat Example 11.8, but for an increase in transmission loss by 6 dB and receiving room absorption to 110 dB.

REFERENCES AND FURTHER READINGS

1. D. Cook and S. K. Das, *Smart Environments: Technologies, Protocols, and Applications*, Wiley-Interscience, Hoboken, NJ, 2005.
2. IEA Report- 2013, Key World Energy Statistics, International Energy Agency, Paris, France, 2013.
3. J. F. Kreider (ed.), *Handbook of Heating, Ventilation, and Air-conditioning*, CRC Press, Boca Raton, FL, 2000.
4. R. W. Haines, *Control Systems for Heating, Ventilating, and Air-conditioning* (4th ed.), Van Nostrand Reinhold, New York, NY, 1987.
5. J. F. Kreider, A. Rabl, and P. S. Curtiss, *Heating and Cooling of Buildings* (2nd ed.), McGraw-Hill, New York, NY, 2001.
6. J. Andreas, *Energy-Efficient Motors: Selection and Application*, Marcel Dekker, New York, 1992. Zumtobel Lighting GmbH, The Lighting Handbook (6th ed.), 6851 Dornbirn, AUSTRIA, 2018. https://www.zumtobel.com/PDB/teaser/EN/Lichthandbuch.pdf
7. G. Rosenquist, M. McNeil, M. Lyer, S. Meyers, and J. McMahon, Energy Efficiency Standards for Equipment: Additional Opportunities in the Residential and Commercial Sectors, LBNL, 2006.
8. ISO 16484-1, Building Automation and Control Systems (BACS)—Part 1: Project specification and implementation, 2010
9. ISO, 2005b. ISO 16484-3: Building Automation and Control Systems (BACS)—Part 3: Functions, International Organization for Standardization, Switzerland, 2005.

10. ISO, 2004. ISO 16484-2: Building Automation and Control Systems (BACS)—Part 2: Hardware, International Organization for Standardization, Switzerland, 2004.
11. ISO 16484-5, Building Automation and Control Systems (BACS)—Part 5: Data communication protocols, 2017.
12. ISO/IEC 14543-3-1, Information Technology - Home Electronic Systems (HES) Architecture—Part 3-1: Communication layers - Application layer for network based control of HES Class 1, 2006.
13. ISO/IEC 14908-1, Information Technology - Control Network Protocol—Part 1, 2012.
14. P. Domingues, P. Carreira, R. Vieira, and W. Kastner, Building automation systems: Concepts and technology review, *Comput. Stand. Interf.*, Vol. 45, pp. 1–12, 2016.
15. IEA 2001. IEA Task 31 - Daylighting Buildings in the 21st Century, 2001; http://www.iea-shc.org/task31/.
16. NBI 2003. *Advanced Lighting Guidelines*, New Building Institute Inc., 2003.
17. O. Gassmann and H. Meixner, *Sensors in Intelligent Buildings*, Wiley-VCH Verlag GmbH, Weinheim, Germany, 2001.
18. UL 521, *Standard for Safety Heat Detectors for Fire Protective Signaling Systems*, Underwriters Laboratories Inc., Northbrook, IL, 1993.
19. NFPA 72®, National Fire Alarm Code®, National Fire Protection Association, Quincy, MA, 2007.
20. F. Kreith and D. Y. Goswami (eds.), *Energy Management and Conservation and Handbook*, CRC Press, Boca Raton, FL, 2008.
21. R. Montgomery and R. McDowall, *ASHRAE Fundamentals of HVAC Control*, American Society of Heating, Refrigerating and Air-Conditioning Engineers, Inc., Atlanta, 2011.
22. D. Mumovic and M. Santamouris, *A Handbook of Sustainable Building Design & Engineering, An Integrated Approach to Energy, Health and Operational Performance*, Earthscan, 2009.
23. NFPA 75, Standard for the Fire Protection of Information Technology Equipment, National Fire Protection Association, Quincy, MA, 2011.
24. A. Sumper and A. Baggini, *Electrical Energy Efficiency: Technologies and Applications*, Wiley, 2012.
25. G. Petrecca, *Energy Conversion and Management - Principles and Applications*, Springer, Cham Heidelberg New York Dordrecht London, 2014.
26. G. W. Gupton, Jr., *HVAC Controls Operation and Maintenance* (3rd ed.), Fairmont Press, Lilburn, GA, 2002.
27. F. C. McQuiston, J. D. Parker, and J. D. Spitler, *Heating, Ventilation, and Air-conditioning: Analysis and Design* (6th ed.), Wiley, 2006.
28. E. G. Pita, *Air-conditioning Principles and Systems* (4th ed.), Prentice Hall-India, 2012.
29. R. N. Helms and M. C. Belcher, *Lighting for Energy Efficient Luminous Environments*, Prentice Hall, 1991.
30. R. F. Barron, *Industrial Noise Control and Acoustics*, Marcel Dekker, Inc., 2001.
31. Zumtobel Staff, *The Lighting Handbook*, Austria, 2004, http://www.zumtobelstaff.com (accessed 2005).
32. G. Rosenquist, M. McNeil, M. Lyer, S. Meyers, and J. McMahon, *Energy Efficiency Standards for Equipment: Additional Opportunities in the Residential and Commercial Sectors*, LBNL, 2006.
33. W. C. Turner and S. Doty (eds.), *Energy Management Handbook* (6th ed.), The Fairmont Press, 2007.
34. W. C. Turner, *Energy Management Handbook*, John Wiley and Sons, 2017.
35. P. Sansoni, L. Mercatelli, and A. Farini, *Sustainable Indoor Lighting*, Springer, 2015.

12 Energy Management, Energy Conservation, and Efficiency

12.1 INTRODUCTION

A typical building is designed for a 40-year to 60-year or even longer economic life, which implies that the existing inventory of buildings, with all their good and bad features, is turned over very slowly. In modern society, it is critical to design a new building to be cost effective, with a high degree of energy efficiency, because the savings on operating and maintenance costs will repay the higher initial investment many times over the building lifetime. In the last three decades, many technological advances have resulted in striking reductions in the energy required to operate buildings safely and comfortably. An added benefit of these developments is the reduction in air pollution, which has occurred as a result of generating less electricity. There are large varieties, even hundreds, of building types, and the buildings can be categorized in several ways: by use and designation, type of construction, size, structure, thermal characteristics, etc. In general, based on their use, the two main designations are: residential and nonresidential. The residential category includes features common to single-family dwellings, apartments, and hotels, while the nonresidential category mainly emphasizes office buildings, but also includes features common to retail stores, hospitals, restaurants, and laundries. Most of this space is contained in buildings larger than 10,000 sq. ft. Industrial facilities include several types with large varieties in terms of size, structure, electromechanical systems, etc. The extension to other types is either obvious, or can be pursued by referring to the literature. The industrial sectors in the United States, and any developed or developing countries, are highly diverse, consisting of manufacturing, mining, power, agriculture, and construction activities. In the United States, for example, the industrial sector consumes one-third of the nation's primary energy use, at an annual cost of over $100 billion. Industry energy-efficiency improvements and energy conservation require emphasis on energy-management activities, making capital investments in new plant and industrial processes, equipment, and facility improvements. Reducing the energy cost per unit of manufactured product is one way that the United States can become more competitive in the global market.

Electricity is an intermediate product, generated by using primary energy sources (oil, natural gas, coal, hydro-power, or nuclear energy), converted to electrical energy and transported via power transmission and distribution networks to consumers. Established electrical power systems, developed over the past century, feed electrical energy from large generation units through step-up transformers and high-voltage interconnected electrical transmission systems, substations, and power subtransmission and distribution networks to consumers or end users. Consumers purchase the electricity as an intermediate step toward final non-electric products. Large-scale electricity usage is due to factors such as: easy to generate and transmit over long distances at very high efficiency, and easy to distribute to consumers at any time, any amount, and almost anywhere, for a large variety of uses (heat, light, electrical motors, computers, communication, etc.). In order to meet the power operation criteria (safety, reliability, quality, economy, and supply security) within grid operations, the following tasks are performed: maintain accurate load-generation and reactive power balances to control the voltage profile, maintain an optimum generation schedule to control costs and the environmental impacts, and ensure the network's security against credible contingencies, requiring network protection against reasonable outages, equipment failure, or unauthorized interventions. In terms of electricity uses, the single most significant residential end-use of electricity is space conditioning, with heating and hot water using over 50% of the electricity, and lighting counting for about 10%. The combination of other uses, such as entertainment systems, personal computers, printers, etc., is also substantial, accounting for one-quarter of residential electricity use. In the commercial sector, space conditioning, the heating, ventilating, and air-conditioning (HVAC), uses about 40% of the electricity. Space cooling accounts for the majority of space conditioning electricity use; indeed, by itself, space cooling represents one-quarter of all commercial electricity use. The next two largest end users of electricity in the commercial sector are lighting with about 25% and office equipment at 20% (Kreith and Goswami, 2008; Sumper and Baggini, 2012).

Energy, an important company or organization resource, must be managed and controlled by a systematic method in collaboration with the management of other resources. Minimizing the energy uses in buildings and industrial processes is an important aspect of sustainability. Energy management in buildings is concerned with maximizing the use of energy resources while providing the desired environmental conditions and services inside the building at the least cost. Energy management can also be applied to industrial processes, particularly those requiring the use of steam or heat, such as textile manufacture and steel making, but this chapter is mainly concerned with the use of energy in buildings to provide thermal, visual, and acoustical comfort for the occupants. There are opportunities for savings by reducing the unit price of purchased energy, and by improving the efficiency and reducing the use of energy-consuming systems. Energy management procedures must be as simple, specific, and direct as possible. However, very large and complex facilities, processes, or buildings, such as hospitals or university campuses, industrial complexes, or large office buildings, usually require systematic efforts and processes. Energy used in buildings and facilities composes a significant amount of the total energy used for all purposes, and thus affects energy resources. Energy management in the form of implementing new energy efficiency technologies, new materials, and new manufacturing processes and the use of new technologies in equipment and materials for business and industry is also helping companies improve their productivity and increase their product or service quality. Often, the energy savings is not the main driving factor when companies decide to purchase new equipment, use new processes, and use new high-tech materials. However, the combination of increased productivity, increased quality, reduced environmental emissions, and reduced energy costs provides a powerful incentive for companies and organizations to implement these new technologies. The effectiveness of energy utilization varies with specific industrial operations because of the diversity of the products and the processes required manufacturing them. The organization of personnel and operations involved also vary. Consequently, an effective energy management program should be tailored for each company and its plant operations.

Energy management refers to the administration of all energy forms used in the company or facility making an optimum program of purchasing, generating, and consuming various energy types based on the company or facility overall short-term and long-term management programs, with due consideration of costs, availability, economic factors, etc. The main goal of energy management is to produce goods and provide services with the least costs and environmental effects. Building energy management controls energy use and cost while maintaining indoor environmental conditions to meet comfort and functional needs. Good energy management has the goal of reducing energy expenses to the lowest level possible without affecting the comfort, productivity, or functionality of the users or occupants. Energy efficiency improvements need not sacrifice any facility functionalities. The most important are to save energy, to minimize energy costs and waste without affecting living quality, and finally to minimize environmental effects. Energy management systems control the equipment energy usage, making them operate efficiently and effectively. Energy management systems save 10% or more of overall annual building energy consumption, influencing several aspects of operation and activities including the energy cost profitability, the market competitiveness, national energy supply-demand balance, trade and financial health, environment, occupational safety, loss prevention and waste reduction, productivity, and quality. Energy management in the form of implementing energy-efficient technologies, manufacturing processes, new technologies in equipment and materials also helps companies improve their productivity and increase their product or service quality. Energy management is about reducing the energy costs used by an organization, now with the added spin of minimizing emissions as well. Reducing energy costs has two facets: price and quantity. Energy efficiency can be defined as utilizing minimum amounts of energy for heating, cooling, lighting, and the equipment that is required to maintain facility needed conditions. An important factor impacting energy efficiency is not only the building energy envelope but also the management of energy within the premises. The energy consumed varies depending on the building or facility design, the available electrical systems, and how these systems operate. Heating and cooling systems consume the most energy in a building or facility, and the control system of the building energy management systems can significantly reduce the energy use of these systems. There are large opportunities to improve energy use efficiency by eliminating waste through process optimization. Applying today's computing and control equipment and techniques is one of the most cost-effective and significant opportunities for larger energy users to reduce their energy costs and improve profits. An energy management system (EMS) is an important element of a comprehensive management program, providing relevant information to key individuals and departments, which enables them to improve energy performance and uses.

12.2 BUILDING ENERGY CONSERVATION AND EFFICIENCY

Energy efficiency, energy conservation, and renewable energy technologies are the pillars of a sustainable energy policy. Among the many reasons to pursue energy conservation, efficiency, and renewable energy use are: increasing and diversifying our energy supply, making the existing energy supply last longer, and having a cleaner and less affected environment. Greater energy efficiency and higher levels of energy conservation are critical issues in sustainable development. For example, increased energy efficiency and energy conservation provide significant economic, environmental, and energy security benefits in any country. Higher energy efficiency translates directly into lower costs of energy, products, and services and a higher life standard. Energy efficiency and conservation methods can also prolong the equipment life. The two concepts are closely related, but they have different connotations. Efficiency involves using less energy to achieve the same results, while conservation stresses using less energy even though there may be a compromise on the results. An energy balance is a set of relationships accounting for all the energy that is produced and consumed, and matches inputs and outputs in a system over a given time period. The system can be anything from a whole country to an area to a process in a factory. An energy balance is usually made with reference to a year, though it can also be made for consecutive years to show variations over time. Energy balances provide overviews, and are basic energy-planning tools for analyzing the current and projected energy situation. The overviews aid sustainable resource management, indicating options for energy saving, or for policies of energy pricing and redistribution, etc. An energy-conversion device or equipment, as represented in Figure 12.1, has efficiency defined as the ratio of the energy output to the energy input:

$$\text{Equipment (Device) Efficiency} = \frac{\text{Useful Energy Output}}{\text{Energy Input}} \tag{12.1}$$

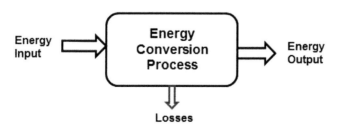

FIGURE 12.1 Schematic diagram of energy-conversion equipment or process.

The meaning of the word *useful* depends strongly on the device or equipment purpose. Energy efficiency of a device, equipment, or process is a quantitatively dimensional value between 0 and 1, with higher values indicating higher efficiency. The concept of efficiency embodies the laws of thermodynamics. On the other hand, even though almost of the energy on our planet comes directly from the sun, we are not yet able to harness large-scale solar energy directly and efficiently. Instead, for the most part, we have to rely on the chemical energy of fossil fuels for most of our energy needs. The problem with chemical energy is that it is a potential energy form and must be converted to other energy forms before it can be used. The only way to exploit stored solar energy is to release it by burning fossil fuels, through the combustion process, and eventually convert it into mechanical, electrical, or other energy forms. The conversion processes of the stored energy involves two or more conversion devices, each with its own efficiency, so it is useful to introduce the concept of *system efficiency*. A system, consisting of two or more devices or subsystems, means a well-defined space in which two or usually more than two energy conversions take place. The efficiency of a system is equal to the product of efficiencies of the individual devices or subsystems, and the concept is exemplified in the following example (Belu, 2018).

Example 12.1

Calculate the overall efficiency of a power plant if the efficiencies of the boiler, steam turbine, generator, and step-up transformer are 90%, 42%, 96%, and 98%, respectively.

SOLUTION

The overall (system) efficiency of the power plant is the product of the conversion subsystem efficiencies.

$$\eta_{Power\ Plant} = \eta_{Boiler} \times \eta_{Turbine} \times \eta_{Generator} \times \eta_{Transformer} = 0.90 \times 0.42 \times 0.96 \times 0.98$$
$$= 0.3556 \text{ or } 35.56\%$$

Note that the system efficiency is lower than any one of the efficiencies of the individual components of the system. In the case of this electric power plant, only 35% of the chemical energy input is converted to electricity, while the rest is lost to the environment, mostly as heat. However, if part of the wasted heat is recovered via combined heat and power (CHP) methods, the overall power plant (system) efficiency can be significantly improved.

The building sector accounts for up to 40% of the total energy uses in the United States, European Union, Canada, and most other developed countries. However, at the same time, building sectors have a documented cost-effective savings potential of up to 80%, which can be effected over the next 40 years. In order to ensure energy conservation and savings, while also using renewable energy in an optimal way, building-integrated energy technologies are required. Energy demands in the commercial, industrial, residential, and utility sectors vary on a daily, weekly, and seasonal basis. These demands are matched by various energy-conversion systems, operating synergistically. Peak hours are the most difficult and expensive to supply, being met by conventional gas turbines or diesel generators, which are reliant on costly oil or natural gas. Energy storage provides alternative ways to supply peak energy demands or to improve the operation of distributed generation (DG), solar, and wind facilities. Energy storage also plays significant roles in energy conservation. In processes where there is a large recoverable wasted energy, the storage can result in savings of premium fuels. Thermal energy storage (TES) is a key technology for energy conservation, being of great importance in energy sustainability, and a well-suited technology for hot water, heating, and cooling thermal applications, which can be easily embedded into DG, contributing significantly to the overall system efficiency improvement. TES deals with the stored energy by cooling, heating, melting, solidifying, or vaporizing a material, while the thermal energy becomes available when the process is reversed. Large TES systems are employed for applications ranging from solar hot water storage to building air-conditioning systems. However, advanced TES technologies have only recently been developed to a point that they have significant impacts on modern industries. TES appears to be an important solution to correcting the mismatch between energy supply and demand. TES can contribute significantly to meet the needs for more efficient, environmentally benign energy use. TES is a key component of many thermal systems, and a good TES should allow very little thermal loss, leading to energy savings, while permitting the highest reasonable extraction efficiency of the stored thermal energy.

Green buildings (green construction structures or sustainable buildings) and green building processes are structures and methods that are environmentally responsible and resource-efficient throughout the building life cycle: from siting to design, construction, operation, maintenance, renovation, and demolition. Such practice expands and complements the conventional building design concerns of economy, utility, durability, and comfort. Although new technologies are constantly being developed to complement current practices in creating greener structures, the common objective is that green buildings are designed to reduce the overall impact of the built environment on human health and the natural environment by: efficiently using energy, water, and other resources; protecting occupant health; improving employee productivity; and reducing waste, pollution, and environmental degradation. A similar concept is natural building, which is usually on a smaller scale and tends to focus on the use of natural materials that are available locally. Other related topics include sustainable design and green architecture. Sustainability may be defined as meeting the needs of present generations without compromising the ability of future generations to meet their needs. Modern sustainability initiatives call for an integrated and synergistic design to both new construction and in the retrofitting of an existing structure. Also known as sustainable design, this approach integrates the building life cycle with each green practice employed with a design purpose to create a synergy among the practices used. Green building brings together a vast array of practices and techniques to reduce and ultimately eliminate the impacts of buildings on the environment and human health. It often emphasizes taking advantage of renewable resources, e.g., using sunlight through passive solar, active solar, and photovoltaic techniques, and using plants and trees through green roofs, rain gardens, and for reduction of rainwater run-off. Green buildings often include measures to reduce energy consumption, both the embodied energy required to extract, process, transport, and install building materials and operating energy to provide services such as heating and power for equipment.

12.2.1   BUILDING APPLICATIONS OF THERMAL ENERGY STORAGE SYSTEMS

Thermal energy storage (TES) is considered one of the most important advanced energy storage technologies, and, recently, increasing attention has been dedicated to the utilization of such an essential technique for thermal applications ranging from heating to cooling, or energy savings, particularly in the building and industrial process sectors. TES is a key ingredient of many successful building and industrial energy systems, minimizing thermal energy losses, enhancing energy conservation and savings, and permitting the highest appropriate extraction efficiency of the stored energy. The design and selection criteria for TES systems are examined in this section. Further, energy-saving techniques and applications are discussed and highlighted with illustrative examples. TES is often considered to be an *advanced energy technology*, and there has been increasing interest in using this essential technology for thermal applications such as hot water, space heating, cooling, air-conditioning, and so on. TES systems have enormous potential for permitting more effective use of thermal energy equipment and for facilitating large-scale energy substitutions. The resulting benefits of such actions are especially significant from an economic perspective. In general, a coordinated set of actions has to be taken in several energy system sectors for the maximum TES potential benefits to be realized. TES appears to be the best means of correcting the mismatch that often occurs between the supply and demand of thermal energy. The first step of a TES project is to determine the energy load profile of the building. Parameters influencing the building demand and load profile are the building use, internal loads, and the climatic conditions. Following steps are to determine the type and amount of storage appropriate for the particular application; the effect of storage on system performance, reliability, and costs; and the storage systems or designs available. It is also useful to characterize the TES types in relation to the storage duration. Short-term storage is used to address peak loads lasting from a few hours to a day in order to reduce the system size, taking advantage of the energy-tariff daily structures. Long-term storage is used when waste heat or seasonal energy loads can be transferred with a delay of a few weeks to several months. Related to the energy storage amount required, it is important to avoid both undersized and oversized systems. Undersizing the systems results in poor performance levels, while oversizing results in higher initial costs and energy waste if more energy is stored than is required. The effect of TES on the overall energy system performance should be evaluated in detail. The economic justification for storage systems requires that the annualized capital and operating costs for TES be less than those required for primary generating equipment supplying the same service loads and periods. TES systems are an important element of many energy-saving programs in a variety of sectors, including residential, commercial, industrial, utility, and transportation. TES can be employed to reduce energy consumption or to transfer an energy load from one period to another. The consumption reduction can be achieved by storing excess thermal energy that would normally be released as waste, such as heat produced by equipment, appliances, lighting, and even by occupants. Energy-load transfer is achieved by storing energy at a given time for later use, and can be applied for either heating or cooling capacity. The consumption of purchased energy can be reduced by storing waste or surplus thermal energy available at certain times for use at other times. The demand for purchased electrical energy can be reduced by storing electrically produced thermal energy during off-peak periods to meet the thermal loads that occur during high-demand periods. There has been an increasing interest in the reduction of peak demand or transfer of energy loads from high- to low-consumption periods. The use of TES can defer the need to purchase additional equipment for heating, cooling, or air-conditioning applications and reduce equipment sizing in new facilities. The relevant equipment is operated when thermal loads are low to charge the TES, and energy is withdrawn from storage to help meet the thermal loads that exceed equipment capacity.

Energy storage density, in terms of the energy per unit of volume or mass, is an important factor for optimizing solar ratio (how much solar radiation is useful for the heating/cooling purposes), efficiency of appliances (solar thermal collectors and absorption chillers), and energy consumption for space heating or cooling room consumption. Therefore, the possibility of using phase-change materials (PCMs) in solar system applications is important to investigate. PCMs might be able to increase the energy density of small-sized water storage tanks, reducing storage volume for a given solar fraction or increasing the solar fraction for a given available volume. It is possible to consider thermal storage on the hot and/or cold side of the building, allowing the hot water storage from the collectors and the auxiliary heater to be supplied to the generator of the absorption chiller (in cooling mode) or directly to the users (in heating mode). The latter allows the storage of cold water produced by the absorption chiller to be supplied to the cooling terminals inside the building. It is usual to identify three situations as "hot," "warm," and "cold," storage based on the different temperature ranges. Typically, a hot tank may work at 80°C to 90°C, a warm tank at 40°C to 50°C, and a cold tank at 7°C to 15°C. While heat storage on the hot side of solar plants is always present because of heating and/or domestic hot water (DHW) production, cold storage is justified in larger plants. Cold storages are used not only to gain economic advantages from lower electricity costs (in the case of electric compression chillers) depending on the time of day but also to lower the cooling power installed and to allow more continuous operation of the chiller. In its early uses, thermal storage could not provide effective backup but did help the system to thermally stabilize. Consequently, thermal storage found use in solar-assisted thermal systems. Since then, increased interest in TES technologies, as well as the usability and effects of both sensible and latent heat storage in numerous applications, has led to a number of reviews. These reviews focused only on one side (cold or hot) or component of the system or one of its integral mechanisms.

Two key issues to consider when one is trying to save energy are: (1) design efficiency reflects maximum efficiency and cannot be achieved in actual production, and (2) energy-saving technology resolves the issue when the *equipment is operating in the lower efficiency zone*, but it is not a remedy for poorly maintained equipment. Equipment maintenance is thus vital and should take place on a regular basis, as per the requirements of its maintenance manual. To enhance energy savings on equipment, managers and engineers should regularly inspect the condition of equipment to avoid energy waste caused by poor maintenance, such as heat transfer losses, blockages in fuel-transfer systems, overloading of lubrication systems, and other issues. One company found a filter in a compressed-air system was blocked, causing a 146 kW overload of the power generator. The air generated by the compressor, meanwhile, amounted to only 18 m/min. Equipment design specifications can sometimes cause lower energy efficiency; if this is the case, consider whether the equipment can be replaced with a type that has better characteristics. As an alternative to replacing equipment, companies can install energy-saving devices. Although doing this saves energy to some extent, it does not reduce costs or overall energy consumption. The example below demonstrates this point. Consider adding a frequency converter to increase the frequency cycle of fans or pumps operating in the allowable efficiency zone, to improve systems with fluctuating pressures, or with compressors running constantly.

As previously stated, the concept of TES is not new and was used a few centuries ago to cool churches using blocks of ice that were stored in the cellar. In recent decades, TES technologies have been effective in reducing the operating costs of cooling plant equipment. By operating the refrigeration equipment during off-peak hours to recharge the storage system and discharging the storage during on-peak hours, a significant fraction of the on-peak electrical demand and energy consumption is shifted to off-peak periods. Cost savings are realized because utility rates favor leveled energy consumption patterns. The variable energy rates reflect the high cost of providing energy during relatively short on-peak periods. Hence, these rates constitute an incentive to reduce or avoid operation of the cooling plant during on-peak periods by the cool storage system. A large differential between on- and off-peak energy and peak consumption rates generally makes cool storage systems economically feasible.

Some electric utility companies actually encourage the use of TES systems to reduce the cost required to generate on-peak electric power. Indeed, the need to build new generation plants in order to meet the demand during on-peak hours can be eliminated by promoting the use of off-peak power. Many utilities have initiated different rate structures to penalize the use of electric power during on-peak periods. In addition to the differential charges for on-peak versus off-peak energy rates, the utilities have imposed demand charges, based on the monthly peak energy demand. These demand charges are intended to recover fixed costs such as investing in new generation plants, and transmission and distribution lines. In the United States, it is estimated that 35% of the electrical peak demand is due to cooling. Therefore, the use of TES systems can be a good alternative to delay the use of chillers to meet space cooling, especially for commercial buildings. TES systems can also be used to reduce the thermal or cooling equipment size and thus the initial cost of the cooling equipment especially in applications where peak loads occur only for a limited period during a year, such as churches, community centers, etc.

Example 12.2

A church has a peak cooling load of 360 kW over a 4-hour period that occurs once a week, Sunday. The local utility charges 20$/kW. What is the saving using a lower-power chiller in 12-hour operation interval? Select two variants of the lower power chiller.

SOLUTION

Instead of using a chiller of 360 kW to operate for 4 hours in order to provide the required 1440 kWh of cooling load, a 144 kW cooling system can be installed. The same cooling load (1440 kWh) is produced by operating the 144 kW cooling system for a longer period, at least a 10-hour period to account for any storage losses. A 210 kW needs to work about 7 hours to provide the required energy. However, the billing savings for the two choices are:

$$\text{Power Charge Saving \#1} = 20 \times (360 - 144) = \$4320.0$$

and, respectively

$$\text{Power Charge Saving \#2} = 20 \times (360 - 210) = \$3000.0$$

Notice that due to the lower compressor efficiency and the TES losses, cooling systems with TES may actually consume more energy than cooling plants without TES systems. However, TES systems if designed and controlled properly can reduce the overall operating costs of cooling plants. To determine the potential operating cost reduction, the auditor should carefully consider the various factors that affect the design and operation of TES systems. In the following sections, an overview of the types of TES systems as well as factors that affect both the design and operation of cooling plants with TES systems is provided. Finally, some simplified calculation examples as well as results of parametric analyses, reported in the literature, are presented to illustrate some typical cost savings incurred due to adequately installing and properly controlling TES systems for space cooling applications.

12.2.2 Energy Conservation, Savings, and Balances

The energy used in residential and commercial buildings provides many services, including weather protection, thermal comfort, communications, facilities for daily living, aesthetics, healthy work environment, etc. Today, people spend most of their time inside buildings, so the quality of the indoor (or built) environment is important to their comfort. Good thermal performance of buildings is important for energy efficiency as well as productivity of workers in commercial buildings. Until recent decades, energy efficiency, conservation, and savings have been relatively low priorities and low perceived opportunities to building sectors. However, with the dramatic increase in awareness of energy-use concerns, new energy regulation requirements, and advances in cost-effective technologies, energy efficiency, conservation, and savings are fast becoming part of building management, industrial processes and facilities management, and operations strategy. Energy saving also constitutes one of the primary measures for the protection of the environment. One method uses the reduction of exchange efflux, which is used to purchase polluting fossil fuels, mainly oil or oil-based fuels. The concepts are also now making significant inroads into domestic residential house-building sectors. For example, for lighting only, the energy savings can be up to 75% of the original circuit load, which represents 5% of the total energy consumption of the residential and commercial sectors. Energy-savings potential from water heating, cooling, or hot water production, can be up to 10%, which represents up to 7% of the total energy consumption of the domestic residential and commercial sectors. Energy efficiency is about getting the same or better service from less energy. This contrasts with energy conservation, which generally involves doing less with less. This latter approach is no longer seen as acceptable practice—life has to go on, but we need to be more careful about how we use resources to get the best results for ourselves and future generations. The critical issue for energy efficiency and energy management is to identify the services that are needed and make sure that these are being provided cost-effectively and for the least energy uses This is a two-pronged approach that means we have to assess what services (and quality of service) we really need, as well as the energy associated with their provision. System predictive control models (SPCM) can offer significant energy-saving potentials in large buildings, facilities, and systems for which response to external factors, disturbances, or control inputs is slow, having high-inertia responses, i.e., on the order of hours or so. SPCM can effectively provide a means to optimize systems dynamically to take advantage of building utilization, local weather patterns, and utility rate structures. Dynamic system models could be used to identify critical control variables (from energy performance and use standpoints), guide facility operation, and generate a predictive control model that reduces energy use. Modern HVAC systems and building hydronic systems are becoming prevalent and robust systems, but often multivariable control methods are lacking and are not used. With the visualization capabilities providing comparison of performance metrics to previous years, building models, and benchmarks, the energy and cost savings of a particular action can be assessed with good certainty.

An energy balance is a set of relationships accounting for all the energy that is produced and consumed, and matches inputs and outputs in a system over a given time period. The system can be anything from a whole country to an area to a process in a factory. An energy balance is usually made with reference to a year, though it can also be made for consecutive years to show variations over time. Energy balances provide comprehensive overviews of energy uses, and are the basic energy planning tools for analyzing the current and projected energy situation. The overviews aid sustainable resource management, indicating options for energy saving, or for policies of energy pricing and redistribution, etc. A number of different types of energy balances can be made, depending on the information you need: the energy commodity account, the energy balance, and the economic balance. The energy commodity account includes all flows of energy carriers, from the point of extraction through conversion to end use, in terms of their original, physical units such as kilotons of coal and GWh of electricity. The energy balance is similar to the energy commodity account, except for the fact that all physical units are converted into a single energy unit (e.g., TJ, ktoe, ktce, etc.). This type of balance uses mean energy content values because of the inevitable variations in fuel composition, especially with coal. This means that there are inherent inaccuracies within the balance, but as long as they are within accepted limits, this is common practice. The conversion ratios used always need to be noted with the balance. In the economic balance, different forms of energy are accounted for in terms of their monetary value. However, variations in currency rate, subsidies, taxes, etc. make it difficult to compare different energy forms, especially between countries.

An important step in preparing an energy balance is the model of the energy chain to trace the energy flows within a system, a building, a facility, or a process, starting from the supply primary source(s) through the conversion, transformation, and transportation processes, to final/delivered energy, and finishing with end use. If the energy balances are constructed in terms of primary energy without taking into account conversions or transformations, then incorrect conclusions may result. The most common example is electricity, a secondary energy form. Electricity is usually included in a primary energy balance either on the basis of the amount of fossil fuel needed to produce it, or, when the electricity is generated from hydro, nuclear, or renewables, an energy equivalent or heat content is used. However, it is not simply a matter of taking a conversion of one single fuel to another, because the conversion efficiency varies with the primary energy source. For example, for hydropower it is around 90% whereas for coal it is around 40%. The amount of energy required to produce the electricity needs to reflect these differences. To produce one unit of electricity would take at least twice the amount of energy (on a joule basis), by using coal as by using hydro. An energy balance should include the following concepts and terms. For commercial energy, it must clearly specify which energy forms are included in commercial and in noncommercial energy categories. Noncommercial energy usually includes the biomass (e.g., woody and agricultural residues) and other bioenergy forms. Despite their important contribution to the energy supply, particularly in rural areas, these sources do not usually appear in national energy balances. They are difficult to quantify physically because they are traded and used in nonstandard units and also they fall outside of the monetized economy, so their flows are often not monitored. The non-energy products from primary energy carriers such as petrochemicals from crude oil, coal, and natural gas, should be listed separately. The energy imports, exports, bunkers, stock changes, transformation (conversion of one fuel to another), distribution, and conversion losses, as well as, if relevant, self-consumption by the energy industry, should all be included. An energy balance is usually constructed from two sides: (1) from end uses back to total primary energy consumption, and (2) from resource extraction to primary energy supply. A statistical difference is sometimes included to balance for inaccuracies in supply and demand, for example, due to evaluation of losses, while such statistical differences can be as high as 10%.

For many buildings, the envelope (i.e., walls, roofs, floors, windows, and doors) can have important impacts on the energy used to condition the facility, affecting the energy conservation and efficiency. During an energy audit, it is critical to determine the actual building envelope characteristics. During the survey, a description of the building envelope should be established to include information such as construction materials (for instance, the level of insulation in walls, floors, and roofs), the area, and the number of building envelope assemblies (for instance, the type and the number of panes for the windows). A major cause of energy wastage is air entering or leaving a home. Unintentional air transfer toward the inside is referred to as infiltration, and unintentional air transfer toward the outside is referred to as exfiltration. However, infiltration is often used to imply air leakage both into and out of a home, and this is the terminology used in this chapter. Infiltration also affects concentrations of indoor pollutants and can cause uncomfortable drafts. Air can infiltrate through numerous cracks and spaces created during building construction, such as those associated with electrical outlets, pipes, ducts, windows, doors, and gaps between ceilings, walls, floors, and so on. Infiltration results from temperature and pressure differences between the inside and outside of a home caused by wind, natural convection, and other forces. Major sources of air leakage are attic bypasses (paths within walls that connect conditioned spaces with the attic), fireplaces without dampers, leaky ductwork, window and door frames, and holes drilled in framing members for plumbing, electrical, and HVAC equipment. In addition, comments on the repair needs and recent replacements should be noted during the survey and included in the audit report. Some commonly recommended energy-conservation measures to improve the thermal performance of the building envelope include the following measures: *Addition of thermal insulation* for building surfaces without any thermal insulation; this measure can be very cost-effective. *Replacement of windows*, because windows represent a significant portion of the exposed building surfaces, using more energy-efficient windows (e.g., high R-value, low-emissivity glazing, airtight) can be beneficial in both reducing the energy use and improving the indoor comfort level. *Reduction of air leakage*, because the infiltration load is significant, the leakage area of the building envelope can be reduced by simple and inexpensive weather-stripping techniques. The energy audit of the envelope is especially important for residential buildings, because the energy use from residential buildings is dominated by weather inasmuch as heat gain or loss from direct conduction of heat or from air infiltration/exfiltration through building surfaces accounts for a major portion (50% to 80%) of the energy consumption. For commercial buildings, improvements to the building envelope are often not very cost-effective due to the fact that the building envelope modifications (e.g., replacing windows, adding thermal insulation in walls) are typically considerably more expensive than in the case of residential buildings. However, it is recommended to audit the envelope components systematically not only to determine the potential for energy savings but also to ensure the integrity of the envelope's overall conditions. For instance, thermal bridges can have significant impacts on heat transfer, and can lead to a heat transfer increase and to moisture condensation. The moisture condensation is often more damaging and costly than the increase in heat transfer because it can affect the structural integrity of the building envelope.

12.3   BUILDING ENERGY MANAGEMENT

Energy management systems are at the heart of all infrastructures, from communications, economy, and society's transportation to the society, making systems more complex and more interdependent.

Energy management offers the largest and most cost-effective opportunity for both industrialized and developing nations to limit the enormous financial, health, and environmental costs associated with burning fossil fuels. The increasing number of disturbances occurring in the system has raised the priority of EMS infrastructure, which has been improved with the aid of technology and investments. Modern industrial and commercial facilities operate complex and interrelated power systems. Energy conservation and facilities/equipment are only part of the approach to improve energy efficiency. Most energy efficiency in industry is achieved through changes in how energy is managed in a facility, rather than through installation of new technologies. Systematic management and the behavior approach have become the core efforts to improve energy efficiency today. An EMS provides methods and procedures for integrating energy efficiency into existing industrial or commercial management systems for continuous improvements in the use of the energy. Energy management is defined as a system, methods, and procedures for effective and optimum energy use in industrial processes and in the operation of residential, commercial, and industrial facilities to maximize profits and to enhance competitive positions through organizational measures and optimization of energy efficiency in the industrial and commercial processes (Kreith and Goswami, 2008; Belu, 2018). Profit maximization is also achieved by lowering the energy costs during each productive and operational phase, while the three most important operational costs are for materials, labor, and energy (fuels, electricity, and thermal energy). Moreover, the improvement of competitiveness is not limited to the reduction of sensible costs, but can be achieved with good energy cost management, which can increase the flexibility and compliance to the changes of market and environmental regulations. Energy management is a well-structured process, both technical and managerial in nature. In this chapter, we discuss the structure, methods, and techniques used in energy management, as well as new approaches and developments in the field. A rich, comprehensive, and up-to-date list of references regarding energy management topics and problems is also included at the end of chapter for professionals, engineers, students, and interested readers.

In the past, most of the manufacturing, industrial, and large commercial organizations and facilities have lacked complex energy monitoring and control systems when compared to those frequently found in electrical generating plants. The reason for this is that the primary goal of any manufacturing organization is to cost effectively produce high-quality products for its customer and thus energy management often takes a somewhat secondary role to production objectives. However, with increasing energy prices and market competition, manufacturing firms are taking a closer look at energy-saving opportunities and/or the potential for on-site electrical generation from alternative energy sources. Manufacturing firms are now re-examining the energy management functional requirements and internal power networks to increase efficiency and integration with the future power grid. Energy efficiency, conservation, and cost are top priorities all over the world, in particular for heavy energy consumers. Electrical energy costs are about 50% or so of total energy in many manufacturing industries, motivating companies to employ various strategies to control and reduce the rising energy costs. Certainly many of our environmental problems today arise from the types of energy we use, and increased burning of fossil fuels will accelerate climate change. Energy conservation technology and facilities or equipment are only part of the approach to improve energy efficiency. Systematic management and the behavior approach have become the core efforts to improve energy efficiency today. Energy management represents a significant opportunity for organizations to reduce their energy use while maintaining or boosting productivity. Industrial and commercial sectors jointly account for approximately 60% of global energy use. Organizations in these sectors can reduce their energy use 10% to 40% by effectively implementing an EMS. Fossil fuels are currently the major source of energy in the world. However, as the world is considering more economical and environmentally friendly alternative energy-generation systems, the global energy mix is becoming more complex. Factors forcing these considerations are the increasing demand for electric power by both developed and developing countries; many developing countries lacking the resources to build power plants and distribution networks; some industrialized countries facing insufficient power generation; and pollutant emission and climate change concerns. Renewable energy sources such as wind turbines, photovoltaic solar systems, solar-thermal power, biomass power plants, fuel cells, gas microturbines, hydropower turbines, combined heat and power (CHP), and hybrid power systems are now part of power generation systems.

The modern manufacturing plant has a large set of IT equipment including EMS and manufacturing execution systems (MES), which are used to control processes and infrastructure. MESs are single or multiple software applications that perform such roles as equipment scheduling and monitoring, inventory and product tracking, quality monitoring, maintenance dispatching, and operating allocation and planning. These systems serving to manage aspects of manufacturing operations are not always well-connected nor integrated despite their common efficiency objectives, which include improving energy efficiency. With the increased availability of all types of real-time information from plant equipment, quality, reliability, and energy consumption monitoring, there is growing interest to take advantage of this valuable data to improve not only the productivity of the manufacturing operations but also simultaneously of the energy performances. However, due to the large amount of data and complex nature of most manufacturing systems, extracting meaning and understanding from this rich source of data is often difficult. To complicate matters further, manufacturing systems are highly dynamic environments where events such as machine breakdowns, operator shortages, and quality problems are commonplace. These events, whether deterministic or stochastic, require effective management responses. Without an event clear integrated view, it is difficult to take appropriate actions to ensure efficient operations from both a productivity and energy perspective. The key question for energy management is how to provide the best case for successful energy management within an organization, achieve the desired buy-in at top management level, and implement a successful EMS. The purpose of energy management is to provide an organizational framework to integrate energy efficiency into management practices, including fine-tuning production processes and improving the system energy efficiency. Energy management seeks to apply to the energy use the same culture of continual improvement that has been successfully used by companies to improve quality and safety practices. Its guidelines recommend that companies need to track energy consumption, benchmark, set goals, action-plan design, evaluate progress and performances, and create energy awareness throughout the organization, enabling the integration of energy efficiency and conservation into existing management system for continuous improvements and to reach the requirements for high efficient companies. Ultimately these efforts should reduce costs and increase revenue. Among the high efficient company requirements are: efficiency is a core strategy, leadership and organizational support is real and sustained, the company has energy-efficiency goals, there is in place a robust tracking and measurement system, substantial resources are dedicated to energy efficiency, the energy-efficiency strategy is working, and results are communicated. An energy management standard is needed to influence how energy is managed in an industrial facility, thus realizing immediate energy-use reduction through changes in operational practices, as well as creating a favorable environment and setting for the adoption of more capital-intensive energy-efficiency measures, practices, and technologies. Efficient energy management requires the identification of where energy is used, where it is wasted, and where any energy-saving measures will have the most effect. The key feature of a successful EMS is that it is owned and fully integrated as an embedded management process within an organization or company. Energy management implications should be considered at all stages of the development process of new projects, and these implications are part of any change control process. Standards should lead to reductions in energy cost and pollutant emissions, and minimize negative environmental impacts. A change in the organizational culture is needed in order to realize industrial energy-efficiency potential. An EMS standard can provide a supportive organizational framework necessary to move beyond an energy-saving project approach to an energy-efficiency approach that routinely and methodically seeks out opportunities to increase energy efficiency, no matter how large or small.

Businesses that are wasting energy are reducing their profitability and are causing avoidable pollution, primarily through increased carbon emissions. If comparisons are made between industries, corporations, or manufacturing plants, a wide variation in the structure of individual energy programs are found. Even with such variations in scope and implementation, there is a good chance of a successful energy management program, while comparing one company with another results in energy management program variations based on how energy is consumed. A company whose energy use is based primarily in lighting and heating, develops a different program and strategy than a cement manufacturer, with hundreds of electric motors in addition to significant thermal energy use. Even when a successful energy management program is implemented in a complete different way, it will usually feature some common elements. A company with centralized management philosophy may dictate program structure across the company's plants and facilities, while one with decentralized management structure can set only broad program goals, allowing individual units to develop individual programs according to the local circumstances. Making businesses more energy efficient is still a largely untapped solution addressing environmental pollution, energy security, and fossil fuel depletion. As pressures mount on businesses to become more energy efficient, managing resources effectively is proving more essential than ever. In addition, customers are increasingly asking for assurance from organizations that are environmentally responsible. Energy management is the means to controlling and reducing the organization energy consumption, which is important because it reduces production costs, pollutant emissions, and the cost-related implications of the pollution, while reducing the risks. As an organization consumes more energy, greater are the risks that any energy price increases or supply shortages will affect the organization's profitability, or even deem it impossible for the organization to continue. Energy management can reduce such risks by reducing energy demands and controlling the demands to make the energy more predictable. On top of these reasons, it is advisable to have some rather aggressive energy usage reduction targets. Energy audits are the process of detecting operating issues, improving occupant comfort, and optimizing/reducing energy consumption of existing buildings. This is done through a series of inspections, surveys, and an extensive analysis of energy flows throughout different processes. In turn, this helps to achieve a great amount on the total utility cost of the building. This method has been found to be extremely useful in understanding the building energy performance and the energy consumption by different utilities of the building. It helps to analyze the energy balance of the installed system, which elucidates possible improvements for energy efficiency and reduced energy costs.

12.3.1 Identification of Energy Usage Factors and Parameters

With energy costs and the importance attributed to climate change having risen over the past several years, energy efficiency has become paramount. Identifying key energy performance indicators is vital for the planning process, as it provides managers with a clear picture of how their company uses energy and can highlight ways to manage resources better. The energy management plan's first step is to determine the energy consumption *structure* of the company, in other words, what are the needed organization energy resources to run its operations (e.g., natural gas, coal, gasoline, or electricity). In order to maintain and evaluate the effectiveness of an energy management program, energy data must be collected and analyzed. Collecting credible utility data in the plant data history provides tracking capability at quite a low cost. Ultimately, all energy use and energy-saving projects and actions must be collected, preferable in a database format. A management process is required to proactively assess, manage, and measure energy use and conservation. However, many companies and organizations have limited levels of expertise necessary to achieve these reductions and so need guidance, etc. Improvements in energy efficiency will require systems and processes necessary to improve energy performance. Companies and organizations need to manage the way in which energy is used in order to reduce pollutant emissions and other environmental impacts, as well as reduce energy costs and wastage. Energy management plans set the energy-efficiency benchmarks that measure consumption, the energy-consumption assets in operation, and seek opportunities and ways to reduce energy usage. Such benchmarks allow companies to compare their energy management systems against best practices, and to compare the energy efficiency results against national standards or the country's energy-saving technologies policy. Energy management shows

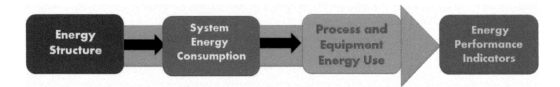

FIGURE 12.2   Energy management flow chart and structure.

the company energy performance ratings and helps to track energy and water consumption, to set targets and indicators and as a result, to improve policies, identify areas that need improvements, and measure the investments required for facility upgrades. Energy management systems aim at reduced energy use and costs, while representing a key element in any company energy management program. Energy management must be based on real-time information obtained from process monitoring and control systems, and on production plans received from production planning systems. It is also useful to compare each unit and site with its own past performance to determine energy savings, but the internal view is not very useful in determining how the process compares with its competitors. Often total and comprehensive solutions include planning and scheduling tools to optimize energy use and supply, energy balance management tools to support the real-time monitoring and control of the energy balance, and reporting tools to evaluate and report energy consumption, costs, efficiency and other energy-related information. Opportunities for cost reduction are greatest when both electricity consumption and prices vary over time, which is common in process industries, and open electricity market environments.

Energy management concepts are built upon the plan-do-check cycle, being used as a basis for many management systems. Figure 12.2 illustrates how continued use of the management action process leads to continuous improvement. An EMS is a collection of procedures, methods, and tools designed to engage staff at all levels within an organization in managing energy use on an ongoing basis. An EMS allows industrial plants, commercial, institutional, and governmental facilities and entire organizations to systematically track, analyze, and plan their energy use, enabling greater control of and continual improvement in energy performance. Organizations may choose to pursue only certain components of energy management, based on their needs. While EMS implementation expertise is typically concentrated in technical and engineering positions within an organization, a variety of nontechnical personnel (e.g., executive staff, accountants, and financial managers) also exert significant influence on energy decision-making. While these nontechnical staff members need not be knowledgeable in all aspects of energy management, they are included in this analysis because they can be critical to the success of an EMS within their organization. Many long-standing professional training or credentialing programs cover some skills or knowledge areas relevant to EMS implementation, but most fall short of providing the entire spectrum of skills and expertise needed to implement an EMS effectively. Energy is one of the management resources of a company, and shall be managed and controlled by a systematic method in harmony with the management of other resources. Energy management involves managing all kinds of energy used in the company by making out an optimum program of purchasing, generating, and consuming various types of energy based on the company's overall short-term and long-term management program, with due consideration of costs, availability, economic factors, and so on. Energy management is necessarily required because it influences a number of aspects of company operation and activities, such as energy costs that affect the company profitability, energy costs that affect competitiveness in the world market, national energy supply-demand balance, national trade and financial balance, local and global environments, occupational safety and health, loss prevention and waste disposal reduction, improved productivity, and quality. Energy management in the form of implementing new energy-efficiency technologies, new materials, and new manufacturing processes and the use of new technologies in equipment and materials for business and industry is also helping companies improve their productivity and increase their product or service quality. Often, the energy savings is not the main driving factor when companies decide to purchase new equipment, use new processes, and use new high-tech materials. However, the combination of increased productivity, increased quality, reduced environmental emissions, and reduced energy costs provides a powerful incentive for companies and organizations to implement these new technologies. Energy management is all about reducing the cost of energy used by an organization, now with the added spin of minimizing carbon emissions as well. Reducing energy costs has two facets: price and quantity. Improving energy efficiency is, in part, a technical pursuit with a scientific basis.

The knowledge and skills of the energy management report impart guidance for development, generate opportunities for collaboration among developing or expanding training programs, facilitate greater consistency among existing professional programs, and increase awareness about energy efficiency potentials. Steps that are needed during implementation of an energy management system are:

1. *Initiating an energy management program:* Understanding basic concepts and requirements, getting organization higher management commitment, establishing an energy management team, and developing an energy policy and planning;
2. *Conducting an energy review:* Collecting energy data, analyzing energy consumption and costs, identifying major energy uses, conducting energy assessments, and identifying potential energy-saving opportunities;
3. *Energy management planning:* Setting a baseline, determining performance metrics, evaluating energy-saving opportunities and selecting best projects, and developing action plans;
4. *Implementing energy management:* Obtaining resource commitments, providing training and raising energy saving awareness, communicating to all stakeholders, and executing action plans;
5. *Measurement and verification:* Providing the knowledge and skills required to monitor, measure, verify, track, and document energy use and savings; and
6. *Management review:* Reviewing progress, modifying goals, objectives, and action plans as needed.

These steps are embedded in the *plan-do-check-act* cycle, which in addition to the common knowledge areas, includes the identified ancillary knowledge and skills that enhance the understanding of key energy management topics. To manage the energy, a complete understanding of the company energy use, structure, and demand is critical. This is based on a comprehensive energy review, consisting of the analysis of all energy use and consumption, determining the significant energy uses, and then identifying and prioritizing potential opportunities for improvements and energy savings. An energy review is an essential element of energy management planning and must be performed by personnel with a broad range of knowledge and skills. An energy review requires the collection of energy-consumption data from utility bills, energy meters, and other information sources. The data must be analyzed and interpreted within the context of the sites, facilities, processes, business units, and equipment. For example, in many smaller facilities, lighting and space heating and cooling are often the dominant energy users, while in larger facilities larger portions of energy are used in process or large equipment. These are just a tiny percentage of the costs at large petrochemical sites, where the dominant energy users are associated with moving, transforming, and separating feed and product materials. This analysis requires personnel that not only understand buildings, processes, energy using equipment, and other process or facility factors, but also possess the knowledge and skills necessary to identify viable improvement opportunities. Some of the fundamental tasks involved in conducting an energy review include: data logging and collection, metering, monitoring, measurement, verification, facilitating, and managing the process for identifying energy efficiency and conservation opportunities.

Businesses, industry, and government organizations are all under tremendous economic and environmental pressures in the last decades to reduce energy uses and protect the environment. Being economically competitive in the marketplace and meeting increasing environmental standards are the major driving factors in most of the operational cost and capital cost-investment decisions. Energy management is an important tool to help organizations meet these critical objectives for their short-term survival and long-term success. A common meaning of energy management, often that signifies different things to different people, is the efficient and effective energy use to maximize profits, while minimizing the costs and to enhance competitiveness of the organization. Some goals and objectives of energy management include improving energy efficiency and reducing energy use, reducing costs and pollutant emissions, complaining to the various regulations, and developing and maintaining effective monitoring, reporting, and management strategies for intelligent energy use. Other goals include finding better ways to increase returns from energy investments through research and development, developing interest in and dedication to the energy management program from all employees, and reducing the impacts of curtailments or any interruption in energy supplies. Although energy conservation is an important energy management aspect, it is not the only consideration. In fact, for many organizations, energy management is one of the most promising revenue and profit improvements and cost reduction programs available in today's industries. Energy management programs are vitally needed today. Commercial activity is very diverse, and this leads to greatly varying energy intensities depending on the nature of the facility. Recording energy use in a building or a facility of any kind and providing a history of this use is necessary for the successful implementation of an energy management program. A time record of energy use allows analysis and comparison so that results of energy productivity programs can be determined and evaluated.

The most important organizational step affecting the success of an energy management program is the appointment of one person or a small group that owns the full responsibility for its operation. Preferably that person reports directly to the top company management and has enough authority in directing technical and financial resources within the bounds set by the level of management commitment to implement the energy management plan. It is difficult to stress enough the importance of making the position of plant energy manager a full-time job. Any diversion of interest and attention to other aspects of the business may adversely affect the energy management program. One reason is that the greatest opportunity for energy cost control and energy-efficiency gains is in improved operational and maintenance practices. As the energy management program progresses to the energy audit and beyond, it is necessary to keep all employees informed as to its purposes, its goals, and how its operations are impacting plant or facility operations and employee routines, comfort, and job performances. The education can be delivered through several channels as best benefits the organizational structure. In addition to general education about energy conservation, it may prove worthwhile to offer specialized courses for boiler, mechanical, and electrical equipment operators and other workers whose jobs can affect energy utilization in the plant. Staff training is the second important step and a key to staying on track for energy conservation and success in any energy management program. It is management's responsibility to ensure that technical and operating personnel are trained to operate the equipment safely and in a proper manner. Effective training is not accomplished in a single session that, once completed, may be quickly forgotten. Training must be thorough and continuous to help not only to inform but also to change attitudes. Training allows the staff to explore new ideas, interchange them with experts and with other trainee participants, and feel more comfortable with the role they must fulfill. In turn, trained technical and management staff should be encouraged to provide in-house training to operating and lower-level technical staff. Staff training is the primary tool by which awareness is generated and knowledge is transmitted. As part of the energy management program, there are two major areas for employee training.

The first area is training in the development of new skills in technologies, and training to adopt new attitudes towards energy wastage and reduction of waste. The introduction of new technologies, process equipment, operating and maintenance procedures, and energy documentation methods requires training at many levels. There is a need to train new as well as experienced personnel in energy-efficient operation of company facilities. The need for training in each should be reviewed periodically to assure that all new personnel are properly trained and to refresh the skills of existing personnel. Staff training is typically at three levels: management, engineering (technical level), and supervisory (operators). It is appropriate and even necessary to select an energy-reduction goal for the first year of the energy management program or very early in the program. The purpose is to gain the advantage of the competitive spirit of those employees who can be aroused by a target goal.

12.3.2 ENERGY AUDIT

An energy audit is a systematic study or survey to identify how energy is being used in a building, facility, or plant, and identifies energy savings opportunities. An energy audit is key to developing an energy management program. Although energy audits have various degrees of complexity and can vary widely from one organization to another, every audit typically involves: data collection and review, plant surveys and system measurements, observation and review of operating practices, data analysis, and reporting. In short, the audit is designed to determine where, when, why, and how energy is being used. Energy audits refer to formal energy accounts in terms of energy consumption or use and the associated costs of the energy system, over a certain period, usually on a yearly basis. The energy audit is one of the first tasks in an effective energy cost-control program. An energy audit consists of a detailed examination of how a facility uses energy, what are the energy costs, and finally, recommending programs for changes in operating practices or energy-consuming equipment that will cost effectively save dollars on energy bills. Energy audits (energy surveys or energy analyses) are positive experiences with significant benefits to the business or individual. Using proper methods and equipment, an energy audit provides essential information on how much, where, and how energy is used within an organization (facility, plant, or building). The audit is divided into different phases, the energy audit components. Utility bills or energy application data are collected for specific time periods to allow the evaluation of energy supply patterns, energy demand rate structures, and energy usage profiles. Energy audits are critical to a systematic approach for decision-making in energy management, attempting to balance the total energy inputs with its use, helping to identify all the energy streams in a facility or a building. An energy audit is a periodic examination of the energy system to ensure that energy is used efficient and to quantify the energy uses according to the system functions. An industrial energy audit is an effective tool in defining and pursuing a comprehensive energy management program. Saving money on energy bills is attractive to businesses, industries, and individuals alike. In any industry, the three top operating expenses are the energy (electrical and thermal), labor, and materials. If one were to relate to the manageability of the cost or potential cost savings, the energy invariably emerges as a top ranker, and thus energy management constitutes a strategic area for cost reductions. Energy audits help to explain the ways in which the energy and fuel are used, and to identify areas where waste occurs and room for improvement exists. Energy audits gives an orientation to energy cost reduction, preventive maintenance, and quality-control programs, which are vital for production and utility activities. An energy audit program helps to keep focus on variations occurring in energy costs, energy supply availability, and reliability, to decide on an appropriate energy mix, to identify energy conservation technologies, or to retrofit for energy conservation equipment. In general, an energy audit translates energy conservation ideas into realities, lending technically feasible solutions with economic and other organizational considerations within a specified time frame. The primary energy audit objective is to determine ways to reduce energy use per unit of product and/or to lower operating costs. An energy audit provides a *benchmark* (reference) for managing energy in an organization and also provides the basis for planning more effective energy use throughout the organization. The main part of the energy audit report is energy savings proposals comprising of technical and economic analysis, with an energy audit being ultimately defined as a systematic search for energy conservation opportunities.

In many situations, major cost savings can be achieved through the implementation of no-cost or low-cost measures, such as changing energy tariffs, rescheduling production activities to take advantage of preferential tariffs, adjusting existing controls so that plant operation matches the actual requirements of the building or manufacturing process, implementing good housekeeping policies, in which staff are encouraged to avoid energy-wasteful practices, and investing in small capital items such as thermostats and time switches. Sometimes it is necessary to undertake more capital-intense measures in order to achieve the energy audit objectives. When a substantial fraction of a company's operating costs is due to energy, that organization has stronger motivations to initiate and continue energy cost-control programs and to perform energy audits. The energy audit is one of the first tasks to be performed in an effective energy cost control and reduction program. An energy audit consists of detailed examinations of how a facility uses energy, what the facility is spending for the energy, and, finally, a recommended program for changes in operating practices or energy-consuming equipment that will cost effectively save on energy bills. Energy audits are performed by different groups. Electric and gas utilities offer free residential energy audits, consisting of an analysis of the monthly energy bills, inspection of the building construction, and an inventory of all the energy-consuming appliances in the house. Ceiling and wall insulation is measured, ducts are inspected, and appliances, heaters, air conditioners, water heaters, refrigerators, freezers, and the lighting system are examined and checked. An initial summary of the basic steps involved in conducting a successful energy audit is provided here, and these steps are explained more fully in the sections that follow. This audit description primarily addresses the steps in an industrial or large-scale commercial audit, and not all of the procedures described in this section are required for every type of audit. The audit process starts by collecting information about a facility's operation and about its past record of utility bills. This data is then analyzed to get a picture of how the facility uses, and possibly wastes, the energy, as well as to help the auditor learn what areas to examine to reduce energy costs. Similar to financial audits, the energy account or energy audit is separated into different energy-application sections. Some of the important terms and concepts used in energy audits are:

- *Specific energy consumption* is the energy consumed by the energy system per unit output, e.g., in the case of a building, it can be the energy consumption per unit area of the building.
- *Energy target* is the energy goal that an entity (usually a government) is going to achieve in the future, usually 3, 5, or 10 years ahead of the policy made.
- *Energy benchmark* is the *energy reference, control,* or *baseline energy case* based on past achievements, such as those given by energy regulatory entities.

An energy audit, as a periodic examination of an energy system to ensure that energy is used as efficiently as possible, is a systematic study or survey to identify how energy is being used in a building or plant, and identifying energy savings opportunities. An energy audit, also known as an energy survey, energy analysis, or energy evaluation, examines the ways energy is currently used in that facility and identifies some alternatives for reducing energy costs. The energy audit goals are: to identify the types and costs of energy use, to understand how the energy is being used, and possibly wasted, to identify and analyze alternatives such as improved operational techniques and/or new equipment that substantially reduce the energy costs, and to perform an economic analysis on the alternatives to determine which ones are cost-effective for the organization. Using the proper methods, an energy audit provides essential information on how much, where, and how energy is used, indicating the performance at the overall plant or process level. These performances can be compared against past and future levels for proper energy management. The major parts of the energy audit report are energy savings proposals comprising technical and economic analysis of projects. Looking at the final output, an energy audit can also be defined as a systematic search for energy conservation opportunities. Major cost savings can be often be achieved through the no-cost or low-cost measures, e.g., new energy tariffs, production rescheduling to take advantage of preferential tariffs, adjusting operating controls to match the facility or process requirements, reducing energy wastes, or making investments in new technologies. However, it is often necessary to undertake more capital-intense measures, investments, and changes.

Energy auditing of buildings can range from a short walk-through of the facility to a detailed analysis with hourly computer simulation. The type of energy audit to be performed depends on the function and type of industry, facility, or building, depth to which final audit is needed, and the potential and magnitude of cost reduction desired. Depending on the level of details on the collected information, purpose, and the organization goals and objectives, energy audits might be distinguished into two main types:

1. *Walk-through energy audits* assess the site energy consumption and relevant costs on the basis of energy bills (invoices) and a short on-site autopsy. Housekeeping or minimum capital-investment energy-saving options of direct economic return are determined, and a further list of other energy-saving opportunities (often involving considerable capital) is proposed on a cost-benefit basis.

2. *Detailed (or diagnostic) energy audits* requesting a detailed recording and analysis of energy and other site data. Energy consumption is disaggregated in different end uses (heating, cooling, different processes, lighting, etc.) and the different factors that affect end use are presented and analyzed (production/services capacity, climatic conditions, raw material data, etc.). All the costs and benefits for energy-saving opportunities that meet the criteria and requirements of the site administration are determined. A list for potential capital-intensive energy investments requiring more detailed data acquisition and processing is also provided, with an estimation of the associated costs and benefits. To perform an energy audit, several tasks are typically carried out depending on the type of the energy audit and the size and function of the building, facility, plant, or industrial process. Some of the tasks may have to be repeated, reduced in scope, or even eliminated based on the findings of other tasks.

Energy audits are the first step to improve the energy efficiency of buildings and industrial facilities.

Generally, four types of energy audits can be distinguished, as briefly described below (Krarti, 2011; Kreith and Goswami, 2008). A walk-through audit consists typically of a short on-site visit of the facility to identify areas where simple and inexpensive actions (typically operating and maintenance measures) can provide immediate energy-use and/or operating cost savings. A utility cost analysis includes a careful evaluation of metered energy uses and operating costs of the facility. Typically, the utility data over several years are evaluated to identify the patterns of energy-use, peak demand, weather effects, and potential for energy savings. A standard energy audit consists of a comprehensive energy analysis for the energy systems of the facility. In particular, the standard energy audit includes the development of a baseline for the energy-use of the facility, evaluation of the energy savings, and the cost-effectiveness of appropriately selected energy-conservation measures. A detailed energy audit is the most comprehensive type, but also the most time-consuming and expensive. Specifically, a detailed energy audit includes the use of instruments to measure energy use for the whole building and/or for some energy systems within the building (for instance by end uses such as lighting systems, office equipment, fans, chillers, etc.). In addition, sophisticated computer-simulation programs are typically considered for detailed energy audits to evaluate and recommend energy retrofits for the facility. By making the audits specific rather than general, more energy is saved. Examples of some types of audits that might be considered are: tuning-operation-maintenance, compressed air, motors, lighting, steam and heat systems, water, controls, HVAC, and employee suggestions. By defining individual audits in this manner, it is easy to identify the proper team for the audit. Bringing in outside people such as electric utility and natural gas representatives to be team members is a good practice. Scheduling the audits, then, can contribute to the events that keep the program active.

The energy audit process starts with an examination of the historical and descriptive energy data for the facility. Specific data that should be gathered in this preliminary phase include the energy bills for the past 12 months, descriptive information about the facility such as a plant layout, and a list of each piece of equipment that significantly affects the energy consumption. Often the most effective way to find and begin implementing energy improvements is to get all the interested parties into one room for a structured discussion of their plant. Careful preparation and actively involved stakeholders are critical to success. A good energy review leaves a site with its own plan, developed by its own people and supported by its management. A walk-through or preliminary audit comprises: 1 day or half-day visit to a plant, where the output is a simple report based on observation and walk-through or preliminary audit historical data provided during the visit. The findings are general comments based on rules-of-thumb, energy best practices, or manufacturer or facility data, to establish the quantity and cost of each energy form used in that facility. An overview of the energy-use patterns provides guidance for the energy accounting system and the personnel with perspectives of processes and equipment. The purpose is to identify energy-intensive processes, system components, and equipment, to identify the energy inefficiency, if any, at the stage or phase of a detailed energy survey.

An energy balance is a set of relationships accounting for all the energy that is produced and consumed, and matches inputs and outputs in a system over a given time period. The system can be anything from a whole country to an area to a process in a factory. An energy balance is usually made with reference to a year, though it can also be made for consecutive years to show variations over time. Energy balances provide overviews, and are basic energy-planning tools for analyzing the current and projected energy situation. The overviews aid sustainable resource management, indicating options for energy saving, or for policies of energy pricing and redistribution, etc. A number of different types of energy balance can be made, depending on the information you need: the energy commodity account, the energy balance, and the economic balance. The energy commodity account includes all flows of energy carriers, from the point of extraction through conversion to end use, in terms of their original, physical units such as kilotons of coal and kWh of electricity. Although this report does not analyze such policies, it acknowledges their vital role in improving workforce programs and work quality. In the economic balance, different forms of energy are accounted for in terms of their monetary value. An important part of preparing an energy balance is the construction of an energy chain to trace the flows of energy within an economy or system, starting from the primary source(s) of supply through the processes of conversion, transformation, and transportation, to final/delivered energy and finishing with end use.

12.3.3 Types and Structure of Energy Audits

The energy audit type depends on the function and type of industry, depth to which final audit is needed, and the potential and magnitude of cost reduction desired. Thus energy audits, besides the two main types (e.g., *preliminary energy audit* and *detailed energy audit*), are classified as *walk-through energy audit*, utility cost analysis, *standard energy audit*, and *detailed energy audit*. A walk-through energy audit, referred to sometimes as operating and maintenance (O&M) measures, consists of a short on-site facility visit to identify areas where simple and inexpensive actions can provide immediate energy use or operating-cost savings. Such O&M measures can include setting back heating set-point temperatures, replacing broken windows, insulating exposed hot water or steam pipes, and adjusting boiler fuel-air ratio. The purpose of utility cost analysis is to analyze the facility operating costs. Usually, the utility data over several years is evaluated to identify the energy use patterns, peak demands, weather effects, and potential for energy savings. To perform such analysis, a walk-through survey of the facility and its energy systems and uses is conducted, to get a clear understanding of the utility rate structure that applies to the facility. Its purpose is to check the utility charges and ensure that there are accurate estimates, to determine the main charges in the utility bills, and to identify whether the facility can benefit from using other utility rate structures to purchase the cheapest fuel and reduce its operating costs. This analysis can provide a significant reduction in the utility bills especially with implementation of electrical deregulation and the advent of real-time pricing rate structures. The standard energy audit provides a comprehensive energy analysis for the facility energy systems. In addition to the activities described in the previous two audit types, the standard energy audit includes the baseline development for the facility energy uses and the evaluation of the energy savings and the cost-effectiveness of appropriately selected energy conservation measures. The step-by-step approach of the standard energy audit is similar to that of the detailed energy audit described later. Typically, simplified tools are used in the standard energy audit to develop baseline energy models and to predict the energy savings and conservation measures. The detailed energy audit is the most comprehensive and time-consuming audit type. The detailed energy audit includes the use of instruments to measure energy use for the whole building or for some energy systems within the building (e.g., lighting systems, office equipment, fans, chillers). In addition, sophisticated computer-simulation programs are typically considered for detailed energy audits to evaluate and recommend energy retrofits for the facility. Monitoring a building over a short or even an extended period may be considered necessary in order to validate this information. The equipment required for this purpose includes handheld instruments, data loggers, and a range of sensors. A large range of instruments and measurement devices is now on the market and the measurements required for energy management can be made relatively easily; modern electronic instruments are compact, durable, and inexpensive. The main measurements of interest are: fuel consumption, temperature, electrical power, electric energy, ventilation and air movement, water flow, relative humidity, heating efficiency, and U-value.

The techniques available to perform measurements for a detailed energy audit are diverse. However, a comprehensive energy audit provides a detailed energy project implementation plan for a facility, because it evaluates all major energy-using systems, offering the most accurate estimate of energy savings and costs. It considers the interactive effects of all projects, accounts for the energy use of all major equipment, and includes detailed energy cost-saving calculations and project cost. In a comprehensive audit, one of the key elements is the energy balance. This is based on an inventory of energy-using systems, assumptions of current operating conditions, and calculations of energy use. This estimated use is then compared to utility bill charges. However, the audit structure and procedures are based on industry-to-industry approaches, so the methodology of energy audits needs to be flexible. Notice, a quick overview of energy-use patterns provides guidance for an energy accounting system, personnel with perspectives of processes and equipment, identifying energy-intensive processes and equipment, energy inefficiency, if any, and can set the stage for eventually detailed energy surveys. A preliminary energy report structure consists of an introduction, overview of current systems, how much energy is being consumed, what type of energy is being consumed, the performance of the facility compared with other similar facilities, the characteristic performance of the facility, scope and goals of for energy audit, recommendations and the associated costs and savings, and conclusions. A detailed energy audit is carried out for the energy-savings proposal recommended in the walk-through or preliminary audit. It will provide detailed data on the energy inputs to, and energy flows within, a facility and also technical solution options and economic analysis for the factory management to decide project implementation or priority. A feasibility study is required to determine the viability of each option. Detailed evaluations of the process and equipment energy-use patterns, measurements of energy-use parameters, and a review of operating characteristics are providing their efficiency evaluations, identifying the energy-saving options and measures, and leading to the recommendations for the energy savings and conservation implementation. In detailed energy audits, more rigorous economic evaluations of the energy conservation measures are usually performed. Specifically, the cost-effectiveness of energy retrofits may be determined based on the life-cycle cost (LCC) analysis rather than the simple payback period analysis. Life-cycle cost analysis, described in a later chapter section, takes into account economic parameters such as interest, inflation, and tax rates.

12.3.4 STRUCTURE OF ELECTRICITY AND FUEL RATES

Energy cost is an important part of an organization's economic viability and the effectiveness of the energy conservation measures, making it crucial to understand energy costs. Usually, several utility rate structures exist within the same geographical location; each of them can include several clauses and charges that sometimes make it complicated to decipher the energy billing procedures. The complexity of utility rate structures increases with the deregulation of the electric industry. However, with new electric utility rate structures (e.g., real-time pricing rates), there are more opportunities to reduce facility energy costs. Sources of energy include: coal, natural gas, petroleum products, and electricity. The electricity is generated in power plants fueled from primary energy sources (i.e., coal, natural gas, or fuel oil) or from nuclear power plants or renewable energy sources (e.g., hydroelectric, geothermal, biomass, wind, photovoltaic, and solar thermal systems). In the United States or any other country, the energy consumption fluctuates in response to significant changes in oil, coal, or natural gas prices, economic growth rates, policy changes, and environmental concerns. However, over the last several decades in the United States, coal remains the cheapest energy source, while the cost of electricity is still high relative to the other fuel and energy types (Krarti, 2011; Vanek, 2012).

To generate electricity, the utilities have to consider several operating costs to determine their rates, such as: the generation plant operation costs, the transmission and distribution costs, and fuel and administrative costs (Krarti, 2011; Fehr, 2015). Other factors affecting the electricity cost include the generating capacity of the utility, and the demand/supply ratios at any specific time period (i.e., on-peak and off-peak periods). Utilities allocate the cost of electricity differently by offering various rate schedules depending on the customer type. Three customer types are usually considered by utilities: residential, commercial, and industrial. Each utility may offer several rate structures for each customer type. It is therefore important that the auditor know the various rate structures that can be offered to the audited facility.

Utilities can tailor their rates to customer needs for electricity using several methods. The common rate structures used by U.S. utilities are: block pricing rates, seasonal pricing rates, and innovative rates. In addition to these rates, utilities may provide riders and discounts to their customers. A ride modifies the structure of a rate based on specific qualifications of the customer. For example, a utility can change the summer energy charges of their residential customers using an air-conditioning rider, or can offer a voltage discount when the customer is willing to receive voltages higher than the standard voltage level. However, to benefit from the voltage discount, the customer may have to install and maintain a properly sized transformer, increasing its operation and maintenance costs.

Several utility rate features and concepts must be known and understood in order to be able to interpret and analyze the utility billing procedure correctly and take advantage of them in order to reduce the facility energy costs. The power demand that is billed by the utility is referred to as the billing demand, being often determined from the peak demand obtained for 1 month (or a specific billing cycle). The peak demand, also known as actual demand, is defined as the maximum demand or maximum average measured demand in any 15-minute period in the billing cycle. To better understand the concept of billing demand, you must consider different monthly load profiles, such as: a rugged profile and a flat profile, and further assume that the average demand for each profile coincides, thus the total energy use in kWh for both profiles is the same. Notice that it is not equitable for the utility to bill the same charges for the two profiles, because the utility supplies higher demand for rugged profiles and thus increases its generation capacity for only a short period of time. Meanwhile, the flat profile is ideal for the utility because it does not change over time. Therefore, often utilities charge their customers for the peak demand incurred during the billing period. This demand charge may serve as an incentive to the customers to reduce their peak demand. In some rates, the billing demand is determined based on the utility specifications and may be different from the maximum demand actually measured during a billing period. For instance, ratchet and power factor clauses can change the determination of the billing demand. Moreover, a minimum billing demand can be specified in a contractual agreement between the utility and the customer. This minimum demand is often called the contract demand. The power factor is defined as the ratio of actual active power used by a consumer (kW) to the total apparent power supplied by the util-ity (kVA), as discussed in Chapter 2. For the same actual power consumed by two customers but with different power factor values, the utility has to supply higher total power to the customer with the lower power factor. To penalize customers for low power factors (usually less than 0.85), many utilities use a power factor clause to change the billing demand or to impose new charges (on total power demand or reactive power demand) depending on the power factor. For example, the billing demand increases if the measured power factor (PF) of the customer is less than the base (reference) power factor, expressed as:

$$\text{Billed Demand} = \text{Actual Demand} \left(\frac{PF_{Base(Reference)}}{PF_{Actual}} \right) \qquad (12.2)$$

Example 12.3

A utility has the following monthly billing structure for its industrial customers: the customer charge, the demand charge, and the energy charge are $420/month, $20/kW, and $0.032/kWh, respectively, and a power factor clause as expressed in Equation (12.2) with an 0.80 base (refer-ence) power factor. Calculate the utility bill for an industrial facility with the following energy-use characteristics during a specific month, an actual demand of 350 kW, energy consumption of 50 MWh, and 64% average monthly power factor.

SOLUTION

From Equation (12.2) the billed demand is:

$$\text{Billed Demand} = 350 \times \left(\frac{0.80}{0.64} \right) = 437.5 \text{ kW}$$

Then the customer monthly bill is computed by using the rate structure:

$$\text{Total Monthly Charge} = 420 + 437.5 \times 20 + 50000 \times 0.032 = \$10,770.0$$

Typically, utility charges are billed monthly, particularly the demand charges, which are based on monthly peak demand. However, when the peak demand for one month is significantly higher than for the other months (such as the case for buildings with high cooling loads in the summer months), the utility has to supply the required peak demand and thus operate additional generators only for a couple months. For the rest of the year, the utility has to maintain these additional generators with significantly additional operating costs. To recover some of this maintenance cost and to encour-age demand shaving, some utilities use a *ratchet clause* in the determination of the billed demand. For instance, the billed demand for any given month is a fraction of the highest maximum demand of the previous 6 months (or 12 months) or the actual demand incurred in the month. Specifically, the ratchet clause states that no billing demand shall be considered as less than a percentage of the highest on-peak season maximum demand corrected for the power factor previously determined during the 12 months ending with the current month. Almost all utilities purchase the primary energy sources (fuel oil, natural gas, and coal) to generate electricity. The cost of these fossil fuels (commodities) fluctuates and changes over time, so the utilities impose adjustments to their energy charges to account for the fossil-fuel cost variations. In the rate structure description, utilities provide the formula used to calculate the fuel-cost adjustment. In addition to fuel-cost adjustment, utilities may levy taxes and surcharges to recover imposts required from them by federal or state governments or agencies. In some cases, the fuel-cost adjustment can be a significant proportion of the utility bill. Example 12.4 illustrates the effects of the ratchet clause, the fuel-cost adjustment, and the sales tax on the monthly utility bill of a facility.

Example 12.4

Calculate the utility bill for the industrial facility in Example 12.3, by taking into account a 70% ratchet clause and both the fuel-cost adjustment of 0.013/kWh and the sales tax of 7.5%. Assume that the previous highest demand (during the last 12 months) is 640 kW.

SOLUTION

The minimum billing demand determined by the ratchet clause is calculated as:

$$\text{Ratchet Clause Minimum Billing Demand} = 640 \text{ kW} \times 0.70 = 448 \text{ kW}$$

The 448 kW is selected, being higher than 437.5 kW, power factor clause computed in Example 12.3. Then, the monthly bill can be calculated using the utility rate structure, as:

$$\text{Total Monthly Charge} = 420 + 448 \times 20 + 50000 \times 0.032 = \$10,980.0$$

The fuel-cost adjustment is applied to the energy use, and the sales tax should be applied to the total cost, so the monthly bill for the industrial facility is:

$$\text{Total Monthly Charge} = 420 + 448 \times 20 + 50000 \times (0.032 + 0.013) = \$11,630.0$$

and

$$\text{Total Monthly Charges (Sales Tax)} = 1.075 \times 11630 = \$12,502.25$$

This is an increase in the monthly electricity bill of $12,502.25 − $10,980.0 = $1,522.25.

Utilities often offer several rates to customers depending on the service type, such as different structure rates depending on the voltage level provided to the customers. The higher the delivery voltage level, the cheaper the energy rate is, with utilities offering reduced rates for demand or energy charges to customers that own their own service transformers. This type of rate is described under various clauses of a utility bill including customer-owned transformer, voltage level, or service type. The organization has to determine the cost-effectiveness of owning a transformer and thus receiving a higher voltage level by comparing the utility bill savings against the cost of purchasing, leasing, and maintaining the transformer. Usually, transformer energy losses are charged to the owner, who should also have a standby transformer or make another arrangement (e.g., standby utility service) in case of service transformer breakdown.

In utility *block pricing rates*, the most common used in the United States, the energy price depends on the rate of electricity consumption using either inverted or descending blocks. An inverted-block pricing rate structure increases the energy price as the consumption increases. On the other hand, a descending, or a declining-block rate structure reduces the price as the energy consumption increases. Typically, the rate is referred to as a *flat rate* when the energy price does not vary with the consumption level. Many electric utilities offer *seasonal rate structures* to reflect the monthly variations in their generation capacity and energy cost differences. Usually, utilities provide seasonal rate structures using different energy or demand charges during winter and sum-mer months. The summer charges are usually higher than winter charges for most electric utilities due to higher energy consumption attributed to cooling of buildings. Over half of US electric utili-ties offer seasonal pricing rates for residential customers. Due to the increased focus on integrated resource planning, demand-side management, and the competitive energy market, several innova-tive rates have been implemented by utilities. These innovative rates have the main objective to profitably meet customer needs. Moreover, some utilities foster new technologies through the use of innovative rates to retain their customers. Several categories of rate structures can be considered in the category of the innovative rates, such as: time-of-use rates, real-time pricing rates, end-use rates, specialty rates, financial incentive rates, energy purchase rates, etc. Interested readers are directed to the end-of-chapter references or elsewhere in the literature to find out more about these innova-tive rates. Example 12.5 illustrates the calculation details of the utility bill, based on a block-type seasonal structure rate for demand charges for residential general service.

Example 12.5

Using the utility rate structure given in the table, calculate the utility bill for a residence during a summer month when the energy use is 750 kWh.

Item	Winter	Summer
Customer charge	$7.25	$7.25
Minimum charge	$7.25	$7.25
Fuel-cost adjustment	0.00	0.00
Tax rate	6.75%	6.75%
No. of energy blocks	3	3
Block 1 energy size (kWh)	0	400
Block 2 energy size (kWh)	0	400
Block 3 energy size (kWh)	0	800
Block 1 energy charge ($/kWh)	0.085	0.080
Block 2 energy charge ($/kWh)	0.000	0.121
Block 3 energy charge ($/kWh)	0.000	0.140

SOLUTION

Considering all the charges on the utility rate, as given in the table, with a customer charge of $7.50 and 0.0 fuel-cost adjustment, the monthly bill is computed as:

$$\text{Energy Charge} = 400 \times 0.080 + 350 \times 0.0121 = \$74.35$$

The total, after the sales-tax adjustment is then:

$$\text{Monthly Energy Charge} = (1 + 0.0675) \times 74.35 = \$79.37$$

The rate structures for natural gas are quite similar to those for electricity, but easier to understand and apply. Natural gas providers usually charge for peak demands, but energy charges using block rates or seasonal rates are commonly offered. Examples of natural gas utility rates are illustrated in Table 12.1 for energy block pricing rates and Table 12.2 for energy and demand block pricing rates. Calculation details of monthly utility bills for natural gas are shown in Example 12.6 (GRI, 1993; Krarti, 2011). In addition to energy and demand charges, the natural gas price is determined based on the interruptible priority class, as selected by the customer. A customer with a low priority has a cheaper rate but can be curtailed whenever a shortage in the gas supply is experienced by the

TABLE 12.1
Typical Gas Utility Rate for Residential General Service

Item	Any Month
Customer charge	$4.50
Minimum charge	$4.50
Tax rate	4.00%
No. of energy blocks	2
Block 1 energy size (MMBtu)	2.5
Block 2 energy size (MMBtu)	>2.5
Block 1 energy charge ($/MMBtu)	5.145
Block 2 energy charge ($/MMBtu)	4.033

utility. The utility rate structures for energy sources other than electricity and natural gas (e.g., coal, oil, gasoline) are usually based on a flat rate. For example, crude oil is typically charged per gallon whereas coal is priced on a per-ton basis. The prices of oil products and coal are set by market conditions but may vary within a geographical area depending on local surcharges and tax rates. Moreover, fuel oil or coal can be classified in a number of grades, and the costs are different. The grades of fuel oil depend on the distillation process. In many applications, it can be possible and economically advisable to purchase steam or chilled water to condition buildings rather than using primary fuel to operate boilers and chillers. Steam can be available from large cogeneration plants, while chilled water and steam may also be produced based on the economics of scale in district heating and cooling systems. The steam and chilled water are usually charged on either a flat rate or a block rate structure for both energy and demand. The steam is charged based on pound per hour (for demand charges) or thousands of pounds (for energy charges) or SI equivalents. Meanwhile, the chilled water is charged on the basis of tons (for demand charges) or ton-hours (for energy charges).

TABLE 12.2
Typical Gas Utility Rates for Commercial General Services

Item	Any Month
Customer charge	$0.00
Minimum charge	$4.50
Tax rate	4.00%
No. of energy blocks	3
Block 1 energy size (MMBtu)	10,000
Block 2 energy size (MMBtu)	10,000
Block 3 energy size (MMBtu) >20,000	>20,000
Block 1 energy charge ($/MMBtu)	3.482
Block 2 energy charge ($/MMBtu)	3.412
Block 3 energy charge ($/MMBtu)	3.385
No. of demand blocks	2
Block 1 demand size (MMBtu/day)	10
Block 2 demand size (MMBtu/day)	>10
Block 1 demand charge ($/MMBtu/day)	5.50
Block 2 demand charge ($/MMBtu/day)	4.50

Example 12.6

Calculate the gas utility bill for a residence during a month when the energy use is 10.4 MMBtu, by using the information provided in Table 12.1.

SOLUTION

Considering all the charges as given in Table 12.1, the monthly natural gas bill is:

$$\text{Total Monthly Charge} = (1 + 0.04) \times (4.50 + 2.5 \times 5.145 + 7.9 \times 4.033) = \$57.27$$

12.3.5 DETAILED ENERGY AUDIT STRUCTURE AND PHASES

An energy audit begins with a detailed analysis of the energy bills for previous year or 2 years. This is an important task because the energy bills show the use of each energy source in the overall energy cost, where the energy is used, energy wastes, the total energy costs, and the upper limit of the savings. A complete analysis of the facility or building energy bills requires detailed knowledge of the energy rate structures in effect in order to determine the individual equipment and process energy costs, energy demand charges, and the most accurate estimates of the energy savings for the energy management opportunities, such as high-efficiency equipment, rescheduling of some on-peak electrical, and/or thermal uses and processes. An initial site visit takes 1 day and gives the energy auditor an opportunity to meet the personnel concerned, to familiarize with the site, and to assess the procedures necessary to carry out the energy audit. During the initial site visit the energy auditor or engineer carries out the following actions: discuss with senior management the energy audit aims, as well as the economic guidelines associated with the recommendations of the audit; analyze the major energy consumption data with the relevant personnel; obtain site drawings where available (e.g., building layout, steam distribution, compressed-air distribution, electricity distribution); and finally tour the site accompanied by engineering/production. The main aims of this visit are: structure and finalize the energy audit team, identify the main energy-consuming areas or plant items to be surveyed during the audit and any existing instrumentation/additional metering required, decide whether any meters have to be installed prior to the audit (kWh, steam, oil, or gas meters), identify the instrumentation required for carrying out the energy audit, plan the audit time frame, collect macro data on plant energy resources and major energy consuming centers, and create awareness through meetings and program dissemination.

Depending on the site or facility nature and complexity, a comprehensive audit can take from several weeks to several months to be completed. Detailed studies to establish and investigate energy and material balances for specific plant departments or items of process equipment are carried out. Whenever possible, checks of plant operations are carried out over extended periods of time, at nights and at weekends as well as during normal daytime working hours, to ensure that nothing is overlooked.

The audit report includes a description of energy inputs and product outputs by major department or by major processing function, and will evaluate the efficiency of each step of the manufacturing process. Means of improving these efficiencies will be listed, and at least a preliminary assessment of the cost of the improvements will be made to indicate the expected payback on any capital investment needed. The audit report should conclude with specific recommendations for detailed engineering studies and feasibility analyses, which must then be performed to justify the implementation of those conservation measures that require investments. The information and data collected during the detailed energy audit include energy consumption by type, department, major process equipment, and by end use. Part of this data and information collection includes material balance data (raw materials, intermediate and final products, recycled materials, use of scrap or waste products, production of by-products for re-use in other industries, etc.), energy cost and tariff data, process and material flow diagrams, site power generation and distribution or other site services (compressed air, steam), energy supply sources (e.g., electricity from the grid or self-generation), potential for fuel substitution, process modifications, and the use of cogeneration systems (combined heat and power generation), and last but not least energy management procedures and energy awareness training programs within the establishment. A general procedure can be outlined for most medium-size and large buildings, and commercial and industrial facilities, consisting of:

1. *Building and utility data analysis:* The main purpose is to evaluate the characteristics of the facility energy systems and the energy use patterns. Tasks performed in this step are: collect at least 3 years of utility data to identify historical energy-use patterns and fuel types (electricity, natural gas, oil, etc.) used, in order to determine the fuel type that accounts for the largest energy use, determine patterns of fuel and energy uses to identify the peak energy demands by fuel type, understand the utility rate structure (energy and demand rates) in order to evaluate if the building is penalized for peak demand and if cheaper fuel can be purchased, analyze the effect of local weather patterns on fuel and energy consumptions, and perform utility energy-use analysis by building type and size (building signature is determined from energy use per unit area), in order to compare against typical indices.

2. *Walk-through survey:* From this step, potential energy savings measures can be identi-fied. Tasks involved in this step include: identification of the customer or organization concerns, expectations, and needs; checking the current operating and maintenance proce-dures; determining the existing operating conditions of major energy-use equipment; and estimating the occupancy and the use of equipment and lighting (energy-use density and operation hours).

3. *Baseline for building energy use:* The main purpose is to develop a *base-case model* that represents the building existing energy use and operating conditions. The model is used as a reference to estimate the energy savings due to the selected energy conservation measures. The major tasks performed during this step are: obtain and review architec-tural, mechanical, electrical, and control drawings, inspect, test, and evaluate building equipment for efficiency, performance, and reliability, obtain all occupancy and operating schedules for equipment, develop a building energy-use baseline model, and validate and calibrate the baseline model using the utility data and/or measurement data.

4. *Evaluation of energy-savings measures:* In this phase, cost-effective energy conservation measures and recommendations are determined by using energy savings, new technology, and economic analysis. The following tasks are included: preparing a comprehensive list of energy-conservation measures (using the information collected in previous steps), estimate the energy savings due to the various energy-saving measures pertinent to the building using the baseline energy-use simulation model developed in phase or step 3, estimate the initial costs required to implement the energy-conservation measures, and evaluate the cost-effectiveness of each energy-conservation measure using an economic analysis method (e.g., simple payback or life-cycle cost analysis).

Existing baseline information and reports are useful to get consumption pattern, production cost, and productivity levels in terms of product output per raw material inputs. The audit team should collect the following baseline data: technology, processes used and equipment details, capacity uti-lization, amount and type of input materials used, water, fuels, steam and electricity consumption, other inputs such as compressed air, cooling water, quantity and type of produced wastes, percent-age rejection or reprocessing, efficiencies, or yields. An overview of unit operations, important process steps, material and energy use, and sources of waste generation should be gathered and represented in flowcharts or tables. Existing drawings, records, and shop floor walk-throughs help to analyze the data. Simultaneously, various inputs and output streams at each process step must be identified. It is important to plan additional data gathering carefully, while the measurement systems must be easy-to-use and provide the information to the needed accuracy, not the accuracy that is technically possible, measurement equipment should be inexpensive and easy-to-use, the data quality must be such that the correct conclusions are drawn, define how frequent data collection should be to account for process variations, and do measurement exercises over abnormal workload periods (such as start-up and shutdowns). Design values can be taken where measurements are difficult (cooling water through heat exchanger).

The energy audit report begins with an executive summary that provides the audited organization or facility with a brief synopsis of the total savings available and the highlights of all energy conservation and savings opportunities. The report describes the facility that has been audited, and provides information on the facility operations that are related to its energy costs. The structure and analysis of the energy bills is presented, in the most informative ways, such as tables and plots showing the costs and consumption. Following the section of the energy cost analysis, the recommended measures and energy-saving opportunities are presented, along with the calculations for the costs and benefits, and the cost-effectiveness criterion. Regardless of the audience for the audit report, it must be written in a clear, concise, and easy-to-understand format and style. The executive summary is tailored to nontechnical personnel, and technical jargon must be minimized. A client who understands the report is more likely to implement the recommended measures. An outline for a complete energy audit report consists of: an executive summary that includes a brief summary of the recommendations and cost savings, a table of contents, introduction, the purpose of the energy audit advocating for the needs for a continuing energy cost-control program, the facility description consisting of the product or service descriptions, and materials flow, size, construction, facility layout, and hours of operation, and equipment list, with specifications. In the report is included a comprehensive energy bill analysis and utility rate structures, tables and graphs of energy consumption and costs, a discussion of energy costs and energy bills, as well as energy-conservation opportunities and a listing of potential energy-savings opportunities. In the report are also included cost and savings analysis, economic evaluation, an action plan, recommended measures and energy-saving opportunities, an implementation schedule, the designation of an energy monitor and ongoing program, and a report summary and conclusions. Additional comments not otherwise covered may also be included in more comprehensive reports as needed.

12.3.6 ENERGY CONSERVATION MEASURES

This section discusses some common energy conservation measures (ECMs) that are usually recommended for residential, commercial, and industrial facilities. It should be noted that the list of ECMs presented below is neither exhaustive nor comprehensive. It is provided merely to indicate some of the options that the energy auditor can consider when performing an energy analysis of a residential, commercial, or industrial facility. However, it is strongly advised that the energy auditor keep abreast of any new technologies that can improve building energy efficiency. For many buildings, the envelope (i.e., walls, roofs, floors, windows, and doors) has important impacts on the used energy, so the actual characteristics of the building envelope are critical for energy conservation. Building envelope characteristics include: construction materials (e.g., level of insulation in walls, floors, and roofs), area, and building envelope assemblies (e.g., the type and the number of windows). Commonly recommended energy conservation measures to improve the building envelope thermal performance are: addition of thermal insulation, window replacement, or air leakage reduction. The envelope energy audit is very important for residential buildings. Indeed, the energy use of residential buildings is dominated by weather inasmuch as heat gain or loss from direct heat conduction or from air infiltration/exfiltration through building surfaces, accounts for 50% to 80% of the energy consumption. For commercial buildings, improvements to the building envelope are often not cost-effective due to the fact that modifications to the building envelope (replacing windows, adding thermal insulation in walls) are typically considerably expensive. However, it is recommended to audit the envelope components systematically not only to determine the potential for energy savings but also to ensure the integrity of its overall condition.

For most commercial buildings and a large number of industrial facilities, the electrical energy cost constitutes the dominant part of the utility bill. Lighting, office equipment, and motors are the electrical systems that consume the major part of energy in commercial and industrial buildings. Lighting for a typical office building represents about 40% of the total electricity use. A variety of simple and inexpensive measures can improve the efficiency of lighting systems, such as the use of energy-efficient lighting lamps and ballasts, the addition of reflective devices, de-lamping (when the luminance levels are above the recommended levels by the standards), and the use of daylighting controls. Most lighting measures are especially cost-effective for office buildings for which payback periods are less than 1 year. Office equipment (e.g., computers, fax machines, printers, copiers) constitutes the fastest growing part of the electrical loads in commercial buildings, and the newer systems are very energy-efficient. The energy cost to operate electric motors can be a significant part of the operating budget of any commercial or industrial building. Measures to reduce the energy cost of using motors include reducing operating time (turning off unnecessary equipment), optimizing motor systems, using controls to match motor output with demand, using variable-speed drives for air and water distribution, and installing energy-efficient motors. In addition to the reduction in total facility electrical energy use, retrofits of the electrical systems decrease space cooling loads and therefore further reduce the electrical energy use in the building. These cooling energy reductions as well as possible increases in thermal energy use (for space heating) should be accounted for when evaluating the cost-effectiveness of improvements in lighting and office equipment. The energy use due to HVAC systems can represent 40% of the total energy consumed by a typical commercial building. A large number of measures can be considered to improve the energy performance of both primary and secondary HVAC systems, e.g., setting back thermostat temperatures, retrofitting the HVAC components, or installation of heat recovering systems. It is worth noting that there is a strong interaction among various components of the heating and cooling system. For example, optimizing the energy use of a central cooling plant (which may include chillers, pumps, and cooling towers) is one example of using a whole-system approach to reduce the energy use for heating and cooling buildings. Compressed air has become present in many manufacturing facilities. Its uses range from air-powered hand tools and actuators to sophisticated pneumatic robotics. Unfortunately, large amounts of compressed air are currently wasted in many facilities, with estimates that only up to 25% of input electricity is delivered as useful compressed-air energy. To improve the efficiency of compressed air systems, the factors include: whether compressed air is the right tool for the job (for instance, electric motors are more energy-efficient than air-driven rotary devices), how compressed air is applied (for instance, lower pressures can be used to supply pneumatic tools), how it is delivered and controlled (for instance, the compressed air needs to be turned off when the process is not running), and how the compressed-air system is managed (for each machine or process, the cost of compressed air needs to be known to identify energy and cost savings opportunities). Specialized process equipment must be analyzed on an individual basis because it will vary tremendously depending on the type of industry or manufacturing facility involved. Much of this equipment will utilize electric motors and will be covered in the motor category. Other electrically powered equipment, such as drying ovens, cooking ovens, welders, and laser and plasma cutters, are non-motor electric uses and must be treated separately. Equipment nameplate ratings and hours of use are necessary to compute the energy and demand for these items. Process chillers are other special classes that are somewhat different from the comfort air-conditioning equipment, because the operating hours and loads are driven by the process requirements and not weather patterns and temperatures.

With constant decreases in the cost of computer technology, automated control of a wide range of energy systems within commercial and industrial buildings is becoming increasingly popular and cost-effective. An energy management and control system (EMCS) can be designed to control and reduce the building energy consumption within a facility by continuously monitoring the energy use of various pieces of equipment and making appropriate adjustments. If an EMCS is already installed in the building, it is important to recommend a system tune-up to ensure that the controls are operating properly. For instance, the sensors should be calibrated regularly in accordance with manufacturers' specifications. Poorly calibrated sensors may cause an increase in heating and cooling loads and may reduce occupant comfort. Water and energy savings can be achieved in buildings by using water-saving fixtures instead of conventional fixtures for toilets, faucets, shower heads, dishwashers, and clothes washers. Savings can also be achieved by eliminating leaks in pipes and fixtures. There is a significant potential of energy-use reduction, energy conservation, and costs by implementing and integrating new technologies within the facility and building. Such new technologies include, among others: building envelope technologies and materials, light pipe technologies, advanced HVAC controls, cogeneration, and waste energy recovering. Energy management is often just one element of an integrated building-automation system that regulates security, fire safety, lighting, HVAC systems, and elevators. Advanced integrated BAS systems include logic for interaction among various systems such as HVAC, lighting, and security systems. Indeed, the automated occupancy count information obtained for different spaces in a facility can be used to adjust indoor temperature settings, reduce or turn off lights, and ensure elevator operation. Such systems offer significant opportunities for energy savings and cost reductions, even though the initial capital investments maybe quite large. To control and operate equipment for HVAC, or for lighting and process equipment, an energy monitoring (or management) and control system can be used. A typical EMCS is configured into a network that includes sensors and actuators at the bottom level, microprocessor controllers in the middle, and a computer at the top with a modem to allow remote monitoring and control of the building energy systems. For a typical commercial building, an EMCS system can be cost-effective in reducing energy use for HVAC and lighting systems.

12.4   ECONOMIC METHODS

In all engineering projects and applications, including the renewable energy and distributed generation projects, it is necessary to justify the project development, implementation, and/or installation of new equipment. In weighing the various alternatives, the economic analysis problems can be classified as fixed input type, fixed output type, or situations where neither input nor output is fixed. Whatever the nature of the problem, the proper economic criteria are to optimize the benefit-cost ratio. Keep also in mind that the lifetime of any RES or DG project typically spans over 20 years. Therefore, it is important to compare savings and expenditures of various amounts of money properly over the project or development lifetime. In engineering economics, savings and expenditures of amounts of money during the project are usually called cash flows. To perform a sound economic analysis of an RES or DG development and project the most important economic parameters, data and project technical concepts must be well-defined and known. The important parameters and concepts that significantly affect the economic decision-making on the project and development include:

1. The time value of money and interest rates including simple and compound interests;
2. Inflation rate and composite interest rate;
3. Taxes including sales, local, state, and federal tax charges;
4. Depreciation rate and salvage value;
5. Tax credits and incentives, local, state and federal.

In any project finance, the two major costs are the investment costs and the operating costs. The investment cost relates to the fixed costs of materials and installation to deliver an RES or DG system to the client or stakeholders. The client has to make this payment only once, and the system will last for decades. A number of closely related, commonly used methods evaluate economic performances of a project or a development. These include the LCC method, levelized cost of energy, NB (net present worth) method, benefit-cost (or savings-to-investment) ratio method, internal rate-of-return method, overall rate-of-return method, and payback method. All of these methods are used when the important effects can be measured in dollars. If incommensurable effects are critical to the decision, it is important that they also be taken into account. The life-cycle costing (LCC) method sums, for each investment alternative, the costs of acquisition, maintenance, repair, replacement, energy, and any other monetary costs (less any income amounts, such as salvage value) that are affected by the investment decision. The time value of money must be taken into account for all amounts, and the amounts must be considered over the relevant period. All amounts are usually measured either in present-value or annual-value dollars. The levelized cost of energy (LCOE) is similar to the LCC method, in that it considers all the costs associated with an investment alternative and takes into account the time value of money for the analysis period. Lumpy costs are common to energy system deployments and projects, especially large-scale power systems for utilities. In comparison to a nuclear power plant, the investment in incremental PV modules is fairly fine-grained and smooth. But from the perspective of a residential homeowner, the incremental investment in solar technologies appears to have a high fixed cost, and to be a lumpy cost. RES and DG investments, the total system costs (C_{Sys}), are divided into direct capital costs (C_{dir}) and indirect capital costs (C_{indir}), both of which can have a scale dependency, as shown here:

$$C_{Sys} = N(units) \cdot C_{dir}(\$/unit) + N(units) \cdot C_{indir}(\$/unit) \tag{12.3}$$

From an economics perspective, we would frame the scaled dependency of the size of the project in terms of the unit cost, that is, the fixed and variable costs per characteristic unit of performance or area. Depreciation is a term used in accounting, finance, and economics to spread the cost of an asset or installation over an interval of several years, for equipment or installation usually the lifespan. In simple words, depreciation is the reduction in the asset or good value due to the usage and factor as: time performance degradation, technological changes, depletion, inadequacy, etc. The depreciation method or approach is a useful method for the recovery of an investment through income tax over a period of time. Usually, depreciation is accounted on capital assets such as a house, car, installation, and industrial equipment. Depreciation is not applicable for items such as land, salvage value, and interest amounts. Investment depreciation is defined as the decline in the capital value of the investment using the internal return rate as the discount factor. Tax and accounting depreciation are set mechanically by a specific set of rules. The two types of depreciation are account (book) and tax depreciation. The depreciation methods used in engineering economic analysis are: (1) straight-line method (usually not used for tax purpose), (2) sum-of-year method, (3) declining balance method, and (4) the accelerated cost recovery system. The recovery of depreciation on a specific asset may be related to the use rather than time, in which the depreciation is calculated by unit of production depreciation. This is not applicable to electrical and energy equipment and installation. If the investment amount and the number of years of depreciation are known, then the depreciation per years is calculated as the ratio of the investment over the lifetime, in years. In the sum of years of digits method, a greater depreciation amount is used in earlier years and there is lesser depreciation in the later years compared to the straight-line method. In the declining balance depreciation method, a constant depreciation rate is applied to the account (book) value of the property. This depreciation rate is based on the type of property and when it was acquired. Because the depreciation rate in this method is twice the straight-line rate, the method is called the double declining balance method and the *double declining balance depreciation* in any year is given by twice the book value divided by the number of years, N. The book value equals the cost depreciation to date, assuming depreciation of zero in the first year, as expressed by:

$$\text{Double Declining Depreciation in Any Year} = \frac{2 \cdot (Cost - \text{Depreciation to Date})}{N} \tag{12.4a}$$

Straight-line depreciation is the simplest and most used calculation method, in which the real asset value at the end of the period that it generates revenue can be estimated, by expensing a part of the original cost into equal increments over the period. The real value of a good or asset, which can be zero or even negative, is its value at the time of disposal, removal, or selling, and is expressed by this relationship:

$$\text{Annual Depreciation Expenses} = \frac{\text{Original Cost} - \text{Real Value at the End}}{N}, \; (\text{in \$/year}) \qquad (12.4b)$$

12.4.1 Energy Cost Calculations without Capital Return

In any energy project, a payment of interest is expected for the invested capital, which depends on the economic factors and risks. The energy unit cost estimate without considering return is relatively simple. All costs over the generation system lifetime are added (the total cost) and divided by the operating period in order to calculate the annual cost, and then by dividing the annual cost to the annual generated energy, the cost per unit of energy is obtained. If an energy generation system produces annually, E_{anl}, expressed in kWh with a total annual cost, C_{anl}, then the *specific energy cost*, SE_{Cost} is calculated by:

$$SE_{Cost} = \frac{C_{anl}}{E_{anl}} \; (\$/\text{kWh}) \qquad (12.5)$$

The annual cost is calculated by dividing the total cost C_{total} by N, the number of the power plant or energy generation system operating years, or the plant or system lifespan. The total costs include: initial investment, INV_0, and all payments P_k for every operational year k, for the entire lifespan of the system, N. For the power plant or generation system operating period of N years, the annual cost is calculated by:

$$C_{anl} = \frac{C_{total}}{N} = \frac{INV_0 + \sum_{k=1}^{N} P_k}{N} \qquad (12.6)$$

The annual cost is also called the leveled electricity cost (LEC) in the case of power systems and leveled heat cost (LHC), in the case of thermal or heating systems. Before discussing other issues, let us clarify the difference between *price* and *cost* of a product, often mistakenly considered synonyms. The product price is determined by the supply and demand for the product. The price of a product consists of its cost (production), the profit, taxes, installation, transportation, and maintenance costs, from which are subtracted incentives and tax credits (if any), and the resulting amount may be corrected by a scarcity factor.

Example 12.7

A 1.2 MW wind energy conversion system is generating an average of 2.85 millions of kWh per year. The investment cost is $1450/kW for the installation and an average annual cost of 2.5% of the investment cost for the annual operation cost for this wind turbine. Assuming the wind turbine lifespan of 20 years, calculate the specific energy cost.

SOLUTION

The total cost of the wind energy system is:

$$C_{total} = 1.2 \times 10^3 \cdot 1650 \cdot (1 + 20 \times 0.025) = \$2,610,000$$

The specific energy cost, computed with Equation (12.5) is:

$$SEC = \frac{C_{total}}{20 \times 2.85 \times 10^6} = \frac{2.61 \times 10^4}{5.7 \times 10^7} = 0.045789 \ \$/kWh \approx 4.6 \ Cent/KWh$$

Understanding the most common and needed concepts and definitions used in the fields of economics—cost analysis and management—is critical in the decision-making process for energy projects. The term *interest* can be defined as the money paid for the use of money. It is also referred to as the value or worth of money. Two important terms are *simple interest* and *compound interest*. Simple interest is always computed on the original principal (the borrowed money). The *present worth* is the current value of an amount of money due at a later time and the effect of the applied interest. *Average cost* represents the total of all fixed and variable costs calculated over a period of time, usually 1 year, divided by the total number of units produced, while the *average revenue* is the total revenue over a period of time, usually 1 year, divided by the total number of units produced. Average profit represents the difference between average revenue and average cost. *Fixed costs* are all costs, not affected by the level of business activity or production level, e.g., rents, insurance, property taxes, administrative salaries, and the interest on borrowed capital. *Life-cycle cost (LCC)* is the sum of all fixed and variable costs of a project from its initiation to its end of life. LCCs also include the planning costs, abandonment, disposal, or storage costs. *Marginal or incremental cost* is the one associated with the production of one additional output unit (e.g., product, energy unit), while the *marginal or incremental revenue* is the revenue resulting from the production of one additional output unit. The opportunity to use scarce resources, such as capital, to achieve monetary/financial advantage and the associated costs are the opportunity costs, e.g., an opportunity cost to building a new power plant for a company is not to build and instead invest capital in 7% interest-bearing securities. *Time horizon* is the time (expressed usually in years) from the project initiation to its end, including any disposal or storage of equipment and products. *Variable costs* are costs associated with the level of business activity, production, or output levels, such as fuel cost, materials cost, labor cost, distribution cost, storage cost, etc. Variable costs increase monotonically with the number of units produced. *Project term* is the planning horizon over which the project cash flow is assessed, usually divided by a specific number of years. Initial cost is the one-time expense occurring at the beginning of the project of investment, such as purchasing major assets for an energy project. *Annuity* represents the annual increment of the project cash flows, as opposed to the one-time quantity, such as the initial cost. Annuities can be either positive (e.g., the selling energy annual revenues) or negative (e.g., O&M annual costs). *Salvage value*, usually very small compared to the initial cost, is the one-time positive cash flow at the project planning horizon end, consisting of the assets, equipment, buildings, or business sold in its actual condition at the end of the project. In order to gain a true perspective as to the economic value of renewable energy systems and projects, it is necessary to compare the system technologies to conventional energy technologies on the LCC basis. The LCC method allows the total system cost calculation during a specific time period, usually the system lifespan considering not only the initial investment but also the costs incurred during the useful system life or a specific period. The LCC is the *present value* LCC of the initial investment cost and the long-term costs related to repair, operation, maintenance, transport to the site, and fuel used to run the system. *Present value* is the calculation of expenses that are realized in the future but applied in the present. An LCC analysis gives the total system cost, including all expenses incurred over the system lifetime. The main reasons for an LCC analysis are to compare different energy technologies, and to determine the most cost-effective system designs. An LCC analysis allows the designer to study the effect of using different equipment, and components with different characteristics, performances, and lifetimes. The common LCC relationship applicable for energy projects is:

$$LCC = INV0 + \sum OM_t + \sum FL_t + \sum LRC_t - \sum SV_{sys+parts} \tag{12.7}$$

Here, *INV0* is the initial overall installation costs, consisting of the present value of the capital that will be used to pay for the equipment, system design, engineering, and installation (the initial cost incurred by the user), the ΣOM_t, represents the sum of all yearly O&M (operation and maintenance) costs, the present value of expenses due to operation and maintenance programs (O&M costs include the salary of the operator, site access, guarantees, and maintenance), the ΣFL_t is the energy cost, sum of all yearly fuel costs, an expense that is the cost of fuel consumed by the conventional power or auxiliary equipment (the transport fuel cost to site must be included), the ΣLRC_t is sum of all yearly replacement costs, the present value of the cost of replacement parts anticipated over the life of the system, and the $\Sigma SV_{sys+parts}$, is salvage value, the net worth at the end of the final year, typically up to 10% for energy equipment. Future costs must be discounted because of the time value of money, so the present worth is calculated for costs for each year. The RES lifespans are assumed to range from 20 to 30 years. LCC analysis is the best way of making acquisition decisions. On the LCC analysis, many renewable energy systems are economically viable. The financial evaluation can be done on a yearly basis to obtain cash flow, breakeven point, and payback time, discussed later. Notice that social, environmental, and reliability factors are not included here, but if they are deemed important, they can be included.

Example 12.8

A small hybrid power system (1 kW wind turbine, PV array, and battery bank) was installed at a remote weather station, with the following costs: installed cost of $36,000, the loan for 10 years, with the total interest paid, minus the tax credit of $22,500, the total operation and maintenance of 2.5% of the initial cost, or $900 per year, and the total replacement cost is 3.5% of the investment. If the system salvage value is estimated at 7.5% of the initial coast, and no fuel is used for the system operation, calculate the LCC for the investment period.

SOLUTION

Applying Equation (12.7), the LCC for the hybrid power system is:

$$LCC = 36000 + 10 \times 900 + 0 + 2700 - 1260 = \$46,440.00$$

Capital budgeting is defined as a process in which a business determines whether projects such as building a new power plant, developing wind energy, or investing in a long-term venture are worth pursuing. A *capital investment* is expenditure by an organization in equipment, land, or other assets that are used to carry out the objectives of the organization. Most of the times, a prospective project's lifetime cash inflow and outflows are assessed in order to determine whether the returns generated meet a sufficient target benchmark. Capital budgeting, an essential managerial tool, is also known as *investment appraisal*. Project proposals that scale through the preliminary screening and evaluation phase are further subjected to rigorous financial appraisal to ascertain if they would add value to the organization. This stage is also referred to as *quantitative analysis, economic and financial appraisal, project evaluation* or *simply project analysis*. The financial appraisal of a project may predict the expected future cash flows of the project, analyze the risk associated with those cash flows, develop alternative cash flow forecasts, examine the sensitivity of the results to possible changes in the predicted cash flows, subject the cash flows to simulation, and prepare alternative estimates of the project's net present value. For the clarification and good understanding of the economic evaluation and analysis of renewable energy sources, it is necessary to define basic terms and concepts used in common practice.

The *economic value* is the asset or equipment value expressed through money. Different experts or economic schools explain it differently. There are two basic approaches, the *subjective* and *objective* understanding of the value. Subjective understanding of the economic value is based on individual preferences of an individual, while the objective understanding is the relationship between preferences (individual and collective) and the cost of meeting the needs. *Utility (use value)* represents the ability of an asset, equipment, installation, service, or product to meet specific needs. *Non-use value* or *passive use value* is the utility of good for others (subjective economics). *Environmental (internal) value* is the result of the belief that nature has a positive value for the environment independently of human preferences and direct benefit to humankind. *Discounting* is a concept used to evaluate the present (the costs and benefits) higher than the future (costs and benefits); there is decline in the value. Discounting relies on the premise that the value of money declines over time, therefore future values should be discounted relative to the present. There are two important parameters related to discounted cash flows. *Interest rate* is the investment percent return, or percent charged on the amount of money borrowed that time horizon beginning. Usually interest is compounded at the end of each year, and the one-year unit is referred to as the *compounding period*. The *minimum attractive rate of return (MARR)* represents the minimum interest rate required for project returns to make the project financially attractive, as set by the business, government offices, or other entities and organizations that make decisions about the project and investment. There are three major reasons for demanding the MARR on the foregoing present value of an amount of money invested: inflation (reducing the future money value), possible investment alterations (even if there are incentives for energy projects, still the investors prefer the ones with high returns), and, most important, the risks associated with any investment. Notice that when the MARR equals the inflation rate, the real dollar return is zero, but there is no actual value loss. The *nominal discount rate* is a summary rate for investment or capital that includes inflation. The *real discount rate* is the net discount rate, a nominal rate minus the inflation rate.

In the decision-making process, a short list of ideas/alternatives is identified and selected, a detailed economic evaluation for the projects in the short list of the project alternatives is performed; this evaluation is critical for the project viability. This evaluation takes into account not only the economic aspects of the chosen alternatives and determines the profitability of the projects, but also other issues, which may not be quantified and do not affect materially the cash flow and the profitability of the project, the so-called intangible items, such as public good, national security, environmental issues, etc. However, the project economic analysis treats the projects strictly as investments, with the final decisions based on the project profitability. The critical concept of the time value of money is based on the premise that one today monetary unit is worth more than the same monetary unit 1 year from now, the latter is worth more in 2 years from now, and so on. The time value of the monetary funds is intricately related to the concepts of the capital return, stipulating that capital invested must yield more capital at the investment period end, the *interest (discount) rate*, r, the percentage of additional funds that must be earned for the lending of capital, and the current and expected future inflation, increases the cost of goods in the future. *Money time value* is an important parameter in all economic calculations. However, in some situations, such as short-lifespan investments, the discount impact is quite limited, making it reasonable to use only the value monetary amount, ignoring the discounting adjustments. When capital investments, such as energy production or conservation investments, are appraised, there are inherent risks associated with any investment and all or part of the capital may be lost. The *investment risk* is one of the justifications for the charging of an interest rate and the expectation of higher return on the invested capital, and the higher the investment risk the higher would be the expected return on the capital. Two concepts often used in economic analysis are: the *simple payback* and the *capital recovery factor (CRF)*. In simple payback, the value known as the *net present value (NPV)* is computed by summing all into and out cash flows, such as: the initial costs, annuities, and salvage value amounts. If the NPV is positive, then the project is economically viable. In the case of multiyear projects with positive simple payback, the break-even point (BEP) is the year where the project total annuities equal the initial costs, which have been paid back at this point. CRF is a parameter that is used to measure the relationship between cash flows and investment costs, usually applied for short-term investments, up to 10 years. CRF is the ratio of the *annual capital cost (ACC)* to the NPV, estimated for a period of N years:

$$CRF = \frac{ACC}{NPV}$$ (12.8a)

and

$$ACC = \text{Annuity} - \frac{NPV}{N}$$ (12.8b)

Example 12.9

A utility makes an investment of \$75 million to improve the transmission capacity and reduce the losses, over a period of 10 years with an annuity of \$15 million. What is the CRF in this case?

SOLUTION

The investment net present value is the difference between the total 10-year period annuity and \$75 million:

$$NPV = -75 + N \times \text{Annuity} = -75 + 10 \times 15 = 75 \text{ million}$$

By using Equations (12.8b) and (12.8a) the ACC and CRF are:

$$ACC = \text{Annuity} - \frac{NPV}{N} = 15 - \frac{75}{10} = 7.5 \text{ million}$$

and

$$CRF = \frac{7.5}{75} = 0.1 \text{ or } 10\%$$

This value is less than the maximum 15% CRF, as recommended by the U.S. Electric Power Research Institute (EPRI) for power and energy industries. There are two ways to calculate the total interest charges over a project lifetime. In the *simple interest charges*, the total interest fee, I, is paid at the loan end, N, and is proportional with the interest rate r and the lifetime, N, expressed in the same unit, as:

$$I = N \cdot r \cdot P$$ (12.9)

The total amount paid, TP, due at the end of the loan period, includes the principal and the interest charge:

$$TP = P + I = P \cdot (1 + N \cdot r)$$ (12.10)

In the second method, *compound interest charges*, the loan lifetime (period) is divided into smaller periods, the *interest period* (usually 1 year), and the interest fee is charged at the end of each interest period and accumulated from one interest period to the next. Therefore, the total payment at the end of an interest period is computed with Equation (12.10), P being replaced by total payment at the previous interest-period end. If the principal is P, the total amount due at the loan period end is then expressed by:

$$TP = P \cdot (1 + r)^N$$ (12.11)

Equation (12.11) states that in the second method, the amount of loan payment increases exponentially with N, if the interest charges follow the law of compound interest. For most RES, DG, or energy efficiency and conservation projects, the interest rates are usually constant throughout the project lifetime. Otherwise, a common practice in economic analysis is to use average interest rates.

Example 12.10

A wind-energy developer decided to invest in a wind project to borrow a loan of $2,500,000. One bank offered a 10-year loan with 5% compound interest, while the second bank offered the same 10-year loan with a 6.5% simple fixed interest rate. Which is the more advantages loan for the wind developer?

SOLUTION

If the 6.5% simple interest is paid over a 10-year period, the total amount, Equation (12.10) is:

$$TP_{fix-rate} = 2500000 \cdot (1+10 \cdot 0.065) = 4.125 \times 10^6 \text{ USD}$$

Again for a 10-year period, for the loan with 5% compound interest, the total amount Equation (12.11) is then:

$$TP = 2500000 \cdot (1+0.05)^{10} = 4.07223 \times 10^6 \text{ USD}$$

The loan with 5% compound interest is more advantageous.

The process of calculating the future value of money (future cash flow) in the present value is called present worth. The present worth, PW, is calculated from Equation (12.11), by replacing TP with FV, and solving for PW parameter:

$$PW = FV \cdot \left(\frac{1}{(1+i)^N} \right) \qquad (12.12)$$

where FV is the money value or the cost expected at time, after N years (future value) and i is the interest (discount) rate. If FV, N, and the discount rate i, are known, then the PW can be calculated.

Example 12.11

Assuming an interest rate of 7.5%, what is the present value of $100,000 that will be received after 5 years from today?

SOLUTION

By using Equation (12.12) for the above values, the present value is:

$$PW = \frac{100000}{(1+0.075)^5} = 69,655.86 \text{ USD}$$

Inflation occurs when the good and service costs increase from one period to the next. The interest rate defines the cost of money, and the inflation rate r_{ifl} expresses the cost increase of goods and services; therefore, a future cost of a commodity or asset, FC, is higher than its present cost, PC. The inflation rate is usually considered constant over the energy project life, similar to the interest rate.

Notice also that changes in energy costs, especially increases (the so-called energy cost escalation), must be taken into consideration in any energy project (development) economic and cost analysis. The future commodity or asset cost, FC, estimated from the present commodity or asset cost, PC, and the future value (worth), FV, calculated from the present value (worth), PW, considering the combined effect of the interest rate, r, and the inflation rate, i, over the energy project (development and operation) period, N, are expressed by these relationships:

$$FC = PC(1+i) \tag{12.13}$$

And, respectively for the future value:

$$FV = PW\left(1+\frac{r-i}{1+i}\right)^N \tag{12.14}$$

Example 12.12

A residential complex owner decided to invest \$350,000 in a building-integrated PV system. The money is borrowed through a 10-year loan with an interest rate of 6.5%, and the inflation experienced by the economy is 2.75%. Determine the future value and the total cost of the investment.

SOLUTION

Applying Equations (12.14) and (12.11) the future value and the total investment costs are:

$$FV = 350000 \cdot \left(1+\frac{0.065-0.0275}{1+0.0275}\right)^{10} = \$350,475.83$$

and

$$TP = 350000 \cdot (1+0.065)^{10} = \$656,998.11$$

An important parameter and indicator for the annual cash flow analysis is *future worth* (FV). Suppose that an amount A_{dep} is deposited at the end of every year for N years, and if r is the loan interest rate compounded annually, then the future worth amount FV is given by this relationship:

$$FV = A_{dep} \cdot \left[\frac{(1+r)^N - 1}{r}\right] \tag{12.15a}$$

Or for an annuity stream, and certain time horizon, the future value, FVA of the annuity at the end of the Nth year is then given by:

$$FVA = A \cdot \left[\frac{(1+r)^N - 1}{r}\right] \tag{12.15b}$$

From an investment point of view, the translation of the future money to its present value, the discounted cash flow analysis is important, and for N equal annual amount (annuity) A, and a discount rate, i is calculated as:

$$V_P = A \cdot \left(\frac{(1+r)^N - 1}{r \cdot (1+r)^N}\right) \tag{12.16}$$

Equations (12.11) to (12.16) assume a constant (fixed) annuity value; a nonconstant (irregular) annuity case is considered below. If some parameters are known in these relationships, the other can be computed directly. In the case of nonconstant annuities, the present worth value, Equation (12.12), by treating each annuity as a single payment to be discounted from the future to the present, changes to:

$$PW = \sum_{k=1}^{N} \frac{A_k}{(1+r)^k} \tag{12.17}$$

Here A_k is the yearly predicted annuity in year k from 1 to N (the last project year). Costs varying from year to year are predicted to the net present value in a similar way. If the capital cost is subtracted from the present value of the revenue, the *net present value (NPV)* is obtained, as:

$$NPV = V_P - C_{Capital} \tag{12.18}$$

From the above equation, the *rate of return (ROR)* can be computed. The rate of return, r_{rt}, is the discount rate that makes the net present value zero, and is found by solving numerically the equation:

$$0 = V_P - C_{capital} \Rightarrow C_{capital} = A \frac{1-(1+r_{rt})^{-N}}{r_{rt}} \tag{12.19}$$

This equation is usually solved numerically, e.g., Newton-Raphson iteration, however an approximate value of the rate of return can be obtained with a trial-and-error approach. The approximate method consists of finding two r_{rt} values, for which the NPV is slightly negative and slightly positive, and then by linear interpolation between these two values, the value of the rate of return is determined. Once r_{rt} value is computed for any project alternative, the actual market discount rate or the minimum acceptable rate of return is compared to the found ROR value, and if ROR is larger, the project is cost-effective. From the determined discount rate (ROR), making the NPV zero, the *annual energy cost*, A_{cost} can be computed by:

$$A_{cost} = \frac{C_{capital} \times r_{rt}}{1-(1+r_{rt})^{-N}} \tag{12.20}$$

The cost of energy unit (specific energy), C_{eng}, considering the return rate is computed by dividing A_{cost} by the annual generated energy, E_{anl}, in a similar way as in Equation (12.5).

$$C_{eng} = \frac{A_{cost}}{E_{anl}} \tag{12.21}$$

When electric equipment, DG, and RES installations and systems, such as PV systems, wind turbines, energy-storage units, or transformers, used in a wind farm are purchased, there are several associated component costs. These include the equipment capital cost, installation cost, maintenance cost, salvage value, and the annual returns. In order to evaluate the project economic viability, other factors related to income tax, depreciation, tax credit, property tax, and insurance, must be taken into account. The maintenance cost depends on the design and the installation location. Cost comparisons between different energy sources and electricity generation systems are made by the levelized cost of energy (LCOE). Levelized costs represent the present value of building and operating a power plant or a renewable energy system over an assumed plant or system lifetime, expressed in real terms to remove the effect of inflation. The levelized cost of energy, the levelized cost of electricity, and/or the levelized energy cost are economic assessments of the average total cost to build and operate a power-generating system over its lifetime divided by the total power generated of the system during that lifetime. LCOE is often used as an alternative for the average price that the power generating system receives in a market to break even over its lifetime. It is a first-order cost competitiveness economic assessment of an electricity-generating system that incorporates all costs over its lifetime accounting for the initial investment, O&M cost, fuel cost, and capital cost. LCOE is a metric used to assess the cost of electric generation and the total plant-level impact from technology design changes, which can be used to compare electricity generation costs. There are different methods to calculate LCOE; the ones included here are the most common found in the literature and used by practitioners. LCOE calculations are part of the framework of the annual technology baseline studies that summarize current and projected future cost and performance of primary electricity generation technology in the United States, including renewable energy technologies. The levelized construction and operations costs are then divided by the total energy obtained to allow direct comparisons across different energy sources. Methods available for evaluating the economic efficiency of project options and alternatives include the following: the net present value, payback period method, rate of return method, the benefit-cost analysis, and return on investment. The payback method is commonly used for the appraisal of capital investments and engineering projects despite its theoretical deficiencies. The payback period is the time required for the benefits of an investment to equal the investment costs. The payback method is often used when aspects such as project time risk and liquidity are the focus and where pure profit evaluation is the single criterion. In practice, the maximum acceptable payback period is often chosen as a fixed value, for instance for a certain number of years, e.g., the payback period for the domestic consumer is up to 5 years, often only 2 or 3 years. In some cases, the payback period limit value is chosen in relation to the project economic life, e.g., the payback period could be shorter than half the economic life of the investment. In many companies, the payback period is used as a measure of attractiveness of capital budgeting investments. Most often the payback method is used as a first screening device to sort out obvious cases of profitable and unprofitable investments, leaving only the middle group to be scrutinized by means of more advanced and more time-consuming calculation methods based on discounted cash flows (DCF), such as the internal rate of return (IRR) and net present value (NPV) methods. However, it should be noted that the payback method can be developed to handle cases with varying cash flows, although some of its simplicity is lost in the process. Due to the fact that the decision situations in the evaluation of capital budgeting investment typically are uncertain concerning the time pattern and the duration of cash flows, the use of the simple and more robust payback method can be justified even if there will be time for more advanced analyses or methods. The payback period goes on decreasing for high-capacity system used by commercial consumers if electricity is replaced. The payback period (PBP) method of capital budgeting calculates the time it takes to recover the initial investment cost. There are two approaches, the short-cut (simple) payback method and the unequal cash flow method. Both are based on the calculations of the annual net cash flows (cash outflows minus cash inflows). As is true with NPV and IRR methods, each year may have different cash flow amounts. In the short-cut method, the amounts of annual operating cash flows expected from a potential capital asset acquisition are equal each year, and the short-cut calculation is used to determine the payback period. The simple payback period (SPP) is calculated as:

$$SPP = \frac{\text{Initial Investment (Cost)}}{\text{Net Benefits per Year}} \qquad (12.22)$$

Example 12.13

To improve the power factor, a capacitor back is installed at a 100 kW wind turbine. The cost of the unit is $20,000 and the interest rate is 7.5%, the combined federal and state tax credits are 40%, and the incremental tax rate is 45%. The system-related costs are: the loss factor 0.4, maximum reactive power demand 80 kVAR, maximum demand cost of kW, kVAR, kVA is $3.80/kW/month, the cost of released transformer kVA is $12 per kVA/year, cost of reactive energy is equal to $0.0025/kVARh, the combined feeder and transformer resistance per phase is 0.040 Ω, the current before installation 55 A, and after installation 80 A. Assuming a power factor improving from 0.65 to 0.95 due to the power factor correction unit installation, calculate the simple payback period.

SOLUTION

Savings due to the reduction in the reactive power penalty per year is:

$$C_{PFP} = 3.80 \times \left(100 \frac{\sqrt{1-0.65^2}}{0.65} - 100 \frac{\sqrt{1-0.95^2}}{0.95} \right) \times 12 = \$3832.4$$

The saving due to the reduction in kVA demand transformer is:

$$C_{kVA} = \left(\frac{100}{.65} - \frac{100}{.95} \right) \times 12 = 582.9959 \approx \$583.0$$

Savings due to the transmission line loss reduction is:

$$C_{loss} = 3 \times 0.06 \times 0.040 \times \left(80^2 - 55^2 \right) \times 8760 \times 0.4 \times 10^{-3} = \$85.1$$

Savings due the reduction of the reactive energy cost is:

$$C_{Reactive} = 0.0025 \times \left(100 \frac{\sqrt{1-0.65^2}}{0.65} - 100 \frac{\sqrt{1-0.95^2}}{0.95} \right) \times 8760 = \$1840.6$$

The annual benefit is $6341.2 and the simple payback period is then:

$$SPP = \frac{20000}{6341.2} = 3.15 \text{ years}$$

The payback period indicates how long it takes to recover the investment used to acquire an asset or for equipment installation and purchasing. The payback period is equal to the number of full years plus the final investment recovery year fraction. The simple payback period method ignores the money time value. One way to compare mutually exclusive economic aspects is to use the present time using the present worth method. The present worth of annual benefits is obtained by multiplying the benefits with the present worth factor defined for a discount (interest) rate, r and N years, as:

$$PVF = \frac{1}{(1+r)^N} \tag{12.23}$$

Then the cumulative value of the benefits, for AB, the benefits per year is given by:

$$B_{cum} = AB \left(\frac{(1+r)^N - 1}{r \cdot (1+r)^N} \right) \tag{12.24}$$

Economic analysis is an important aspect and tool of the decision-making process, while economic efficiency is a major factor in the planning of RES and DG projects. The fundamentals of present worth and future worth analyses are very important in the planning and analysis of any engineering project and in the choice of project alternative. Fundamental to finance and invest over long timespans is the time value of money. The asset, money, or cash flow worth present value or their future values are important economic parameters. Usually the solution with the best economic benefits is the choice, while the technical and/or ecological aspects are of secondary importance. The economic analysis aims to find one system out of several alternatives or possible solutions that provides the desired energy form at the lowest cost possible. The cost analysis methods, such as payback period methods, cost-benefit analysis, life cost analysis, or cost components involved in any RES and DG project are presented and discussed in detail. The effects of taxes, depreciation, tax credits, and inflation on economic analysis are also discussed. The payback period approach is analyzed for the simple case, taking into account the present worth value of money, taxes, depreciation, tax credits, and inflation. The decision-making process of energy and building retrofitting projects is very similar to the decision-making processes in any engineering projects, involving the following steps: (1) the identification of the project needs, (2) project alternative solutions are formulated, analyzed, and evaluated, and (3) a decision is made on the best and optimum alternative solution and then the project is identified and fully specified. These steps include feasibility analysis and a detailed economic analysis of all the identified alternatives. In evaluating the projects, all factors must be examined and some of the proposed alternative solutions may be excluded for reasons other than economic or technical considerations. It is worth noting here that the final decision must be specific enough to allow the facility planning, installation, and construction, but also allow sufficient flexibility for the next project stages or future changes and adaptations. The decision-making process for engineering projects and especially for energy projects follows a well-defined need, being accomplished and performed by multistep, structured procedures, simulations, and mathematical techniques. The project or investment economic analysis should be made for a few carefully selected alternative solutions to the problem, with the final decision taking into consideration the economic analysis as well as previous experience, environmental, and social factors as well as engineering input.

12.5 METHODS FOR ENERGY ANALYSIS

In order to fully understand and analyze the energy consumption and estimate the cost-effectiveness of energy conservation measures, specific calculation methods and simulation tools needed to be used. The existing energy analysis methods vary widely in complexity and accuracy. To select the appropriate energy analysis method, several factors, such as speed, cost, versatility, reproducibility, sensitivity, accuracy, and ease of use are considered. Energy analysis tools can be classified into either forward or inverse methods. In the forward approach, the energy predictions are based on a building system physical description, such as geometry, location, construction details, and HVAC system type and operation. In the inverse approaches, the energy analysis model attempts to deduce representative building parameters, such as the building load coefficient, the building base-load, or the building time constant, using existing energy use, weather, and relevant performance data. In general, the inverse models are less complex to formulate than the forward models. However, the flexibility of inverse models is typically limited by the formulation of the representative building parameters and the accuracy of the building performance data. Energy analysis tools can also be classified based on their ability to capture the dynamic behavior of building energy systems. They can use either steady-state or dynamic modeling approaches. In general, the steady-state models are sufficient to analyze seasonal or annual building energy performance. However, dynamic models may be required to assess the transient effects of building energy systems such as those encountered for thermal energy storage systems and optimal start controls.

The most common energy analysis methods used in the United States and European Union can be grouped into three main categories:

1. *Ratio-based methods*, which are pre-audit analysis approaches that rely on building energy/cost densities to quickly evaluate building performance. The ratio-based methods are pre-audit analysis approaches to determine a specific energy or cost indicators for the building. These energy/cost building indicators are then compared to reference performance indices obtained from several other buildings with the same attributes. The end-use energy or consumption ratios can provide useful insights on some potential problems within the

building such as leaky steam pipes, inefficient cooling systems, or high water usage. These building energy densities or ratios can be useful to determine if the building has high energy consumption and to assess if an energy audit of the building would be beneficial, to assess if a preset energy performance target has been achieved for the building. If not, the energy ratio can be used to determine the magnitude of the required energy-use reduction to reach the target, to estimate typical consumption levels for fuel, electricity, and water to be expected for new buildings, or to monitor the evolution of energy consumption of buildings and estimate the effectiveness and profitability of any energy management program carried out following an audit. Notice that in order to estimate meaningful reference ratios, large databases have to be collected. For energy ratios, the variables that are typically used include: total building energy use (i.e., including all end uses), expressed in kWh or MMBtu, building energy consumption by end use (i.e., heating, cooling, and lighting), energy demand (in kW), surface area or space volume (such as heating area or conditioned volume in offices), building users (in collective buildings such as hotels or schools), degree-day (usually with 65°F [18°C] as a base temperature), or units of production (especially for manufacturing facilities or restaurants).

2. *Inverse methods*, which use both steady-state and dynamic modeling approaches and include variable-base degree-day methods. Inverse modeling methods can be useful tools in improving building energy efficiency, being used to detect malfunctions by identifying time periods or specific systems with abnormally high energy consumption, to provide estimates of expected savings from a defined set of energy-conservation measures, and to verify the savings achieved by energy retrofits. Regression analyses are usually used to estimate the representative parameters for the building or its systems (such as building load coefficient or heating system efficiency) using measured data. In general, steady-state inverse models are based on monthly or daily data and include one or more independent variables. Dynamic inverse models are usually developed using hourly or sub-hourly data to capture any significant transient effect such as the case where the building has a high thermal mass to delay cooling or heating loads. Steady-state inverse models attempt to identify the relationship between building energy consumption and selected weather-dependent parameters such as monthly or daily average outdoor temperatures, degree-hours, or degree-days, being characterized by simplicity and flexibility. Steady-state inverse models are especially suitable for measurement and verification of energy savings from energy retrofits.

3. *Forward methods*, which include either a steady-state or a dynamic modeling approach, are often the bases of detailed and complex energy-simulation computer programs. Typically, forward models can be used to determine the energy end uses as well as predict any energy savings incurred from energy-conservation measures. Steady-state energy analysis methods that use the forward modeling approach are generally easy to use because most of the calculations can be performed by hand or using spreadsheet programs. Two types of steady-state forward tools can be distinguished: degree-day methods and bin methods. The degree-day methods use seasonal degree-days computed at a specific set-point temperature (or balance temperature) to predict the energy use for building heating. However, the degree-day methods are not suitable for predicting building cooling loads. In the United States, the traditional degree-day method using a base temperature of 65°F has been replaced by the variable-base degree-day method and is applied mostly to residential buildings; while in Europe 18°C is still in use as the base temperature, for both residential and commercial buildings. Dynamic analytical models use numerical or analytical methods to determine energy transfer among various building systems. These models generally consist of simulation computer programs with hourly or sub-hourly time steps to accurately estimate the effects of thermal inertia, due for instance to energy storage in the building envelope or its heating system. Detailed computer programs require a high level of expertise and are suitable to simulate large buildings with complex HVAC systems and

involved control strategies that are difficult to model using simplified energy analysis tools. Such energy simulation programs require a detailed physical description of the building, including building geometry, building envelope details, HVAC equipment type and operation, and occupancy schedules. Thermal load calculations are based on a wide range of algorithms depending on the complexity and the flexibility of the simulation program.

12.6 STANDARDIZATIONS AND SUMMARY OF IMPORTANT STANDARDS

As stated in the previous sections, energy efficiency and conservation face some barriers to success. Examples of such barriers include: the lack of awareness of the savings potential, inadequate performance efficiency information and metrics, the tendency to focus on the performance of individual components rather than the energy yield, consumption of complete systems, or lower initial cost rather than life cycle cost, and split or insufficient incentives. Standards can help to overcome some of these barriers. Standards, for instance, can provide common measurement and test methods to assess the use of energy and the reductions attained through new technologies and processes, as well as provide a means of codifying best practices and management processes for efficient energy use and conservation. Standards can also provide design checklists and guides that are applied to the design of new systems and the retrofit of existing systems, providing calculation methods or comparisons of alternatives for specific situations, helping with the infrastructure adaptation to integrate new technologies and interoperability. In 2009, ISO and the International Electrotechnical Commission (IEC) created the joint project committee ISO/IEC JPC 2, energy efficiency and renewable energy sources. In 2007, IEC began to establish subsidiary bodies to advise its management board on strategic issues that determine future technical work. Among these was the SG 1, which was established on the specific topic of energy efficiency. SG 1 was established at the beginning of 2007 and was tasked to analyze the status quo in the field of energy efficiency and renewable resources (existing IEC standards, ongoing projects), identify gaps and opportunities for new work in IEC's field of competence, set objectives for electrical energy efficiency in products and systems, and formulate recommendations for further actions. IEC's vision on energy efficiency is outlined in a white paper, "Coping with the Energy Challenge." Developed by the IEC Market Strategy Board, this document maps out global energy needs and potential solutions and the IEC's role in meeting the challenges. IEC considers that a system approach, taking into account all aspects of generating, transporting, and consuming energy, must be considered to cope with the energy efficiency challenges and those measurement procedures and methods of evaluating energy efficiency must be specified in order to assess potential improvements and to optimize technological issues.

ISO (International Organization for Standardization) is the principal organization that develops and publishes voluntary standards that impact specifications for products, services, and good practices globally. ISO 50001: 2011, Energy Management Systems—Requirements with Guidance for Use, adopted in June 2011, is the international standard for energy management systems. There were national or regional standards for energy management prior to the adoption of ISO 50001. Superior Energy Performance (SEP) is an energy management program managed by the US DOE, which extends beyond ISO 50001 by adding a verification component to ensure energy savings in industrial facilities. SEP is a voluntary certification that industrial facilities earn by demonstrating continual improvement in energy efficiency and conformance to ISO 50001. Organizations can use the SEP framework as a roadmap to achieve ongoing energy improvements, even if they are not yet ready to pursue SEP or ISO 50001 certification. SEP builds on ISO 50001 to analyze and prioritize energy use and consumption by tracking progress with energy performance metrics. SEP is accredited by the American National Standards Institute (ANSI) and the ANSI-American Society of Quality (ANSI-ASQ) National Accreditation Board (ANAB). SEP certification requires independent verification of the two requirements by an ANSI-ANAB accredited verification body: ISO 50001 conformance; and energy performance improvement levels corresponding to the ANSI/MSE 50021 standard for SEP. SEP provides a robust protocol for energy performance measurement and third-party verification and also generates reliable data for company management and validation to external stakeholders of continual energy performance improvement. In response to the CEN/CENELEC, BTJWG energy management recommendation, CEN and CENELEC have created a horizontal structure, a Sector Forum Energy Management (SFEM), dedicated to the definition of a common strategy for standardization in the field of energy management and energy efficiency. SFEM is a platform for stakeholders to share information and experiences, and to identify priorities regarding standardization in the energy sector. The CEN–CENELEC standards and projects in the field of energy management and energy efficiency include: EN 16001:2009 (22320) Energy management systems – Requirements with guidance for use, and EN 15900:2010 (22416) Energy efficiency services – Definitions and requirements. The standard EN 15232 provides a list of BACS and TBM functions that can affect the energy performance of buildings and introduces four different BAC efficiency classes: Class A: High energy performance BACS and TBM systems; Class B: Advanced BACS and TBM systems; Class C: Standard BACS; and Class D: Non-energy efficient BACS. These classes refer only to the installed BACS and TBM systems and not to the building as a whole, and they are not correlated to the energy classes defined by the European Standard EN 15217.

Directive 2012/27/EU of The European Parliament and of the Council of 25 October 2012 on the energy efficiency, updates and amends Directives 2009/125/EC and 2010/30/EU and repeals the old Directives 2004/8/EC and 2006/32/EC, Official Journal of the European Union. L 315/1,14.11.2012. Actual consumption or the various requirements relating to the standardized use of a building serves as the measure of energy efficiency. Per EU directive "Energy Performance of Building Directive" (EPBD), the following thermal and electrical forms of energy are considered when determining the energy efficiency of a building: The energy performance of a building means the amount of energy estimated or actually consumed to meet the different needs associated with a standardized use of the building. The standards referring to heating are: EN 15316-1 and EN 15316-4, cooling is EN 15243l, domestic hot water is EN 15316, while ventilation and lighting are covered in EN 1524. EN 15193 covers heating, DHW (domestic hot water), cooling, ventilation, lighting, and auxiliary energy usages. A new European standard EN15232 "Energy performance of buildings - Impact of Building Automation, Control and Building Management" is one of a set of CEN (Comité Européen de Normalization, European Committee for Standardization) standards, which are developed within a standardization project sponsored by the European community. The aim of this project is to support the EPBD to enhance energy performance of buildings in EU member states. Standard EN 15232 specifies methods to assess the impact of building automation and control systems (BACS) and technical building management (TBM) functions on the energy performance of buildings, and a method to define minimum requirements of these functions to be implemented in buildings of different complexities.

12.7 CHAPTER SUMMARY

Energy consumption by buildings is a challenge to control, making it an increasing global concern. The annual contribution of residential and commercial building sectors towards energy consumption is usually between 30% and 40% in developed countries, with closer values in many developing countries. The concept of energy conservation, energy efficiency, or energy optimization has become one of the major concerns of our times. Energy sustainability is the cornerstone to the health, vitality, and competitiveness of industries or organizations that are producing and manufacturing in today's global economy, being more than the process of being environmentally responsible and earning the right to operate as a business. It is the ability to utilize and optimize multiple sources of secure and affordable energy for the enterprise, and then continuously improve utilization through systems analysis and through an organizational drive for continuous improvement as a core principle. Management commitment to ensure the best energy efficiency management in existing process operations, as well as a dedicated pursuit of new system technologies and processes, is the only recipe for excellence. Energy management is the philosophy of more efficient energy use, without compromising production levels, product quality, safety, or environmental standards. Energy accounting, monitoring, and control are the first steps in any energy management program, being of great importance in making any program of energy management a success. The effectiveness of energy utilization varies with specific industrial, facility, or building operations because of the diversity of the products and the processes required for manufacturing goods and services. The organization of personnel and operations involved in organizations also varies greatly. Consequently, an effective energy management program needs to be tailored for each company and its plant operations. An energy audit is a systematic study or survey conducted to identify how the energy is being used in a building, facility, or plant; to identify energy savings opportunities; and to indicate the performance at the overall plant, facility, or process level. An energy audit of residential, commercial, and industrial buildings encompasses a wide variety of tasks and requires expertise in a number of areas to determine the best energy-conservation measures suitable for an existing facility. Energy price-rate structures focusing on US utilities are also outlined in a chapter section with a special emphasis on electricity pricing features. Several examples are presented to illustrate how various components of rate structures will affect monthly utility bills. The main goal is to help energy auditors or engineers to understand the complexities of various utility rate structures, and that significant energy cost savings can be achieved by selecting the energy pricing rate best-suited for the audited facility. The chapter provides comprehensive descriptions of energy management and a systematic approach of energy audit types, and how to perform specific energy audits. The chapter gives also a brief presentation of energy analysis methods, characteristics, and suitability. Using proper and well-designed energy audit methods and equipment, an energy audit provides the energy manager with essential information on how much, where, and how energy is used within an organization (factory, plant, or building). The main objectives and goals of the energy audit report are energy savings proposals comprising technical and economic analysis of projects. Goals and objectives of an energy audit are: clearly identifying the energy use types and cost, understanding the energy use and waste, analyzing and identifying the ways to save or preserve energy, improving operational techniques and equipment, performing an economic analysis of alternatives, and finally determining the most cost-effective one. In summary, energy management means lowering the organization or company costs through eliminating unnecessary energy use and wastes, improving the efficiency of the needed energy, buying energy at the lowest cost possible, and adjusting the operations to allow purchasing the energy at lower prices. The design, construction, and implementation of any engineering system are ultimately decided by economic decisions based on cost and economic analysis. Economic analysis of investments, including any engineering project, is one of the critical steps leading to a decision. Therefore, making informed decisions on the design and implementation of renewable energy systems requires comprehensive economic and life cycle and life cost assessment and analysis.

12.8   QUESTIONS AND PROBLEMS

1. What are the differences between energy conservation and energy efficiency?
2. What is the efficiency of an energy conversion machine if it requires 1200 J of energy input and produces 180 J of unwanted waste energy (losses)?
3. An industrial energy system has power output of 450 kW with an overall efficiency of 58.5%. What is the input energy supplied to this generator per hour? What is the rate of energy input required if the system efficiency increases to 63%?
4. Define energy management.
5. List and briefly describe the objectives of an energy management system.
6. Briefly discuss the energy uses in building sectors.
7. Briefly describe how energy-efficient lighting can reduce the building energy consumption.
8. Explain the role of training and awareness in an energy management program.
9. Define in your own words the energy audit.
10. What is a walk-through energy audit? A utility cost analysis?
11. Briefly describe the standard energy audit.

12. Briefly describe the procedures of a detailed energy audit.
13. What are the steps or phases in the implementation of energy management in an organization or facility?
14. Explain briefly the difference between preliminary and detailed energy audits.
15. What is the significance of knowing the energy costs?
16. What are the benefits of benchmarking energy use?
17. Briefly describe the four energy audit types and their purpose.
18. List the main steps of a building energy audit.
19. List all the costs that are expected with the construction and operation of a geothermal power plant, which is to produce electrical energy for 40 years.
20. Do the same as in the previous problem, but for a photovoltaic power plant producing electrical energy for 25 years.
21. Do the same as in the previous example, but for a wind farm producing energy for 30 years.
22. List the measurement instruments and equipment that are most commonly used in an energy audit.
23. Briefly describe possible energy-conservation strategies with thermal energy storage.
24. Briefly describe the energy balance concept and its importance.
25. Determine the cost savings achieved for the month in Example 12.3, if the facility power factor is improved to be always above 85%.
26. Calculate the utility bill for the industrial facility considered in Example 12.3 taking into account the ratchet clause. Assume that the previous highest demand (during the last 12 months) is 630 kW and the ratchet clause is 67%.
27. Calculate the utility bill for the industrial facility considered in the above problem, taking into account both a fuel-cost adjustment of 0.015/kWh and a sales tax of 8%.
28. The industrial facility of Example 12.3 has an option of owning and operating its service transformer with the advantage of a reduced rate structure as described below: customer charge of $650/month, demand charge of $15/kW, and energy charge of $0.025/kWh. Calculate the reduction in the utility bill during the month for which the energy use characteristics are given in Example 12.3.
29. Repeat the monthly electricity bill estimate of Example 12.5, but for a winter month and for an electricity consumption of 850 kWh.
30. By using the utility rate structure given in the table below, calculate the utility bill for a commercial facility during a winter month when the energy use is 72,000 kWh, the billing demand is 400 kW, and the average reactive demand is 200 kVAR.

Item	Winter	Summer
Customer charge	$0.00	$0.00
Minimum charge	$25.0	$25.0
Fuel-cost adjustment	0.015	0.0125
Tax rate	0.00%	0.00%
No. of energy blocks	3	3
Block 1 energy size (kWh)	40,000	40,000
Block 2 energy size (kWh)	60,000	60,000
Block 3 energy size (kWh)	>100,000	>100,000
Block 1 energy charge ($/kWh)	0.060	0.065
Block 2 energy charge ($/kWh)	0.040	0.050
Block 3 energy charge ($/kWh)	0.030	0.40
No. of demand blocks	2	2
Block 1 demand size (kW)	50	50
Block 2 demand size (kW)	>50	>50
Block 1 demand charge ($/kW)	12.5	13.5
Block 2 demand charge ($/kW)	11.5	12.5
Reactive demand charge ($/kVAR)	0.20	0.20

31. Repeat the calculation of the above problem, but for a summer month in which energy use is 60,000 kWh, the billing demand is 400 kW, and the average reactive demand is 120 kVAR.

32. By using the utility rate structure given in Table 12.2, calculate the natural gas utility bill for a commercial facility during a month when the energy use is 17,500 MMBtu and the billing demand is 540 MMBtu/day.

33. A wind energy company is considering adding eight new wind turbines to its wind farm. Each turbine, plus the installation cost is $1,350,000 USD. A bank is offering a loan covering the investment for 15 years with a fixed rate of 7%. How much is the loan total amount paid at the end of the loan period?

34. Calculate the future worth, FW, for a deposit of $60,000. The number of investment years is 6, and the interest rate is 4.5%.

35. If an amount of $7500 annual deposit is made for 5 years, what is the future worth value of those deposits? The interest rate is 3.85%. If a cash flow consists of an annuity of $175,000 over an 8-year period at 6% interest rate, calculate its future value at the end of this period.

36. An investor decided to invest $1.5 million, borrowed with 7.5% interest, in a geophysical district heat, having a lifespan of 20 years. Determine the cost of the money. Assuming the O&M cost per year of 8.5% of the initial investment and 3.5% of the initial investment, in governmental tax credit per year, and electricity saving per year of $9.575 million of kWh at an average cost of 4 cent per kWh, calculate the annual revenue of the development.

37. Briefly describe the life cycle costs (LCC) for a renewable energy system and the main factors that affect the LCC.

38. Briefly describe the common energy conservation measures (methods).

39. Calculate the monthly utility bill for an industrial facility, having an actual demand of 500 kW, energy consumption 90,000 kWh, and average power factor 0.72. The facility is subject to the following rate structure: customer charge is $160/month, billed demand charge is $10/kW, and the energy charge is $0.025/kWh. The rate structure includes a power factor clause that states that the billing demand is determined as the actual demand multiplied by the ratio of 85 and the actual average power factor expressed in percent.

40. List and briefly describe the most common energy analysis methods.

REFERENCES AND FURTHER READINGS

1. U.S. Department of Energy, Energy Use, Loss and Opportunities Analysis: U.S. Manufacturing and Mining, U.S. Department of Energy, Office of Industrial Technologies, Industrial Technology Program, Washington, DC, 2004.

2. U.S. DOE Energy Information Agency, Industrial Sector Energy Price and Expenditure Estimates for 2001. U.S. Department of Energy, Office of Industrial Technologies, Industrial Technology Program, Washington, DC, 2005.

3. GRI. Electric and Gas Rates for the Residential, Commercial, and Industrial Sectors, Volumes 1 and 2, GRI-93/0368.1 and GRI-93/0368.2 by L. J. Whitem, C.M. McVicker, and E. Stiles, Gas Technology Institute, formerly Gas Research Institute, Des Plaines, IL, 1993.

4. M. A. El-Sharkawi, *Electric Energy: An Introduction* (2nd ed.), CRC Press, 2009.

5. W. C. Turner, *Energy Management Handbook*, John Wiley and Sons, 2017.

6. B. L. Capehart (ed.), *Information Technology for Energy Managers: Understanding Web Based Energy Information and Control Systems*, Fairmont Press, Atlanta, GA, 2004.

7. B. L. Capehart and L. C. Capehart (eds.), *Web Based Energy Information and Control Systems: Case Studies and Applications*, Fairmont Press, Atlanta, GA, 2005.

8. G. Rosenquist, M. McNeil, M. Lyer, S. Meyers, and J. McMahon, Energy Efficiency Standards for Equipment: Additional Opportunities in the Residential and Commercial Sectors, LBNL-2006.

9. W. C. Turner and S. Doty (eds.), *Energy Management Handbook* (6th ed.), The Fairmont Press, 2007.

10. M. Krarti, *Energy Audit of Building Systems: An Engineering Approach* (2nd ed.), CRC Press, Boca Raton, FL, 2011.

11. F. Kreith and D. Y. Goswami (eds.), *Energy Management and Conservation and Handbook*, CRC Press, Boca Raton, FL, 2008.
12. I. Dincer and M. A. Rosen, *Thermal Energy Storage Systems and Applications*, Wiley, London, 2001.
13. T. Greadel and B. Allenby, An introduction to LCA, In *Industrial Ecology* (2nd ed.), Prentice Hall, Upper Saddle River, NJ, 2003.
14. ISO 14040 (2006): Environmental management – Life cycle assessment – Principles and framework, International Organization for Standardization (ISO), Geneva, Switzerland, 2006.
15. ISO 14044 (2006): Environmental management – Life cycle assessment – Requirements and guidelines, International Organization for Standardization (ISO), Geneva, 2006.
16. M. Kaltshmitt, W. Streicher, and A. Weise (eds.), *Renewable Energy – Technology, Economics and Environment*, Springer, 2007.
17. F. Kreith and D. Y. Goswami (eds.), *Handbook of Energy Efficiency and Renewable Energy*, CRC Press, Boca Raton, FL, 2007.
18. D. Mumovic and M. Santamouris, *A Handbook of Sustainable Building Design & Engineering, An Integrated Approach to Energy, Health and Operational Performance*, Earthscan, 2009.
19. G. Petrecca, *Energy Conversion and Management - Principles and Applications*, Springer, Cham Heidelberg New York Dordrecht London, 2014.
20. A. Sumper and A. Baggini, *Electrical Energy Efficiency: Technologies and Applications*, Wiley, 2012.
21. P. Sansoni, L. Mercatelli, and A. Farini, *Sustainable Indoor Lighting*, Springer, 2015.
22. N. B. Jacobs, *Energy Policy: Economic Effects, Security Aspects and Environmental Issues*, Nova Science Publishers Inc., 2009.
23. L. Blank and A. Tarquin, *Engineering Economics* (9th ed.), McGraw-Hill, 2012.
24. E. E. Michaelides (ed.), *Alternative Energy Sources*, Springer, 2012.
25. F. M. Vanek, L. D. Albright, and L. T. Angenent, *Energy Systems Engineering – Evaluation and Implementation* (2nd ed.), McGraw-Hill, 2012.
26. IEA Report- 2013, Key World Energy Statistics, International Energy Agency, Paris, France, 2013.
27. A. P. Rossiter and B. P. Jones (eds.), *Energy Management Efficiency for the Process Industries*, Wiley, 2015.
28. R. Fehr, *Industrial Power Distribution* (2nd ed.), Wiley, 2015, ISBN: 978-1-119-06334-6.
29. R. Belu and L. I. Cioca, Energy management systems in *Encyclopedia of Energy Engineering & Technology* (Online) (eds. S. Anwar, R. Belu), CRC Press/Taylor and Francis, 2015.
30. R. Bansal (ed.), *Handbook of Distributed Generation - Electric Power Technologies, Economics and Environmental Impacts*, Springer, 2017.
31. C. J. Corbett and D. A. Kirsch, *International diffusion of ISO 14000 certification, Productions and Operations Management*, Vol. 10(3), pp. 327–342, 2001.
32. European Technical Standard EN 15217, Energy performance of Buildings – Methods for expressing energy performance and for the energy certification of buildings, first ed., CEN, Brussels, Belgium, 2007.
33. W. Kahlenborn, S. Kabisch, J. Klein, I. Richter, S. Schürmann, *Energy Management Systems in Practice, ISO 50001: A Guide for Companies and Organizations*, German Federal Ministry for the Environment, Nature Conservation and Nuclear Safety (BMU), Berlin, 2012.
34. U.S. Department of Energy DOE Guide for ISO 50001, 2015. Available at https://ecenter.ee.doe.gov/ Pages/default.aspx (accessed April 2018).
35. B. L. Capehart, *Guide to Energy Management* (7th ed.), Fairmont Press, 2011.
36. R. Belu, *Industrial Power Systems with Distributed and Embedded Generation*, The IET Press, Stevenage, UK, 2018.

Appendix A: Common Parameters, Units, and Conversion Factors

TABLE A.1
Physical Constants in SI Units

Quantity	Symbol	Value
Avogadro constant	N	$6.022169 \cdot 10^{26}$ kmol^{-1}
Boltzmann	k	$1.380622 \cdot 10^{-23}$ J/K
First radiation constant	$C_1 = 2 \cdot \pi \cdot h \cdot c$	$3.741844 \cdot 10^{-16}$ Wm^2
Gas constant	R	$8.31434 \cdot 10^3$ J/kmol K
Planck constant	h	$6.626196 \cdot 10^{-34}$ Js
Second radiation constant	$C_2 = hc/k$	$1.438833 \cdot 10^{-2}$ mK
Speed of light in a vacuum	c	$2.997925 \cdot 10^8$ m/s
Stefan-Boltzmann constant	σ	$5.66961 \cdot 10^{-8}$ W/m^{2}K^4
Speed of light	c	299,792.458 m/s
Elementary charge	e	$1.602176 \cdot 10^{-19}$ C

TABLE A.2
Multiplication Factors

Multiplication Factor	Prefix	Symbol
10^{12}	Terra	T
10^9	Giga	G
10^6	Mega	M
10^3	Kilo	K
10^2	Hecto	H
10	Deka	da
1	N/A	—
0.1	Deci	d
0.01	Centi	c
10^{-3}	Milli	m
10^{-6}	Micro	μ
10^{-9}	Nano	n
10^{-12}	Pico	p
10^{-15}	Femto	f
10^{-18}	Atto	a

TABLE A.3
System of Units and Conversion Factors

US Unit	Abbreviation	SI Unit	Abbreviation	Conversion Factor
Foot	ft	Meter	m	0.3048
Mile	mi	Kilometer	km	1.6093
Inch	in.	Centimeter	cm	2.54
Square foot	ft^2	Square meter	m^2	0.0903
Acre	acre	Hectare	ha	0.405
Circular mil	cmil	μm^2	—	506.7
Cubic feet	ft^3	Cubic meter	m^3	0.02831
Gallon (US)	gal (US)	Liter	l	3.785
Gallon (UK)	gal (UK)	Liter	l	4.445
Cubic feet	ft^3	Liter	l	28.3
Pound	lb	Kilogram	kg	0.45359
Ounce	oz	Gram	g	28.35
US ton	ton (US)	Metric ton	ton (metric)	0.907
Mile/Hour	mi/h	Meter/second	m/s	0.447
Flow rate	ft^3/h	Flow rate	m^3/s	0.02831
Density	lb/ft^3	Density	kg/m^3	16.020
lb-force	lbf	Force	N	4.4482
Pressure	lb/in^2	Pressure	kPa	6.8948
Pressure	bar	Pressure	Pa	10^5
Torque	lb.force ft	Torque	Nm	1.3558
Power	ft.lb/s	Power	W	1.3558
Power (horsepower)	HP	Power	W	745.7
Energy	ft.lb-force	Energy	J	1.3558
Energy (British Thermal Unit)	Btu	Energy	kWh	3412

TABLE A.4
Summary of Radiometric and Photometric Units

Quantity	Symbol	Unit
Wavelength	λ	m
Foot-candela	fc	lm·ft^{-2}
Radiant (luminous) energy	Q	W·s
Radiant (luminous) energy (photometry)	Q_V	lm·s
Radiant (luminous) energy density	We	W·s·m^{-2}
Radiant (luminous) energy density (photometry)	W_V	lm·s·m^{-2}
Radiant (luminous) radiant energy density (per unit of volume)	$u_0 = \dfrac{dW_e}{dv}$	J·m^{-3}
Energy flux (power)	Φ	W
Radiant (luminous) energy flux (power) (photometry)	Φ_V	lm
Radiant exitance (radiant flux per unit of source area)	M_0	W·m^{-2}
Irradiance (radiant flux per unit of target area)	I_0	W·m^{-2}
Irradiance and illuminance	E	W·m^{-2} (or W·cm^{-2})
Irradiance and illuminance (photometry)	E_V	lx (fc)
Radiance, intensity, and luminance	L	W·m^{-2}/sr (steradian)
Radiance, intensity, and luminance (photometry)	L_1	lm·m^{-2}/sr (steradian)
Radiant and luminous intensity	1	W/sr
Radiant and luminous intensity (photometry)	l_V	cd (lm/sr)

TABLE A.5
Common Energy Conversion Factors

Energy Unit	SI Equivalent
1 electron volt (eV)	$1.6021 \cdot 10^{-19}$ J
1 erg (erg)	10^{-7} J
1 calorie (cal)	4.184 J
1 British thermal unit (Btu)	1055.6 J
1 Q (Q)	10^{18} Btu (exact)
1 quad (q)	10^{15} Btu (exact)
1 tons oil equivalent (toe)	$4.19 \cdot 10^{10}$ J
1 barrels oil equivalent (bbl)	$5.74 \cdot 10^{9}$ J
1 tons coal equivalent (tce)	$2.93 \cdot 10^{10}$ J
1 m^3 of natural gas	$3.4 \cdot 10^{7}$ J
1 liter of gasoline	$3.2 \cdot 10^{7}$ J
1 kWh	$3.6 \cdot 10^{6}$ J
1 ft^3 of natural gas (1000 Btu)	1055 kJ
1 gal. of gasoline (125,000 Btu)	131.8875 kJ

TABLE A.6
Refractive Index of Some of the Common Materials (at 20°C)

Material	Index of Refraction	Material	Index of Refraction
Diamond	2.419	Benzene	1.501
Fluorite	1.434	Carbon disulfide	1.628
Fused quartz	1.458	Carbon tetrachloride	1.461
Glass (crown)	1.520	Ethyl alcohol	1.361
Glass (flint)	1.660	Glycerin	1.473
Ice	1.309	Oil, turpentine	1.470
Polystyrene	1.590	Paraffin (liquid)	1.480
Salt ($NaCl_2$)	1.544	Water	1.333
Teflon	1.380	Air (0°C, 1 atm)	1.000293
Zircon	1.923	Carbon dioxide (0°C, 1 atm)	1.00045

TABLE A.7
Properties of Dry Air

Temperature (K)	ρ (kg/m³)	C_P (kJ/kg K)	k (W/m K)
293	1.2040	1.0056	0.02568
300	1.1774	1.0057	0.02624
350	0.9980	1.0090	0.03003
400	0.8826	1.0140	0.03365
450	0.7833	1.0207	0.03707
500	0.7048	1.0295	0.04038
600	0.5879	1.0551	0.04659
700	0.5030	1.0752	0.05230
800	0.4405	1.0978	0.05779
900	0.3925	1.1212	0.06279
1000	0.3525	1.1417	0.06752

Symbols: T absolute temperature, degrees Kelvin; ρ density; C_P specific heat capacity; μ viscosity; $v = \mu/\rho$; and k = thermal conductivity. The values of ρ, μ, v, k, and C_P are not strongly pressure-dependent and may be used over a fairly wide range of pressures.

Source: Adapted from US National Bureau Standards (US), 1955.

TABLE A.8
Properties of Water

Temperature (°C)	ρ (kg/m³)	C_P (kJ/kg K)	k (W/m K)
0.00	999.8	4.225	0.566
4.44	999.8	4.208	0.575
10.0	999.2	4.195	0.585
20.0	997.8	4.179	0.604
25.0	994.7	4.174	0.625
100.0	954.3	4.219	0.684

TABLE A.9
Temperature Conversion Formulas

	Degree Celsius (°C)	Degree Fahrenheit (°F)	Kelvin Unit (K)
Degree Celsius (°C)	—	$\frac{9}{5} \cdot °C + 32$	$K - 273.15$
Degree Fahrenheit (°F)	$\frac{5}{9}(°F - 32)$	—	$1.8 \times K - 459.67$
Kelvin unit (K)	$°C + 273.15$	$(459.67 + °F)/1.8$	—

Appendix B: Design Parameters, Values, and Data

TABLE B.1-I
Geometric Characteristics of Conduits (Adapted from NEC)

Conduit Trade Size Designator: English (Metric)	Internal Diameter in. (mm)	Cross-Sectional Area in.² (mm²)
1/2 (16)	0.62 (15.7)	0.30 (195)
3/4 (21)	0.82 (20.9)	0.53 (345)
1 (27)	1.05 (26.6)	0.87 (559)
1 1/4 (35)	1.38 (35.1)	1.51 (973)
1 1/2 (41)	1.61 (40.9)	12.05 (1.322)
2 (53)	2.07 (52.5)	3.39 (2.177)
2 1/2 (63)	2.47 (62.7)	4.82 (3.106)
3 (78)	3.07 (77.9)	7.45 (4.794)
3 1/2 (91)	3.55 (90.1)	9.96 (6.413)
4 (103)	4.03 (102.3)	12.83 (8.268)
5 (129)	5.05 (128.2)	20.15 (12.984)
6 (155)	6.07 (154.1)	29.11 (18.760)

TABLE B.1-II
Maximum Occupancy Recommended for Conduits (Adapted from NEC)

Conduit Trade Size Designator: English (Metric)	1 Cable = 53% Fill in.² (mm²)	2 Cables = 31% Fill in.² (mm²)	3+ Cables = 40% Fill in.² (mm²)
1/2 (16)	0.16 (103)	0.09 (60)	0.120 (78)
3/4 (21)	0.28 (183)	0.16 (107)	0.21 (138)
1 (27)	0.46 (296)	0.27 (173)	0.35 (224)
1 1/4 (35)	0.80 (516)	0.47 (302)	0.60 (389)
1 1/2 (41)	1.09 (701)	0.64 (410)	0.82 (529)
2 (53)	1.80 (1.154)	1.05 (675)	1.36 (871)
2 1/2 (63)	2.56 (1.646)	1.49 (963)	1.93 (1.242)
3 (78)	3.95 (2.541)	2.31 (1.486)	2.98 (1.918)
3 1/2 (91)	5.28 (3.399)	3.09 (1.988)	3.98 (2.565)
4 (103)	6.80 (4.382)	3.98 (2.563)	5.13 (3.307)
5 (129)	10.68 (6.882)	6.25 (4.025)	8.06 (5.194)
6 (155)	15.43 (9.943)	9.02 (5.816)	11.64 (7.504)

TABLE B.1-III
Minimum Radius of Bends of Conduits (Adapted from NEC)

Conduit Trade Size Designator: English (Metric)	Layers of Steel within Sheath in. (mm)	Other Sheath in. (mm)
1/2 (16)	6 (160)	4 (100)
3/4 (21)	8 (210)	5 (130)
1 (27)	11 (270)	6 (160)
1 1/4 (35)	14 (350)	8 (210)
1 1/2 (41)	16 (410)	10 (250)
2 (53)	21 (530)	12 (320)
2 1/2 (63)	25 (630)	25 (630)
3 (78)	31 (780)	31 (780)
3 1/2 (91)	36 (900)	36 (900)
4 (103)	40 (1.020)	40 (1.020)
5 (129)	50 (1.280)	50 (1.280)
6 (155)	60 (1.540)	60 (1.540)

TABLE B.2-I
Commonly Used Conductor Cross-Sectional Area

Size	Approximate Cross-Sectional Area Square Inches/ in²				
AWG/kcmil	RHH/RHW With Cover	RHH/RHW Without Cover	TW or THW	THHN THWN	BARE Stranded Conductors
14	0.0293	0.0209	0.0139	0.0097	0.0042
12	0.0353	0.0260	0.0181	0.0133	0.0066
10	0.0437	0.0333	0.0243	0.0211	0.0106
8	0.0835	0.0556	0.0437	0.0366	0.0167
6	0.1041	0.0726	0.0590	0.0507	0.0266
4	0.1333	0.0973	0.0814	0.0824	0.0423
3	0.1521	0.1134	0.0962	0.0973	0.0531
2	0.1750	0.1333	0.1146	0.1158	0.0670
1	0.2660	0.1901	0.1534	0.1562	0.0845
1/0	0.3039	0.2223	0.1825	0.1855	0.1064
2/0	0.3505	0.2624	0.2190	0.2233	0.1379
3/0	0.4072	0.3117	0.2642	0.2679	0.1691
4/0	0.4754	0.3718	0.3197	0.3237	0.2190
250	0.6291	0.4596	0.3904	0.3970	0.2597
300	0.7088	0.5281	0.4536	0.4608	0.3107
350	0.7870	0.5958	0.5166	0.5242	0.3610
400	0.8626	0.6619	0.5782	0.5863	0.4140
500	1.0082	0.7901	0.6984	0.7073	0.5191
600	1.2135	0.9729	0.8709	0.8676	0.6235
750	1.4272	1.1652	1.0532	1.0496	0.7823
1000	1.7719	1.4784	1.3519	1.3478	1.0423

TABLE B.2-II
Conductor Properties Size, Area, and Resistance per Length (Adapted from NEC Chapter 9)

Size	Area (Circular)		Resistance					
AWG or kcmil	mm²	mils	Uncoated Copper		Coated Copper		Aluminum	
			Ω/km	Ω/kFT	Ω/km	Ω/kFT	Ω/km	Ω/kFT
18	0.823	1620	25.5	7.77	26.5	8.08	42.0	12.8
16	1.31	2580	16.0	4.89	16.7	5.08	26.4	8.05
14	2.08	4110	10.1	3.07	10.4	3.19	16.6	5.06
12	3.31	6530	6.34	1.93	6.57	2.01	10.45	3.18
10	5.261	10380	3.984	1.21	4.148	1.26	6.651	2.00
8	8.367	16510	2.506	0.764	2.579	0.786	4.125	1.26
6	13.30	26240	1.608	0.491	1.671	0.510	2.652	0.808
4	21.15	41740	1.010	0.308	1.053	0.321	1.666	0.508
3	26.67	52620	0.802	0.245	0.833	0.254	1.320	0.403
2	33.62	66360	0.634	0.194	0.661	0.201	1.045	0.319
1	42.41	83690	0.505	0.154	0.524	0.160	0.829	0.253
1/0	53.59	105600	0.399	0.122	0.415	0.127	0.660	0.201
2/0	67.43	133100	0.1370	0.0967	0.329	0.101	0.523	0.159
3/0	85.01	167800	0.2512	0.0766	0.261	0.0797	0.413	0.126
4/0	107.2	211600	0.1996	0.0608	0.205	0.0626	0.328	0.100
250	127	-	0.1687	0.0515	0.1753	0.0535	0.2778	0.0847
300	152	-	0.1409	0.0429	0.1463	0.0446	0.2318	0.0707
350	177	-	0.1205	0.0367	0.1252	0.0382	0.1984	0.0605
400	203	-	0.1053	0.0321	0.1084	0.0331	0.1737	0.0529
500	253	-	0.0845	0.0258	0.0869	0.0265	0.1391	0.0424
600	304	-	0.0704	0.0214	0.0732	0.0223	0.1159	0.0353
700	355	-	0.0603	0.0184	0.0622	0.0189	0.0994	0.0303
750	380	-	0.0563	0.0171	0.0579	0.0176	0.0927	0.0282
800	405	-	0.0528	0.0161	0.0544	0.0166	0.0868	0.0265
900	456	-	0.0470	0.0143	0.0481	0.0147	0.0770	0.0235
1000	507	-	0.0423	0.0129	0.0434	0.0132	0.0695	0.0212
1250	633	-	0.0338	0.0103	0.0347	0.0106	0.0554	0.0169

TABLE B.3
Average Luminance of Common Light Sources

Light Source	Comment	Average (approximate) Luminance (cd·m⁻²)
Sun (at earth's surface)	At meridian	$1.60 \cdot 10^9$
Sun (at earth's surface)	Near horizon	$6.0 \cdot 10^4$
Moon (at earth's surface)	Bright spot	$2.50 \cdot 10^3$
Clear sky	Average luminance	$8.0 \cdot 10^3$

(Continued)

TABLE B.3 *(Continued)*
Average Luminance of Common Light Sources

Light Source	Comment	Average (approximate) Luminance (cd·m⁻²)
Overcast sky	—	$2.0 \cdot 10^3$
60 W incandescent lamp	—	$1.20 \cdot 10^6$
Tungsten-halogen lamp (3000 K CCT)	—	$1.30 \cdot 10^7$
Tungsten-halogen lamp (3400 K CCT)	—	$3.90 \cdot 10^7$
CFL	36 W twin tube	$3.0 \cdot 10^4$
T-5 Fluorescent lamp	14–35 W	$2.0 \cdot 10^4$
T-8 Fluorescent lamp	36 W	$1.0 \cdot 10^4$
T-12 Fluorescent Lamp	Cool white 800 mA	$1.0 \cdot 10^4$
High-pressure mercury lamp	1000 W	$2.0 \cdot 10^8$
Xenon short-arc lamp	1000 W	$6.0 \cdot 10^8$

TABLE B.4-I
Percent Effective Ceiling (Floor Cavity Reflectance) for Various Reflectance Combinations (Part I)

% Ceiling or Floor Reflectance (ρ_{cc}, ρ_{fc})	90				80				70		
% Wall reflectance (ρ_w)	90	70	50	30	90	70	50	30	70	50	30
Room cavity ratio (RCR)											
0.2	89	88	86	85	78	78	77	76	68	67	66
0.4	88	86	84	81	77	76	74	72	67	65	63
0.6	87	84	80	77	76	75	71	68	65	63	59
0.8	87	82	77	73	75	73	69	65	64	60	56
1.0	86	80	75	69	74	72	67	62	62	58	53
1.2	85	78	72	66	73	70	64	58	61	57	50
1.4	85	77	69	62	72	68	62	55	60	55	47
1.6	84	75	67	59	71	67	60	53	59	53	45
1.8	83	73	64	56	70	66	58	50	58	51	42
2.0	83	72	62	53	69	64	56	48	56	49	40
2.2	82	70	59	50	68	63	54	45	55	48	38
2.4	82	69	58	48	67	61	52	43	54	46	37
2.6	81	67	56	46	66	60	50	41	54	45	35
2.8	81	66	54	44	65	59	48	39	53	43	33
3.0	80	64	52	42	65	58	47	37	52	42	32
3.2	79	63	50	40	65	57	45	35	51	40	31
3.4	79	62	48	38	64	56	44	34	50	39	29
3.6	78	61	47	36	63	54	43	32	49	38	28
3.8	78	60	45	35	62	53	41	31	49	37	27
4.0	77	58	44	33	61	53	40	30	48	36	26
4.2	77	57	43	32	60	52	39	29	47	35	25
4.4	76	56	42	31	60	51	38	28	46	34	24
4.6	76	55	40	30	59	50	37	27	45	33	24
4.8	75	54	39	28	58	49	36	26	45	32	23
5.0	75	53	38	28	58	48	35	25	44	31	22

TABLE B.4-II
Percent Effective Ceiling (Floor Cavity Reflectance) for Various Reflectance Combinations (Part II)

% Ceiling or Floor Reflectance (ρ_{cc}, ρ_{fc})	50				30			10		
% Wall reflectance (ρ_w)	70	50	30	10	50	30	10	50	30	10
Room cavity ratio (RCR)										
0.2	49	48	47	30	29	29	28	10	10	09
0.4	48	47	45	30	29	28	26	11	10	09
0.6	47	45	43	30	28	26	25	11	10	08
0.8	47	44	40	30	28	25	23	11	10	08
1.0	46	43	38	30	27	24	22	12	10	08
1.2	45	41	36	30	27	23	21	12	10	07
1.4	45	40	35	30	26	22	19	12	10	07
1.6	44	39	33	29	25	22	18	12	09	07
1.8	43	38	31	29	25	21	17	13	09	06
2.0	43	37	30	29	24	20	16	13	09	06
2.2	42	36	29	29	24	19	15	13	09	06
2.4	42	35	27	29	24	19	14	13	09	06
2.6	41	34	26	29	23	18	14	13	09	06
2.8	41	33	25	29	23	17	13	13	09	05
3.0	40	32	24	29	22	17	12	13	09	05
3.2	39	31	23	29	22	16	12	13	09	05
3.4	39	30	22	29	22	16	11	13	09	05
3.6	39	29	21	29	21	15	10	13	09	04
3.8	38	29	21	28	21	15	10	14	09	04
4.0	38	28	20	28	21	14	09	14	09	04
4.2	37	28	20	28	21	14	09	14	08	04
4.4	37	27	19	28	21	14	09	14	08	04
4.6	36	26	18	28	20	13	08	14	08	04
4.8	36	26	18	28	20	13	08	14	08	04
5.0	35	25	17	28	19	13	08	14	08	04

TABLE B.4-III
Adjustment Coefficients for 30% and 10% Floor Cavity Reflectance (Part III)

% Ceiling or Floor Reflectance (ρ_{cc}, ρ_{fc})	80				50			30		
% Wall reflectance (ρ_w)	70	50	30	10	50	30	10	50	30	10
Room cavity ratio (RCR)				Effective floor reflectance ($\rho_{fc} = 30\%$) ($20\% = 1.00$)						
1	1.092	1.082	1.075	1.068	1.049	1.044	1.040	1.028	1.026	1.023
2	1.079	1.066	1.055	1.047	1.041	1.033	1.027	1.026	1.021	1.017
3	1.070	1.054	1.042	1.033	1.034	1.027	1.020	1.024	1.017	1.012
4	1.062	1.045	1.033	1.024	1.030	1.022	1.015	1.022	1.015	1.010
5	1.056	1. 038	1.026	1.018	1.027	1.018	1.012	1.020	1.013	1.008
6	1.052	1.033	1.021	1.014	1.024	1.015	1.009	1.019	1.012	1.006
7	1.047	1.029	1.018	1.011	1.022	1.013	1.007	1.018	1.010	1.005
8	1.044	1.026	1.015	1.009	1.020	1.012	1.006	1.017	1.009	1.004
9	1.040	1.024	1.014	1.007	1.019	1.011	1.005	1.016	1.009	1.004
10	1.037	1.022	1.012	1.006	1.017	1.010	1.004	1.015	1.009	1.003

(*Continued*)

TABLE B.4-III *(Continued)*

Adjustment Coefficients for 30% and 10% Floor Cavity Reflectance (Part III)

% Ceiling or Floor Reflectance (ρ_{cc}, ρ_{fc})	80				50			30		
% Wall reflectance (ρ_w)	70	50	30	10	50	30	10	50	30	10
Room cavity ratio (RCR)				Effective Floor Reflectance (ρ_{fc} = 10%)(20% = 1.00)						
1	0.923	0.929	0.935	0.940	0.956	0.960	0.956	0.973	0.976	0.979
2	0.931	0.942	0.950	0.958	0.962	0.968	0.974	0.976	0.980	0.985
3	0.939	0.951	0.961	0.969	0.967	0.975	0.981	0.978	0.983	0.988
4	0.944	0.958	0.969	0.978	0.972	0.980	0.986	0.980	0.986	0.991
5	0.949	0.964	0.976	0.983	0.975	0.983	0.989	0.981	0.988	0.993
6	0.953	0.969	0.980	0.986	0.977	0.985	0.902	0.982	0.989	0.995
7	0.957	0.973	0.983	0.991	0.979	0.987	0.994	0.983	0.990	0.996
8	0.960	0.976	0.986	0.993	9.981	0.988	0.995	0.984	0.991	0.997
9	0.963	0.978	0.987	0.994	0.983	0.990	0.996	0.985	0.992	0.998
10	0.965	0.980	0.965	0.998	0.984	0.991	0.997	0.986	0.993	0.998

TABLE B.4-IV

Coefficient of Utilization (CU)

Effective floor reflectance (ρ_{fc})	20	20	20	20	20	20	20	20	20
Effective ceiling reflectance (ρ_{cc})	80	80	80	70	70	70	50	50	50
Effective wall reflectance (ρ_w)	50	30	10	50	30	10	50	30	10
Room cavity ratio (RCR)									
0	0.83	0.83	0.83	0.72	0.72	0.72	0.50	0.50	0.50
1	0.72	0.69	0.66	0.62	0.60	0.57	0.43	0.42	0.40
2	0.63	0.58	0.54	0.54	0.50	0.47	0.38	0.35	0.33
3	0.55	0.49	0.45	0.47	0.43	0.39	0.33	0.30	0.29
4	0.48	0.42	0.37	0.42	0.37	0.33	0.29	0.26	0.23
5	0.43	0.36	0.32	0.37	0.32	0.28	0.26	0.23	0.20
6	0.38	0.32	0.27	0.33	0.28	0.24	0.23	0.20	0.17
7	0.34	0.28	0.23	0.30	0.24	0.21	0.21	0.17	0.15
8	0.31	0.25	0.20	0.27	0.21	0.18	0.19	0.15	0.13
9	0.28	0.22	0.18	0.24	0.19	0.16	0.17	0.14	0.11
10	0.25	0.20	0.16	0.22	0.17	0.14	0.16	0.12	0.10

TABLE B.5
Power and Efficiency of Common Light Sources

Light Source	Power (W)	Lamp Efficiency (lm/W)
Incandescent lamp	100	17
Linear tungsten-halogen lamp	300	20
T-5 Fluorescent lamp (4 ft.)	28	100
T-8 Fluorescent lamp (4 ft.)	32	90
CFL	26	70
Mercury vapor lamp	175	45
Metal-halide, low-voltage	100	80
Metal-halide, high-voltage	400	90
High-pressure mercury lamp	1000	50
Xenon short-arc lamp	1000	30
High-pressure sodium, low-voltage	70	90
High-pressure sodium, high-voltage	250	100
Low-pressure sodium, U-type	180	180

TABLE B.6
Color Temperature

Color Temperature	Kelvin Units
Candle	1800 K
Sun on the horizon	2000 K
Sodium vapor lamp	2200 K
Incandescent light bulb	2400 K – 2700 K
Warm white fluorescent tube	2700 K – 3000 K
Metallic halogen lamp	3000 K – 4200 K
Halogen lamp	3000 K – 3200 K
Neutral white fluorescent tube	3900 K – 4200 K
Midday sunshine (cloudless sky)	5500 K – 5800 K
Solar spectrum AM 0	5900 K
Daylight fluorescent tube	5400 K – 6100 K
Electronic flash	5000 K – 6500 K
Cloudy sky	7000 K – 9000 K

Note: Color temperature value describes the apparent color of the light source, which varies from the orange red of a candle flame (1800 K) to the bluish white of an electronic flash (between 5000 K and 6500 K, depending on the manufacturer).

TABLE B.7
Lamp Voltage Factor (VF)

Lamp Type	−5	−4	−3	−2	0	+2
			Rated Lamp Voltage Deviation (%)			
Fluorescent (magnetic ballast)	95	96	97	90	100	102
Fluorescent (electronic ballast)	—	—	—	—	100	—
Mercury (ballast)	88	90	92	95	100	105
Halogen	80	84	88	92	100	108
Incandescent	83	86	89	94	100	106
Mercury (constant power)	97	98	99	99	100	102
Metal halide	91	93	95	97	100	—
High-pressure sodium	—	—	—	—	100	—

TABLE B.8-I
Typical Mounting Height for Standard Luminaires

Luminaire Type/Wattage	Type of Road	Preferred Mounting Height (m)	Minimum Mounting Height (m)
CFL32	Minor	7.5	5.5
2xLF14	Minor	7.5	5.5
2xLF24	Minor	7.5	5.5
S70	Minor	7.5	5.5
S100	Major	9.0	7.5
S150	Major	10.5	9.0
S250	Major	12.0	10.0
S400		15.0	12.0

Symbols: CFL = compact fluorescent light, LF = linear fluorescent, S = high-pressure sodium

TABLE B.8-II
Luminaire Maintenance Factors (LMF) – Adapted from CIE Publication 97.2:2005 "Guide on the Maintenance of Indoor Electric Lighting Systems"

Environmental Conditions	Maintenance Intervals				
	1 Year	2 Years	3 Years	4 Years	5 Years
Very Clean	0.96	0.94	0.92	0.90	0.87
Clean	0.93	0.89	0.85	0.82	0.77
Normal	0.89	0.84	0.79	0.75	0.67
Dirty	0.83	0.78	0.73	0.69	0.62

TABLE B.9

Outdoor and Parking Luminaire Luminous Intensity Distribution

Vertical Angle, θ	Intensity (cd) at Lateral Angle, ϕ						
	0°	10°	20°	30°	45°	60°	75°
0.0°	2886	2886	2886	2886	2886	2886	2886
5.0°	2846	2892	2929	2955	2963	2963	2963
10.0°	2779	2802	2825	2854	2864	2871	2879
15.0°	2678	2695	2740	2755	2780	2812	2820
20.0°	2578	2620	2651	2803	2854	2829	2804
25.0°	2544	2583	2676	2753	2804	2879	2762
30.0°	2427	2448	2501	2720	2904	2963	2820
35.0°	2712	2678	2929	3080	3326	3078	2996
40.0°	3565	2952	3436	4051	4168	3550	3565
45.0°	4118	3359	4067	5000	4812	3944	4109
50.0°	3749	3525	4530	5038	4926	3891	3691
60.0°	1741	2224	3281	4861	4385	3808	3448
75.0°	418	489	1188	3021	2063	2092	1364
90.0°	0	0	0	0	0	0	0

TABLE B.10

Thermal Conductivities of Selected Materials

Solids–Metals	Temp (K)	k (Btu/h ft^2 °F/ft.)	k (W/mK)
Aluminum	573	133	230
Cadmium	291	54	94
Copper	373	218	377
Iron (wrought)	291	35	61
Iron (cast)	326	27.6	48
Lead	373	19	33
Nickel	373	33	57
Silver	373	238	412
Steel 1% C	291	26	45
Tantalum	291	32	55
Admiralty metal	303	65	113
Bronze	—	109	189
Stainless steel	293	9.2	16

Solids–Nonmetals	Temp (K)	k (Btu/h ft^2 °F/ft.)	k (W/mK)
Asbestos sheet	323	0.096	0.17
Asbestos	273	0.09	0.16
Asbestos	373	0.11	0.19
Asbestos	473	0.12	0.21
Bricks (alumina)	703	1.8	3.1
Bricks (building)	293	0.4	0.69
Magnesite	473	2.2	3.8
Cotton wool	303	0.029	0.050
Glass	303	0.63	1.09
Mica	323	0.25	0.43
Rubber (hard)	273	0.087	0.15

(Continued)

TABLE B.10 *(Continued)*
Thermal Conductivities of Selected Materials

Solids–Nonmetals	Temp (K)	k (Btu/h ft^2 °F/ft.)	k (W/mK)
Sawdust	293	0.03	0.052
Cork	303	0.025	0.043
Glass wool	—	0.024	0.041
85% Magnesia	—	0.04	0.070
Graphite	273	87	151

Liquids	Temp (K)	k (Btu/h ft^2 °F/ft.)	k (W/mK)
Acetic acid 50%	293	0.20	0.35
Acetone	303	0.10	0.17
Aniline	273-293	0.1	0.17
Benzene	303	0.09	0.16
Calcium chloride brine 30%	303	0.32	0.55
Ethyl alcohol 80%	293	0.137	0.24
Glycerol 60%	293	0.22	0.38
Glycerol 40%	293	0.26	0.45
n-Heptane	303	0.08	0.14
Mercury	301	4.83	8.36
Sulfuric acid 90%	303	0.21	0.36
Sulfuric acid 60%	303	0.25	0.43
Water	303	0.356	0.62
Water	333	0.381	0.66

Gases	Temp (K)	k(Btu/h ft^2 °F/ft.)	k (W/mK)
Hydrogen	273	0.10	0.17
Carbon dioxide	273	0.0085	0.015
Air	273	0.014	0.024
Air	373	0.018	0.031
Methane	273	0.017	0.029
Water vapor	373	0.0145	0.025
Nitrogen	273	0.0138	0.024
Ethylene	273	0.0097	0.017
Oxygen	273	0.0141	0.024
Ethane	273	0.0106	0.018

TABLE B.11
Standard Thermodynamic Properties for Selected Substances

Substance	Symbol	ΔH_f^o (kJ·mol^{-1})	ΔG_f^o (kJ·mol^{-1})	S_{298}^o (J K^{-1} mol^{-1})
Aluminum	Al(s)	0	0	28.3
	Al(g)	324.4	285.7	164.54
Antimony	Sb(s)	0	0	45.69
	Sb(g)	262.34	222.17	180.16
Arsenic	As(s)	0	0	35.1
	As(g)	302.5	261.0	174.21
Bromine	Br(l)	0	0	152.23
	Br2(g)	30.91	3.142	245.5
	Br(g)	111.88	82.429	175.0
Cadmium	Cd(s)	0	0	51.76
	Cd(g)	178.2	144.3	167.75

(Continued)

TABLE B.11 *(Continued)*
Standard Thermodynamic Properties for Selected Substances

Substance	Symbol	ΔH_f^o (kJ· mol^{-1})	ΔG_f^o (kJ·mol^{-1})	S_{298}^o (J K^{-1} mol^{-1})
Carbon	C(s)/Graphite	0	0	5.74
	C(s)/Diamond	1.89	2.90	2.38
	C(g)	716.681	671.2	158.1
	CO(g)	−110.52	−137.15	197.7
	CO_2(g)	−393.51	−394.36	213.8
	CH_4(g)	−74.6	−50.5	186.3
	CH_3OH(l)	−239.2	−166.6	126.8
	CH_3OH(g)	−201.0	−162.3	239.9
	C_2H_2(g)	227.4	209.2	200.9
	C_2H_4(g)	52.4	68.4	219.3
	C_2H_6(g)	−84.0	−32.0	229.2
	C_2H_5OH(l)	−277.6	−174.8	160.7
	C_2H_5OH(g)	−234.8	−167.9	281.6
	C_6H_6(g)	82.927	129.66	269.2
	C_6H_6(l)	49.1	124.50	173.4
	CH_2Cl_2(l)	−124.2	−63.2	177.8
	CH_2Cl_2(g)	−95.4	−65.90	270.2
Chlorine	Cl_2(g)	0	0	223.1
	Cl(g)	121.3	105.70	165.2
Chromium	Cr(s)	0	0	23.77
	Cr(g)	396.6	351.8	174.50
Copper	Cu(s)	0	0	33.15
	Cu(g)	338.32	298.58	166.38
	$CuSO_4$(s)	−771.36	−662.2	120.9
Fluorine	F_2(g)	0	0	202.8
	F(g)	79.4	62.3	158.8
	HF(g)	−273.3	−275.4	173.8
Hydrogen	H2(g)	0	0	130.7
	H(g)	217.97	203.26	114.7
	H^+(aq)	0	0	0
	OH^-(aq)	−230.0	−157.2	−10.75
Iron	Fe(s)	0	0	27.3
	Fe(g)	416.3	370.7	180.5
	Fe_2^+(aq)	−89.1	−78.90	−137.7
Lead	Pb(s)	0	0	64.81
	Pb(g)	195.2	162.	175.4
	Pb_2^+(aq)	−1.7	−24.43	10.5
Lithium	Li(s)	0	0	29.1
	Li(g)	159.3	126.6	138.8
	Li+(aq)	−278.5	−293.3	13.4
Magnesium	Mg2+(aq)	−466.9	−454.8	−138.1
Mercury	Hg(l)	0	0	75.9
	Hg(g)	61.4	31.8	175.0
Nickel	Ni_2^+(aq)	−64.0	−46.4	−159

(Continued)

TABLE B.11 *(Continued)*
Standard Thermodynamic Properties for Selected Substances

Substance	Symbol	ΔH_f^o (kJ· mol⁻¹)	ΔG_f^o (kJ·mol⁻¹)	S_{298}^o (J K⁻¹ mol⁻¹)
Nitrogen	$N_2(g)$	0	0	191.6
	$N(g)$	472.704	455.5	153.3
	$NO(g)$	90.25	87.6	210.8
	$NO_2(g)$	33.2	51.30	240.1
	$NO_3^-(aq)$	−205.0	−108.7	355.7
	$NH_3(g)$	−45.9	−16.5	192.8
	$NH_4^+(aq)$	−132.5	−79.31	113.4
Oxygen	$O_2(g)$	0	0	205.2
	$O(g)$	249.17	231.7	161.1
	$O_3(g)$	142.7	163.2	238.9
Phosphorus	$P_4(s)$	0	0	164.4
	$P_4(g)$	58.91	24.4	280.0
	$P(g)$	314.64	278.25	163.19
Potassium	$K(s)$	0	0	64.7
	$K(g)$	89.0	60.5	160.3
	$K+(aq)$	−252.4	−283.3	102.5
Silicon	$Si(s)$	0	0	18.8
	$Si(g)$	450.0	405.5	168.0
	$SiO2(s)$	−910.7	−856.3	41.5
	$SiH4(g)$	34.3	56.9	204.6
Silver	$Ag(s)$	0	0	42.55
	$Ag(g)$	284.9	246.0	172.89
	$Ag+(aq)$	105.6	77.11	72.68
Sodium	$Na(s)$	0	0	51.3
	$Na(g)$	107.5	77.0	153.7
	$Na+(aq)$	−240.1	−261.9	59
Sulfur	$S8(s)$ (rhombic)	0	0	256.8
	$S(g)$	278.81	238.25	167.82
	$S_2^-(aq)$	41.8	83.7	22
Tin	$Sn(s)$	0	0	51.2
	$Sn(g)$	301.2	266.2	168.5
Water	$H_2O(l)$	−285.83	−237.1	70.0
	$H_2O(g)$	−241.82	−228.59	188.8
	$H_2O_2(l)$	−187.78	−120.35	109.6
	$H_2O_2(g)$	−136.3	−105.6	232.7
Zinc	$Zn(s)$	0	0	41.6
	$Zn(g)$	130.73	95.14	160.98
	$Zn2+(aq)$	−153.9	−147.1	−112.1

Note: (g) gas; (l) liquid; (aq) aqueous solution; and (s) solid
Source: Adapted from U.S. National Bureau Standards (U.S.), Tables of Thermodynamic and Transport Properties, Circular 564, 1955.

TABLE B.12
Thermal Capacities of Common TES Materials at 20°C

Material	Density (kg/m³)	Specific Heat (J/kg K)	Volumetric Thermal Capacity (10⁶ J/m³ K)
Clay	1458	879	1.28
Brick	1800	837	1.51
Sandstone	2200	712	1.57
Wood	700	2390	1.67
Concrete	2000	880	1.76
Glass	2710	837	2.27
Aluminum	2710	896	2.43
Iron	7900	452	3.57
Steel	7840	465	3.68
Gravely	2050	1840	3.77
Earth Magnetite	5177	752	3.89
Water	988	4182	4.17

TABLE B.13
Thermodynamic Properties of Water

Temperature (K)	$\widehat{G}(T)$ (kJ·mol⁻¹)	$\widehat{H}(T)$ (kJ·mol⁻¹)	$\widehat{S}(T)$ (J·K⁻¹ mol⁻¹)	$C_P(T)$ (J·K⁻¹ mol⁻¹)
273	−305.1	−287.73	63.28	76.10
298.15	−306.69	−285.83	69.95	75.37
320	−308.27	−284.18	75.28	75.27
340	−309.82	−282.68	79.85	75.41
360	−311.46	−281.17	84.16	75.72
373	−312.58	−280.18	86.85	75.99

TABLE B.14
Common Units and Conversion Factors

Type	Name	Symbol	Approximate Value
Pressure	Atmosphere	atm	1.013×10^5 Pa
Pressure	Bar	bar	10^5 Pa
Pressure	Pounds per square inch	psi	6890 Pa
Mass	Ton (metric)	t	10^3 kg
Mass	Pound	lb	0.436 kg
Mass	Ounce	oz	0.02835 kg
Length	Ångström	Å	10^{-10} m
Length	Inch	in	0.0254 m
Length	Foot	ft	0.3048 m
Length	Mile (statute)	mi	1609 m
Volume	Liter	l	10^{-3} m³
Volume	Gallon (U.S.)	gal	3.785×10^{-3} m³

Index